Practical Computer-Aided

Lens Design

Gregory Hallock Smith

Practical Computer-Aided
Lens Design

Gregory Hallock Smith

Published By

Willmann-Bell, Inc.

P.O. Box 35025 • Richmond, Virginia 23235 USA • ☎ (804) 320- 7016
www.willbell.com

Published by Willmann-Bell, Inc.
P.O. Box 35025, Richmond, Virginia 23235

First English Edition

Printed in the United States of America

Library of Congress Cataloging-in-Publication Data.

Smith, Gregory Hallock,

Practical computer-aided lens design / GregoryHallock Smith.

p. cm.

Includes bibliographical references and index.

ISBN 0-943396-57-3

1. Lenses–Design and construction–Data processing. 2. Computer -aided design. I. Title.

QC385.2.D47S6 1998 98-31043

681'.423–dc21 CIP

Gregory Hallock Smith and Willmann-Bell, Inc. hereby acknowledge the following trademarks:

ACCOS by Optikos
AMTIR by Amorphous Materials
APART and ASAP by Breault Research Organization
Astrositall by SRC Technology, Russia
Biogon, Planar, Protar, Tessar, and Topogon by Carl Zeiss, Oberkochen and Jena
Cer-Vit by Owens-Illinois
Cleartran by Morton Advanced Materials
CODE V by Optical Research Associates
Dagor and Hypergon by Goerz
Elmar and Leica by Ernst Leitz, Wetzlar
GUERAP by Lambda Research
Hoya by Hoya Corp., Japan
Kodak and Wratten by Eastman Kodak
LensVIEW by Optical Data Solutions
Metrogon by Bausch & Lomb
Nikon by Nippon Kogaku, Japan
Ohara by Ohara, Inc., Japan
OSLO by Sinclair Optics
Pentium by Intel
Pyrex and ULE by Corning Glass Works
Retrofocus by Angénieux
Schott and Zerodur by Schott Glass Technologies, Germany and USA
Super-Angulon by Schneider, Kreuznach
ZEMAX by Focus Software

The glass maps illustrated in Chapter A.10 are used with permission of Schott Glass Technologies. Readers are cautioned that these maps were recreated from printed Schott maps by Willmann-Bell, Inc. and while they have been carefully checked and no known errors exist, they should be used with caution if they are used beyond the purpose of this book, which is solely illustrative. Accordingly, on projects that have economic implications, glass availability and specifications should be checked directly with Schott.

98 99 00 01 02 03 04 05 9 8 7 6 5 4 3 2

Dedicated to the memory of
Clinton W. Hough,
friend, mentor, inspiration.

Preface

This book is the summation of my more than forty years' study of optics. It began in the Spring of 1954 when an eighth-grade general science class sparked my interest in astronomy, telescopes, photography, and cameras. It continued through school to a Ph.D. in 1972 from the Optical Sciences Center of the University of Arizona. This was followed by several optical jobs in research, defense, and most recently, teaching. As time progressed, my interest in optics only grew. And I have never stopped learning. It has all come together here.

This book is dedicated to the memory of Clinton W. Hough, a retired engineer and neighbor when I was in high school. During those years, I spent many enjoyable and stimulating evenings and weekends at his home, where he had a telescope, Leica cameras, darkroom, machine shop, library, and lots of time for a youngster first exploring his mind. By word and deed, he pointed the way to my professional career. He was truly a gentleman of the old school, and he is greatly missed.

I must also acknowledge the debt I owe to my professors at Occidental College, the University of California at Berkeley, the University of California at Los Angeles, and the University of Arizona. I especially remember Walter Wallin, Daniel Popper, A. Papoulis, Robert Noble, Philip Slater, Roland Shack, Helmut Abt, and Richard Cromwell.

While a student, I was privileged to have had eye-opening summer jobs at Caltech and at the Jet Propulsion Laboratory (JPL). These included the opportunity to observe at Mt. Wilson Observatory, an amazing place.

Since graduation, the several professional positions I have held have provided plenty of experience in the rapidly evolving field of computer-aided lens design. I have been both witness and participant in these developments.

I owe a great debt of gratitude to all the friends and colleagues who have helped me over the years. I especially thank those who agreed to referee the manuscript of this book for errors and other lapses. Some of their reviews were long and detailed. All required much time and effort.

Aden Meinel, who was Director of the Optical Sciences Center when I was a student there, and Marjorie Meinel both reviewed the manuscript in detail. They made many valuable suggestions and contributed the three-bar resolution target. Aden Meinel has been a mentor and inspiration for me for over 30 years, and my admiration continues.

Kenneth Moore, who writes the ZEMAX program at Focus Software, reviewed the manuscript in detail. Without Ken's amazing software and personal help during the writing, this book would be very different and much poorer.

I thank the people at Optical Research Associates (ORA) in Pasadena. Much of what I know about lens design I learned by using their CODE V program (starting in 1977). Thomas Harris, Darryl Gustafson, and David Hasenauer of ORA all carefully reviewed the manuscript. In addition, several other people at ORA have contributed ideas and corrected my thinking, both recently and over the past 20 years.

Richard Buchroeder, in addition to reviewing the manuscript in detail, has been a personal friend and mentor in my lens design efforts since we were both optics students nearly 30 years ago. When I get stuck, I often call Dick and he helps me out. I am greatly in debt to him for his interest, advice, and example.

Lawrence Scherr and I have worked together as optical engineers, and he also audited my lens design classes. I have had the opportunity to try out on Larry many of my ideas about presenting the concepts of lens design. He has been through this book several times during its gestation. His advice has always been generous and accurate.

Harold Suiter, who wrote *Star Testing Astronomical Telescopes* (also published by Willmann-Bell), contributed the pictorial images of point spread functions. He also reviewed the manuscript from an optical physicist's viewpoint. His suggestions, especially about diffraction, have been most appreciated.

William Swantner (another friend from my student days), Graham Brewis, and John Gregory all carefully reviewed the manuscript and made valuable suggestions. An early partial version of the book was reviewed by the late Robert Chambers, who was very encouraging and helpful. And my lens design students at the University of La Verne successfully struggled through another early partial version. Thus, this book has been student-tested.

I am extremely grateful to all of these reviewers for both their praise and their criticisms. I am very fortunate to have such good friends. Any errors that remain are, of course, strictly my fault. I am also extremely grateful to my publisher, who agreed without hesitation to publish my book. His enthusiastic support has made this book possible.

Finally, I acknowledge the debt I owe to the existence of the electronic digital computer and modern lens design software. If I had to design lenses the way it was done years ago without these tools, I would never have become a lens designer and would now be in a different line of work.

Gregory Hallock Smith
Pasadena, California
January, 1998

Table of Contents

Part A

Optical Concepts and Techniques

Chapter A.1

Introduction

In his novel *A Tale of Two Cities,* Charles Dickens begins by saying "It was the best of times, it was the worst of times." Sometimes it may seem like the worst of times, but if you are a lens designer, today is the best of times.

Since about 1940, there have been major advances in the techniques of designing and fabricating optics. These advances are continuing today at an ever-increasing pace. The most revolutionary advance has been the introduction of the electronic digital computer, which has changed nearly everything. Computers have made the job of the lens designer less difficult and arcane, less time-consuming and costly, and much less tedious. Special software with optimization, evaluation, and tolerancing capabilities has made lens design more effective and more innovative. New lens designs are produced today with higher performance than ever before.

Other areas of optics have been advancing too. New optical glasses with exotic properties allow better aberration control. Ultraviolet and infrared materials allow a wider range of wavelengths to be exploited. Stray light can now be analyzed. Anti-reflection coatings make practical complicated lenses that previously would have been useless. Both improved and totally new types of photodetectors are now available that have performance scarcely imagined in previous decades. For the optician, there is computer-aided fabrication, testing, and alignment. For the optomechanical engineer, there is finite element analysis. Active optics allow continuous alignment correction in large systems. Adaptive optics allow real-time correction of atmospheric turbulence or seeing. And the laser has had a major impact on both optical testing/alignment and on the types of systems designed.

All this yields vast new possibilities for lens design creativity. You never had it so good!

Lens design, also called optical design and optical system design, is the science or art of developing optical systems to image, direct, analyze, or measure light. These systems include: camera lenses, telescopes, microscopes, scanners, photometers, spectrographs, interferometers, and so forth. In most cases, it is desirable that these systems be as free of geometrical optical errors as possible. These geometrical errors are called aberrations. Correcting and controlling aberrations is one of the main tasks of the lens designer.

Note that the term lens is meant to include systems with mirrors as well. Note too that design includes performance evaluation and fabrication/tolerancing issues.

This book is primarily intended for people who wish to do actual lens design, although it may also be useful to those who just wish to understand lenses and optical design techniques. The discussion has in mind three groups of readers. The first group is college students at the upper division and graduate level who are studying optics and physics. The second group is non-optical professionals who find that they must design or understand optics as part of their work. They include: physicists, astronomers, chemists, biologists, medical researchers, geologists, and engineers, and their managers. The third group is advanced hobbyists and optical enthusiasts, such as photographers and amateur telescope makers.

This book is not a first introduction. It is assumed that the reader is already somewhat familiar with the basic concepts of geometrical and physical optics. It is also assumed that the reader wishing to do lens design has access to a computer equipped with an optimizing lens design program. In today's world, this latter assumption is no longer unreasonable.

An attempt has been made to make this book compatible with most modern lens design programs. Fundamentally, these programs are all attempting to do nearly the same thing. Only the specific details are different. For a description of a program's features and a discussion of how to use it, the reader is referred to the program's user's manual.

At the present time (1998), the most prominent commercial lens design computer programs are CODE V, OSLO, and ZEMAX. There are also several other excellent but less widely used programs, including some developed by optical companies for their own use. Because software capability and availability change so rapidly, the reader is referred to published reports and advertisements in optics trade publications for the latest developments. The computer-aided examples in this book have all been done with ZEMAX.

The goal of this book is to assist the reader in acquiring an understanding of how lenses work and how lenses are designed today using computers and numerical methods. With the exception of first-order ray tracing, traditional pre-computer methods are not covered.

An intuitive and practical approach is emphasized. The discussion is more descriptive than analytical, and more geometrical than algebraic. There is some mathematics, but derivations and rigor are peripheral. Today, the lens designer can let the computer handle the mathematics while he concentrates on the optics.

The book is divided into two parts. Part A is a discussion of optical concepts and techniques. Part B is a selection of worked design examples. The second part applies and expands on the concepts and techniques presented in the first part. At the end is a list of references and supporting texts.

Chapter A.2

A Brief History of Lens Design

Since about 1960, the way lenses are designed has changed profoundly as a result of the introduction of electronic digital computers and numerical optimizing methods. Nevertheless, many of the older techniques remain valid. The lens designer still encounters terminology and methods that were developed even in previous centuries. Furthermore, the new methods often have a strong classical heritage. Thus, it is appropriate to examine, at least briefly, a history of how the techniques of lens design have evolved.

A.2.1 Two Approaches to Optical Design

The equations describing the aberrations of a lens are very nonlinear functions of the lens constructional parameters (surface curvatures, thicknesses, glass indices and dispersions, etc.). Boundary conditions and other constraints further complicate the situation. Thus, there are only a few optical systems whose configurations can be derived mathematically in an exact closed-form solution, and these are all very simple. Examples are the classical reflecting telescopes.

This predicament has produced two separate and quite different approaches to the practical task of designing lenses. These are the analytical approach and the numerical approach. Historically the analytical dominated at first, but the numerical now prevails.

Neither approach is sufficient unto itself. A lens designed analytically using aberration theory requires a numerical ray trace to evaluate its actual performance. In addition, an analytically designed lens can often benefit significantly from a final numerical optimization. Conversely, a lens designed numerically cannot be properly understood and evaluated without the insight provided by aberration theory.

A.2.2 Analytical Design Methods

The first lenses made in quantity were spectacle lenses (after about 1285). Later (after 1608), singlet lenses began to be made in quantity for telescopes and microscopes. Throughout the seventeenth and eighteenth centuries, optical instruments were designed primarily by trial and error. As might be expected, optical flaws or aberrations remained. Note that aberrations are fundamental design shortcomings, not fabrication errors. Eventually it became clear that understanding and correcting aberrations required greater physical understanding and a more rigorous analytical approach.

At first, progress was slow and the methods largely empirical. Later, mathematical methods were introduced, and these were much more effective. The most outstanding early work on optical theory was done by Newton in 1666. Among the somewhat later pioneers were Fraunhofer, Wollaston, Coddington, Hamilton, and Gauss.

A major advance was made by Petzval in 1840. Petzval was a mathematician, and he was the first to apply mathematics to the general problem of designing a lens with a sizable speed and field for a camera. The techniques he devised were new and fundamental. His treatment of field curvature based on the Petzval sum is still used today. Just as unprecedented, he was able to completely design his very successful Petzval Portrait lens on paper before it was made.

In 1856, Seidel published the first complete mathematical treatment of geometrical imagery, or what we now call aberration theory. The five primary or third-order monochromatic aberrations are thus known today as the Seidel aberrations. They are:

1. Spherical aberration
2. Coma
3. Astigmatism
4. Field curvature
5. Distortion.

There are also two primary chromatic aberrations. These are wavelength-dependent variations of first-order properties, and they are often included with the Seidel aberrations. They are:

6. Longitudinal chromatic aberration
7. Lateral chromatic aberration.

Petzval, Seidel, and many others in subsequent years have now put aberration theory and analytical lens design on a firm theoretical basis.[1]

Until about 1960, the only way to design lenses was by an analytical approach based on aberration theory. Unfortunately, by its nature, aberration theory gives only a series of progressively better approximations to the real world. Thus, the optical designs derived from aberration theory are themselves approximate and usually must be modified to account for the limitations in the process.

Today, most lenses are designed, not with analytical methods, but with computer-aided numerical methods. Nevertheless, the analytical methods remain extremely valuable for deriving or identifying potentially useful optical configurations that can serve as starting points for further numerical optimization. Even more important, aberration theory can *explain* what is happening. It is only through aberration theory that a lens designer can understand the underlying operation of lenses.

A.2.3 Numerical Evaluation Methods

Part of the job of designing a lens is evaluating its performance as the design evolves. And of course, the performance of the final design must be thoroughly characterized. Aberration theory is useful in giving approximate indications, but a rigorous image evaluation requires a different, exact approach.

Note that unless or until a prototype model is made, the design exists only on paper. Thus, to evaluate the paper design, a mathematical procedure is necessary. The most exact mathematical evaluation procedure is numerical and assumes only trigonometry and Snell's law.

[1] For the reader interested in analytical lens design methods, see A. E. Conrady, *Applied Optics and Optical Design,* Vols. 1 and 2, and Rudolf Kingslake, *Lens Design Fundamentals.*

Snell's law for refraction was discovered experimentally by Snell in 1621 and states that for a ray incident on and refracted by lens surface i (even subscripts for surfaces, odd subscripts for spaces),

$$n_{i-1} \sin \varphi_{i-1} = n_{i+1} \sin \varphi_{i+1} \tag{A.2.1}$$

where n_{i-1} and n_{i+1} are the refractive indicies of the bounding media, and φ_{i-1} and φ_{i+1} are the angles of incidence and refraction in the ray plane. The law of reflection for mirrors was known by the ancient Greeks, and is a special case of Snell's law if n_{i+1} equals $-n_{i-1}$.

Evaluating a lens numerically involves tracing many real (or trigonometric) geometrical rays through the system from the object to the image. For each ray, Snell's law is applied as the ray encounters each lens surface in turn. The calculations are repeated again and again at surface after surface for ray after ray. The locations of the piercing points of these rays on the image surface are then used to calculate various measures of image quality.

At first, logarithms were used to do the calculations. After the introduction of mechanical desk calculators around 1930, direct trig tables were used. With a desk calculator, it took an experienced person about five minutes to trace one meridional ray through one spherical surface (assuming no errors). The time to trace a skew ray, which lay out of the meridional plane, was more than twice as long, and thus tracing skew rays was rarely done in those days. Often a prototype model was indeed made, so great was the computational burden and tedium (you can view a prototype as an analog computer).

With the introduction and development of electronic digital computers in the late 1940s, this manual approach to ray tracing began to change. One of the first jobs given to these new machines was trigonometric ray tracing. But the early computers were hard to get time on, hard to program, expensive, and not all that fast. Even as late as the early 1960s, a company doing lens design would have to make the economic decision whether it was cost-effective to buy time on one of the big computers, or better to hire someone to trace rays by hand with a desk calculator and seven- or eight-place trig tables.

That did not last much longer. The growth of the capabilities of computers has been explosive since about 1960, as has their availability. Soon, computers completely eliminated manual ray tracing. By 1998, a fast Pentium-Pro personal computer with the ZEMAX or similar program could trace about 600,000 skew ray surfaces per second. And by the time you read this, 600,000 will be ancient history.

The author had his first course in lens design in 1963. The professor, Walter Wallin, who also ran his own lens design company, related that he was once asked in all seriousness, "But sir, did you specialize in this from choice?" As with dentistry, absolutely no one today gets nostalgic for the "good old days" of lens design.

A.2.4 Optical Design Using Computer-Aided Numerical Optimization

The advent of electronic digital computers did much more than allow rays to be easily traced. Since the mid-1950s, a few pioneers had been working on new numerical algorithms to do what was then called by the misnomer automatic lens design. These methods, which we now call computer-aided lens design, became widely known in 1963.[2] Commercial computer programs using these methods, such as ACCOS (Automatic Correction of Centered Optical Systems), became available soon after. Thus, starting in the mid-1960s, lens designers could use computers, not just to evaluate a lens, but to *change* lens parameters to improve optical performance.

[2]Donald P. Feder, "Automatic Optical Design," *Applied Optics*, Vol. 2, No. 3, pp. 1209–1226, December 1963.

This was truly a revolution. Lens designers used to struggle with a design until image quality was "good enough." Now, when given a starting lens configuration, the computer can by an iterative process *optimize* the lens. After optimization, image quality is the best that the lens can produce under the constraints of basic configuration, required focal length, f/number, field of view, wavelengths, and so forth. Furthermore, the preferred criteria for good image quality are based on trigonometrically traced real rays. Thus, computer optimization is as exact as ray tracing allows.

The first benefit from computer optimization was that many older designs were recomputed to achieve major performance gains. Complex designs benefited most; the simpler ones were already fairly well optimized. Older designs were also simplified to ease production, use fewer elements, use fewer types of glass, be smaller and lighter, and cost less.

Even more interesting, the new optimizing techniques allow the development of new design forms. These may be extensions of older forms, but they also may be wholly new forms discovered by the computer in its quest for better solutions. Fast wide-angle lenses and sharp wide-range zoom lenses are only two examples of current lens types that were virtually unknown in 1960.

The numerical method of designing lenses does have a limitation, however. Although the software writers are very skillful and their optical programs have amazing capabilities, the computer's basic design approach is still only a sophisticated search algorithm. In particular, the computer has no true optical understanding or intelligence. This intelligence must be supplied by the designer through his selection of the starting optical configuration, through his control of the computer program, and especially through his understanding of the underlying optical theory.

But it is exactly this human intelligence that today's fast computers and interactive software exploit. It is now a relatively easy matter for the lens designer to try out different optical ideas inside the computer. In only a short time, the designer can determine which of his ideas are the better ones that should be pursued.

Thus, the computer does not make the human designer obsolete. Rather, the computer plus optimizing software change the way the work of the lens designer is done and the quality of the final results. The computer removes the drudgery and becomes a powerful new tool to be used by the designer for new lens design creativity.

Chapter A.3

Light and Imaging Systems

Lens design, like any other technical discipline, is concerned with procedures not commonly encountered in everyday life. In this chapter, some of the basic concepts and terminology used in lens design are introduced.

A.3.1 The Nature of Light

The wave nature of light was first proposed in the mid-1600s and was further developed by Huygens (ca. 1678). Conclusive confirmation was obtained by Young in 1801 with his famous double-slit interference experiment. By 1864 Maxwell had developed a general theory of electromagnetism that described light as electromagnetic waves. By the turn of the twentieth century, most physicists considered Maxwell's wave theory to be the last word and beyond question.

There was one problem, however. Maxwell's theory could not explain the Photoelectric Effect; that is, the specifics of the photoemission of electrons. In 1905, Einstein succeeded in doing so by proposing the radical concept that light consisted of particles or photons. This insight revolutionized physics and earned Einstein the Nobel Prize (it was not for relativity).

But how could light be both a wave and a particle? The answer came through the subsequent development of quantum theory. This theory has been widely tested and found to perfectly describe to many significant digits all the observed properties of light (and many other things). But quantum theory is very abstract and hard to use.

Quantum theory is the only theory that explains emission, absorption, and photoelectric effects, but not all optical effects require this level of completeness and sophistication. The older electromagnetic theory can be viewed as a subset or simplification of quantum theory, and electromagnetic theory is quite capable of explaining reflection, refraction, polarization, diffraction, and interference. Similarly, scalar wave theory is a simplification of the full electromagnetic theory, and scalar wave theory is also capable of handling diffraction and interference. These theories and the optical effects they address are the province of physical optics. Any further simplification of the theory of light leads to geometrical optics.

Geometrical optics is a considerable simplification of reality. Here light is reduced to geometrical wavefronts and to geometrical light rays. This is the theory of light used by lens designers to calculate geometrical aberrations. To calculate aberrations, light rays are traced trigonometrically through a lens from the object to the image.

Often, however, pure trigonometric ray tracing does not offer much insight into where in a lens certain aberrations originate, nor does it allow generalizations. To provide these, geometrical aberration theory has been developed. But aberration theory is yet another approximation, and the results of aberration theory in describing images are usually not as exact as those of trigonometric ray tracing.

The final optical simplification yields first-order ray tracing and first-order image properties. Here the trigonometric equations are reduced to their simplest form, and these give perfect, aberration-free images. It is first-order theory that is used to calculate all the basic optical properties of an imaging system, such as: entrance pupil diameter, focal length, focal ratio or f/number, field of view, magnification, pupil locations, image locations, and image orientation.

Note the relationship that the higher the level of the theory, the more correct it is, but also the more abstract and hard to use it is. Conversely, the lower the level of the theory, the more incorrect or incomplete features it contains, but also the more useful and intuitive it is. If at any time the inherent lack of correctness or completeness of whatever theory you are using becomes a problem, you can always switch to one of the higher-level theories as needed. This is what the lens designer does when he adds the effects of diffraction to the usual geometrical aberrations in optical performance evaluations.

A.3.2 Spectral Regions

The concept of light can be generalized to include electromagnetic radiation of all wavelengths. The electromagnetic spectrum extends from radio waves with the longest wavelengths, through the microwave region, the infrared, visible light, the ultraviolet, x-rays, and finally to gamma-rays with the shortest wavelengths. Although there are exceptions, the lens designer is usually concerned only with the ultraviolet, visible, and infrared regions.

In the visible region, the wavelength of light determines its perceived color. High light levels produce the photopic visual response curve to various wavelengths. The range of photopic response extends from about 0.400 μm (micrometers or microns) at the blue end of the spectrum to about 0.700 μm at the red end. The wavelength of peak photopic sensitivity is near 0.555 μm (or 555 nm or 5550Å); this wavelength appears yellow-green. By no coincidence, 0.555 μm is also near the peak of the sun's output. Low light levels produce the scotopic visual response curve. The range of scotopic response extends from about 0.400 μm to about 0.650 μm. Peak scotopic sensitivity is near 0.510 μm.

Incidentally, in the language of spectroscopy, blue may just indicate shorter wavelengths, and red may indicate longer wavelengths, even if you are working in the ultraviolet or infrared. An example is the term red shift.

The energy E of a photon of light is inversely proportional to its associated wavelength. If λ is wavelength, ν its corresponding frequency, and c the speed of light (3×10^8 meters/sec), then:

$$E = h\nu = h\frac{c}{\lambda} \tag{A.3.1}$$

where h is Planck's constant. Many optical properties are wavelength-dependent. For the lens designer, the most important wavelength-dependent property is the index of refraction of transmitting materials such as glass.

A.3.3 Objects, Light Rays, and Wavefronts

The lens designer assumes that objects of interest consist of self-luminous (or radiant) points (even if the light is really being reflected), and that extended objects are accumulations of such points. Furthermore, it is (usually) assumed that light from various different object points does not interact or interfere; that is, the objects are not mutually coherent.

Photons emanate from each of these points, and the trajectories of the photons through an optical system define the light rays. Light rays are thus physically real and not just a mathematical abstraction. If diffraction is neglected, then physical rays become geometrical rays.

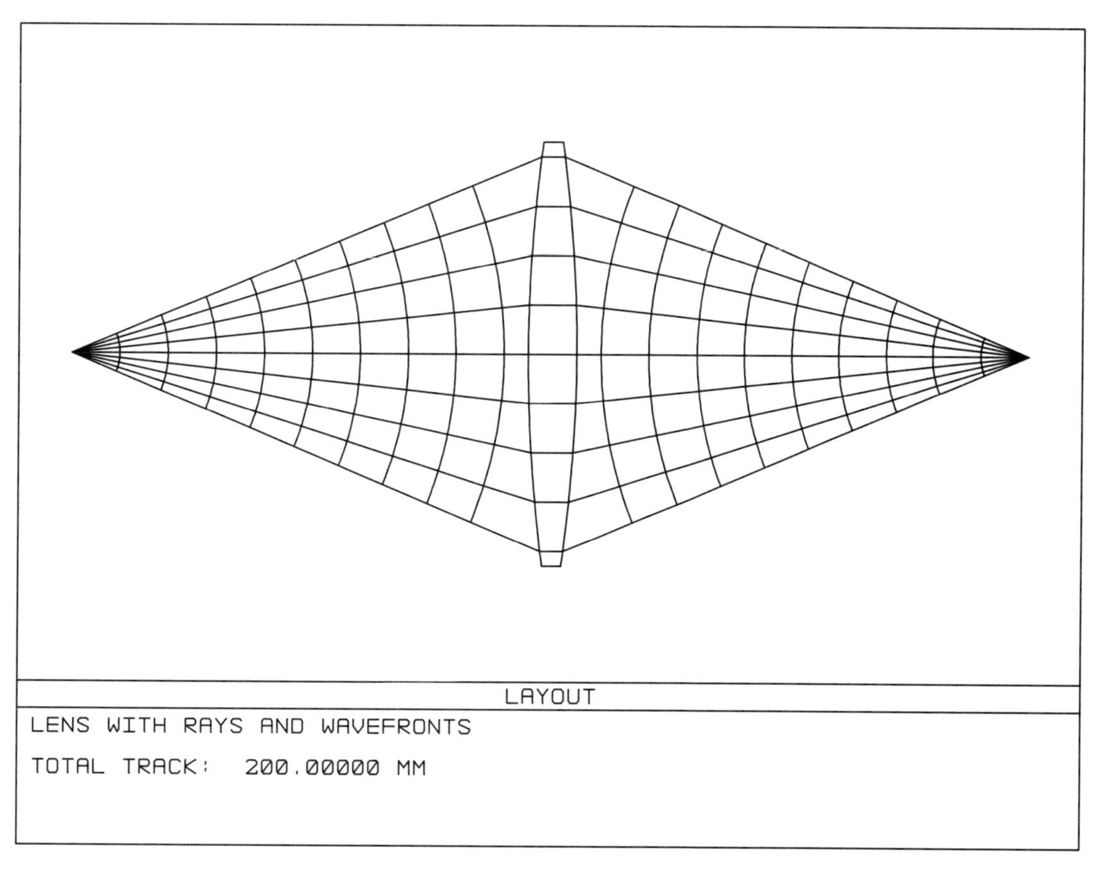

Figure A.3.1.

Wavefronts are surfaces normal to rays. For a given wavefront, the light travel time from the object point to the wavefront along all rays is the same. As with rays, if diffraction is neglected, then physical wavefronts become geometrical wavefronts. Geometrical wavefronts are a good approximation to physical wavefronts except near wavefront boundaries or edges where diffraction is greatest.

For an object point in a homogeneous medium of constant refractive index, light rays are straight lines diverging from the object point, and thus the corresponding wavefronts are spherically shaped with their centers of curvature at the object. If the medium is not homogeneous, then the rays are not straight and the wavefronts are not perfectly spherical, but rays still originate at and diverge from object points.

A.3.4 Images and Imaging Systems

The purpose of an imaging system is to collect a portion of the rays emanating from all object points in the field of view, and then to redirect these rays so they are reunited at their corresponding image points.

Equivalently, an imaging system takes and reshapes parts of the spherical wavefronts emanating from all object points in the field of view. The reshaped wavefronts are again spherical, but the new centers of curvature are at their corresponding image points to which the wavefronts collapse.

Figure A.3.1 illustrates the concepts of objects, images, rays, and wavefronts. The light originates at a point object on the left, passes through a lens, and is focused to an image on the right. The rays and wavefronts also travel from left to right, with the wavefronts normal to the rays. Note that in practice this ray reunification or wavefront reshaping process can never be done perfectly. Errors are caused by geometrical aberrations, diffraction, and other sources.

If the object is at infinity, then the spherical wavefronts entering an optical system are effectively plane or flat. Thus, the rays from a given object point are all parallel to each other; that is, the rays are collimated. Note that although rays from a single object point may be parallel to each other, rays from any two *different* object points are not parallel to each other. If the image is at infinity (for example, a telescope with eyepiece) and aberrations, diffraction, etc. are neglected, then similar statements can be made for rays leaving an optical system.

If light rays physically originate at an object point, the object is called real. If light rays are physically reunited at an image point, the image is called real. The test of a real image is whether it can be viewed on a screen. If the rays do not physically pass through an object or image, but only appear to do so due to auxiliary optics, then the object or image is called virtual.

For example, the light reflected from your face is from a real object, and the image formed of your face by a camera is a real image. But if you look at your reflection in a plane mirror, you see a virtual image of yourself behind the mirror because no light is really present behind the mirror. This virtual image acts as a virtual object for your eyes, which in turn form real images on your retinas.

The space in which the object is located is called object space. The space in which the image is located is called image space. Object space can be extended to include all space if the virtual images seen looking from the object into the front of the lens are included. Similarly, image space can be extended to include all space if the virtual images seen looking from the image into the back of the lens are included. Thus, object space and image space overlap and are three-dimensional images of each other; that is, one space maps into the other.

Most imaging systems consist of volumes of homogeneous media separated by abrupt physical boundaries, such as lenses in air. If an imaging system contains inhomogeneous media, then some of the details are slightly different, but the basic ideas are the same.

A.3.5 The Optical Axis

By far the majority of optical systems are collections of spherical surfaces whose centers of curvature are all located along a common axis. This axis is called the optical axis, and these systems are said to consist of centered spherical surfaces. Plane surfaces are considered spherical with an infinite radius of curvature. The point where the axis intersects a surface is called the surface vertex. The object point on the axis is called the axial or on-axis object; object points not on the axis are called off-axis objects. Similarly, the image on the axis is called the axial or on-axis image, and images not on the axis are called off-axis images.

Systems arrayed along an axis are not restricted to spherical surfaces. All types of rotationally symmetric surfaces are allowed, provided that their individual axes coincide with the optical axis. Familiar examples are surfaces of revolution of ellipses, parabolas, hyperbolas, and polynomials. As in analytical geometry, the symmetry point is again called the vertex.

For an axially centered, rotationally symmetric optical system, a longitudinal cross-section containing the optical axis defines a meridional plane. Because of axial symmetry, all meridional planes are equivalent. Rays lying in a meridional plane are called meridional rays. Rays lying out of a meridional plane are called skew rays.

The concept of an axially centered, rotationally symmetric optical system can be extended to include systems with plane reflections that only redirect the light. These bent systems are still *effectively* axially centered and rotationally symmetric.

An axially centered system need not even be rotationally symmetric. It is possible to employ, for example, surfaces shaped like cylinders or toroids. Note that in these systems, all meridional planes are not equivalent.

But systems with surface tilts and decenters no longer have a single optical axis.

These systems are fundamentally different from axially centered systems and must be handled in special ways. Spectrographs with wavelength-dispersing prisms or gratings are an old and venerable example. Non-axial or so-called off-axis systems are another example. In an off-axis system, the central ray (from the center of the object area through the center of the stop to the center of the image area) is called the gut ray, not the axis. Note that it is possible to have an eccentric or transversely decentered entrance pupil in an otherwise axially centered system; conventional reflecting telescopes with central obscurations are often equipped with auxiliary eccentric stops.

Most of the discussion in this book is concerned with only axially centered, rotationally symmetric systems, and unless otherwise specified, this configuration is assumed.

A.3.6 Stops and Pupils

Somewhere in every optical system, there is (or should be) only one physical aperture that limits the extent of the wavefront or ray bundle transmitted through the system to the on-axis image point. This defining aperture is called the aperture stop (or just the stop of the system). This term is used because the stop determines which rays get through to the image and which are stopped. The surface containing the stop is called the stop surface. In many lenses, a variable iris diaphragm is placed at the stop surface to allow its diameter to be easily changed. For general, non-axial systems, the idea is the same for light reaching the image point in the center of the field of view.

If the diameters of all the system optical elements are large enough, then this same aperture stop also limits the wavefronts or ray bundles for off-axis image points. However, if some of the element diameters are not large enough, then the off-axis beams are further clipped by these undersized apertures. This beam clipping is called mechanical vignetting (pronounced vin-yetting). Mechanical vignetting usually increases with off-axis distance, and thus image irradiance is a maximum in the center of the field and progressively decreases toward the edges. Although the effects look similar, mechanical vignetting is not to be confused with cosine-fourth vignetting, to be discussed in Chapter A.9. The term derives from the old-fashioned photographs called vignettes that deliberately shade off gradually toward the edges.

Note that selecting the size and location of the physical stop are two of the most important decisions that the lens designer makes. This is because the stop does more than influence the radiometric properties of the system. Image aberrations are also strongly influenced by stop size and location.

An image of the object is not the only image an optical system can form. The stop can be imaged too. The image of the stop as seen in object space is called the entrance pupil. The image of the stop as seen in image space is called the exit pupil. In most cases, the entrance and exit pupils are virtual images. In complex optical systems, the object and/or stop may be imaged and reimaged any number of times. These systems can have any combination of real and virtual images of the object and stop. Parts and locations within an optical system that are corresponding objects and images are said to be conjugate to each other.

Instead of specifying the size of the stop opening itself, usually the entrance pupil diameter (EPD) is given. Alternatively, EPD can be specified by giving the focal ratio for a known focal length. Focal ratio is also called f/number and f-stop. For an object at infinity, focal ratio is the ratio of focal length to EPD. In other words, focal ratio is the number by which you divide focal length, f, to get EPD. This division of f is why focal ratios are often written with an f and a slash. Thus, an $f/5.6$ lens opening means that f divided by 5.6 equals EPD.

For a given diaphragm opening, a lens has only one true focal ratio, which is based on an object at infinity. However, the concept can be extended to objects at finite distances. Focal ratio then loses its simple connection to EPD, and instead becomes a measure of the aspect ratio of a focused cone of light. For example, a 4:1 aspect cone can be said to be an $f/4$ cone. When the object is not at infinity, aspect ratio gives an effective or working $f/$number. By further extension, you can even informally speak of the $f/$number of the cone of light collected from an object point at a finite distance.

A quantity related to focal ratio is numerical aperture (NA). NA also describes a cone of light and includes the effect of the index of refraction. Numerical aperture is given by:

$$NA = n \sin u$$

where n is the index of refraction of the medium, and u is the half-angle of the cone of light. NA is most commonly encountered when the cone of light collected by a microscope objective is specified.

If the medium is air (index equals unity), then numerical aperture and effective $f/$number are approximately related by:

$$f/\text{number} \approx \frac{1}{2NA}.$$

If the object is at infinity and if spherical aberration and coma are corrected (as they would be in a good lens), then this relationship is exact for the image.

For an extended object (but not a point object), image brightness (irradiance or flux density) varies directly as the square of the image NA, and inversely as the square of the effective image $f/$number.

A.3.7 Marginal and Chief Rays

In the analysis of optical systems, there are two rays that are especially useful. These are the marginal ray and the chief ray. These are shown schematically in Figure A.3.2. The marginal ray originates at the object point on the axis, and goes to the edge (or margin) of the defining aperture or stop of the system. From there, the marginal ray is focused back to the axis. Thus, for the marginal ray, it is center-edge-center.

The chief ray, also called the principal ray, originates at the object point at the edge of the field of view, and goes to the center of the defining aperture or stop of the system. At the stop the chief ray crosses the axis. From the center of the stop, the chief ray continues to the edge of the image field. Thus, for the chief ray, it is edge-center-edge.

The axial height (the transverse distance away from the axis) of the marginal ray is zero at the object and at all images of the object. At these locations, the axial height of the chief ray determines the size (semi-diameter) of the object and its images. Similarly, the axial height of the chief ray is zero at the stop and at all images of the stop (pupils). At these locations, the axial height of the marginal ray determines the size of the stop opening and its images. Throughout lens design, the marginal and chief rays are used as convenient reference rays.

Actually, a ray from *any* off-axis object point that passes through the center of the stop is a chief ray. Similarly, *any* ray from an on-axis or off-axis object point that passes through the edge of the stop opening is a marginal ray. But when lens designers speak of *the* chief ray, they usually mean the chief ray from the extreme edge of the field. And *the* marginal ray usually means the on-axis marginal ray. How the terms chief ray and marginal ray are being used can be determined from the context.

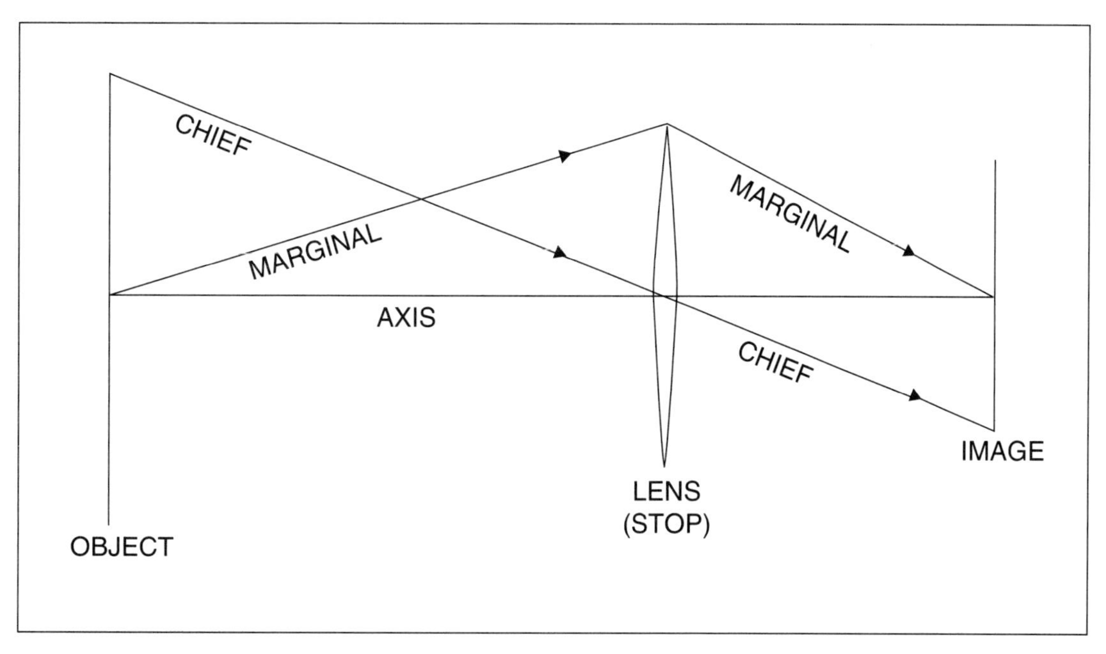

Figure A.3.2. Marginal and Chief Rays. *The marginal ray originates at the center of the object field (on the axis) and travels first to the edge (or margin) of the stop aperture. After passing through the lens, the ray travels to the center of the image field (again on the axis). The chief (or principal) ray originates at the edge of the object field and travels first to the center of the stop aperture. After crossing the axis at the stop, the chief ray continues on to the edge of the image field. Thus, the marginal ray shows the size of the stop aperture, and the chief ray shows the size of the field of view.*

A.3.8 Perfect Imagery

Maxwell, in addition to studying electromagnetism, also studied imagery. He proposed that, independent of the constructional details, any axially centered optical system must satisfy the following three criteria in order to achieve perfect imagery:

1. From any object point, all rays that pass through the system must converge to a single image point, or points are imaged as points;
2. If the object surface is a plane perpendicular to the optical axis, then the images of any and all object points must also lie on a plane perpendicular to the axis, or the system has a flat field;
3. Straight lines on the object plane must be imaged as straight lines on the image plane, or the system has no geometrical distortion.

Except for plane mirrors, no real optical system completely satisfies Maxwell's criteria, although some come close. The concept of perfect imagery can also be extended to general, non-axial systems.

A.3.9 Causes of Image Quality Degradation

There are six main causes of image quality degradation. Four are related to image sharpness.

Of the four, the first is design error residuals, or geometrical aberrations. Except in a very few highly restricted special cases, the lens designer is never able to completely eliminate all aberrations in a real system. The geometrical light rays are (almost) never perfectly reunited. If image quality is limited by aberrations, the system is said to be aberration limited.

The second cause of poor images is errors in optical and mechanical fabrication and alignment. This area is the province of the optics shop, the machine shop, and the glass manufacturer. Predicting, analyzing, restricting and compensating for these real-world errors is done during tolerancing. In addition, many optical systems degrade somewhat with time; for example, reflecting telescopes must occasionally be recollimated. In some large and complex systems, active optics can be used to maintain alignment.

The third cause of imperfect images is diffraction of light by aperture and obscuration edges in the system. Diffraction is a consequence of the wave nature of light and is thus impossible to eliminate. Diffraction spreads out the light in the images of point objects and rules out the possibility of a perfect point image of a point object. In other words, diffraction creates the fundamental limit to image quality beyond which it is impossible to go. In some systems, the effects of diffraction may be small relative to other sources of image degradation. But in highly corrected systems, diffraction is the main source of image degradation, and these systems are said to be diffraction limited.

The fourth cause of unsharp images is atmospheric turbulence, or what astronomers call atmospheric seeing. An extreme case of seeing is the shimmering that is apparent when looking at a distant object in the desert in the summer heat. Seeing is always an important effect for ground-based telescopes that attempt to achieve high angular resolution while viewing through the dynamic and inhomogeneous air. Note that seeing is related to but not the same as the scintillation or twinkling of stars. In very special systems, adaptive optics can partially remove the effects of seeing in real time. Optical systems whose image quality is limited by atmospheric turbulence are said to be seeing limited.

A.3.10 The Point Spread Function

Because it is impossible to image a point object as a mathematically perfect point image, the concept of point spread function, or PSF, has been introduced. The PSF gives the physically correct light distribution in the image of a point object due to the effects of aberrations and diffraction.

Neither fabrication and alignment errors nor atmospheric seeing are included in a standard PSF analysis. These errors are handled separately because they are of a statistical rather than a deterministic nature. Although these sources of error are often important, the PSF due only to aberrations and diffraction is very useful for the lens designer because it provides a measure of the best image quality a nominal design is capable of achieving. Most lens design computer programs can calculate the PSF and display the results either numerically or graphically.

For a two-dimensional extended object, the light distribution in the image is derived mathematically by a two-dimensional convolution of the object (or its perfect image) with the PSF. Image processing and Fourier optics are based on this relationship, and they are career specializations in their own right.

A.3.11 Image Motion

The fifth cause of image quality degradation is image motion. Image motion can be produced by a linear translation or an angular rotation of an optical system. The classic example is camera vibration, although there are many other sources. In addition, sometimes the object itself is in motion. In either case, unless precautions are taken, the result is image smearing. Although image motion is often important in the design and use of optical systems, it is only rarely a factor in lens design.

A.3.12 Stray Light

Stray light, the sixth cause of image quality degradation, is caused by unwanted light scattered and/or reflected by optical and mechanical surfaces within an optical system. Stray light can take two forms. The first form is a roughly uniform veiling glare across the image that reduces contrast. The second form is ghost images. Ghost images can also take two forms. The first is a complete duplicate image, often inverted and in fairly good focus. The second consists of discrete flare spots, often highly aberrated, corresponding to bright light sources within or close to the field of view.

Excessive stray light can disqualify a potential optical design as readily as excessive aberrations. Stray light is a particularly difficult problem for complex systems that either have many lens surfaces or that have large mechanical distances between surfaces (including the image surface). It is important for the optical designer to be aware of stray-light considerations from the beginning of the design process. Waiting until a design is nearly finished and then trying to correct stray-light problems is usually doomed to failure.

Since about 1940, there have been two advances that have revolutionized the control of stray light. The first advance is thin-film anti-reflection lens coatings. Uncoated glass reflects about 4% of the incident light per surface, with higher-index glasses reflecting somewhat more than lower-index glasses. For a multi-element lens with many air-to-glass surfaces, the total of this reflected light can be substantial. The big problem is not the loss of image light; the big problem is the increase in stray light. This reflected light must go somewhere, and much of it invariably ends up on the image surface as stray light. A coated lens surface, however, typically reflects less than 1%, and often much less. Thus, stray light is reduced. Coated lenses make practical many complex systems that otherwise would be useless.

The second advance is the development of computer programs that analyze stray light, especially light reflected by interior mechanical surfaces such as baffles. The results from this capability have pointed the way toward great improvements in high-performance optical systems such as orbiting telescopes. At the present time, the major stray-light computer programs are APART, ASAP, and GUERAP. Note that these programs analyze stray light, but they do not design baffles. Baffles must still be designed by hand. Thin-film coatings and stray-light suppression are each career specializations in their own right.

A.3.13 Focal and Afocal Systems

If either the object or the image is located at a finite distance from the lens, then the lens is called focal. There are three different cases of focal systems. In the first case, the object is at infinity and the image is at a finite distance. An example is an ordinary camera lens focused on a distant object. In the second case, the object is at a finite distance and the image is at infinity. Examples are a hand magnifier and a compound microscope, both focused for the eye. In the third case, both the object and image are at finite distances. An example is a macro camera lens focused up close.

However, if both the object and the image are at infinity, then the system is called afocal. The classic example of an afocal system is a telescope with an eyepiece focused for the eye. A laser beam expander is also afocal.

A.3.14 Fast and Slow Lenses and Detectors

The term fast applied to an optical system originated historically to describe a lens giving a relatively bright (high irradiance or illuminance) image that would expose a photographic plate quickly. A fast lens illuminates an image point with

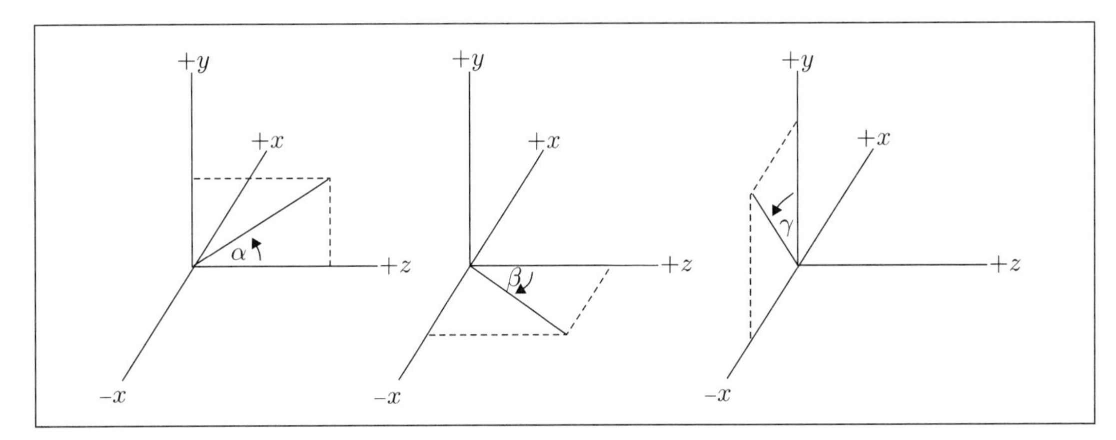

Figure A.3.3. Coordinate and Sign Conventions. *The local coordinates at a given lens surface form a rectangular, Cartesian, or xyz system. Coordinate tilts are by rotation angles α, β, and γ about the x, y, and z axes, respectively. The tilts must be applied sequentially. For CODE V, all tilt angles are shown positive. For ZEMAX, the α and β tilts are shown negative.*

a large solid angle of light; that is, it is a lens with a large image numerical aperture or a small focal ratio or f/number. Similarly, a fast photodetector, such as photographic film, is one requiring little exposure to record the light. As might be expected, slow means the opposite of fast, and the speed of an optical system or detector refers to how fast or slow it is.

A.3.15 Coordinate Systems and Sign Conventions

Coordinate systems and sign conventions are subjects about which there is no universal agreement. Different lens design programs may use different systems, and this lack of standardization may lead to confusion. In particular, CODE V and ZEMAX do not use the same system, although the two systems are similar. Thus, first the CODE V system will be described, followed by the ZEMAX system. For other programs, the user must determine his own specific conventions.

Figure A.3.3 illustrates the CODE V system. The coordinate axes comprise a rectangular, Cartesian, or xyz coordinate system. The origin is at the vertex of whatever surface is under consideration; that is, these are local coordinates, not global coordinates. The z axis is the optical axis (or sometimes just a mechanical reference axis) with positive values of z to the right of the origin. Light originates from somewhere along the negative z axis, and initially moves in the direction of the positive z axis from left to right. A reflection reverses this direction of light travel but not the orientation of the coordinate system. The y axis is orthogonal to the z axis and runs up and down, with positive values of y above the origin. Thus the yz plane is the plane of the paper. The x axis is orthogonal to both the y and z axes and runs in and out of the paper, with positive values of x into the paper. This sign convention creates a right-handed xyz system.

Because local coordinates are used, the origin of the coordinate system moves successively from the vertex of one surface to the vertex of the next as you progress through a lens. As many as seven parameters govern this coordinate transformation. One is the thickness associated with the present surface. Others are the decenters and tilts associated with the next surface. Up to three decenters and three tilts can be specified for each surface.

The order in which these parameters are applied during the transformation is important; changing the order may give different results. First the coordinate system is moved down the optical axis by the amount specified by the thickness parameter. Then the decenters associated with the new surface are applied in the order of x, then y, and then z. Actually, the order so far does not matter. Also, note that the thickness parameter and the z decenter parameter have the same effect. Finally, the tilts associated with the new surface are applied, and here the order does indeed matter.

The three rotation or tilt angles are α, β, and γ. In CODE V, a positive α tilt is a left-handed rotation of coordinates about the positive x-axis. A positive β tilt is a left-handed rotation of coordinates about the positive y-axis. And a positive γ tilt is a right-handed rotation of coordinates about the positive z-axis. Thus, the CODE V coordinate system is not completely right-handed. For this sign convention, refer again to Figure A.3.3, where the rotation angles are all shown positive.

To apply the tilts, first the α tilt is applied to give a new intermediate coordinate system. Then in this intermediate coordinate system, the β tilt is applied to give a second new intermediate coordinate system. Lastly, in the second intermediate coordinate system, the γ tilt is applied to give the final coordinate system.

Of course, some of these seven parameters can be zero. In particular, to avoid confusion, specifying only one nonzero tilt angle at a time is recommended (use multiple single-angle transformations as needed). In addition, note carefully that if you wish to reverse or undo a coordinate transformation, you cannot merely reverse the signs of the parameters; you must also completely reverse the order that the parameters are applied. CODE V has a command that allows this to be done easily.

In ZEMAX, coordinates, decenters, and tilts are handled the same way with three exceptions. First, in ZEMAX there is no z decenter; this function must be handled by the thickness parameter. Second, a positive α tilt is a *right*-handed rotation about the positive x-axis. And third, a positive β tilt is a *right*-handed rotation about the positive y-axis. Thus, the ZEMAX system is completely right-handed, and the α and β angles shown in Figure A.3.3 now become negative.

There are more sign conventions. If the lens surface is curved, a positive curvature has the center of curvature to the right of the surface, and a negative curvature has the center of curvature to the left of the surface. For a lens as a whole, a positive lens causes rays to increase convergence, while a negative lens causes rays to decrease convergence (or increase divergence). Convex lenses are positive and concave lenses are negative, but concave mirrors are positive. Still more sign conventions will be introduced later.

Finally, recall that for rotationally symmetric, axially centered systems, any plane containing the optical axis is a meridional plane, and all meridional planes are equivalent. Thus, with no loss of generality, the yz meridional plane is usually selected, and objects and images are located along the y-axis.

A.3.16 Optical Prescriptions

In order to be able to enter a lens into a computer for optimization or evaluation, the optical prescription is required. An optical prescription contains all the information needed to specify the nominal lens configuration. Tolerances are usually not included and are handled separately.

Because most lenses consist of collections of surfaces through which the light passes sequentially, the main part of a conventional prescription is a table listing the surface parameters. The surfaces are numbered and tabulated in order starting with the object surface (surface 0) and ending with the image surface. A minimum of three surfaces is required: the object surface, the stop surface, and the image surface. The stop cannot be on the object or image surfaces. The details of how an optical prescription is presented depend very much on your particular lens design program.

For each surface, the parameters include: surface number, radius of curvature, thickness to the next surface, glass type in the next medium, any aspheric data, any apertures and obscurations, any tilt and decenter data, and any solves. One surface is designated the stop surface.

Instead of radius of curvature, the reciprocal of radius of curvature, known simply as the curvature, is often substituted. Curvature has the advantages that a greater absolute value denotes a greater effect and that a flat surface does not need to be described with an infinite quantity. However, curvature does require the inconvenient use of tiny numbers (of course, the computer does not mind).

Thickness is vertex (not edge) separation along the optical axis from one surface to the next. Glass type can be any optical glass as well as a crystal, water, air, vacuum, and so forth. A special glass designation is mirror, where the light is reflected back into the previous medium. Solves are special operations that adjust lens parameters to satisfy conditions; solves are discussed in Chapter A.13.

A multiple configuration lens is a special situation. Here there are two or more discrete lens configurations, with one or more lens parameters changing between configurations. One of the most common multiple configuration lens types is a zoom lens. In a zoom lens, the lens elements move longitudinally relative to each other to vary the focal length. But multiple configuration lenses are not limited to zoom lenses. Other examples include lenses with scanning mirrors, multiple wavelength passbands, beam splitters, and so forth. In Chapter B.4, the lens is given multiple entrance pupil diameters. In a lens prescription, the parameters that change between configurations are listed separately along with their values in each configuration. Alternatively, more than one lens table is given.

Another special situation is unusual characteristics, such as surfaces with diffractive properties or gradient index media. Again, these parameters are often listed separately.

Three other parameters or groups of parameters are also part of the optical prescription. The first is the set of wavelengths used. The set may consist of only one wavelength for a monochromatic or an all-reflecting system. For a color-corrected refracting system, two or more wavelengths are necessary. The extreme wavelengths define the wavelength passband, and usually a central wavelength is designated as the reference, principal, or primary wavelength for calculating first-order properties.

The second parameter group specifies the positions of the object points in the field of view. Either angular position as seen from the entrance pupil or linear position on the object surface can be used. Three or four object points are commonly specified. One object point is usually on the optical axis, and one is usually at the edge of the field. Half-field angles or distances relative to the optical axis are given; a full-field of $40°$ corresponds to a half-field of $\pm 20°$. For asymmetric or off-axis systems, many object points arrayed in two dimensions across the object surface may be needed.

The third parameter controls the diameter of the on-axis beam transmitted by the system. Most commonly, entrance pupil diameter (EPD) is specified. Alternatively, the diameter of the aperture on the stop surface can be specified. Another alternative is to specify the f/number or numerical aperture of the image. For finite conjugate lenses, the numerical aperture of the object is often used. Other variations are also possible, but the idea is the same.

Finally, an optical prescription includes a title, any apodization (nonuniform partial transmission across the entrance pupil), any vignetting factors (a way of specifying the sizes of off-axis beams), any special controls for the program (such as ray aiming), and the units used. Millimeters are becoming the world standard for optics; use inches if you must; avoid centimeters.

For convenience, when listing a prescription, your software may append several derived quantities. These may include: focal length, focal ratio, back focal length, overall length, maximum image height, entrance and exit pupil diameters and locations, clear apertures, and so forth. Be sure to differentiate between specified and derived quantities. The derived quantities are technically not part of the prescription.

A.3.17 Aspheric Surfaces

Optical surfaces are most often spherically shaped, but general conics, polynomials, and other shapes can also be used. For the more common rotationally symmetric surfaces (including spheres), the departure of a surface from a plane, also called the sag of a surface parallel to the optical axis, is given by:

$$z = \frac{ch^2}{1 + [1 - (1+k)c^2h^2]^{1/2}} + Ah^4 + Bh^6 + Ch^8 + Dh^{10} \tag{A.3.2}$$

where:

$h^2 = x^2 + y^2$ is the axial height on the surface (transverse distance from the axis),

c is the curvature of the surface at its vertex,

k is the conic constant of the surface, and

A, B, C, D, are the 4th-, 6th-, 8th-, and 10th-order polynomial surface deformation coefficients, respectively.

For the conic constant,

$$k = -e^2$$

where e is the conic eccentricity from analytical geometry, and:

$k = 0$ gives a sphere,

$-1 < k < 0$ gives a prolate ellipsoid (like the pointed end of a football),

$k = -1$ gives a paraboloid,

$k < -1$ gives a hyperboloid, and

$k > 0$ gives an oblate ellipsoid (like a door knob or the top of a spinning planet, such as the Earth or especially Jupiter).

Note that by definition, *any* surface that is not a pure sphere is an asphere. As previously mentioned, a plane surface is considered a sphere with an infinite radius of curvature. Spheres are distinguished from all other surface shapes because spheres (and flats) are the only surfaces that can be made relatively easily in the optical shop. Any asphere is hard to make.

A.3.18 Thin Lenses

For a real singlet lens with a nonzero thickness, the paraxial focal length is

$$\frac{1}{f} = (n-1)\left(\frac{1}{r_1} - \frac{1}{r_2} + \frac{(n-1)t}{nr_1r_2}\right)$$

where f is the focal length, n is the refractive index of the glass, r_1 and r_2 are respectively the front and rear radii of curvature, and t is the axial thickness. If the thickness of the lens is small relative to its other dimensions, then this equation reduces to the lensmaker's equation:

$$\frac{1}{f} = (n-1)\left(\frac{1}{r_1} - \frac{1}{r_2}\right).$$

The concept of a thin lens is encountered again and again in lens design. A thin lens has such small curvatures and center and edge thicknesses that all of its refractive power is effectively concentrated in a plane. Of course, truly thin lenses do not exist in the real world, and the concept is an approximation. Nevertheless, for many applications, the errors are small and can be neglected, at least initially.

Note that at the center of a thin lens, the two lens surfaces are parallel to each other and are effectively in contact. Thus, a ray passing through the center of a thin lens is undeviated.

Many useful optical formulas are thin-lens formulas. Perhaps the most useful is the thin-lens equation:

$$\frac{1}{f} = -\frac{1}{s} + \frac{1}{s'} \tag{A.3.3}$$

where f is the focal length, s is the object distance, and s' is the image distance. The sign conventions are as given above.

Note that the focal length of a thin lens is equal to the distance from the lens to the image when viewing a distant object.

Among lens designers, there is an additional and somewhat nonrigorous way of using the term thin lens. If the thickness of a lens element is not too thick, and if the thickness does not make much difference in the design, then the element can be regarded as conceptually thin. For example, the elements of an achromatic doublet telescope lens are thin by this terminology.

A.3.19 The Pinhole Camera Example

As an example of an imaging system, consider the pinhole camera. The pinhole camera is the simplest optical system capable of forming an image. No lenses or mirrors are required. Only an enclosed box is needed with a small pinhole in one side and, say, a sheet of photographic film on the opposite side. For this discussion, assume the box to be square or rectangular with the pinhole in the center of one of the sides. The optical axis (z-axis) passes through the pinhole and is normal to the plane of the pinhole and to the object and image planes. Such an arrangement is illustrated by the meridional layout in Figure A.3.4.

From a given object point, a small (very small) bundle of rays is admitted by the pinhole. According to geometrical optics, these rays should pass without deviation through the pinhole and subsequently fall on the film as a small spot of light. The size of the spot should be about the same as the size of the pinhole. But this is one of those cases where geometrical optics fails to correctly predict what really happens. Diffraction at the edge of the pinhole actually does deviate the rays slightly, and the light in the spot on the film is thus spread out by some amount. If the diameter of the pinhole is increased to reduce diffraction (and admit more light), then the geometrical image spot is enlarged. But if the pinhole is made smaller, then diffraction is increased and the spot is again enlarged. These two effects combine to limit the image sharpness of a pinhole camera. Nevertheless, pinhole cameras do work, and they also demonstrate many of the basic properties of an imaging system.

Note in Figure A.3.4 that surfaces are numbered with even numbers, 0, 2, 4, 6, etc., and that spaces or thicknesses are numbered with odd numbers, 1, 3, 5, etc. This scheme of numbering surfaces and spaces separately is not the usual way; usually the space following a surface has the same number as the surface. But for the purposes of hand layouts and hand ray traces, numbering separately is convenient and avoids confusion. Just be aware that this separation is not usually done in other situations, such as when optimizing a lens with computer software. You should be able to switch back and forth.

In Figure A.3.4, the (extreme) chief ray is drawn. The height of the object is H_0 (shown positive). The thickness between the object and pinhole is t_1 (shown positive). The angle of the ray in object space is U_1 (shown negative with a downward slope). The thickness between the pinhole and the image is t_3 (shown positive). The angle of the ray in image space is U_3, which equals U_1 because there is nothing in the pinhole to bend the ray (neglecting diffraction). And the height of the image is H_4 (shown negative).

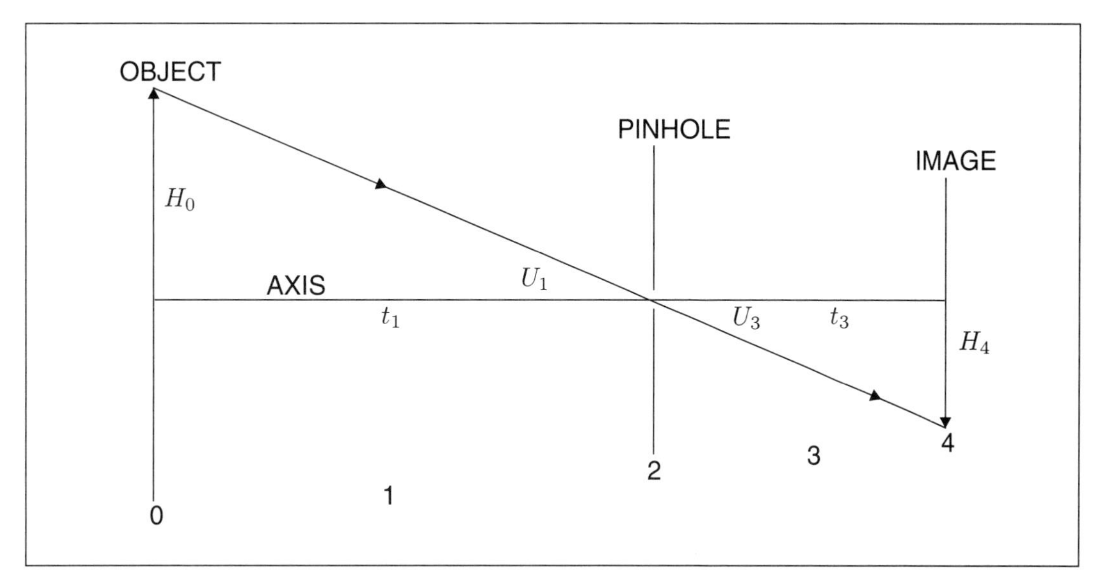

Figure A.3.4. Pinhole Camera. *A meridional layout of a pinhole camera is shown. Surfaces are given even number subscripts, and thicknesses are given odd number subscripts. Many of the fundamental constituents and properties of any imaging system are revealed by studying the pinhole camera.*

Note first that the object and image heights have opposite signs. This sign reversal is also true for rays in the xz plane. Top and bottom, and right and left, have each been interchanged. But this image orientation could have also been obtained by a 180° rotation about the optical axis. Because the image as a whole would be rotated, handedness would not be changed (right hands remain right hands). Such an image orientation is very different from a so-called mirror image where only right and left are interchanged; in a mirror image, right hands become left hands.

Note second the two similar triangles made by the optical axis, the chief ray, the object height, and the image height. In these triangles,

$$\frac{H_0}{t_1} = -\tan U_1 = -\tan U_3 = -\frac{H_4}{t_3}.$$

If m is linear magnification, then

$$m = \frac{H_4}{H_0} = -\frac{t_3}{t_1} < 0.$$

But t_1 and t_3 are constant, and thus magnification is a constant for all values of H_0. A constant magnification regardless of field height means that there is no distortion. Thus, if the object is a square grid of straight lines (like a piece of graph paper) on a flat surface, then the image also looks like a square grid of straight lines on a flat surface. Two of Maxwell's criteria for perfect imagery are satisfied: the object and image are both planes, and straight lines are imaged as straight lines. Only the finite size of the pinhole and diffraction prevent the pinhole camera from being a perfect imaging device.

Pinhole cameras have no focal length in the strict sense of the term. But the height of the chief ray on the image surface of a pinhole camera can be compared to the height of the chief ray on the image surface of a lens. In other words, image scales can be compared.

Thus, for an object at infinity, the effective focal length of a pinhole camera can be said to equal the focal length of a thin lens that gives the same image scale. In addition, because a ray passing through the center of either a thin lens or a pinhole is undeviated, it follows that the pinhole and its equivalent thin lens are both the same distance from their image surfaces. This common image distance is their common focal length.

Now, suppose you have a complex lens whose focal length you wish to determine. For an object at infinity, again the effective focal length is equal to the focal length of a thin lens that gives the same image scale. And the effective focal length of the complex lens is also equal to the length of a pinhole camera that gives the same image scale. This connection of focal length to image scale is fundamental and is even useful in understanding aberrations, as is discussed in Chapter A.6.

Chapter A.4

First-Order, Paraxial, and Gaussian Optics

In this chapter, first-order, paraxial, and Gaussian optics are discussed. Paraxial and Gaussian optics are similar and related to each other, but they are not identical. Both, however, are first-order geometrical optical descriptions where aberrations are absent.

A.4.1 Snell's Law to First Order

Recall that Snell's law for refraction at surface i is given by Equation A.2.1 as:

$$n_{i-1} \sin \varphi_{i-1} = n_{i+1} \sin \varphi_{i+1}$$

where n is the index of refraction and φ is the angle of incidence or refraction. If the sine function is expressed as a power series, then

$$\sin \varphi = \varphi - \frac{\varphi^3}{3!} + \frac{\varphi^5}{5!} - \frac{\varphi^7}{7!} + \cdots \tag{A.4.1}$$

where φ is measured in radians. If φ is small, then the series may be truncated at its first term, and

$$\sin \varphi \approx \varphi.$$

Thus, in the limiting case for small angles, Snell's law reduces to

$$n_{i-1}\varphi_{i-1} = n_{i+1}\varphi_{i+1}. \tag{A.4.2}$$

This equation is Snell's law to first order. Note that the equation is linear in terms of its variables, contains no transcendental trigonometric functions, and can be manipulated by simple algebra. Applying this linear equation to optical systems yields first-order geometrical optics with no aberrations.

It is of interest that if the third-order term is included in the sine expansion, then the third-order Seidel aberrations result. This is why the Seidel aberrations are called third-order. Similarly, adding the fifth-order term to the sine expansion adds the fifth-order aberrations, and so forth for higher orders.

A.4.2 Paraxial Optics

A few optical designs with aspheric surfaces do succeed, on paper, in producing a mathematically perfect geometrical image *on axis*. But so far, no one has succeeded in designing an optical system with perfect geometrical images over an extended field of view. There is one lens configuration, however, that does come close. This is the paraxial region of an axially centered system.

Paraxial means very close to, but not restricted to, the axis. Thus, the paraxial region of an axially centered optical system can be viewed as a tiny channel surrounding the axis. Rays lying everywhere within this channel are called paraxial rays. Paraxial ray heights and ray angles are very small and approach zero in the limit. Thus, the first-order form of Snell's law is valid for tracing paraxial rays. This is a major simplification; the first-order form is much easier to apply than the full trigonometric form or even the third-order form.

But the first-order form of Snell's law gives something more. When used to trace paraxial rays, it can be shown that all rays from an object point on the axis are perfectly reunited at an image point on the axis. There are no aberrations. For a paraxial off-axis object point and paraxial rays, a perfect point image is similarly achieved. Furthermore, this close to the axis, there is no field curvature and distortion. Even off-axis, there are no aberrations. If diffraction is ignored, then geometrically and within the paraxial region, all three of Maxwell's criteria are satisfied.

Paraxial optics is first-order optics and yields *perfect geometrical imagery*. It is important to note that the paraxial description is a limiting case, not an approximation. It really happens.

A.4.3 Usefulness of Paraxial Optics

At first it may seem that paraxial optics has at best a limited usefulness due to its restricted region of applicability near the axis. It is true that only infinitesimal distances are involved transverse to the axis, but these infinitesimals have relative size and sign, and ratios can be taken that are not infinitesimal. Equally important, longitudinal distances are not infinitesimals. If the paraxial equivalent of the marginal and chief rays are traced through an optical system, then the ray heights and axial crossings determine the location, magnification, and orientation of images and pupils. Furthermore, the sizes of images and pupils can be estimated by an extension of the paraxial description to finite transverse distances, as described later in this chapter. By relegating aberrations to a separate issue, the mathematics is simplified, basic system parameters can be readily calculated, and the optics are made more comprehensible and intuitive. Paraxial optics (or, more generally, first-order optics) is surely one of the most valuable tools of the lens designer.

A.4.4 Principal Planes and Cardinal Points

The first person to correctly analyze paraxial optics was Gauss in 1840. At that time, he introduced the concept of two principal planes and six cardinal points.

In Gauss's treatment of paraxial rays from a distant object, all refraction by even a complex lens with many surfaces is effectively replaced by refraction at a single plane normal to the axis. This plane is called the principal plane. The point on the axis where the axis intersects the principal plane is the principal point. The point on the axis where the light focuses is the focal point. The distance from the principal point to the focal point is the focal length. Often, the focal length of a complex lens is termed the effective focal length. The adjective effective refers to an equivalent thin lens and its focal length.

Actually, two principal planes are defined. The first is for collimated light incident from the right, and the second is for collimated light incident from the left. This also yields two principal points and two focal points (and thus two focal lengths). These four points are four of Gauss's six cardinal points, and they are useful in deriving several important paraxial relationships.[1]

The remaining cardinal points are the two nodal points. If a paraxial ray incident on a lens passes through the first nodal point, then the ray emerges from the lens

[1]See Rudolf Kingslake, *Lens Design Fundamentals*, pp. 48–54.

through the second nodal point. Furthermore, the slope of the ray is unchanged. If the lens is in air (index of unity), then the nodal points coincide with the principal points.

A.4.5 Collinear Mapping and Gaussian Optics

In his 1840 treatise, Gauss only considered true paraxial rays. Thus, he did not address several basic optical properties related to finite apertures and fields. These include: stop and pupil diameters, object and image diameters, focal ratio, and so forth. These quantities are subject to aberrations, but for many applications only first-order estimates are required. Thus, the concept of first-order paraxial optics has been extended to cover these nonparaxial system properties.

To accommodate finite apertures and fields, the paraxial region around the optical axis is transversely scaled up by some extremely large number so that all transverse infinitesimal distances become finite in size (there are ways to handle this mathematically). No changes are made to longitudinal positions and distances. Thus, along the axis, the locations of surface vertices and centers of curvature are unchanged. All principal planes, cardinal points, and focal lengths are unchanged. And the locations, magnifications, and orientations of all paraxial objects, images, and pupils are unchanged. However, all curved optical surfaces become so stretched out that they resemble planes tangent to the vertices of the unstretched surfaces. Nevertheless, these "planes" retain their optical power to focus light rays to form images.

When lens designers speak of tracing paraxial rays through an optical system, it is actually this modified first-order configuration, free of infinitesimals, that is used. If angles are handled properly, then the paraxial ray tracing equations still apply. Recall that when tracing paraxial rays, sines of angles are replaced by the angles themselves in radians. Such an angle is measured by the ratio of an infinitesimal transverse distance to a finite longitudinal distance. Now in the stretched version, the corresponding angle is measured by the ratio of a finite transverse distance to a finite longitudinal distance. Thus, sines of angles are replaced by angles, and then both are replaced by *tangents* of angles. In other words, ray angles are replaced by ray slopes. The techniques of tracing first-order rays are covered in the next chapter.

The imaging properties of this stretched paraxial configuration were developed by Maxwell and Abbe and are properly called a collinear mapping. The rays traced are properly called collinear rays. However, over the years, collinear optics has become widely known as Gaussian optics, even though Gauss himself never considered this case. Although it is historically inaccurate to speak of collinear optics as Gaussian, the terms Gaussian optics and Gaussian rays will be adopted here as labels for these concepts to conform with common practice.

Because the Gaussian description is merely a huge transverse scaling of the first-order paraxial description, the Gaussian description also yields perfect geometrical images; that is, all of Maxwell's criteria are satisfied (neglecting diffraction). But the Gaussian description of an optical system is clearly not physically correct. The real lens surfaces are not planes with power, and the real images are not geometrically perfect. Thus, the Gaussian description is an approximation, not a limiting case.

It follows that the first-order Gaussian description is an aberration-free ideal, and that actual departures from this ideal imagery are aberrations. In other words, for a point object and a well-corrected lens, the real rays tend to go to the Gaussian image. Thus, during the optimization of a lens, one goal becomes making the real images of point objects approach the first-order Gaussian images by minimizing higher-order effects.

In summary, both the paraxial and Gaussian descriptions are first-order descriptions with perfect geometrical imagery (no aberrations), but they are not identical.

The paraxial description applies only to infinitesimal transverse distances from the axis and is an accurate limiting case. The Gaussian (collinear) description is an extension of the paraxial description, includes the paraxial description, uses finite transverse distances, and is an approximation. Gaussian ray tracing, not true paraxial ray tracing, is actually done in practice. People sometimes incorrectly use the terms first order, paraxial, and Gaussian interchangeably. If you understand the two concepts, you can recognize whether the limiting case or the approximation is being considered.

A.4.6 Where First-Order Optics Do Not Work

In the general case of systems with tilted or decentered surfaces, there is no single axis of symmetry, and thus no single axis with which to be paraxial. In these systems, first-order, paraxial, and Gaussian analyses must be used very carefully or not at all. Lens design programs are usually so generally written that when calculating first-order properties, any tilts or decenters are simply ignored. For most tilted or decentered systems, the resulting first-order quantities are meaningless and must be disregarded. An exception is an otherwise axial system that is folded or bent only by plane reflections (mirrors or prisms). Here, first-order, paraxial, and Gaussian descriptions do apply because the system is still effectively axially centered.

A second case where first-order theory must be used with care is anamorphic optics (where image scale is different in different directions). Anamorphic optics often contain axially centered cylindrical, toroidal, or other surfaces without rotational symmetry. A third case with potential problems is any system with a curved object or image surface.

A.4.7 Paraxial Properties of Surfaces

All first-order optics, both paraxial and Gaussian, are concerned with only the region of an optical surface very near the vertex. But from Equation A.3.2, any aspheric departure from a sphere approaches zero near the vertex where h approaches zero. Thus, to first order, all rotationally symmetric surfaces look alike and look like spheres. All have equal first-order properties for equal vertex curvatures.

Chapter A.5

First-Order Ray Tracing

One of the most important optical tools of the lens designer is first-order, paraxial, and/or Gaussian ray tracing. Not only does a first-order geometrical analysis reveal the basic nature of an optical system, but first-order techniques are very valuable during the initial design stage. If the system does not work to first order, it does not work at all.

First-order ray tracing is also one of those exceptions, in this computer age, where doing an old-fashioned hand calculation is often the best way to understand the trade-offs necessary when designing a complex system. It is certainly a technique with which all lens designers should be very familiar.

In first-order ray tracing, only axially centered systems are considered and only meridional rays are used (no skew rays). In an axially centered system, all meridional planes are equivalent. Thus, distances transverse to the axis can be labeled h (for height) rather than x or y. The most common rays traced in a first-order analysis are the first-order versions of the marginal and chief rays.

There are several ways to set up a first-order ray trace. The method described here is called the h-theta method, which the author first learned from Walter Wallin in 1963 and still prefers. The h-theta method has an especially convenient way of incorporating the Lagrange invariant.

A.5.1 Recursion Formulas for Surfaces

There are two recursion formulas, used over and over again, to trace a first-order ray through a lens. The first formula is the transfer equation that transfers the ray from one surface to the next. The second formula is the refraction equation that refracts (or reflects) the ray through this next surface. These equations are derived below.

A.5.2 Transfer Equation

In Figure A.5.1, a ray is to be transferred from surface $i-1$ to surface $i+1$ through space i (as before, surfaces have even numbers and spaces have odd numbers). In the Gaussian approximation, all surfaces with optical power look like planes, even though they are really curved. Also, in the Gaussian approximation, $\sin\varphi \approx \varphi \approx \tan\varphi$, or sines and angles (in radians) are replaced by their tangents or slopes. When you see the word angle, think slope.

The angle of the ray relative to the axis is u_i. Relative to the axis, the ray heights on surfaces $i-1$ and $i+1$ are h_{i-1} and h_{i+1}. The axial separation between the surfaces is t_i with positive thickness if surface $i+1$ is to the right of surface $i-1$. The refractive index of the medium between the surfaces is n_i.

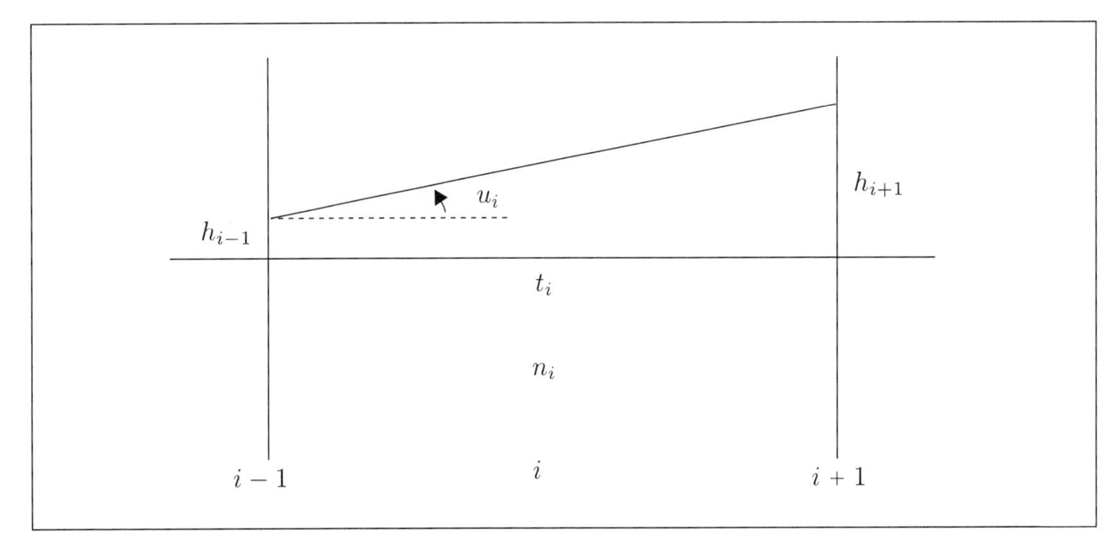

Figure A.5.1. Geometry of the Transfer Equation. *A Gaussian ray is transferred through space i from surface $i-1$ to surface $i+1$.*

Then,

$$u_i = \frac{h_{i+1} - h_{i-1}}{t_i}.$$

Clearing fractions,

$$h_{i+1} = h_{i-1} + t_i u_i.$$

Define the optical ray angle θ_i to be:

$$\theta_i = n_i u_i.$$

Also, define the air-equivalent thickness δ_i to be:

$$\delta_i = \frac{t_i}{n_i}.$$

Then, the transfer equation becomes:

$$h_{i+1} = h_{i-1} + \delta_i \theta_i. \tag{A.5.1}$$

A.5.3 Refraction Equation

The optical power of a surface is a function of two variables, the change in refractive index across the surface, and the vertex radius of curvature of the surface. As mentioned earlier, the radius of curvature is often replaced by the reciprocal of the radius of curvature, called simply the curvature. Curvature has the advantage that larger numbers (absolute values) give more surface optical power. Radius of curvature has the advantage of avoiding the use of a lot of very small numbers. As the radius goes to infinity, the curvature goes to zero.

In Figure A.5.2, a ray is to be refracted at point P on surface i. The surface is spherical with center of curvature at C and radius of curvature r_i. The Gaussian approximation of stretching out curved surfaces into planes looks worse here than in Figure A.5.1, but to make things seem better, imagine everything compressed down close to the axis. The surface normal, as with any spherical surface, is along the line connecting the surface intersection point P with the center of curvature C.

The angle of the surface normal relative to the axis is φ_i. The angles of incidence and refraction of the ray relative to the surface normal are I_{i-1} and I_{i+1}. The media refractive indices and the ray angles and height relative to the axis are all handled as with the transfer equation. Then,

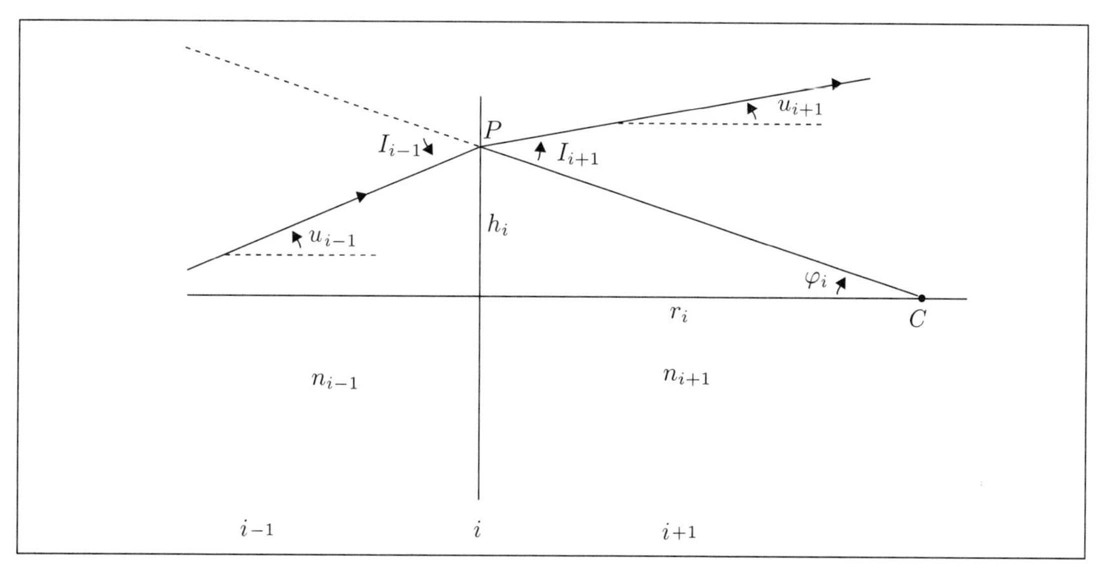

Figure A.5.2. Geometry of the Refraction Equation. *A Gaussian ray is refracted at surface i from medium $i-1$ to medium $i+1$.*

$$I_{i-1} = \varphi_i + u_{i-1} = \frac{h_i}{r_i} + u_{i-1}$$

and

$$I_{i+1} = \varphi_i + u_{i+1} = \frac{h_i}{r_i} + u_{i+1}.$$

Recalling Snell's law for small angles:

$$n_{i+1} I_{i+1} = n_{i-1} I_{i-1}.$$

Replacing the angles of incidence and refraction with their equivalent expressions,

$$n_{i+1}\left(\frac{h_i}{r_i} + u_{i+1}\right) = n_{i-1}\left(\frac{h_i}{r_i} + u_{i-1}\right).$$

Rearranging,

$$\left(\frac{n_{i+1}}{r_i} - \frac{n_{i-1}}{r_i}\right) h_i + n_{i+1} u_{i+1} = n_{i-1} u_{i-1}.$$

Define surface refractive power d_i to be:

$$d_i = \frac{n_{i+1} - n_{i-1}}{r_i}.$$

Recall the definition of optical ray angle given above:

$$\theta_i = n_i u_i.$$

The refraction equation then becomes:

$$\theta_{i+1} = \theta_{i-1} - d_i h_i. \tag{A.5.2}$$

If the surface is a mirror surface instead of a lens surface, then the direction of light travel reverses after reflection. To apply the recursion formulas to a mirror, reverse the signs of all the refractive indices following the reflection. For more than one mirror, the signs change again after each reflection.

Repeating, the two recursion formulas are:

$$h_{i+1} = h_{i-1} + \delta_i \theta_i, \tag{A.5.1}$$

$$\theta_{i+1} = \theta_{i-1} - d_i h_i. \tag{A.5.2}$$

A.5.4 Recursion Formulas for Thin Lenses

If a system is complex and has, for example, relayed images of the object and stop, then it may be helpful or necessary to do a preliminary first-order analysis by hand to just get light through to the final image surface. In this rough analysis, the individual lens elements are usually modeled as thin lenses. This procedure works quite well, even though many of the lenses may in reality be decidedly not thin. Each thin lens is modeled as a transmitting plane with optical power and is assigned only one surface number in the first-order ray trace.

If the definition of optical power is slightly modified, then the two recursion formulas given above can be used with thin lenses. Instead of surface refracting power, element power is used; that is,

$$d_i = \frac{1}{f_i}$$

where f_i is thin-lens focal length. The reason the symbol d is used for optical power is that the reciprocal of lens focal length in meters is lens power measured in diopters.

When working with systems having mirrors, it is often convenient to conceptually unfold the system during a first-order analysis. Mirrors with power are replaced by equal-focal-length thin lenses, and plane mirrors are replaced by transmitting plane dummy surfaces. Dummy surfaces are surfaces with no optical power and media of the same refractive index on both sides.

A.5.5 Reduced Thickness

Many optical systems contain thick filters or thick nondispersing prisms. These filters and prisms are modeled in a first-order ray trace by thick plane-parallel glass plates placed normal to the optical axis. However, the calculations can be simplified if each of these glass plates is conceptually replaced (only in this analysis) by its air-equivalent thickness. In this application, the air-equivalent thickness is called the reduced thickness and is given by the same equation used previously:

$$\delta_i = \frac{t_i}{n_i}.$$

The ray heights on the surfaces bounding the reduced thickness of air are the same as on the surfaces bounding the actual thickness of glass, and the incident and exiting ray angles are also unchanged.

If the reduced thicknesses are used in a thin-lens first-order analysis, then the only spaces are airspaces whose refractive indices are unity. In this case, the optical angles and the actual angles become equal; that is,

$$\theta_i = n_i u_i = u_i.$$

Note that it is customary in lens design to assign to air (not to vacuum) a refractive index of exactly unity. More will be said about this convention later.

A.5.6 The Lagrange Invariant

There is a very powerful relationship called the Lagrange invariant (also called the optical invariant and the Smith-Helmholtz invariant) that relates the heights and angles of the marginal and chief Gaussian rays anywhere throughout an optical system. For any space i and relative to the axis, the angle of the marginal ray is θ_i, and the heights on the previous and following surfaces are h_{i-1} and h_{i+1}. For the chief ray, the corresponding quantities are Θ_i, H_{i-1}, and H_{i+1}. For these two rays, the two transfer equations are:

$$h_{i+1} = h_{i-1} + \delta_i \theta_i$$

and

$$H_{i+1} = H_{i-1} + \delta_i \Theta_i.$$

Multiplying the first equation by Θ_i and the second by θ_i, they become:

$$h_{i+1}\Theta_i = h_{i-1}\Theta_i + \delta_i \theta_i \Theta_i$$

and

$$H_{i+1}\theta_i = H_{i-1}\theta_i + \delta_i \theta_i \Theta_i.$$

The Lagrange invariant is obtained by subtracting the first equation from the second:

$$L = H_{i+1}\theta_i - h_{i+1}\Theta_i = H_{i-1}\theta_i - h_{i-1}\Theta_i. \tag{A.5.3.1}$$

This relationship can be rewritten in determinant form as:

$$L = -\begin{vmatrix} h_{i-1} & H_{i-1} \\ \theta_i & \Theta_i \end{vmatrix} = \begin{vmatrix} \theta_i & \Theta_i \\ h_{i+1} & H_{i+1} \end{vmatrix} \tag{A.5.3.2}$$

The reason for using determinant form and the positions of the several terms will become clear when an example is worked below.

At the object surface, h_0 is zero, and

$$L = H_0 \theta_1.$$

Thus, the value of the Lagrange invariant is equal to the height of the object times the collected angle of the light emitted from the axial object point (neglecting the question of signs and the question of full versus half heights and angles). By similar reasoning, at the image surface, the value of the Lagrange invariant is equal to the height of the image times the angle of the light incident on the axial image point.

In addition, at the system stop surface, surface i, H_i is zero, and

$$L = h_i \Theta_{i-1} = h_i \Theta_{i+1}.$$

Thus, the value of the Lagrange invariant is also equal to the height of the stop times the convergence or divergence angle of light to or from the axial point in the stop.

The concept can be extended to more complex systems. The Lagrange invariant holds at the object and all images of the object, and at the stop and all images of the stop (pupils). And, most important, the value of the Lagrange invariant as given by the determinant remains constant even at surfaces that are not located at images and pupils. Thus, the Lagrange invariant connects together the basic first-order properties throughout an optical system.

A.5.7 Physical Significance of the Lagrange Invariant

The Lagrange invariant can be generalized to three dimensions by using areas and solid angles. If there is no mechanical vignetting (beam clipping), then the following four products are all equal:

1. The area of the object times the solid angle subtended by the entrance pupil (the stop as seen from the object);
2. The area of the image times the solid angle subtended by the exit pupil (the stop as seen from the image);

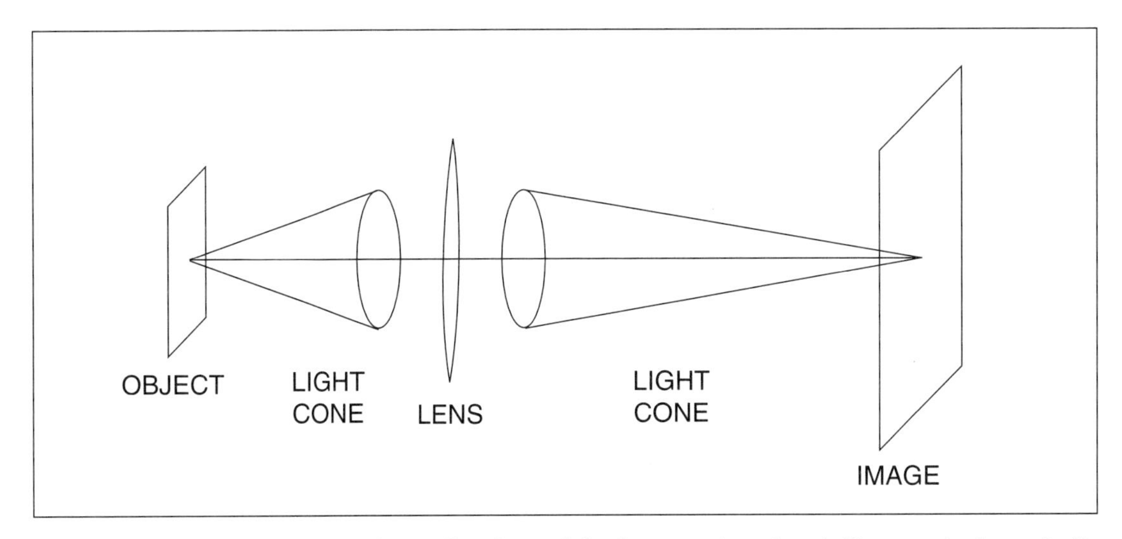

Figure A.5.3. Lagrange Invariant. *One form of the Lagrange invariant is illustrated schematically. The object area times the solid angle collected from an object point equals the image area times the solid angle incident on an image point.*

3. The area of the stop opening times the solid angle subtended by the object as seen from the stop;

4. The area of the stop opening times the solid angle subtended by the image as seen from the stop.

If the system has reimaging, the Lagrange invariant further maintains a constant product of area and solid angle at all images of the object and at all images of the stop. Figure A.5.3 illustrates the three-dimensional Lagrange invariant for just the object and final image.

In radiometry, an image area times the solid angle illuminating any point in the image gives a quantity proportional to the flux throughput, or what the French call étendue. Thus, the Lagrange invariant is a statement of the constancy of flux throughput from one end of the system to the other. Neglecting losses, there is the same amount of light (photons/second) passing through each of the various surfaces in the system. In other words, the Lagrange invariant is a statement of conservation of energy. This connection to conservation of energy is why the Lagrange invariant is so fundamental and so useful.

A.5.8 First-Order Ray Trace Used to Design a Projector

As an example to illustrate first-order ray tracing methods, a thin-lens design of a conventional slide projector is given below. Only first-order ray tracing of the marginal and chief rays, the Lagrange invariant, and the thin-lens equation (Equation A.3.3) are used.

The first-order layout of the projector is shown in Figure A.5.4. Surface 0 represents a coiled-filament lamp whose filament size is to be determined. Surface 2 represents a condenser lens whose diameter and focal length are to be determined. Do not be concerned now that it is highly unrealistic to model the thick condenser as a thin lens. Surface 4 is a dummy surface representing the slide or transparency to be projected. For a standard 35 mm slide, the diagonal length is 40 mm giving a semi-height of 20 mm. Note that the field covered by an axially centered system is circular, and field diameter must be at least as large as the *diagonal* of the rectangular slide. Surface 6 represents the projector lens; a focal length of 125 mm and a semi-diameter of 15 mm are arbitrarily chosen to give a focal ratio of about $f/4$. Surface 8 represents the screen.

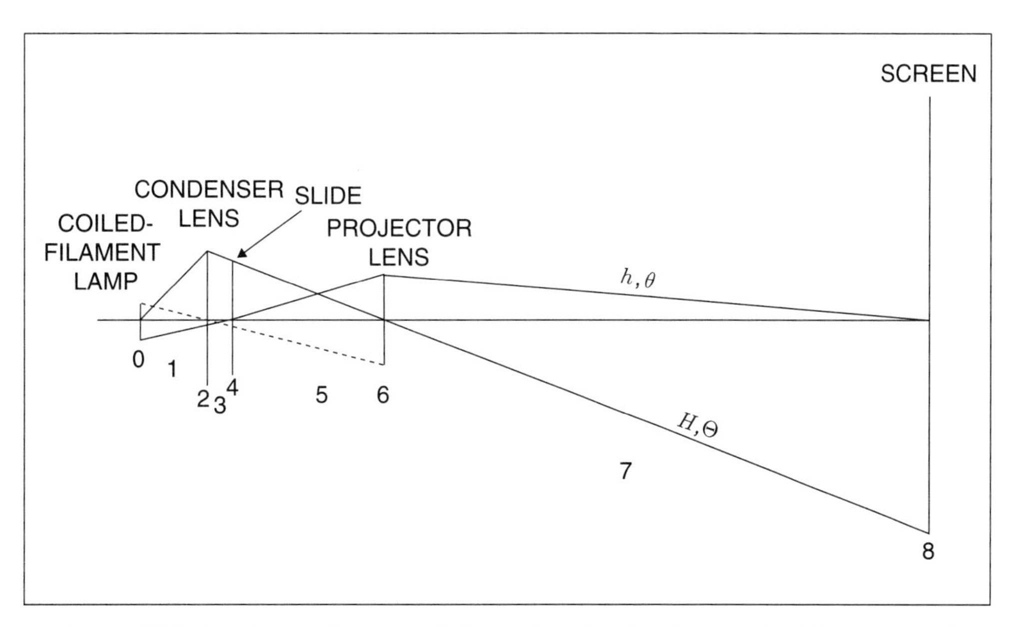

Figure A.5.4. Slide Projector Layout. *A first-order, thin-lens layout of a slide projector illustrates the relationship of the several components and shows the marginal and chief Gaussian rays.*

If the slide diagonal is 40 mm, then the diameter of the condenser must be a bit larger, say about 50 mm. If the solid angle of light collected from the axial point in the lamp is to be slightly less than an $f/1$ cone (a realistic value), then adopt 70 mm for space 1. To allow some room for the actual condenser thickness, adopt 20 mm for space 3. Space 5 is calculated based on the projector lens focal length and screen distance. Space 7, the screen distance, is chosen to be exactly 5 meters or 5000 mm from the lens.

The function of the projector lens is to sharply image the slide on the screen. The marginal ray for the slide-lens-screen system is labeled h, θ. By the definition of a marginal ray, h_4 and h_8 are both zero. The marginal ray height on the projector lens h_6 has been specified as 15 mm. The chief ray has been labeled H, Θ, and from the definition of a chief ray, H_6 is zero. The chief ray height on the slide H_4 has been specified as 20 mm.

If the chief ray is extended backwards from surface 6 to surface 4 and then further, it must eventually run into the lamp. If not, the ray will not get any light. The function of the condenser is to bend the chief ray down to the center of the lamp so that H_0 is zero. But if H_0 and H_6 are both zero, this satisfies the condition that surface 0 is imaged on surface 6. Thus, the condenser forms an image of the lamp filament on the projector lens. This really happens, and if you (carefully) look sideways into a real projector, you can see the coils of the lamp in focus on the dust on the interior surfaces of the projector lens. This imaging of one optical element onto another to get light through an optical system is the basic function of a field lens (a condenser lens is a field lens). Field lenses are crucial for the operation of many instruments from projectors to periscopes to photometers.

The designation of the two rays as marginal and chief seems to reverse itself in the region between the lamp and the slide. This effect is very common in systems where objects and pupils are interchangeable depending on how you look at things. Both descriptions are equally correct. Study the schematic layout (which is not to scale) in Figure A.5.4 until you see these relations.

The first-order ray trace of the marginal and chief rays is set up as shown in Table A.5.1. Enter initially only the specified and adopted values.

Table A.5.1
First-Order Ray Trace, Initial Entries

Number	f, t	d, δ	h, θ	H, Θ
0	—	—		0
1	70	70		
2				
3	20	20		
4	∞	0	0	20
5				
6	125	0.008000	15	0
7	5000	5000		
8	—	—	0	

Now that the basic system values have been entered, the next step is to fill in the blanks with calculated ray heights, ray angles, focal lengths, and spaces. There is more than one way to do this. The ray tracing equations could be used exclusively, but the author prefers to use whatever method or equation is easiest or most intuitive.

The angle of the marginal ray in space 7 is

$$\theta_7 = u_7 = -\frac{h_6}{t_7} = -\frac{15}{5000} = -0.003000.$$

The lens equation yields space 5,

$$-\frac{1}{s} = \frac{1}{f} - \frac{1}{s'} = \frac{1}{f_6} - \frac{1}{t_7} = \frac{1}{125} - \frac{1}{5000} = -\frac{1}{-128.2051282}.$$

Or, space 5 between surface 4 and surface 6 is 128.2051282 mm. Surface 4 is a dummy surface; no refraction occurs there. Thus, the angle of the marginal ray in both space 3 and space 5 is the same and is

$$\theta_3 = \theta_5 = u_5 = \frac{h_6}{t_5} = \frac{15}{128.2051282} = 0.117000.$$

Similarly, the angle of the chief ray in space 3 and space 5 is

$$\Theta_3 = \Theta_5 = U_5 = -\frac{H_4}{t_5} = -\frac{20}{128.2051282} = -0.156000.$$

Because a ray passing through the center of a thin lens is undeviated, the angle of the chief ray in space 7 also equals its angle in space 5.

Finally, there is the condenser. The focal length is

$$\frac{1}{f} = -\frac{1}{-t_1} + \frac{1}{t_3 + t_5} = -\frac{1}{-70} + \frac{1}{20 + 128.2051282} = \frac{1}{47.54406578}.$$

Note the second chief ray drawn dashed in Figure A.5.4. This ray passes through the center of the condenser lens and is also undeviated. The filament size can be derived from the equal ray angles

$$\frac{h_0}{t_1} = \frac{-h_6}{t_3 + t_5}$$

or

$$h_0 = \frac{-t_1 h_6}{t_3 + t_5} = \frac{-(70)(15)}{(20 + 128.2051282)} = -7.084775087.$$

It will be interesting to see later whether the ray trace gives the same answer.

Now the values derived above are added to the ray trace table.

Table A.5.2
First-Order Ray Trace, Added Entries

Number	f, t	d, δ	h, θ	H, Θ
0	—	—	−7.084775087	0
1	70	70		
2	47.54406578	0.02103311914		
3	20	20	0.117000	−0.156000
4	∞	0	0	20
5	128.2051282	128.2051282	0.117000	−0.156000
6	125	0.008000	15	0
7	5000	5000	−0.003000	−0.156000
8	—	—	0	

Note how the structure of the ray trace columns on the right side of the table agrees with the determinant form of the Lagrange invariant. If you look only at these two columns and make a determinant with any two adjacent rows, the value of the determinant is the Lagrange invariant. If you shift up or down one row and make a new determinant, you get the same answer, except the sign is reversed. You can repeat this all up and down the table.

This procedure is useful in several ways. First, once you know the Lagrange invariant for your system, you can fill in some of the missing entries in your table. Second, you do not need to know all the details of a system to fill in widely spaced entries, even if intervening parts of the system are still unknown. Third, it is an excellent way to check for errors.

The value of the Lagrange invariant can now be determined. Look at rows 4 and 5. The marginal ray height h_4 is zero, and for the moment it does not matter what Θ_5 is. Thus,

$$L = h_4\Theta_5 - H_4\theta_5 = 0 - (20)(0.1170000) = -2.340000.$$

Note the negative value. Now look at rows 5 and 6. Here, the chief ray height H_6 is zero, and,

$$L = \theta_5 H_6 - \Theta_5 h_6 = 0 - (-0.156000)(15) = +2.340000.$$

Note the positive sign. Actually, the sign of the Lagrange invariant is not important. The things to note are the absolute value and the fact that the signs of the determinants flip back and forth from one pair of adjacent rows to the next.

As an example of using the Lagrange invariant,

$$L = 2.340000 = \theta_7 H_8 - \Theta_7 h_8 = (-0.003000)(H_8) - 0$$

and

$$H_8 = \frac{2.340000}{-0.003000} = -780.000000.$$

Note in Table A.5.2 that all the focal lengths and spaces are now known. Thus, it is now easy to do a first-order ray trace to fill in the remaining blanks. The author has programmed the two recursion formulas into a programmable pocket calculator with the option to trace either forward or backward. The reader should do the same if he really wants to understand this analytical method and plans to do this type of work later. The ray traces yield the final results given below.

Table A.5.3
First-Order Ray Trace, Final Results

Number	f, t	d, δ	h, θ	H, Θ
0	—	—	−7.084775087	0
1	70	70	0.067782501	0.330285715
2	47.54406578	0.02103311914	−2.340000	23.120000
3	20	20	0.117000	−0.156000
4	∞	0	0	20
5	128.2051282	128.2051282	0.117000	−0.156000
6	125	0.008000	15	0
7	5000	5000	−0.003000	−0.156000
8	—	—	0	−780.000000

The values for h_0 and H_8 do agree to within a couple of digits in the last place. Incidentally, the reader may have thought it inappropriate to carry ten significant digits in a calculation that is clearly such an approximation to the real world. Physically, three significant digits is more than enough. But in a calculation like this where propagation of round-off error builds up, carrying all ten of the digits available in most pocket calculators is good practice. Not only is propagation of error reduced, but mistakes are more readily spotted. The final answers can be rounded off later.

Looking at the results, the size of the screen and lamp can be determined by inspection because either the height of the marginal ray or the height of the chief ray is zero. Doubling ray height to get full diameter, the image on the screen is 1560 mm across diagonally, which is a good-sized image for home viewing. The lamp filament is about 14 mm across diagonally, which means a practical 10x10 mm square array of coils.

In the more general case where neither ray height is zero, the element semi-diameter required to let all the light pass without vignetting (beam clipping) is equal to the sum of the absolute values of the marginal and chief ray heights. For the condenser lens, the marginal ray height is 2.34 mm, the chief ray height is 23.12 mm, and the sum is 25.46 mm. Thus the condenser must have a clear diameter of twice this value, or 50.92 mm (the initial guess of 50 mm was about right).

The f/number (aspect ratio) of the collected solid-angle light cone from the filament (which affects system efficiency) is $f/1.51$, which is not too stressing for one of the available molded glass aspheric condenser lenses. Note that the f/number of the collected light cone from the filament is not the f/number of the lens. The lens f/number is calculated with the object at infinity, and the present arrangement uses finite conjugates. The actual collected solid angle is what counts here.

As long as system efficiency has been mentioned, there is an additional optical element that is usually added to projectors that nearly doubles light output. If a spherical mirror is placed behind the lamp with the filament at the center of curvature, an image of the filament is formed back on itself. Light that would otherwise be lost to the back of the projector is now used to effectively double the number of coils in the filament. Sometimes the real filament coils partially block the reflected image, but the filament can be designed with carefully chosen gaps that are neatly filled with imaged coils.

This completes the first-order design of the projector. Enough information is here to allow selection of the condenser lens, lamp, and spherical mirror. A setup on an optical bench is probably all that is necessary to adjust the details to make a workable system. Alternatively, you could use this first-order design as a starting place for a realistic trigonometric computer optimization, with thick and/or multi-element lenses substituted for the thin lenses.

Chapter A.6

Basic Optical Analysis

Before discussing aberrations, more optical concepts must be presented. Included is an introduction to the three most important analytical tools for identifying and evaluating optical problems and image defects. These tools are: the layout, the spot diagram, and the transverse ray fan plot.

A.6.1 Gaussian and True Entrance Pupils

As related earlier, the entrance pupil is the image of the stop opening as seen by the object. Many basic lens parameters are defined with reference to the entrance pupil. Like any other image, the entrance pupil is subject to aberrations. On-axis, these aberrations usually have little effect. The difference between the true entrance pupil (as defined by trigonometric rays) and the aberration-free Gaussian entrance pupil is usually quite small.

Off-axis, however, the true entrance pupil may have significant aberrations, especially in wide-angle lenses. These aberrations cause two major effects. The first is an asymmetric transverse expansion or contraction in pupil size as a function of off-axis field angle or distance. The second is a longitudinal shift in pupil location, again as a function of off-axis field angle or distance.

Different lens design programs handle the entrance pupil in different ways. Consult your program's user's manual to determine how your program works, and whether there are any special controls that you as the user must exercise. Much more will be said about the entrance pupil, both later in this chapter and in subsequent chapters.

A.6.2 Effective Refracting Surface

Assume a distant on-axis object and collimated paraxial rays incident from the left. If these rays are traced through a lens, then from paraxial theory, all refraction can be conceptually replaced by refraction at only the second principal plane.

If these paraxial rays are now replaced by real trigonometric rays arrayed over the finite entrance pupil, and if the entering and exiting rays are extended forward or backward until they meet, then all refraction again appears to effectively take place at a single surface. This trigonometric analog of the second principal plane still intersects the axis at the second principal point, but now the surface is usually curved, not flat. This curved surface, illustrated in Figure A.6.1, is called either the effective refracting surface or the equivalent refracting locus.

For off-axis objects and objects at finite distances, similar constructions can be made.

There is a special case where the effective refracting surface is known without tracing rays. In a lens, if both spherical aberration and coma are corrected to zero, then the lens is said to be aplanatic. For an on-axis object point at infinity and

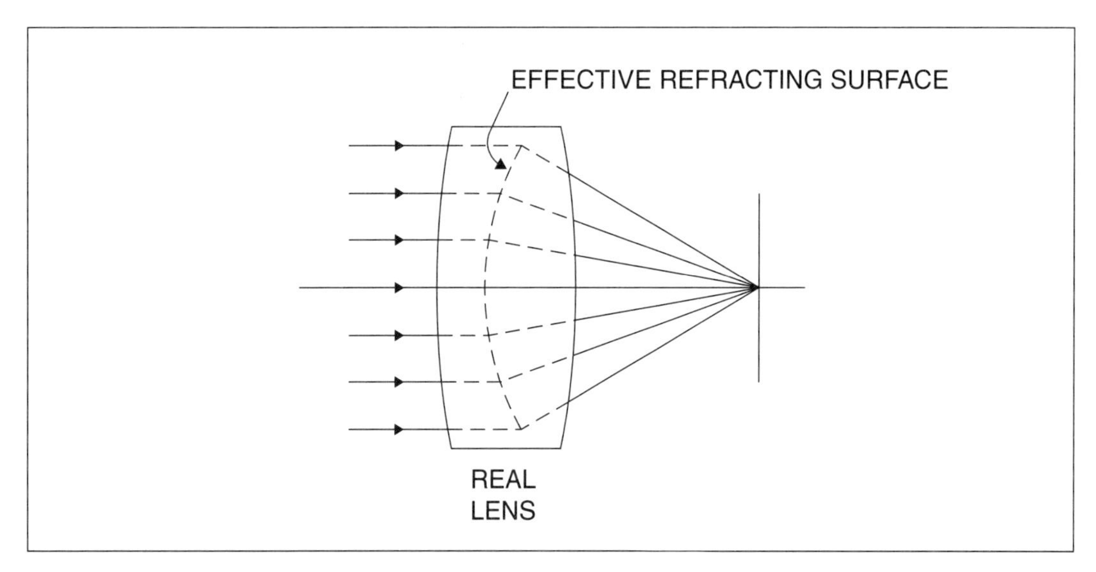

Figure A.6.1. Effective Refracting Surface. *The trigonometric analog of the second principal plane is the generally curved effective refracting surface, where all refraction appears to take place.*

a lens that is aplanatic for this object distance, the effective refracting surface is spherical with its center of curvature at the image.[1]

This shape of the effective refracting surface has an interesting consequence. For an aplanat to have the maximum possible theoretical lens speed, the spherical effective refracting surface is extended into a hemisphere surrounding the image. Thus, the image point is illuminated by a full 2π steradian solid-angle cone of rays. Entrance pupil diameter is now equal to the diameter of the sphere, and focal length is equal to its radius. Thus, the fastest you can conceive of an aplanat is $f/0.5$, which gives a numerical aperture (NA) of 1.0 in air.

In practice, the optical speed limit is closer to $f/0.65$. Note that oil-immersion microscope objectives can have a NA greater than 1.0. This super speed is not the result of gathering a solid angle greater than 2π steradians, but is due to the index of oil being greater than unity.

A.6.3 Zones

In axially centered, rotationally symmetric optical systems, zones on surfaces are annular regions of constant distance from the optical axis. In systems without a single axis, the concept of surface zones can be extended to include regions concentric about an arbitrarily chosen central reference point. The term zone can be applied to any surface in a lens, including: stops and pupils, objects and images, the effective refracting surface, and physical lens surfaces.

A.6.4 Bending a Lens

The shape of a singlet lens is determined by the signs and relative magnitudes of the curvatures on its two surfaces. For a positive singlet, the terms double-convex, plano-convex, and meniscus are used. For a negative singlet, the equivalent terms are double-concave, plano-concave, and meniscus.

Often, it is desired to change the shape of a singlet lens without changing its power or focal length. This process is called bending the lens. The shape or bending of a singlet lens has a major effect on its aberrations. When singlets are grouped

[1]For a proof, see the discussion of Abbe's sine condition in Rudolf Kingslake, *Lens Design Fundamentals*, pp. 157–166.

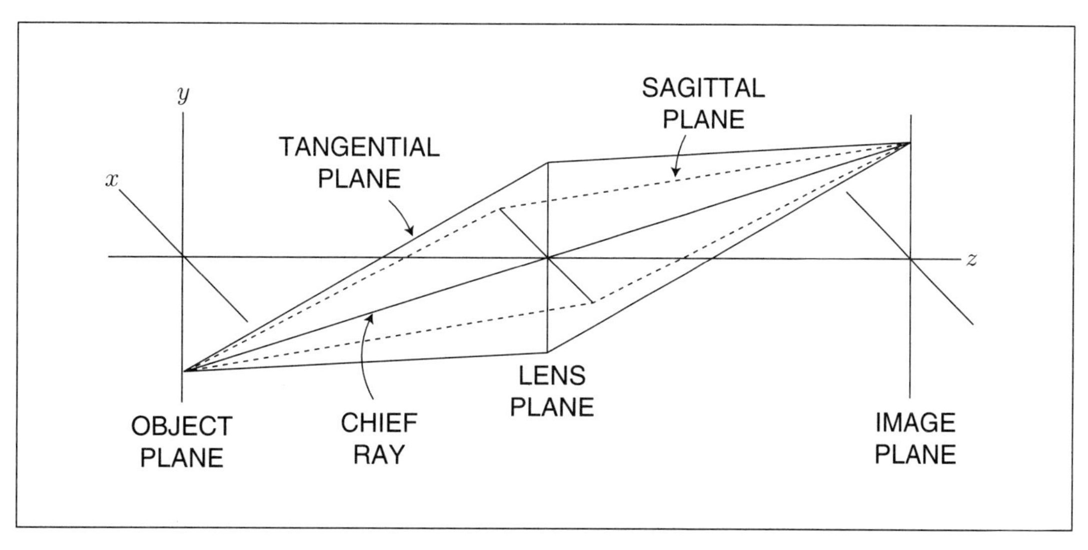

Figure A.6.2. Tangential and Sagittal Planes. *A three-dimensional representation of a thin-lens layout is shown. The object is off-axis and the stop is at the lens. The tangential plane is identical to the meridional plane, the plane of the paper. The sagittal plane is orthogonal to the tangential plane and crosses the tangential plane along the chief ray. Note the right-left bilateral symmetry about the meridional plane.*

together to make a compound lens, the powers and bendings of the various elements are two of the most effective types of optical variables (or degrees of freedom) for controlling aberrations during optimization. For most systems, the only other types of optical variables of comparable effectiveness are air (and sometimes glass) thickness and glass type (index and dispersion).

A.6.5 Tangential and Sagittal Planes

The concept of tangential and sagittal planes for off-axis objects in an axially centered system is illustrated by the three-dimensional drawing in Figure A.6.2. The object plane, lens plane, and image plane, all normal to the optical axis, are indicated. The lens here is assumed to be thin with the stop at the lens. Thus, the chief ray, also indicated, passes directly through the center of the lens without deviation.

The object is located in the yz meridional plane (the plane of the paper). Because all meridional planes are equivalent, there is no loss of generality if off-axis object points are located only in the yz meridional plane.

The tangential plane is identical to the meridional plane. Rays lying in the tangential plane are called tangential rays and all tangential rays are meridional rays.

The sagittal plane is orthogonal to the tangential plane and intersects the tangential plane along the chief ray. Except for the chief ray, rays lying in the sagittal plane are called sagittal rays. Because sagittal rays lie out of the meridional plane, all sagittal rays are skew rays.

Note that this geometry has right-left bilateral symmetry about the meridional plane. Thus, sagittal rays passing through the lens at equal positive and negative x-direction heights are equivalent and redundant.

For a thick or complex lens, the concept of tangential and sagittal planes is nearly identical. The difference is that now the chief ray is bent as it passes through each successive optical surface. The chief ray is a meridional ray, and a meridional ray always remains a meridional ray. Thus, the tangential plane remains the same plane throughout the system. The sagittal plane, however, is not the same plane throughout the system. The sagittal plane changes its tilt after each optical surface to follow the deflection of the chief ray.

The names tangential and sagittal are historical and are explained in Chapter A.8 in the discussion on astigmatism.

A.6.6 Back Focal Length and Effective Focal Length

Back focal length or BFL, is often called back focal distance or just back focus. BFL specifies where the light is focused relative to the mechanical lens. As usually applied, BFL is considered to be a paraxial quantity. As such, BFL is measured as the distance along the axis from the vertex of the rear lens surface to the on-axis paraxial focus for an object at infinity. Thus, BFL gives the longitudinal *location* of the focus.

Sometimes a related quantity called the flange focal length is used instead. This alternate term measures the location of the focus from the plane of some convenient mechanical flange on the lens mount.

Note that the image surface (what we used to call the film plane) is not necessarily located at the focus specified by the BFL. Any departure causes the image to be defocused.

When the focal length of a complex lens is compared to the focal length of an equivalent thin lens, then the focal length of the complex lens is usually called the effective focal length or EFL. As usually applied, EFL is considered to be a paraxial quantity. As such, EFL is measured as the distance along the axis from the vertex of the effective refracting surface to the on-axis paraxial focus for an object at infinity. For a distant off-axis object, EFL controls the height of the chief ray on the image surface. Recall the example of the pinhole camera in Chapter A.3. Thus, EFL governs the transverse *image scale* of an extended image.

More generally, the concepts of BFL and EFL can be expanded to include the analogous quantities based on real trigonometric rays (still from a distant object). These rays are not confined to the paraxial region. They can originate at any object height, and they can pass anywhere through the pupil. And these real rays can also have different wavelengths. BFL is still measured along the optical axis, even for off-axis objects. But now EFL is measured as the distance from the effective refracting surface to the focus along a ray bundle, not along the optical axis.

Note that the location of the effective refracting surface relative to the mechanical lens can change as a function of wavelength, pupil zone, and field zone. Note too that the shape of the effective refracting surface can also change as a function of wavelength, pupil zone, and field zone.

A.6.7 Telephoto and Retrofocus Lenses

To illustrate the concepts of effective focal length and back focal length, consider the telephoto and retrofocus lens types.

Contrary to much popular usage, a telephoto lens is not merely a lens having a relatively long focal length and narrow field of view. A true telephoto lens has negative power in its rear section to create a more compact and convenient system whose physical length is shorter than its EFL. This unsymmetrical configuration is shown in Figure A.6.3.1. Note that the rear negative element causes the chief ray to increase its outward divergence, thus increasing image size. Astronomers use this same telephoto principle when they add a negative Barlow lens in front of focus to increase the focal length of their telescopes.

Conversely, a retrofocus lens (originally a trade name by Angénieux) is a backwards telephoto with negative power in front to create a system whose BFL is greater than its EFL. This unsymmetrical configuration is shown in Figure A.6.3.2. The retrofocus principle is most often used in wide-angle lenses that must provide extra clearance for a single-lens-reflex mirror. Retrofocus lenses are big and bulky

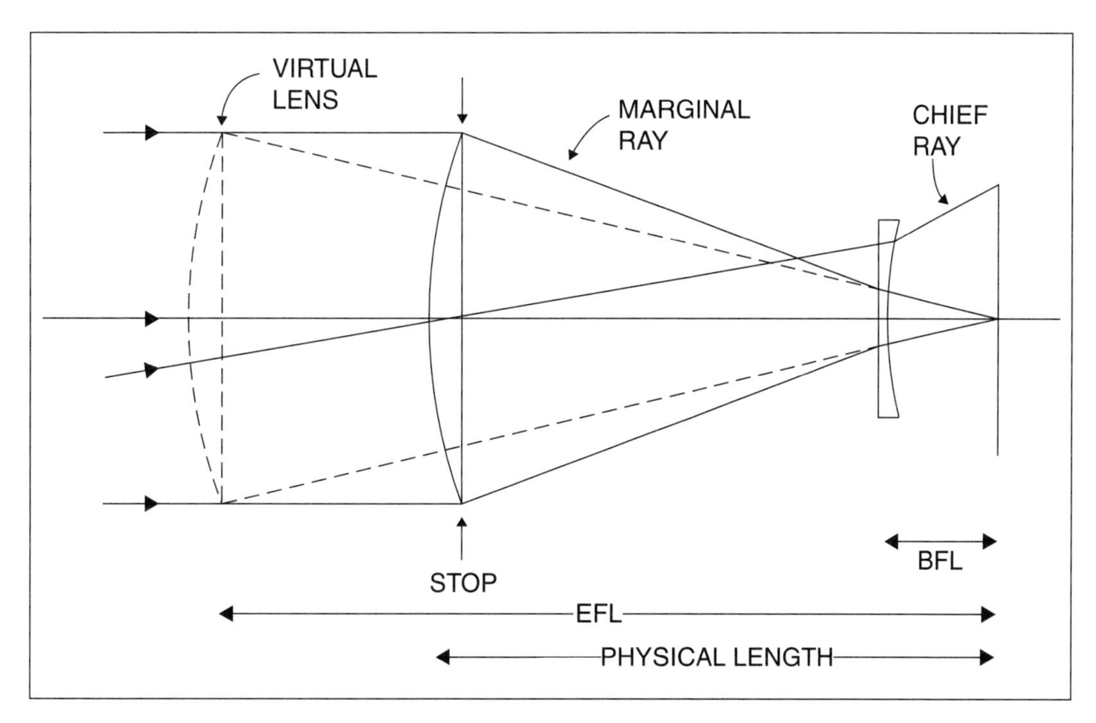

Figure A.6.3.1. Telephoto Lens. *A schematic layout of a telephoto lens is shown. The separated positive and negative elements together simulate the system depicted by the dashed lines. The effective refracting surface is located out in front in the virtual lens.*

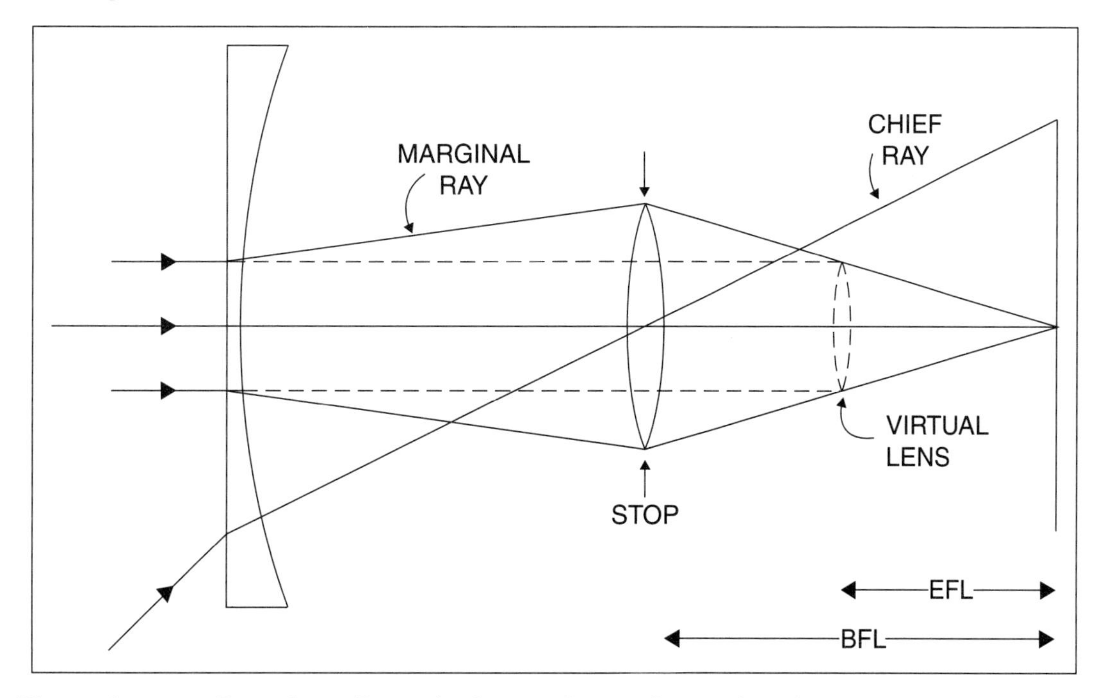

Figure A.6.3.2. Retrofocus Lens. *A schematic layout of a retrofocus lens is shown. The separated negative and positive elements together simulate the system depicted by the dashed lines. The effective refracting surface is located toward the rear in the virtual lens.*

for their focal length, but because wide-angle lenses have short focal lengths, this relatively large size is still practical. Note in Figure A.6.3.2 that the front negative element reduces the chief ray angle both inside the lens and on the image, thus allowing the rear part of the wide-angle lens to work with an effectively narrower field of view.

A.6.8 BFL, EFL, and Aberrations

For a lens and a distant object, much insight can be gained about the basic seven aberrations by analyzing how the generalized forms of BFL and EFL change with wavelength, pupil zone, and field zone. For BFL:

1. A change in BFL with wavelength causes longitudinal chromatic aberration;
2. A change in BFL with pupil zone causes spherical aberration;

3 & 4. A change in BFL with field zone causes astigmatism and field curvature; that is, tangential and sagittal field curvature.

Similarly for EFL:

5. A change in EFL with wavelength causes lateral chromatic aberration;
6. A change in EFL with pupil zone causes coma;
7. A change in EFL with field zone causes distortion.

The first four aberrations, those related to BFL, are called the longitudinal aberrations because they cause image ray errors that are parallel to the axis. The last three aberrations, those related to EFL, are called the transverse aberrations because they cause image ray errors that are perpendicular to the axis.

For many purposes, it is better to replace the term EFL above with either image scale, image height, or magnification. This is especially so when the concepts are further extended to include objects not at infinity, such as with microscope objectives.

Thus, when designing a lens and controlling its aberrations, the two crucial questions about the image become "Where is it?" and "How big is it?"

A.6.9 Sign Conventions for Aberrations

A word on aberration sign conventions is necessary at this point. As with other conventions, there is no universal agreement on this subject. Before the Second World War, it was customary to regard the undercorrected (usually that means *not* corrected) aberrations of a single positive lens with spherical surfaces as all being positive. However, most lens designers today usually say that undercorrected aberrations are negative and overcorrected aberrations are positive. The situation is further complicated by whether you are speaking of the Seidel aberration coefficients, transverse ray aberrations, longitudinal ray aberrations, wavefront aberrations, and so forth. In case of doubt, run a known case with your lens design software and see what you get.

A.6.10 Three Basic Analytical Tools

In computer-aided lens design and evaluation, there are many analytical tools that are useful. Three of the most important are:

1. The meridional-plane, cross-sectional layout,
2. The spot diagram, and
3. The transverse ray-intercept ray fan plot.

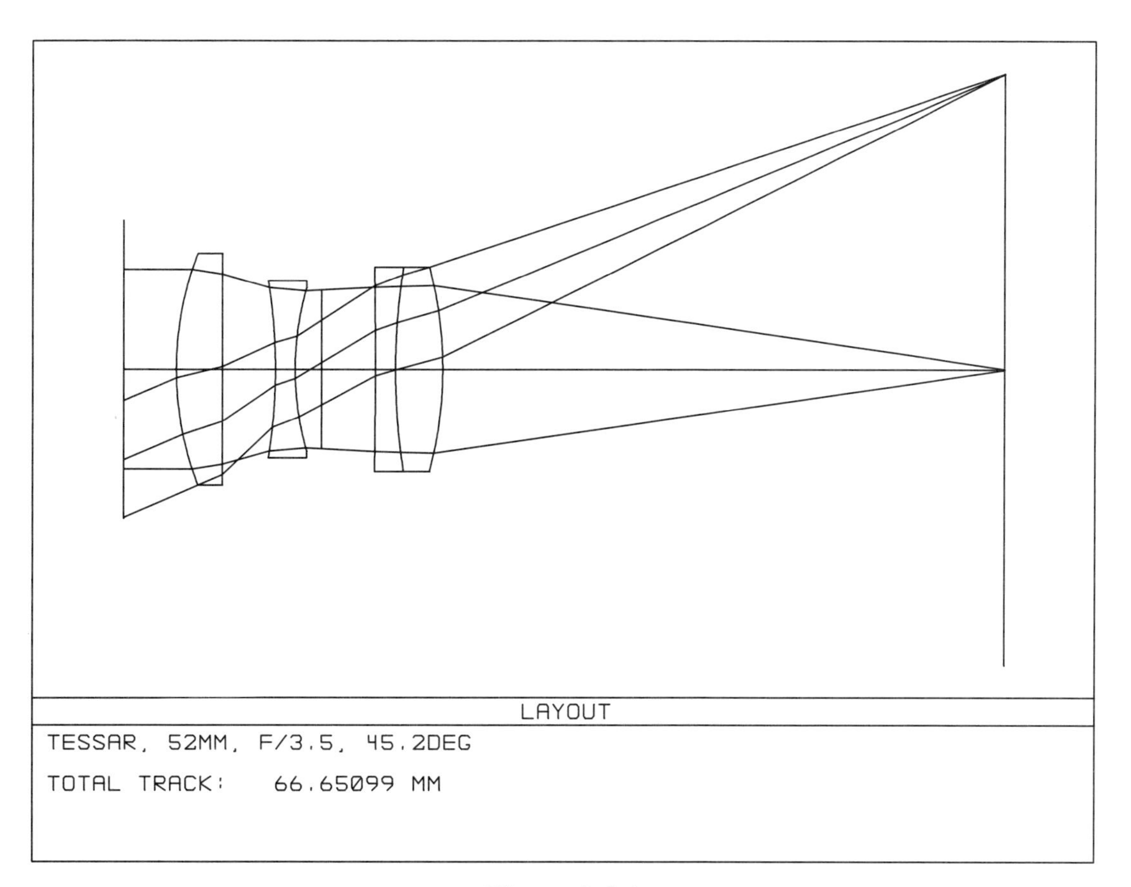

Figure A.6.4.

Some of the other analytical tools that are also useful are: the OPD ray fan plot, the field curvature plot, the distortion plot, the diffraction point spread function, the modulation transfer function, the encircled energy plot, and the surface aberration contribution table. But for now, the first three will suffice.

A.6.11 Layout

For an axially centered optical system, a meridional-plane, cross-sectional layout (or simply a layout) is a diagram showing what the system would look like if it were sliced in half along the optical axis. The lens designer usually does a layout as the first step in evaluating any lens.

As an example, a layout of a four-element Tessar camera lens is shown in Figure A.6.4. Although the lens prescription specifies four object points, to avoid too many confusing lines, only the rays from the on-axis and extreme off-axis objects have been drawn. For the purposes of a layout, these are the most important field positions.

For an off-axis or non-axial optical system, various cross-sections and projections can be made. However, great care is needed in interpretation, and the programming details in your software are probably important. Unless otherwise stated, axially centered systems will be assumed here.

After entering a lens by hand into the computer, it is certainly a good idea to do a layout as a check for mistakes such as sign errors. In addition, when generating a starting point for an optimization run, the designer can often enter a very preliminary lens configuration, look at the layout, make changes by hand, and look at the layout again until he (often quite quickly) converges on something the computer can accept and optimize. It may sound funny to someone who has never done it, but merely ensuring that the initial design transmits light to the image can be a prime consideration.

Once an optimization run has been made, a layout is again the first thing to do. At a glance, the designer can see what the lens looks like and how it is operating. A layout is a particularly good sanity check to reveal practical problems such as: excessive surface curvatures, inappropriate glass or air thicknesses (such as excessive center or negative edge thicknesses), inappropriate overall length or back focal clearance, inappropriate element diameters, grazing-incidence rays, and rays suffering total internal reflection within an element. A closer look reveals the approximate distribution of optical power among the several elements. Elements that are relatively too strong or too weak are often signs of a poor design.

A layout is also an excellent way of checking mechanical vignetting. In Figure A.6.4, note that the off-axis beam is vignetted (clipped) by apertures on the front and rear lens surfaces, and thus the off-axis beam does not fill the stop opening. Incidentally, the central off-axis ray as drawn in Figure A.6.4 is not the chief ray but is merely the ray in the middle of the beam. The true chief ray would, by definition, cross the axis as it passes through the stop surface.

Finally, a layout shows whether the new design is what the designer intended, or whether the computer has gone off on a tangent and come up with a ridiculous solution to be immediately discarded. In the latter case, the operands in the merit function that control the optimization procedure would need to be adjusted. Of course, the computer may also come up with an unexpected solution that is good. Throughout the design process, the designer will find that layouts reveal much about his system.

A.6.12 Spot Diagram

A spot diagram is an analog of the geometrical point spread function (PSF). Diffraction effects are ignored. Thus, a spot diagram illustrates the geometrical image blur corresponding to a point object, such as a star. A spot diagram is one of the best ways to view the effects of aberrations.

To construct a spot diagram, start with a single object point that emits a cone of monochromatic rays. These rays are aimed into and uniformly fill the entrance pupil. Then trace these rays trigonometrically through the lens and on to the image surface. The aggregate of the piercing points of the rays on the image surface is a spot diagram.

In other words, if light rays are the trajectories of photons, and if the entrance pupil is uniformly illuminated by a single monochromatic object point, and if diffraction is neglected, then a spot diagram is a map of the impact points of the photons on the image surface.

Most optical systems use polychromatic white light and contain glass whose index of refraction varies with wavelength. In these systems, several spots for a single object point must be generated, each using a different wavelength. Either the several different monochromatic spots can be displayed separately in a matrix spot diagram, or the spots can be superimposed to obtain a single polychromatic spot diagram. Polychromatic spot diagrams are by far the more common of the two.

For a system with an extended field, the above procedures are repeated for several object points covering the field of view.

Figure A.6.5.1 shows a matrix spot diagram for the Tessar lens in Figure A.6.4. Five wavelengths and four field angles are used. By inspection, it is clear that image quality varies strongly with field angle, and to a lesser degree with wavelength. Figure A.6.5.2 shows the equivalent polychromatic spot diagram.

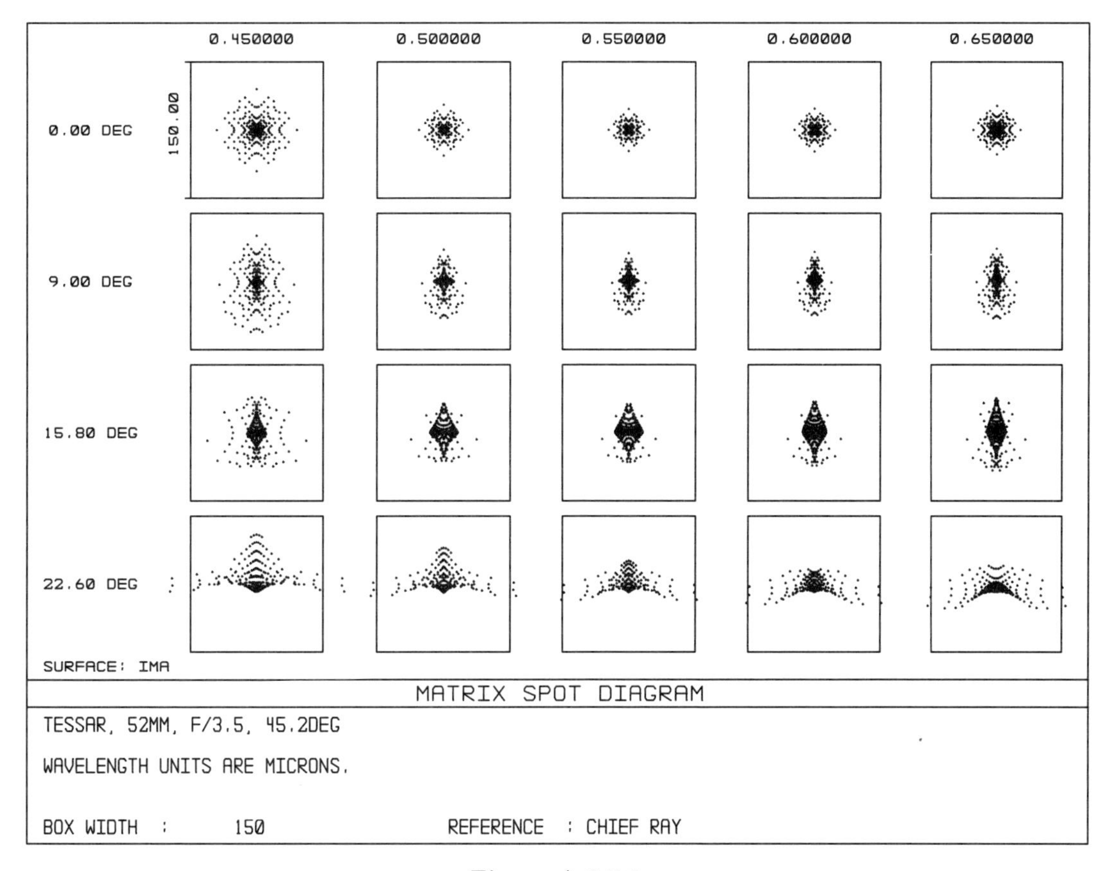

Figure A.6.5.1.

A.6.13 Filling the Lens with Rays

The process of making a spot diagram requires that the lens opening be filled with rays. This is done by projecting a uniform ray array into the entrance pupil. There are three common types of ray arrays, and some programs allow you to choose from more than one. The first array type is a rectangular grid and is illustrated in Figure A.6.6.1. If no vignetting factors are used, then the array is actually square. The second array type is a hexapolar array and is illustrated in Figure A.6.6.2. A hexapolar array is created by a series of concentric circular rings of rays. The radial spacing of the rings is equal, and the number of rays per ring is adjusted to give a roughly uniform ray density. The third array type is a dithered or pseudo-random array and is illustrated in Figure A.6.6.3. A dithered array can be created by taking either of the previous arrays and randomly displacing the positions of the rays entering the lens. In Figure A.6.6.3, a hexapolar array was dithered. Rays displaced outside the lens opening are deleted.

Each of these array types yields a similarly sized but different looking spot diagram. For rectangular and hexapolar arrays, the systematic pattern of the rays entering the lens may cause a corresponding pattern of artifacts to appear in the image spot. A spot made from a dithered array looks more realistic and is usually free of artifacts. However, the pseudo-random process involved in creating a dithered array is not reproducible, thereby causing a slightly different spot to be produced each time the process is repeated. In lens design, rectangular arrays are by far the most common. Hexapolar arrays are most useful when analyzing a spot diagram for third-order aberrations such as coma and astigmatism, and in some instances also fifth-order aberrations such as oblique spherical aberration. Dithered arrays are rarely encountered. Figures A.6.5.1 and A.6.5.2 use a rectangular array.

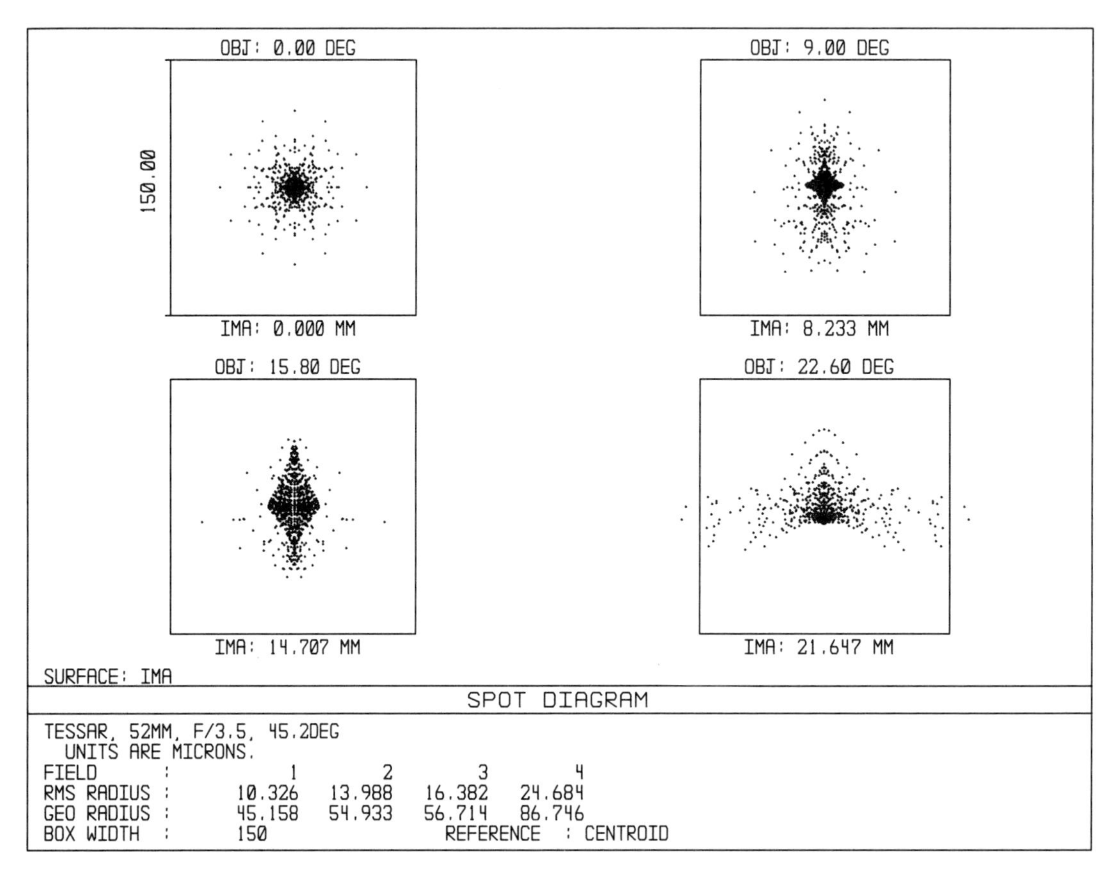

Figure A.6.5.2.

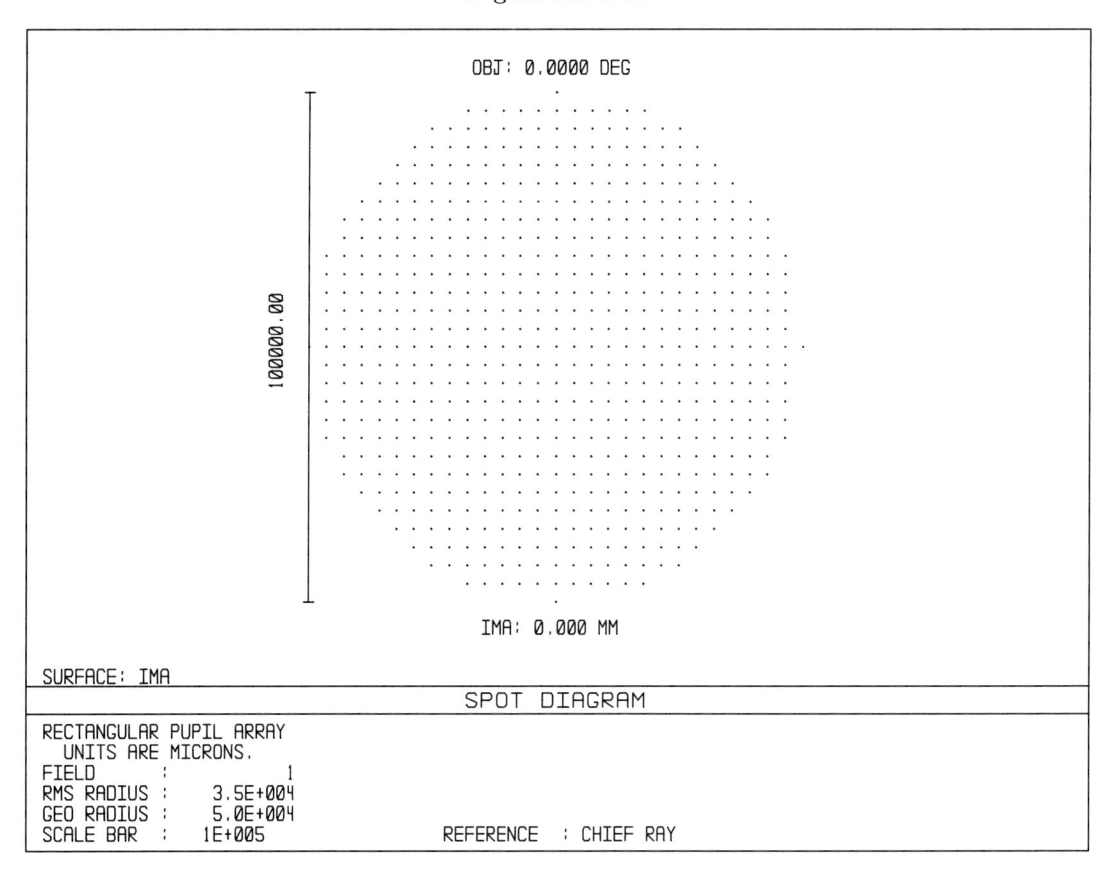

Figure A.6.6.1.

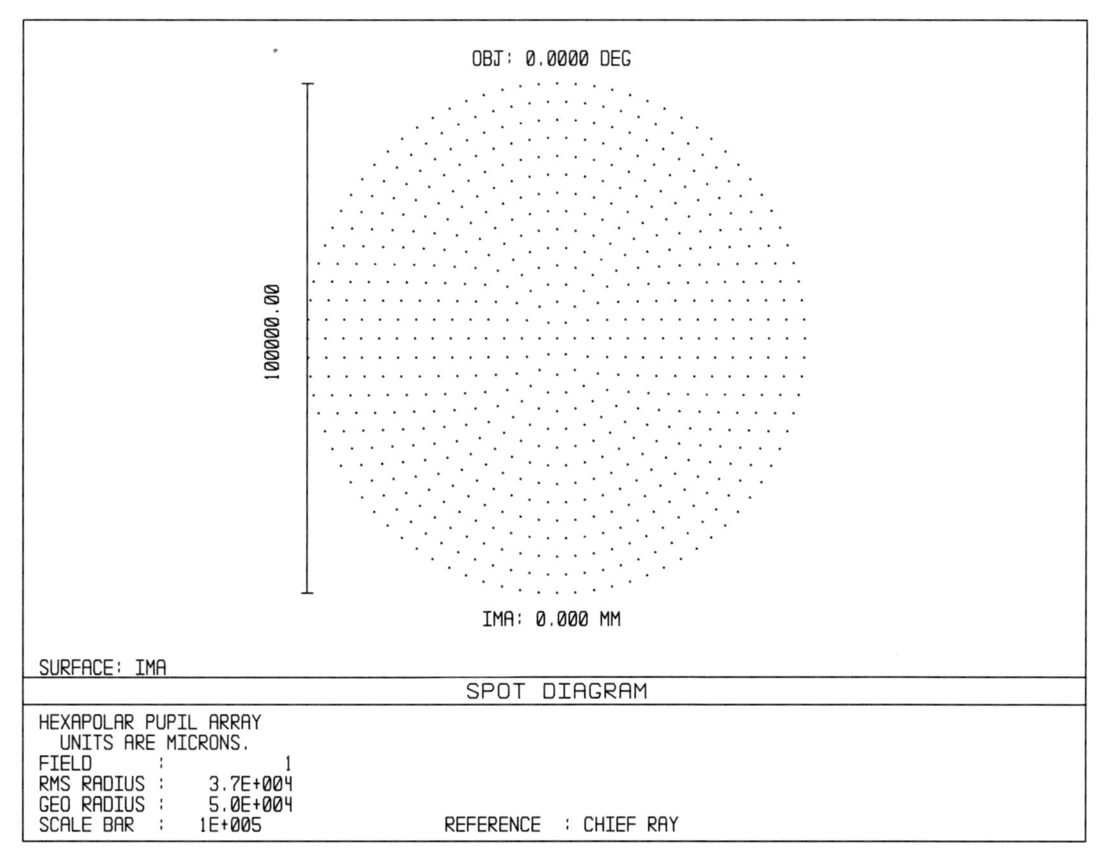

Figure A.6.6.2.

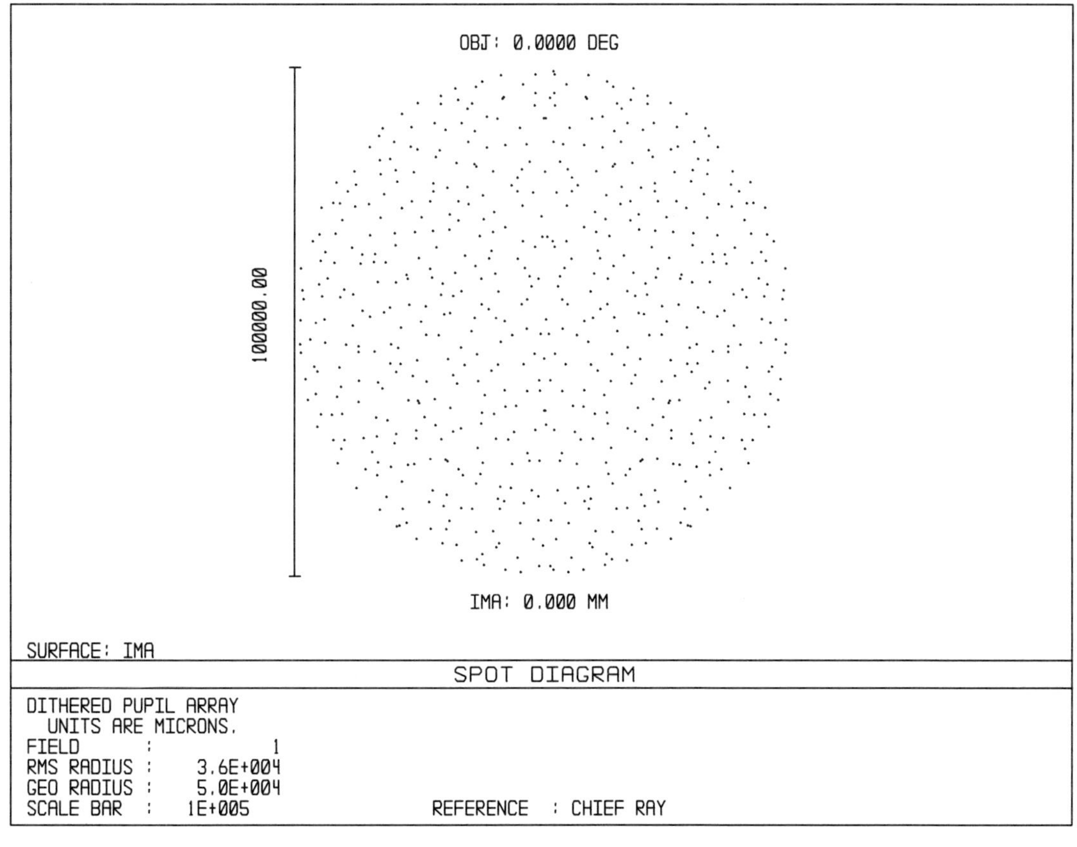

Figure A.6.6.3.

The way the entrance pupil is filled with rays depends on more than your choice of a rectangular, hexapolar, or dithered array. Recall that for off-axis objects, entrance pupil aberrations can cause the true pupil to be significantly shifted and reshaped relative to the unaberrated Gaussian pupil. If you use the Gaussian pupil, you risk including rays that the actual lens does not transmit, and omitting rays that the lens does transmit. Your results could be seriously in error.

There are two ways to handle this problem. The first way is to use the true trigonometric entrance pupil. The ray array is defined on the stop surface, not the Gaussian pupil. Ensuring that the rays pass correctly through the stop opening requires an iterative ray aiming procedure, but with today's computers, these computations are not much of a burden. The second way is to modify the Gaussian entrance pupil to more closely match the true pupil. This shifting and reshaping uses vignetting factors. Using vignetting factors is faster computationally, but using the true pupil more accurately models the situation. Your approach may depend on the features in your lens design program.

Actually, as the name implies, the original purpose of vignetting factors was to account for mechanical vignetting. For vignetting, the pupil is again modified, this time to account for off-axis beam clipping by apertures other than the stop opening. Note that in this application, vignetting factors can be used with either the true entrance pupil or the Gaussian entrance pupil. Alternatively, vignetting can be handled by completely filling the stop opening with rays (using the true pupil without vignetting factors), and by setting hard apertures on the various lens surfaces. Off-axis rays blocked by the hard apertures are deleted from the ray set. Once again, the approach without vignetting factors more accurately models the situation. However, the approaches with vignetting factors may be completely adequate. Much more will be said about mechanical vignetting in Chapters B.3 and B.4.

A.6.14 Transverse Ray-Intercept Ray Fan Plot

Sometimes, the artifacts in a rectangular or hexapolar spot diagram are useful in diagnosing the types of aberrations present. But as Figures A.6.5.1 and A.6.5.2 illustrate, often it is hard to glean much more from spot diagrams than trends and the basic image sizes and shapes. Little information is presented about where in the entrance pupil the rays passed on their way to various parts of an image spot. That diagnostic information, however, is presented in a transverse ray-intercept ray fan plot. For brevity, these plots are also called transverse ray fan plots, ray-intercept ray fan plots, or, simply ray fan plots. In addition, they are sometimes called rimray plots, but this is a misnomer because a rimray plot is actually something else entirely.

Ray fan plots are one of the most important tools that the lens designer has available for diagnosing the nature of aberrations in optical systems. A set of ray fan plots plus a set of spot diagrams together give an excellent indication of the types and magnitudes of aberrations present.

A ray fan is a collection of rays from one object point that all lie in one plane. For a ray fan plot, this plane is made to pass through the center of the entrance pupil, with the fan extending from one side of the pupil to the other.

As a given ray in the fan passes through the lens on its way to the image surface, the ray passes through the entrance pupil at a particular zone height. Let P be this pupil zone height.

When the ray intercepts the image surface, it generally falls some small but nonzero distance away from the chief ray. This transverse distance from the chief ray is the ray height error or aberration, Δh, corresponding to pupil zone height P.

A ray fan plot graphically presents transverse ray height errors on the image surface as a function of corresponding pupil zone height. In other words, Δh on the vertical axis is plotted versus P on the horizontal axis. Along the horizontal axis, negative values of pupil height are on the left, and positive values of pupil height are on the right. The center of the pupil, where the optical axis passes, is at the origin where P equals zero.

It is customary when doing ray fan plots to trace the rays and plot the errors for two specific fans of rays from a given object point. These fans are the tangential (meridional) fan and the sagittal fan, with rays intersecting the entrance pupil along the y- and x-axes, respectively (refer again to Figure A.6.2). These two fans thus give pupil heights P_y and P_x.

It is also customary to consider only the components of Δh, Δh_y and Δh_x, in the y- and x-directions, respectively. For the tangential fan, Δh_y in the image is plotted versus P_y in the pupil. For axially centered systems, meridional rays remain in the meridional plane, and Δh_x is zero.

For the sagittal fan, Δh_x in the image is plotted versus P_x in the pupil. Note that in general for the sagittal fan, aberrations cause the sagittal rays to slightly depart from the sagittal plane in image space, and thus Δh_y is usually not zero. Thus, it is possible to plot Δh_y versus P_x for the sagittal fan, but this is rarely done in practice. Unless specifically stated otherwise, the conventional sagittal ray fan plot is assumed.

Recall that an axially centered optical system is right-left symmetrical about the meridional plane. Thus, the positive and negative sides of a sagittal ray fan plot are also symmetrical; you can obtain one side from the other side. To avoid showing this redundancy, some lens design programs display only the positive side of a sagittal ray fan plot. Of course, for a system with tilts or decenters, symmetry is lost. In this case, tangential and sagittal must be replaced by y-fan and x-fan and both sides of the x-fan plot must be displayed.

In the special case where the object point is on the axis of an axially centered system, the tangential and sagittal ray fan plots become identical. Now the curves are multiply redundant because only one side of either of the plots gives all the information. Nevertheless, most lens design programs still give you two complete plots because of the general way these programs are written. Once again, for a system with tilts or decenters, symmetry is lost, and the complete y-fan and x-fan plots are both required.

Note that there is a second type of ray fan plot called an OPD ray fan plot. OPD stands for optical path difference. An OPD is the linear distance along a ray between (1) the actual aberrated wavefront in the exit pupil and (2) the ideal (aberration-free) wavefront. OPDs are measured relative to the ideal wavefront and are customarily expressed in units of wavelength. An OPD ray fan plot (or simply an OPD plot) graphs wavefront OPD errors in the exit pupil as a function of entrance pupil zone height, P_y and P_x. There should be no confusion between transverse ray-intercept ray fan plots and OPD ray fan plots. OPD plots are discussed further in Chapter A.11.

A.6.15 Example of a Ray Fan Plot

The concepts governing transverse ray fan plots are illustrated in Figures A.6.7.1 and A.6.7.2. Figure A.6.7.1 is a layout of an aberration-free thin lens producing a perfect on-axis focus. The image surface is deliberately located outside or beyond focus, and thus the image is out of focus. Figure A.6.7.2 is the corresponding ray fan plot of image ray height error Δh_y versus pupil ray height P_y. In this special on-axis case where the chief ray coincides with the axis, Δh_y can be replaced by just h_y.

The ray fan plot in this example is an inclined straight line with negative slope. The reason the line is straight can be seen by noting in the layout the two similar right triangles defined by (1) the optical axis (z-axis), (2) any ray from the lens

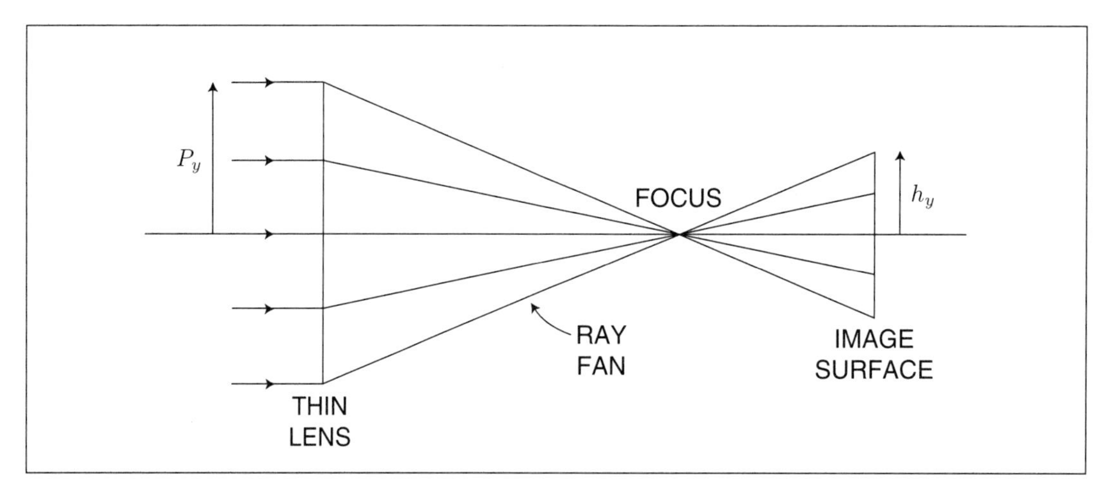

Figure A.6.7.1. Defocus Ray Fan Plot Layout. *An aberration-free thin lens perfectly focuses a meridional fan of rays. The image surface is deliberately located outside or beyond focus, and thus the image is out of focus. The pupil and image heights are indicated.*

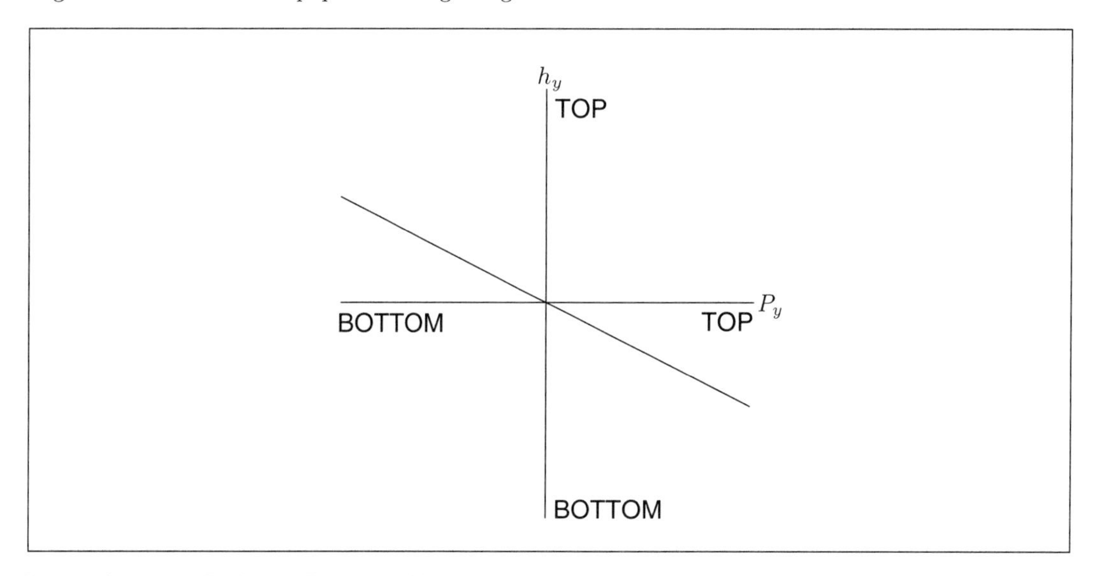

Figure A.6.7.2. Defocus Ray Fan Plot. *The plot of pupil height, P_y, on the horizontal axis versus transverse image height error, h_y, on the vertical axis is shown. For pure defocus, the curve is an inclined straight line. For the image surface located outside of focus, the slope is negative.*

through the focus to the image surface, (3) the y-axis on the lens surface, and (4) the y-axis on the image surface. Because the legs of the triangles along the optical axis are identical for all pupil heights, h_y is linearly proportional to P_y, and the plot is a straight line. The negative slope is caused by the ray crossing the axis before reaching the image surface; that is, P_y and h_y have opposite signs.

Note three things. First, if defocus is in the other direction (inside focus toward the lens), then the slope of the ray fan plot is reversed (positive). Second, if the lens is not thin and its effective refracting surface is curved, or if the image surface is curved, then the out-of-focus ray fan plot is no longer perfectly straight (this effect is discussed further in the next chapter). And third, if the object point is not on the optical axis, then the idea is still the same, except that the chief ray is used instead of the optical axis, image ray height differences from the chief ray, Δh_y, are used instead of total values, h_y, and the similar triangles are not right triangles.

If defocus in this example were reduced to zero, yielding a perfect in-focus image, then h_y would become identically zero and the ray fan plot would coincide with

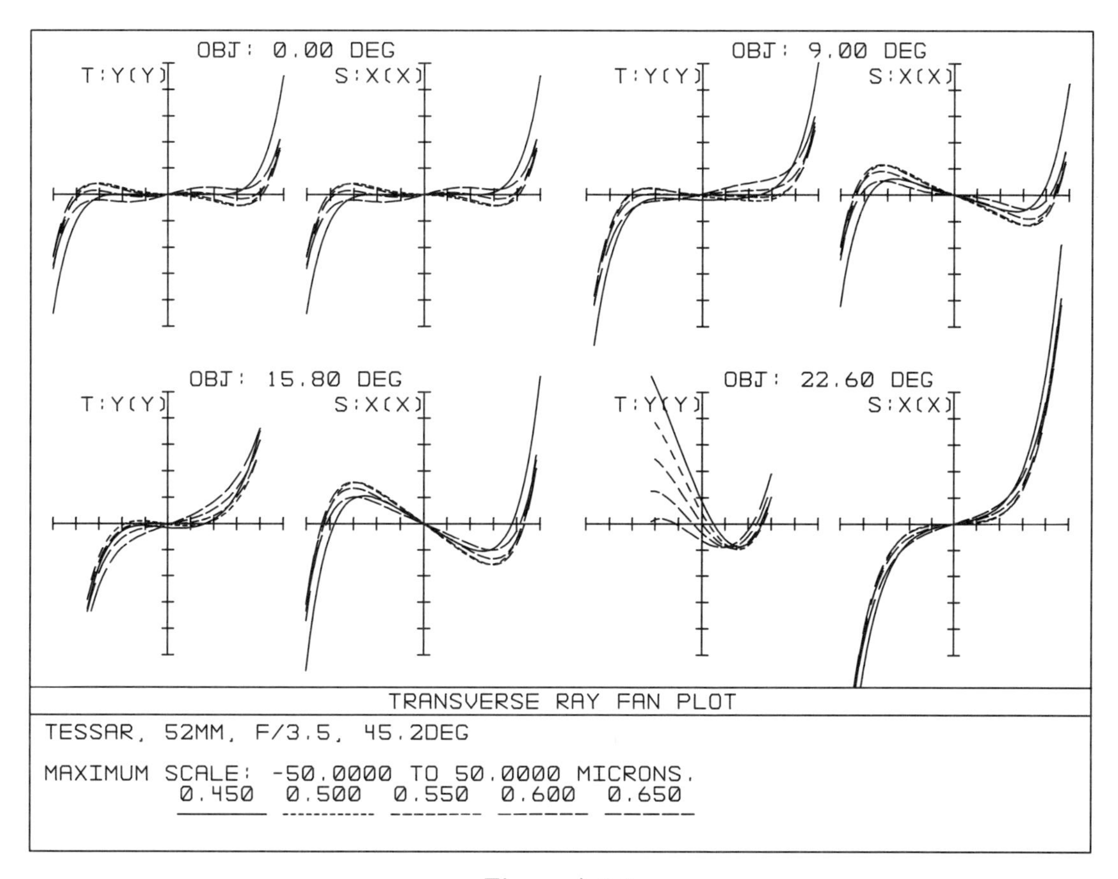

Figure A.6.8.

the horizontal axis. Clearly, perfect imagery giving no image errors and a ray fan plot lying on the horizontal axis are always the ideal.

A.6.16 Use of Ray Fan Plots

Although the slope of the ray fan plot indicates the rate of increase or decrease of image ray error as a function of pupil zone height, slope tells you more than that. As was just demonstrated, an inclined ray fan plot implies defocus. This is true in general, even for complex plots. Thus, if a ray fan plot has an overall inclination, then an improvement in image quality might be achieved merely by refocusing. Because defocus causes geometrical image degradation, defocus can be classified as a type of aberration.

A related use of a ray fan plot is as an indication of whether the system is near the paraxial focus. Look at the ray fan plot of an on-axis image. If the region surrounding the origin where the paraxial rays are plotted has a slope, then the system is not at the paraxial focus. A similar effect can be seen in off-axis images. If the part of the ray fan plot surrounding the origin has a slope, then the rays near the chief ray are not focusing on the image surface.

But the most important use of ray fan plots is in *identifying* aberrations. Each of the Seidel and other aberrations has its own characteristic appearance in a ray fan plot. The designer learns to recognize these different curve shapes so he can diagnose at a glance which aberrations are present. This skill should extend to both transverse ray fan plots and OPD ray fan plots.

An example of a complete ray fan plot analysis is given in Figure A.6.8. The lens is the same Tessar that is used for the layout in Figure A.6.4 and spot diagrams in Figures A.6.5.1 and A.6.5.2. To the uninitiated, however, the meaning of these complex curves is far from clear. Therefore, in the next two chapters, the classical on-axis and off-axis aberrations are discussed with emphasis on building insight and technique.

Chapter A.7

On-Axis Geometrical Aberrations

In the notes for his 1963 course on lens design, Walter Wallin made the following astute observation on the subject of optical aberrations. To paraphrase slightly:

"The individual who has grown up knowing that that thing on the front of a camera forms a perfectly acceptable image is usually tempted, sometime in a course in geometrical optics, to ask, 'Why do lenses have aberrations?' The student so inclined is advised to look more closely at the results of trigonometric ray tracing and then consider whether a more reasonable question might not be, 'If that is the way light rays pass through a lens, how can a lens possibly form an image?' "

That second question is no joke. In this and the next chapter, the classical on-axis and off-axis geometrical aberrations are described and discussed.

It should be noted before proceeding that the examples given are not just schematic illustrations or cartoons. With the help of the ZEMAX optical design computer program, what follows are all the results of real trigonometric ray tracing of actual lens designs carefully chosen to demonstrate the points under consideration. Granted, the examples may be extreme, but the idea is to make the aberrations visible in the layouts.

Note too that, although the several different types of aberrations can be listed and analyzed separately, in most real lenses they appear together in a mixture.

A.7.1 Plane Surfaces

Before discussing aberrations, it should be noted that a reflection by a first-surface plane mirror merely redirects the light and changes the handedness of the image, but no aberrations are introduced. Likewise, no aberrations are introduced into a collimated beam passing normally through a plane-parallel glass plate or (equivalently) through a folding (not dispersing) prism. If the collimated beam is inclined to the plane-parallel plate, then the plate shifts the whole beam laterally (the shift is a function of wavelength), but there is no other change. However, a transmitting plane-parallel plate anywhere in a converging or diverging light beam introduces the whole zoo of aberrations.

A.7.2 Correcting Versus Controlling Aberrations

When an optical designer speaks of correcting an aberration, he usually means reducing the aberration to exactly zero. When he speaks of controlling an aberration, he usually means deliberately adjusting the aberration to some nonzero value for the purpose of balancing or cancelling other aberrations to yield a best overall compromise. Less rigorously, however, designers often refer to the optimization process as the process of correcting aberrations rather than controlling aberrations. They also speak of a well-corrected lens to describe a lens with excellent performance but whose aberrations are actually well controlled rather than corrected to

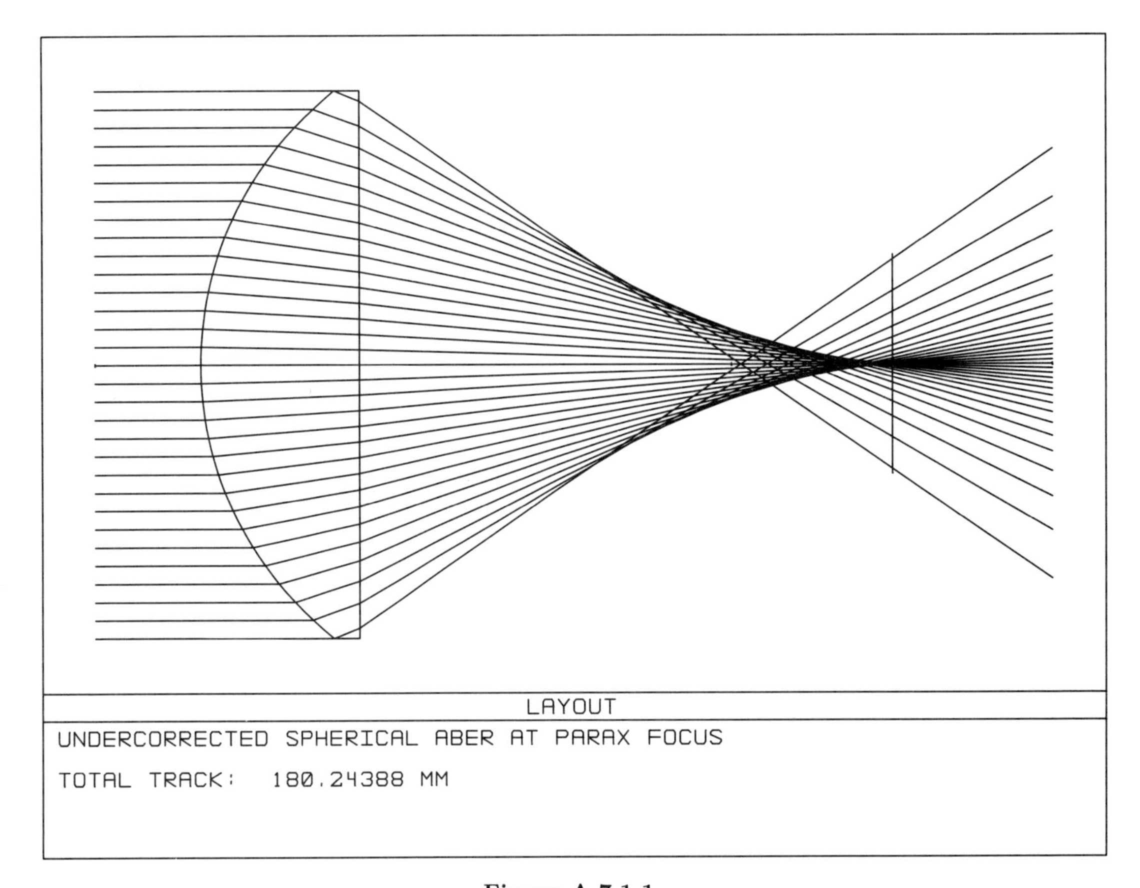

Figure A.7.1.1.

zero. And they use the terms undercorrected and overcorrected. Thus, the words correct and control are frequently confused, and the intended meaning must often be determined by context.

A.7.3 Undercorrected Spherical Aberration at Paraxial Focus

In this and the next several sections, spherical aberration is discussed step-by-step, including the effects of orders, signs, and refocusing.

Figure A.7.1.1 is the layout of an actual plano-convex singlet lens. The diameter of the lens is 100 mm, its paraxial focal length is 120 mm, and thus its Gaussian focal ratio is $f/1.2$. The front curved lens surface is purely spherical, and the glass is ordinary Schott BK7. To avoid the complications of chromatic aberration, only one wavelength is used: 0.550 μm. The object is at infinity, and only the rays from the on-axis object point are shown in the layout. A vertical line is drawn to indicate the image surface, which in Figure A.7.1.1 is located at the paraxial focus. The rays have been extended past the image surface to illustrate where they would have gone had they not been intercepted.

Ideally, of course, you might hope that all of the rays passing through the lens would focus together to a single tiny image point on the image surface. However, although the paraxial rays near the optical axis focus on the image surface, none of the other rays do. The rays passing through the outer zones of the lens all focus closer to the lens than the paraxial rays. Thus, BFL is a function of pupil zone. This is classical undercorrected (not corrected) spherical aberration. The result in the present example is a very poor image indeed. Clearly, this lens does not satisfy the Maxwell criterion that says that point objects should be imaged as point images.

Note that the rays drawn in the layout show a concentration in the center of the image blur. This concentration is again visible in the spot diagrams. Figure A.7.1.2 is the rectangular (square) array spot diagram; Figure A.7.1.3 is the hexapolar array spot diagram; Figure A.7.1.4 is the dithered array spot diagram. All three spots show a concentrated image core surrounded by extended image wings.

In addition, note the different qualitative appearance of the three spots and the geometrical artifacts in the rectangular and hexapolar spots. For most work, the optical designer asks the computer to make only one of the three types of spot diagrams. Usually, the rectangular spots are preferred. All three have been included here for illustrative purposes.

One of the most important functions of a spot diagram is to provide a measure of spot size; that is, the size of the geometric PSF. Total diameter can be used, but total diameter may be misleading if the spot has broad low-intensity wings. For most purposes, the preferred measure is RMS spot diameter. On the rectangular spot diagram in Figure A.7.1.2, the RMS spot *diameter* is 15600 μm (twice the RMS spot *radius* given in the plot caption).

Figure A.7.1.5 is the transverse ray fan plot. The object point is on the axis of an axially centered system, and thus the plot is multiply redundant. Note that around the origin, the curve is near the horizontal axis, indicating small paraxial ray errors. Furthermore, around the origin, the slope of the curve is near zero, indicating an in-focus paraxial image. These two observations from the ray fan plot confirm that the image surface really is at the paraxial focus, as specified.

Note too that away from the origin, both the departure of the curve from the horizontal axis and the slope of the curve become progressively greater. These trends indicate increasing ray errors and an increasing rate of change of ray errors for outer pupil zones. This curve shape is typical of spherical aberration.

One further observation from Figure A.7.1.5 is that the whole curve is significantly inclined with a negative overall slope. This inclination indicates general defocus with the image surface outside best focus (just like the example in Chapter A.6). It also suggests that the lens designer should try refocusing.

A.7.4 Undercorrected Spherical Aberration at Best Focus

Figure A.7.2.1 is a layout of the same lens as in Figure A.7.1.1, except that the image surface has been shifted away from the paraxial focus and toward the lens. The longer vertical line indicates the location of the new image surface, and, for reference, the shorter vertical line indicates the location of the paraxial focus (this convention will continue to be used in this chapter). The position of the new image surface was selected by using the optimization feature in ZEMAX; the program was asked to refocus to find the minimum RMS spot size. This is not the only possible "best" image criterion, but it is one of the common ones and will serve here.

Note immediately on the layout that the total extent of the image blur has been significantly reduced. Figure A.7.2.2 is the rectangular array spot diagram. Note the changed image scale and reduced spot size relative to Figure A.7.1.2. The new RMS spot diameter is 5200 μm, only a third of the RMS spot diameter at paraxial focus. Note also that there has been a major change in the qualitative nature of the light distribution in the image. In particular, there is no longer a concentrated core. Whether you prefer the refocused image or the paraxial-focus image depends on your application.

Figure A.7.2.3 is the ray fan plot. First, note the different scale from Figure A.7.1.5 and the reduced magnitude of the image ray errors. Second, note that the overall inclination of the curve has been removed and that the slope of the curve is nonzero at the origin, confirming that the image surface is no longer at the paraxial focus. Third and most important, note how the effects of defocus and spherical aberration are balancing each other to yield a compromise over the whole pupil. In the center of the pupil, defocus predominates. As you move to outer zones

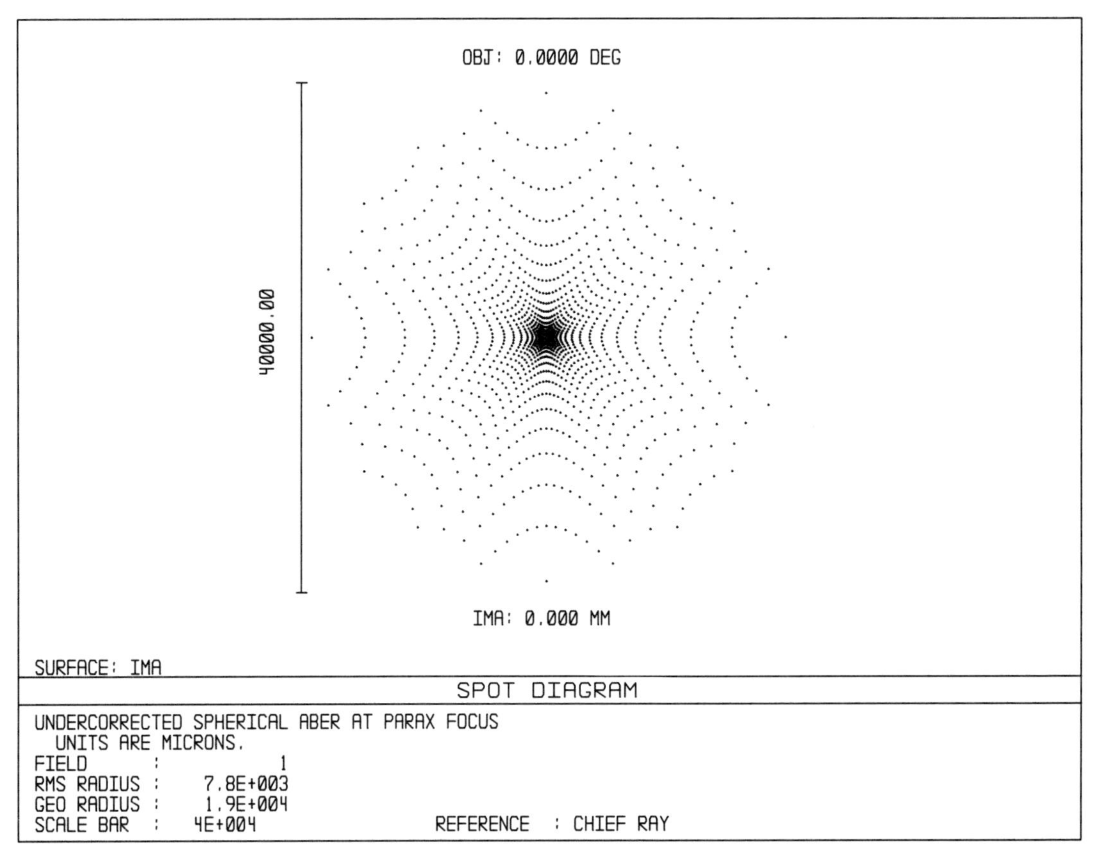

Figure A.7.1.2.

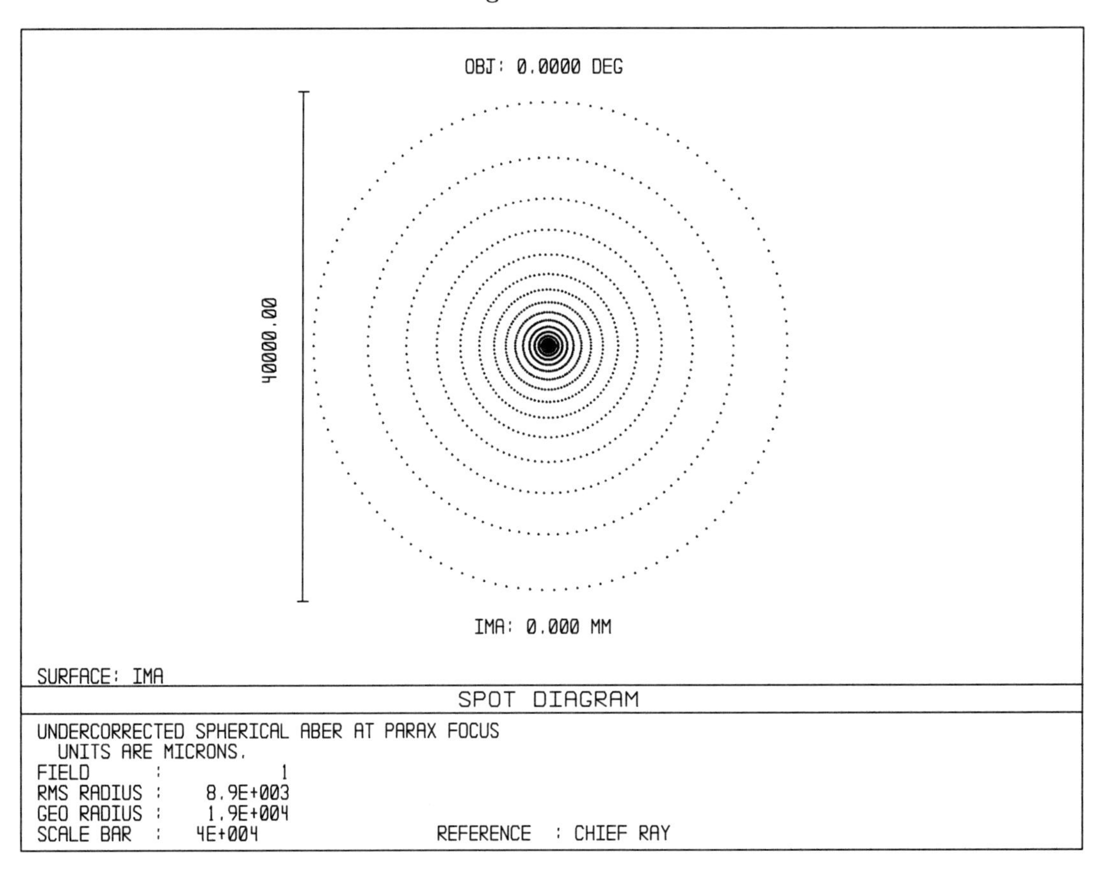

Figure A.7.1.3.

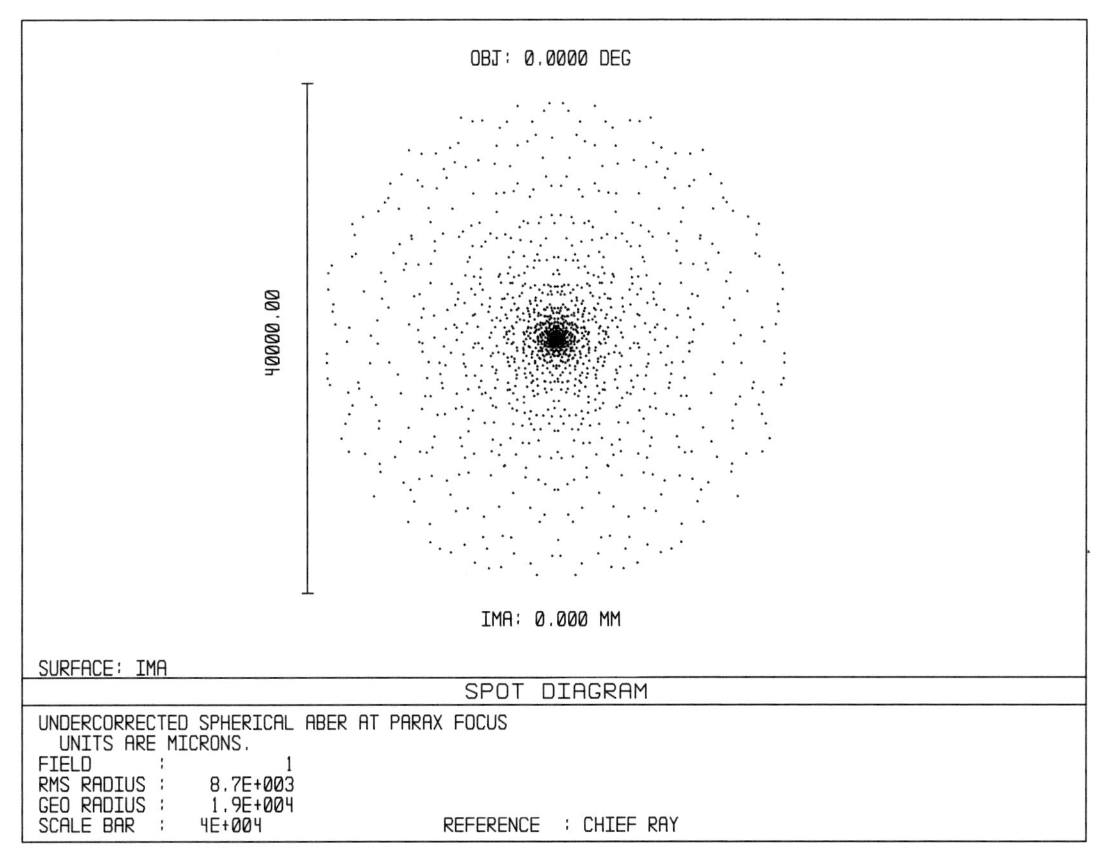

Figure A.7.1.4.

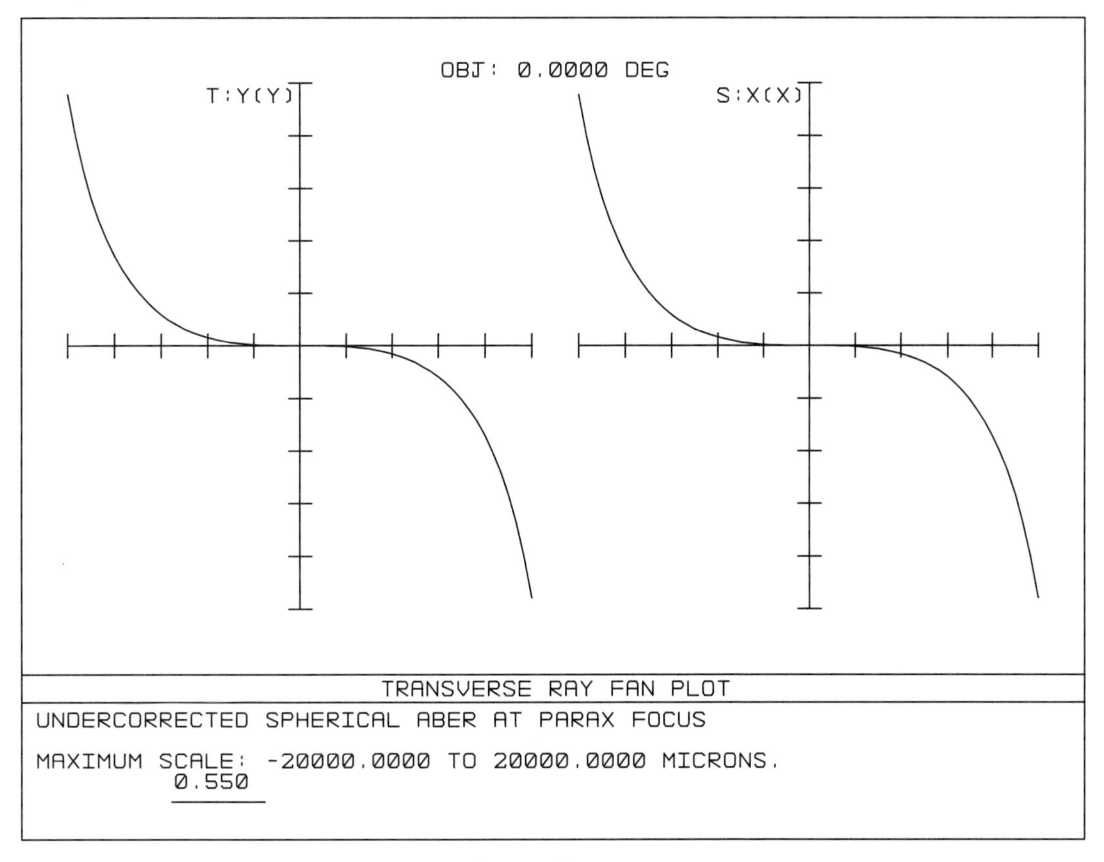

Figure A.7.1.5.

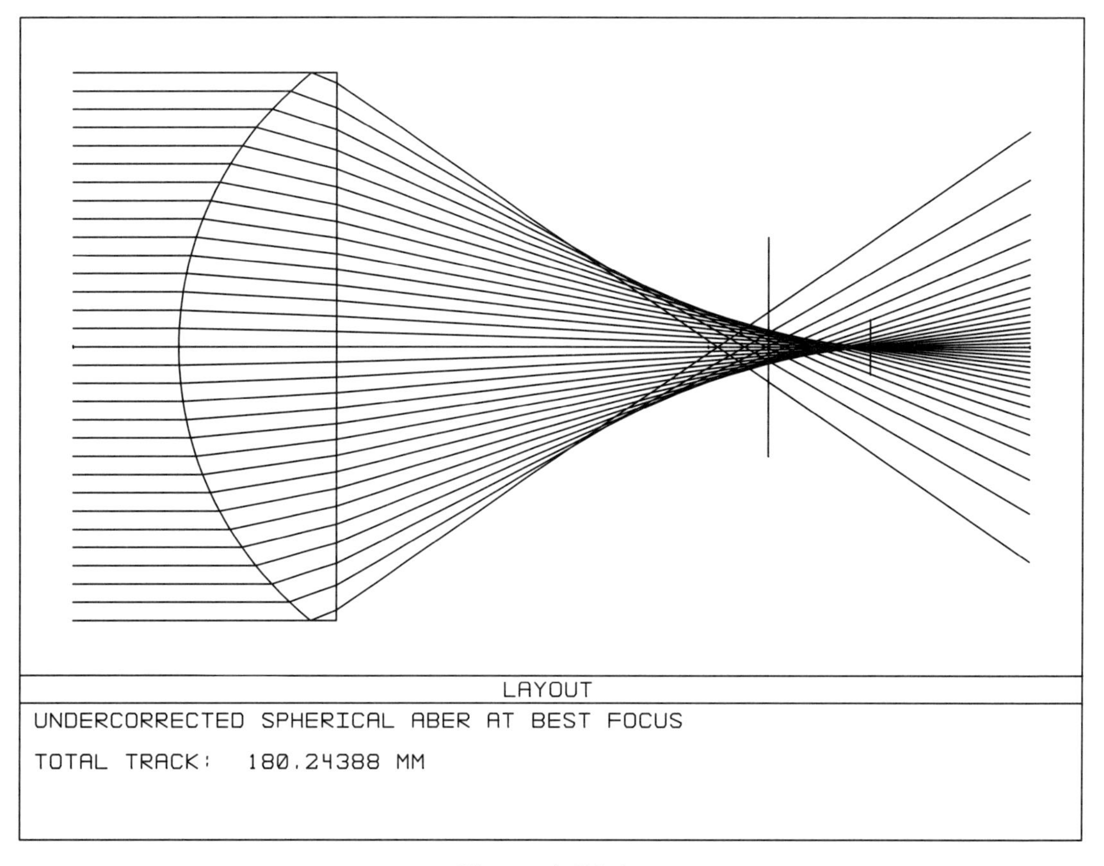

Figure A.7.2.1.

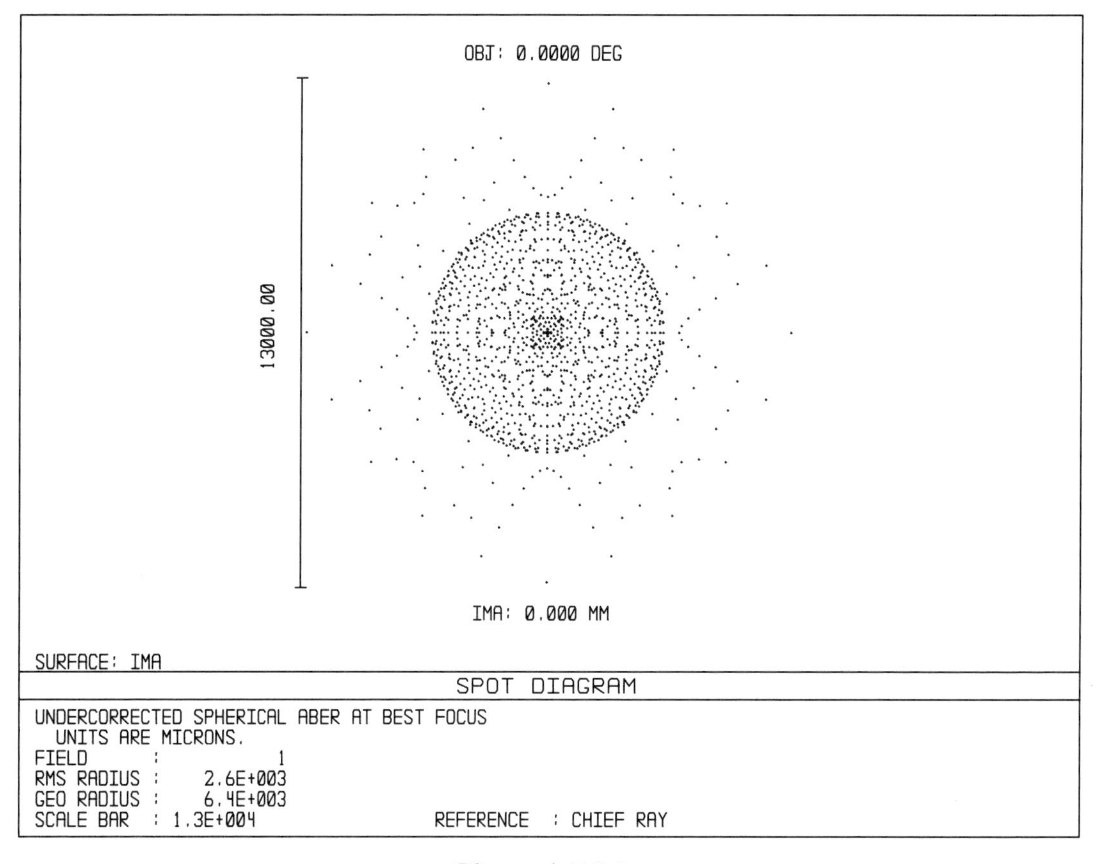

Figure A.7.2.2.

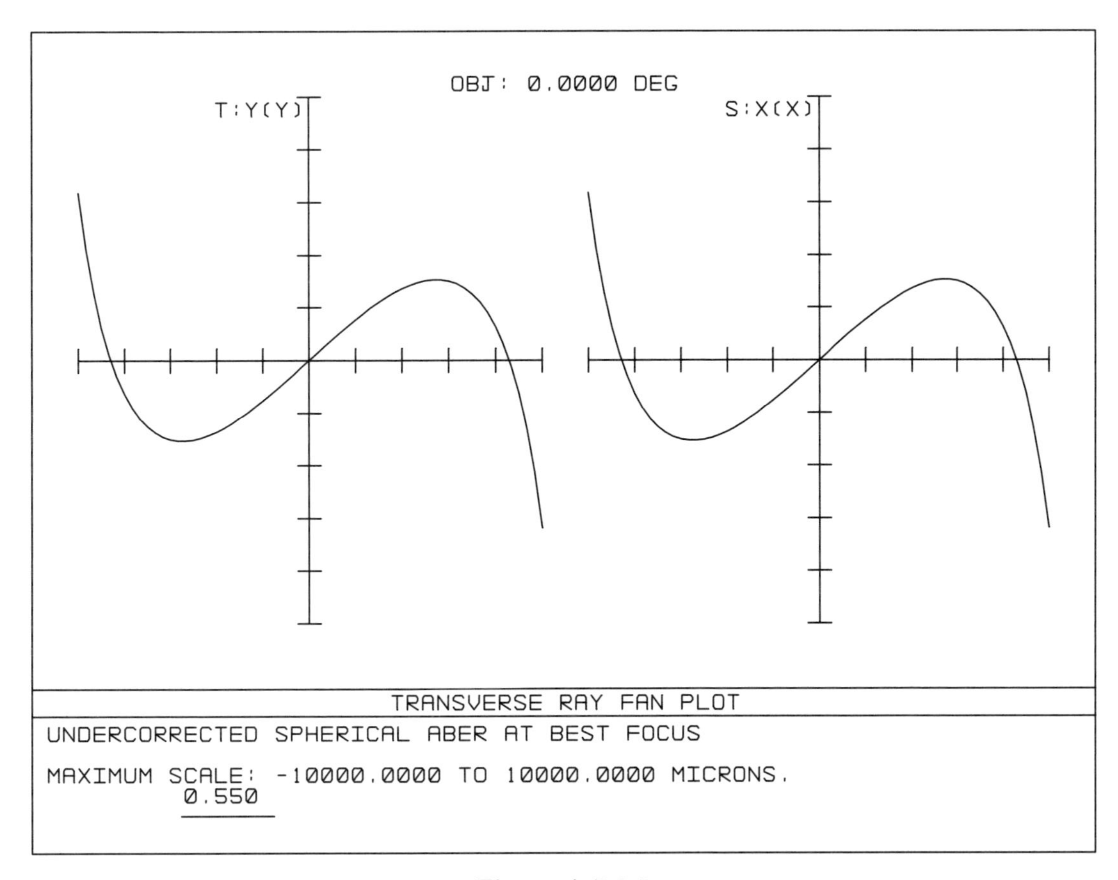

Figure A.7.2.3.

in the pupil, spherical aberration becomes progressively more important until at a little over half way to the pupil edge, spherical aberration halts the increase in image blur size caused by defocus. At about the 0.85 pupil zone, the combined effects of spherical aberration and defocus exactly balance each other to yield zero ray error. Finally, at the pupil edge, undercorrected spherical aberration predominates.

A.7.5 Overcorrected Spherical Aberration at Paraxial Focus

You need not have a spherical surface to get spherical aberration, although historically, lenses with spherical surfaces were the first to be seen with the problem. Figure A.7.3.1 is a layout of a lens similar to the lens in Figures A.7.1.1 and A.7.2.1, except that the front spherical surface has now been replaced by a hyperboloid (with the same vertex curvature) to purposely give spherical aberration of the opposite sign; that is, overcorrected spherical aberration. All the first-order optical properties remain unchanged. The conic constant of the hyperboloid was manually adjusted to give a ray distribution in the layout that clearly illustrates the effect. The image surface is located at the paraxial focus. Figure A.7.3.2 is the ray fan plot. Compare the shape of this ray fan plot with that in Figure A.7.1.5.

In a positive singlet lens, the only way to produce overcorrected (or zero) spherical aberration is with an aspheric surface. However, in more complex multi-element designs, overcorrected spherical aberration can be produced with all-spherical surfaces.

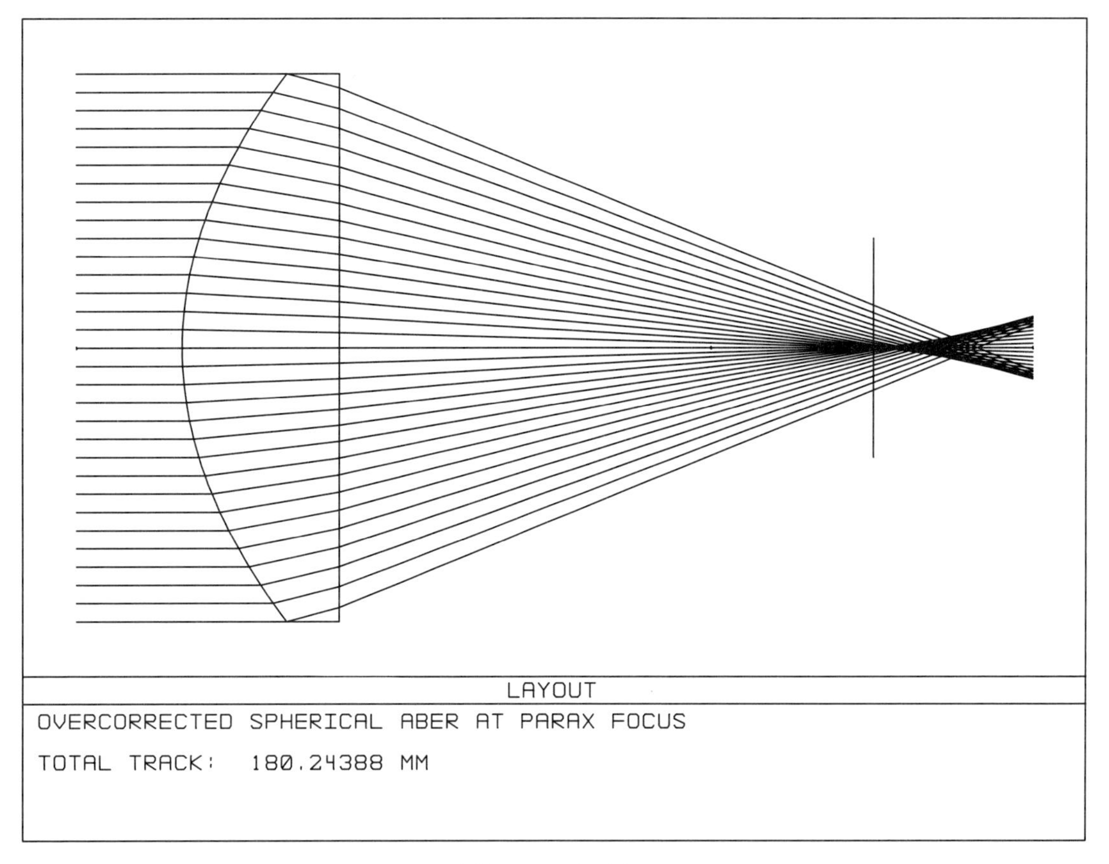

Figure A.7.3.1.

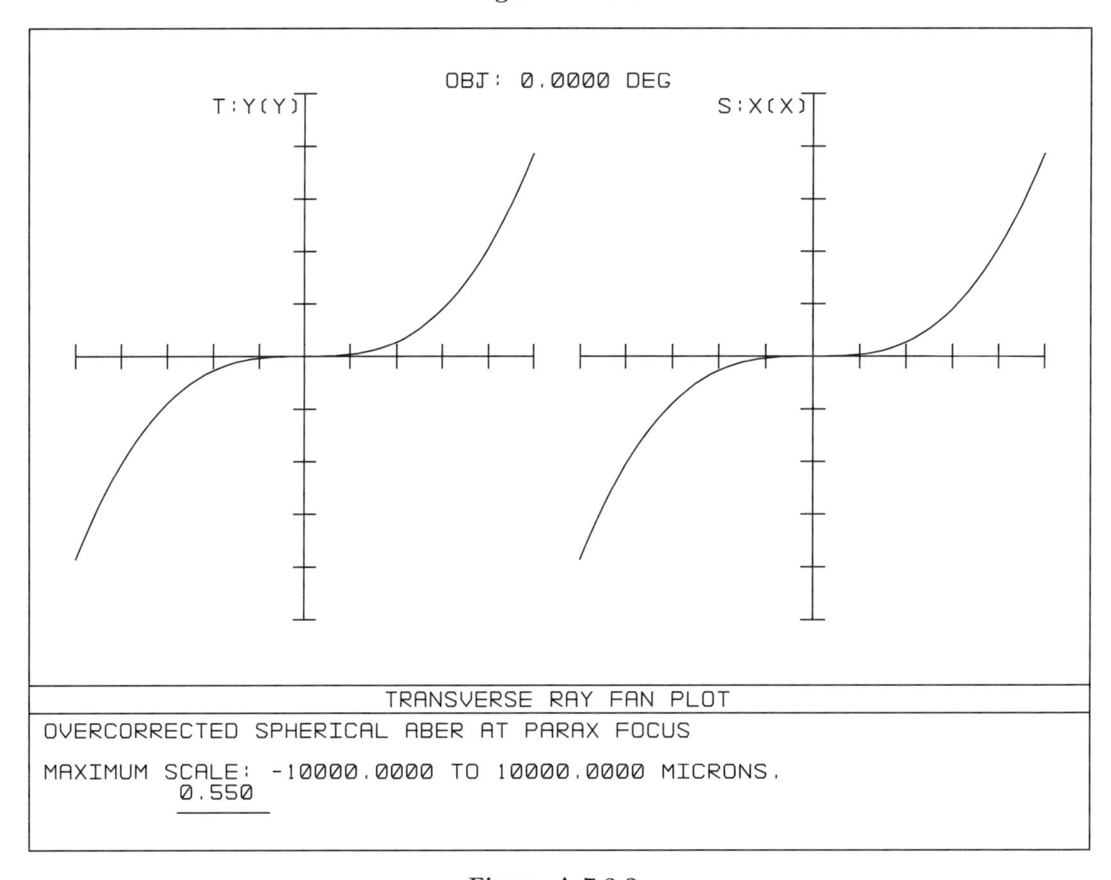

Figure A.7.3.2.

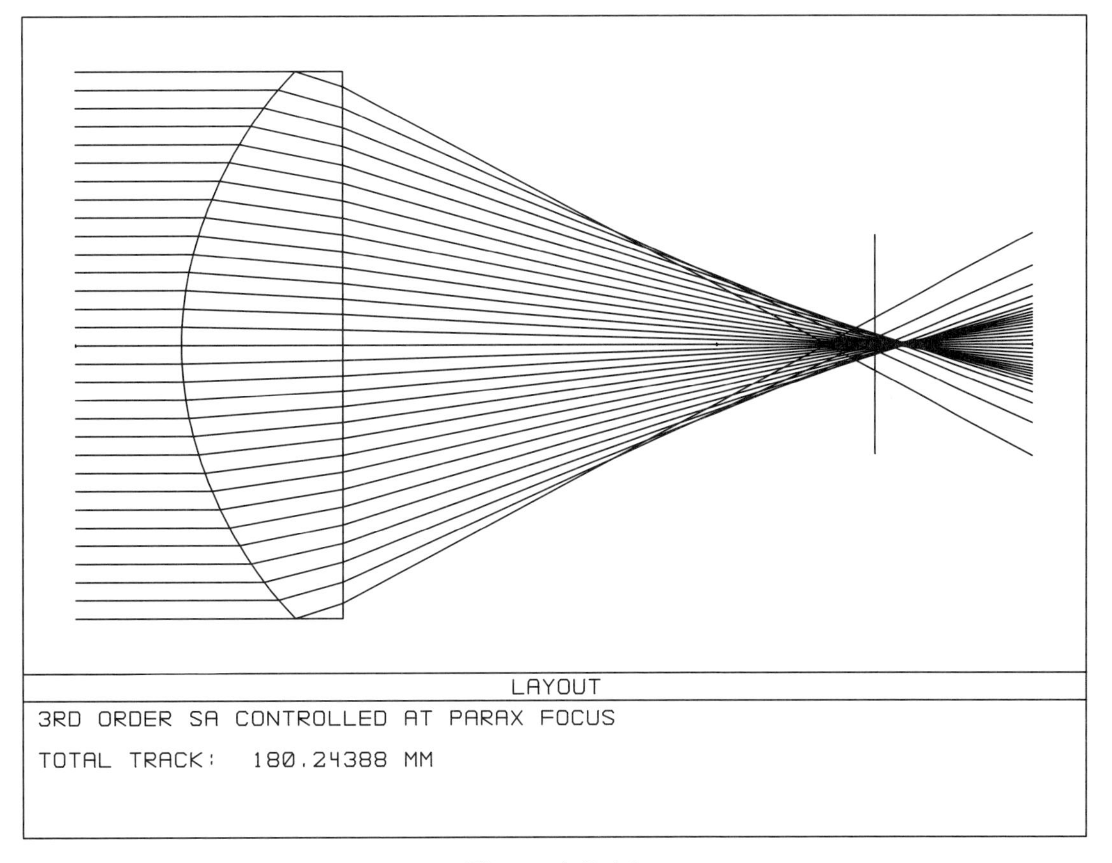

Figure A.7.4.1.

A.7.6 Third-Order Spherical Aberration Controlled at Paraxial Focus

Recall Equation A.4.1 where the sine function was expressed as a power series expansion containing terms of various orders. The other trigonometric functions can also be expressed by similar power series expansions. The classical optical aberrations are all based on these power series expansions, so it follows that the classical aberrations also come in various orders. For example, there is third-order spherical aberration, fifth-order spherical aberration, seventh-order spherical aberration, and so forth. As an example of controlling aberrations, the lens introduced in Figure A.7.1.1, which contains all orders of spherical aberration, will now be aspherized to control third-order spherical to balance higher orders.

Figure A.7.4.1 is the layout of the lens as optimized for minimum RMS spot size. All the first-order properties are the same as those of the lens in Figure A.7.1.1, and the image surface is at the paraxial focus. The difference is that a fourth-order polynomial deformation has been superimposed on the spherical front surface. A fourth-order polynomial has the ability to control third-order spherical aberration, but not higher orders. Controlling higher aberration orders requires higher orders of polynomial deformation. Nevertheless, controlling just a third-order aberration can be very effective because the third-order contributions are usually the largest.

Figure A.7.4.2 is the spot diagram. RMS spot diameter is 3000 μm. Figure A.7.4.3 is the ray fan plot. First, note the scale. Second, note that the curve is flat around the origin, indicating paraxial focus. Third, note that the curve initially goes in the opposite direction from the curve in Figure A.7.1.5 and in the same direction as the curve in Figure A.7.3.2, indicating overcorrected third-order spherical aberration for inner pupil zones. Fourth, note that at about the 0.9 pupil zone, higher orders of undercorrected spherical aberration overcome the overcorrected third-order spherical to yield a total ray error

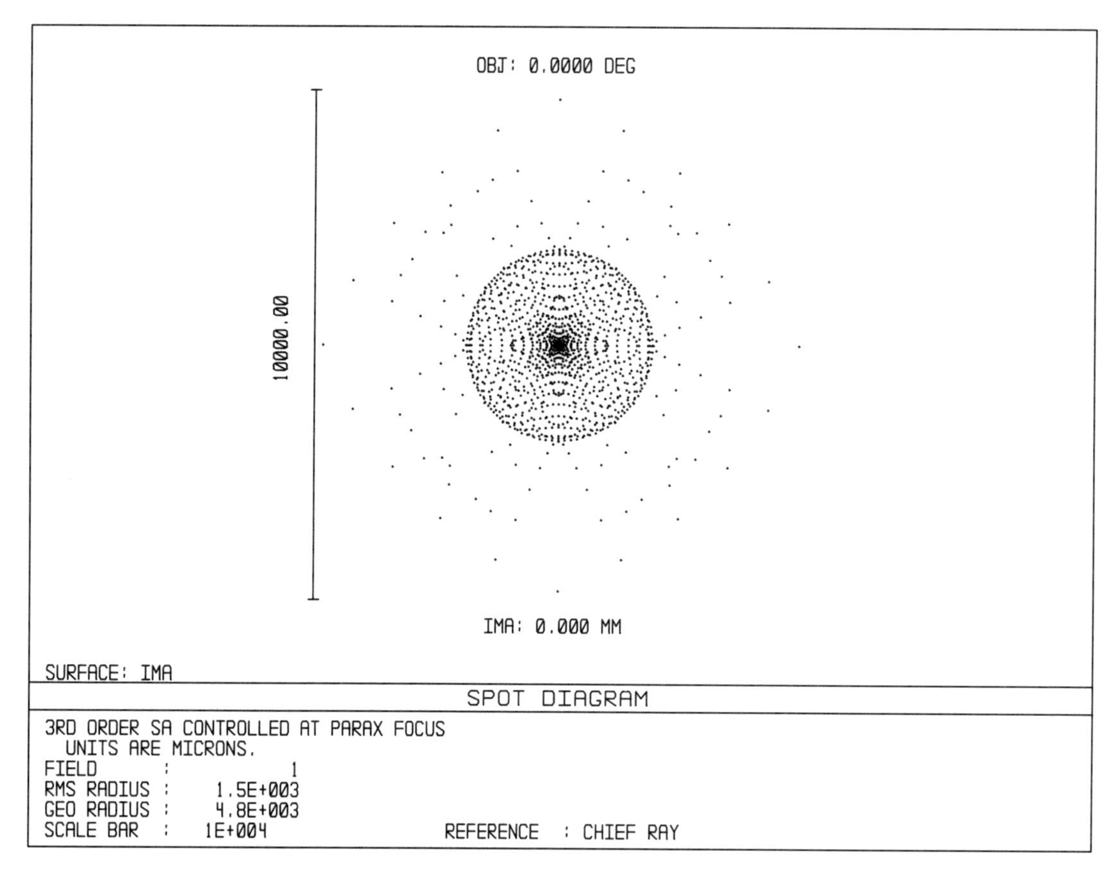

Figure A.7.4.2.

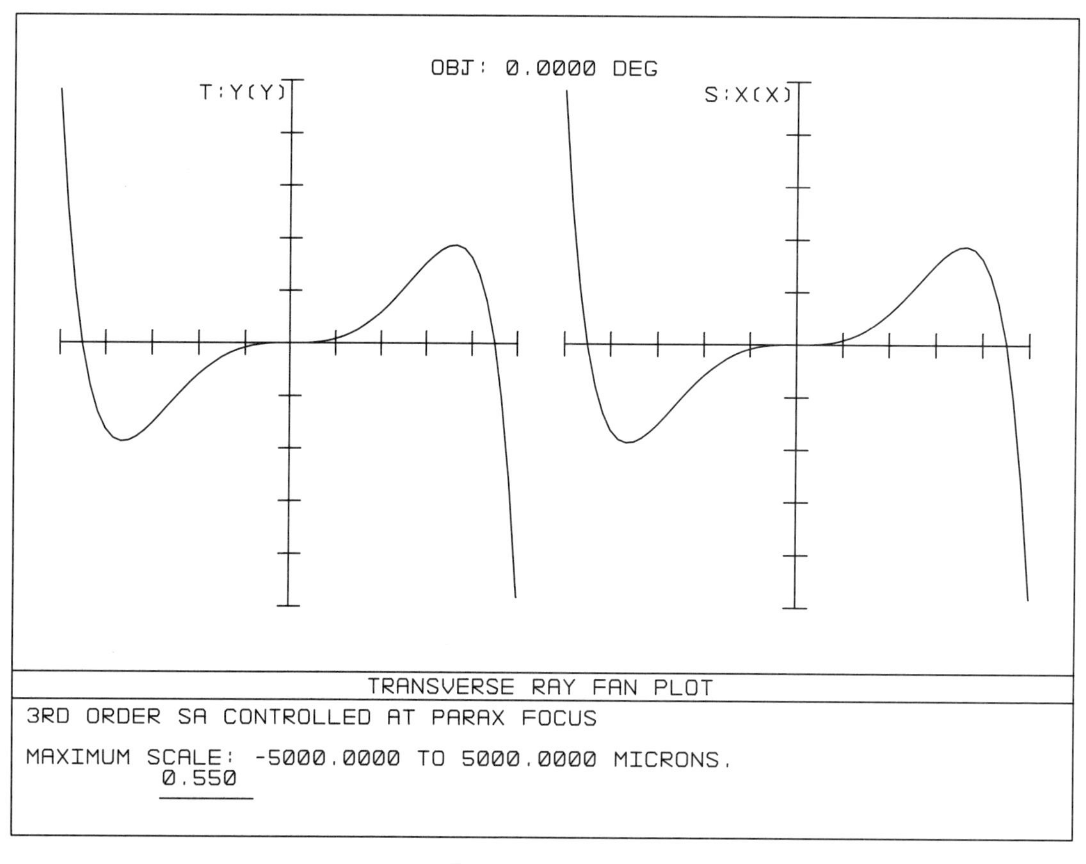

Figure A.7.4.3.

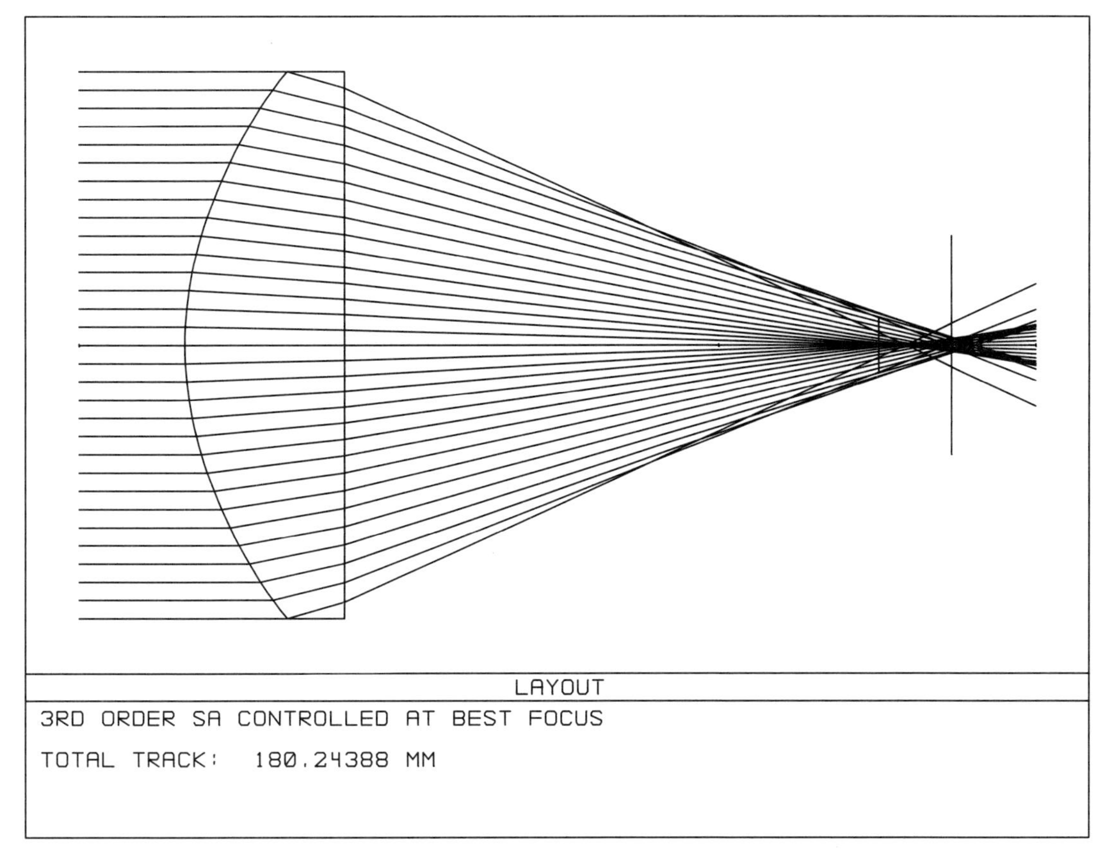

Figure A.7.5.1.

of zero. Total spherical aberration is thus *corrected* to zero at only this one zone. The residual spherical at other zones is known as zonal spherical aberration. Zonal spherical can be very important; in some designs it is the limiting aberration. Fifth, note that in the outermost pupil zones, undercorrected spherical predominates.

This type of spherical performance is very common, and often you know ahead of time that the optimized ray fan plot will resemble Figure A.7.4.3. Spherical aberration was controlled here by minimizing RMS spot size. But in this case, the same result could have been obtained by taking a single ray passing through the 0.9 pupil zone and asking the computer to optimize the lens to make the height of this ray zero on the image surface. This second approach is simpler and takes much less computing. However, if you are not sure what the ray fan plot will look like (as in most of the examples in this chapter), then controlling spherical by minimizing RMS spot size is the safer way to optimize a lens. In practice, lens designers use both approaches, the choice depending on circumstances.

A.7.7 Third-Order Spherical Aberration Controlled at Best Focus

The lens of Figure A.7.1.1 had only one free variable, front vertex curvature, and this was used to control paraxial focal length. In Figure A.7.2.1, a second degree of freedom, defocus, was added. After optimization, the result was a smaller image blur. In Figure A.7.4.1, defocus was replaced by a different second degree of freedom, fourth-order polynomial surface deformation. After optimization, this lens yielded even better imagery. The question now is what would be the result if both of these additional degrees of freedom were allowed to vary *simultaneously* during optimization.

The result for minimum RMS spot size is shown in Figure A.7.5.1. Note that this

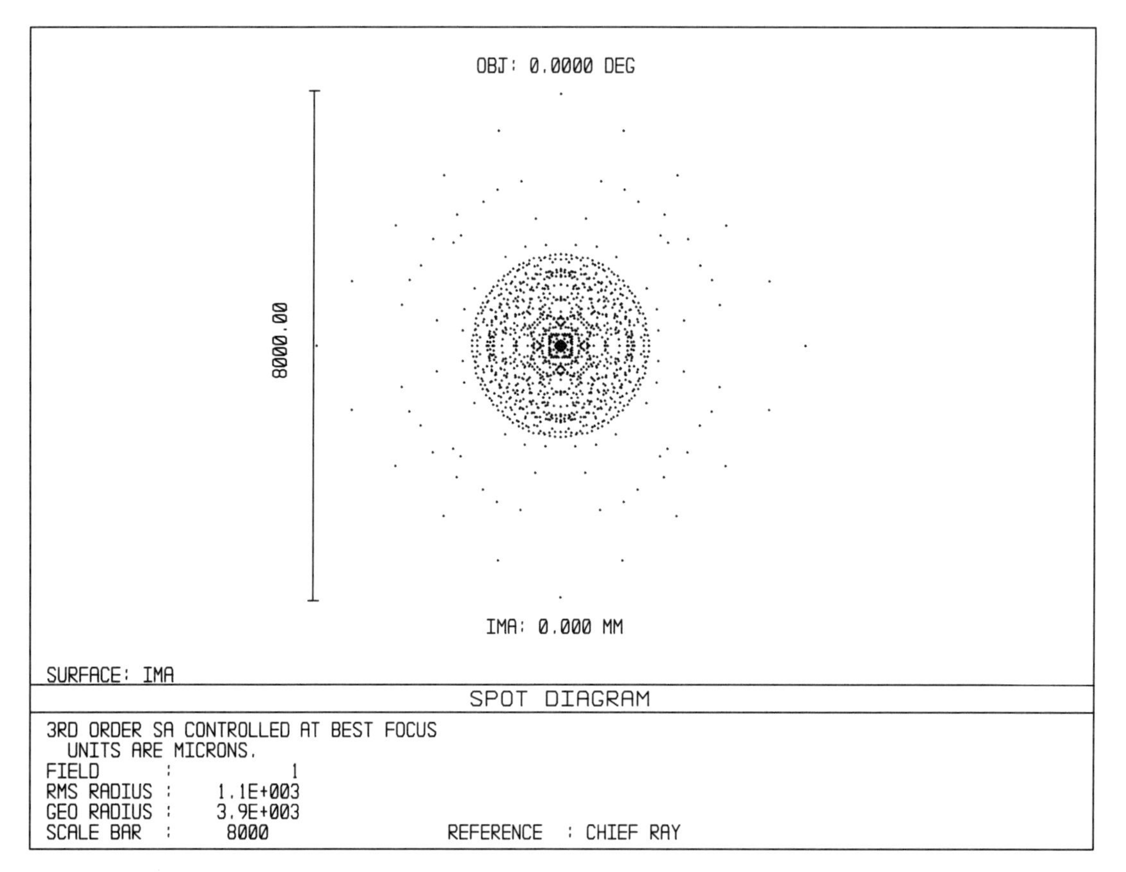

Figure A.7.5.2.

lens is not merely a refocused version of the lens in Figure A.7.4.1. The lens in Figure A.7.5.1 is a complete reoptimization.

Figure A.7.5.2 is the spot diagram. RMS spot diameter is 2200 μm. Figure A.7.5.3 is the ray fan plot. First, note the scale, which is the same as in Figure A.7.4.3. Second, note that the curve is not flat around the origin, indicating that the image surface is not at the paraxial focus. Third, note that for innermost pupil zones, defocus predominates; for intermediate zones, overcorrected third-order spherical aberration predominates; and for outermost zones, undercorrected higher-order spherical aberration predominates. Thus, there is a compromise over the pupil. Fourth, note that the total ray error is zero for two pupil zone heights, near 0.6 and 0.9. But between the two zero crossings, there are still residual zonal ray errors or zonal spherical.

A.7.8 Third- and Fifth-Order Spherical Aberration Controlled at Paraxial Focus

If controlling third-order spherical aberration yielded improved imagery, it might be expected that controlling both third- and fifth-order spherical aberration would be even more effective. In the singlet lens under consideration, third-order spherical is controlled by a fourth-order polynomial deformation superimposed on the spherical front surface. Similarly, fifth-order spherical is controlled by a sixth-order deformation. Such an optimized lens is shown in Figure A.7.6.1. The first-order properties remain as before, and the image surface is at the paraxial focus.

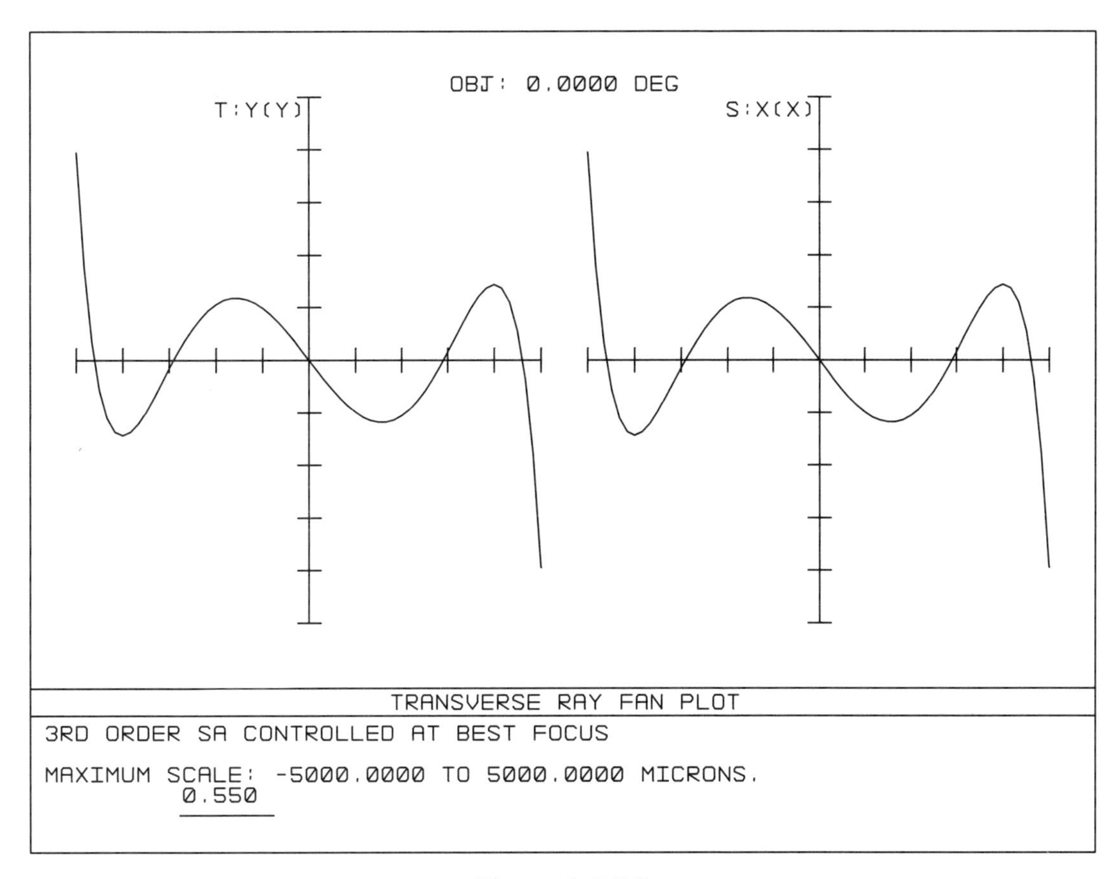

Figure A.7.5.3.

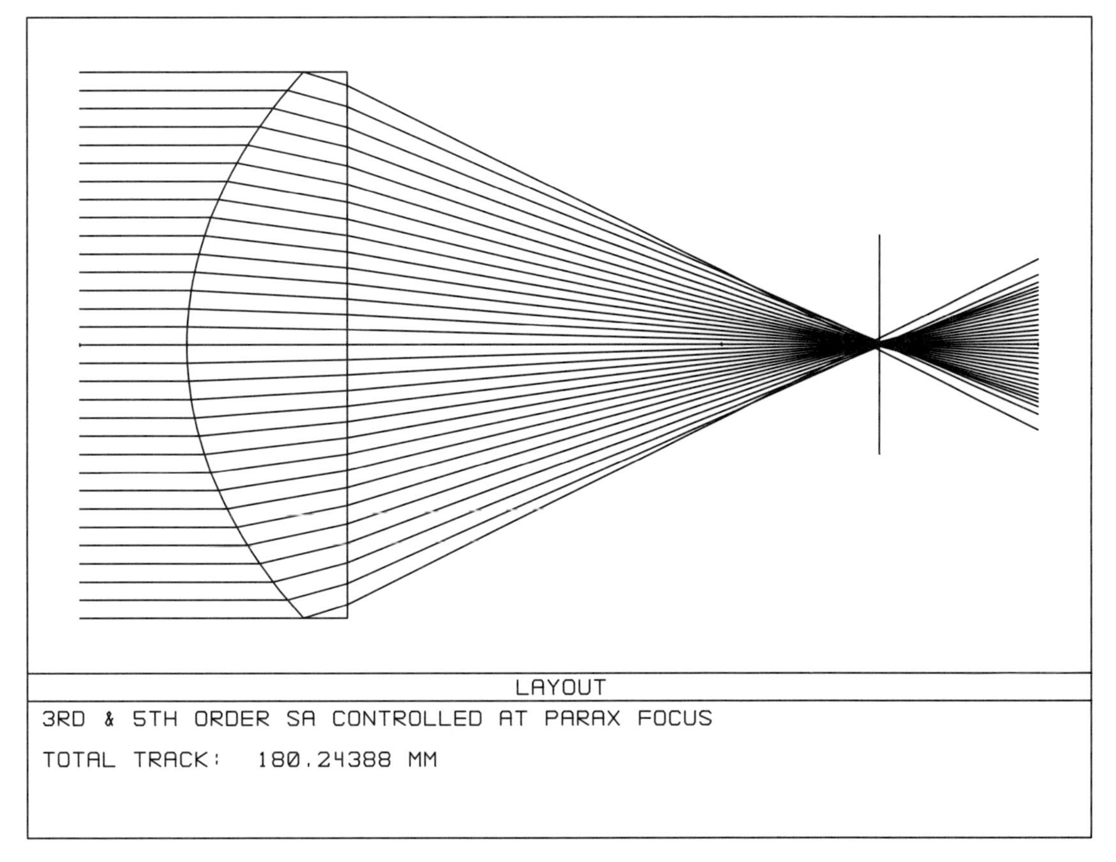

Figure A.7.6.1.

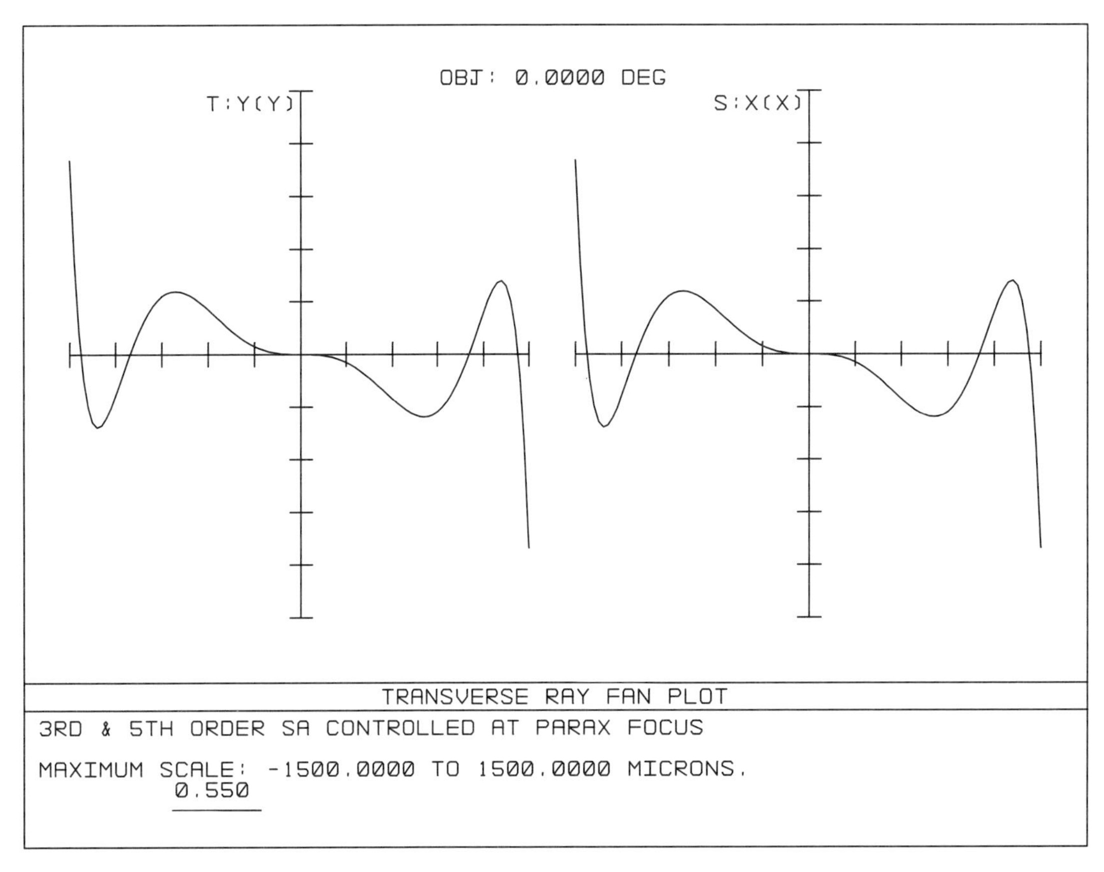

Figure A.7.6.2.

The spot diagrams are now beginning to all look the same except for spot size. Little diagnostic information is conveyed. Thus, the spot diagram is not shown, but RMS spot diameter is 594 μm.

Figure A.7.6.2 is the ray fan plot, and here the lens designer finds the diagnostic information he needs. First, note the scale. Second, note that the curve is flat around the origin, indicating paraxial focus. Third, note that the curve initially goes in the same direction as the curve in Figure A.7.1.5, and in the opposite direction from the curve in Figure A.7.3.2, indicating undercorrected third-order spherical aberration for inner pupil zones. Fourth, note that farther out in the pupil, the curve reverses direction, indicating that here overcorrected fifth-order spherical has become dominant. Fifth, in the outermost pupil zones, the curve reverses direction still again, indicating that higher orders of undercorrected spherical finally predominate. Sixth, note that there are two zero crossings where the total error goes to zero.

A.7.9 Third- and Fifth-Order Spherical Aberration Controlled at Best Focus

If adding defocus helped the lens in Figure A.7.4.1, then it might be expected that defocus would similarly help the lens in Figure A.7.6.1. The reoptimized lens is shown in Figure A.7.7.1.

Again the spot diagram is not shown, but RMS spot diameter is 407 μm. Figure A.7.7.2 is the ray fan plot. Most of what was said about Figure A.7.6.2 applies here too. One difference is that the slope of the curve is no longer zero around the origin, indicating a departure from paraxial focus. As you go outward in the pupil, the dominant aberrations are: defocus near the center, then undercorrected third-order spherical, then overcorrected fifth-order spherical, and finally undercorrected higher

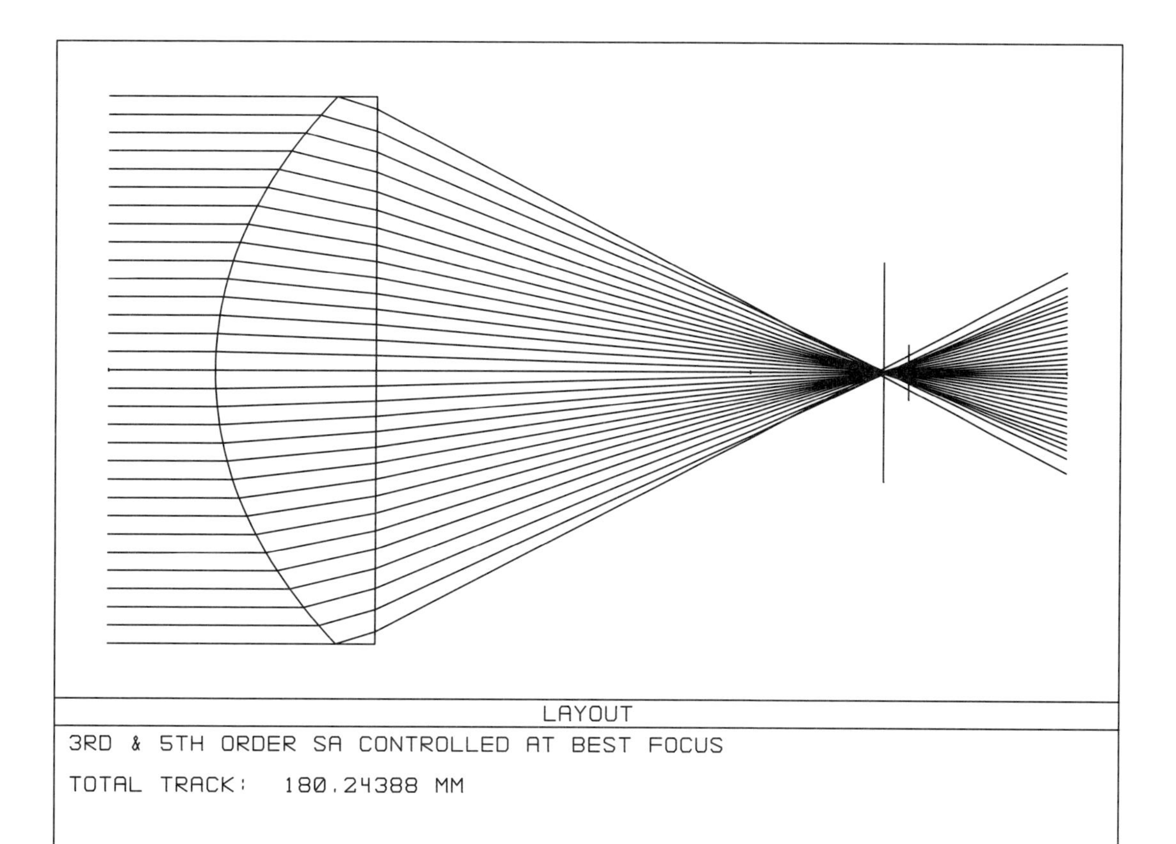

Figure A.7.7.1.

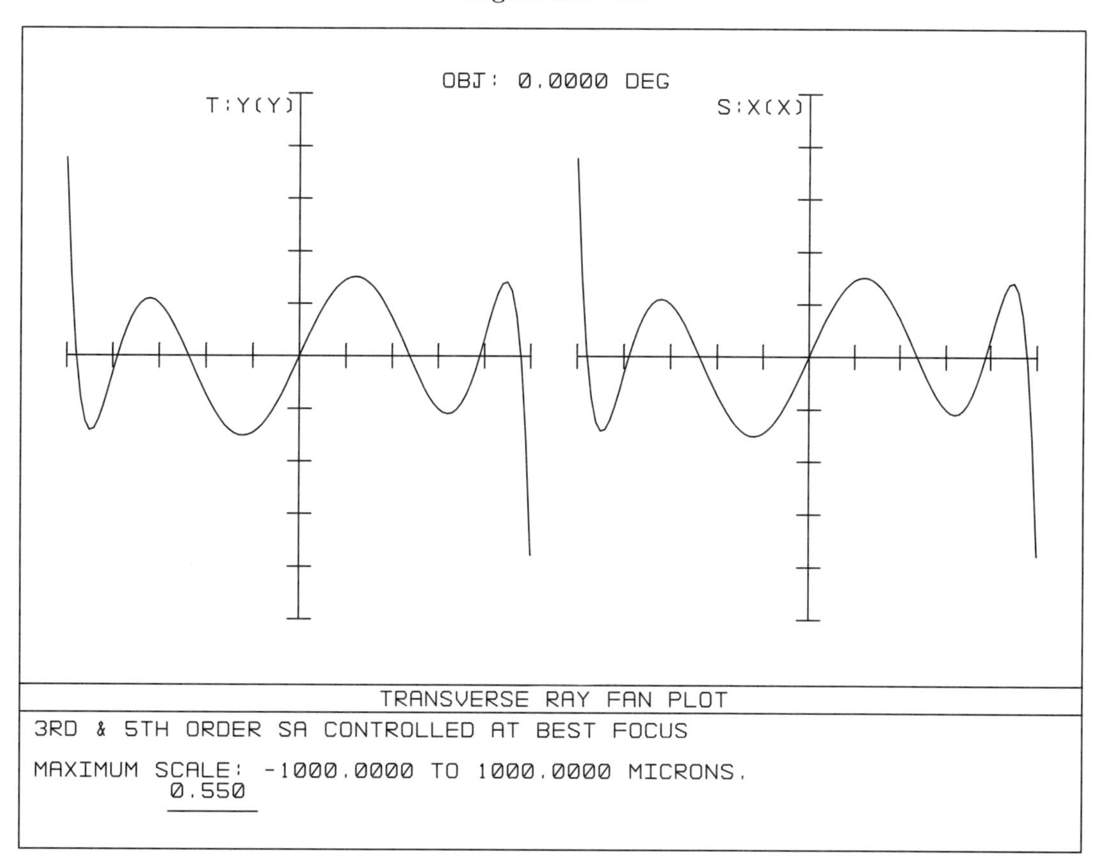

Figure A.7.7.2.

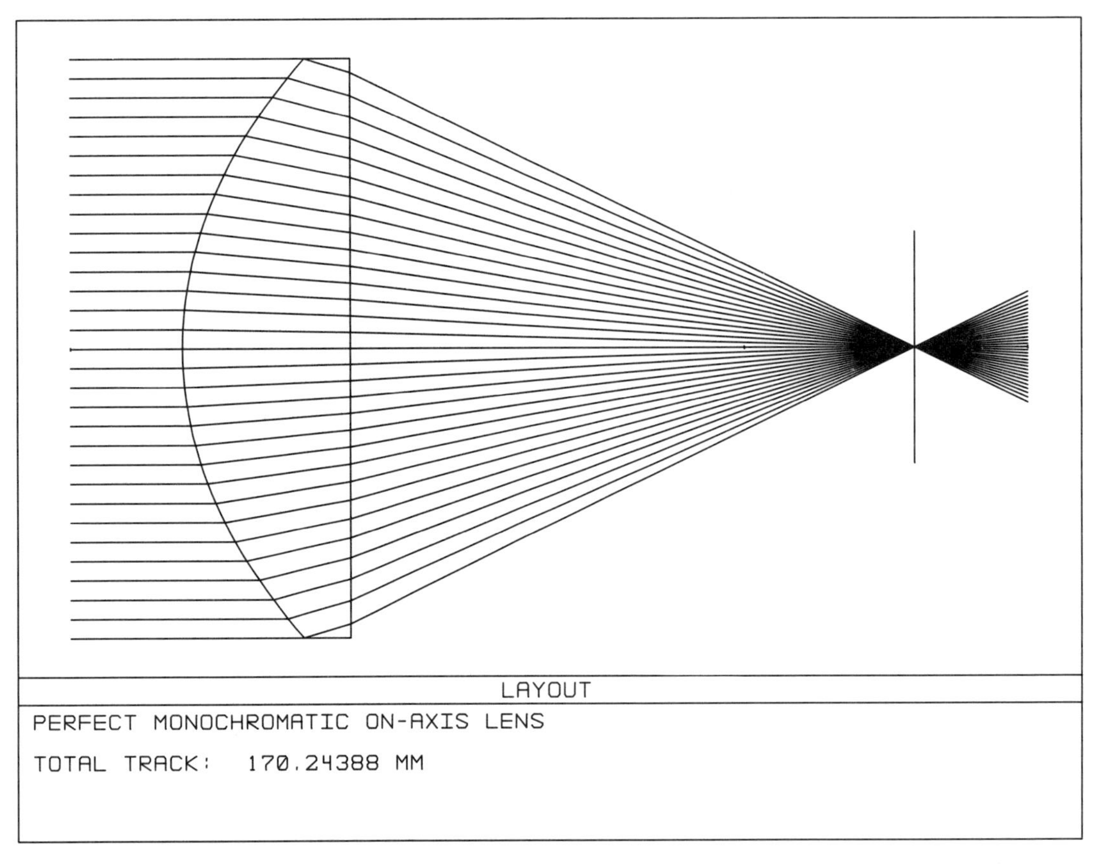

Figure A.7.8.1.

orders of spherical at the pupil edge. A second difference from Figure A.7.6.2 is that now there are three zero crossings.

A.7.10 A Perfect Monochromatic On-Axis Lens

The above procedure can be continued, in theory, for as long as you wish. Higher and higher orders of polynomial deformations can be added to control higher and higher orders of spherical aberration. If continued infinitely, the front lens surface would approach the ideal shape that causes all geometrical rays to be perfectly united at the focus.

For a finite number of polynomial deformation terms, the residual image errors are not zero. But with enough terms, these errors are significantly less than the image spreading caused by diffraction. Any further improvement is inconsequential, and the lens is effectively perfect. However, note that this perfection applies only to one wavelength of monochromatic light and only to the on-axis image. A layout of the final "perfect" lens is shown in Figure A.7.8.1.

It may be of interest that for a plano-convex singlet lens with the curved surface forward and the object at infinity, the ideal shape of the curved surface for no spherical aberration can be approximated by a prolate ellipsoid. The approximation is not exact, but an ellipsoid is certainly much better than a sphere. Polynomial deformations can be added to the ellipsoid to further reduce residual errors. It is significant that with the ellipsoid rather than the sphere as the base curve, many fewer polynomial terms are required for the same level of aberration control.

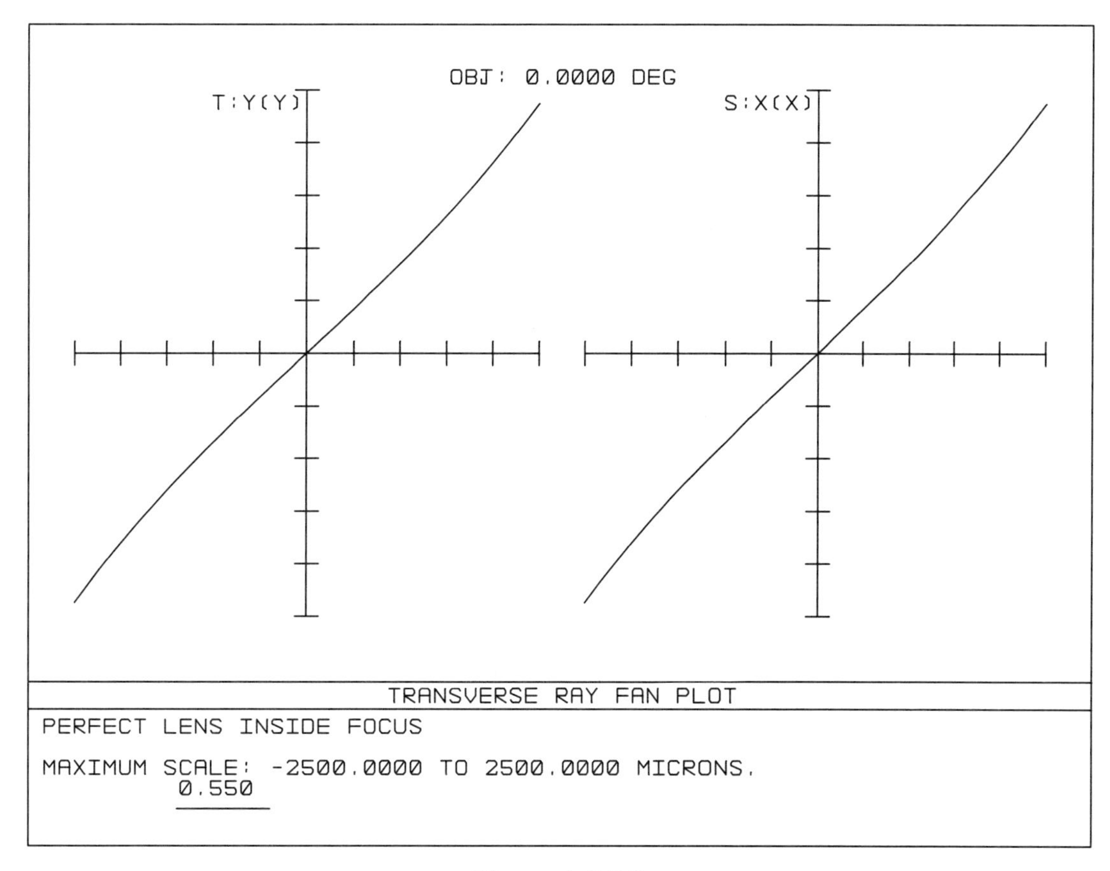

Figure A.7.8.2.

A.7.11 A Defocused Perfect Lens

If the image surface of the lens in Figure A.7.8.1 is displaced longitudinally away from focus, then pure defocus is introduced into an otherwise perfect image. For pure defocus, the relative ray distribution in the spot diagram is nearly identical to the ray distribution in the pupil. Of course, the scale is different. Refer again to Figure A.6.6.1.

Figure A.7.8.2 is the ray fan plot for the image surface shifted 5.0 mm toward the lens or inside focus. Based on the discussion of defocus in the previous chapter, a straight line with positive slope would be expected. But the line in Figure A.7.8.2 is not perfectly straight. At first, the curve seems to indicate defocus plus a little spherical aberration. But this lens has no spherical aberration. Thus, the departure of the ray fan plot from a straight line must be caused by a curved effective refracting surface. In fact, a lens of this shape has very little coma, and with spherical aberration corrected, the lens is nearly aplanatic. Thus, the effective refracting surface is nearly spherical with the center of curvature at the image.

If the image surface of the lens in Figure A.7.8.1 is displaced 5.0 mm in the other direction (away from the lens or outside focus), then the ray fan plot should look very similar to Figure A.7.8.2 but with a negative slope. Figure A.7.8.3 shows this plot.

A.7.12 Balancing Aberrations in Multi-Element Lenses

In the above examples, spherical aberration was controlled by aspherizing a lens surface. This technique does work, and it illustrates in a simple way the concept of aberration balancing. However, in most cases in practical optics, spherical

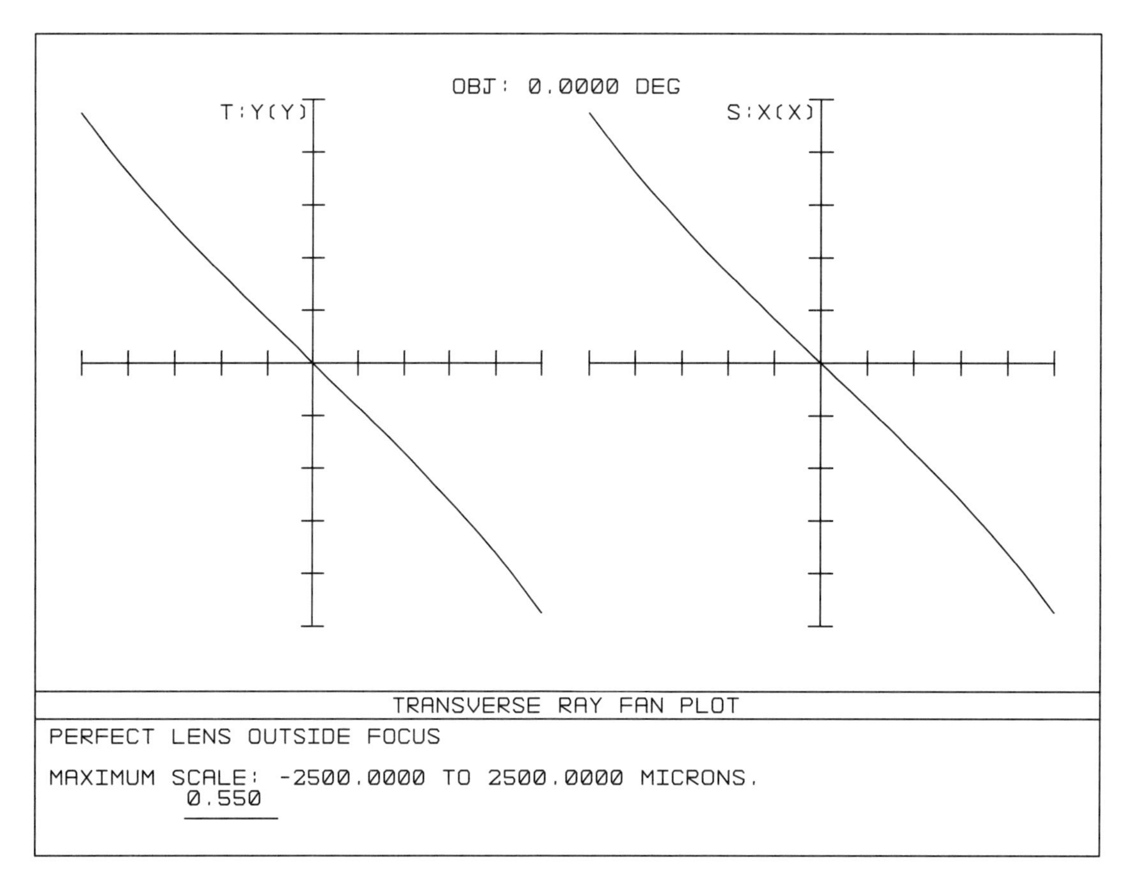

Figure A.7.8.3.

aberration is *not* controlled with aspherics because of their fabrication difficulty and cost. Instead, multi-element lenses with all-spherical surfaces are used. In these compound lenses, the amounts and signs of the spherical aberration contributions from the several different lens surfaces are adjusted during optimization to balance or cancel each other and thus give a small overall sum.

The principle of aberration balancing or cancelling in multi-element lenses can be applied to all types of aberrations, not just to spherical aberration. Furthermore, the principle can be applied during optimization to several different aberrations *simultaneously*. This is very important because this is the principle on which nearly all well-corrected optical systems operate.

A.7.13 Longitudinal Chromatic Aberration

Figure A.7.9.1 again shows the BK7 singlet lens of Figure A.7.8.1. In this case, however, the light is no longer monochromatic. Polychromatic light is used at three wavelengths: 0.300 μm in the ultraviolet, 0.550 μm in the visible, and 2.50 μm in the infrared. Only the marginal rays have been drawn to avoid confusion in the layout. Note the significant changes in focal position (BFL) between the three wavelengths. Index of refraction is greater for shorter wavelengths. Thus, for this singlet lens, the ultraviolet light focuses nearest to the lens, and the infrared light focuses farthest from the lens.

This wavelength-dependent change in BFL is called longitudinal chromatic aberration. Alternative terms are axial chromatic aberration and primary chromatic aberration. For simplicity, the term chromatic aberration is often replaced by the word color, as in longitudinal color.

Figure A.7.9.2 is the ray fan plot. The different wavelengths are distinguished by different dashed lines identified in the caption. The image surface has been

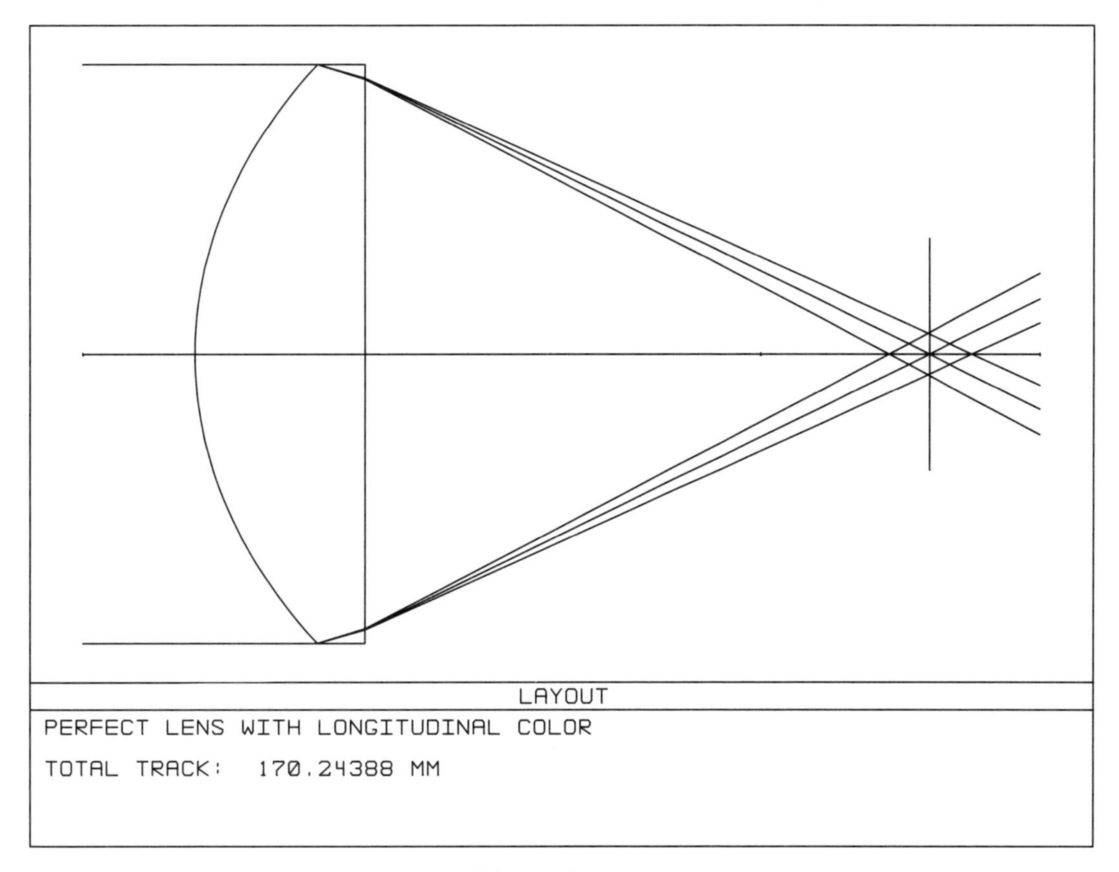

Figure A.7.9.1.

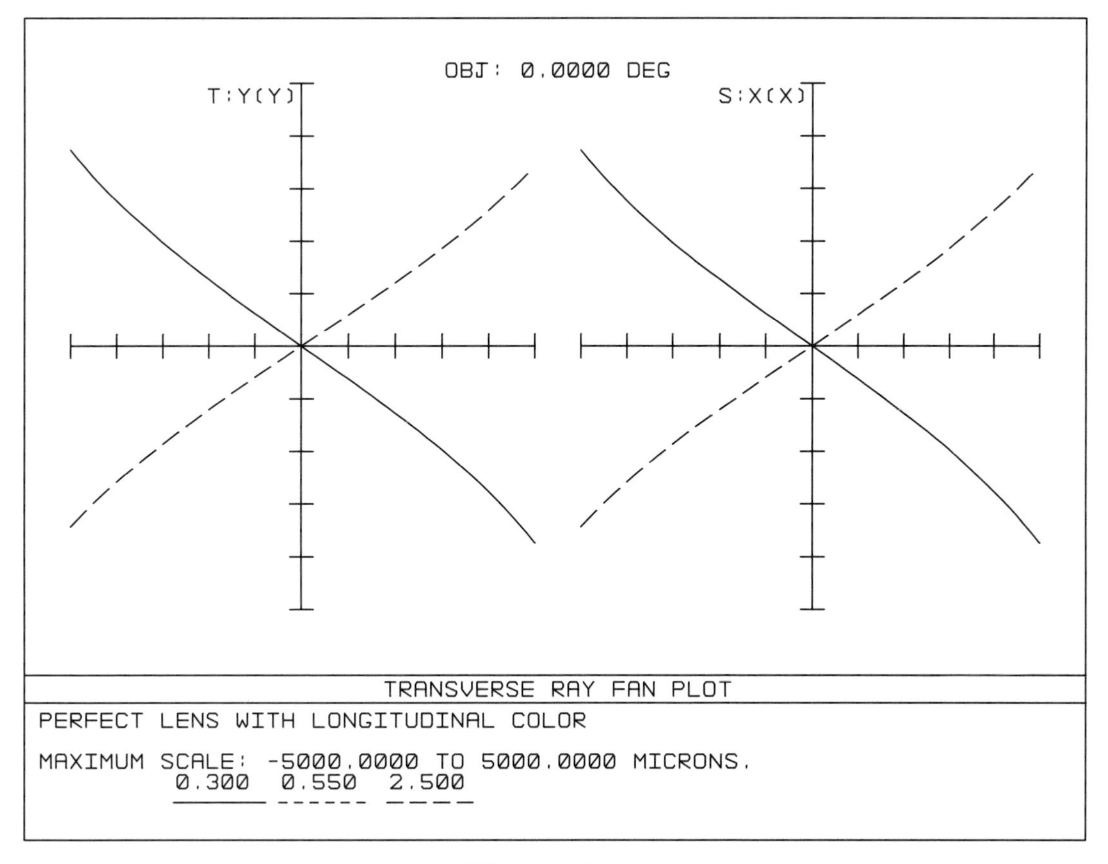

Figure A.7.9.2.

deliberately placed at the focus of the visible light. Because the lens was designed to produce a perfect image at 0.55 μm, the curve for this wavelength on the ray fan plot lies on top of the horizontal axis and cannot be separately seen. The curve for the ultraviolet light shows a negative slope, indicating that, for this wavelength, the image surface lies outside focus (the focus lies closer to the lens than the image surface). Similarly, the curve for the infrared light shows a positive slope, indicating that, for this wavelength, the image surface lies inside focus (the focus lies beyond the image surface). This is just what is seen in the layout.

The process of controlling color is called achromatization (meaning no-color), and the resultant lens is called an achromat. To control longitudinal color, the BFLs for two different wavelengths are made equal during optimization. To achieve this, it is necessary to have a compound lens made of two or more elements. The various elements are made of at least two different glass types, each type having a different dispersion.

Note that an achromat brings two colors, but not all colors, to the same focus. The residual BFL errors for other wavelengths are called secondary longitudinal chromatic aberration, secondary color, or, traditionally, secondary spectrum. By choosing special glasses, lenses can be designed that bring three colors to a common focus. Such lenses are called apochromats. Much more about controlling longitudinal color will be said in Chapter B.1.

A.7.14 Other Chromatic Aberrations

As was just explained, BFL can change with wavelength, and this causes a wavelength-dependent change in focus; that is, longitudinal chromatic aberration. EFL can also change with wavelength, and this causes a wavelength-dependent change in transverse image scale; that is, lateral chromatic aberration. When unspecified, the term chromatic aberration is usually understood to mean the longitudinal type. In practice, lateral chromatic aberration is identified as lateral. More will be said about lateral color in the next chapter. Note that these changes of BFL and EFL are both wavelength-dependent variations of *first*-order properties.

There may also be wavelength-dependent variations of the monochromatic third- and higher-order aberrations. For example, there is nothing to guarantee that if you optimize a lens for zero spherical aberration at one wavelength, then other wavelengths also have zero spherical. The variation of spherical aberration with wavelength is called spherochromatism. Similarly, all the off-axis monochromatic aberrations can vary with wavelength.

The designer must always be vigilant against all types of chromatic image defects. Because reflection is wavelength independent, only all-reflecting systems are completely free of all chromatic aberrations.

A.7.15 Defocus with an On-Axis Paraboloidal Mirror

Figure A.7.10.1 shows the layout of a very fast concave paraboloidal (or parabolic) mirror imaging a distant object located on the optical axis. The mirror focuses rays into nearly a hemisphere or 2π steradians; that is, nearly the maximum possible solid angle for a system where light is not allowed to approach the image surface from behind. Focal length is 100 mm, entrance pupil diameter is 300 mm, and focal ratio is $f/0.33$. Such a mirror, in somewhat slower form, is the heart of the classical reflecting telescope.

A paraboloidal mirror imaging a distant on-axis object is a very special imaging system. It is one of the few known examples of a "lens" (photographers even use the term mirror-lens) that can form at least one mathematically perfect geometrical image completely free of *all* aberrations. First, the system is all-reflecting,

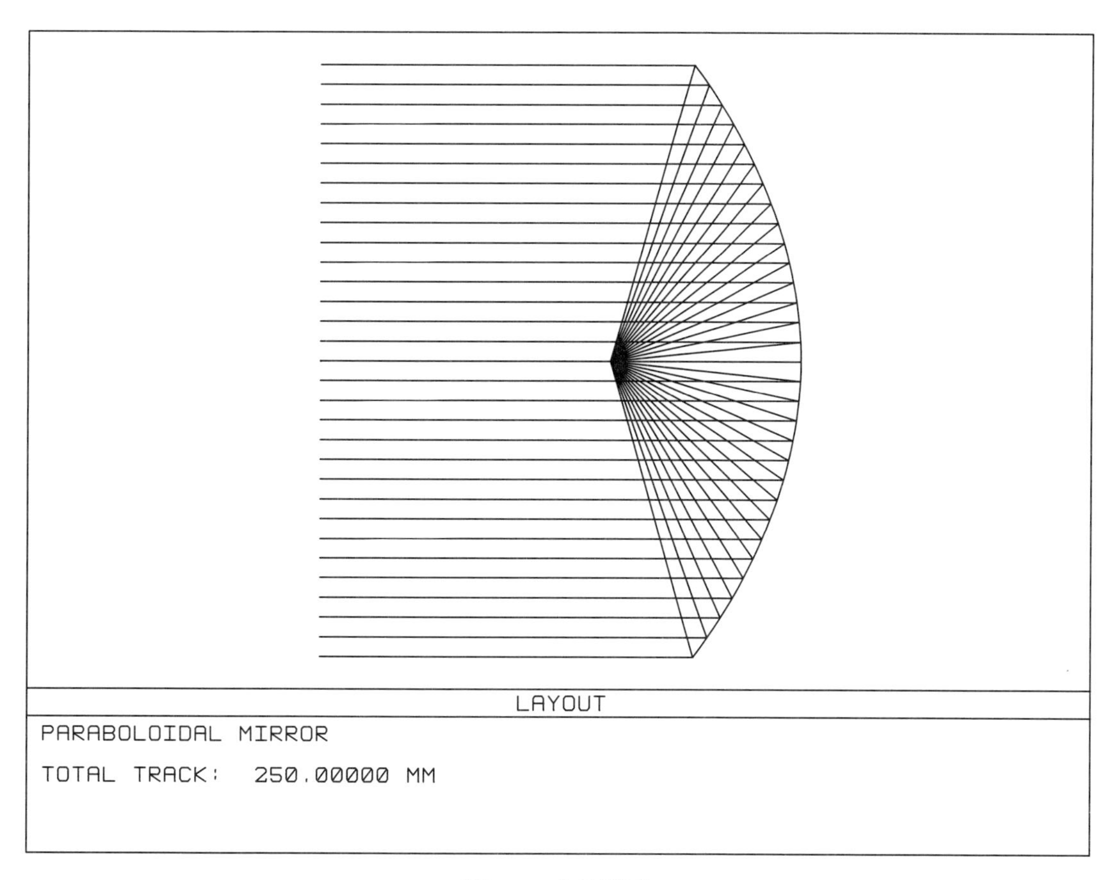

Figure A.7.10.1.

and thus there are no chromatic aberrations. Second, spherical aberration is the only monochromatic aberration (except defocus and fabrication errors) that can be present on-axis. Third, all orders of spherical aberration are identically zero; this is especially unusual. Thus, when in focus, perfect geometrical imagery is obtained in theory even with polychromatic white light. But this perfect imagery only occurs exactly on the optical axis. Going the slightest distance off-axis introduces coma to the image. And, of course, diffraction ruins even the on-axis perfection, but diffraction is ignored here.

This perfect image yields a spot diagram that is a single point and a ray fan plot that is coincident with the horizontal axis. These match the ideal. Figure A.7.10.2, however, illustrates what happens if you deliberately defocus a short distance outside focus. Figure A.7.10.2 is a partial layout showing the rays diverging from the focus toward a plane image surface. Focus is on the right, and the rays are traveling toward the left (recall that the direction of travel is reversed by the reflection at the mirror). The ray distribution in the image has now become finite because defocus is an aberration.

In Figure A.7.10.1, the rays entering the system are equally spaced and a perfect focus is formed. But in Figure A.7.10.2, the ray piercing points on the image surface can be seen to be not equally spaced. Examining Figure A.7.10.1 reveals that the effective "refracting" surface of the system is the reflecting surface of the mirror itself. The nonuniform ray spacing on the image surface is caused by the effective refracting surface being curved. In fact, if the paraboloid were made only a little faster (400 mm aperture), then the surface of the concave mirror would extend forward enough to intersect the extension of the image surface. The rays reflected from the edge of the mirror would then graze the image surface and not intersect at all; that is, the ray spacing would go to infinity.

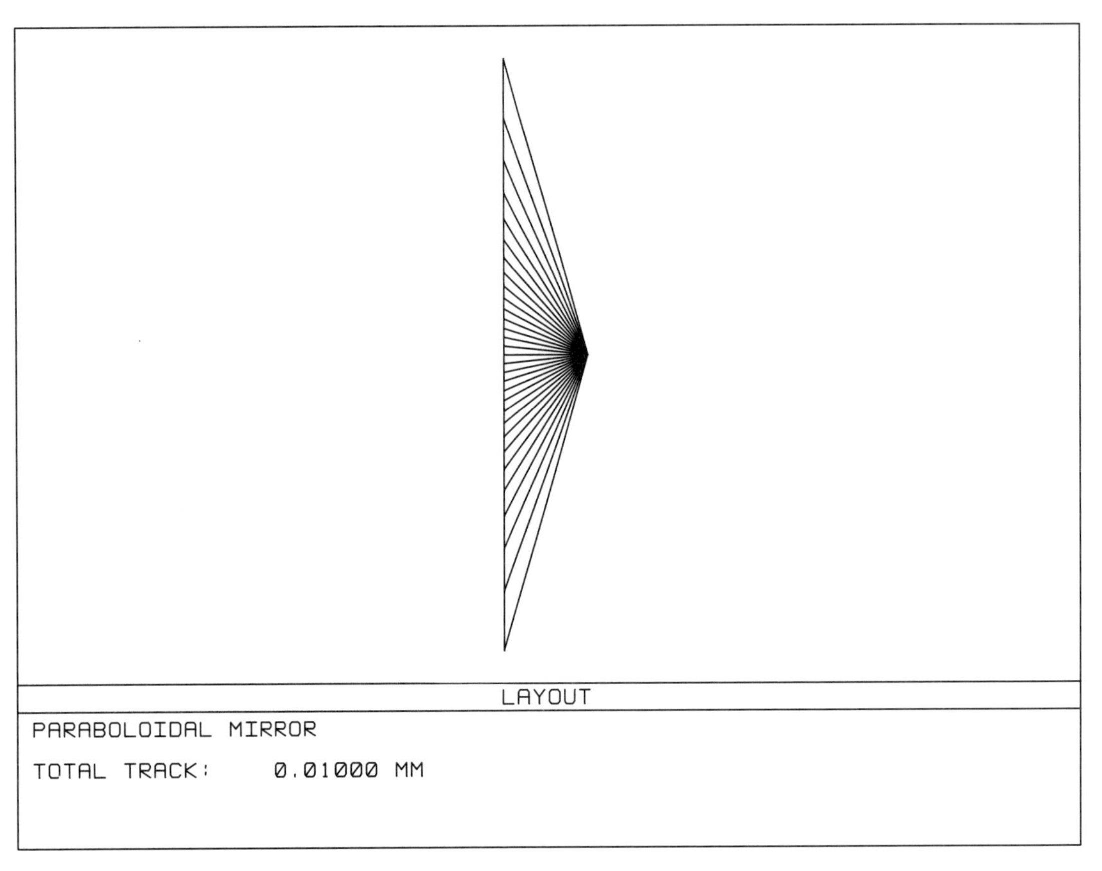

Figure A.7.10.2.

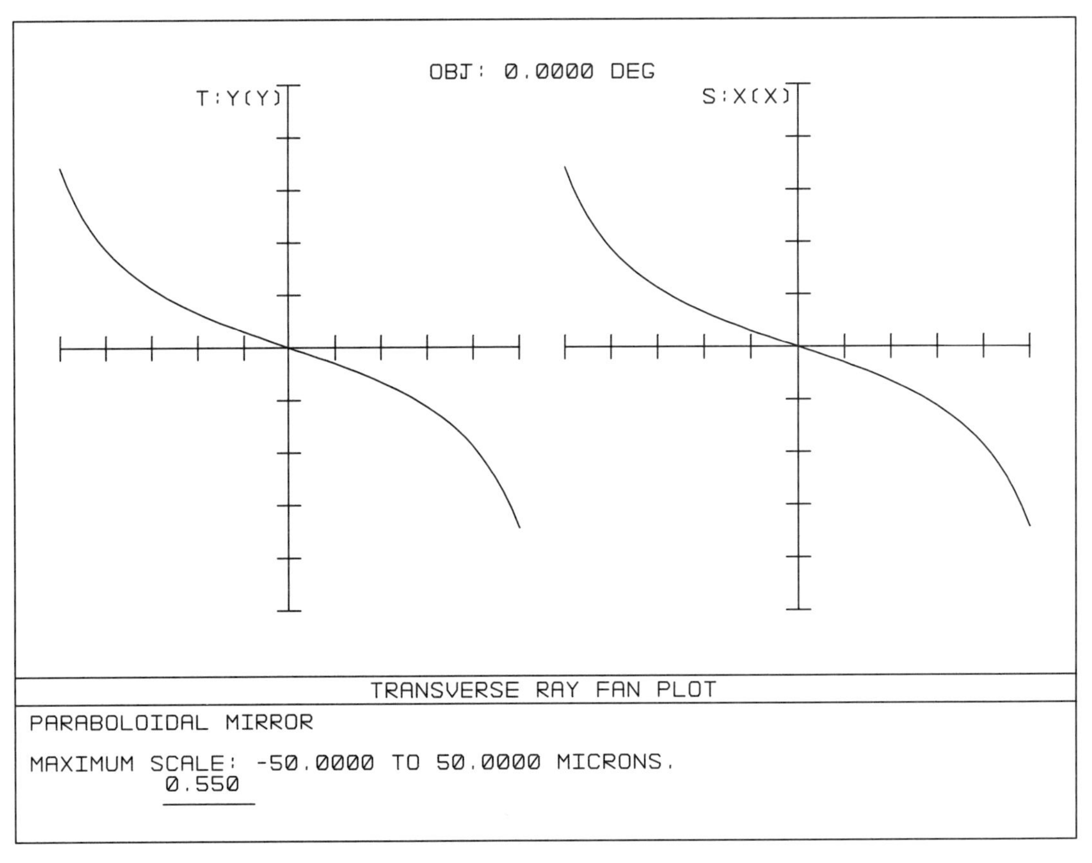

Figure A.7.10.3.

Figure A.7.10.3 is the ray fan plot. Note the nonzero negative slope around the origin (the image surface is not at paraxial focus) and the strong curvature of the line. This curvature cannot be caused by spherical aberration because there is none. Rather, this curvature is again caused by the curved effective refracting surface. Recall that this same effect was seen earlier to a lesser extent in Figures A.7.8.2 and A.7.8.3 for the singlet lens. Fortunately, this confusion with spherical aberration on the ray fan plots of defocused images is only encountered rarely, and then only for very good and very fast systems. For slower, more normal systems, pure defocus yields a virtually straight, inclined ray fan plot.

A fast paraboloidal mirror also illustrates another interesting situation. In Chapter A.6, the theoretical maximum speed of an aplanatic system is given as $f/0.50$. But for the paraboloidal mirror, the maximum aperture of 400 mm and a focal length of 100 mm yields a focal ratio of $f/0.25$. How can this be? Recall that the effective refracting surface of an aplanat is spherical, and the effective refracting surface of the mirror is paraboloidal (which is definitely not a sphere). The paraboloidal mirror is thus definitely not aplanatic. Far from having a constant EFL with pupil zone, the marginal EFL at the edge of an $f/0.25$ paraboloid is twice the paraxial EFL at its center. This shape of the effective refracting surface is why a paraboloidal mirror has such a large amount of coma.

This concludes the introduction to image aberrations found on the axis of axially centered optical systems. The next chapter continues the discussion by introducing the off-axis aberrations.

Chapter A.8

Off-Axis Geometrical Aberrations

In the previous chapter, the on-axis aberrations were introduced. These aberrations, however, are not confined to just the axial region in the center of the image; if they are present, they are equally present all across the field. On-axis aberrations were discussed separately because they can be isolated from the off-axis aberrations that appear *only* off-axis.

In this chapter, the off-axis aberrations are introduced. The only first-order off-axis aberration is lateral chromatic aberration. The monochromatic third-order Seidel off-axis aberrations are: coma, astigmatism, field curvature, and distortion. There are many more higher-order off-axis aberrations, such as fifth- and higher-order versions of the Seidel aberrations. There are also some uniquely higher-order off-axis aberrations, such as fifth-order oblique spherical aberration. In addition, all these monochromatic aberrations have chromatic variations.

As in the previous chapter, the examples are actual designs and not cartoons. Although each of these examples has been optimized to isolate the particular aberration under consideration, small amounts of other aberrations do creep in. In the real world, image blurs are usually the result of many aberrations appearing simultaneously.

A.8.1 Lateral Chromatic Aberration

If the EFL of a lens varies with wavelength, then the size of the image, the transverse image scale, or chief ray height also varies with wavelength. Depending on the sign of the aberration, the red image can be either larger or smaller than the blue image. This image defect is called lateral chromatic aberration, lateral color, or, traditionally, chromatic difference of magnification.

If a lens has lateral color and is used to image a field of point objects such as stars, then the images of the stars are stretched out radially from the axis like a shell burst. Each image looks like a small spectrum whose length depends linearly on the off-axis angle or distance.

Lateral color is seen on a polychromatic spot diagram as relative transverse displacements of the image spots for the various colors. On a ray fan plot, lateral color is seen as vertical shifting of the curves. Usually, the curve for the reference wavelength is drawn exactly through the origin, with the curves for other wavelengths crossing the vertical axis above or below the origin. Although these crossing points along the vertical axis only reveal lateral color for the polychromatic chief rays, often this is the most useful indicator. For outer pupil zones, too many other effects can confuse the issue.

To control lateral color, the design approach is often similar to the way longitudinal color is controlled; that is, a compound lens made of different glasses with different dispersions is used. During optimization, the heights of the chief rays for two different wavelengths are made equal on the image surface. Other wave-

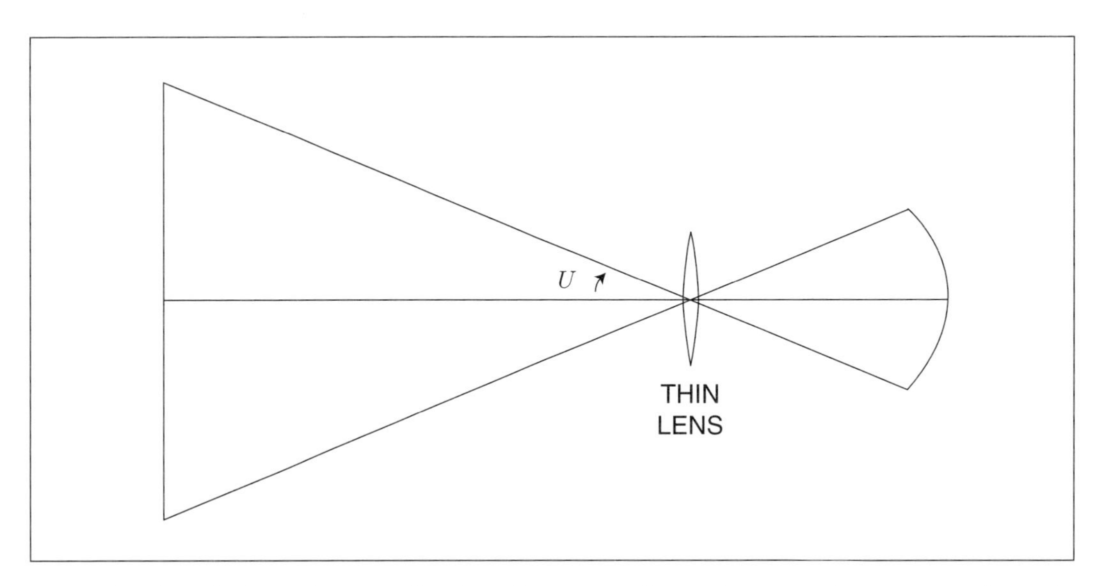

Figure A.8.1.1. Natural Curved Field. *A thin lens with the stop at the lens images a flat object surface. Off-axis object points are farther from the lens than the on-axis object point. From the thin-lens equation, the off-axis images are closer to the lens than the on-axis image, thereby producing a curved field.*

lengths have small residual errors, and these are called secondary lateral chromatic aberration or secondary lateral color.

Note that unlike longitudinal color, controlling lateral color does not *necessarily* require different glass types. There are some lens designs, for example, the Huygens eyepiece, that control lateral color with only one glass type. This control is a result of the use of a symmetrical optical system. Symmetry works by pairing aberrations in two halves of a lens. The aberrations have opposite signs and cancel. The lens need not be perfectly symmetrical for this principle to be effective. Symmetry, which is also useful in cancelling coma and distortion, is further discussed in Chapter A.9.

Finally, note too that the lateral color of a thin singlet lens with the stop at the lens is always zero. In this thin-lens approximation, the chief rays of all wavelengths pass straight through the lens center with no deviation.

A.8.2 Field Curvature

In a lens imaging an extended field of view, if the best image surface is flat, then the lens is said to have a flat field. If the best image surface is not flat, then the lens is said to have a curved field or field curvature. The major component of field curvature is usually a field-dependent defocus; that is, BFL changes with field zone.

A curved image surface is usually a practical disadvantage because most two-dimensional image detectors, such as photographic plates/films and CCD arrays, are flat. If an optical system cannot be designed to have a flat field, then either the image detector must be built or warped into a curve, or progressively degraded off-axis images must be tolerated.

There is nothing in optics that says that lenses intrinsically want to have a flat field. In fact, a flat image of a flat object is not the natural state of affairs. Figure A.8.1.1 illustrates the problem. A thin lens with the stop at the lens views a flat object surface. The distance from the lens to an off-axis object point varies as $1/\cos U$, where U is the off-axis angle. Thus, off-axis objects are farther from the lens than the on-axis object. It follows from this increasing object distance plus the thin-lens equation, Equation A.3.3, that off-axis images should focus closer to the lens than the on-axis image, thus giving a curved field.

By the same reasoning, if the image surface is also to be flat, as shown with the

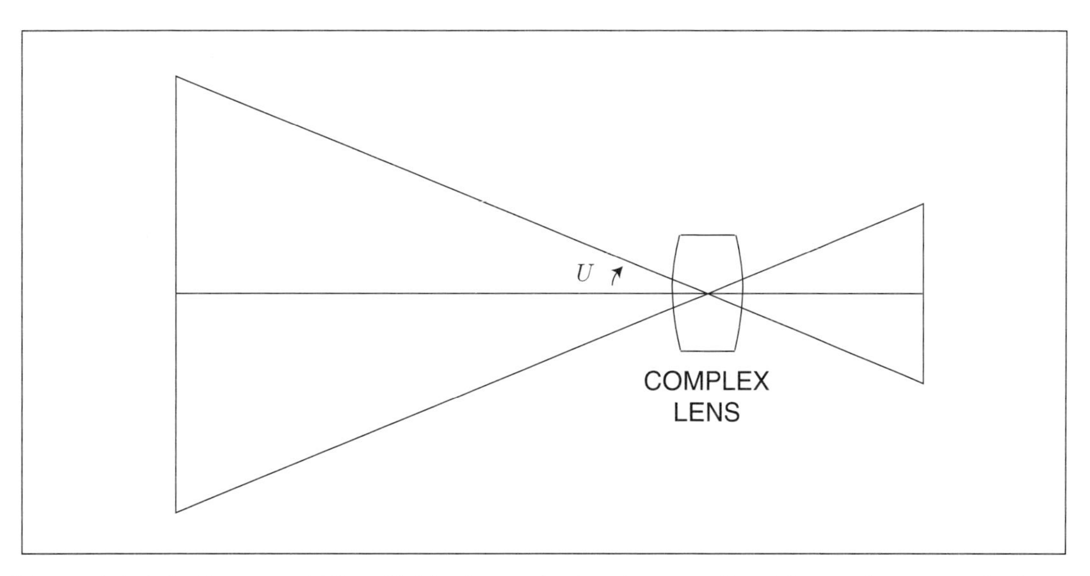

Figure A.8.1.2. Corrected Flat Field. *A complex lens images a flat object surface onto a flat image surface. Not only are the off-axis objects farther from the lens, but so too are the off-axis images. Clearly, the optical properties of the complex lens are significantly different from those of the thin lens.*

complex lens in Figure A.8.1.2, then off-axis images are farther from the lens than the on-axis image, not closer. Clearly, the complex lens must possess optical properties quite different from those of the thin lens. Achieving this flat-field performance has been one of the driving concerns of lens designers since the invention of the first flat-field lens, the Wollaston Landscape lens, in 1812.

Field curvature is usually discussed in connection with astigmatism. Certainly, field curvature and astigmatism are closely related, and field curvature will be further discussed in the section on astigmatism later in this chapter. However, too much emphasis on this connection may be misleading. Field curvature or field-dependent defocus affects all off-axis aberrations in the same way that axial defocus affects all on-axis aberrations.

A.8.3 Coma

Perhaps the most serious and most subjectively disturbing off-axis aberration is coma. Coma causes off-axis object points to be imaged as asymmetrically flared out (or in) image blurs. The aberration coma was named after the word comet. Comets are diffuse objects that appear now and then in the night sky. Often, comets have long tails that spread out in one direction from a brighter nucleus. Comatic images resemble these comets with their tails. It may be of further interest that the word comet derives from the Greek word kome, meaning hair, because comets were thought to look like hairy stars.

Figure A.8.2.1 is the layout of a lens specially designed to demonstrate coma. This lens is to be used monochromatically and has relatively little spherical aberration. The stop has been shifted to remove the astigmatism so that off-axis the only significant aberration is a large amount of nearly pure positive third-order coma. There are two object points on the distant object surface: the on-axis object point and an object point 5° off-axis. Note that the image surface is slightly curved and that the upper and lower marginal meridional rays focus together farther from the axis than the chief ray. It is possible to have optical systems with negative coma, where the marginal rays focus closer to the axis than the chief ray.

Figure A.8.2.2 gives hexapolar spot diagrams. Recall that in making a hexapolar spot, the rays are incident on the entrance pupil in a series of concentric rings. A hexapolar ray array is particularly effective in demonstrating coma.

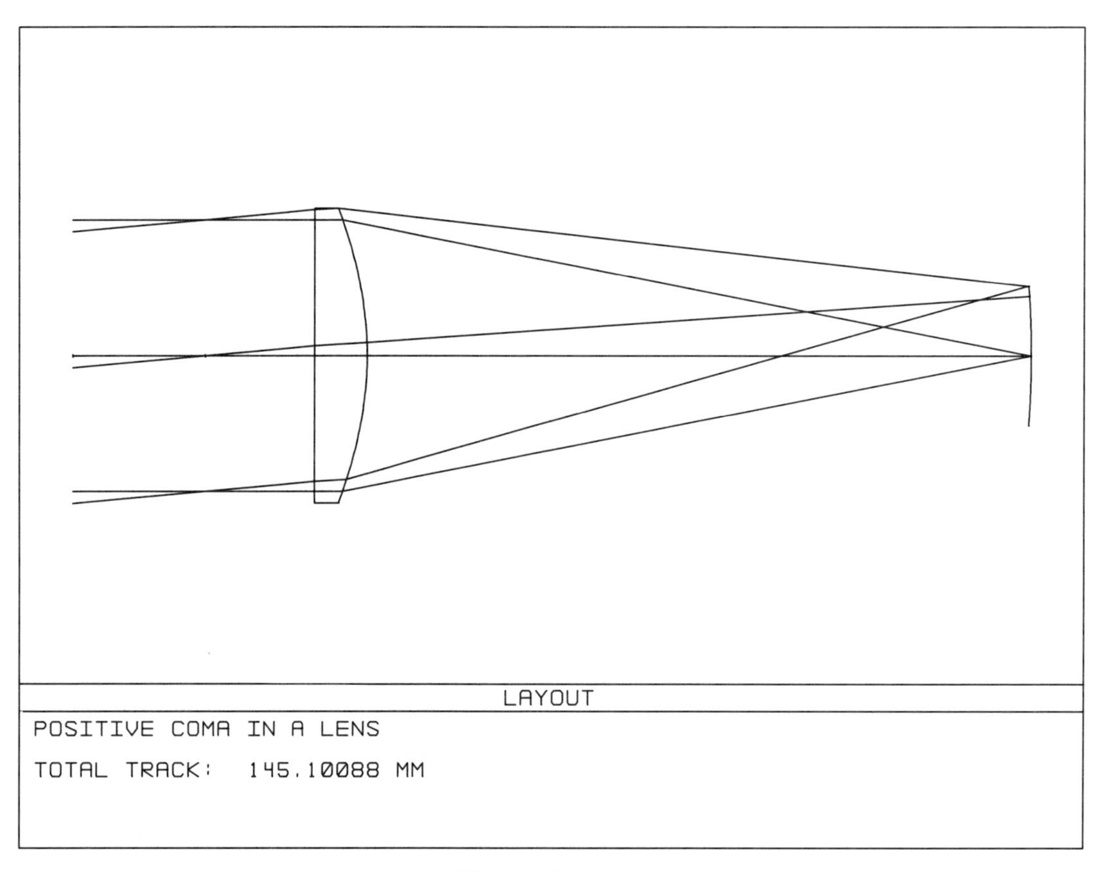

Figure A.8.2.1.

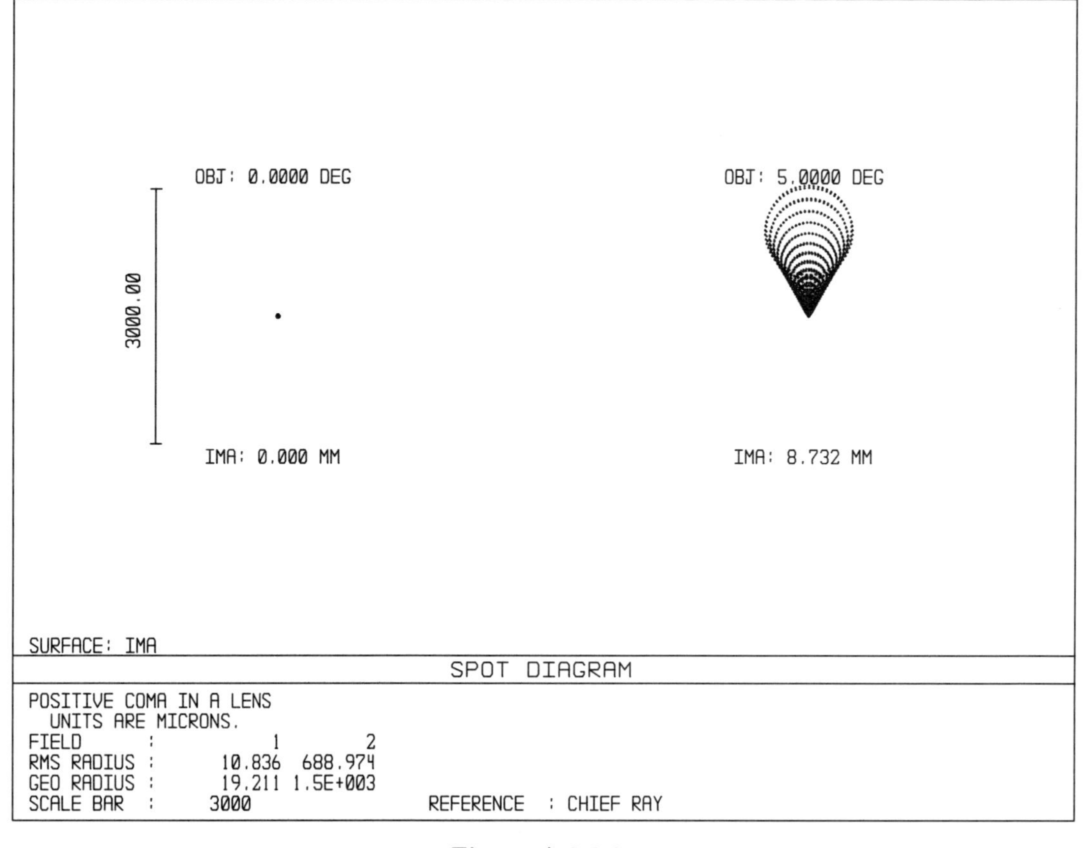

Figure A.8.2.2.

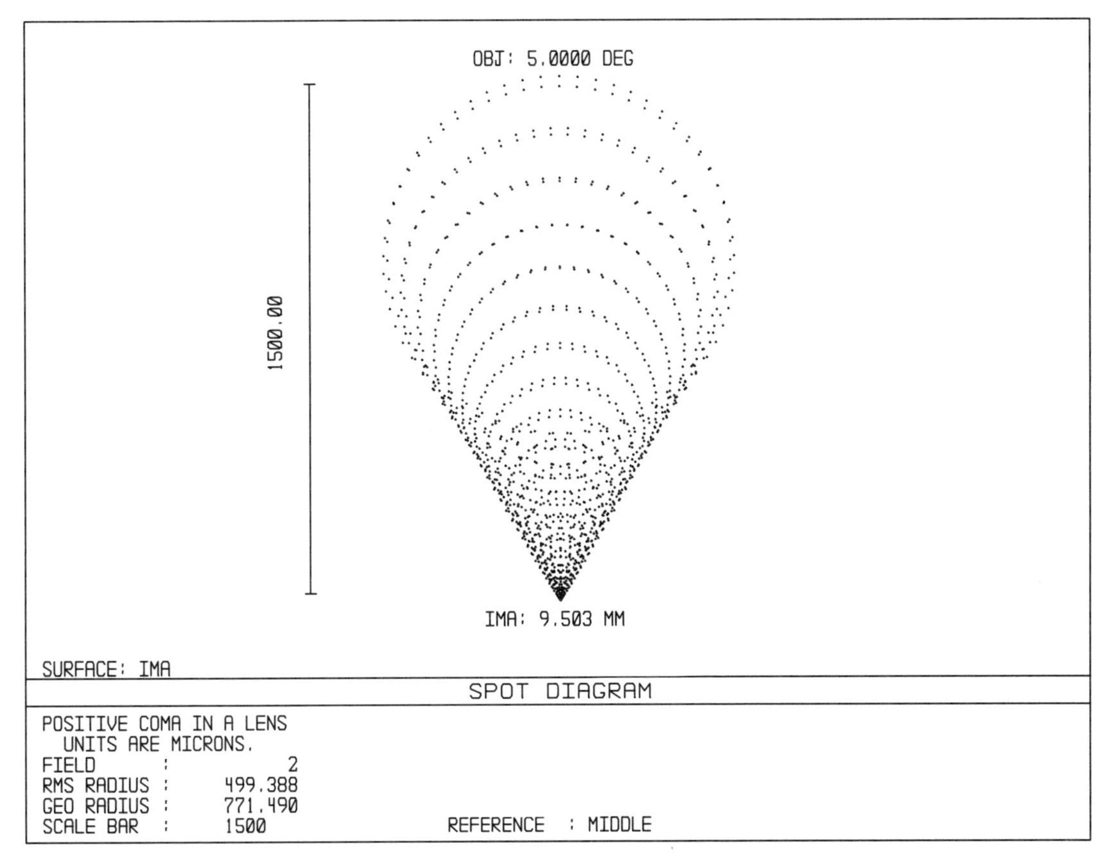

Figure A.8.2.3.

The axial image is virtually perfect at this scale. The off-axis image shows comatic flare spreading away from the chief ray, which is at the tip of the V-shaped pattern. This flare also spreads radially away from the center of the field. Figure A.8.2.3 is an enlargement of just the off-axis spot to show its structure better.

Coma is the result of a change in EFL with pupil zone height. There are two consequences. The first is that as EFL changes, image height changes, and this causes a change in image scale or magnification with pupil zone. The second is that as EFL changes, there is an accompanying effect that somewhat resembles a change in defocus with pupil zone. The two effects together produce the comatic flare.

The hexapolar spot in Figure A.8.2.3 shows how this works. If you trace a ring of rays all around any one pupil zone, the piercing points of the rays on the image surface also fall in a ring. The diameters of the image rings increase as pupil zone height increases. The positions of the centers of the image rings also increasingly shift radially as pupil zone height increases. The direction or sign of the shift depends on the sign of the coma. The amount of change in diameter and amount of radial shift are such that, for pure third-order coma, one side of a given image ring is three times as far from the chief ray as the other side. The image rings are all tangent to two straight lines at a 60° angle to each other that intersect at the chief ray. The result is the characteristic V-shaped appearance of coma. All this can be seen in Figure A.8.2.3.

Actually, one circuit around a pupil zone takes you *two* circuits around an image ring. For pure third-order coma, both image circuits are on top of each other. But in the real example in Figure A.8.2.3, small amounts of other aberrations slightly perturb the image rings and allow both circuits to be separately seen. If the image were deliberately defocused a small amount, then a double image ring would be changed into a closed double loop with one loop larger than the other. This double loop behavior is clearly not the result of simple defocus as a function of pupil zone,

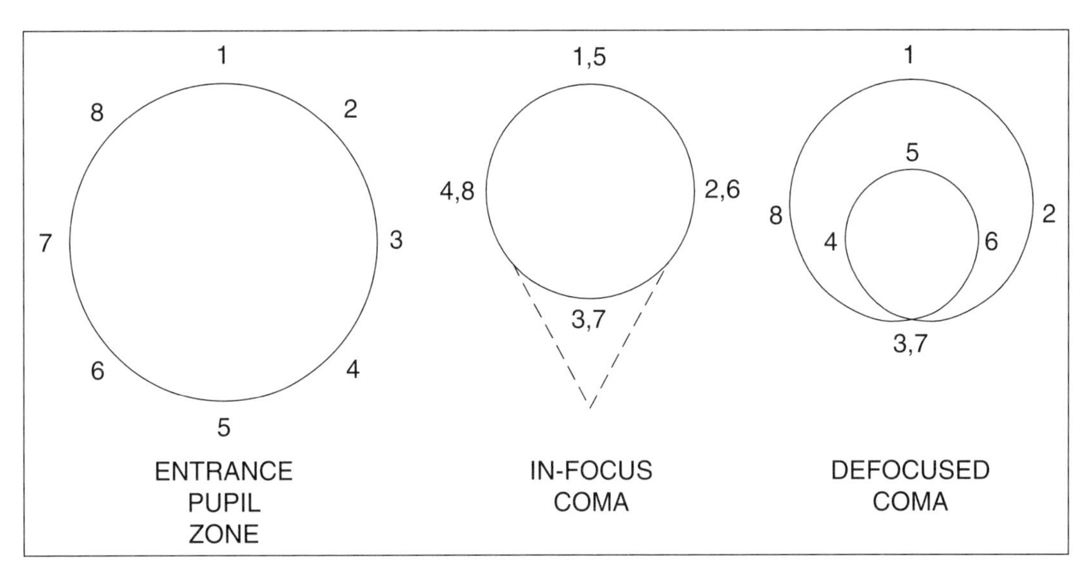

Figure A.8.2.4. Coma Loops. *The mapping of a circular zone in the entrance pupil onto the comatic image, both in-focus and defocused, is illustrated.*

as is the case with spherical aberration. The double coma loop is illustrated in Figure A.8.2.4.

This complex behavior has led to two ways to measure coma. These measures use rays in either the tangential plane or the sagittal plane in object space.

Take the case of an image with only pure third-order coma and no additional defocus (both loops the same size), such as was attempted in Figure A.8.2.3. Consider a single pupil zone and two tangential or meridional rays. One ray from the object passes through the top of the pupil zone (at a position angle of 0°), and the second ray passes through the bottom of the pupil zone (at a position angle of 180°). These are the two off-axis marginal rays shown in image space in Figure A.8.2.1.

For this or any axially centered system, meridional rays always remain in the meridional plane. This fact plus the double comatic image circuit means that the ray at 0° position angle in the pupil zone falls at 0° position angle in the image ring, and that the ray at 180° position angle in the pupil zone falls at 360° position angle in the image ring. But 0° and 360° yield the same image location. Thus the two rays focus together. This focal point on the comatic image ring is the point farthest from the chief ray and is at the top in Figure A.8.2.3. Coma measured by these two tangential rays is thus known as tangential coma.

Now consider the same comatic image, the same pupil zone, and two sagittal skew rays. One ray from the object passes through the right side of the pupil zone (at a position angle of 90°), and the second ray passes through the left side of the pupil zone (at a position angle of 270°). The double comatic image circuit means that the ray at 90° in the pupil zone falls at 180° in the image ring, and that the ray at 270° in the pupil zone falls at 540° in the image ring. But 180° and 540° yield the same image location. Thus, these two sagittal rays also focus together, and they do so on the opposite side of the image ring from the 0° position angle where the tangential rays focus. Therefore, the sagittal focal point on the comatic image ring is the point nearest to the chief ray. Coma measured by these two sagittal rays is thus known as sagittal coma. Actually, because the system is right-left symmetrical about the meridional plane, only one of the two sagittal rays is necessary to describe sagittal coma.

Recall that for pure third-order coma, the far side of a comatic image ring is exactly three times the distance from the chief ray as the near side. Thus, third-order tangential coma is exactly three times as large as third-order sagittal coma. It

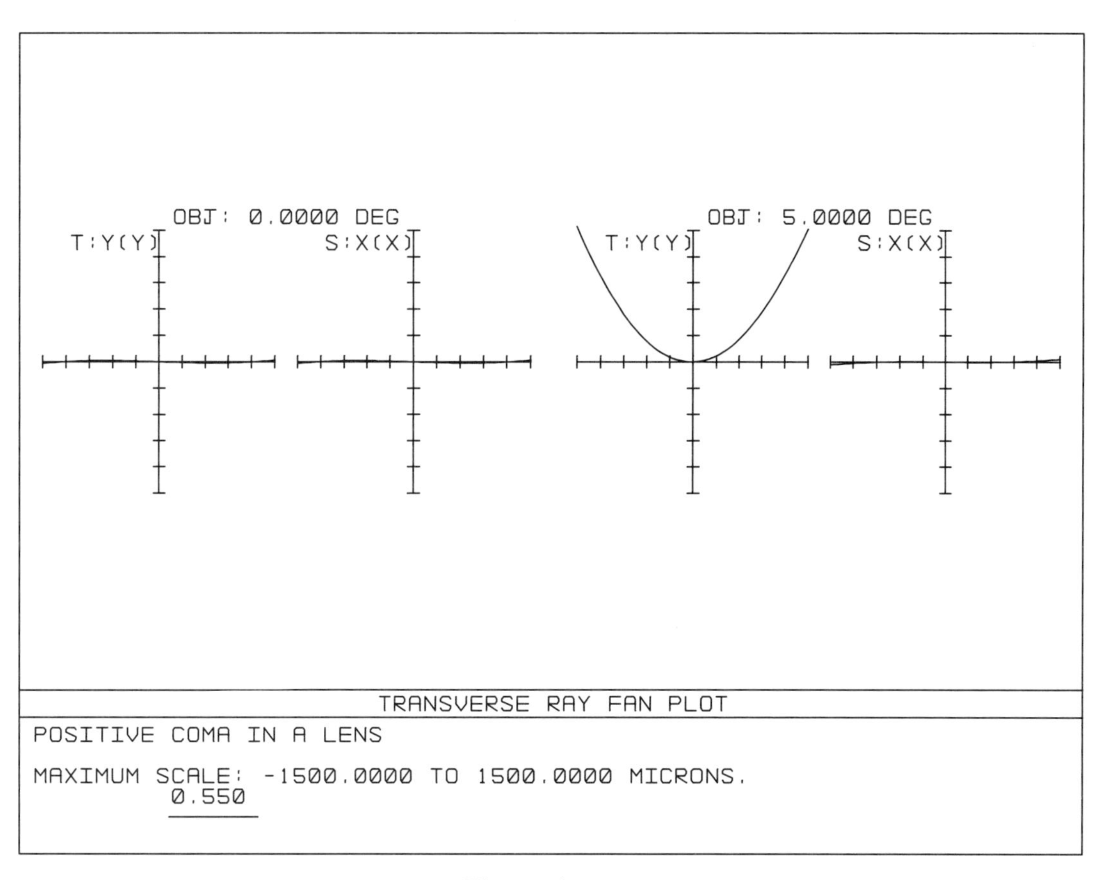

Figure A.8.2.5.

may seem that these two quantities are nearly redundant, but by convention, both types of third-order coma are often listed separately in lens evaluations.

Figure A.8.2.5 gives the ray fan plots. Recall that in conventional ray fan plots, the y-direction tangential ray errors and the x-direction sagittal ray errors are plotted. But coma is characterized by y-direction tangential and y-direction sagittal ray errors, but not by x-direction sagittal ray errors. Thus, the tangential ray fan plot easily reveals the presence of coma, but the sagittal ray fan plot is insensitive to coma and reveals only defocus and other aberrations.

For the off-axis object in Figure A.8.2.5, the tangential ray fan gives a U-shaped error curve characteristic of coma. Because coma is the only significant aberration and defocus is nearly zero, the sagittal ray fan shows almost no image errors. In general, a U-shaped tangential ray fan curve, even if distorted by additional aberrations, is always a quick indication of the presence of coma.

To correct tangential coma during optimization, the average height on the image surface of the upper and lower tangential rays is made equal to the chief ray height. To correct sagittal coma during optimization, the y-direction height on the image surface of either of the two sagittal rays is made equal to the chief ray height. These coma controls work even if the image is defocused. The effect is to force a symmetrical image blur about the chief ray. There are applications where it is imperative that the point spread function (PSF) be symmetrical.

As with lateral color, coma (and distortion) can be cancelled or controlled during the design process by the use of a symmetrical optical system. Also, note that the coma of a singlet lens can be made zero. This is done by properly adjusting the curvatures on the two lens surfaces; that is, by bending the lens. Bending the lens also has a strong effect on spherical aberration. Although the spherical of a singlet never goes to zero (without aspherization), it does reach a minimum for one bending. For a positive singlet with a distant object, the shape for minimum spherical is nearly the same as the shape for zero coma if the stop is at the lens. This shape is very close to plano-convex with the curved surface forward.

A.8.4 Astigmatism and Field Curvature

In Greek, the word stigma means point. A stigmatic lens produces point images. The word astigmatism is derived from not-a-point. A lens with astigmatism does not produce point images. If a lens is free of astigmatism and has a flat field, then the lens is called anastigmatic (not-not-a-point).

Astigmatism and field curvature are usually discussed together. This is because a lens with astigmatism has two focal surfaces, one or both of which are curved. For off-axis objects, tangential rays focus on the tangential focal surface, and sagittal rays focus on the sagittal focal surface. Thus, astigmatism and field curvature together can be equivalently described as tangential field curvature and sagittal field curvature. Similarly, astigmatism and field curvature can also be seen as a change in BFL with off-axis field angle or distance, with the effect being different in the tangential and sagittal directions. If astigmatism is zero, then there is no difference between the tangential and sagittal focal surfaces, and this mutual focal surface can be either flat or curved.

To illustrate astigmatic images, imagine that you have a lens whose only significant aberrations are astigmatism and field curvature. With this lens you make two different photographs of a field of point objects, such as stars. For the first picture, you warp the film to conform to the tangential focal surface; the resultant picture would look like Figure A.8.3.1. For the second picture, you warp the film differently to conform to the sagittal focal surface; the resultant picture would look like Figure A.8.3.2. In both cases, the images of the stars have been stretched out into lines whose lengths increase as the square of the object distance off-axis. On the tangential focal surface, the lines are all tangent to circles centered on the axis; this is the origin of the term tangential. On the sagittal focal surface, the lines have been rotated 90° and are all oriented pointing radially outward like a shell burst. Sagittal astigmatism is, for this reason, also called radial astigmatism. The term sagittal comes from the Latin word sagitta, meaning arrow.

In between the tangential and sagittal focal surfaces is a third surface called the medial surface. On the medial surface, the image of a point object is a compromise between the two astigmatic line foci. Geometrically, the medial image is a round disk known as the circle of least confusion.

Subjectively, none of these three types of images looks very good, although perhaps the medial is preferable followed by the tangential. But the sagittal image is usually avoided because it gives the perception that the image is blowing up. All these image blurs, however, have the advantage of being symmetrical about the chief ray. The shape of all the focal surfaces is paraboloidal to third order, but with different vertex curvatures.

One way to conceptually visualize the cause of astigmatism is to take a relatively thin singlet lens and tilt it so that you look into it off-axis. In one direction, the lens width appears foreshortened and reduced; in the other orthogonal direction, the lens width remains unchanged. In the foreshortened direction, lens surfaces are effectively compressed, so that lens curvatures are effectively increased. The increased effective curvatures increase lens power or shorten focal length. Thus, a lens with astigmatism has two focal lengths, one for each of the two directions. This produces the two separated astigmatic images. The further you go off-axis, the greater the effect.

This (somewhat nonrigorous) argument also explains why cylindrical and toroidal lenses (and sometimes people's eyes) are said to have astigmatism. Cylindrical and toroidal lenses actually do have two different curvatures in two orthogonal directions. Thus, such a lens has two focal lengths and produces two separated line

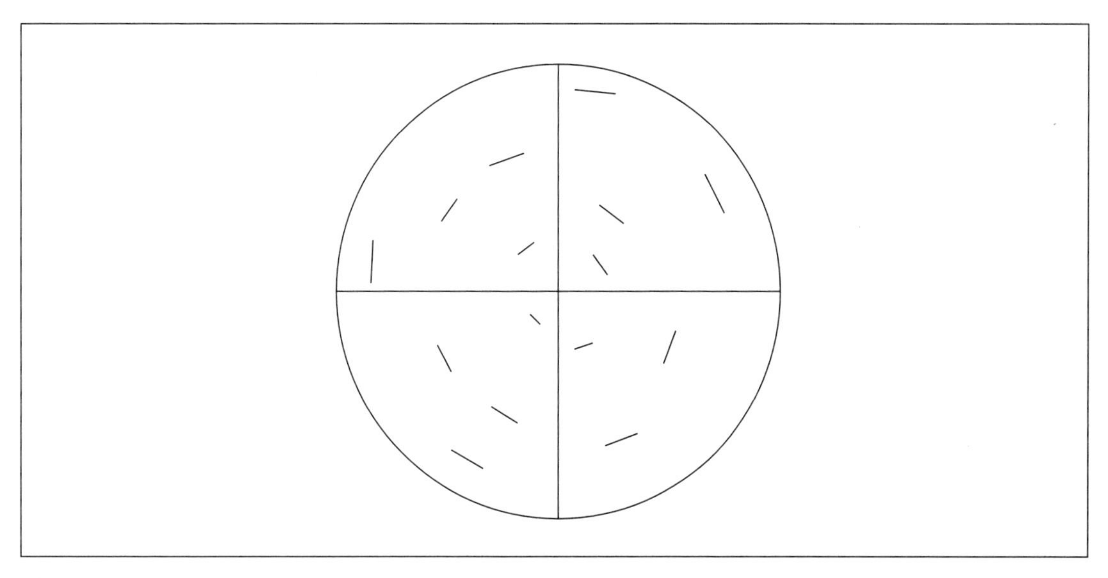

Figure A.8.3.1. Tangential Astigmatism. *A field of stars is photographed with a lens having astigmatism. The film has been warped to conform to the tangential image surface. The star images appear as lines tangent to circles centered on the field center. The lengths of the lines actually increase as the square of the off-axis distance.*

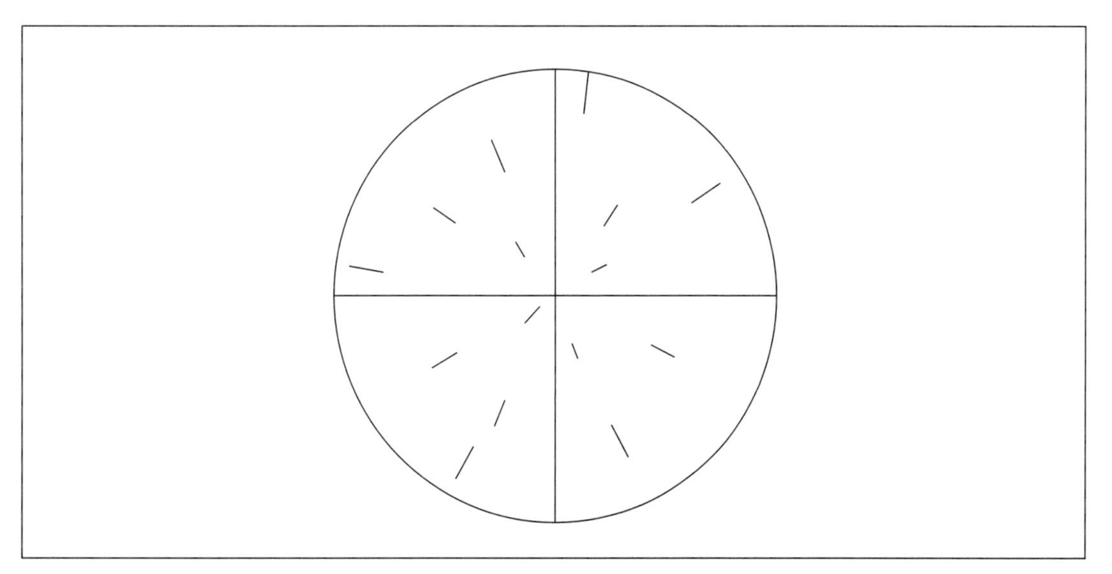

Figure A.8.3.2. Sagittal Astigmatism. *The same lens is used as in Figure A.8.3.1. The film has now been warped to conform to the sagittal image surface. The line images have become reoriented radially outward from the field center. Once again, the lengths of the lines actually increase as the square of the off-axis distance.*

foci. However, the effect is not the same as off-axis Seidel astigmatism in a rotationally symmetric, axially centered lens. In cylindrical and toroidal lenses, all the line images are oriented parallel to each other, rather than pointing radially outward or being tangent to circles. Also in cylindrical and toroidal lenses, the line images are all the same length, even in the center of the field, rather than increasing with off-axis angle or distance. For these reasons, the terms cylinder power and cylindrical aberration are sometimes introduced. As an example, a prescription for eyeglasses lists two types of lens power, sphere and cylinder. Whatever you call them, the important idea is that both types of astigmatism look similar and have causes that are related.

Figure A.8.4.1 is the layout of a lens specially designed to demonstrate astigma-

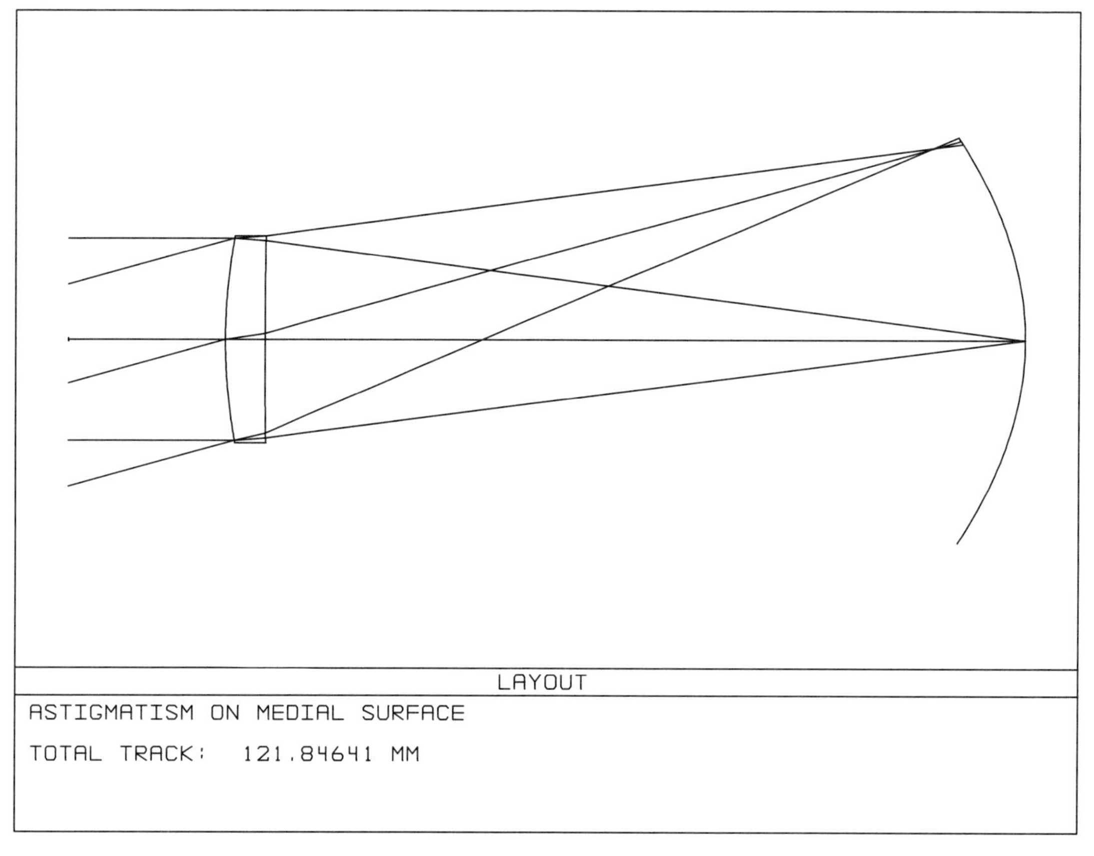

Figure A.8.4.1.

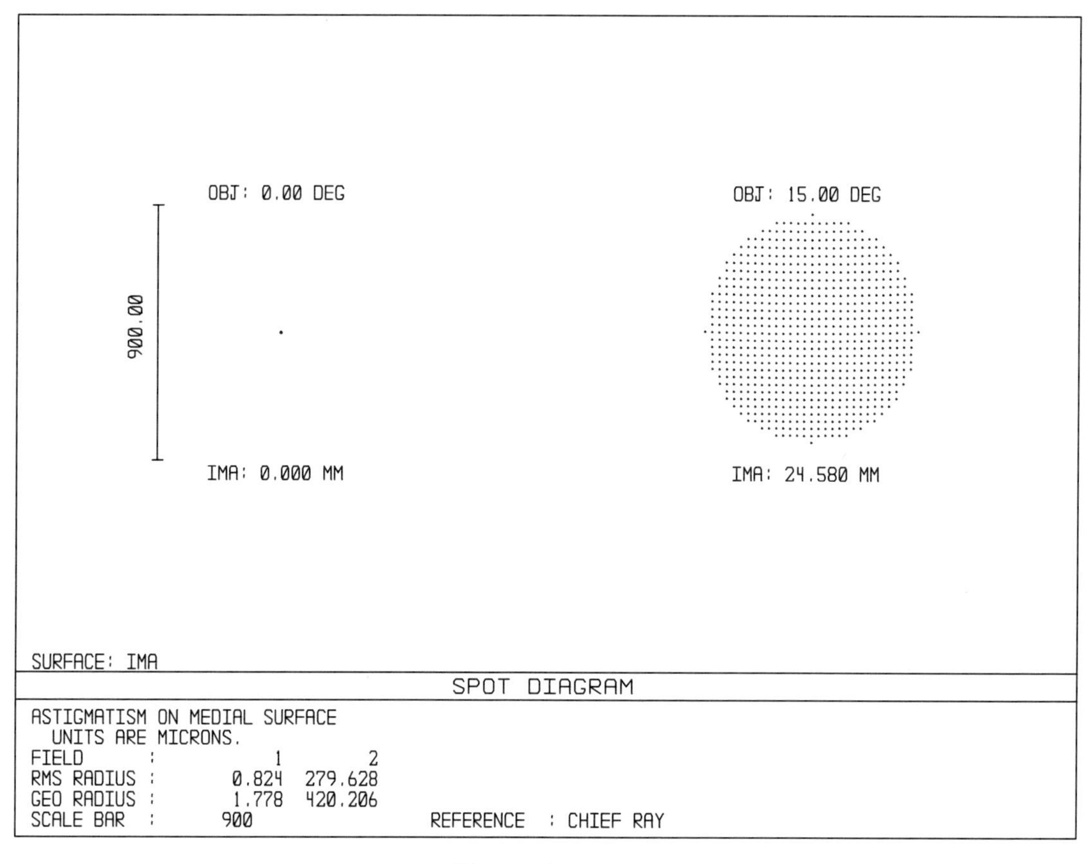

Figure A.8.4.2.

tism. This lens is to be used monochromatically and has relatively little spherical aberration. The lens has been bent to remove the coma so that off-axis the only significant aberrations are large amounts of nearly pure undercorrected third-order astigmatism and field curvature. The medial image surface is selected. There are two object points on the distant object surface: the on-axis object point and an object point 15° off-axis. Note that the off-axis image does not look in focus on the layout.

Figure A.8.4.2 gives the spot diagrams. The on-axis image appears virtually perfect at this scale. The off-axis image is not good and looks out of focus.

However, the ray fan plots in Figure A.8.4.3 reveal something more. The off-axis plots do show inclined straight lines indicating defocus, but the slopes of the tangential and sagittal curves have *opposite* signs. The medial surface truly is located in between two different foci; it is inside one focus and outside the other.

Figure A.8.5.1 shows the same lens as in Figure A.8.4.1, except that the tangential focal surface is selected. Note that field curvature is more extreme, and that the off-axis image now looks in focus on the layout. Figure A.8.5.2 gives the spot diagrams. The off-axis image is a thin line tangent to a circle centered on the axis and perpendicular to the radial direction. The tangential rays (those visible in the layout) are indeed in focus, but the sagittal rays are not. The ray fan plots in Figure A.8.5.3 tell the same story.

Figure A.8.6.1 again shows the same lens as in Figures A.8.4.1 and A.8.5.1, but now the sagittal focal surface is selected. Note that field curvature is much less than in either of the two previous cases, and that the off-axis image looks farther out of focus on the layout. The spot diagrams in Figure A.8.6.2 look very much like the spot diagrams in Figure A.8.5.2, except that the off-axis line image is rotated by 90° to be parallel to the radial direction. Now the sagittal rays are in focus, but the tangential rays are not. This conclusion is confirmed by the ray fan plots in Figure A.8.6.3. Note that the slopes of the off-axis curves in Figures A.8.5.3 and A.8.6.3 have opposite signs. In this example, the tangential focus is inside the sagittal focus, or conversely, the sagittal focus is outside the tangential focus. These relative positions are reversed if the sign of the astigmatism is reversed.

For technical and historical reasons, astigmatism is actually defined only for rays very close to the chief ray, and not for rays across the full pupil width. For off-axis object points, the chief ray now acts as an off-axis axis. Astigmatism is calculated using the Coddington equations (ca. 1829), and these use only parabasal or quasi-paraxial rays closely surrounding the chief ray. All lens design computer programs have a feature that plots tangential and sagittal field curvature as a function of field angle or distance using these quasi-paraxial rays. But these field curvature plots must be used with caution. They are incomplete in the same sense that the paraxial description of on-axis images is incomplete.

An example of a field curvature plot is given on the left in Figure A.8.7 for the same Tessar lens used for Figures A.6.4, A.6.5.1, A.6.5.2, and A.6.8. Here there are both third- and fifth-order field curvature and astigmatism. Tangential and sagittal field curvature plots are often complex curves that may (or may not) cross at one or more field angles. If the curves do cross, then astigmatism is corrected to zero at that field angle. For other field angles, residual field-zonal astigmatism remains. In Figure A.8.7, astigmatism is zero near the 0.9 field zone.

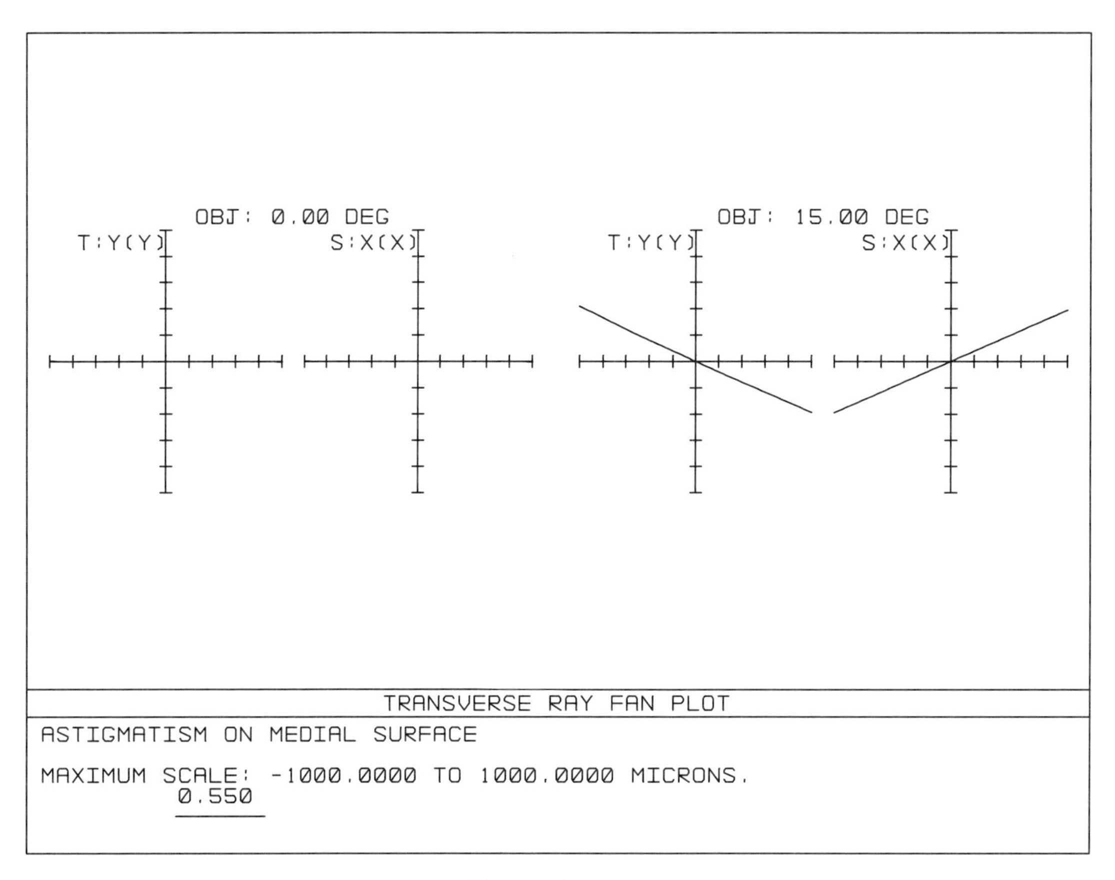

Figure A.8.4.3.

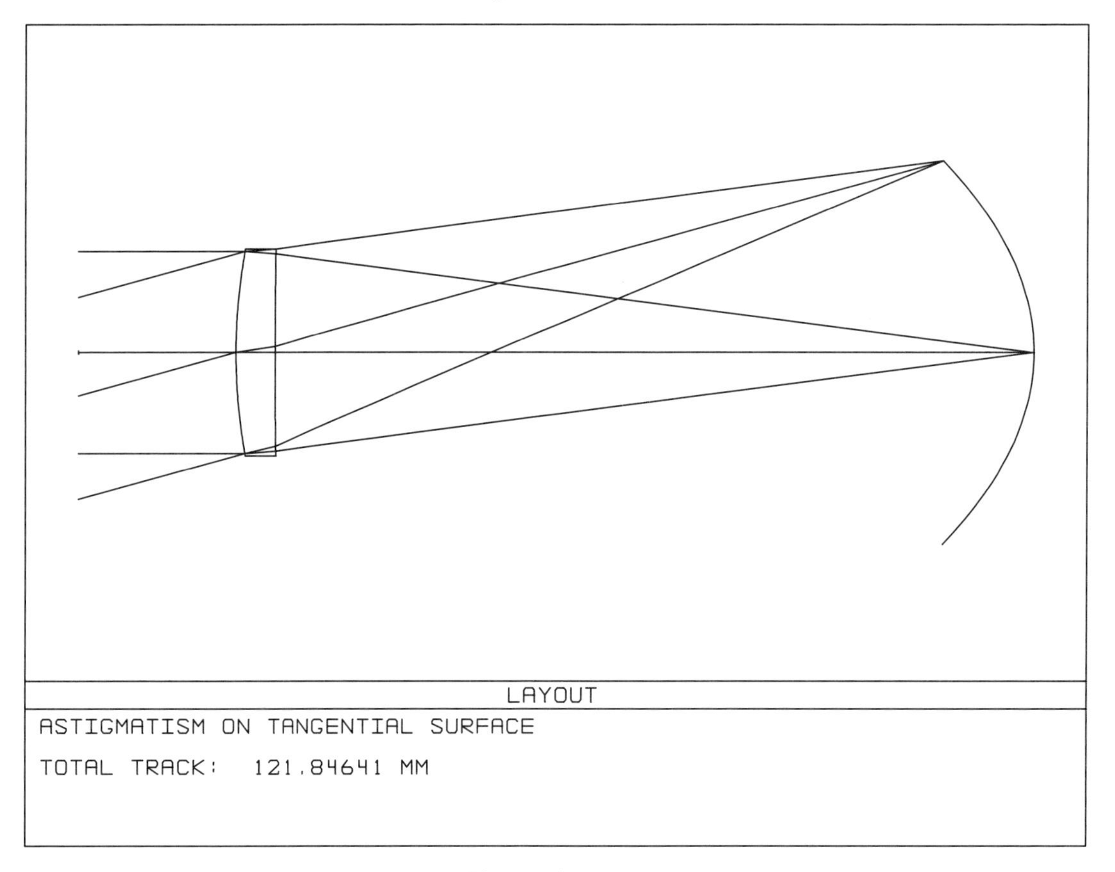

Figure A.8.5.1.

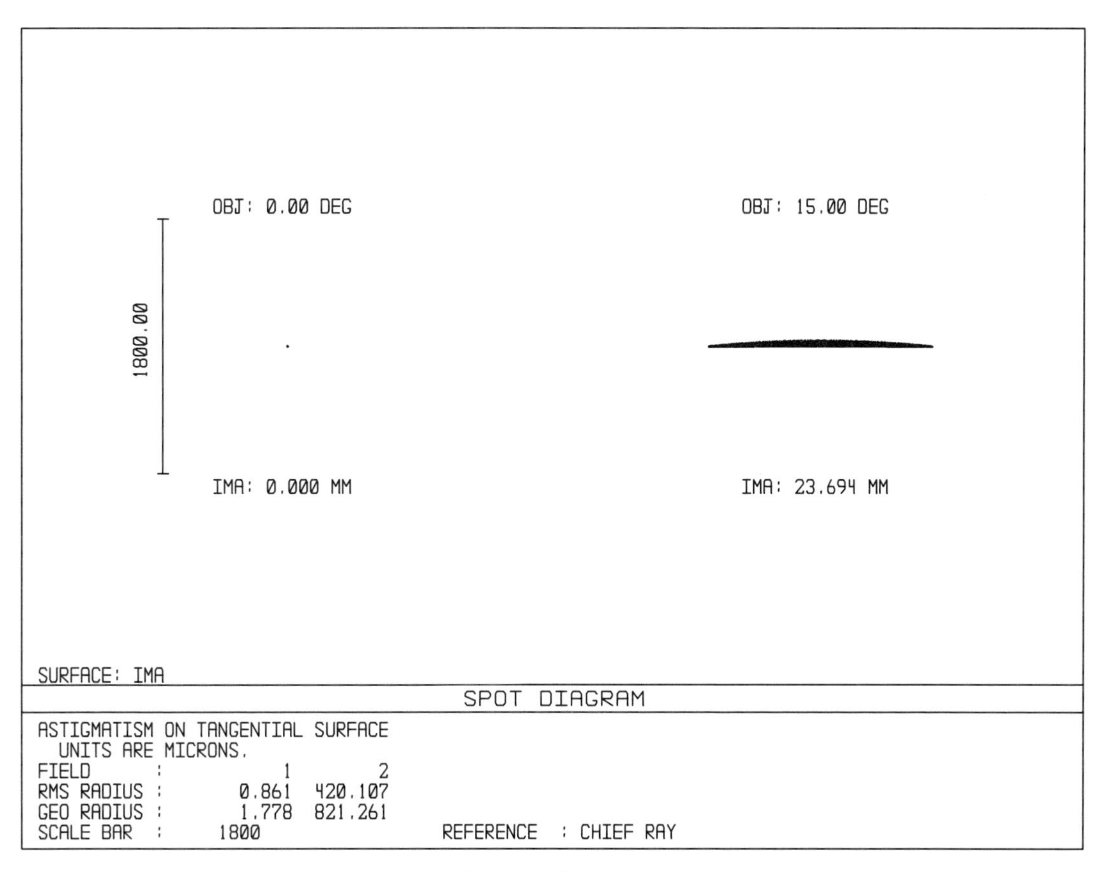

Figure A.8.5.2.

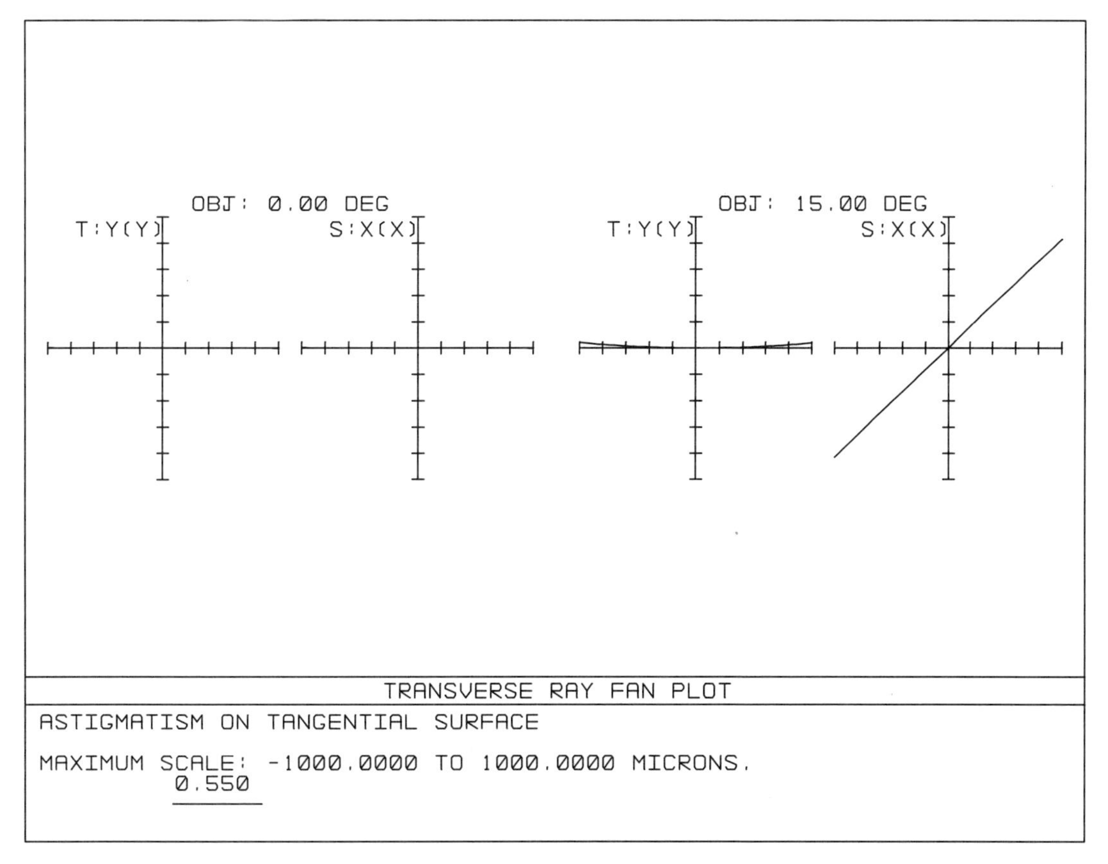

Figure A.8.5.3.

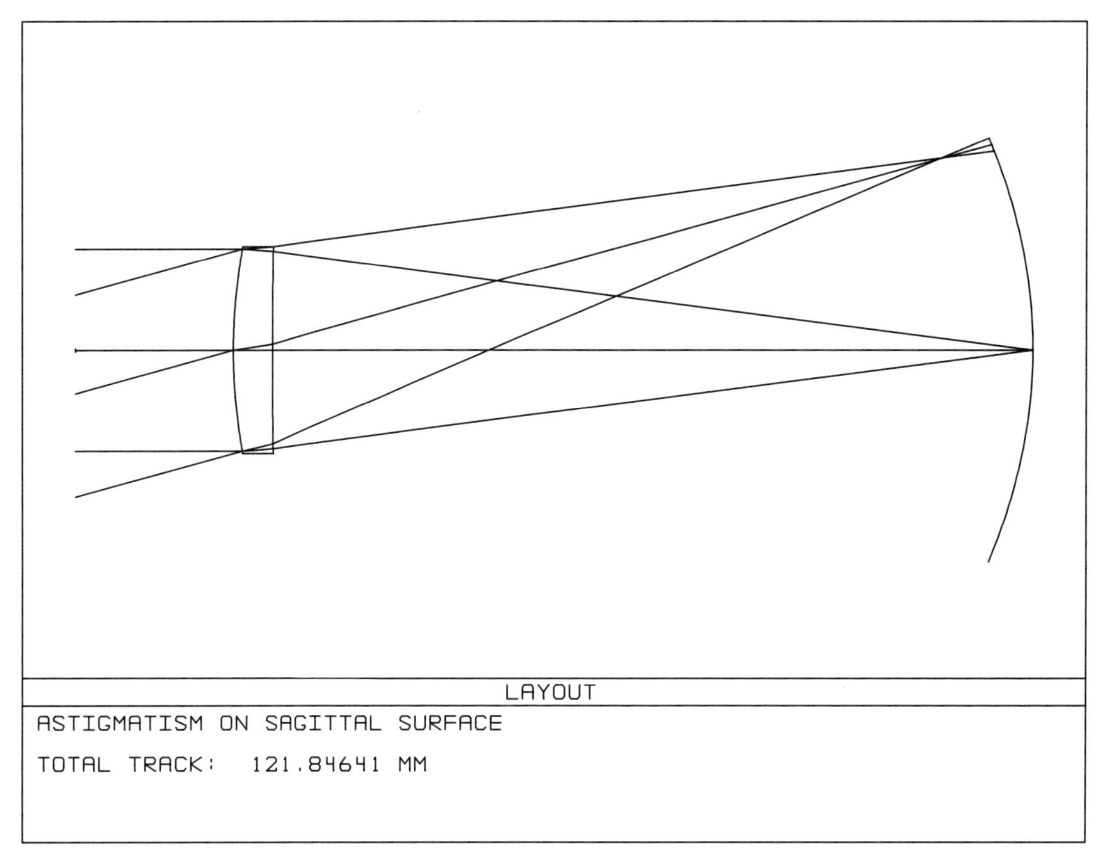

Figure A.8.6.1.

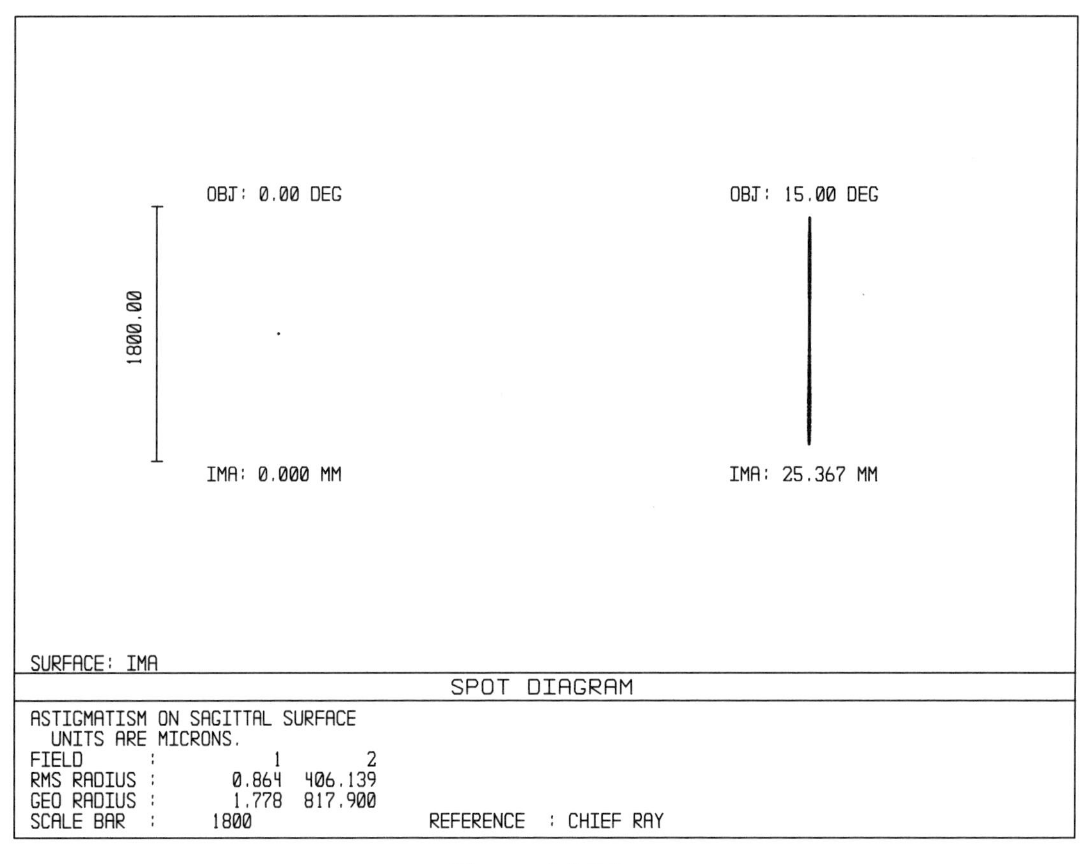

Figure A.8.6.2.

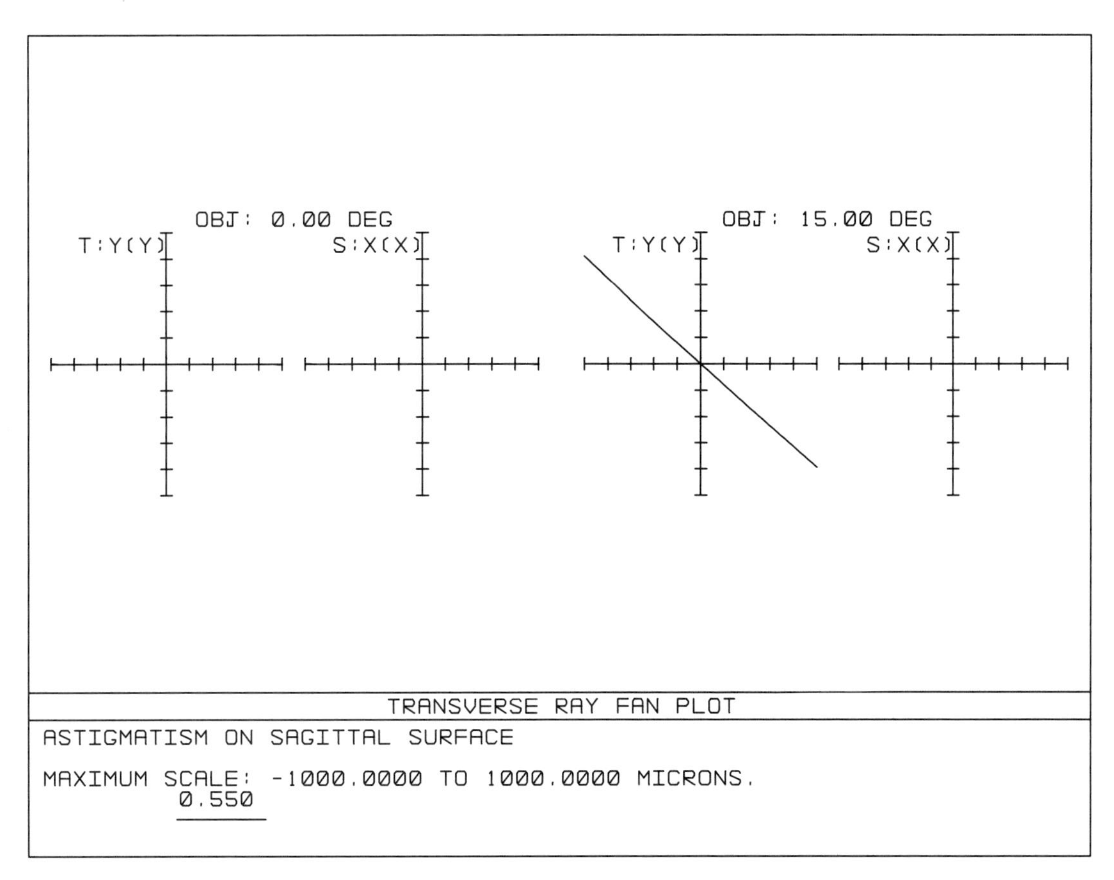

Figure A.8.6.3.

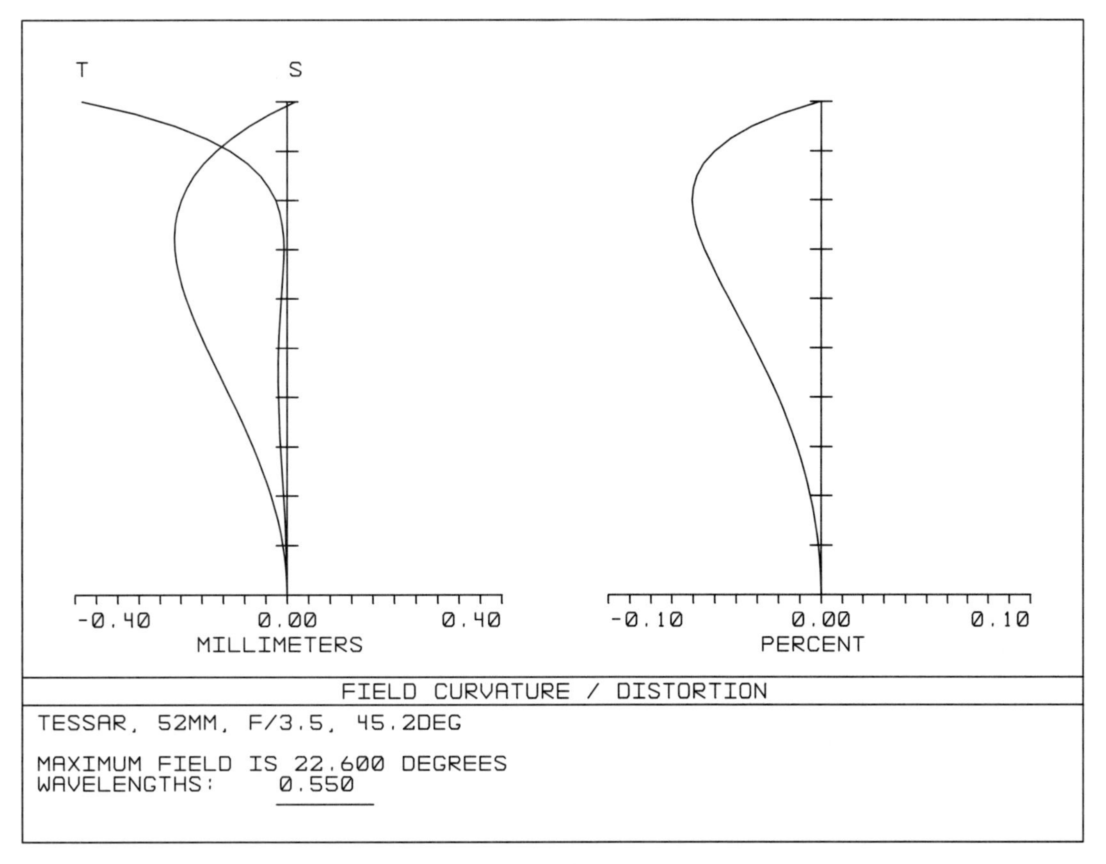

Figure A.8.7.

Unlike field curvature plots, ray fan plots do use the full pupil width. Thus, ray fan plots are more general and more useful. Figures A.8.4.3, A.8.5.3, and A.8.6.3 showing astigmatism completely across the pupil are possible only because almost no other aberrations are present. To isolate astigmatism in complex ray fan plots, look at the origins of the tangential and sagittal plots. If the curves at the two origins have different slopes, then there is a relative difference in focus for rays passing near the pupil center, and this is astigmatism.

Astigmatism and field curvature can be specifically controlled during optimization. However, this direct approach may have limited success due to the limited applicability of quasi-paraxial rays. Often, a better approach is to optimize using rays covering the entire pupil; that is, by minimizing RMS spot size or RMS wavefront errors.

If astigmatism is zero, then the tangential and sagittal focal surfaces are coincident. Furthermore, the tangential and sagittal surfaces both collapse onto another surface called the Petzval surface. In other words, the Petzval surface is very close to the best image surface in the absence of astigmatism. Thus, the curvature of the Petzval surface is the basic underlying field curvature of the system. The curvature of the Petzval surface is given by a quantity called the Petzval sum. If a flat image surface is required, then the Petzval sum must be made small during optimization. Obtaining a small Petzval sum may present difficult design problems. Much more will be said about the Petzval surface and the Petzval sum in later chapters.

A.8.5 Distortion

Not all aberrations are concerned with unsharp or blurry images. Some are concerned with the *location* of images. Field curvature, in the absence of other aberrations, can be thought of as a field-dependent longitudinal error in image location. Similarly, distortion is a field-dependent lateral or transverse error in image location. You can have a perfectly sharp image of an extended field where the parts of the image are simply in the wrong places.

Distortion was noticed early in the history of photography. Because the first photographic process, the daguerreotype (ca. 1839), was very slow and required long exposures, favorite subjects for photography were architectural and city views that did not move. But photographers immediately noticed that the lenses available at the time imaged straight lines, such as building edges, as curves. There was thus much demand for better lenses free of rectilinear distortion.[1]

Distortion is the Seidel aberration that relates how a flat object surface normal to the optical axis is projected or mapped onto a flat image surface also normal to the optical axis. If the object surface looks like a flat rectilinear grid or a big sheet of graph paper, then the ideal undistorted image likewise looks like a flat rectilinear grid. The only difference is an overall change of scale.

However, if the image is distorted, then the off-axis grid lines are curved. This inaccurate image mapping is caused by a variation in transverse image scale or magnification as a function of off-axis field angle or distance; that is, by a variation in EFL with field zone.

Distortion comes in two forms depending on its sign. Positive distortion, illustrated in Figure A.8.8.1, causes the position of an image point to be erroneously given an incremental transverse displacement radially outward. The magnitude of this position error increases as the cube of off-axis angle or distance. From the appearance of images with this error, positive distortion is called pincushion dis-

[1] People unfamiliar with optical terminology sometimes incorrectly say that an image is distorted when they mean unsharp. The term distortion in optics is properly used only to describe errors in image geometry. Distortion, at least monochromatically, has nothing to do with image sharpness.

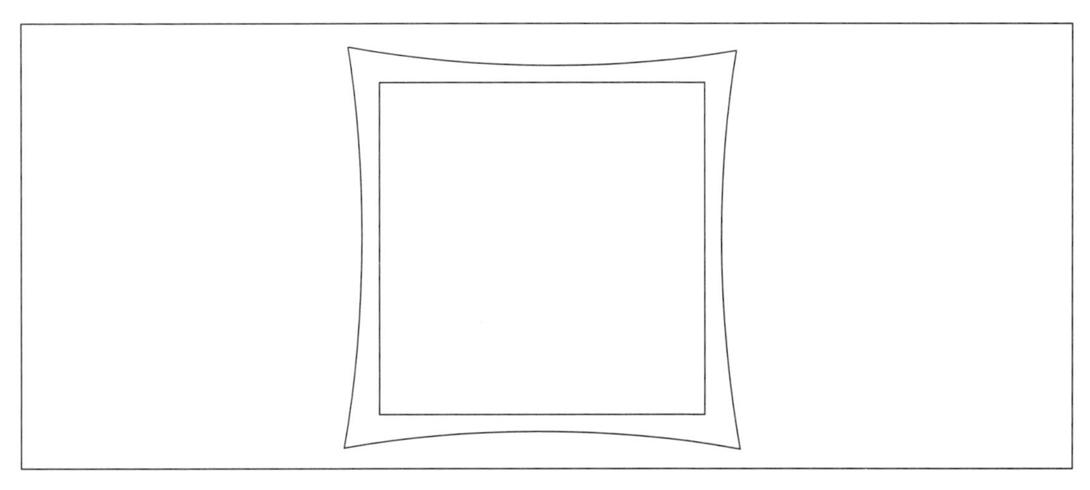

Figure A.8.8.1. Pincushion Distortion. *A square centered about the field center is imaged by a lens having positive or pincushion distortion. For comparison, both the distorted and undistorted images are shown.*

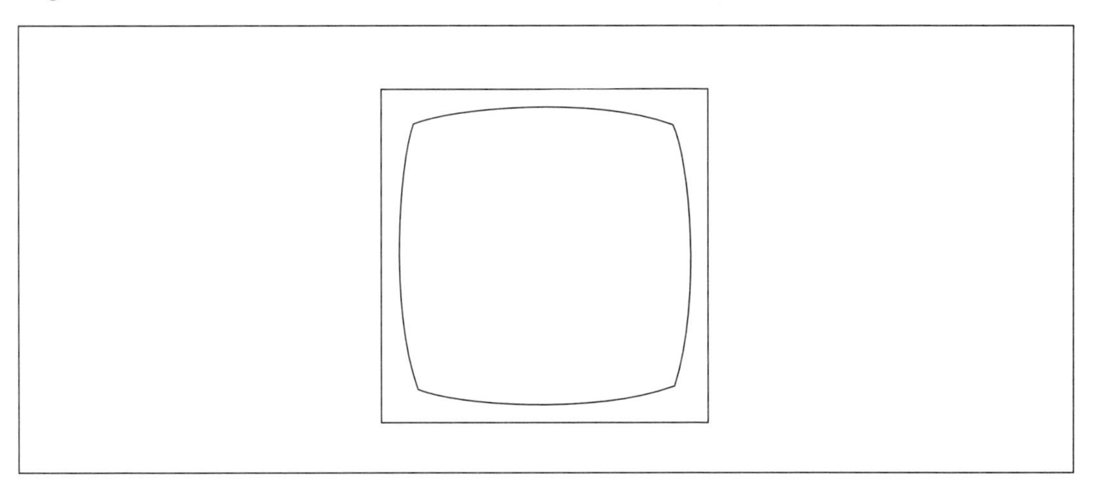

Figure A.8.8.2. Barrel Distortion. *The same square as in Figure A.8.8.1 is now imaged by a lens having negative or barrel distortion. Once again, both the distorted and undistorted images are shown.*

tortion. Negative distortion, illustrated in Figure A.8.8.2, causes the position of an image point to be displaced radially inward; this is called barrel distortion.

Linear distortion is measured as the difference between the trigonometric and Gaussian chief ray heights on the image surface. Relative distortion is measured as the linear distortion divided by the Gaussian chief ray height, and is usually given in percent. All lens design computer programs have a feature that plots percent distortion versus field angle or distance. An example is given on the right in Figure A.8.7 for the same Tessar lens used previously.

During optimization, distortion is corrected by making the trigonometric and Gaussian chief ray heights equal. But there is a problem with this approach if the image surface is curved. The trigonometric ray height is measured on the true curved image surface, whereas the Gaussian ray height is measured on a different surface, the Gaussian image plane. Two dissimilar things are being compared, and the result is not very meaningful. In this case, the above measure of distortion fails. In other words, rectilinear optical distortion is not *defined* on a curved image surface. However, most lens design programs are written so generally that they do calculate distortion on a curved image surface. When this happens, disregard it.

On a more fundamental level, if either the object or image surface is curved and the other is flat, or if both are curved, then any map maker will tell you that it is impossible to project one surface onto the other without distortion. But this distortion is different from conventional optical distortion. To handle this general distortion, a special object-to-image mapping function is required.

Some lenses, especially highly asymmetric lenses, can have higher orders of distortion. Even the Tessar of Figure A.8.7 shows both third- and fifth-order distortion, with the two cancelling at the edge of the field. For extreme cases, the designer may have to control distortion simultaneously for more than one field angle or distance. Chromatic variation of distortion is also possible, and this looks like higher-order lateral color. As with lateral color and coma, distortion is more easily controlled with a symmetrical optical system. Also like lateral color, distortion is always zero for a thin lens with the stop at the lens.

Rarely is it required that a lens be perfectly corrected for zero distortion. A small residual amount of distortion is often permitted in even the best lenses to allow better control of other aberrations. Therefore, during optimization, a constraint is often used to bound the absolute value of distortion at the edge of the field to be less than some number, typically about 1%, which the eye can only rarely notice. However, if there is a choice, it is preferable that the residual distortion be of the barrel type, which is subjectively less disturbing. Pincushion distortion makes the image look like it is coming apart. Barrel distortion also somewhat counteracts vignetting by compressing the off-axis image, thereby artificially raising image irradiance.

If a lens is found to have distortion of one sign when traced with light going in a particular direction, it will be found to have distortion of the opposite sign when traced in reverse. When designing a lens, it is more common to configure it with the light going from the long conjugate to the short conjugate; that is, the object distance is greater than the image distance. For lenses such as microscope objectives, eyepieces, and projector lenses that are easier to design in reverse, the actual distortion when used will be of the opposite sign to what the computer program reports.

A distortion-free lens maps the tangent of the off-axis field angle onto a flat image surface. The equation is:

$$h = f \tan\theta$$

where h is image height, f is focal length (or image distance for the object not at infinity), and θ is off-axis field angle. However, there are other types of lenses that are specifically designed to map the off-axis angle itself onto a flat image surface. These lenses are called fisheye lenses and are used for extremely wide-angle coverage. Fisheye lenses can have a total angular coverage of 180° (±90° from the axis) or even more. However, fisheye lenses have a huge amount of barrel distortion.

Incidentally, the name fisheye comes from the fact that most fisheye lenses are adjusted to give a round image on the film, and that such an image is what a fish in the water would see when looking up out of the water (index of 1.33) toward the sky (index of 1.0). The 2π steradian hemispheric view in the air is converted into a smaller conical solid angle in the water. For angles greater than this cone angle, the fish sees back into the water (where it is dark) due to total internal reflection.

Lenses corrected for rectilinear distortion may not have barrel or pincushion distortion, but they do have another type of distortion called elliptical distortion. Elliptical distortion is especially noticeable in wide-angle lenses, and is even present in pinhole cameras. The problem relates back to the projection or mapping geometry. A distortion-corrected lens is designed to make small two-dimensional circles drawn near the edge of a flat object surface project as circles on a flat image surface. But if your lens does this, then small three-dimensional objects located near the edge of the field appear in the image to be stretched; that is, spheres look elliptical, hence the name. However, these spheres do project as circles with a fisheye lens. You cannot have it both ways.

A.8.6 Higher-Order Off-Axis Aberrations

The third-order aberrations are by no means the only monochromatic aberrations that have a major impact on off-axis images. There are fifth- and higher-order aberrations that can be just as important.

Each of the third-order aberrations has its fifth-order counterpart. There are fifth-order versions of spherical aberration, coma, astigmatism, field curvature (Petzval), and distortion. Similarly, there are seventh- and higher-order counterparts too. In a great many lenses, the fifth-order aberrations are crucial. However, the seventh-order versions are usually less prominent, except when deliberately introduced to balance aberrations in high-performance lenses.

There are also two aberrations that only begin with the fifth order and have no third-order precursors. These are fifth-order oblique spherical aberration and fifth-order elliptical coma.

Oblique spherical aberration is very common. In many lenses, it is the most serious off-axis aberration. Oblique spherical actually has two fifth-order components, and the amounts of both of these can be different in the tangential and sagittal directions. On a transverse or an OPD ray fan plot, oblique spherical aberration appears only off-axis, and the curves resemble those for ordinary spherical aberration except that they are usually different in the tangential and sagittal directions. On a spot diagram, however, the patterns usually look quite different from those of ordinary spherical. The spots can variously resemble an elliptical spherical aberration blur, an hour glass (two lobes), a four-leaf clover (four lobes), or a four-pointed star. Adding other aberrations such as defocus from field curvature complicates things even more.

Elliptical coma also has two fifth-order components whose amounts can be different in the tangential and sagittal directions. In most cases, the spot diagram resembles ordinary coma except that circular zones in the pupil map onto the in-focus image as ellipses rather than circles. In addition, the elliptical coma blur flares out from the chief ray at an angle that is no longer necessarily 60° (it can be greater or smaller). Depending on the orientation of the ellipse and the flare angle, the result can range from a long, skinny coma blur to a short, stubby coma blur.

In theory, this analysis can be extended indefinitely to aberrations beginning with the seventh, ninth, and still higher orders, but these aberrations are not encountered very often in practice.

Chapter A.9

Analytical Relationships for Imagery

There are some analytical relationships governing image formation from aberration theory and radiometry that the lens designer should understand to better know what to expect and how to proceed. The basic first-order chromatic and third-order Seidel aberrations were introduced in the previous chapters. This chapter discusses a few more theoretical relationships that are both useful and commonly encountered.

A.9.1 Petzval Surface and Petzval Sum

One of the first questions that a lens designer asks a potential customer is whether he wishes his lens to have a flat field; that is, a flat image surface. Usually the answer is yes. To satisfy this constraint, the lens designer must be able to control the curvature of a special theoretical surface called the Petzval surface.

Astigmatism is present if the tangential and sagittal focal surfaces are not coincident. If astigmatism is made zero, not only do the tangential and sagittal focal surfaces collapse onto one another, but both collapse onto the Petzval surface. In other words, the Petzval surface is the best (or close to the best) image surface in the absence of astigmatism. Thus, for a well-corrected system with a flat field, the curvature of the Petzval surface, or Petzval curvature, must be near zero. This is a fundamental requirement.

If the image detector is flat, then any field curvature causes a field-dependent defocus. Recall that axial defocus can be used to balance on-axis aberrations. In a similar way, field-dependent off-axis defocus can be used to balance off-axis aberrations. Thus, the lens designer may wish to purposely leave in a small amount of Petzval curvature to achieve the best overall solution. The important thing is that Petzval curvature must be controllable during optimization so that it can be reduced to the desired and usually small value.

If the object surface is flat, then the curvature (reciprocal of radius of curvature) of the Petzval surface is given by the Petzval sum over all lens surfaces in a compound lens

$$P = -\sum \frac{\phi}{nn'}$$

where ϕ is the optical power of a surface given by

$$\phi = \frac{n' - n}{r}$$

where r is the radius of curvature of the surface, and n and n' are the preceding and following refractive indices on the two sides of the surface. For a mirror, the indices go from n to $n' = -n$. If the lens is made of several separated thin elements, the Petzval sum over all elements reduces to

$$P = -\sum \frac{\phi}{n}$$

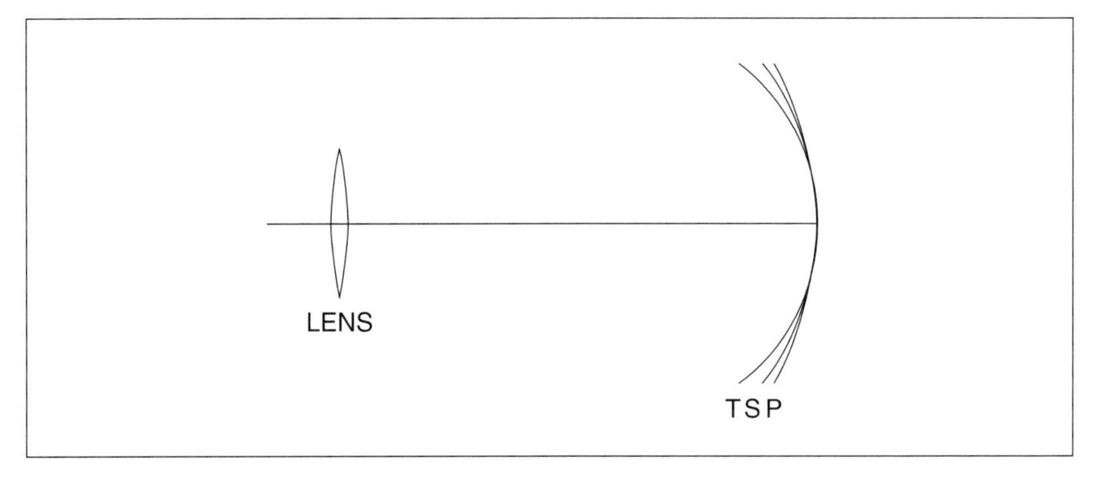

Figure A.9.1.1. Undercorrected Astigmatism. *The shapes of the tangential (T), sagittal (S), and Petzval (P) surfaces are shown for a lens with undercorrected third-order astigmatism and an inward curving Petzval surface. The curves are all parabolas. The tangential and sagittal surfaces lie on the same side of the Petzval surface. The sag distance of the tangential surface from the Petzval surface is three times the sag distance of the sagittal surface from the Petzval surface. If the medial surface had been shown, it would be halfway between the tangential and sagittal surfaces.*

where ϕ is now the power of an element (reciprocal of element focal length) and n is the refractive index of the element's glass.

The sign convention for Petzval curvature is different from the sign convention for curvatures of lens surfaces. A positive Petzval sum means that light is incident on a convex Petzval surface, whereas a negative Petzval sum means that light is incident on a concave Petzval surface. This convention is true regardless of which way the light is traveling. Thus, for a simple singlet positive lens, Petzval sum is negative, indicating that light is incident on a Petzval surface concave toward the lens and toward the incident light. For a single concave first-surface mirror, optical power is again positive but now the Petzval sum is positive, indicating that light is incident on a Petzval surface convex toward the mirror and toward the incident light.

Note that the Petzval sum does not depend on lens aperture, field of view, glass or air thicknesses, stop position, or conjugate distances. This is remarkable. Note too that elements with higher-index glass contribute less to the Petzval sum. This fact has led to the development of higher-index crown glasses to help reduce the absolute value of the Petzval sum in multi-element lenses.

A second way to reduce the Petzval sum is to use large glass thicknesses or large airspaces to separate surfaces or elements having power. The spaces themselves do not reduce Petzval, but in the example of an overall positive lens, large spaces allow the marginal ray heights to be reduced on surfaces or elements having negative power relative to surfaces or elements having positive power. These differences in the marginal ray heights allow the power of negative surfaces or elements to be increased relative to what they would be if marginal ray heights were nearly constant. More negative power means a greater positive contribution to the Petzval sum. Thus the overall Petzval sum, which is naturally negative for a positive lens, can be reduced and the field flattened. The situation is reversed for an overall negative lens.

If a lens has astigmatism, then the tangential and sagittal focal surfaces depart from the Petzval surface. There are two cases, depending on the sign of the astigmatism. Figure A.9.1.1 shows undercorrected, negative, or inward curving astigmatism. The example used in the section on astigmatism in the previous chapter also has this type of astigmatism. For third-order astigmatism and for surface sags (longitudinal departures) measured relative to the Petzval surface, the sag of the

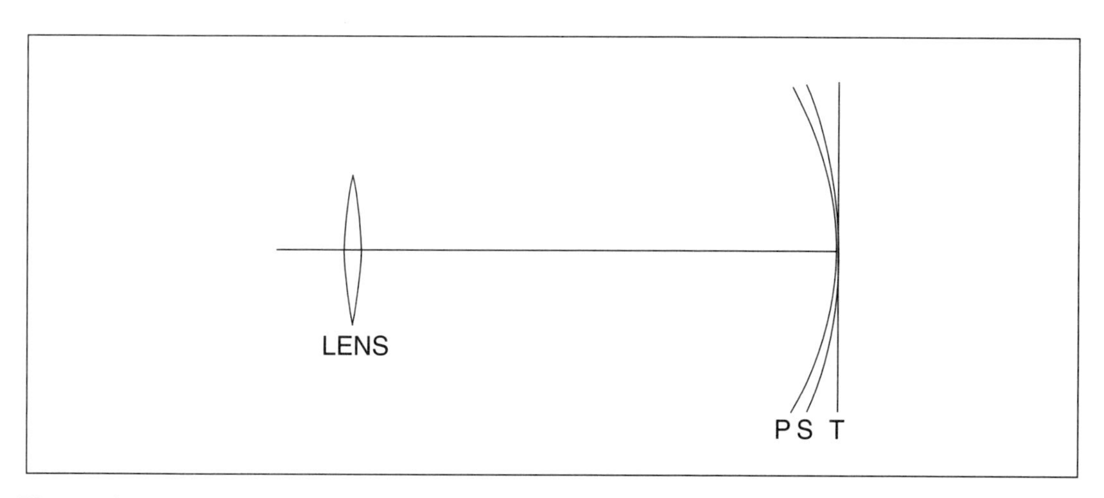

Figure A.9.1.2. Overcorrected Astigmatism. *The shapes of the tangential, sagittal, and Petzval surfaces are shown for a lens with overcorrected third-order astigmatism but the same inward curving Petzval surface as in Figure A.9.1.1. The tangential and sagittal surfaces now both lie on the other side of the Petzval surface. The amount of astigmatism has been adjusted to make the tangential surface flat. Once again, the sag distance of the tangential surface from the Petzval surface is three times the sag distance of the sagittal surface from the Petzval surface.*

tangential surface is exactly three times the sag of the sagittal surface. This constant three-to-one ratio is the reason why the tangential and sagittal surfaces must collapse onto the Petzval surface when astigmatism is made zero.

The medial surface is halfway between the tangential and sagittal surfaces; that is, medial sag is twice sagittal sag. Note that the tangential, medial, and sagittal surfaces are all on the same side of the Petzval surface. All four of these surfaces are actually paraboloidal in shape for third-order astigmatism. The curvatures calculated for these surfaces are vertex curvatures.

Figure A.9.1.2 shows overcorrected, positive, or outward curving astigmatism, as might be obtained with a complex lens. Petzval curvature, however, has not been changed. Now the tangential, medial, and sagittal focal surfaces are all on the other side of the Petzval surface. Note that the tangential surface has been made flat. Flattening the tangential image surface was a common way to artificially flatten the field in the days before techniques were developed to reduce the Petzval sum.

Lenses with a flat field and no astigmatism are called anastigmatic lenses. Lenses with no spherical aberration, no coma, no astigmatism, and a flat field are both aplanatic and anastigmatic.

A.9.2 Aberration Dependence on Aperture and Field

When designing a lens, it is often useful to know how the different aberrations, as measured by transverse ray errors on the image surface, vary as functions of aperture and field. For the seven basic geometrical aberrations, Table A.9.1 summarizes the exponent dependence of blur size on entrance pupil semi-diameter (aperture) and off-axis image distance (field). The image surface is assumed to be flat and at the paraxial focus.

For distortion-corrected systems (except for telecentric lenses where the exit pupil is at infinity and the chief ray is parallel to the axis), off-axis image distance is proportional to the tangent of off-axis image angle. However, for conceptual estimates, there is little error introduced by using the angle itself rather than its tangent. Thus, the terms off-axis angle and off-axis distance are often used interchangeably.

Note that for most systems, real image blur sizes are affected by combinations of the basic seven aberrations plus higher-order aberrations. Thus, the dependences in

Table A.9.1
Blur Size Dependence on Aperture and Field

	Aperture Exponent	Field Exponent
Longitudinal Color	1	0
Lateral Color	0	1
Spherical Aberration	3	0
Coma	2	1
Astigmatism	1	2
Field Curvature	1	2
Distortion (linear, not percent)	0	3

the table are most valuable as indications of trends. If an accurate value of geometric blur size is needed, enter the lens into the computer and do a spot diagram.

From the table, longitudinal color blur varies linearly with aperture and is independent of field. This linear dependence on aperture is just what is shown graphically in Figure A.7.9.2. Conversely, lateral color is independent of aperture and varies linearly with field. The sum of the two exponents for the two color aberrations is always unity because these are first-order aberrations.

Similarly, spherical aberration blur varies as the cube of the aperture, as is shown in Figure A.7.1.5. Comatic blur varies as the square of the aperture, as is shown in Figure A.8.2.5. Astigmatic blur varies linearly with aperture, as is shown in Figures A.8.4.3, A.8.5.3, and A.8.6.3. Field curvature blur, which represents defocus, also varies linearly with aperture. And distortion is independent of aperture. There are similar relationships for field. Note that here the sum of the aperture and field exponents is always three, because these are third-order aberrations.

These are relationships that the lens designer (and lens user) should memorize. If your lens has a lot of spherical aberration, stopping down the lens just a bit can make a big difference. But if your lens has distortion, stopping down the lens makes no difference at all. For a wide-angle lens, astigmatism may be harder to control than coma, and so forth.

For the fifth-order aberrations, Table A.9.2 summarizes the exponent dependence of blur size on aperture and field. Note how the two new aberrations, oblique spherical aberration and elliptical coma, fit in with the fifth-order versions of the Seidel aberrations. Note too that here the sum of the aperture and field exponents is always five.

Table A.9.2
Fifth-Order Blur Size Dependence on Aperture and Field

	Aperture Exponent	Field Exponent
Fifth-Order Spherical Aberration	5	0
Fifth-Order Coma	4	1
Oblique Spherical Aberration	3	2
Elliptical Coma	2	3
Fifth-Order Astigmatism	1	4
Fifth-Order Field Curvature	1	4
Fifth-Order Distortion	0	5

There is one final relationship that is of supreme value. If all linear (not angular) dimensions of an optical system are scaled up or down by the same factor, then all geometrical aberrations, both longitudinal and transverse, are also linearly scaled up or down by the same factor. Everything gets enlarged or reduced together while maintaining the same relative geometry. This is why a design for a lens for a 35 mm camera cannot just be scaled up to be used on an 8x10 view camera. The resolving power of the film remains the same, but the aberrations have been scaled up and are now unacceptably large. Of course, when a lens is scaled, the wavelength of light is not scaled, so diffraction effects are not scaled.

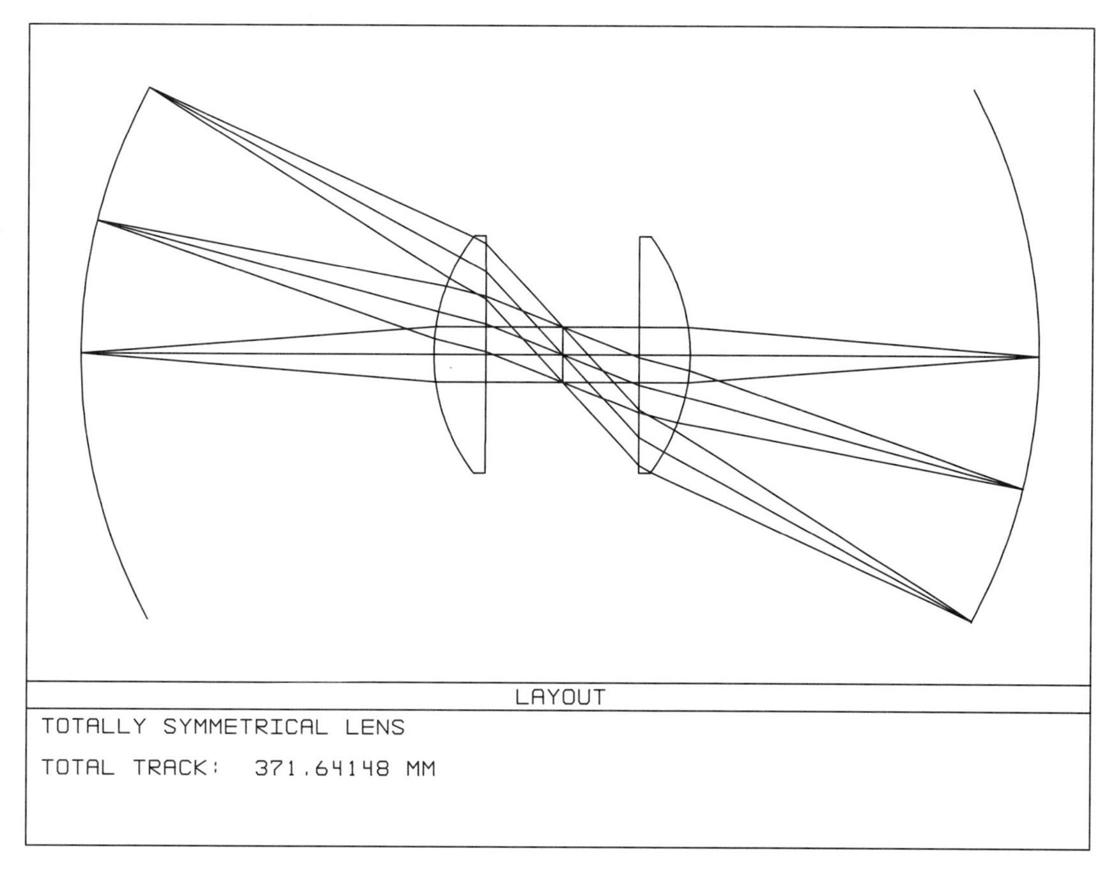

Figure A.9.2.

A.9.3 Use of Symmetry in Controlling Transverse Aberrations

The use of symmetry has been mentioned in previous chapters. Symmetrical optical systems often have advantages because they exploit the fact that there are aberrations whose surface contributions depend on the *sign* of the chief ray height on a surface. These are the transverse (rather than longitudinal) aberrations: lateral color, coma, and distortion. These three are also the aberrations in Table A.9.1 that depend on even powers of the aperture and odd powers of the field.

Consider a lens that is symmetrical about a central stop. Each half of the lens resembles a mirror image of the other half reflected about the stop surface. Even the object and image distances and curvatures are symmetrical (neglecting a fine point concerning the different tangential and sagittal image surfaces). In this lens, each surface has its equivalent surface in the other half of the lens; that is, the lens surfaces come in pairs. Such a lens is illustrated in Figure A.9.2.

Although the lens may be mirror-image symmetrical, the path taken by light through the lens is not. In particular, the chief ray, which runs diagonally through the lens and crosses the axis at the stop surface, is inverted symmetrical. The chief ray surface heights still come in pairs, but with opposite signs. Thus, so too do the surface contributions for the three transverse aberrations. Consequently, each of these surface contributions is automatically cancelled by the corresponding contribution from the other side of the lens to yield a zero overall sum. The use of symmetry is thus a very effective way of controlling lateral color, coma, and distortion.

Note that for the longitudinal aberrations, the surface aberration contributions do not change sign when the chief ray height changes sign, and thus the contributions in the two halves of the lens add rather than cancel. Thus, symmetry cannot control longitudinal color, spherical aberration, astigmatism, and field curvature.

The symmetry principle can also be applied to a lens with an object at infinity and a flat field. In this configuration, transverse aberration cancellation is no longer perfect, but the residuals are not large. If, in addition, the design of the lens is allowed to depart somewhat from symmetry, then the resulting quasi-symmetrical lens can reduce the transverse aberrations to very small values. A great many camera lens types, including the Cooke Triplet and the Double-Gauss, are quasi-symmetrical. Controlling the transverse aberrations in a highly asymmetric lens is very much harder to do.

A.9.4 Effect of a Stop Shift

In the absence of mechanical vignetting, the stop is the aperture that selects which light rays get through the lens, and which are blocked or stopped. As photographers know, opening up or stopping down the iris diaphragm of a lens changes aberrations as well as image brightness. But the lens designer has another degree of freedom with the stop. While designing the lens, he can move the stop back and forth along the axis to further select the rays that pass through the lens. This is called a stop shift.

For an on-axis object point, a stop shift has no effect. The same rays get through no matter where the stop is located, provided that the stop diameter is adjusted to maintain a constant entrance pupil diameter or focal ratio. But for off-axis object points, the rays are inclined to the axis. Shifting the stop location strongly affects which of these off-axis rays get through and which are stopped. The ray pattern, or footprint, on a given element will move up or down as the stop is shifted. Look again at Figure A.9.2, and see how the stop location determines where the off-axis beams fall on the lens surfaces.

The consequence of a stop shift for off-axis aberrations is to select which part of a larger aberrated ray bundle gets through to form the image. By carefully clipping some of the more aberrated rays, image quality can be dramatically improved. Conversely, if all the rays in the larger off-axis ray bundle would form a perfect aberration-free image, then any clipping by the stop will not change the perfect imagery of a ray subset. In other words, a stop shift only has an effect if there are aberrations to modify.

Thus, for the on-axis aberrations of longitudinal color and spherical aberration, a stop shift has no effect. And, as was discussed above, a stop shift has no effect on the Petzval sum and Petzval curvature. For the off-axis aberrations and to third-order accuracy, the stop shift equations available in standard texts[1] yield the following rules. A stop shift has no effect on lateral color if there is no longitudinal color. A stop shift has no effect on coma if there is no spherical aberration. For a stop shift to have an effect on astigmatism, there must be present either spherical aberration or coma, or both. Finally, for a stop shift to have an effect on distortion, there must be present one or more of spherical aberration, coma, astigmatism, or Petzval curvature.

A.9.5 Vignetting and the Cosine-Fourth Law

If an extended flat object surface of constant radiance is imaged without rectilinear distortion onto a flat image surface, then the resulting image irradiance will be seen to progressively decrease toward the edge of the field. This effect is called vignetting.

There are two different causes of vignetting. The first cause, which has been mentioned previously, is clipping of off-axis beams by apertures other than the

[1] See, for example, Warren J. Smith, *Modern Optical Engineering*, second edition, pp. 70–74 and pp. 314–319.

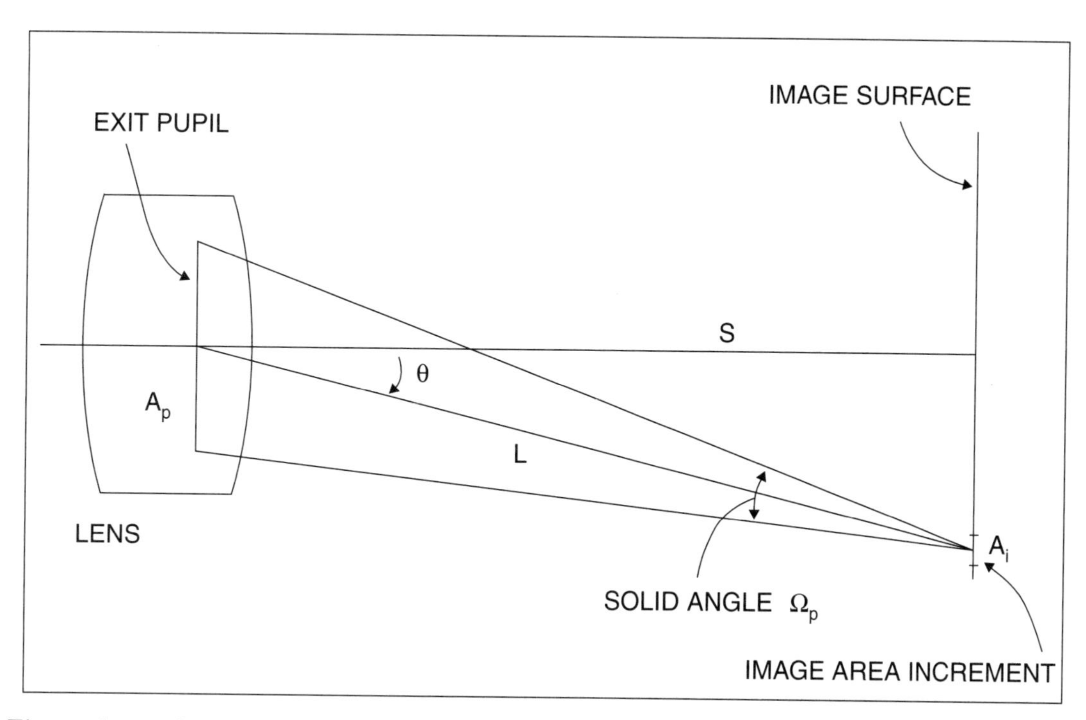

Figure A.9.3. Cosine-Fourth Vignetting. *A schematic layout of the image space of a lens is shown to illustrate cosine-fourth vignetting.*

stop. This type of vignetting by undersized optical elements is called mechanical vignetting and is not discussed in this section.

The second cause of vignetting is a fundamental radiometric property of all distortion-free imaging systems. In most cases, the image irradiance decreases roughly as the fourth power of the cosine of the angle of incidence of the light on the image surface (this angle is the same as the corresponding chief ray angle in image space). Therefore, this type of vignetting is usually called either cosine-fourth vignetting or cosine-fourth falloff. Cosine-fourth vignetting is present even in a pinhole camera, where the decrease goes exactly as cosine-fourth of the incidence angle. If there is mechanical vignetting, it is in addition to cosine-fourth vignetting.

To explain cosine-fourth vignetting, Figure A.9.3 shows a schematic layout of the light leaving the exit pupil and incident on the image surface. The object and image surfaces are flat, there is no distortion (magnification is constant across the field), and there is no mechanical beam clipping. The radiance of the object surface (flux emitted per unit area per unit solid angle) is constant both with view angle (Lambertian) and with position, and it has a value of R. For the moment, assume that the exit pupil is not subject to aberrations. In this case, the exit pupil is circular, is normal to the optical axis, and remains in a fixed location as you go off-axis.

An object area increment, such as a small square of fixed size, is located anywhere on the object surface. For no distortion, the corresponding image area increment is also a tiny square, and it has an area A_i, which is the object area times the square of the system magnification. The axial distance of the image surface from the exit pupil is S. Relative to the axis, the angle of the chief ray for the image area increment is θ. The distance from the center of the exit pupil to the image area increment is L. The area of the exit pupil is A_p. The solid angle subtended by the exit pupil as seen by the image is Ω_p.

Assume that there are no losses due to absorption and surface reflections. In such a loss-free imaging system, radiance is conserved. Thus, the radiance of the exit pupil as seen by any point on the image surface is the same as the radiance of the object surface as seen by any point in the stop opening or entrance pupil. The light (flux, lumens, watts, or photons/second) incident on the image area increment is W. From Figure A.9.3,

$$W = R\,\Omega_p A_i \cos\theta.$$

The cosine factor is due to the image area increment being inclined and foreshortened along the line of sight. The solid angle is

$$\Omega_p = \frac{A_p \cos\theta}{L^2}.$$

This cosine factor is due to the exit pupil being inclined and foreshortened along the line of sight (the simple cosine dependence is an approximation for slow systems). Substituting $L = S/\cos\theta$,

$$\Omega_p = \frac{A_p \cos^3\theta}{S^2}.$$

Thus, the light falling on the image area increment becomes

$$W = \frac{R A_p A_i \cos^4\theta}{S^2}.$$

But R, A_p, A_i, and S are all constants. Thus, W/A_i, or image irradiance, decreases as the fourth power of the cosine of the image incidence angle. This result for a slow lens with no exit pupil aberrations is called the cosine-fourth law.

However, this "law" is not always rigorously enforced. There are many distortion-free lenses where the fourth-power relationship does not hold very well. In practice, the falloff can be more severe or less severe than cosine-fourth.

Most lenses whose falloff dependence is significantly different from cosine-fourth have aberrations in the entrance or exit pupils or both. Because these pupils are each an image of the stop opening, they are subject to aberrations like any other image, especially off-axis. The pupil aberrations of greatest concern are those that (1) enlarge or shrink its apparent size, (2) shift its apparent location or line-of-sight direction, and (3) control its apparent inclination along the line-of-sight and thus its foreshortening.

A change in the apparent size of an off-axis entrance or exit pupil is caused by a change in the magnification of the stop by the intervening optics as a function of the oblique angle of view. In forming the pupil, what is called the object, stop, and image must be redefined. The new object is the old stop opening. The new image is the pupil. And the extent of the light rays forming the pupil is limited by the old object or image area, which now acts as a substitute stop. Changing the zone height on this substitute stop changes the oblique angle of view. If magnification changes with oblique angle, then the oblique effective focal length of the optics is also changing. But a change in oblique EFL as a function of stop zone is coma (refer again to Section A.6.8). In this case, it is coma of the pupil.

For ordinary image coma, you are interested in the effect integrated over the whole stop area. But for pupil coma, you are only interested in what is seen from one point at a time in the substitute stop (the original object or image area). Thus, pupil coma does not blur the apparent pupil; instead the oblique magnification change merely resizes the pupil. Although the aberration is pupil coma, its effect on the pupil resembles distortion. In addition, the change in oblique magnification is often different in the tangential and sagittal directions, an effect that can cause the apparent off-axis pupil to become elongated and deformed.

A change in the apparent location or line-of-sight direction to an off-axis pupil is caused by the chief rays from all field zones not intersecting at one point, and this is spherical aberration of the pupil.

A change in the apparent inclination or tilt of an off-axis pupil is caused by the internal chief ray angle on the stop becoming different than the external chief ray angle outside the lens. Recall that in the derivation of the cosine-fourth law, an unaberrated off-axis pupil was inclined relative to the line-of-sight by the external chief ray angle (this applies to both the entrance and exit pupils). But more generally, an off-axis pupil is the projected image of the stop as viewed along the external chief ray. When the chief ray is traced to the stop surface, however, it is this internal chief ray angle that determines how the light passes through the stop aperture. Therefore, as viewed along the external chief ray, the apparent inclination of the pupil is actually controlled by the internal chief ray angle on the stop. For this reason, a smaller chief ray angle on the stop is beneficial in reducing apparent pupil foreshortening.

Of course, when a pupil is the stop opening itself with no intervening optics, then there are no pupil aberrations.

If the exit pupil has aberrations, then the previous derivation of cosine-fourth vignetting must be modified. In particular, if the off-axis exit pupil grows, then the image solid angle increases and falloff thereby decreases (even though the chief ray angle on the image remains about the same). However, to satisfy conservation of energy in this case, additional light must be collected from the off-axis object. To do this, the off-axis entrance pupil must grow to match. Thus, both pupils become aberrated. For ultra-wide-angle coverage without distortion, the reduction in vignetting from pupil growth can make a big improvement.

Examples of lenses that can thus beat the "cosine-fourth" law are the nearly symmetrical Roossinov wide-angle lens and its variations such as the Biogon and Super-Angulon. These lenses have strong negative power in both the front and rear elements and positive power in the middle. This optical construction produces both a decreased chief ray angle internally on the stop (thereby reducing pupil foreshortening) and considerable off-axis pupil enlargement (from pupil coma).

In contrast, examples of lenses with falloff worse than cosine-fourth are the old-style symmetrical wide-angle lenses, such as the Hypergon, Topogon, and Metrogon. These lenses have strong positive power in the front and rear elements and an increased chief ray angle internally on the stop.

There is a second group of distortion-free lenses that also achieves a reduced off-axis light falloff. These are the highly unsymmetrical retrofocus (reverse telephoto) wide-angle lenses, which were illustrated previously in Figure A.6.3.2. In a retrofocus lens, a large amount of negative power in the front part, but not in the rear part, causes the angle of the chief ray to be greatly reduced upon passing through the lens. This reduction applies both internally on the stop and externally on the final image.

Here you do not actually beat the cosine-fourth law. Rather than progressively increasing the off-axis solid angle incident on the image, you decrease the chief ray angle instead. Of course, whenever you change the chief ray angle, you must also ask what is happening simultaneously to the exit pupil as a consequence.

In a retrofocus lens, the exit pupil has been moved to a position further away from the image than the system focal length. Compare this configuration with that of a more conventional lens having the same focal length and image size, but with no chief ray angle reduction and no pupil shift. The retrofocus lens has a smaller chief ray angle on the exit pupil (and only minor exit pupil aberrations), it has a smaller chief ray angle on the image, and it has a smaller relative increase in the oblique pupil-to-image distance (the L parameter above). All of these smaller parameters combine, according to the cosine-fourth law, to yield less off-axis falloff.

Another comparison is with a "standard" lens of the same focal ratio. A wide-angle retrofocus lens has a short focal length and an exit pupil distance that is longer than its focal length. A narrower-coverage standard lens has a longer focal length and an exit pupil distance that is similar to its focal length and to the exit pupil distance of the retrofocus lens. Although the focal lengths and object field coverages are different, the two lenses have nearly the same exit pupil and image chief ray angles, nearly the same exit pupil sizes and inclinations, and nearly the same pupil-to-image distances. Thus, as far as image irradiance is concerned, both lenses behave in nearly the same way. But a standard lens with its narrower coverage has relatively little cosine-fourth vignetting, and so too must the retrofocus lens.

As with the Biogon-type lenses, to achieve this more uniform image irradiance and also satisfy conservation of energy, the off-axis entrance pupil of a retrofocus lens must grow to collect more light to fill the stop opening. This pupil growth is again caused by pupil coma and reduced foreshortening. Thus, the aberrations of the entrance pupil are large even though the aberrations of the exit pupil are small (remember, the lens is unsymmetrical).

Examples of retrofocus lenses are the many wide-angle lenses for 35 mm single-lens-reflex cameras. In addition to their favorable radiometric properties, these lenses have the sizable back focal clearance required for the swinging reflex mirror. Historically, achieving this clearance was the original reason for shifting the exit pupil.

In some Biogon and retrofocus lenses, pupil aberrations can be quite noticeable, even startling. If you look into the front of one of the more extreme examples, and if you tip the lens off-axis, you will clearly see the entrance pupil enlarge and deform, its line-of-sight shift, and its inclination almost seem to follow you.

All of the above discussion assumes that there is absolutely no final image distortion. If some distortion is allowed, then the cosine-fourth relationship is again modified. If positive (pincushion) distortion is present, the outer parts of the image are stretched, and this reduces image irradiance and increases vignetting. Thus, pincushion distortion is doubly bad. If negative (barrel) distortion is present, the outer parts of the image are compressed, and this increases image irradiance and decreases vignetting. Thus, barrel distortion can be beneficial, at least for vignetting. In fisheye lenses with gigantic fields of view ($\pm 90^\circ$ or more) and enormous barrel distortion, it is the image distortion along with large amounts of spherical aberration, coma, and tilt of the entrance pupil (fisheyes are all of retrofocus construction) that prevent vignetting from becoming impossibly large.

If a lens has vignetting that is excessive, there are ways to reduce the effects. If the system has electronic image detection, such as a CCD camera, then the gain of the video amplifier can be varied during the image readout to compensate for vignetting, or the image can be processed later in a procedure called flat-fielding. In photographic systems, an anti-vignetting neutral-density filter that is darker in the middle can be placed some distance in front of the lens to reduce the irradiance in the otherwise brighter image center. Such a graded-density filter throws away light and is a brute force method, but sometimes it is the only practical solution.[2]

[2]For their assistance during the development of the ideas in this section, the author is indebted to Richard Buchroeder, David Hasenauer, Donald Koch, Kenneth Moore, and John Rogers.

Chapter A.10

Optical Glass

The elements or components of most lenses are made of optical glass. There are exceptions, however. There are unusual amorphous materials, such as fused quartz and several infrared transmitting glasses. And there are optical crystals, either single crystal or polycrystaline, such as calcite, fluorite, silicon, and many others. But when a lens designer speaks of glass, he implicitly includes all optical materials, except perhaps mirrors, which are a separate issue.

Good optical glass has many requirements to satisfy. First, of course, it must transmit light in the wavelength region of interest. Second, it must, as far as possible, be homogeneous and free of scattering centers, striae, bubbles, internal stress, and other flaws. Third, it must be mechanically, thermally, and chemically stable. Chemical stability also includes resistance to weathering and staining. And fourth, it must be practical to work in the optical shop. Some optical materials do not satisfy all of these requirements very well, but they are used nonetheless because of their other unique and desirable properties.

To control chromatic aberrations, compound lenses combining glasses of two or more types are usually required. To control higher orders of chromatic aberrations, special attention to glass selection is crucial. The only optical systems completely free of chromatic aberrations are all-reflecting systems composed entirely of mirrors.

This chapter discusses glass from the perspective of the lens designer. Glass selection is one of the more difficult aspects of lens design. The process is perhaps the one least amenable to automation during computer optimization. Often, glass selection is strongly influenced by non-optical considerations. In nearly all cases, the final glass choices are made by hand by the designer.

A.10.1 Index of Refraction

For a transmitting optical material, the index of refraction or refractive index, n, as defined in physics, is given by

$$n = \frac{c}{v}$$

where c is the speed of light in vacuum and v is the speed of light in the material. For a common glass, such as Schott BK7, the index is about 1.52. Similarly, the index of air is about 1.0003, and the index of vacuum is exactly 1.0. These values of the index of refraction are called absolute indices.

For historical reasons, lens designers rarely use absolute indices. Most lenses are used in air, and in the days before computers, it was a lot of extra work to maintain an index for air slightly greater than unity. Thus, lens designers redefined the refractive index of air at standard temperature and pressure as exactly unity. To do this and maintain the validity of Snell's law, both sides of the equation for Snell's law are divided by the absolute index of air. The new relative indices (relative to air) are thus equal to the old absolute indices divided by the absolute index of air.

Today, with computers, it is no longer necessary to continue this fiction, but nevertheless the practice remains. The published values of index of refraction are all relative indices as determined by measuring the deflection of a beam of light by a prism in air and applying Snell's law.

The use of relative indices may at times cause confusion or anomalies. When doing environmental analyses where temperatures and pressures are changed, some computer programs convert all indices into absolute indices. Glass manufacturers publish both relative and absolute values of temperature coefficient of refraction, $\Delta n/\Delta T$. And if an optical system is to be used in vacuum, the relative index of refraction of vacuum is about 0.9997. All this must seem rather strange to a physicist.

A.10.2 Dispersion

The index of refraction of all transmitting optical materials varies with wavelength. For wavelengths of interest to lens designers, index is always higher for shorter wavelengths (the blue end of the spectrum) and lower for longer wavelengths (the red end of the spectrum). Because it is index variation that causes a prism to disperse white light into spectral colors, chromatic variation of index is called dispersion.

Figures A.10.1.1 and A.10.1.2 show plots of index versus wavelength for two different glass types, Schott BK7 and SF2. Note that the curves are nonlinear and quite different. In particular, the index increments on the vertical axis of the SF2 plot are twice as large as those on the BK7 plot. Plots such as these are called dispersion plots, although the term is a misnomer because they are really refractive index plots. In general, every glass has its own unique dispersion plot.

Dispersion is often expressed by the refractive properties of a prism or lens; that is, by the ratio of polychromatic differential refraction relative to average total refraction. Accordingly, the resulting dispersion function is (1) directly proportional to the rate of change of index (to the slope or derivative of the dispersion curve), and is (2) inversely proportional to the material's index above that of air. However, this function varies from wavelength to wavelength and from glass to glass. To simplify glass comparisons, a measure of average dispersion over a range of wavelengths is needed. Thus, if three separated wavelengths are arranged in order of increasing wavelength, and if the waveband is defined by the first and third wavelengths, then the following formula expresses as a single number the overall dispersion, ω:

$$\omega = \frac{n_1 - n_3}{n_2 - 1}.$$

Note that as the waveband is reduced and approaches zero, overall dispersion converges to wavelength-dependent dispersion (when both are normalized to unit waveband).

To avoid using small numbers to express dispersion, Abbe introduced reciprocal dispersion. Thus, reciprocal dispersion is called the Abbe number. Originally, the symbol ν (nu) was used, but more recently, at least in the United States, the symbol V has been adopted. Thus,

$$V \equiv \frac{1}{\omega} = \frac{n_2 - 1}{n_1 - n_3}.$$

Traditionally in the visible region, wavelength 1 is the blue hydrogen Fraunhofer F line (Balmer beta) at 0.4861 μm. Wavelength 2 is the yellow sodium Fraunhofer D line at 0.5893 μm. Wavelength 3 is the red hydrogen Fraunhofer C line (Balmer alpha) at 0.6563 μm. Thus, in the visible,

$$V_D = \frac{n_D - 1}{n_F - n_C}.$$

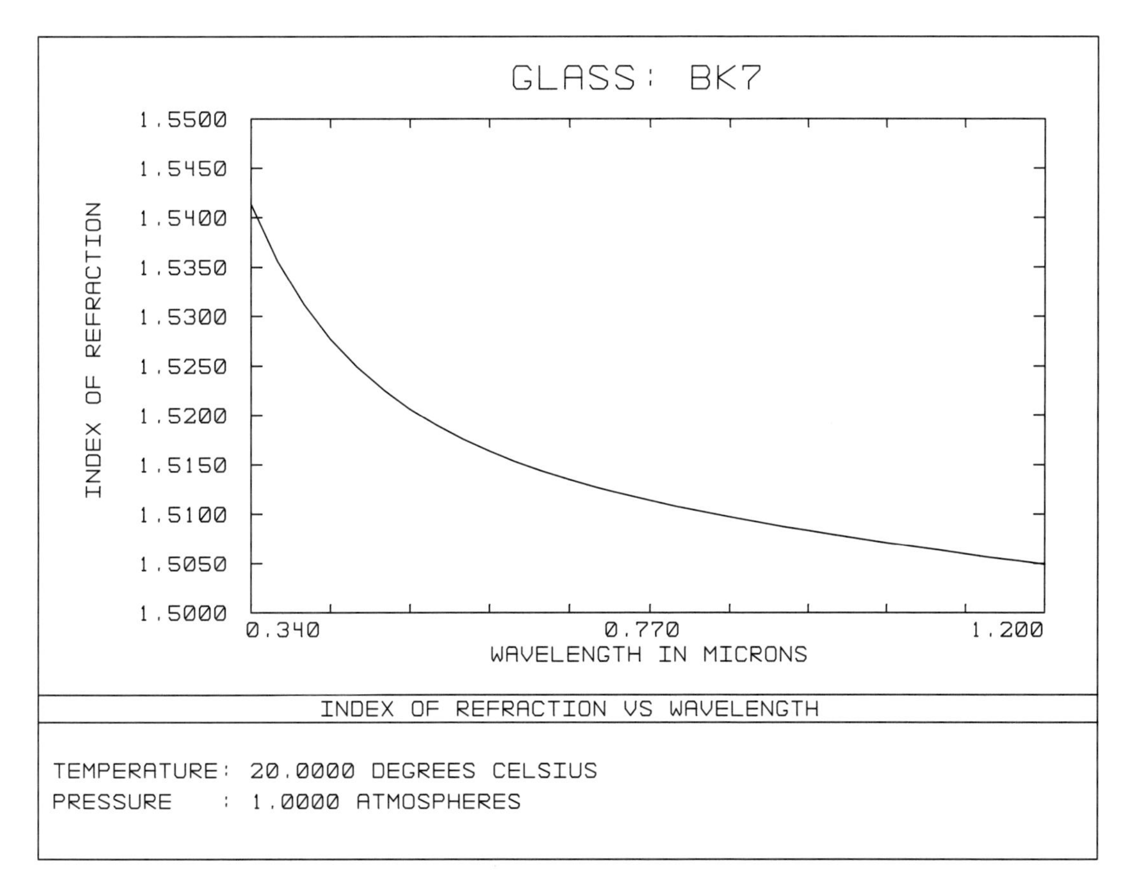

Figure A.10.1.1.

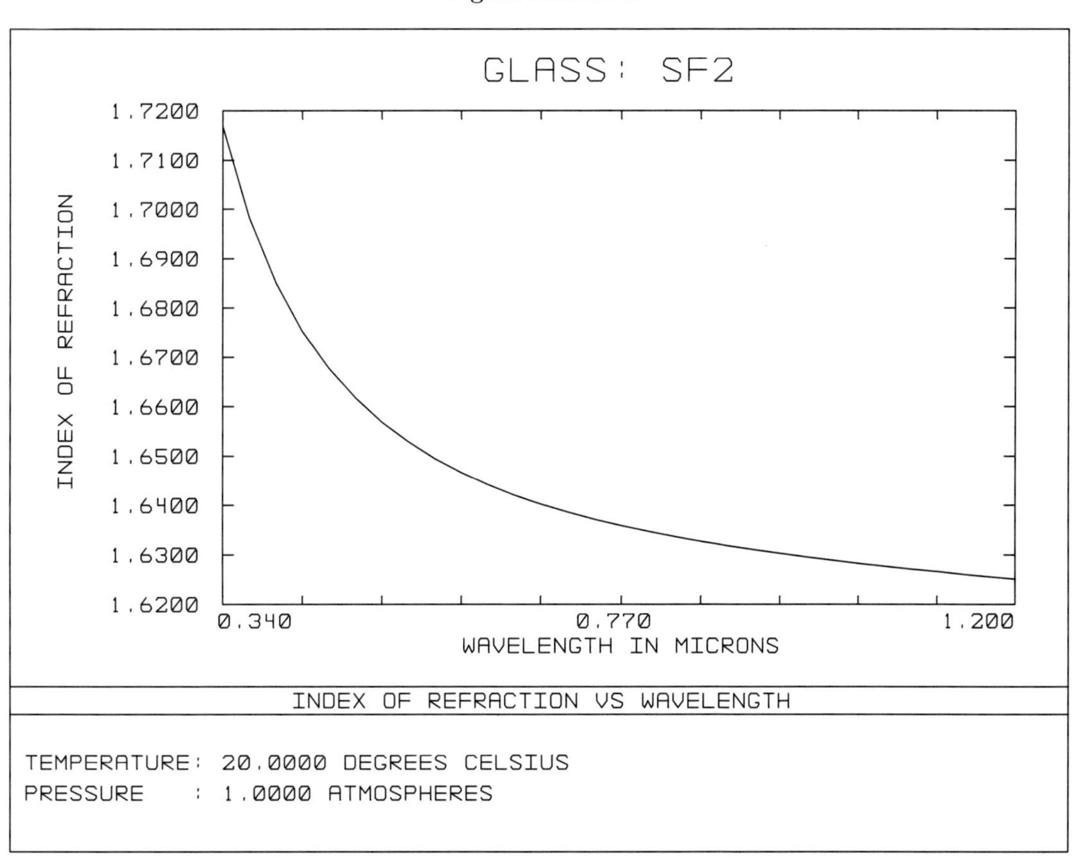

Figure A.10.1.2.

The sodium D line is actually a close double line, so in recent years the yellow helium d line at 0.5876 μm has been substituted. For BK7 glass, V_d is 64.2; for SF2 glass, V_d is 33.9. In addition, reciprocal dispersion is often quoted based on the blue cadmium line, green mercury line, and red cadmium line. In general, and especially in the ultraviolet and infrared, any three wavelengths can be used.

In addition to its name as given by the manufacturer, a glass can be identified by its index and dispersion. This identification is usually expressed in a six digit shorthand notation. For example, Schott BK7 is 517642, which translates to mean that n_d is 1.517 and V_d is 64.2.

A.10.3 Crown and Flint Glasses

Glasses for use in or near the visible wavelength region are divided into two groups based on their dispersions (not indices). These two groups are called crown glasses and flint glasses. The names are historical. Crown glasses have relatively low dispersions, and flint glasses have relatively high dispersions. If made into a prism for a spectrograph, a crown prism forms a relatively compressed spectrum, whereas a flint prism with the same average deviation forms a relatively extended spectrum. For this reason, most prism spectrographs use flint prisms. Similarly, a crown singlet lens has less longitudinal chromatic aberration than a flint lens of the same focal length. For the two glasses mentioned above, BK7 is a crown glass, and SF2 is a flint glass.

In unusual spectral regions, such as the ultraviolet and infrared, crown and flint are merely relative terms. Lens designers speak of one glass as being "crowny" or "flinty" relative to another. If wavelength regions change, the same material can even switch from a crown material to a flint material. When doing glass selection in these circumstances, the designer may have to calculate the dispersion of candidate glasses by hand with a pocket calculator. Contrary to tradition, the author prefers the direct dispersion, ω, rather than the reciprocal dispersion, V, because the numbers vary in a more intuitively obvious way. If you do not like comparing small numbers, just multiply all dispersions by 100 or 1000, and tabulate those numbers.

A.10.4 Partial Dispersion

Overall dispersion, expressed as a single number within a waveband, does not describe how dispersion varies within the waveband. Thus, a second dispersion function has been introduced to express this variation as another single number. This second function is called partial dispersion. As with overall dispersion, partial dispersion simplifies the comparison of different glasses.

Refer again to the dispersion plots in Figures A.10.1.1 and A.10.1.2 for BK7 and SF2 glass. These curves are typical for glasses in the visible, near ultraviolet, and near infrared, but not necessarily in the thermal infrared. In both curves, the slope in the blue part of the spectrum is considerably greater than the slope in the red part of the spectrum. Thus, for both glasses, blue dispersion is greater than red dispersion (even when the $n-1$ factor in the denominator of the dispersion formula is taken into account).

But there is a second-order effect. The *shapes* of the two curves are different, and thus the ratio of blue slope to red slope is not the same for both glasses. For SF2, the ratio is greater. Thus, the dispersion in the blue part of the spectrum relative to the red part of the spectrum is greater for SF2 than for BK7 (again taking the $n - 1$ factor into account).

Imagine now that you make a prism spectrograph with interchangeable prisms of different glasses. For prisms with different dispersion curve shapes, the spectral features will fall in different places relative to each other. Normalized to the total length of the spectrum, if a prism is used with relatively greater blue dispersion, then the spectrum in the blue will be relatively longer. Similarly, if a prism is used with relatively less blue dispersion (but still much more than in the red), then the spectrum in the blue will be relatively shorter. Thus, if a glass produces a longer blue spectrum, it is called a long glass. And if a glass produces a shorter blue spectrum, it is called a short glass. Note carefully that the terms long and short do *not* refer to overall dispersion.

In addition to having less overall dispersion, most crown glasses are also short glasses. In addition to having more overall dispersion, most flint glasses are also long glasses. But there is no general linkage and there are exceptions. There are a few long crowns and short flints. These unusual glasses are called abnormal-dispersion glasses.

Recall that in general, every glass type has its own dispersion curve. In particular, the shape of each of these curves is unique. Partial dispersion is a measure of this shape. For any three separated wavelengths arranged in order of increasing wavelength, partial dispersion is given by:

$$P_{1,2} = \frac{n_1 - n_2}{n_1 - n_3}.$$

Also,

$$P_{2,3} = \frac{n_2 - n_3}{n_1 - n_3}.$$

And it follows that

$$P_{1,2} + P_{2,3} = 1.0.$$

Thus, partial dispersion, defined this way, expresses the linear fractional position of an intermediate wavelength relative to two extreme wavelengths in the spectrum formed by a prism spectrograph. In other words, the partial dispersion of a glass is a measure of its "longness" or "shortness." There are also similar expressions for partial dispersion using other wavelength combinations, but the idea of expressing the shape of the dispersion curve is the same.

For nonstandard wavelength regions, especially the ultraviolet and infrared, the lens designer may find it necessary, as with overall dispersions, to calculate partial dispersions by hand with a pocket calculator.

A.10.5 Glass Maps

As an aid to glass selection in the visible region, it is customary to make a plot of available glasses, where V_d is plotted on the horizontal axis and n_d is plotted on the vertical axis. This plot is called a glass map. As an example, the glass map for the Schott glasses is shown in Figure A.10.2. The n_d values range from about 1.45 at the bottom of the glass map to about 2.05 at the top. For crown glasses, the V_d values range from a very crowny 90 at the far left on the glass map to between 50 and 55 in the middle. For flint glasses, the V_d values range from between 50 and 55 to a very flinty 20 at the far right. The dividing line between crown and flint glasses is fairly arbitrary and not straight. Once again, these dispersions are reciprocal dispersions. A V_d value of 90 represents low dispersion, and a V_d value of 20 represents high dispersion.

Not all of the area on the glass map is populated by available glasses. In fact, much of the area on the glass map is blank. The available glasses appear mostly in a diagonal swath running from the lower left to the upper right.

The most common crown and flint glasses fall within a curved region along the bottom and right edges of the swath. This region is called the old glass line. For centuries, the only glasses available fell on this line and ranged from the K glasses to the F glasses. The old crown glasses, which are high quality versions of window glass, are composed mostly of silicon dioxide (silica or quartz) with a few added compounds

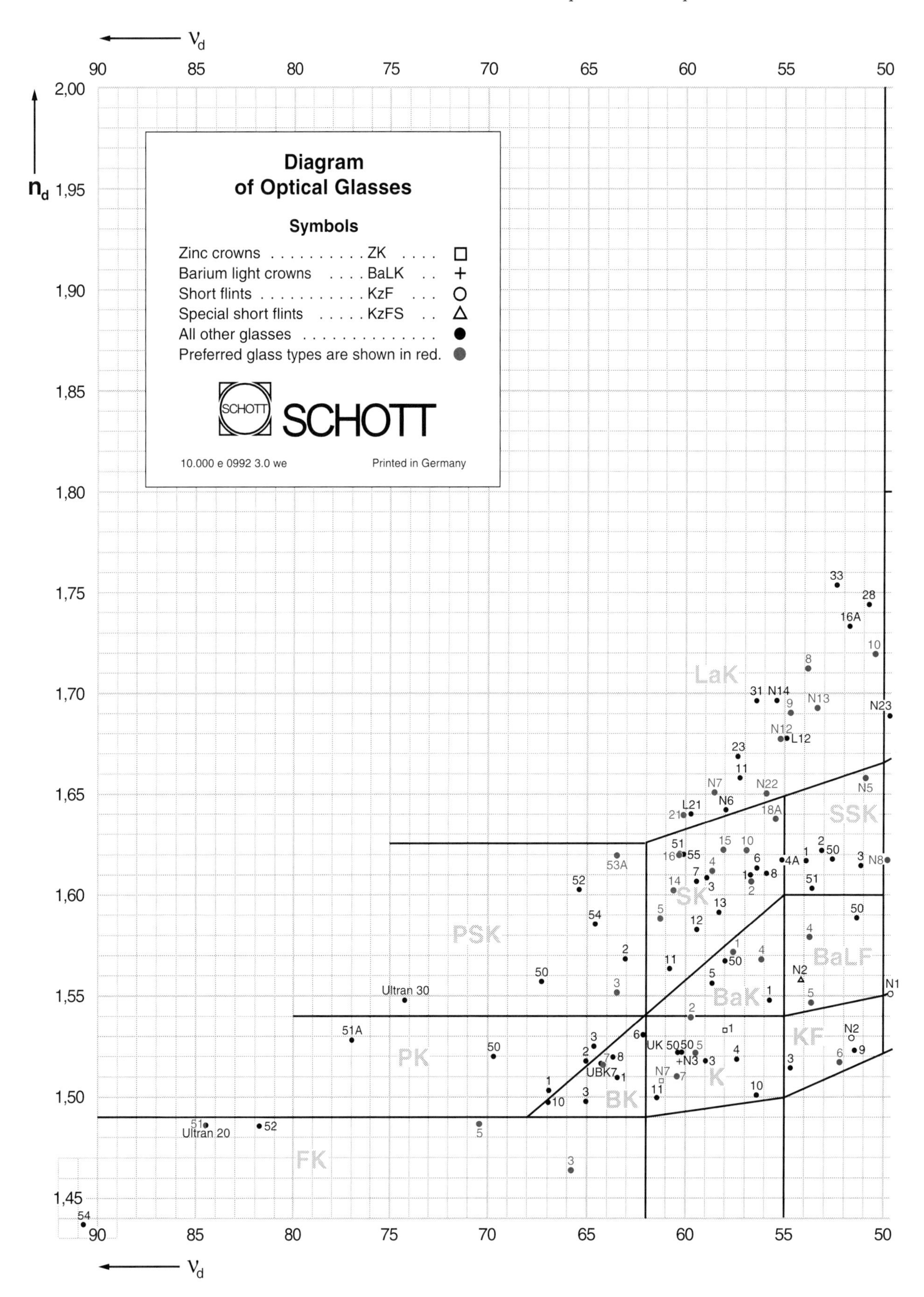

Figure A.10.2. Left

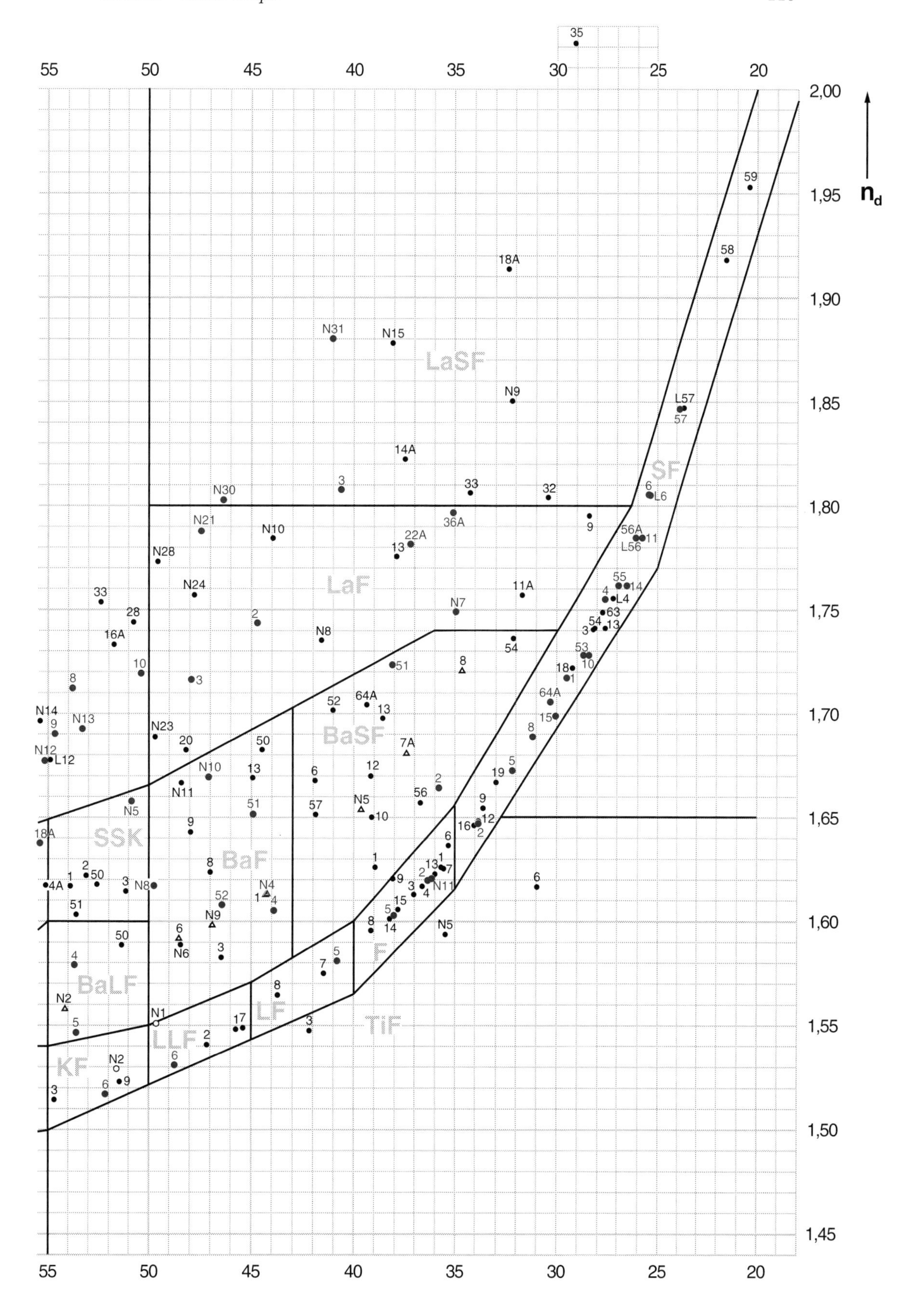

Figure A.10.2. Right

to help the glass melt at practical temperatures during manufacture (thus the name soda-lime glass). The old flint glasses are composed mostly of silicon dioxide and lead oxide. More recently, the old glass line has been extended to range from the BK glasses to the SF glasses.

In 1886, Abbe and Schott introduced the new glasses. Most noteworthy among these were the high-index barium crowns (BaK and SK glasses) that for the first time allowed anastigmatic lenses to be designed. In the 1930s, Eastman Kodak introduced the rare earth glasses that contain lanthanum (La glasses). Optical glass is now made using a wide range of materials, and new types are continuously being developed. But glass chemistry has its limits, and some areas of the glass map will probably remain blank.

There is another type of glass map that is often seen. This is a plot of overall dispersion, V_d, on the horizontal axis versus partial dispersion on the vertical axis. Two types of partial dispersions are commonly used for these plots. The first, $P_{g,F}$, relates the shape of the dispersion curve for blue wavelengths, and the second, $P_{C,s}$, relates the shape of the dispersion curve for red wavelengths. Thus,

$$P_{g,F} = \frac{n_g - n_F}{n_F - n_C}$$

and

$$P_{C,s} = \frac{n_C - n_s}{n_F - n_C}$$

where the wavelengths are: g = 0.4358 μm, F = 0.4861 μm, d = 0.5876 μm, C = 0.6563 μm, and s = 0.8521 μm. Examples for the Schott glasses are shown in Figures A.10.3.1 and A.10.3.2.

What is most interesting about the partial dispersion glass maps is that nearly all glasses fall close to a single straight or slightly curved line called the normal glass line. Thus, for the vast majority of glasses for use in the visible, partial dispersion is approximately linearly related to overall dispersion. This is a more quantitative way of saying that most crown glasses are short glasses and most flint glasses are long glasses. The few exceptions that fall significantly away from the normal glass line are the abnormal-dispersion glasses. These abnormal-dispersion glasses are very useful in controlling secondary chromatic aberrations.

A.10.6 Ultraviolet and Infrared Glasses

Most optical glass is transparent only in or near the visible region of the spectrum. If a lens is to be used in the ultraviolet or infrared, special optical materials are often required.

In the ultraviolet, the common materials are optical crystals, a few plastics, and a few special glasses. These materials are all transparent in the visible, and they remain transparent down to some ultraviolet cutoff wavelength. The optical material that transmits farthest into the ultraviolet is lithium fluoride, whose cutoff wavelength is 0.105 μm. For ultraviolet wavelengths shorter than 0.105 μm, nothing transmits, and all-reflecting systems become mandatory. Still farther into the ultraviolet and into the x-ray region, reflective films also fail, and here grazing-incidence mirrors must be used. Table A.10.1 summarizes the cutoff wavelengths of several useful ultraviolet materials.

Note that air has a cutoff wavelength of about 0.185 μm. Wavelengths shorter than about 0.185 μm are said to be in the vacuum ultraviolet. The situation is even worse if you view up out of the atmosphere through the ozone layer. For these astronomical observations, the ultraviolet cutoff is about 0.300 μm.

In the infrared, the common materials are again optical crystals (single crystal or polycrystalline) and a few special glasses. Many of these materials are opaque in the visible and transmit a band in the infrared. The exact passband depends on specimen purity, percent transmission criterion, and thickness. Table A.10.2 is a guide to the useful wavelength passbands of some of the more practical infrared materials.

All of these materials are used, even diamond (which is available relatively cheaply in small thin sections). Rocksalt is widely used, but must be protected and slightly heated to avoid being etched by moisture in the air. Fused silica (or fused quartz) is very popular. But ordinary (slightly impure) fused silica has a strong water absorption band at about 2.7 μm, and even in water-free form it only transmits to about 3.5 μm. Calcium fluoride is also popular, but it must not be subjected to thermal shock or it will fracture. For longer wavelength infrared work, perhaps the best materials are silicon, germanium, zinc sulfide, and zinc selenide. These four materials are all very practical, relatively inexpensive, non-hygroscopic, and available in large pieces.

Table A.10.1
Ultraviolet Cutoff Wavelengths

LiF	0.105 μm
MgF_2	0.115 μm
CaF_2 (fluorite)	0.125 μm
BaF_2	0.135 μm
Al_2O_3 (sapphire)	0.150 μm
SiO_2 (fused silica)	0.165 μm
NaCl (rocksalt)	0.200 μm
$CaCO_3$ (calcite)	0.210 μm
Diamond	0.230 μm
Schott UBK7 glass	0.280 μm
PMMA (acrylic)	0.300 μm

Lens designers accustomed to working in the visible find infrared materials somewhat surprising. Refractive indices are available ranging from less than 1.5 to more than 4.0. Both overall dispersions and partial dispersions can be highly unusual. There are even a few infrared materials that are so unusual that dispersion is greater, not less, for longer wavelengths. High indices are capable of reducing monochromatic aberrations, and unusual dispersions can yield lenses with superb color correction.

Table A.10.2
Passbands for Infrared Materials

Schott BK7 glass	0.30 to 2.5 μm
$CaCO_3$ (calcite)	0.21 to 3.0 μm
SiO_2 (fused silica)	0.165 to 3.5 μm
Al_2O_3 (sapphire)	0.150 to 6.0 μm
MgF_2	0.115 to 7.0 μm
MgO	0.25 to 8.0 μm
Silicon	1.2 to 8.5 μm
CaF_2 (fluorite)	0.125 to 9.0 μm
BaF_2	0.135 to 11.5 μm
As_2S_3 glass	0.65 to 11.5 μm
Germanium	1.8 to 11.5 μm
Cleartran ZnS	0.40 to 12 μm
ZnS	0.55 to 12 μm
AMTIR-1	0.8 to 12 μm
GaAs	1.0 to 16 μm
NaCl (rocksalt)	0.20 to 17 μm
ZnSe	0.55 to 18 μm
CdTe	1.0 to 30 μm
KRS-5	0.6 to 50 μm
Diamond	0.23 to $>$80 μm

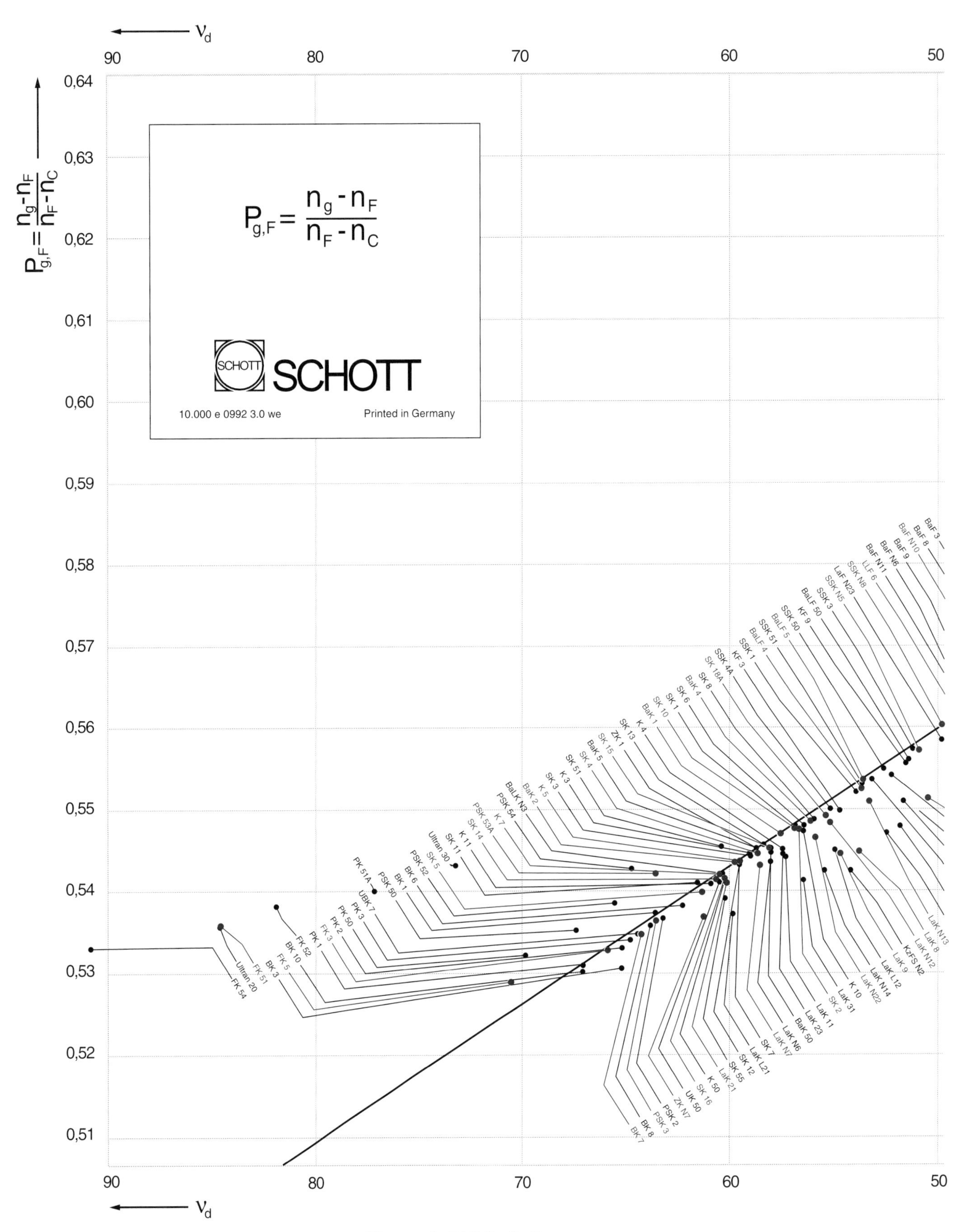

Figure A.10.3.1. Left

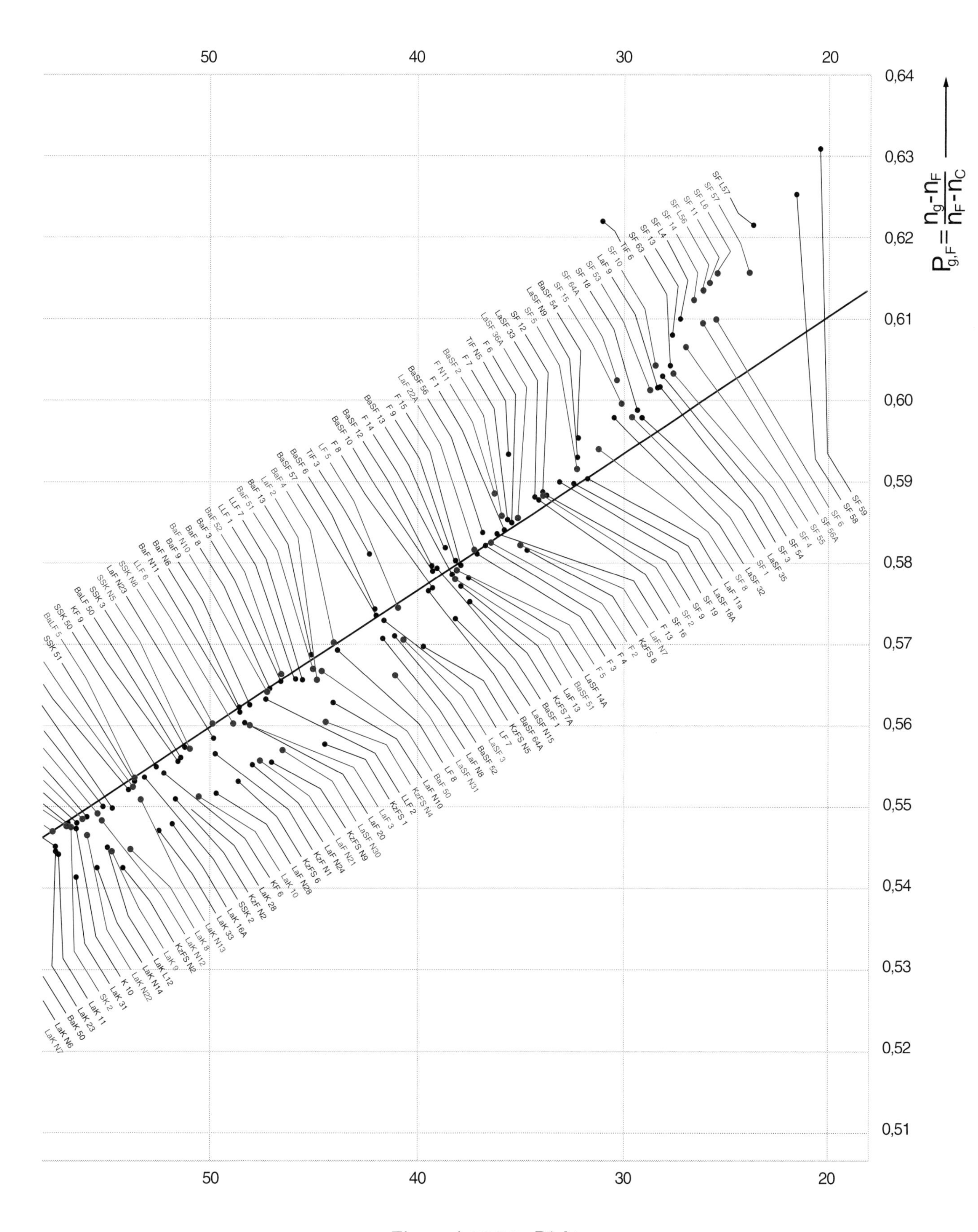

Figure A.10.3.1. Right

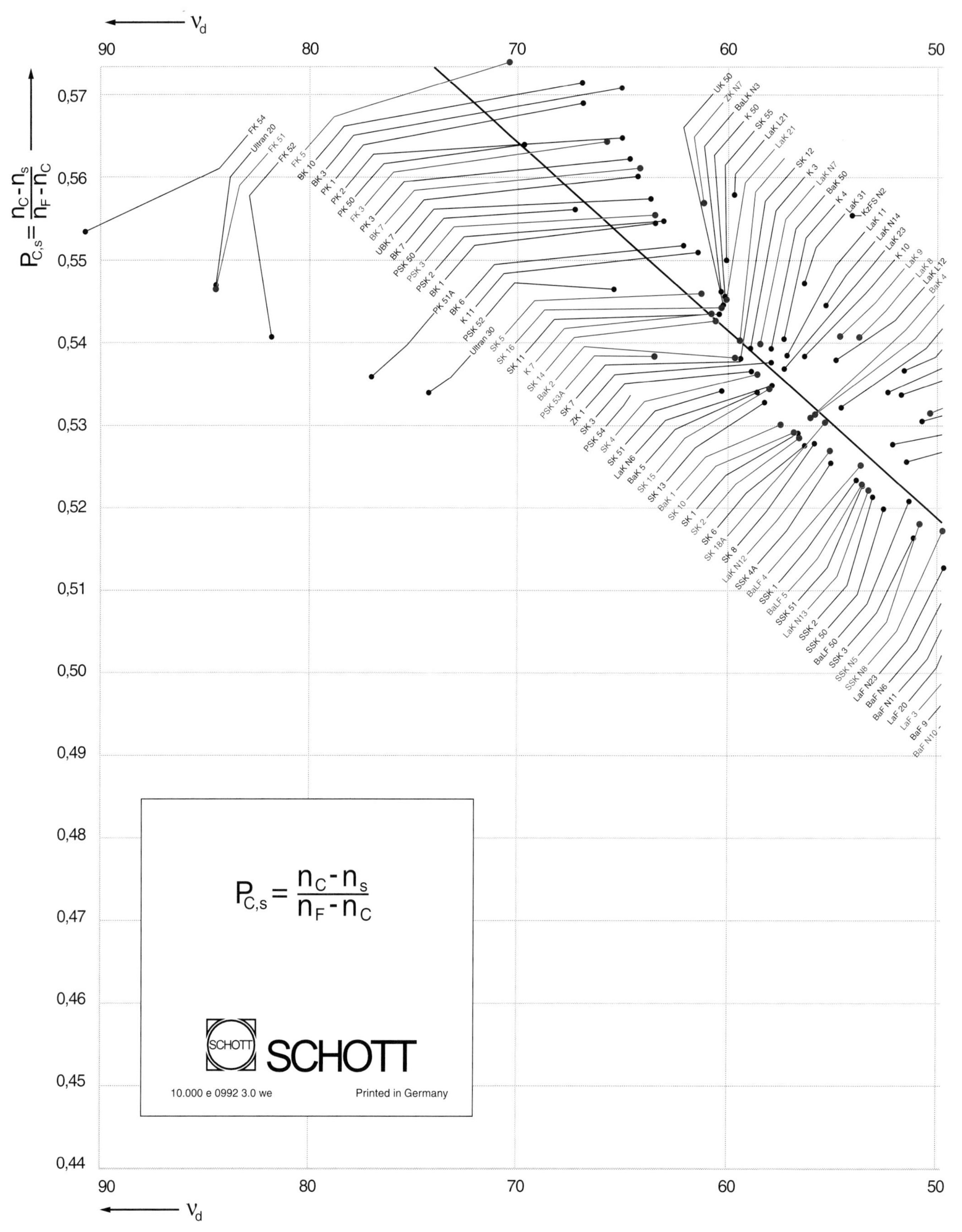

Figure A.10.3.2. Left

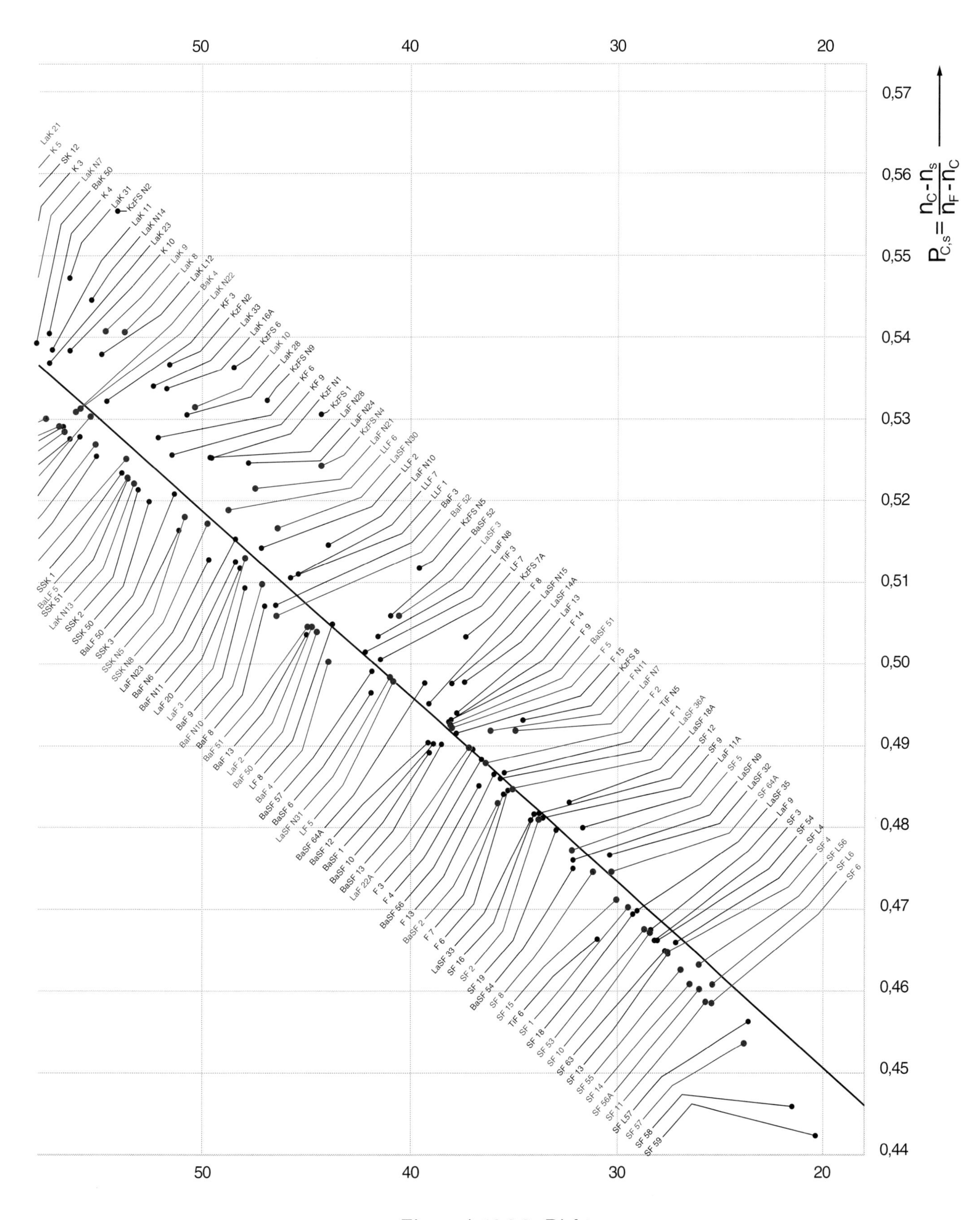

Figure A.10.3.2. Right

A.10.7 Glass Selection

There are so many different glasses on the glass map that it may seem that selecting the right glass types for a compound lens would be very difficult. Actually, the multiplicity of glass types is a legacy from the days when it was easier to order up a new glass to fit a design, rather than redesign the lens to utilize existing glasses. Today, most lenses can be designed using relatively few glass types. An exception is where the exact index and/or dispersion change at a cemented lens interface makes a big difference. A second exception is where a good match is needed for one of the abnormal-dispersion glasses.

For most work in the visible region, the designer can often restrict himself to a few favorite glasses scattered about the glass map. Along the old glass line, they might include BK7, LLF1, F2, SF2, and several of the higher-index SF glasses out to SF57. An interesting barium glass is SK16. Lanthanum glasses might include LaK8 and LaFN21. An excellent short flint is KzFSN4. For a long crown, crystal calcium fluoride is very effective and increasingly widely used. In the ultraviolet and/or infrared, all the glasses and crystals listed earlier should be considered because these materials have such widely varying properties.

When designing a lens, a starting configuration or geometry is always needed. Thus, initial glasses must be specified. The computer program has an algorithm that can model index and dispersion for normal-dispersion glasses. Therefore, during optimization, normal glasses can be made variable. The computer will then tell you which region of the glass map is best for a given element. Often, the computer will change the glasses to extreme cases. These answers are best taken only as indicators.

To convert variable glasses into real glasses, the designer manually changes model glasses into non-variable real glasses, one element at a time, and reoptimizes. After each reoptimization, it will be seen that the remaining variable glasses have shifted somewhat to accommodate the real glasses. Often the best approach is to convert the crowns first because there are so few good high-index crowns available. When it comes time to convert the flints, the designer will have an abundance of choices.

After all the glasses have been converted, the designer may sometimes manually try out other similar glasses, reoptimize, and compare performance. For example, if a high-index flint is indicated, it may not matter much if SF56, SF57, or SF58 is used. The lower-index glass would have many practical advantages. If a higher index does make a big difference, then you know.

If abnormal-dispersion glasses are to be used, then you specify them at the beginning and allow only the remaining normal glasses to be variable. Note that in this case, the normal glass model is not totally reliable. After the computer has indicated its choices for the normal glasses, you should manually try out all the glasses in the vicinity of each indicated glass to double-check. Second-order effects are very important here. Also, be sure to use enough wavelengths to adequately sample the waveband.

A.10.8 Melt Sheets

No two batches or melts of optical glass are the same. Even if the glass type is very common and has been made for decades, the index of refraction will vary to some degree from batch to batch. When an optician buys glass blanks to be made into lenses, he is furnished by the manufacturer with a certificate called a melt sheet. For the specific glass batch, a melt sheet contains a listing of actual measured (relative) indices and index differences for specified wavelengths.

Thus, the lens designer may use nominal values of the refractive indices for initial design studies. But when the final design is being computed, the actual measured values from the melt sheets should be substituted for the catalog values.

A.10.9 Non-Optical Glass Considerations

Often, glass selection is influenced by non-optical considerations such as cost, availability, weight, ease of fabrication, resistance to environmental deterioration and staining, mechanical hardness, thermal properties, and many other factors. The designer is wise to keep these factors in mind because they can make or break a candidate design.

Cost is doubly important. First, of course, cost is of direct importance. But cost tells the designer something more. If a glass is expensive, it is for one of two reasons. The raw materials may be expensive, such as the lanthanum used in most of the rare earth glasses. However, the more interesting cost factor is the difficulty in making the glass. A glass that is hard to make costs more, usually because the yield is low. These difficult glasses, even when nominally acceptable, may still have problems, such as inhomogeneity, bubbles, or striae. A factor related to cost is melt frequency. If they make a lot of a certain type of glass, they get good at it and the price drops. Inexpensive glass is usually good glass. Keep a copy of a glass price list handy. An especially convenient form of price list is a table of relative glass prices compared to BK7 at 1.0.

Glasses that are readily available in large amounts are called preferred glasses. Preferred glasses are usually specially designated on a manufacturer's glass map, such as by a red dot rather than a black dot. If you choose to use a non-preferred glass, you had better have a good reason.

A.10.10 Glass Manufacturers

Optical glass is made all over the world. The de facto standard of optical glass has been set by Schott Glass Technologies Inc. of Mainz, Germany and Duryea, Pennsylvania, USA. The commonly accepted designations for the various types of optical glass are Schott names.

There are two major optical glass manufacturers in Japan, Hoya Corp. and Ohara Inc. There are also major optical glass manufacturers in the United States, Britain, France, Russia, and other countries. In addition, there are numerous manufacturers of special optical glasses and crystals for all wavelength regions from 0.105 μm to the far infrared. Finally, there are manufacturers of mirror substrate materials.

The student of optical design should, as soon as possible, obtain a glass catalog from one or more of the major glass manufacturers. Two versions are usually available, the big catalog and a small pocket catalog. These catalogs contain a huge amount of information about all aspects of glass selection.

A.10.11 Mirror Substrate Materials

Common mirrors have the reflecting surface on the back side for the reflecting surface's protection. These are called second-surface mirrors. In the design of optical systems, second-surface (Mangin) mirrors are occasionally used. But nearly always, optical mirrors have the reflecting surface on the front side. These mirrors are called first-surface mirrors.

For a first-surface mirror, the mirror substrate merely serves as a mechanical support for a reflecting surface or thin reflecting film. In particular, the substrate does not have to be of optical quality. It can be full of striae, bubbles, and inclusions (not too near the surface), and it can even be opaque. The important criteria for a mirror substrate become mechanical, thermal, and chemical stability, and the ability to be ground, polished, and figured to a smooth and correct shape. For many applications, weight is also a factor.

When Isaac Newton made his first reflecting telescope in 1668, he made his mirrors of uncoated speculum metal. Although the formula for speculum metal varied over the years, Newton's formula was six parts copper, two parts tin, and one part arsenic. Speculum mirrors, however, tarnish and have to be periodically repolished and thus also refigured. This is a very great disadvantage.

This problem was solved in 1856 when Liebig developed a method of chemically depositing a thin layer of silver on glass. In the same year, Steinheil and Foucault independently applied Liebig's silvering process to glass telescope mirrors with great success. Although the silver also tarnished, it could easily be removed and replaced without having to repolish and refigure the mirror. Later, in 1932, Strong and others developed a method of evaporating a thin film of aluminum on a glass mirror in a vacuum. Aluminized mirrors do not tarnish and the coating lasts for years (how long depends mostly on weathering and cleaning). Most reflecting telescopes today still use aluminized mirrors. For infrared work, mirrors are sometimes gold coated.

The first glass telescope mirrors were made of ordinary soda-lime glass of the same type used in windows and bottles. There are stories at Mt. Wilson of how the 60- and 100-inch telescopes have mirrors (cast in 1896 and 1908) of French wine bottle glass. For the Palomar 200-inch telescope, greater resistance to thermal distortion was needed. Thus, the 200-inch mirror (cast in 1934) was made of high-silica, low-expansion Corning Pyrex glass. By the 1960s, pure fused silica or fused quartz (amorphous silicon dioxide), with an even lower coefficient of thermal expansion, had become the mirror material of choice. For some applications, fused silica is still the best material today. And low-expansion glass is staging a comeback. The University of Arizona is spin casting very large, lightweighted mirror blanks using Ohara low expansion E6 glass.

Incidentally, lightweighted mirrors, pioneered with the 200-inch, are not solid disks, but rather have large openings or voids inside. The maximum practical lightweighting is about 80%, which means that 20% of the material of a solid disk is left. Further lightweighting produces a mirror that is too delicate.

By the mid-1960s, new glass-making technologies yielded two different types of mirror materials with virtually zero coefficients of thermal expansion. One of these was Cer-Vit (Ceramic-Vitreous), developed by Owens-Illinois. Although Owens-Illinois later stopped making Cer-Vit, very similar materials are now available as Zerodur from Schott in Germany and as Astrositall from Russia. Zerodur and similar materials are semi-transparent and have a color that can range from pale yellow through orange to brown. Recent developments by Schott are yielding excellent lightweighted cast Zerodur mirrors.

During manufacture, Zerodur starts out as a molten glass that can be readily stirred. But Zerodur has an unusual chemical composition and is given a special heat-treating process during cooling. The result is a final material that is a mixture of tiny micro-crystals in an amorphous glass matrix. The thermal properties of Zerodur are obtained because the coefficients of thermal expansion of the micro-crystals and glass matrix have opposite signs. One constituent expands while the other contracts, and the two changes cancel out each other.

Note that there is a potential problem with this arrangement. On a micro-crystal level, large internal stresses can develop, and these stresses can lead to long-term mechanical instability or creep. In practice, however, many successful large telescope mirrors have been made of Cer-Vit, Zerodur, and Astrositall.

The second of these zero-coefficient mirror materials is ULE (Ultra Low Expansion), developed by Corning. ULE has no micro-crystals and requires no special heat-treating process. Instead, ULE is an amorphous homogeneous mixture of approximately 92.5% silicon dioxide and 7.5% titanium dioxide. Different percentages are sometimes used during manufacture depending on the anticipated temperature at which the final material will be used, with less titanium for lower temperatures.

Like more conventional fused silica, during manufacture, ULE cannot be melted and stirred at practical temperatures (any container would melt first). Consequently, ULE is manufactured by a flame hydrolysis deposition process; molten material rains down onto a substrate where it accumulates. Unfortunately, even when carefully done, this process allows the final bulk material to have small layered inhomogeneities caused by minor variations in the titanium content. These variations and the internal stresses they cause are visible from a side cut, although not when viewed normal to the slab. Thus, although it looks almost completely clear, ULE is not suitable for making transmitting optics such as lenses and windows.

The basic principle of operation of ULE is similar to that of Zerodur; the two constituents have coefficients of thermal expansion with opposite signs. The crucial difference is that the cancellation in ULE takes place on a molecular scale rather than on a micro-crystal scale. Thus, unlike in Zerodur, in ULE no large internal stresses are developed between two different and physically separated constituents. Thus, long-term creep is greatly reduced.

Another advantage of ULE is that, because no special heat-treating process is required, separate pieces of ULE can be welded together to make lightweight cellular or egg-crate mirrors. The heat from such welding would destroy the thermal properties of Zerodur. These lightweight ULE mirrors have been very successful in space applications. The disadvantage of ULE is that it is more expensive than Zerodur.

Finally, much effort was spent during the 1980s to develop spacecraft mirrors made of beryllium metal. Beryllium has the advantages of having low density yet being very stiff. Thus, we have now come full circle back to metal mirrors.

Chapter A.11

Wavefronts and Diffraction

There are many optical effects that geometrical optics cannot explain. To explain color, polarization, interference, and diffraction, it is necessary to invoke physical optics where light is more correctly modeled either as electromagnetic waves or as photon wave packets. In this chapter, wavefronts and diffraction are discussed from the point of view of the lens designer.

A.11.1 Diffraction by Aperture Edges

Diffraction can be explained in two ways. The first uses nineteenth century electromagnetic theory; the second uses twentieth century quantum theory.

According to electromagnetic theory, a light source emits into space a continuous, unbounded, outward-flowing stream of electromagnetic waves. The speed of propagation is the speed of light. If the light source is of very limited spatial or angular extent, then the source is effectively a point source. At any instant of time, all light of a given wavelength emitted by a point source is in phase; that is, the source is coherent. Consequently, on surfaces of constant light travel time from the point source, all of the waves are still in phase; that is, the monochromatic waves from point to point on these surfaces are spatially coherent. These surfaces of constant phase are called wavefronts. If the source is immersed in a single medium of constant index of refraction, then these wavefronts are spherically shaped and centered on the source.

One way to describe the propagation of wavefronts is by Huygens' principle. Huygens' principle asserts that each point on a wavefront acts as a source of new wavelets that combine to form the next wavefront according to the laws of interference. It is important to note that Huygens' principle reveals that perfectly coherent wavefronts continue to propagate as perfectly coherent wavefronts only if the wavefronts are unbounded with no edges.

The entrance pupil of an optical system intercepts at some location a limited portion of the moving stream of wavefronts. At an aperture edge, such as the stop or an obscuration, the Huygens wavelets on one side of the edge are blocked. Thus, in image space beyond the exit pupil, the interference of the remaining wavelets is different from what it would be if all the wavelets were present. Beyond the exit pupil, the stream of now bounded wavefronts loses its perfect coherence; that is, the phases become partially smeared. The effect is greatest near wavefront edges. One way to think of this loss of spatial coherence is to visualize the wavefronts as becoming frayed around the edges. This concept is illustrated in Figure A.11.1.

The physical result of bounding the wavefronts is called diffraction. Diffraction causes an angular dispersing or spraying out of the light that in an imaging system degrades image sharpness. But diffraction is not to be confused with aberrations, which also degrade image sharpness. Diffraction is an interference phenomenon that can only be explained by physical optics. Aberrations are a geometrical phenomenon

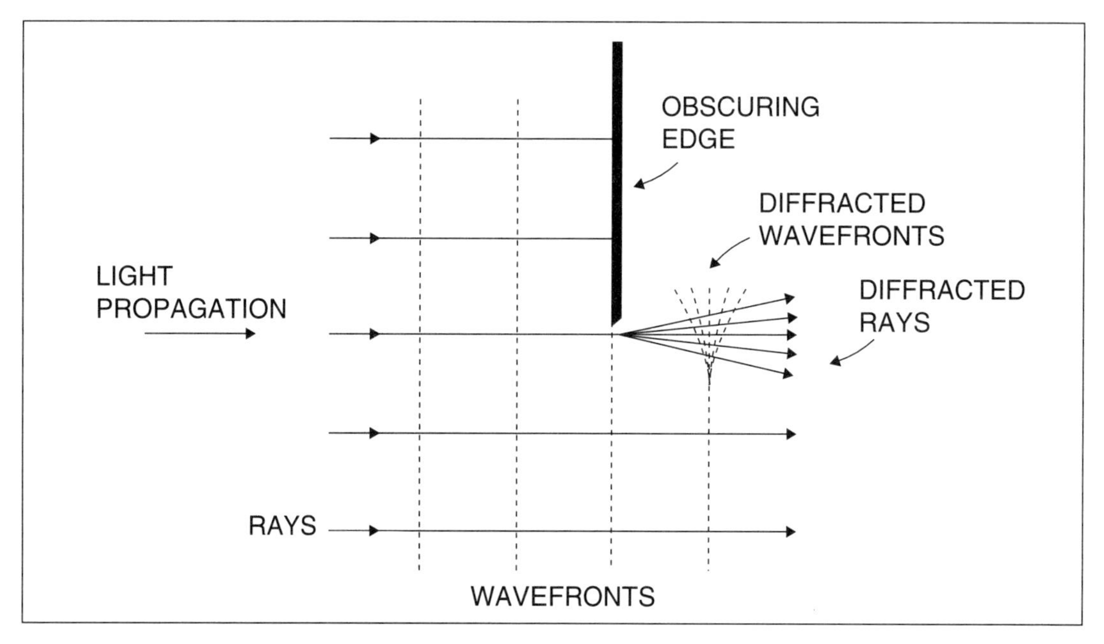

Figure A.11.1. Diffraction Near an Obscuring Edge. *Parallel rays and plane wavefronts are incident on an opaque obscuring edge. The rays near the edge are diffracted in various directions. The wavefronts, always normal to the rays, are thus no longer coherent (unique). Instead, the wavefronts are smeared out or* frayed around the edges.

where the wavefront is deformed without changing its coherence. Note carefully that if diffraction acted as a type of aberration, then in principle it would be possible to add an optical element to restore the correct wavefront shape. But this correction is not possible. Diffraction and aberrations are two separate phenomena that together conspire to enlarge the image point spread function (PSF).

Diffraction is a process that begins at an aperture edge and continues downstream as the waves propagate. Right at the edge, there is little change in the wavefront other than blocking part of it. Just past the edge, the effects of truncating the wavefront become significant and evolve with increasing distance. This region of changing effects just past the edge is called the near field. At greater distances from the edge, diffraction effects cease evolving and stabilize. This region of constant effects farther from the edge is called the far field.

In lens design, most images are formed in the far field of the defining aperture (the stop opening) and its image (the exit pupil). In the far field of the exit pupil, all diffraction effects can be considered to happen at the exit pupil alone rather than throughout the system. Thus, the exit pupil is in practice the only aperture for which diffraction effects need to be accounted. If there is mechanical vignetting or a central obscuration, then the vignetted or obscured exit pupil is used.

Recall that in geometrical optics, the spot diagram is the measure of the PSF, and that by tracing rays the exit pupil is uniquely mapped onto the spot diagram. Now with diffraction, the spot diagram must be replaced by a more accurate diffraction PSF. Furthermore, the irradiance of light at any one point in the diffraction PSF is the result of the interference of the Huygens wavelets originating over the entire exit pupil. Thus, there is no longer a one-to-one mapping of the exit pupil onto the PSF.

In the alternative quantum theory explanation of diffraction, light is modeled as wave packets or photons. Light rays are the trajectories of photons, and these trajectories are normal to the wavefronts.

A photon passing close to an aperture edge suffers a random change in its direction of motion, and this is diffraction. For many similar photons, the combination of these random angular deflections is a spraying out of the rays; that is, after passing

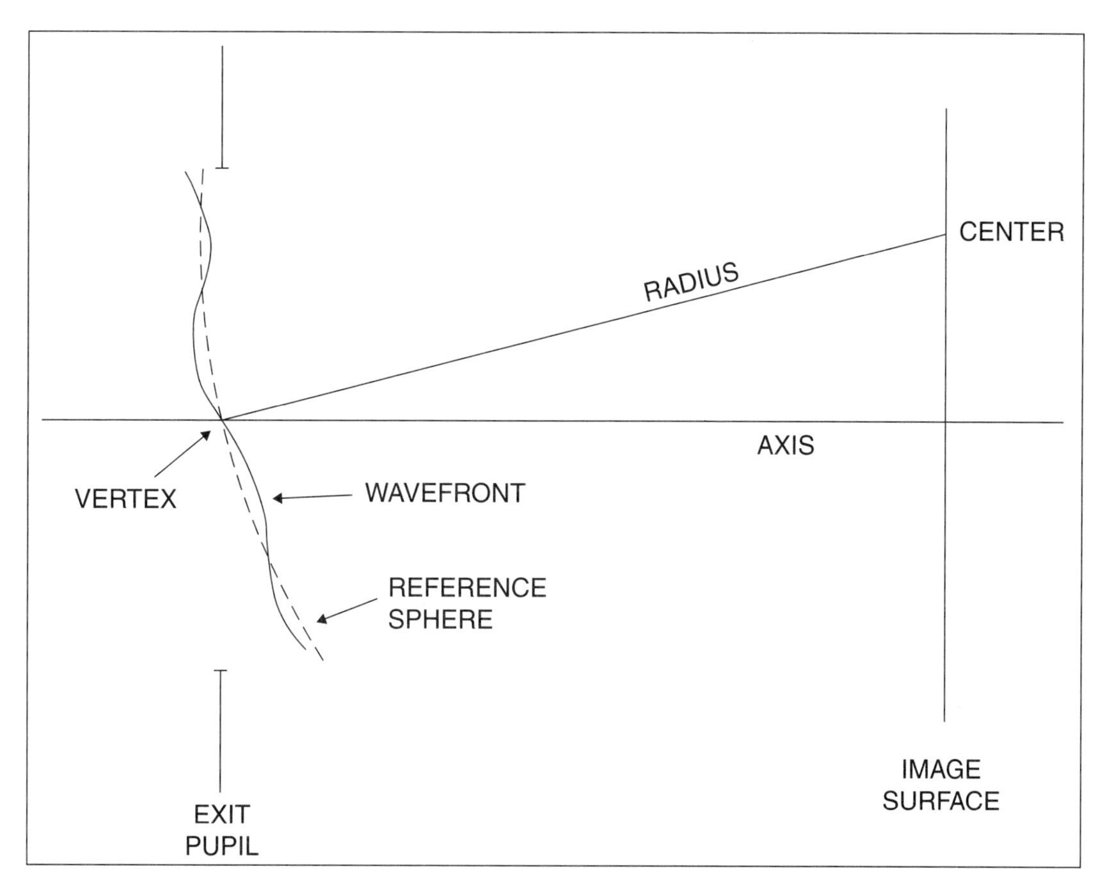

Figure A.11.2. Exit Pupil Wavefront and Reference Sphere. *The meridional cross-section of the actual aberrated wavefront and the ideal unaberrated wavefront (reference sphere) are shown in the exit pupil for a general off-axis image. The wavefront errors have been greatly exaggerated. The center of curvature (located at the image) and radius of curvature of the reference sphere are noted. In this example, the wavefront passes through the vertex of the exit pupil, and thus no wavefront piston has been removed. Optical path differences (OPDs) are positive when the actual wavefront is closer to the image (to the right in this case) than the reference sphere.*

by the edge, one ray becomes an aggregate of many diverging rays. This divergence causes the smearing (loss of point-to-point spatial coherence) of the corresponding wavefronts, which are still normal to the rays. Again refer to Figure A.11.1.

Some deflections are more likely than others, with the probabilities given by a statistical distribution function. Note that the diffraction of photons is a statistical process, whereas tracing geometrical rays is deterministic. Again, diffraction and aberrations are separate phenomena. Furthermore, although most photons cluster about their geometrical rays, there is a finite probability that a photon at any point in the PSF could have come from anywhere in the exit pupil. Thus, once again, there is no one-to-one mapping of the exit pupil onto the PSF.

If the aberrations in an optical system are large, then the effects of diffraction can be neglected, and the optical description given by geometrical optics is reliable. However, if aberrations are small enough that diffraction is significant, then a complete evaluation of optical performance must include the effects of diffraction as well as aberrations.

A.11.2 Geometrical Wavefronts

Again, rays are lines constructed normal to wavefronts, or wavefronts are surfaces constructed normal to rays. Refer again to Figure A.3.1. If, as an approximation, diffraction is ignored so that truncated wavefronts remain coherent and the path of a photon is determinate, then the rays and wavefronts are unique. In this case, the light downstream from the exit pupil can be modeled as a stream of bounded wavefronts with no edge effects. These diffraction-free wavefronts are called geometrical wavefronts. An imaging system with no aberrations produces a stream of spherically shaped geometrical wavefronts collapsing onto each image point.

To simplify the analysis of geometrical wavefronts, you can stop the propagation motion of light by stopping time. For a further simplification, you can consider only one wavefront from each object point by conceptually selecting only one light travel elapsed time from the source. There are advantages in selecting this single wavefront to be the wavefront located in the stop opening. Here, diffraction has not yet had a chance to significantly alter the coherence of the wavefront and fray its edges. Thus, the physical and geometrical wavefronts in the stop opening are identical for practical lens design purposes. This is also true at all images of the stop, including the exit pupil. Thus, describing and controlling aberrations on the wavefront in the exit pupil is justified both geometrically and physically.

A.11.3 Aberrations Measured by Optical Path Differences

Previously, geometrical aberrations were described in terms of transverse ray-intercept errors on the image surface. Controlling these ray errors yields small image spot sizes. Now, a second way of describing aberrations is introduced that uses optical path difference (OPD) errors in the exit pupil wavefront. Although the two methods are related, they are not equivalent.

Ray-intercept errors in the image are defined by rays constructed normal to the geometrical wavefronts leaving the last surface of the lens. To avoid physical wavefront uncertainties caused by diffraction, these rays are extended back to the trigonometric exit pupil where diffraction effects can be disregarded. Recall that the exit pupil is the image of the stop as seen by the image. You can then think of the rays forming the geometrical image as being launched normal to the exit pupil wavefront toward the image surface. Thus, any errors in the directions of these rays are proportional to errors in the slopes or derivatives of the exit pupil wavefront.

On the other hand, wavefront OPD errors are distance or path length errors measured longitudinally along rays. The distances are the separations between (1) the actual aberrated wavefront in the exit pupil and (2) an aberration-free spherical wavefront in the exit pupil.

The aberration-free wavefront is called the reference sphere. The center of curvature of the reference sphere is located on the image surface, and the vertex of the reference sphere is located at the vertex of the exit pupil surface. Figure A.11.2 illustrates the concepts of an aberrated wavefront in the exit pupil and a reference sphere.

Although the sign convention for OPDs is arbitrary, the more common convention is that an OPD is positive if the actual wavefront leads the reference sphere; that is, if the actual wavefront is closer to the image than the reference sphere.

Path lengths in optics are often not expressed in direct linear measure, such as millimeters. Instead, the number of wavelengths in the distance is given. This quantity is called the optical path length. Optical path length is the linear path length, multiplied by the index of refraction, divided by the wavelength in air.

Note that using optical path lengths rather than linear path lengths results in a changing scale with wavelength. For different wavelengths, two identical wavefront errors when measured in millimeters translate into very different numbers when expressed as optical path lengths. The shorter wavelength appears to give the larger error.

Clearly, wavefront slope errors and wavefront OPD errors are not the same thing. It is also clear that for good imagery, it is desirable that both be made as small as possible. Although you can attempt to optimize an optical system both for small wavefront slope errors and for small wavefront OPD errors, usually this approach is asking too much. The design solution that minimizes one type of error cannot simultaneously minimize the other type of error unless all aberrations are reduced to zero and the image becomes geometrically perfect.

Fortunately, if wavefront slope errors are minimized, then wavefront OPD errors are usually also small. However, the converse is not necessarily true. Small amplitude but high spatial frequency wavefront OPD errors can produce large wavefront slope errors. This might be called the phonograph groove effect.

A.11.4 Specifying the Amount of OPD Aberrations

When the amount of a wavefront OPD aberration in the exit pupil is specified, the meaning may be uncertain or ambiguous. Not only are there different ways to specify the wavefront departures from the reference sphere, but different wavefronts and different reference spheres can be used.

Four common ways to specify wavefront departures are:

1. Total peak-to-valley OPD range (from the extreme positive wavefront departure to the extreme negative wavefront departure);

2. Maximum absolute-value wavefront departure from the reference sphere (the sign of the OPD error does not matter);

3. RMS (root-mean-square) wavefront departure from the reference sphere evaluated over the whole exit pupil; and

4. Wavefront departures from the reference sphere expressed in terms of Zernike polynomials.

Two common ways to select the wavefront are:

1. The wavefront is chosen that passes through or is tangent to the vertex of the reference sphere (making the OPD of the chief ray zero); and

2. The wavefront is chosen that most nearly fits the reference sphere (this means a possible time or phase shift of the wavefront relative to the previous way).

The vertex of the reference sphere is always located at the vertex of the exit pupil. However, there are five common ways to specify the center of curvature of the reference sphere on the image surface:

1. Center on the Gaussian image point;

2. Center on the piercing point of the chief ray;

3. Center on any arbitrary point in the region of the image;

4. Center on the image point that minimizes the RMS wavefront departure; and

5. Center on the peak of the diffraction PSF.

Clearly, when specifying the amount of a wave aberration, it is imperative that the assumptions be explicitly stated.

Each of the above options has its advantages and disadvantages, depending on the application.

For visualizing the wavefront, either the maximum absolute-value departure or the peak-to-valley OPD range is often preferred. Because the absolute values of the extreme positive and negative departures are not necessarily equal, the peak-to-valley wavefront range is usually more convenient and unambiguous. A peak-to-valley measure is also not affected by exactly which wavefront is selected; that is, peak-to-valley is insensitive to wavefront piston (see below). Note that, instead of peak-to-valley (abbreviated P-V), some optical engineers use the term peak-to-peak (abbreviated P-P) to mean the same thing; the two peaks are the extreme positive and negative "peaks."

But for analyzing aberrations and the resultant PSF, conceptually and physically it is not just the worst-case departures that matter. Instead it is the distribution of the wavefront departures over the whole exit pupil relative to the reference sphere that matters. In this case, the relevant criterion is RMS wavefront departure.

For the various third-order aberrations, the value of the RMS wavefront departure is about a factor of 3.5 to 5 less than the peak-to-valley value.

For selecting the wavefront, the wavefront that gives an OPD of zero for the chief ray is often advantageous or convenient. But this wavefront may not be a good fit to the reference sphere; that is, there may be a relative longitudinal overall displacement. Such a displacement is called wavefront piston. If the wavefront has piston, then the average (mean) OPD is not zero (across the exit pupil, the OPDs of one sign are generally larger than the OPDs of the opposite sign).

Except in unusual lenses for use in interferometers, wavefront piston has no direct physical consequence. However, there is an indirect consequence when considering the PSF. The PSF depends on the wavefront variance, which is the square of the wavefront standard deviation. But the standard deviation is the RMS of the wavefront departures calculated relative to the wavefront mean, not relative to the reference sphere. If the wavefront has piston, then its standard deviation is not equal to the RMS of its OPDs relative to the reference sphere (its RMS OPD is greater). Only if any piston is removed will the wavefront mean coincide with the reference sphere. Thus, to get the variance from the OPDs, piston must be removed by shifting the wavefront relative to the reference sphere so that the mean OPD is zero (this also minimizes the RMS OPD). Note that now the OPD of the chief ray may be non-zero.

For selecting the reference sphere, placing the center of curvature of the reference sphere at either the Gaussian or the chief ray image points can be useful (especially during optimization). Using the chief ray has the advantage of not being affected by distortion.

But for analyzing the wavefront to determine the irradiance at a specific point in the PSF, the center of curvature of the reference sphere must be at that point. Then, if aberrations are small (the system is near the diffraction limit), the irradiance at that point is inversely related to the corresponding wavefront variance. This is true regardless of the type of aberration.

When the reference sphere is not centered on the chief ray, the effect is to take the system wavefront aberrations and to add or subtract an amount of wavefront tilt. If the center of curvature of the reference sphere is carefully chosen to remove the maximum amount of wavefront tilt, then the resulting wavefront variance is the minimum, and the reference sphere center is at the peak of the PSF. In other words, this is a two-way minimum; the RMS OPD is simultaneously minimized as a function of two variables, tilt and piston. For asymmetrical aberrations such as coma, the minimum-variance image point can be considerably shifted from the chief ray image point.

Note that if aberrations are not small (the system is far from the diffraction limit), then the minimum-variance image point may not give the PSF peak. In these cases, the PSF peak may actually be closer to the chief ray than to the minimum-variance point. When optimizing such a lens using OPDs, it may be beneficial to try the optimization three different ways: with reference to the chief ray, with only piston removed, and with both piston and tilt removed. All three designs are saved and evaluated, and the best one is selected.

Another way to visualize OPD aberrations is to think how a computer might be programmed to calculate them. First, trace a grid or array of rays covering the entrance pupil from the object point to the image surface. For an infinite object distance, begin the trace at a plane dummy surface located just in front of the lens that is perpendicular to the incoming beam. Of course, the chief ray is traced too. The location where each ray hits the image surface is now ignored; instead, for each ray, the total optical path length is found.

To get the optical path differences referenced to the chief ray, take the optical path length of the chief ray, and from this value subtract the optical path length of each of the other rays in turn. Of course, this makes the OPD of the chief ray zero.

If at some place in the exit pupil the actual aberrated wavefront leads the reference sphere, then the optical path length of the corresponding ray from the object to the image surface is less than the optical path length of the chief ray from the object to the image surface. In this case, the difference (chief ray minus general ray) is positive, and so is the OPD.

When using the chief ray as a reference, the resulting OPDs may have piston and tilt. To remove wavefront piston, find the average (mean) optical path length for all rays in the pupil, and from this value subtract the optical path length of each of the specific rays to get a revised set of OPDs. These OPDs are said to be referenced to the mean. The RMS of these OPDs is now the wavefront standard deviation, whose square is the variance.

To remove wavefront tilt, determine whether there is any systematic linear variation of the OPDs across the exit pupil (both in the x- and y-directions). If there is, then this linear variation is subtracted out of the OPDs while simultaneously maintaining piston removal. Subtracting out a linear OPD variation is equivalent to shifting the center of curvature of the reference sphere to the minimum-variance image point.

For brevity in ZEMAX, if both piston and tilt have been removed from the wavefront, then the OPDs are said to be referenced to the image centroid. Although not strictly accurate, this label will be retained here.

Note that many authors say that piston and tilt are quantities that must added to the basic wavefront (whose chief ray has an OPD of zero, and whose departures are measured relative to the reference sphere with center of curvature at the chief ray or at the Gaussian point on the image surface). This is in contrast to the present way of viewing piston and tilt as quantities that must be subtracted from the wavefront by shifting the wavefront and by shifting the center of curvature of the reference sphere. The two approaches are complementary.

A.11.5 OPD Ray Fan Plots

Like transverse ray-intercept errors, longitudinal OPD ray errors can also be displayed using a ray fan plot. This plot is called an OPD ray fan plot or simply an OPD plot. As before, pupil zone height is plotted on the horizontal axis, but now the corresponding OPD error is plotted on the vertical axis. If the wavefront is chosen that passes through the vertex of the reference sphere at the exit pupil center (the conventional mode), then the OPD error is zero at the plot origin.

Note that because OPDs are longitudinal, they do not depend on the separate transverse x- and y-directions in the image. Thus, there is only one type of sagittal OPD plot. This is in contrast to ray-intercept errors where two different types of sagittal plots are needed, although only the x-direction sagittal plot is commonly given.

Because transverse errors in the image are proportional to slope errors in the wavefront, a transverse ray fan plot and its corresponding OPD ray fan plot are related. For meridional rays in the tangential plane, the function displayed in the tangential ray-intercept plot is proportional to the derivative (slope) of the function displayed in the tangential OPD plot. Conversely, the function displayed in the tangential OPD plot is proportional to the integral (area under the curve evaluated from the pupil center outward) of the function displayed in the tangential ray-intercept plot. For skew rays in the sagittal plane, total derivatives must be replaced by a pair of partial derivatives evaluated in the x- and y-directions, but the idea is the same.

A.11.6 The Diffraction-Limited PSF

If all geometrical wavefront aberrations (OPDs) in an optical system are made much less than the wavelength of light, then aberrations become negligible and the physical point spread function is determined wholly by diffraction (not considering fabrication errors, stray light, and atmospheric seeing). This aberration-free PSF represents the ultimate imaging performance that the system can achieve. Such a system is said to be diffraction limited.

The diffraction PSF for a diffraction-limited system with a circular pupil and monochromatic light is shown pictorially in Figure A.11.3.1.[1] Figure A.11.3.2 shows this same pattern in a three-dimensional graphical representation, where PSF irradiance (what physicists call light intensity) is shown as a function of position on the two-dimensional image surface. A cross-section through the middle of this PSF is given in Figure A.11.3.3; to show the details, the scale is now logarithmic. All of these figures reveal a large central peak surrounded by a series of concentric rings of decreasing irradiance. Between the bright rings, there appear to be dark rings. These dark rings are where the irradiance goes to zero at specific distances from the center (with higher resolution, the minima really would go to zero in Figure A.11.3.3). This diffraction-limited light distribution is called the Airy pattern and the central peak is called the Airy disk, both named after G.B. Airy who first derived the distribution theoretically in 1835.

Another way of displaying the light distribution in a PSF is by an encircled energy plot. Here, the integrated PSF light enclosed inside a circle is plotted as a function of circle radius. The encircled energy values are usually given as a fraction or percent of total light; that is, relative to the encircled energy for a circle of very large radius. If the PSF is symmetrical, then the centers of the circles are at the symmetry point, which is at the chief ray. For unsymmetrical PSFs, the centers of the circles can be either at the chief ray or, preferably, at the image centroid.

Figure A.11.3.4 is the encircled energy plot for the PSF shown in Figures A.11.3.1 through A.11.3.3. For reference and as a check, the theoretical diffraction-limited curve is also drawn for a lens having the same first-order properties. Because the image formed by the real system is nearly free of aberrations, the two curves lie on top of each other.

The irradiance distribution of the Airy pattern is given mathematically by:

$$I(r) = I(0)\left[\frac{2J_1(x)}{x}\right]^2 \tag{A.11.1}$$

where J_1 is the Bessel function of order 1 and $I(0)$ is the peak irradiance, usually normalized to unity or 100%. The argument of the Bessel function is:

$$x = \frac{\pi r}{\lambda F} \tag{A.11.2}$$

where r is the linear radial distance from the center of the peak, λ is the wavelength of light in the same units as r, and F is the system focal ratio (dimensionless). The linear radius of the Airy disk, which is equal to the radius of the first dark ring, is:

$$r_{\text{Airy}} = 1.22\lambda F. \tag{A.11.3}$$

[1]The author thanks H.R. Suiter for supplying this and the other pictorial images of diffraction patterns in this chapter.

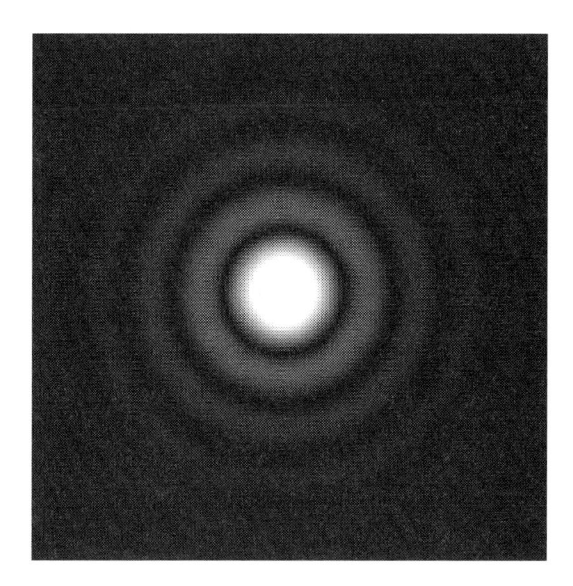

Figure A.11.3.1.

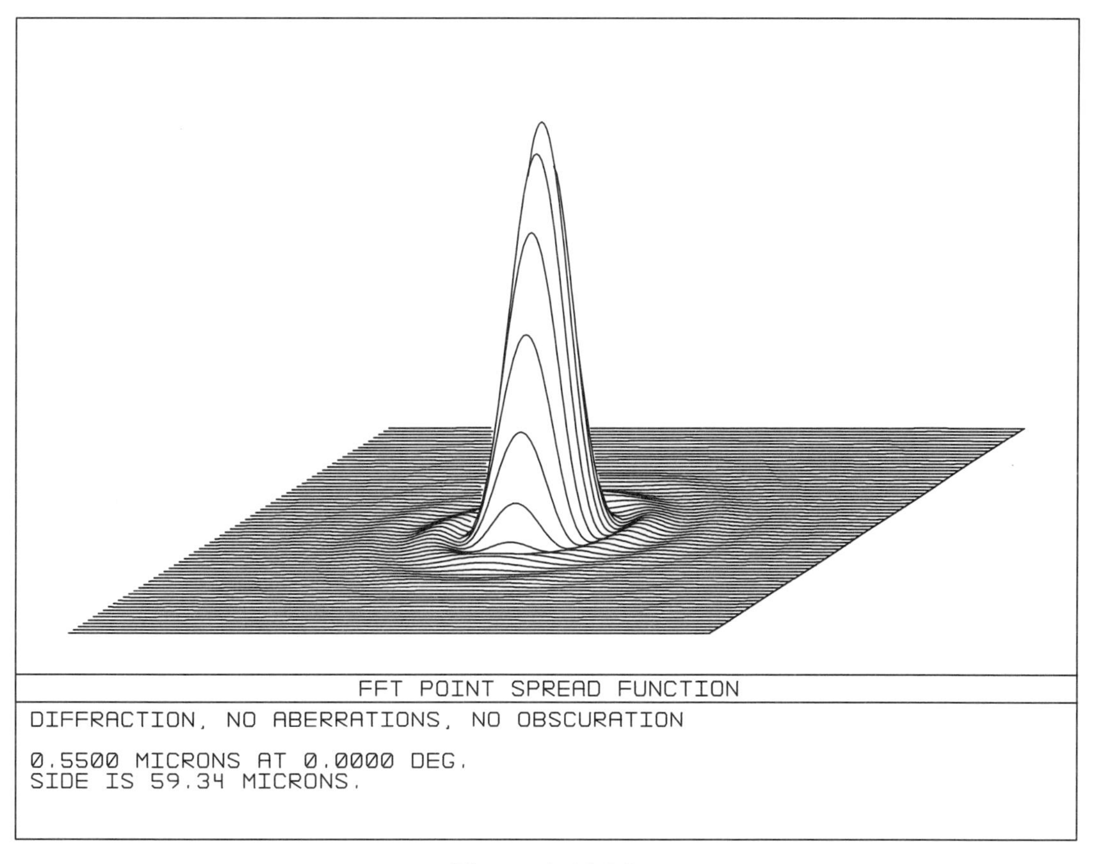

Figure A.11.3.2.

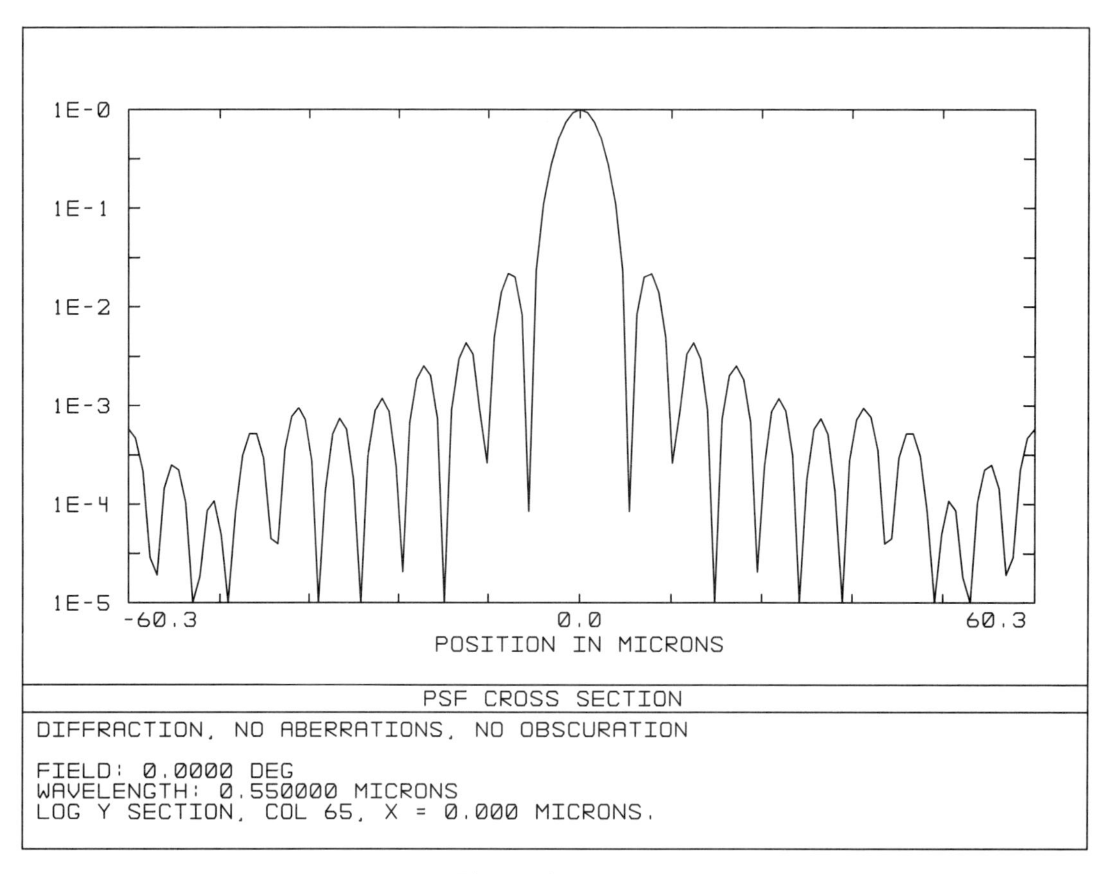

Figure A.11.3.3.

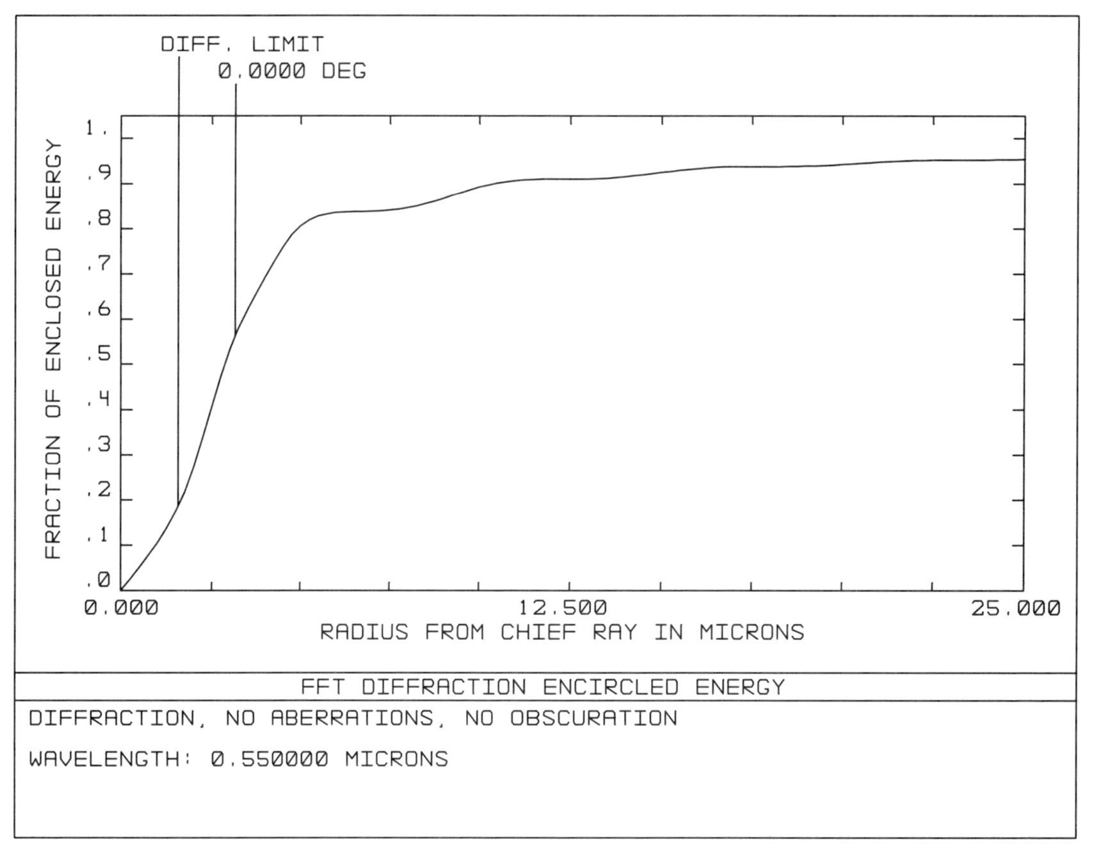

Figure A.11.3.4.

Note that for many applications, the diameter, rather than the radius, of the Airy disk is required. Thus:

$$d_{\text{Airy}} = 2.44\lambda F. \tag{A.11.4}$$

For an annular pupil (a circular pupil with a circular central obscuration or obstruction), the irradiance distribution in the diffraction pattern is modified. The fraction of the total light in the central Airy disk is reduced, the fraction of the total light in the ring system is correspondingly increased, the relative distribution of light in the various rings is changed, the diameter of the Airy disk is reduced, and the diameters of the rings are changed.

Figures A.11.4.1 and A.11.4.2 illustrate the PSF for a diffraction-limited system with a 30% diameter central obscuration. Figure A.11.4.3 is the logarithmic cross-section plot. Figure A.11.4.4 is the encircled energy plot. Compare these plots with Figures A.11.3.1 through A.11.3.4. Note how much brighter the first ring is in the system with the central obscuration. As they did in Figure A.11.3.4, the curve for the actual lens and the curve for the diffraction-limited lens (with the same obscuration) lie on top of each other in Figure A.11.4.4.

The irradiance distribution of the Airy pattern for an annular pupil is given mathematically by a modified form of Equation A.11.1:

$$I(r) = \frac{I(0)}{(1-\varepsilon^2)^2}\left[\frac{2J_1(x)}{x} - \varepsilon^2\frac{2J_1(\varepsilon x)}{\varepsilon x}\right]^2 \tag{A.11.5}$$

where ε is the fractional diameter obscuration.

Incidentally, when speaking of central obscurations, be sure to specify whether diameter or area is being discussed. A 30% diameter central obscuration means that 9% of the pupil area is obscured. Also, it can be argued that the terms Airy pattern and Airy disk should only be used for the unobscured case, because the theory for the obscured case was first worked out by Lord Rayleigh in 1881. Nevertheless, nearly everyone speaks of Airy pattern and Airy disk for the obscured case too.

Table A.11.1 gives a more quantitative description of the diffraction-limited diffraction patterns of systems with different central obscurations.[2] All pupils and obscurations are circular. The outer radii of the Airy disk and first five bright rings are given; this is the same as the radii of the minima of the first six dark rings. The values of the radii are all relative to the radius of the first dark ring of the pattern for no obscuration, whose linear radius is given by Equation A.11.3. Maximum irradiance in each bright ring is given normalized to 100 for the peak of the Airy disk of each pattern. The percent of total light in the disk and various bright rings is given relative to the light passing the pupil; that is, the reduction caused by the central obscuration is not included. Obscurations range from zero to 90% by diameter.

A.11.7 Diffraction Plus Aberrations

The lens with no central obscuration of Figures A.11.3.1 through A.11.3.4 is now used slightly defocused. Figure A.11.5.1 shows the OPD ray fan plot for the on-axis image. The amount of defocus has been adjusted to give a maximum OPD range, measured in wavelengths, of 0.5 waves peak-to-valley.

Recall that an axially centered optical system has right-left bilateral symmetry about the meridional plane, and that consequently sagittal transverse ray fan plots are also symmetrical. For the same reason, sagittal OPD ray fan plots are symmetrical too. In addition, recall that on-axis transverse ray fan plots become multiply redundant. Again, on-axis OPD ray fan plots become multiply redundant too. This multiple redundancy is seen in Figure A.11.5.1.

[2]This table has been adapted from Virendra N. Mahajan, *Aberration Theory Made Simple*, pp. 116–117.

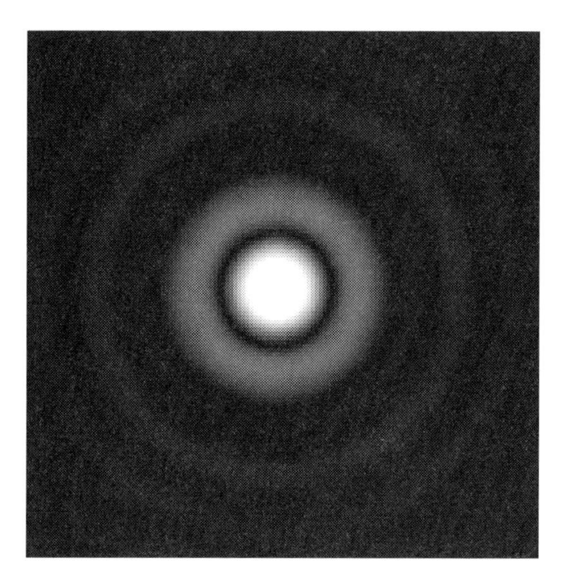

Figure A.11.4.1.

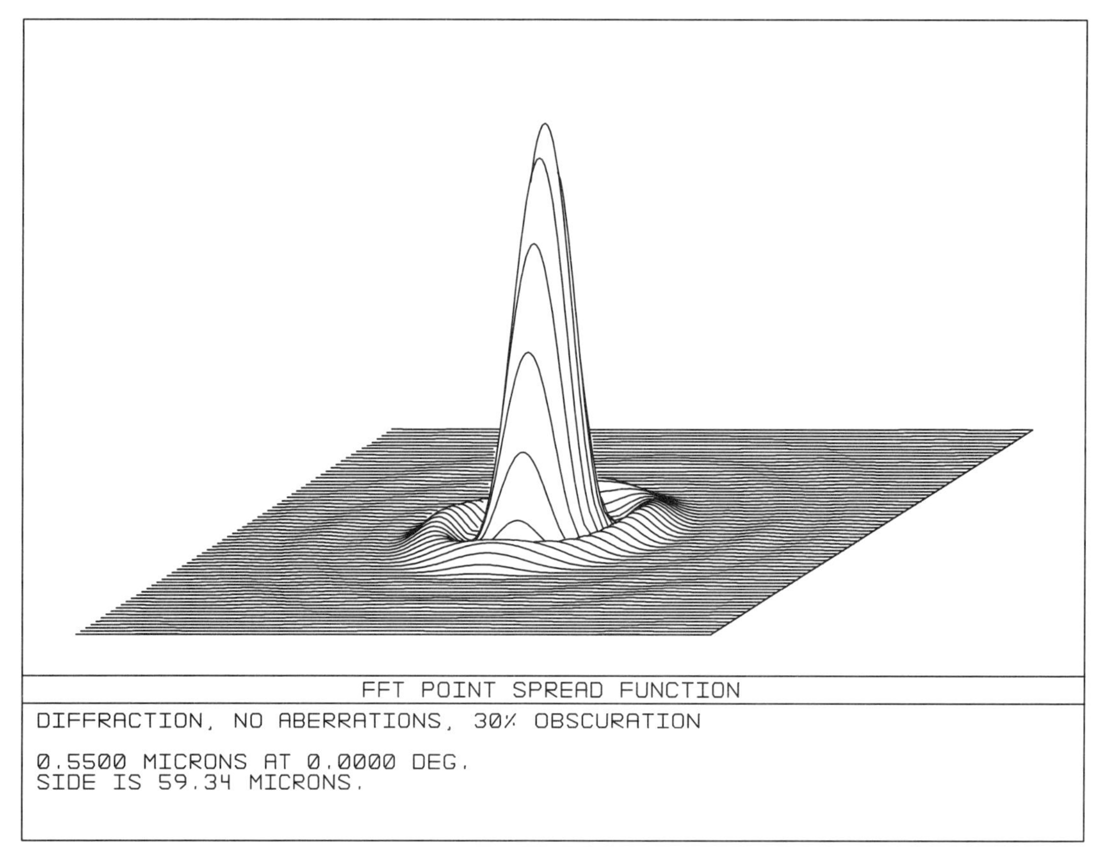

Figure A.11.4.2.

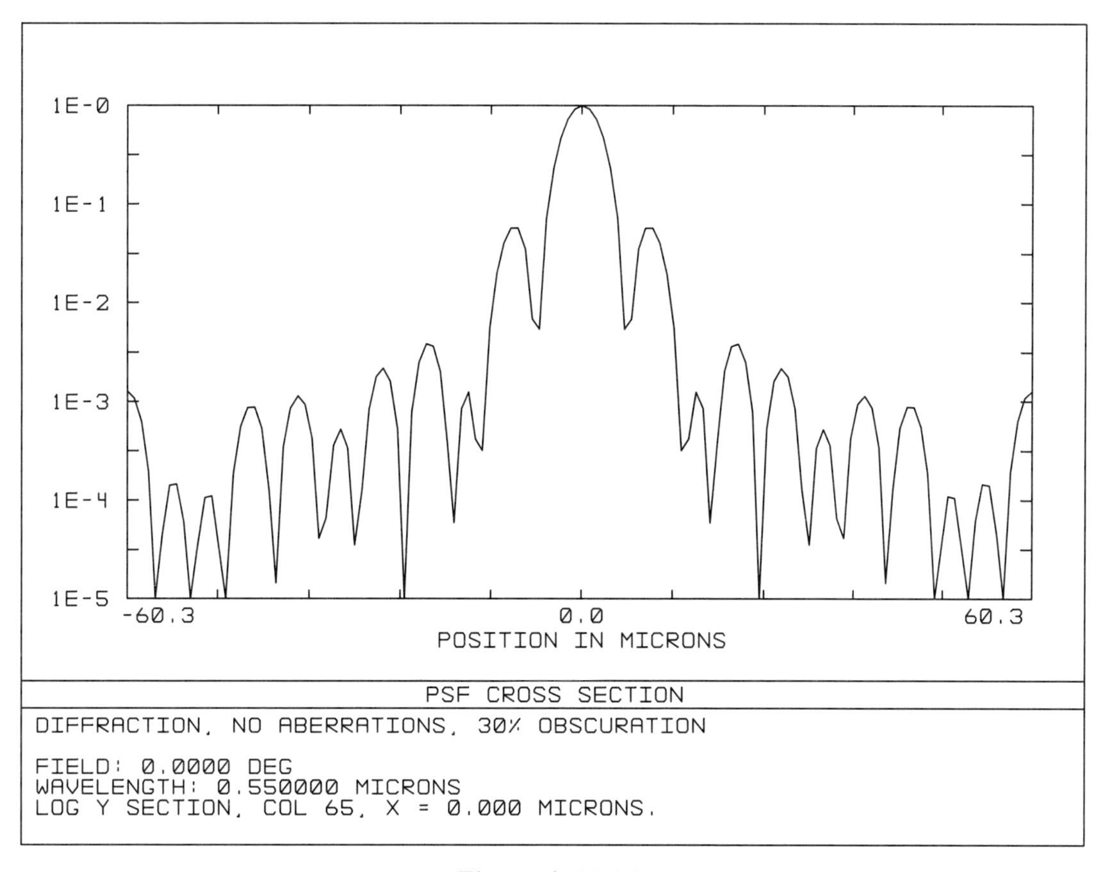

Figure A.11.4.3.

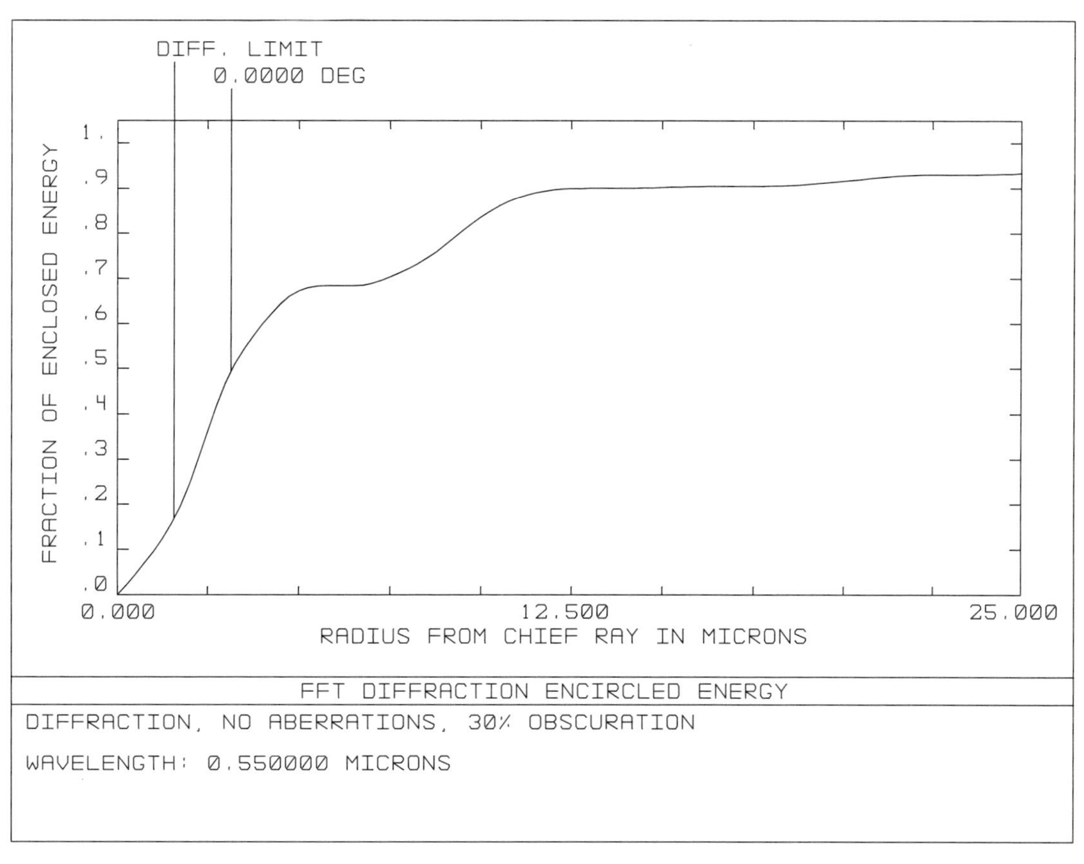

Figure A.11.4.4.

Table A.11.1
Diffraction-Limited PSF for Circular Pupil
with
Central Obscuration from 0% to 90%

		Ring					
	Airy Disk	1	2	3	4	5	>5
0% Obscuration							
Relative Outer Radius	1.00	1.83	2.66	3.48	4.30	5.11	-
Relative Irradiance (Max.)	100	1.75	0.42	0.16	0.08	0.04	-
Total Light (%)	83.8	7.2	2.8	1.4	0.9	0.7	3.2
10% Obscuration							
Relative Outer Radius	0.99	1.86	2.61	3.54	4.22	5.20	-
Relative Irradiance (Max.)	100	2.06	0.31	0.24	0.04	0.08	-
Total Light (%)	81.8	8.8	1.9	2.4	0.4	1.2	3.5
20% Obscuration							
Relative Outer Radius	0.96	1.93	2.53	3.58	4.23	5.11	-
Relative Irradiance (Max.)	100	3.04	0.15	0.37	0.04	0.06	-
Total Light (%)	76.4	13.6	0.8	3.9	0.4	0.8	4.1
30% Obscuration							
Relative Outer Radius	0.91	1.98	2.54	3.46	4.44	4.98	-
Relative Irradiance (Max.)	100	4.75	0.11	0.28	0.16	0.01	-
Total Light (%)	68.2	21.7	0.5	2.5	2.0	0.1	5.0
40% Obscuration							
Relative Outer Radius	0.87	1.96	2.70	3.31	4.35	5.27	-
Relative Irradiance (Max.)	100	7.07	0.33	0.07	0.28	0.08	-
Total Light (%)	58.4	30.1	1.8	0.4	3.2	1.1	5.0
50% Obscuration							
Relative Outer Radius	0.82	1.88	2.86	3.38	4.14	5.16	-
Relative Irradiance (Max.)	100	9.63	1.24	0.04	0.09	0.22	-
Total Light (%)	47.9	35.0	7.2	0.2	0.7	2.8	6.2
60% Obscuration							
Relative Outer Radius	0.78	1.78	2.78	3.70	4.19	4.92	-
Relative Irradiance (Max.)	100	12.03	3.06	0.45	0.01	0.04	-
Total Light (%)	37.2	34.5	15.6	2.9	0.1	0.3	9.4
70% Obscuration							
Relative Outer Radius	0.74	1.69	2.64	3.59	4.52	5.30	-
Relative Irradiance (Max.)	100	13.95	5.33	1.92	0.50	0.05	-
Total Light (%)	26.9	29.1	20.1	10.4	3.4	0.4	9.7
80% Obscuration							
Relative Outer Radius	0.70	1.60	2.50	3.41	4.32	5.22	-
Relative Irradiance (Max.)	100	15.27	7.34	4.01	2.18	1.10	-
Total Light (%)	17.2	20.4	17.8	14.1	10.0	6.2	14.3
90% Obscuration							
Relative Outer Radius	0.66	1.52	2.38	3.24	4.10	4.96	-
Relative Irradiance (Max.)	100	16.00	8.61	5.66	4.04	2.99	-
Total Light (%)	8.2	10.2	10.0	9.5	8.9	8.1	45.1

Note that the curves in Figure A.11.5.1 are parabolas; that is, the OPDs for defocus vary as the square of pupil zone height. This is in contrast to the transverse ray fan curves for defocus that are linear with pupil zone height, as illustrated in Figures A.7.8.2 and A.7.8.3. In general, the order of a transverse ray fan curve is one less than the order of the corresponding OPD ray fan curve. This relationship is the result of the ray-intercept errors being a function of the derivative of the wavefront shape.

Figures A.11.5.2 and A.11.5.3 show the corresponding diffraction PSF. Compare Figure A.11.5.3 with Figure A.11.3.2. Both figures have the same horizontal scale and both are normalized to a peak of unity. In the defocused PSF, the central maximum is broadened by the expanded geometric spot, and there is more diffracted light in the rings.

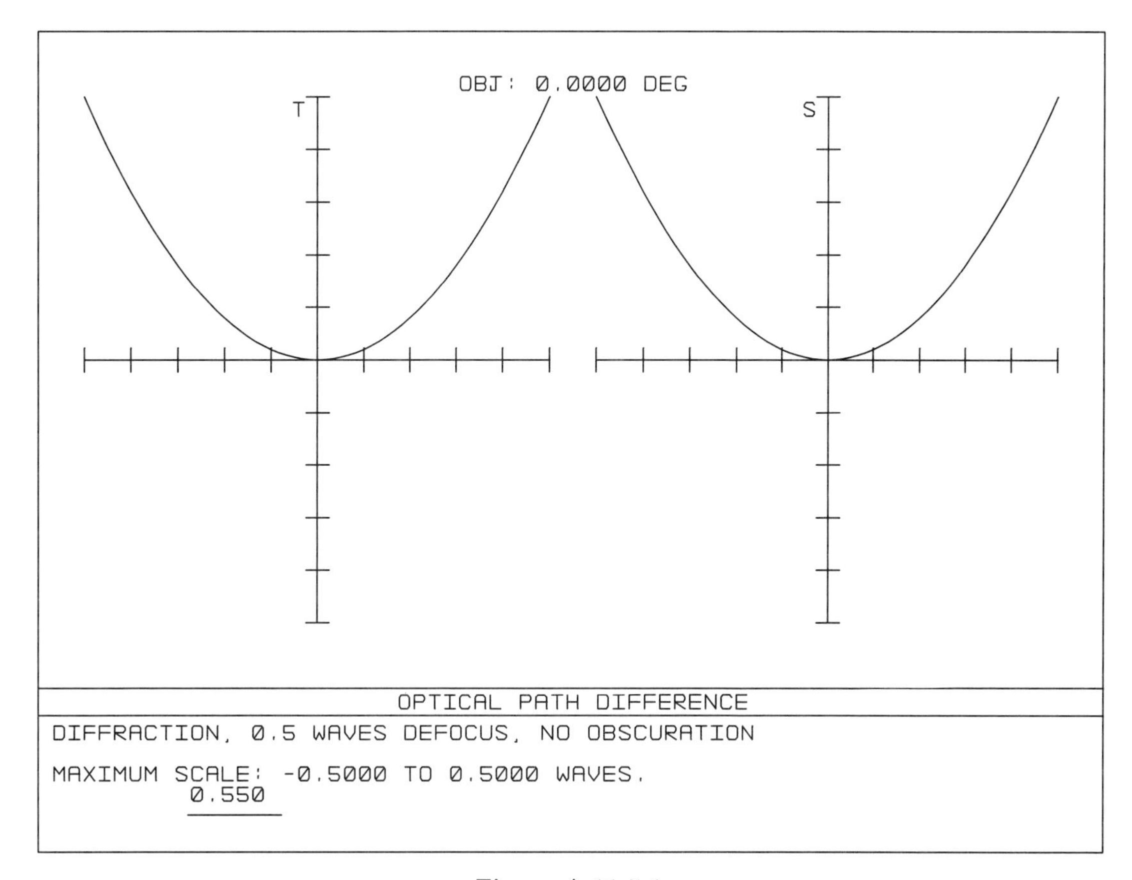

Figure A.11.5.1.

However, normalizing the PSF peaks gives an incorrect picture of the image radiometry. In a true PSF, the three-dimensional volume of the PSF is proportional to the total amount of light. For the lenses of Figures A.11.3.2 and A.11.5.3, the entrance pupil areas (light gathering powers) are the same, thus giving the same true PSF volumes. Because the PSF of the defocused lens is broader, the peak must be lower.

Figure A.11.5.4 is the encircled energy plot for the defocused lens. Now, with the image degraded by defocus, the actual curve is not the same as the diffraction-limited curve.

Figure A.11.6.1 is the OPD ray fan plot for an unobscured lens similar to the lens of Figure A.11.5.1, but now there is 0.5 waves peak-to-valley of spherical aberration and no paraxial defocus. Again, the on-axis object is used, giving multiple redundancy. Note that the OPDs vary as the fourth power of pupil zone height. Contrast this with the third-power variation for spherical aberration in the transverse ray fan plot in Figure A.7.1.5.

Figures A.11.6.2 and A.11.6.3 show the diffraction PSF. Relative to Figure A.11.3.2, note that again the central maximum is broadened by the expanded geometric spot, and there is more diffracted light in the rings. Comparing Figures A.11.5.3 and A.11.6.3, note the different shapes of the inner parts or cores of the PSFs. These differences are caused by the dissimilar structures of the geometric spots. Figure A.11.6.4 is the encircled energy plot.

Figure A.11.7.1 is the OPD ray fan plot for a lens similar to the previous unobscured lenses, except now the object is off-axis and the wavefront has 2.0 waves peak-to-valley of tangential coma relative to the reference sphere centered on the chief ray. Note that the tangential and sagittal curves are no longer identical. The tangential OPD errors vary as the cube of pupil zone height, and the sagittal OPD errors are nearly zero. Contrast this with the second-power variation for coma

Figure A.11.5.2.

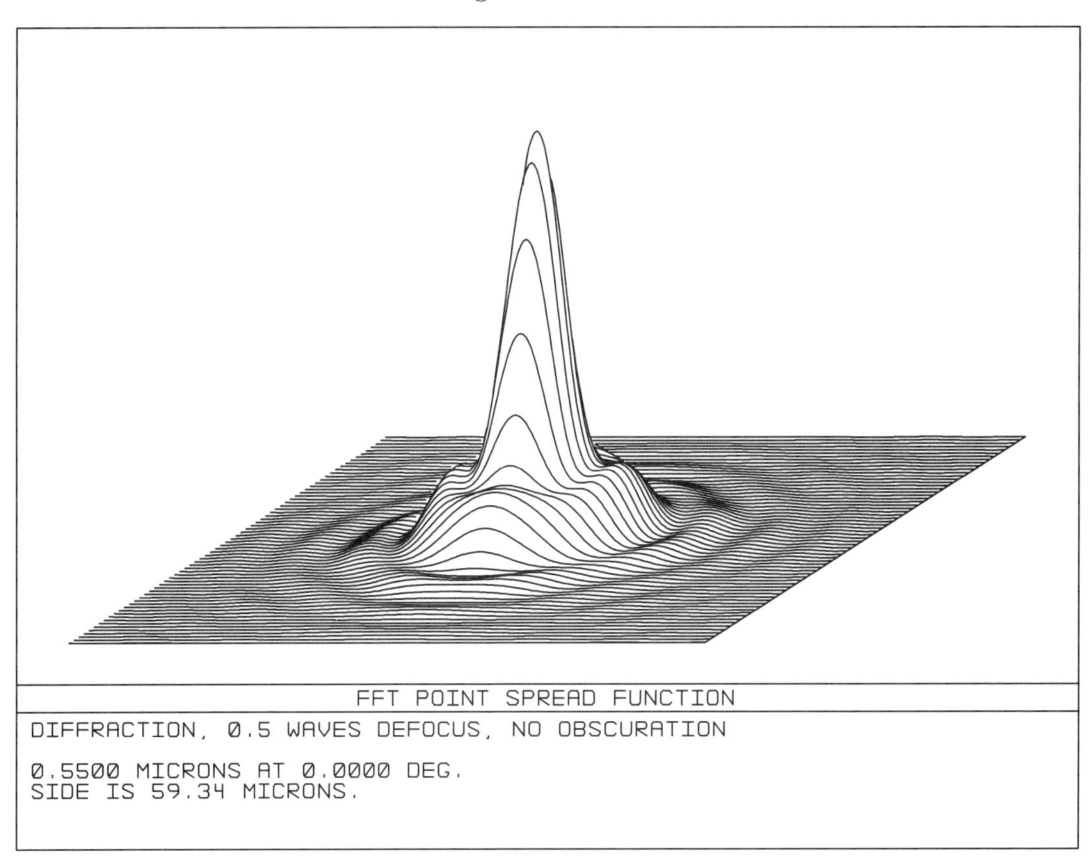

Figure A.11.5.3.

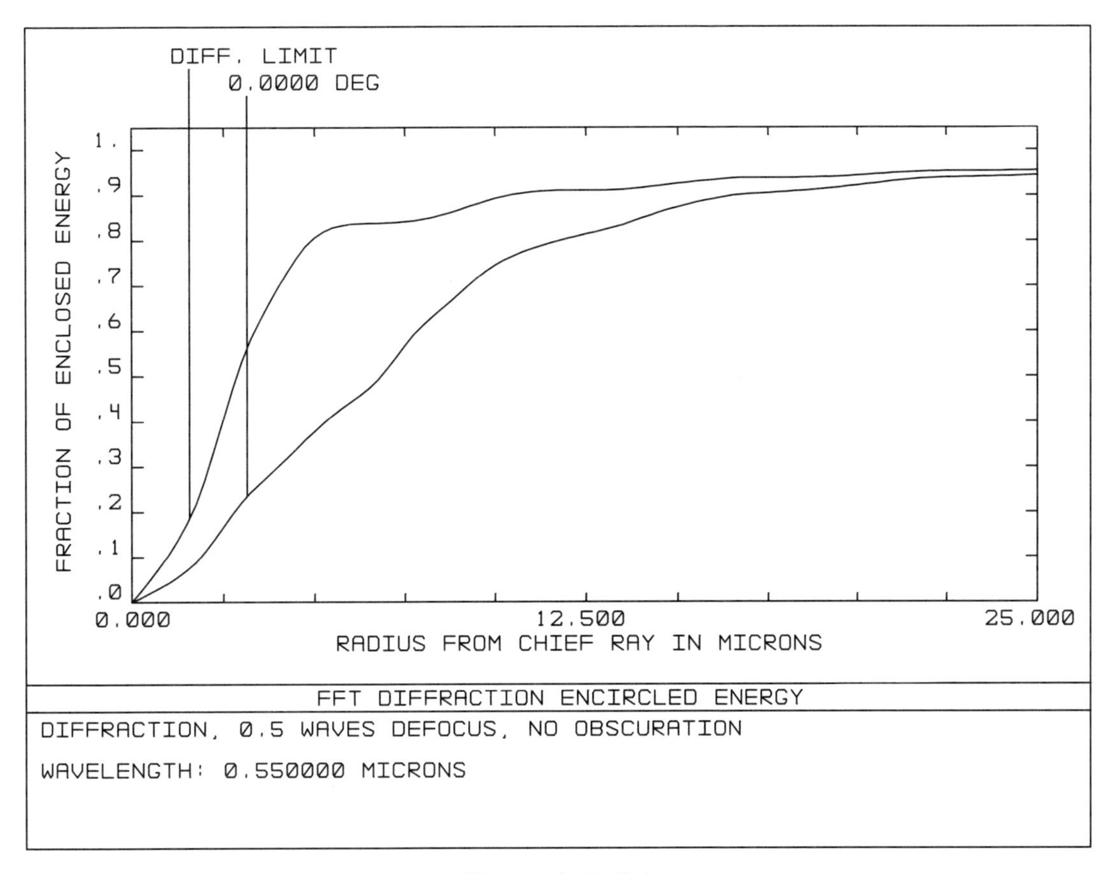

Figure A.11.5.4.

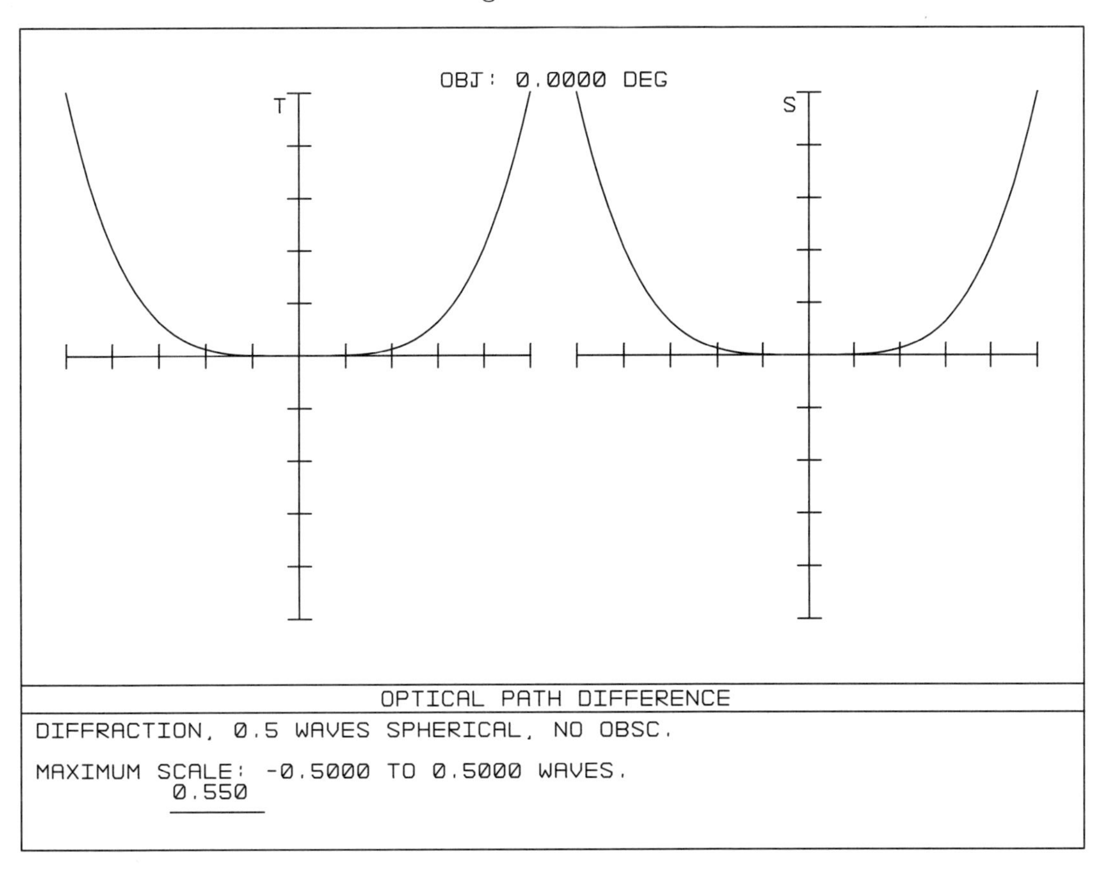

Figure A.11.6.1.

Figure A.11.6.2.

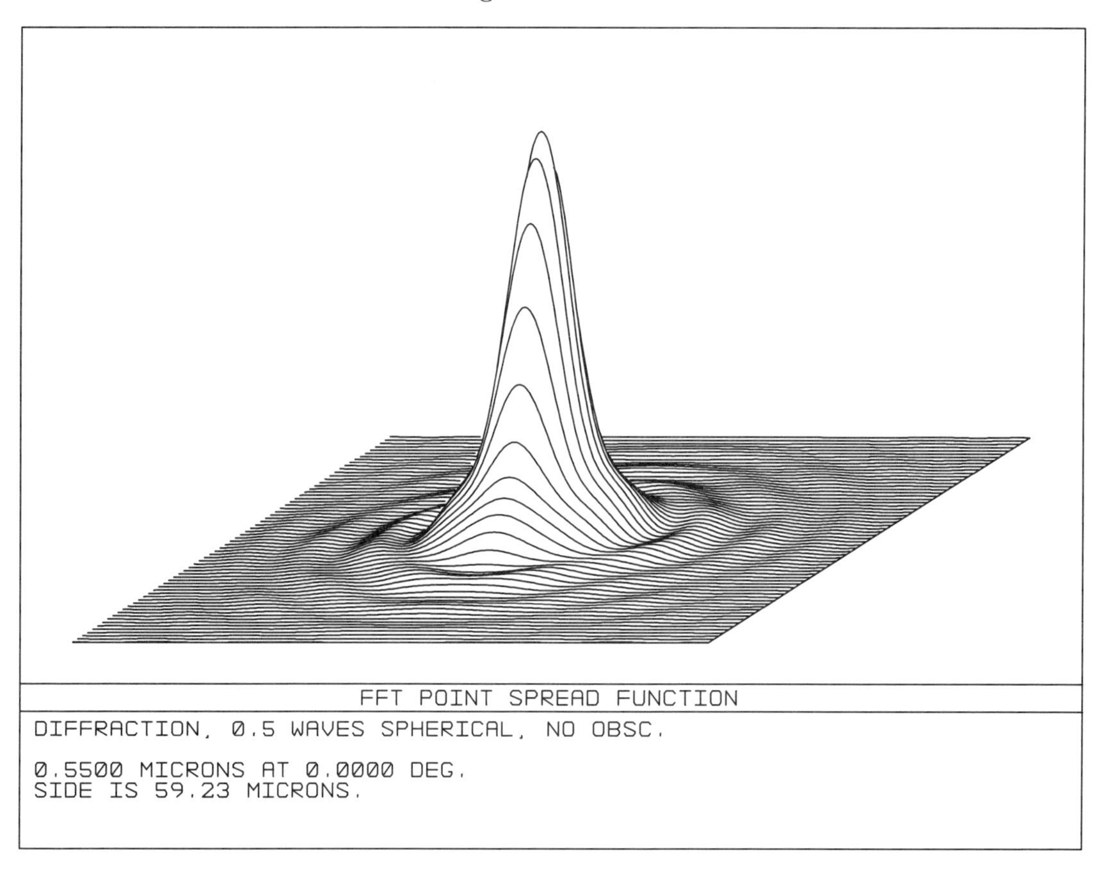

Figure A.11.6.3.

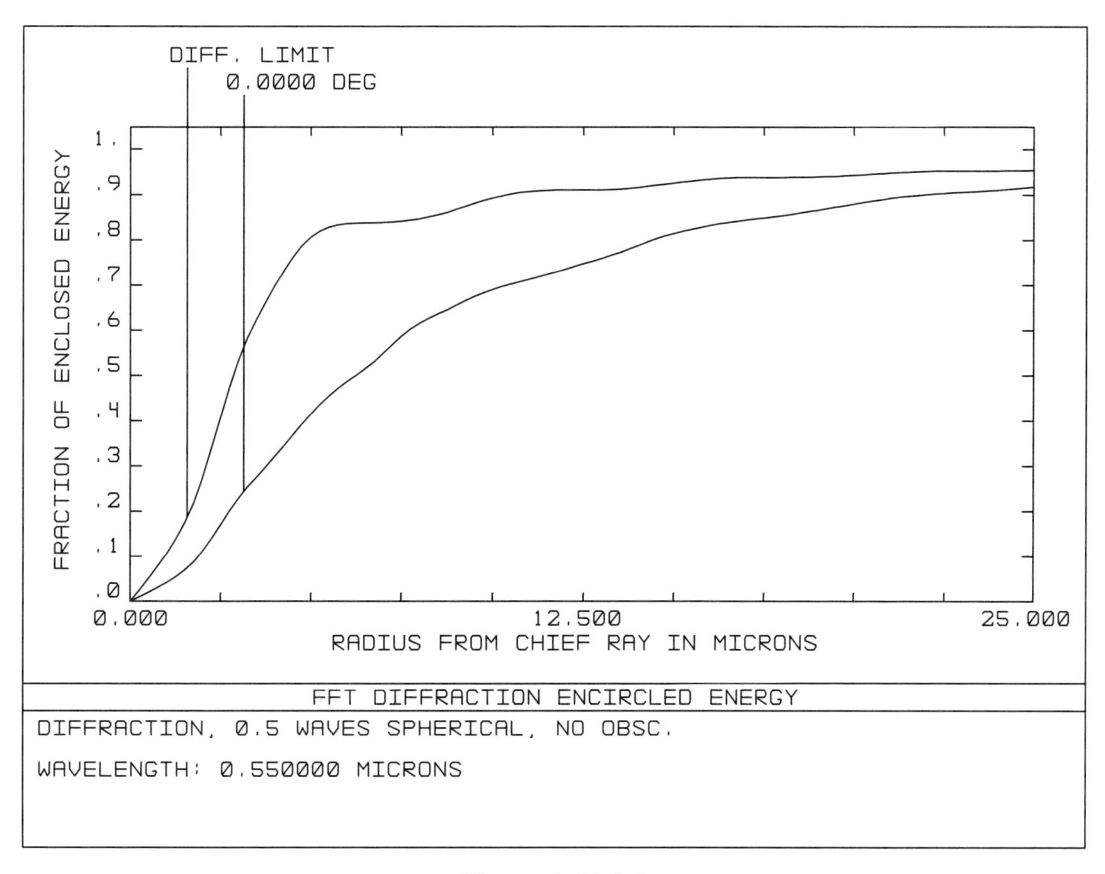

Figure A.11.6.4.

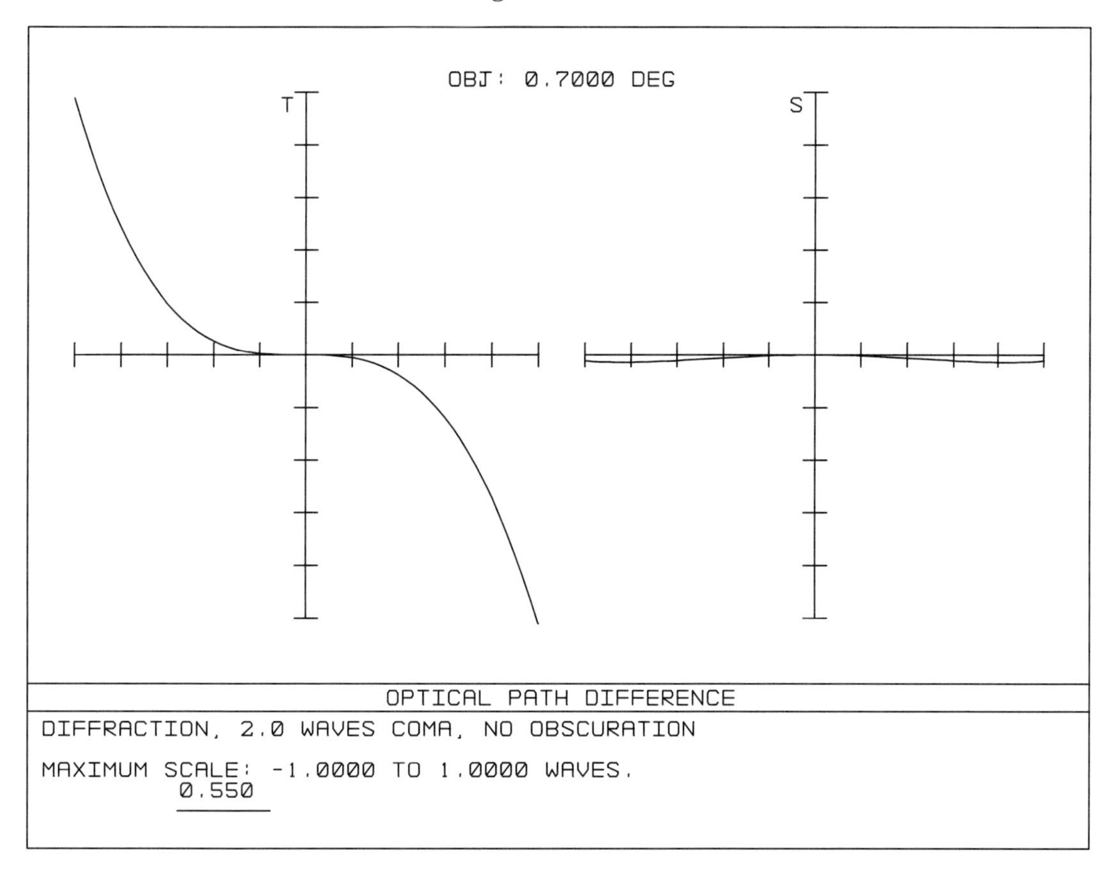

Figure A.11.7.1.

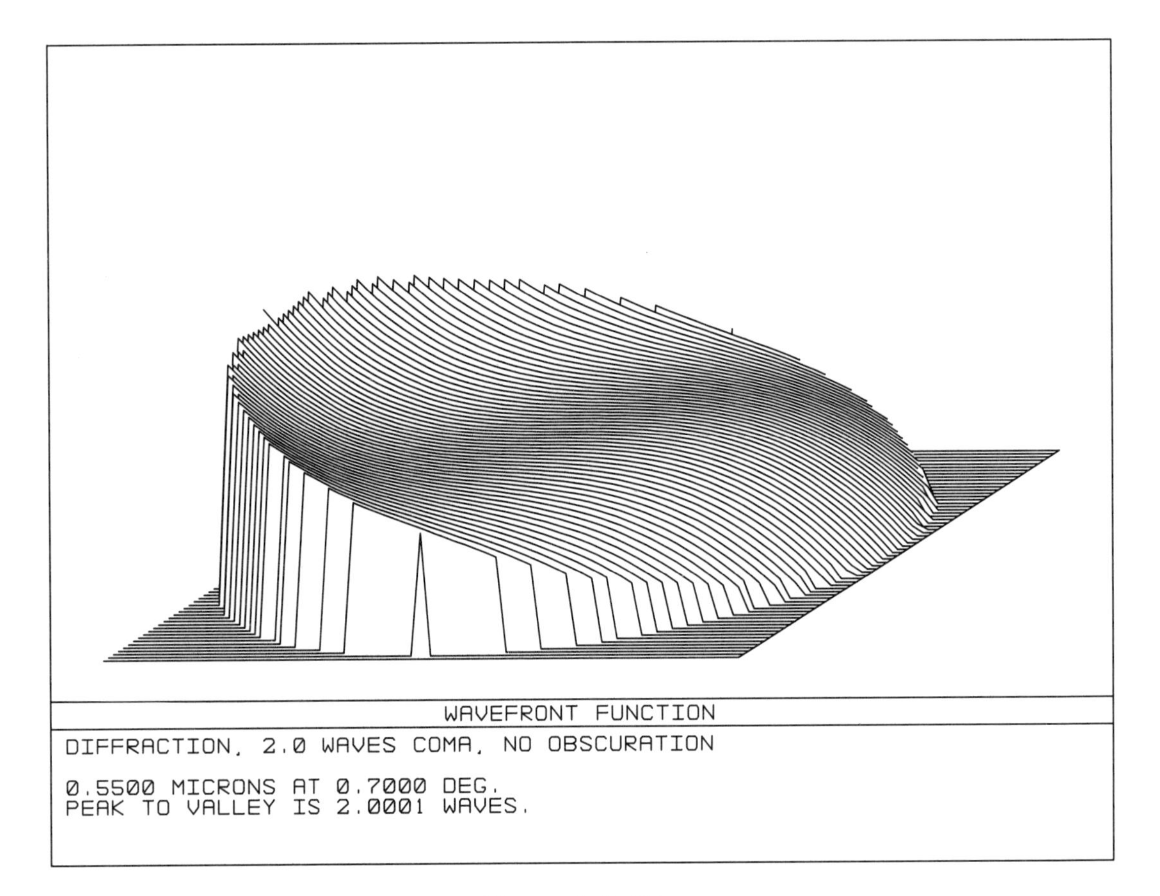

Figure A.11.7.2.

in the transverse ray fan plot in Figure A.8.2.5.

The tangential and sagittal OPD ray fan plots are actually two orthogonal cuts through a three-dimensional wavefront OPD function representing the OPD errors across the two-dimensional exit pupil. For rotationally symmetric OPD functions, all cuts are the same and the OPD ray fan plots tell everything. But for unsymmetrical OPD functions, the OPD ray fan plots are incomplete, and the entire OPD function may be of interest. A practical suggestion is that if you plot the full wavefront function, you might also plot the OPD ray fan plots; these latter are much easier to read quantitatively.

Figure A.11.7.2 is a three-dimensional representation of the OPD function across the exit pupil of the lens for Figure A.11.7.1. Apparently, coma causes the OPD error surface to be shaped somewhat like a tilted potato chip. However, if the intrinsic overall tilt is subtracted out of the wavefront function by shifting the center of the reference sphere to the image centroid, then the remaining wavefront error is only a third as much, or 0.667 waves peak-to-valley. Note that this wavefront tilt removal only concerns the general location of the image and does not change the relative light distribution in the PSF.

Figures A.11.7.3 and A.11.7.4 show the corresponding diffraction PSF. Note the asymmetry of the light distribution; the image centroid (PSF peak) is not on the chief ray. This is the same asymmetry that was revealed in the geometrical analysis of coma in Chapter A.8.

Figure A.11.8.1 is the off-axis OPD ray fan plot for a lens similar to the lens of Figure A.11.7.1, except now the lens has astigmatism instead of coma. The image surface is the medial image surface halfway between the two astigmatic foci. Like the defocused lens of Figure A.11.5.1, the tangential and sagittal curves each show a maximum of 0.5 waves of an OPD error that varies as the square of pupil zone height. But there is a difference here; the tangential and sagittal errors have

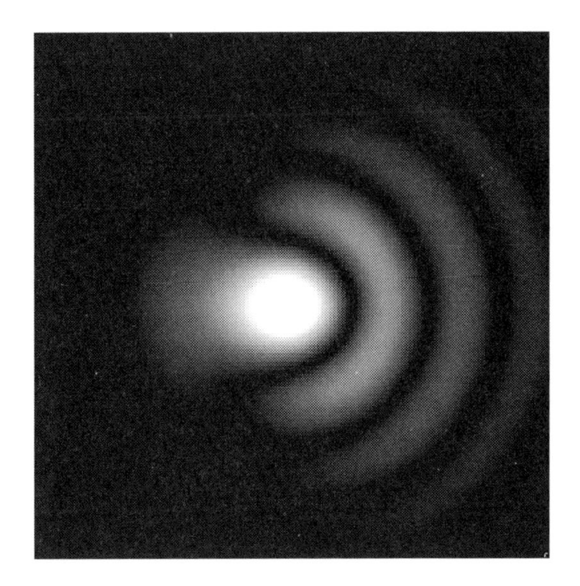

Figure A.11.7.3.

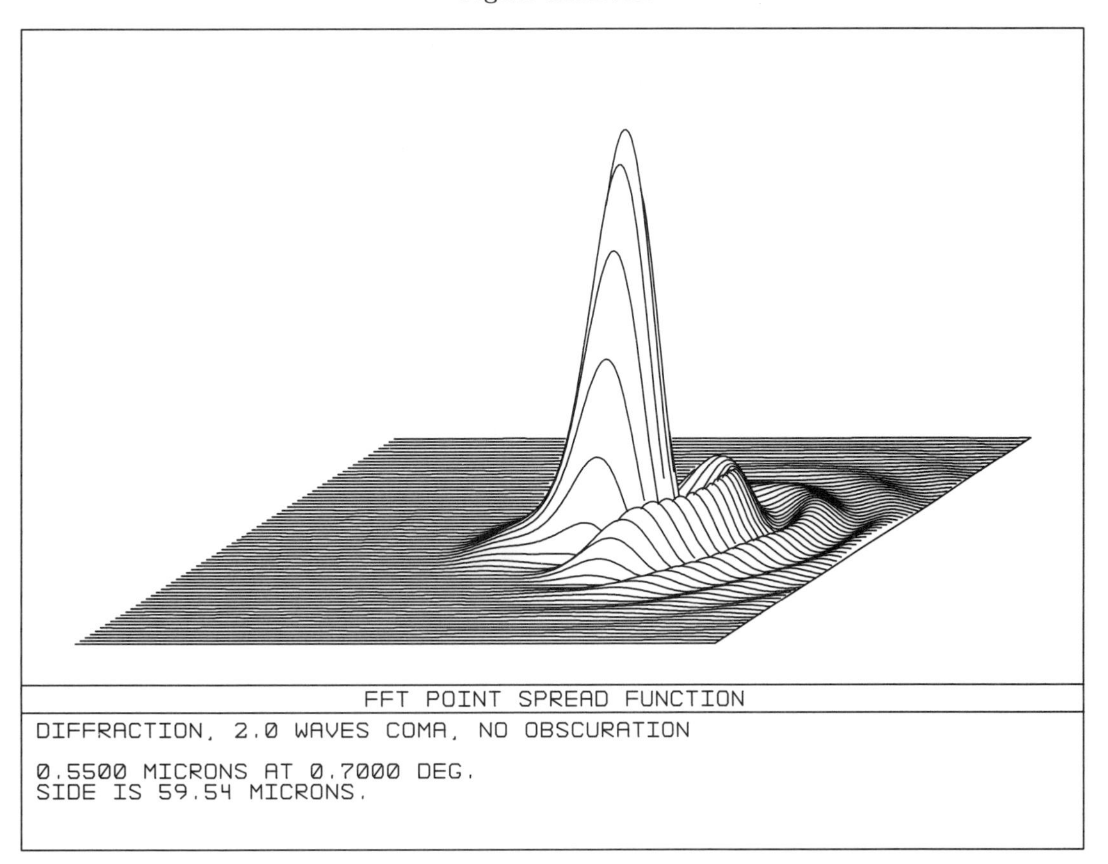

Figure A.11.7.4.

opposite signs. Thus, the total OPD error now becomes 1.0 waves peak-to-valley. Recall that a second-power OPD relationship indicates defocus, and a difference in defocus between the tangential and sagittal ray fans indicates astigmatism. Contrast Figure A.11.8.1 with Figure A.8.4.3, which shows a linear relationship with opposite tangential and sagittal slopes for medial astigmatism in a transverse ray fan plot.

Figure A.11.8.2 shows the three-dimensional OPD function in the exit pupil. Apparently, medial astigmatism causes the OPD error surface to be shaped like a saddle. Unlike coma, medial astigmatism has no overall wavefront tilt to remove. Unlike defocus and spherical aberration, medial astigmatism has no wavefront piston to remove either.

Figures A.11.8.3 and A.11.8.4 show the diffraction PSF. Although the image surface is at the geometric "circle" of least confusion, the actual light distribution resembles something between a cross and a diamond. This pattern shows the influence of the two orthogonal astigmatic line foci, and is seen only when diffraction is included with aberrations.[3]

A.11.8 OPD Plots for Chromatic Aberrations

Longitudinal color is just defocus as a function of wavelength. Thus, on OPD plots, longitudinal color appears like the curves in Figure A.11.5.1, except that now there is a different curve for each wavelength.

Lateral color is a difference in chief ray height on the image surface as a function of wavelength. On the wavefronts in the exit pupil, the chromatic chief rays must be launched in different vertical directions to produce lateral color. Thus, on OPD plots, lateral color appears as a wavelength-dependent change in the slope of the tangential OPD curve at the origin.

Of course, all of the monochromatic aberrations also have wavelength-dependent variations. These appear on OPD plots as chromatic variations of the curves discussed in the previous section.

A.11.9 Full Width at Half Power

When a single number is needed to describe the overall size of an aberrated diffraction PSF, a reliable measure is the diameter of the circle enclosing half of the PSF energy (or power). This number can be obtained from an encircled energy plot. It is customary here to use diameter and not radius. This criterion is called the full width at half power (or FWHP). However, there is another criterion that is often confused with FWHP; this is full width at half maximum (or FWHM). FWHM gives the diameter of the circle that encloses the PSF at a level that is half of the peak level. Unfortunately, FWHM is not a reliable measure of the overall size of the PSF. This failing is worst in cases where the PSF has a narrow central peak containing a small fraction of the total energy, and extended low-level wings containing a large fraction of the total energy (an all-too-common situation). FWHM is unrealistic and misleading here because it only measures the size of the central peak and ignores the fact that most of the energy in the PSF is elsewhere. The original Hubble Space Telescope with its huge spherical aberration had excellent FWHM but terrible FWHP. To be safe, use FWHP.

A.11.10 Diffraction-Limited Resolution

If two separate object point sources are close together, there is a limit to how close these two points can be and still have a diffraction-limited optical system

[3]Excellent references for the appearance of additional diffraction effects are Michel Cagnet, Maurice Françon, and Jean Claude Thrierr, *Atlas of Optical Phenomena*, and Michel Cagnet, Maurice Françon, and Shamlal Mallick, *Atlas of Optical Phenomena Supplement*.

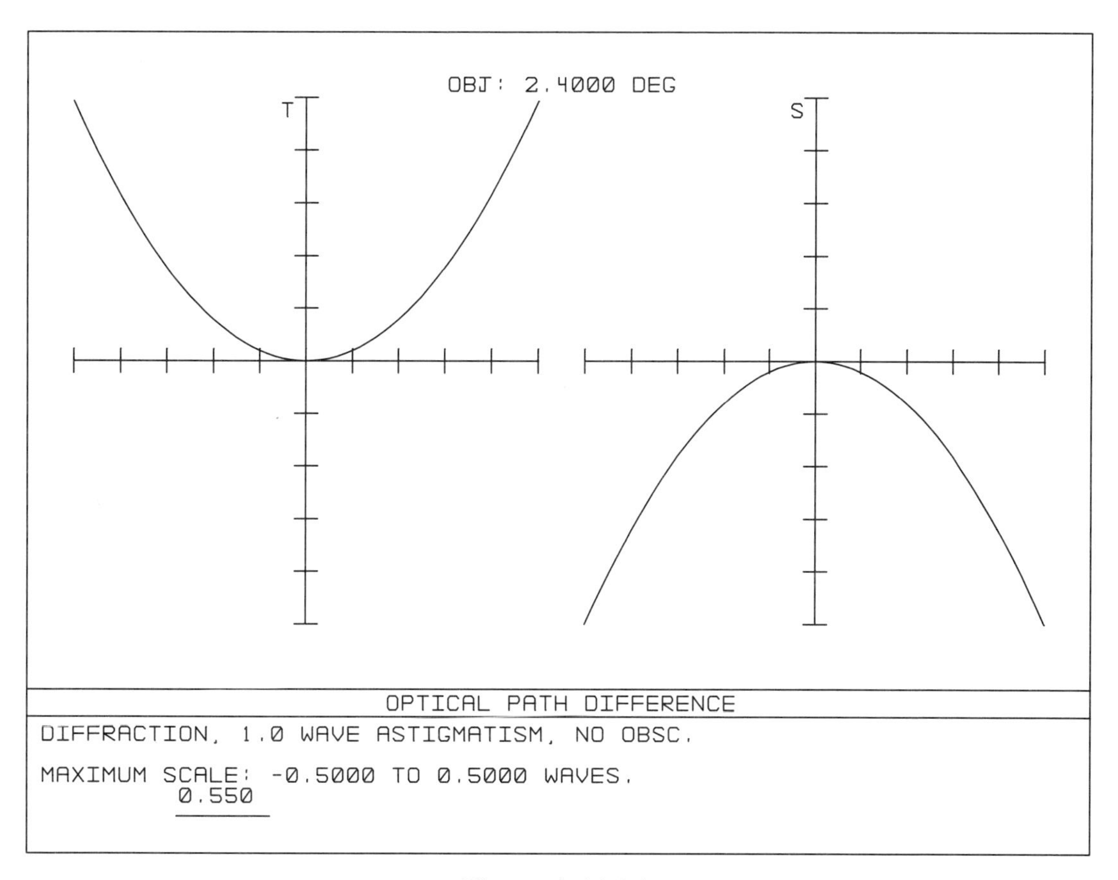

Figure A.11.8.1.

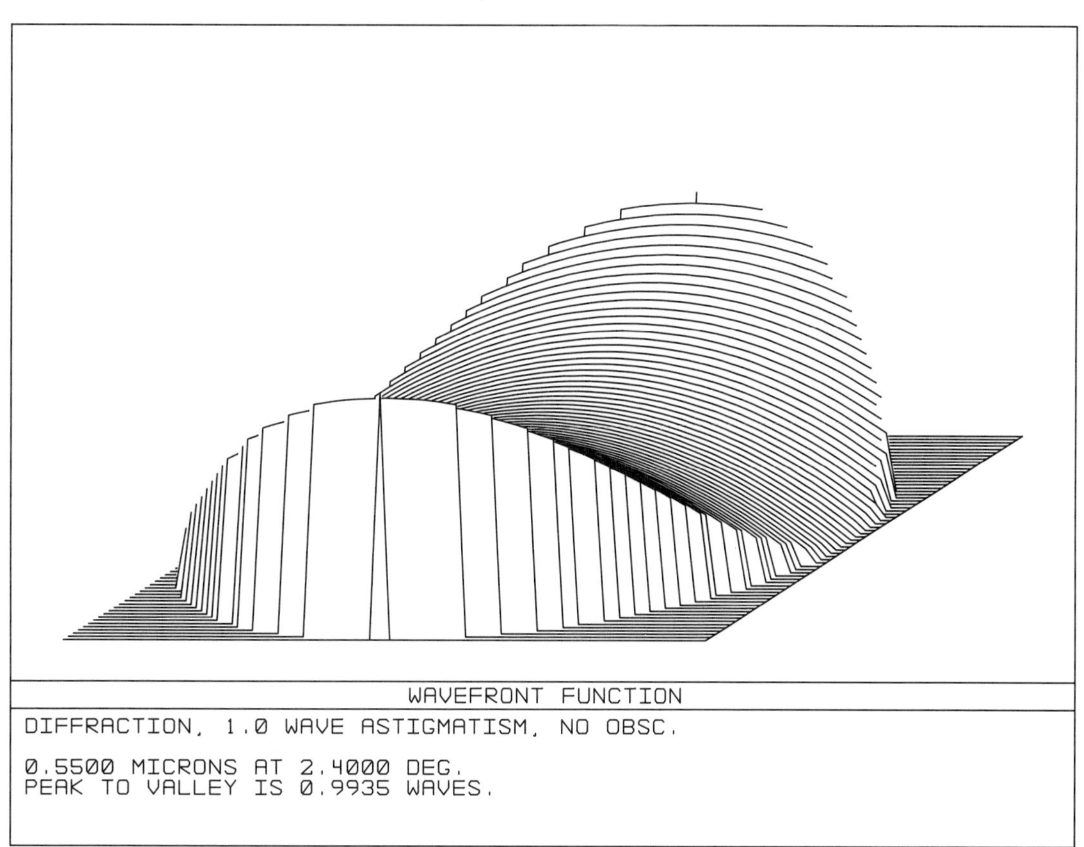

Figure A.11.8.2.

Figure A.11.8.3.

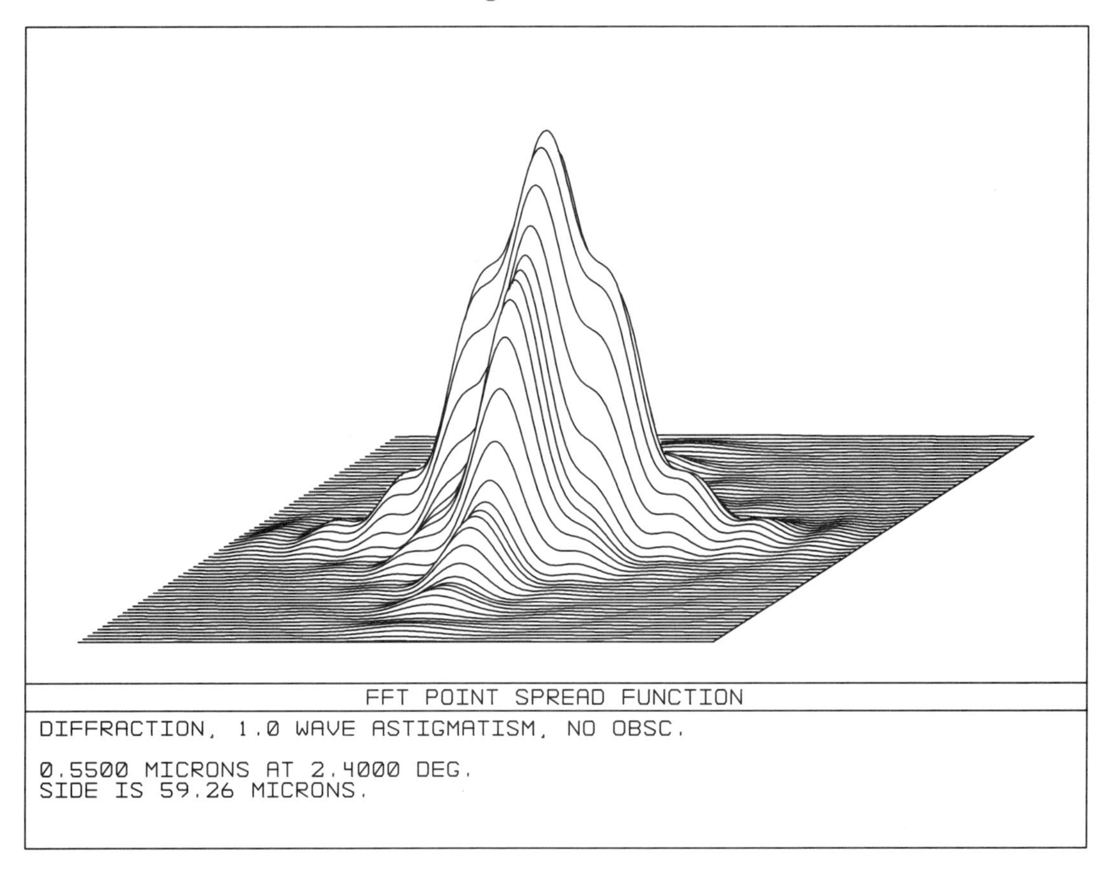

Figure A.11.8.4.

recognize them as two. The classic example of this situation is resolving the two components of a binary or double star.

There are several different but similar criteria for saying that two nearby objects are resolved. The most widely accepted standard is called the Rayleigh criterion. The Rayleigh criterion says that two equally bright point sources will be resolved if the peak of the Airy disk of one source falls on the first dark ring of the Airy pattern of the other source (and vice versa). In this case, the combined diffraction pattern has two separate peaks with a small (26%) dip in between.

If the Rayleigh criterion is adopted, then the linear resolving power of an optical system becomes the radius (not diameter) of the Airy disk as given by Equation A.11.3:

$$r_{\mathrm{Airy}} = 1.22\lambda F = 1.22\lambda\frac{f}{d}$$

where focal ratio, F, is replaced by focal length, f, divided by entrance pupil diameter, d. To get angular resolving power, the linear resolving power is divided by focal length to give:

$$\theta_{\mathrm{Airy}} = 1.22\frac{\lambda}{d} \tag{A.11.6}$$

where θ_{Airy} is in radians.

For some applications, the small irradiance dip between the two peaks is not necessary. If the 1.22 factor is replaced by 0.95, a smaller separation is allowed, the dip disappears, but the net effect is still an elongated image that reveals the presence of two sources. This standard is Sparrow's criterion and gives a resolving limit that agrees approximately with Dawes' empirically determined limit.

For a continuous extended object, resolution is more complex. An extended object can be regarded as the aggregate of a large number of individual object points. The image of each of these object points is subject to the system point spread function. The result of this spreading is a blurring of the image and a loss of resolution and contrast. In general, the light distribution in an extended image is derived mathematically by the two-dimensional convolution of the object (or perfect image) with the PSF.

If the smaller details in an extended object are resolved, then the relative sizes of these details in the image are approximately correct relative to the sizes of larger parts of the image. However, small extended object details are often detected in the image but not resolved. For example, the presence of a narrow but finite bright or dark line on a contrasting background may be detected, but its width may not be resolved. Thus in an aerial photograph, a road may be resolved, but the painted line down the middle may only be detected. Unresolved details appear relatively larger than they really are due to image spreading. Be vigilant against this detection versus resolution source of error.

Although many resolution criteria for extended images are possible, the limit based on the Rayleigh criterion remains a good indicator of ultimate diffraction-limited resolution. Of course, actual resolution may be considerably worse than the diffraction limit if aberrations, fabrication errors, stray light, or atmospheric seeing are present.

A.11.11 Strehl Ratio and the Quarter-Wave Rule

To get diffraction-limited images, it is not necessary that a lens have absolutely perfect geometrical performance; that is, truly zero aberrations. If the size of the spot in a spot diagram is considerably smaller than the Airy disk, then diffraction overwhelms aberrations, and images of point objects are practically indistinguishable from perfect Airy patterns. Similarly, if the OPD errors in the exit pupil are much smaller than the wavelength of light, then aberrations are again negligible. If diffraction-limited images are the goal, the question then becomes one of determining when geometrical aberrations are small enough.

If a lens has zero aberrations, then the resultant Airy pattern is the most compact image the lens can produce. The irradiance at the center of the Airy disk is also the maximum possible. If a lens has aberrations, then the light is spread out to some extent, and the peak irradiance of the broadened PSF is always reduced.

The ratio of the peak irradiance of an aberrated PSF to the peak irradiance of the corresponding aberration-free PSF is called the Strehl ratio. Note that the two peaks are not necessarily at the same place. If the pupil has central obscurations or mechanical vignetting, these are included in the calculation of both terms of the Strehl ratio.

For practical diffraction-limited images, the allowable departure of the Strehl ratio from unity or 100% depends on the instrument and its application. But for many applications, a Strehl ratio of 80% has been found to be an acceptable value. In 1879, Lord Rayleigh showed that if the only aberration in a lens is third-order spherical, then a reduction in the peak of the diffraction pattern of 20% (a Strehl ratio of 80%) corresponds in the exit pupil to a peak-to-valley OPD wavefront error of a quarter wavelength, or $\lambda/4$.[4] More recently, others such as Maréchal have shown that after any overall wavefront tilt has been removed, a Strehl ratio of 80% corresponds roughly to a peak-to-valley error of about a quarter-wave in the wavefront for other aberrations as well. This OPD guide to diffraction-limited images has thus become known as the Rayleigh quarter-wave rule (which is not to be confused with the other standard named after Rayleigh, the Rayleigh double star resolution criterion yielding a resolution limit of $1.22\lambda F$ as given in Equation A.11.3).

More precisely, a Strehl ratio of 0.8 actually corresponds to somewhat different amounts of wavefront peak-to-valley OPDs, depending on the aberration type. Nevertheless, regardless of the aberration type, for a Strehl ratio of 0.8, the minimum wavefront variance (both piston and tilt removed) is always 0.00565 waves-squared. This variance value yields an RMS wavefront departure (standard deviation in this case) equal to the square root of 0.00565, or about $1/14$ wave.

To determine the specific peak-to-valley wavefront errors for several common aberrations that produce a Strehl ratio of 0.8, lenses with the right amounts of these aberrations were devised and analyzed using ZEMAX. The results are given below.

For pure defocus, peak-to-valley OPD is 0.26 waves.

For pure third-order spherical aberration at the paraxial focus, peak-to-valley OPD is 0.25 waves.

For defocus and third-order spherical aberration balancing each other (the OPD errors at the exit pupil center and edge are made equal), peak-to-valley OPD is also 0.25 waves. Although this value is the same as for pure spherical aberration, the amount of spherical involved is four times as much as without compensating defocus.

For pure third-order coma with the reference sphere centered on the chief ray, peak-to-valley OPD is 1.28 waves. Because no wavefront piston needs to be removed with coma, the corresponding extreme wavefront departures are each half this amount, or plus and minus 0.64 waves. When the center of the reference sphere is shifted to the image centroid to remove overall wavefront tilt, the resulting wavefront errors are reduced by a factor of three, to 0.43 waves peak-to-valley, or plus and minus 0.21 waves.

For pure third-order medial astigmatism, peak-to-valley OPD is 0.37 waves. Because no wavefront piston or tilt needs to be removed with medial astigmatism, the corresponding extreme wavefront departures are each half this amount, or plus and minus 0.18 waves.

The reason why it takes almost twice as much peak-to-valley coma as peak-to-valley spherical aberration to produce the same Strehl ratio is that the coma

[4]See Max Born and Emil Wolf, *Principles of Optics*, fourth edition, p. 468.

wavefront errors are mostly in one dimension across the exit pupil (like a warped or bent disk), whereas the spherical aberration wavefront errors are distributed over two dimensions (like a bowl). Medial astigmatism is somewhere in between (like a ruffled disk).

It is also of interest to examine the examples given earlier to determine their Strehl ratios.

For Figures A.11.5.1 through A.11.5.4, 0.5 waves peak-to-valley of pure defocus yields a Strehl ratio of 0.44.

For Figures A.11.6.1 through A.11.6.4, 0.5 waves peak-to valley of pure third-order spherical aberration at the paraxial focus yields a Strehl ratio of 0.42.

For Figures A.11.7.1 through A.11.7.4, 2.0 waves peak-to-valley of pure coma relative to the reference sphere centered on the chief ray yields a Strehl ratio of 0.58.

And for Figures A.11.8.1 through A.11.8.4, 1.0 waves peak-to-valley of pure medial astigmatism yields a Strehl ratio of 0.20.

Note that all the values of Strehl ratio given here were calculated by ZEMAX using an approximation that assumes small aberrations. However, a system with a Strehl ratio as low as 0.20 is getting away from the diffraction limit, and thus this value may not be very accurate (although it is still indicative). Any approximately calculated Strehl ratio less than 0.1 would be untrustworthy.

From the user's and fabricator's perspectives, it is important to remember that the Rayleigh quarter-wave rule is only a guide. In many cases, a quarter-wave peak-to-valley of spherical aberration is too much aberration. A good example is a well-corrected telescope objective lens or mirror, where virtually perfect diffraction-limited performance requires something like $\lambda/10$ peak-to-valley in the wavefront or less. Nevertheless, the Rayleigh $\lambda/4$ rule is the source of the $\lambda/8$ surface fabrication error specification found on innumerable telescope mirrors (the factor of two results from light traveling a double pass on reflection). A better fabrication error specification would be a smooth surface with a peak-to-valley surface error of $\lambda/20$.

The mention of making a smooth surface follows from the fact that the distribution of light in the diffraction PSF actually depends in a complicated way on many factors. Not only are RMS and peak-to-valley OPD errors important, but so too is geometrical image size (spot size). Geometrical image size in turn depends on wavefront slope errors (both peak and RMS) in the exit pupil, and thus on smoothness. Both the lens designer and the lens maker (optician) should always keep in mind the importance of all these factors.

The quarter-wave rule has one further benefit for the lens designer. When designing a lens, you need not calculate the diffraction PSF for intermediate results to determine whether your lens is approaching the diffraction limit. Rather, one glance at an easily calculated OPD ray fan plot supplies this information. If your lens has peak-to-valley OPDs of several waves, then you immediately know that aberrations are large relative to diffraction. Conversely, if your lens has OPDs less than $\lambda/4$, then you know that your lens is getting pretty good and may be approaching the point of diminishing returns. Only at major milestones in the design process is it necessary to do a complete diffraction evaluation of image quality to see what you really have.

A.11.12 Scaling the Lens

If all linear dimensions of a lens are scaled up or down by the same factor, then the lens is enlarged or reduced like the image in a photographic enlarger. By this means, an initial lens design can be converted to any desired focal length. However, this process does not change relative dimensions; that is, ratios of dimensions. Thus, focal ratio, percent distortion, and all angular geometrical properties remain unchanged by scaling.

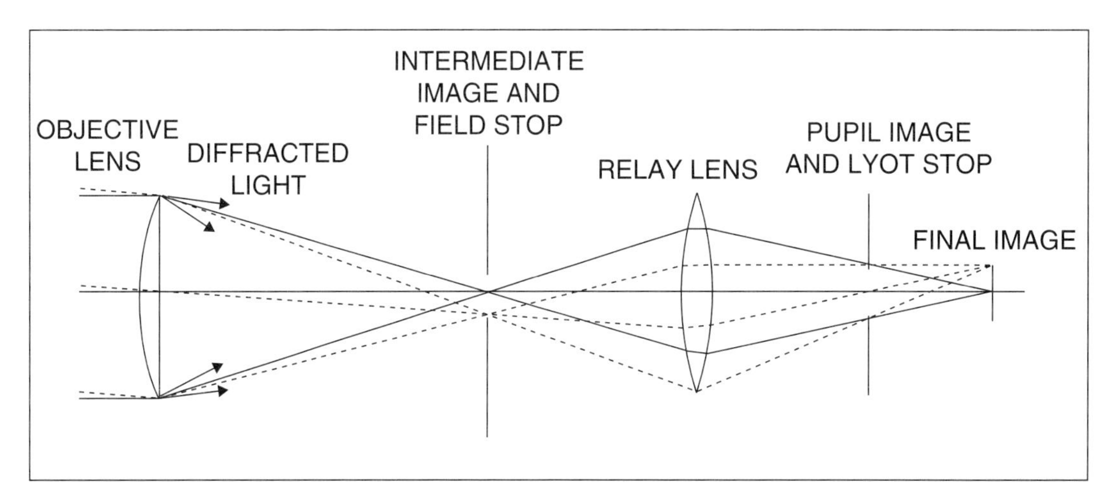

Figure A.11.9. Lyot Stop. *In this thin-lens schematic layout, the relay lens reimages the intermediate image onto the final image. The relay lens also images the objective lens onto the Lyot stop. A bright light source is just outside the field of view as defined by the field stop at the intermediate image. The light from the bright source that is diffracted around the edge of the objective lens is blocked by the Lyot stop, which is slightly smaller than the image of the objective.*

It is of particular interest that when a lens is scaled, all linear geometrical aberrations also scale with the lens. This is true whether the aberrations are expressed as transverse or longitudinal ray errors in the image, or as OPD errors in the exit pupil wavefront (all expressed in linear measure such as millimeters).

There is one length, however, that does not change when a lens is scaled. This exception is the wavelength of light. There are two consequences. The first is that OPD errors, when measured in constant wavelengths, still scale with the lens. The second is that diffraction effects do not scale with the lens. When a lens is scaled, all diffraction effects must be recalculated. For example, if a lens is scaled up, the angular size of the Airy disk is reduced (not increased), but the linear size remains constant. It is very important to keep separate in your mind the geometrical and physical properties of an optical system.

A.11.13 The Lyot Stop

Suppose you wish to view a faint object on a dark background that is located close to a very bright object. An example is a star near the full moon. In most optical systems, there is so much stray light from the bright object that the faint object is lost in the glare. This is true even with clean optics and excellent conventional baffles.

This problem was first solved by Bernard Lyot (pronounced Leo) in 1930 with his invention of the coronagraph.[5] Lyot wished to observe the corona of the sun without an eclipse. But the corona is roughly a million times fainter than the disk of the sun. Obviously, extreme control of stray light is necessary. The heart of Lyot's invention is now called a Lyot stop. A Lyot stop is a device that controls stray light by reducing the effects of diffraction. Note that it is the effects that are reduced, not the diffraction itself.

Figure A.11.9 is a schematic thin-lens layout of a system incorporating a Lyot stop. Rays from on-axis and off-axis object points are shown as solid and dashed lines, respectively. A bright object (not shown) is located just outside the field of view.

[5] Bernard Lyot, "A Study of the Solar Corona and Prominences without Eclipses," *Monthly Notices of the Royal Astronomical Society*, Vol. 99, pp. 580–594, 1939; excerpted as "Invention of the Coronagraph," in Harlow Shapley, ed., *Source Book in Astronomy 1900–1950*, pp. 17–22.

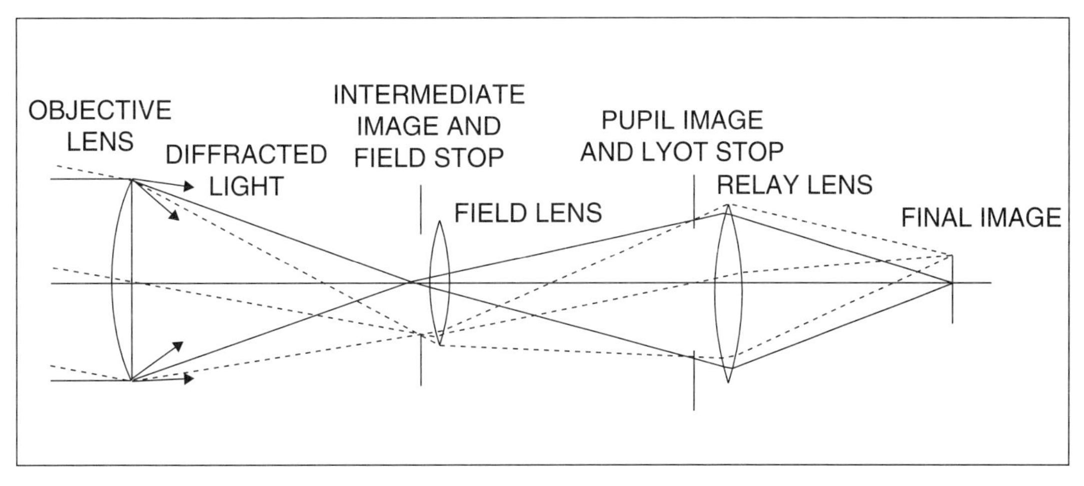

Figure A.11.10. Lyot Stop with Field Lens. *The field lens images the objective lens near the relay lens to reduce the required diameter of the relay lens. Otherwise, the principle is the same as in Figure A.11.9.*

An objective lens with the stop at the lens images the object at the intermediate image. Beyond the intermediate image is a relay lens that reimages the intermediate image at the final image. Thus, the marginal ray crosses the axis at the intermediate image and again at the final image, as seen in Figure A.11.9.

But the relay lens images more than just the intermediate image. The relay lens also images the objective lens; that is, it forms a pupil image. Thus, the chief ray crosses the axis at the objective lens and again at the pupil image, as also seen in Figure A.11.9. If you place a viewing screen at the pupil image, you will see all the dust particles on the objective lens in sharp focus.

A field stop at the intermediate image defines and limits the field of view. In particular, the field stop isolates and transmits the region of interest around the faint object, but it blocks the light from the bright nearby object outside the field of view. Thus, direct light from the bright object is prevented from entering the rear part of the system.

However, the bright object still illuminates the objective lens and causes light to be diffracted from all around the edge of the lens mount. If you could place your eye at the center of the intermediate image, you would see a bright ring around the edge of the objective lens. This diffracted light is indicated in Figure A.11.9. Some of this diffracted light passes through the field stop and, unless something is done, reaches the final image to produce unacceptable glare.

It would be nice if you could just mask off the region around the edge of the objective lens, thereby removing the diffracted light. But the mask would also be illuminated by the bright light, its edge would also diffract, and you would be back where you started.

But the situation is very different at the image of the objective lens, the pupil image. Because the field stop blocks the bright light and prevents it from entering the rear part of the system, you can mask off the diffracted light at the pupil image without introducing new diffracted light.

This mask at the pupil image is the Lyot stop. By removing the diffracted light caused by the bright object, the Lyot stop suppresses glare and allows the faint object to be seen.

Because the diameter of the Lyot stop is slightly less than the diameter of the image of the objective lens, overall system speed is slightly reduced. If necessary, the objective lens can be made slightly oversized to compensate. In any case, it is the Lyot stop, not the objective lens, that is the true system aperture stop for radiometric purposes.

Note that a Lyot stop only works as long as the bright object continues to be kept out of the field of view. If the bright object should enter the field of view, then bright light enters the rear part of the system and the glare returns.

Incidentally, in an actual coronagraph, the field of view is annular. The sun is placed in the middle of the field, and an opaque central occulting disk at the intermediate image blocks the bright light. At the final image, the faint corona can be seen surrounding the occulting disk. The result looks very much like a total solar eclipse, which is what the system is trying to simulate.

A.11.14 A Lyot Stop Plus a Field Lens

Note in Figure A.11.9 that the off-axis rays do not pass through the center of the relay lens. Thus, the diameter of the relay lens must be enlarged to pass these rays. If the field is not too large, this extra diameter does not matter. But for substantial fields, an additional lens, a field lens, is necessary to prevent the relay lens from getting too big.

Figure A.11.10 is a schematic layout of an optical system with both a Lyot stop and a field lens. The field lens is located just aft of the intermediate image. The purpose of the field lens is to redirect the diverging light cones from the intermediate image so they pass more nearly through the center of the relay lens and minimize its required diameter. In other words, the field lens forms a pupil image of the objective lens near the relay lens. The Lyot stop, located close to the relay lens, is exactly at the pupil image where the chief ray crosses the axis.

Using field lenses to control the location of relayed pupil images is important in the design of many complex optical systems. The concept is not restricted to systems with a Lyot stop. Field lenses are used in systems ranging from eyepieces to periscopes to photometers to spectrographs. In projectors and enlargers, a field lens is called a condenser lens (see the example in Chapter A.5). Incidentally, do not place a field lens too close to an intermediate image; if you do, then all of the dust particles on the field lens will be in focus in the final image.

Chapter A.12

Modulation Transfer Function

Modulation transfer function, or MTF, is a direct measure of how well the various details in the object are reproduced in the image. MTF is also known as spatial frequency response and sine-wave response. MTF values are given from zero to 100%. Since the 1960s, MTF has become the most widely accepted criterion for specifying and judging image quality. This acceptance is based on firm theoretical ground from information theory and Fourier analysis. This chapter discusses MTF for both the design and evaluation of optical systems.

A.12.1 Frequency Response

The concepts of spatial frequency and optical frequency response may seem a bit strange at first. But very similar concepts are well known to electrical engineers. Every hi-fi enthusiast knows about frequency response, and he rates his electronic equipment by how uniform the response curves are throughout the audible range, such as flat to within ± 3 db from 20 to 20,000 cycles/second or Hz. These frequencies are functions of time; that is, they are temporal frequencies.

The situation is much the same in optics, except that objects and images are two-dimensional, and frequency content is measured in cycles/mm. These frequencies are now functions of length; that is, they are spatial frequencies.

Although he may not think of it as such, spatial frequency response is familiar to every photographer who has tested his camera lenses by photographing a resolution test target. An example, the widely used U.S. Air Force three-bar target, is shown in Figure A.12.1. The target consists of a series of similar but progressively smaller patterns, each formed by a number of light and dark stripes of equal width. The effect is a series of radiance square waves of varying spatial frequency. When photographed, the finest pattern that can be resolved gives the limiting square-wave spatial frequency response, or limiting resolution. The result is given in line pairs/mm on the film; a line pair is one bright stripe and one dark stripe. Performance reviews of photographic equipment in camera magazines also quote limiting resolution, such as 64 line pairs/mm. More recently, contrast transfer at a specific square-wave frequency has also begun to be quoted, such as 40% at 30 line pairs/mm. In a similar way, television engineers rate their equipment in TV scan lines, where it takes two TV scan lines to make a line pair.

A.12.2 Fourier Analysis

The general study of the decomposition of functions into their frequency components is called Fourier analysis. Any function can be resolved into its constituent sine-wave frequency components by a mathematical procedure called a Fourier transform. Many operations with functions are much easier to do if the functions

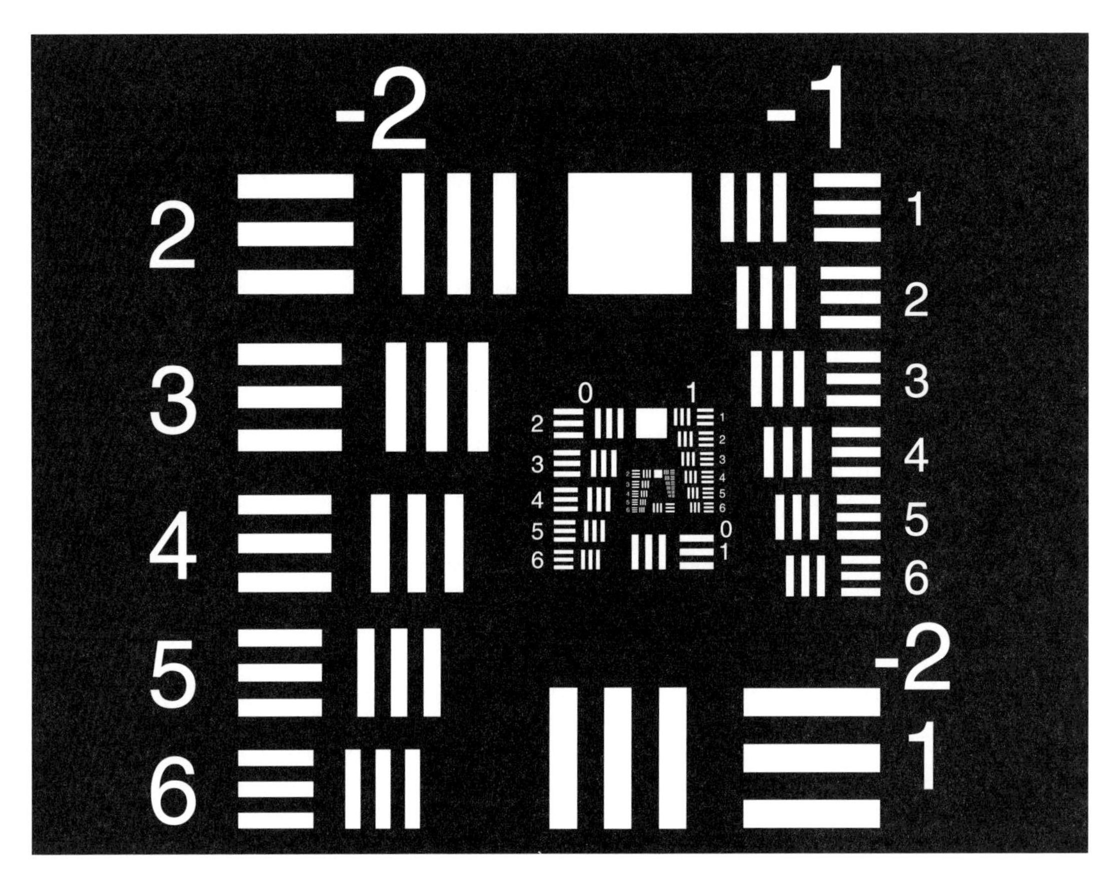

Figure A.12.1. U.S. Air Force 1951 Three-Bar Resolution Test Target. *Both positive and negative versions are available; shown here is the negative version with a dark background. An element consists of two patterns at right angles to test both tangential and sagittal resolution. Each (negative) pattern consists of three bright lines and two dark lines of equal width, with line length five times the width. Element size decreases geometrically as the sixth root of two, causing the spatial frequency to double for six element size reductions. At the scale of the original target, the spatial frequencies range from 0.25 line pairs/mm to 228 line pairs/mm (a line pair consists of one bright line and one dark line). Such a target is used to assess how much information can be extracted from a scene. Three-bar targets have been variously viewed and photographed, and for special tests they have been painted on roofs, painted on the underside of airplane wings, and so forth.*

are first transformed into the frequency domain, operated upon, and then inverse transformed to get the final answer. A good example is the convolution of two functions; convolution is replaced in the frequency domain by a simple multiplication of the transforms. The techniques of Fourier analysis are a field of specialization in itself.

In optics, the study of image formation by means of spatial frequencies is called Fourier optics. Recall that in the presence of diffraction and other image degrading effects, all real lenses have point spread functions (PSFs) of greater than zero extent and thus limited resolving capabilities. More precisely, the light distribution in the image is the two-dimensional convolution of the object (or perfect image) and the PSF. The convolution process causes fine details in the object to be smeared out and lost in the image. In frequency terms, higher spatial frequencies in the object are partially or completely attenuated in the image. Because lower spatial frequencies are preferentially passed, and higher spatial frequencies are preferentially attenuated, a lens can be described, using electronic terminology, as a low-pass filter. Thus, the spatial sine-wave response or MTF of a lens strongly influences its imaging capabilities.

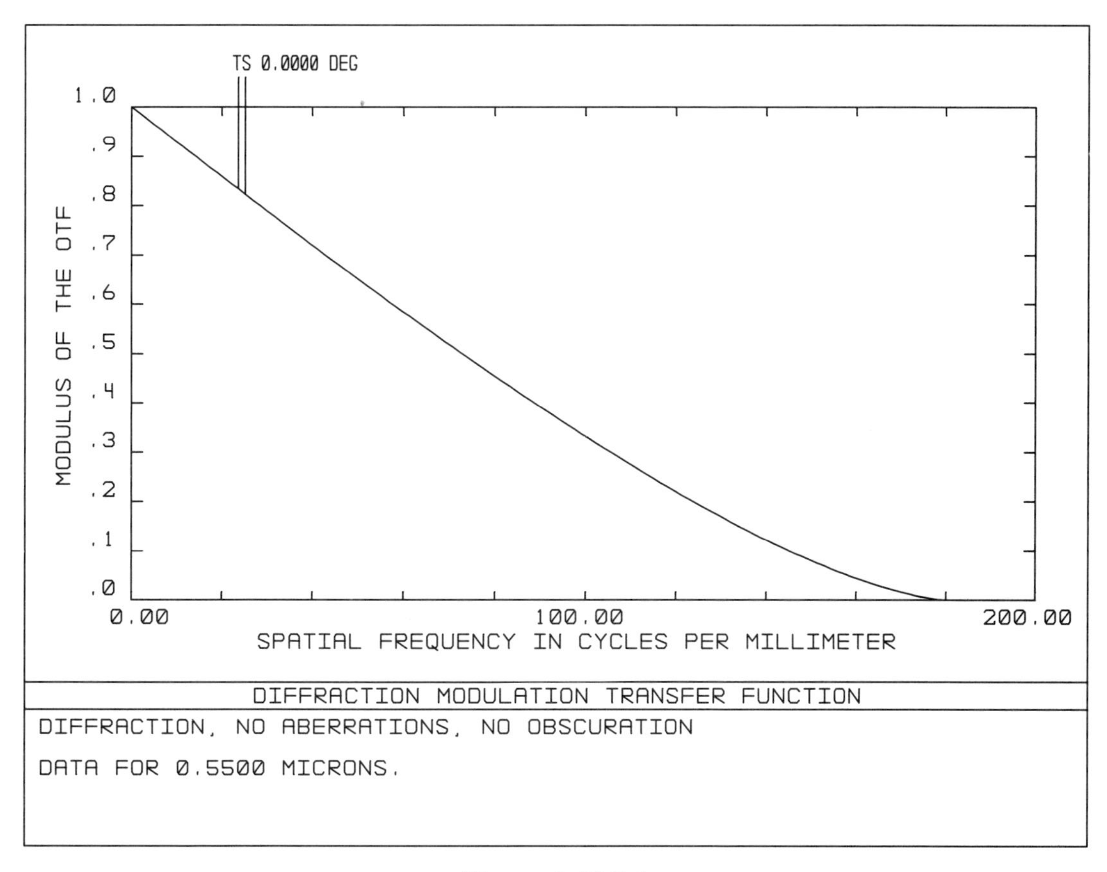

Figure A.12.2.1.

Figure A.12.2.1 is an example of an MTF plot. The lens is the same on-axis diffraction-limited lens with no central obscuration used for Figures A.11.3.1 through A.11.3.4. On the MTF plot, note the curve shape. This classic MTF curve is unity (or 100%) at zero spatial frequency and drops off nearly linearly with increasing frequency except for a rounded toe.

Incidentally, in the ZEMAX-generated Figure A.12.2.1, the TS 0.0000 DEG notation at the top and the accompanying two vertical lines refer and point to the tangential and sagittal MTF curves for the 0.0000° field position. In this case, because the field is on-axis, the tangential and sagittal curves are identical and lie on top of each other. However, on other MTF plots later in this book, off-axis fields will yield different tangential and sagittal curves.

Actually, the Fourier analysis of images is more complicated than this. Not only are various spatial frequencies attenuated, but they can also be shifted relative to each other if the PSF is unsymmetrical. This positional shift is a spatial phase shift. Thus, the complete Fourier description of an image requires the complex optical transfer function or OTF. The modulus or absolute value of the OTF is the MTF, and the phase of the OTF is the phase transfer function or PTF.

A.12.3 Measuring MTF

For an existing real lens, one way to measure (not calculate) the sine-wave response is to examine how well the lens images a special MTF test target. The test target is similar to the standard photographer's resolution target mentioned above, but sinusoidal radiance patterns replace the previous square waves. The modulation (or contrast) M of a spatial frequency ν is given by

$$M(\nu) = \frac{I_{\text{max}} - I_{\text{min}}}{I_{\text{max}} + I_{\text{min}}} \tag{A.12.1}$$

where $I_{\max}$ and $I_{\min}$ are the maximum and minimum radiances. MTF is the ratio of image modulation to object modulation as a function of spatial frequency, or

$$MTF(\nu) = \frac{M_{\text{image}}(\nu)}{M_{\text{object}}(\nu)}. \tag{A.12.2}$$

The plot of MTF as a function of frequency gives the shape of the low-pass filter passband.

Note that this procedure gives the MTF, not the OTF. It is possible to also measure spatial phase shifts with test targets, but large errors are likely.

There are other ways besides a test target to create spatial sine waves of different frequencies. One commercial MTF measuring instrument uses two synchronized counter-rotating glass disks, each imprinted with a pattern of radial stripes that get progressively narrower and closer as you go around the disk. As the disks rotate, a moiré pattern is created whose spatial frequency varies continuously and repeatedly with time from zero to a maximum value. Other MTF instruments use the line spread function or the edge spread function, both of which are related to the point spread function. Here, the measured distribution of light in the image must be analyzed using various mathematical techniques, all of which include a Fourier transform. However, line and edge spread techniques suffer from errors caused by noise problems, and do not always agree with more direct methods.

Note that in addition to spatial frequency, MTF depends in general on the angular orientation of the test pattern. Thus, a three-dimensional polar plot is required to display MTFs for all pattern orientations. Cross-sections at various angles through the MTF surface give the more conventional MTF curves. Because the test pattern looks the same when rotated by 180°, the three-dimensional MTF is an even function; that is, positive and negative frequencies give equal MTFs. Thus, only half of a given cross-section need be shown.

As an example, Figure A.12.2.2 is the three-dimensional MTF plot corresponding to Figure A.12.2.1. Because the object is on-axis, the PSF is rotationally symmetrical, and all test pattern orientations give the same MTF curve. Actually, this example was calculated, not measured; a measured example would not look so flawless.

For axially centered systems, only two orthogonal cross-sections through the three-dimensional MTF surface are usually given. The angles are in the tangential and sagittal directions. Note the convention that tangential spatial frequencies are modulations obtained by scanning in the y-direction (in the radial or sagittal direction) across a test pattern with sinusoidal stripes oriented in the tangential direction. Similarly, sagittal spatial frequencies are modulations obtained by scanning in the x-direction (along tangential circles) across radial or sagittal stripes. This somewhat confusing situation is illustrated in Figure A.12.2.3.

A.12.4 Calculating the Diffraction MTF by Autocorrelation

For a lens modeled in a computer, there are two equivalent ways to calculate (not measure) the MTF with diffraction effects included. Both require knowledge of the pupil function. The pupil function is complex and is defined for coordinates ξ and η across the pupil as

$$G(\xi,\eta) = K\tau(\xi,\eta)e^{ikW(\xi,\eta)} \tag{A.12.3}$$

where $G(\xi,\eta)$ is the pupil function, $\tau(\xi,\eta)$ is the amplitude (not intensity) pupil transmission function (intensity is the square of amplitude), $W(\xi,\eta)$ is the pupil OPD wave aberration function, and K and k are normalizing factors. Thus, the modulus of the pupil function is proportional to transmission, and the complex phase is proportional to OPD in wavelengths.

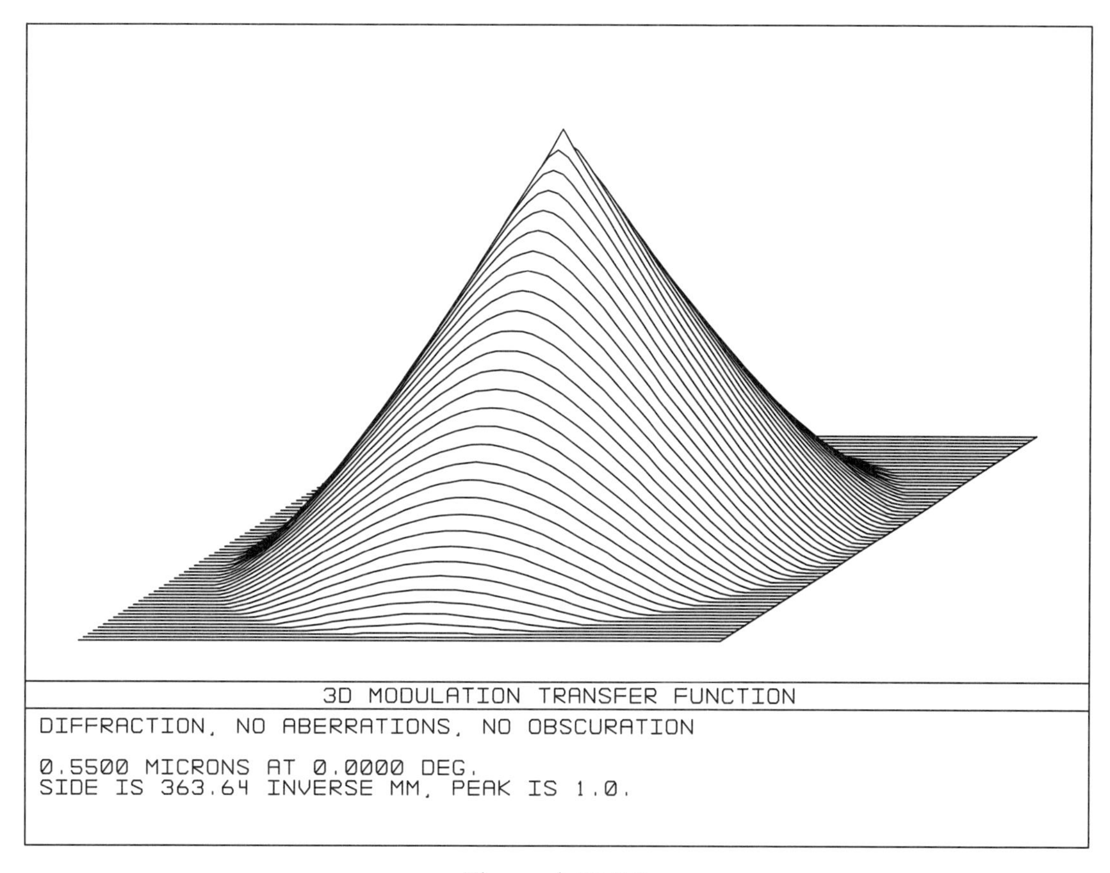

Figure A.12.2.2.

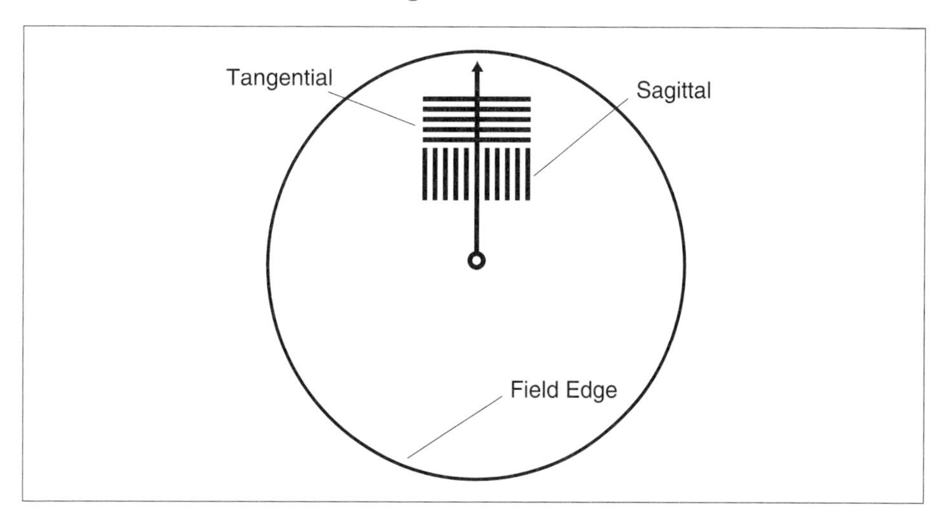

Figure A.12.2.3. Tangential and sagittal spatial frequencies.

The first and more direct method for calculating the diffraction MTF performs a two-dimensional autocorrelation of the pupil function. The result is the OTF, which combines both the MTF and the PTF.

To see how this works, take the case of a perfectly diffraction-limited lens (zero aberrations). Because $e^{i0} = 1$, the pupil function becomes real (no longer complex) and is simply the transmission function, which is three-dimensional (transmission plus the two pupil coordinates). For a circular pupil with perfect transmission, the transmission function resembles a top hat. If there are absorptions, then the hat is rounded (like a derby) or rippled. If there are obscurations, then transmission is zero in the obscured regions as well as beyond the pupil edges. Note that the pupil

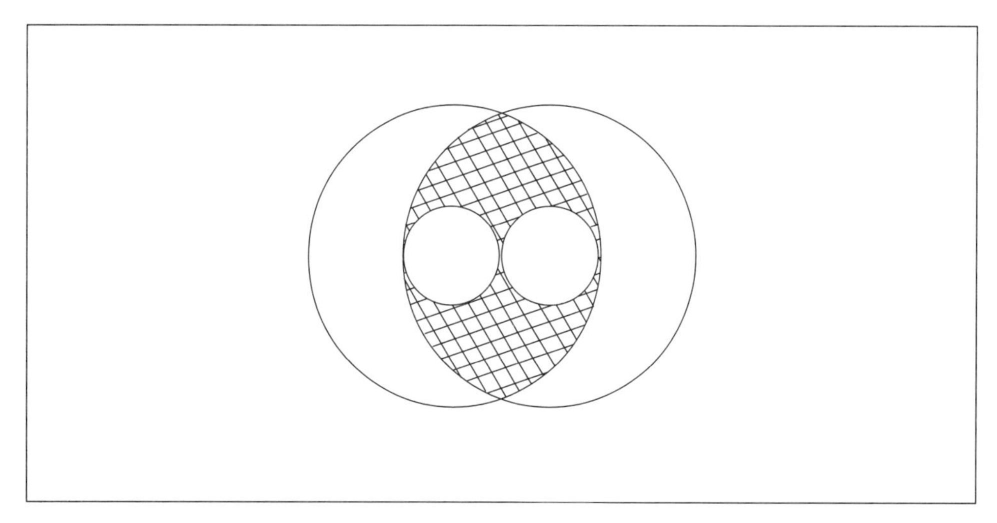

Figure A.12.3.1. MTF by Autocorrelation. *Spatial frequency is 1/3 of cutoff.*

need not be circular.

The diffraction-limited OTF is then the easily calculated autocorrelation of this real transmission function. Because the PTF is zero for a diffraction-limited system, the OTF here equals the MTF.

To do this autocorrelation, two copies of the pupil are sheared or translated in opposite directions relative to each other. For each value of shear, the two pupils overlap to some extent. The two corresponding transmission functions are multiplied, and in the region of overlap where the product is nonzero, another three-dimensional function is obtained. The volume of this product function is obtained by integration. The volumes for all shears are then normalized to unity for zero shear.

A specific amount of shear corresponds to a specific spatial frequency. The maximum shear is where the two functions cease to overlap and the product volume goes to zero. This limiting shear corresponds to the MTF cutoff frequency, and is given by

$$\nu_{\text{cutoff}} = \frac{1}{\lambda F} \tag{A.12.4}$$

where λ is wavelength (in millimeters to get cycles/mm) and F is working focal ratio. Note that even with aberrations to make the pupil function complex, cutoff frequency still depends only on wavelength and the size of the pupil that can be sheared. In all of the examples in this chapter, λ is 0.00055 mm, F is 10, and thus cutoff frequency is 181.8 cycles/mm.

As an example of autocorrelation, take the case of a diffraction-limited lens with a perfectly transmitting circular pupil except for a 30% central obscuration. This is the same optical configuration used for Figures A.11.4.1 through A.11.4.4. Because transmission is either all or nothing, the autocorrelation process now reduces to a two-dimensional shearing of the transmitting pupil *area*. The autocorrelation values are represented by the overlapping areas. Figure A.12.3.1 shows the two copies of the pupil sheared by one-third of the cutoff frequency. Figure A.12.3.2 shows the same two pupils sheared by two-thirds of the cutoff frequency. The overlapping areas are shaded. Any areas occupied by the central obscuration are not shaded.

The resulting MTF curve is shown in Figure A.12.3.3. Compare this curve with the curve in Figure A.12.2.1 for a similar lens with no obscuration. The obscuration causes a significant drop in mid-range frequency response and a very slight rise in frequency response for higher frequencies (if this sounds like hi-fi jargon, it is no coincidence). The improvement at higher spatial frequencies is caused by the slight reduction in the diameter of the Airy disk for an obscured pupil.

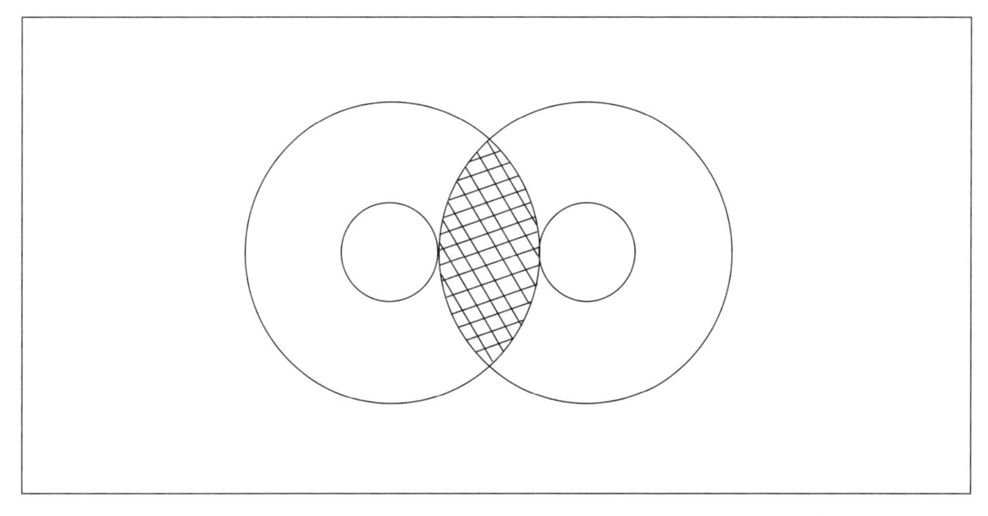

Figure A.12.3.2. MTF by Autocorrelation. *Spatial frequency is 2/3 of cutoff.*

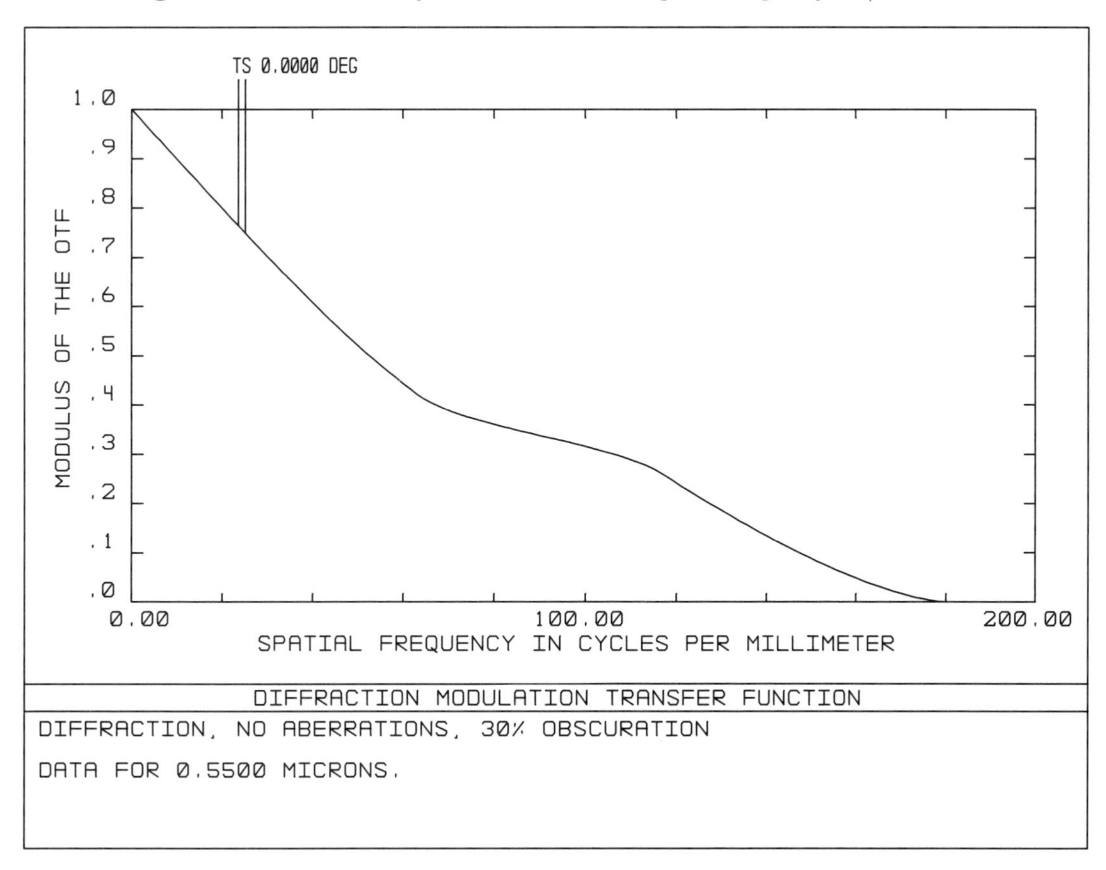

Figure A.12.3.3.

Note that for an arbitrarily shaped pupil, the cutoff frequency as well as the shape of the MTF curve depend on the angular direction of the shear. Thus, as was mentioned above, a complete description of MTF requires a three-dimensional polar plot whose radial cross-sections are the two-dimensional MTF curves.

Unfortunately, when a lens has aberrations and the pupil function is complex, the autocorrelation method becomes quite cumbersome mathematically. Thus, for this general case, autocorrelation is rarely used in practice to get the MTF (although with modern computers, this problem is less important). Fortunately, there is another way.

A.12.5 Calculating the Diffraction MTF by Fourier Transforms

For a lens with aberrations, the second and easier way of calculating the diffraction MTF uses Fourier transforms.

The Fourier transform method is done in three steps. First, a two-dimensional Fourier transform of the pupil function is performed. Second, this transform is squared by multiplying by its complex conjugate. The result of the second step is the image intensity (irradiance) point spread function (PSF). Third, a two-dimensional Fourier transform of the PSF is performed to get the OTF. Make a special note that the Fourier transform of the PSF gives the OTF.

In both methods, the modulus (absolute value) of the OTF is the MTF. All of these mathematical calculations are handled routinely by any professional lens design software package, and the lens designer normally need not be concerned with the details.

A.12.6 Consequences for Optical Design

There is more, however, to the second method, the one using Fourier transforms, than mathematical manipulations. There are important consequences for the basic approach to designing optics.

Look at the definition of the pupil function (Equation A.12.3). The pupil function is closely related to wavefront OPD errors in the exit pupil. Recall that wavefront slope errors (ray aberrations) are proportional to the derivative of wavefront OPD errors, and that wavefront OPD errors are proportional to the integral of wavefront slope errors. Thus, the pupil function is only distantly related through an integral to the wavefront slope errors and to the corresponding image transverse ray-intercept errors.

Now look at how the PSF and MTF are calculated. PSF and MTF are derived from the pupil function, and thus PSF and MTF are also more closely related to OPD errors than to ray-intercept errors. Although it may seem counterintuitive, it follows that during optimization, better PSF and MTF results are obtained if OPD errors rather than ray-intercept errors (as with spot size) are minimized.

More theoretically, recall that Strehl ratio is the ratio of the peak irradiance of the actual aberrated PSF (including diffraction) relative to the peak irradiance of the corresponding diffraction-limited PSF. A high Strehl ratio is very desirable in an optical system. It can be shown that for small aberrations, Strehl ratio is independent of the details of the aberrations and depends only on the statistical variance (square of the standard deviation) of the wavefront OPD errors evaluated over the whole pupil.[1] The greater the variance, the lower the Strehl ratio. Once again, optimizing with OPDs is indicated.

Consequently, lenses with small aberrations that were previously optimized using spots can often be subsequently reoptimized to advantage using OPDs. In fact, reoptimizing using OPDs can be beneficial even for lenses with large aberrations. However, as is so often the case in lens design, no rule is absolute. Sometimes in the end, the solution derived from transverse ray errors is better or is otherwise preferable to an OPD solution. A common design approach is to optimize both ways, compare the results, and select the one you prefer.

A.12.7 MTF in the Presence of Aberrations

Up until now, the MTF examples have been for lenses that are diffraction limited. But if the same unobscured lens used for Figures A.11.3.1 through A.11.3.4 and Figures A.12.2.1 and A.12.2.2 is used slightly out of focus, then the images are degraded. Figure A.12.4 shows the MTF curve for 0.5 waves peak-to-valley of

[1] For the proof, see Max Born and Emil Wolf, *Principles of Optics*, fourth edition, pp. 463–464.

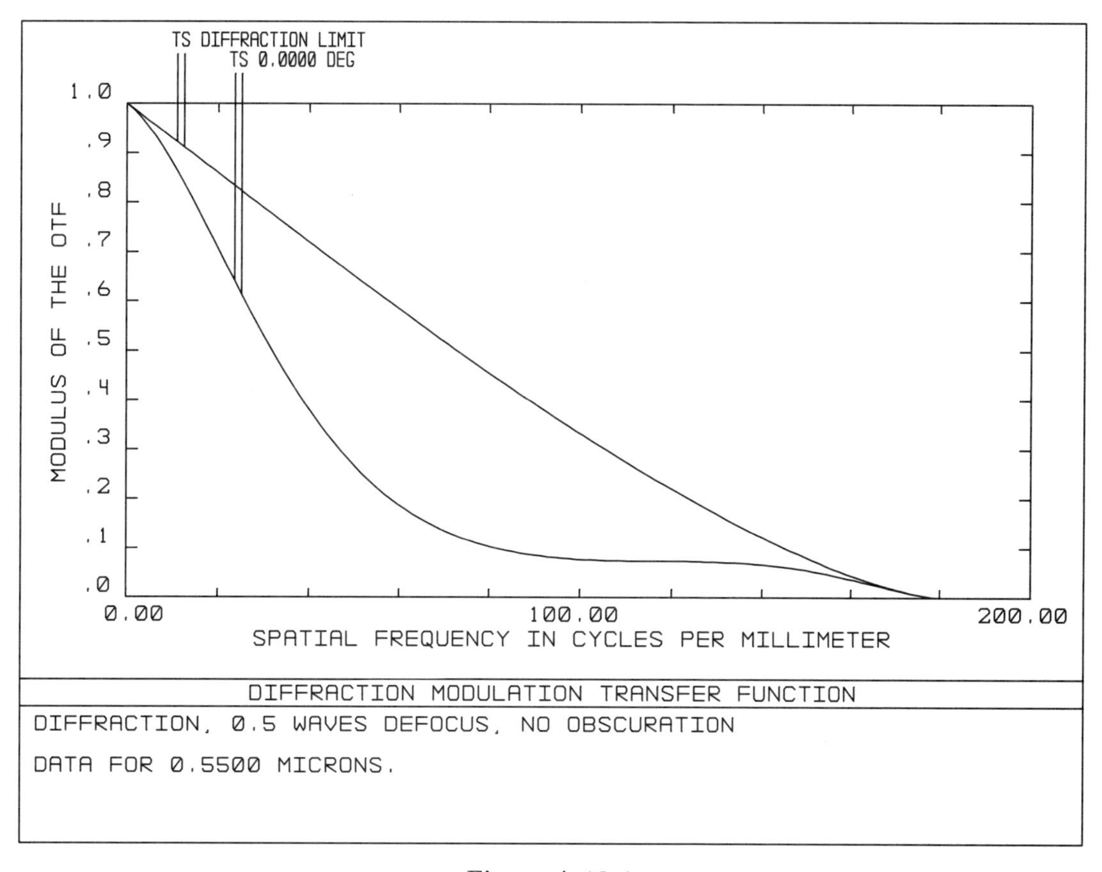

Figure A.12.4.

defocus, making the optical configuration the same as used for Figures A.11.5.1 through A.11.5.4. As is often done, for comparison the theoretical MTF curve for a similar diffraction-limited system has also been plotted on Figure A.12.4. The departure of the actual lens from diffraction-limited performance is clear.

Similarly, Figure A.12.5 is the MTF plot for the unobscured lens with 0.5 waves peak-to-valley of spherical aberration used for Figures A.11.6.1 through A.11.6.4. The diffraction-limited curve has again been included, and the reduction in MTF is again clear.

If an off-axis object point is used and if the lens has a non-rotationally symmetric aberration such as coma, then the tangential and sagittal MTF curves are no longer identical. Such an MTF plot is shown in Figure A.12.6.1 for the same unobscured lens with 2.0 waves peak-to-valley of coma used for Figures A.11.7.1 through A.11.7.4 (reference sphere centered at the chief ray, equivalent to 0.667 waves peak-to-valley with the reference sphere centered on the image centroid). Note that the tangential MTF curve is worse than the sagittal curve because the image spread from coma flare is worse in the radial direction, and scanning in the radial (sagittal) direction yields tangential MTF. This difference is confirmed by the three-dimensional MTF plot in Figure A.12.6.2. Note the dimple on the right side; there is a partially hidden symmetrical dimple on the left side too.

A second example of off-axis MTF is astigmatism on the medial image surface; that is, at the "circle" of least confusion. Figure A.12.7 shows the three-dimensional MTF surface for the lens with 1.0 waves peak-to-valley of medial astigmatism used for Figures A.11.8.1 through A.11.8.4. Here, the tangential and sagittal MTF curves are identical while the MTF curves at other angular orientations are different. This behavior would not be apparent on a conventional MTF plot.

Note that the MTF surfaces in Figures A.12.6.2 and A.12.7 reveal that the coma example has better performance than the astigmatism example. But this would be

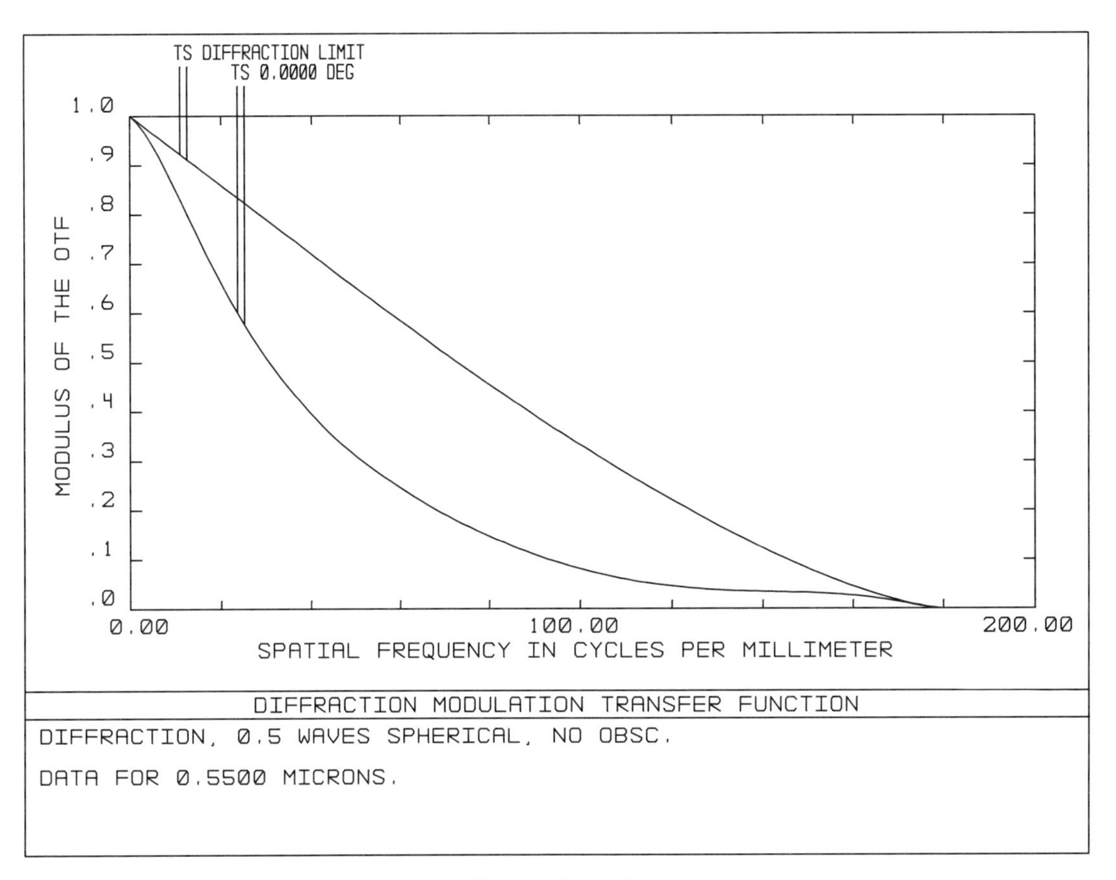

Figure A.12.5.

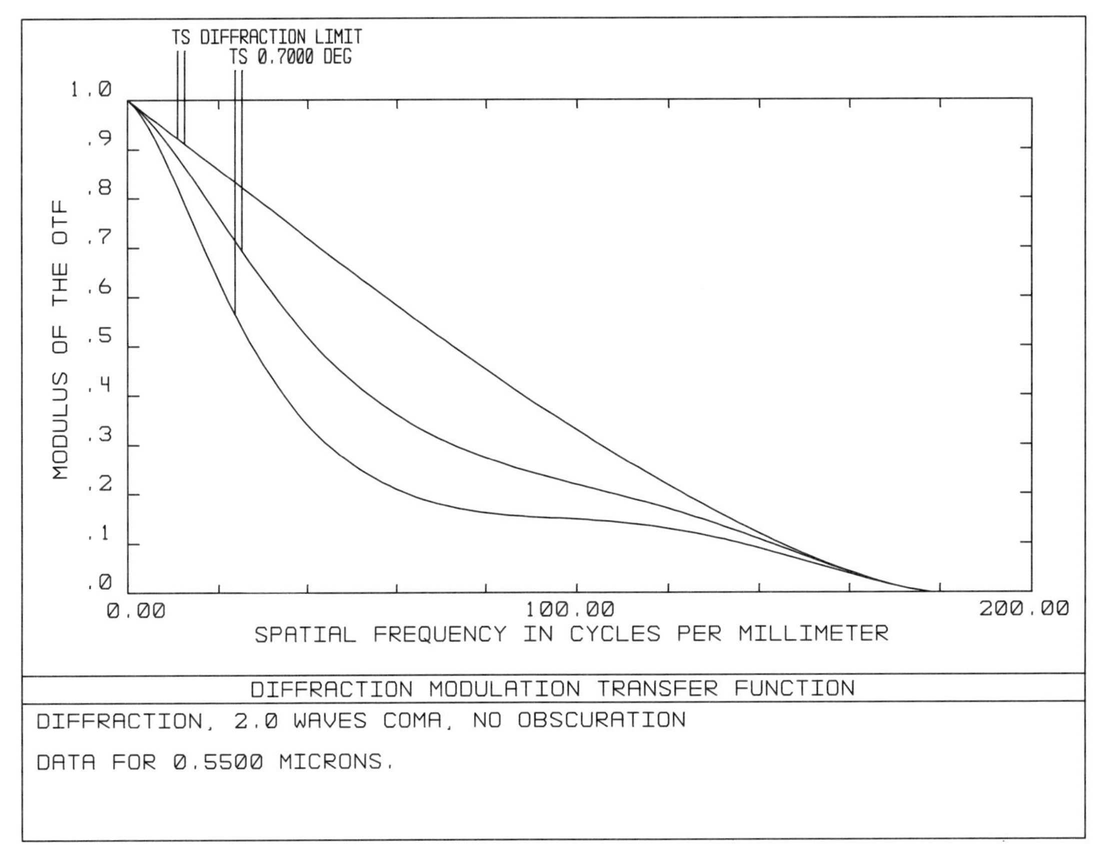

Figure A.12.6.1.

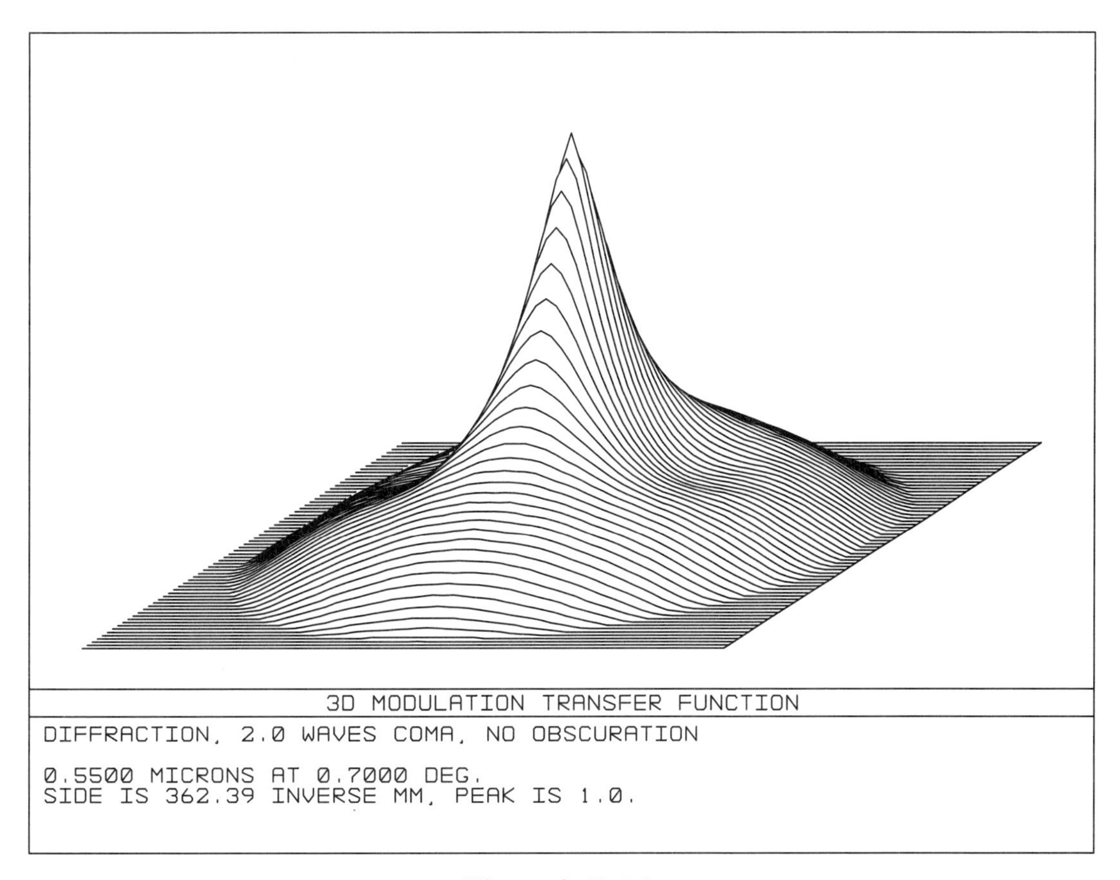

Figure A.12.6.2.

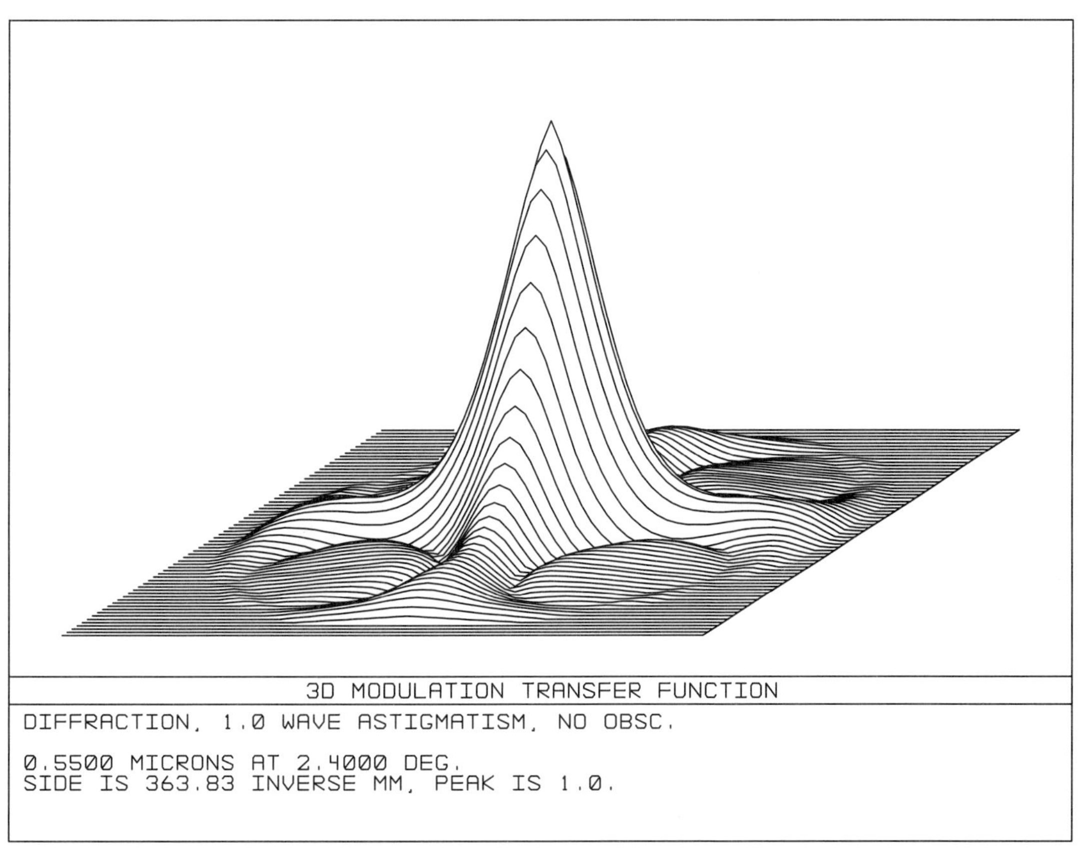

Figure A.12.7.

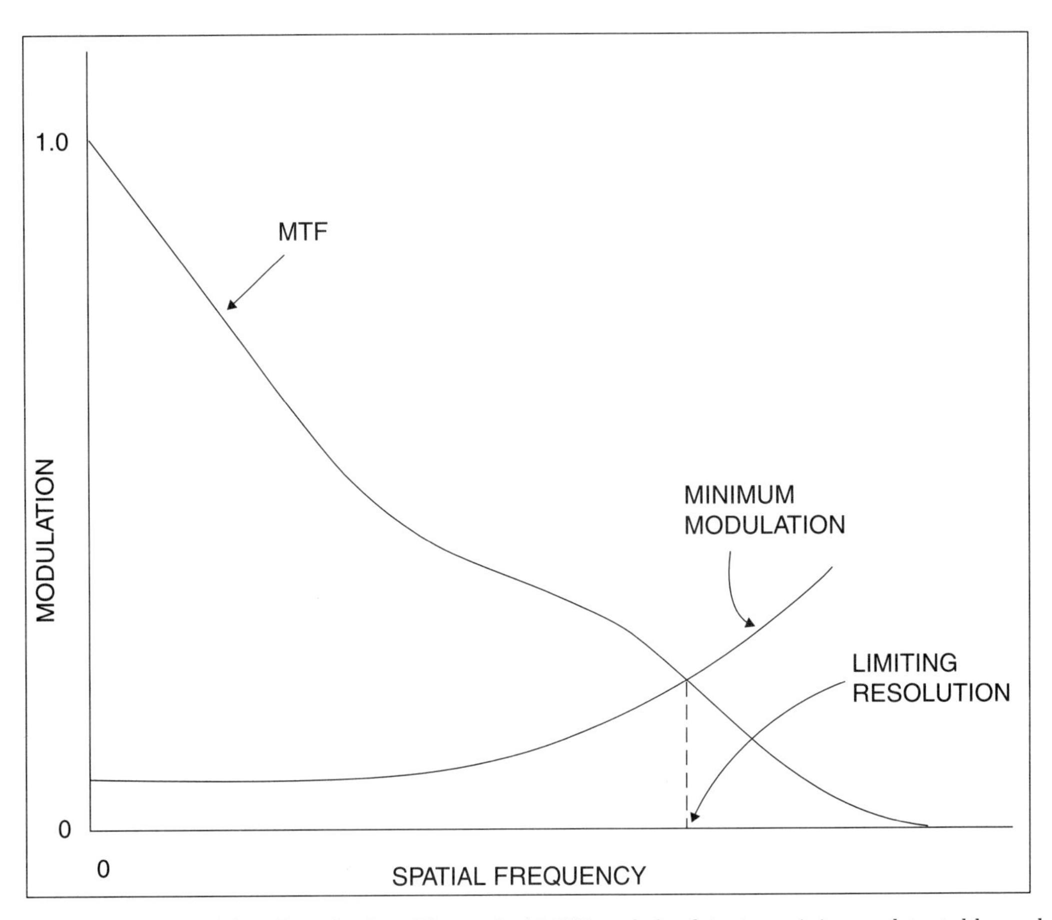

Figure A.12.8. Limiting Resolution. *The optical MTF and the detector minimum detectable modulation are plotted on the same graph. The point where the two curves cross is at the maximum resolved spatial frequency or limiting system resolving power.*

expected based on their Strehl ratios as given in the previous chapter. The coma example has a Strehl ratio of 0.58, whereas the astigmatism example has a Strehl ratio of only 0.20.

A.12.8 Minimum Detectable Modulation and Limiting Resolving Power

Actual image detectors cannot respond to extremely small values of image modulation. Some minimum modulation must be present for detection. This minimum modulation is very much a function of the detector and how it is being used. In addition, the minimum detectable modulation varies with spatial frequency. In general, higher spatial frequencies require greater modulations for detection.

If the detector's minimum detectable modulation is plotted as a function of spatial frequency, and if the optical MTF curve is plotted on the same graph, then the point where the two curves cross is located at the highest spatial frequency that the system can record. This system cutoff frequency is called the limiting resolving power of the lens-detector combination, and is illustrated in Figure A.12.8.

Note that limiting resolving power is always somewhat less than the diffraction-limited MTF cutoff frequency given by Equation A.12.4. In many cases, aberrations, detector properties, and other practical problems cause the actual limiting resolution to be much less.

When designing a lens, sometimes the designer is given the minimum detectable modulation curve or can determine it. More often, only an estimate can be made. In the absence of specific knowledge, a good rule of thumb is that the cutoff frequency is near the point where the MTF curve first drops below 10%.

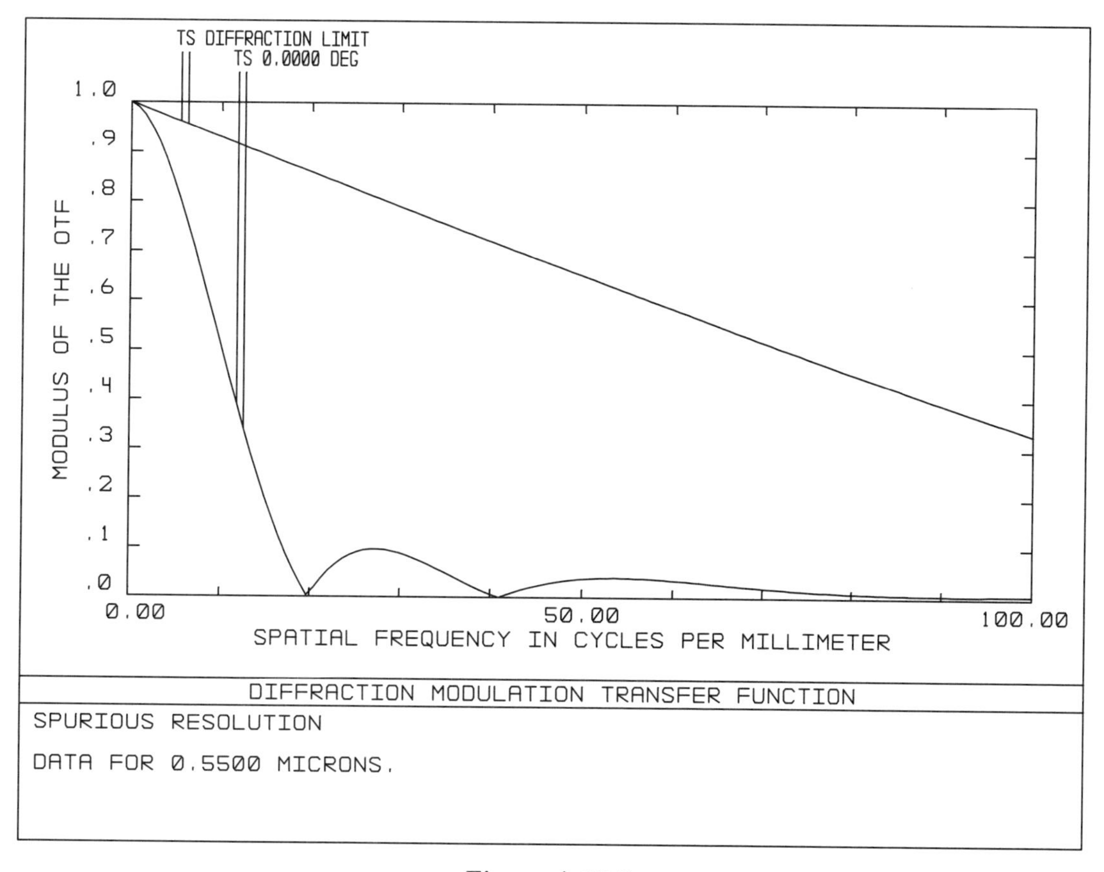

Figure A.12.9.

A.12.9 Spurious Resolution

The lens designer will often find that he obtains MTF curves that descend to zero and then rise again to substantial modulation values at higher spatial frequencies. Sometimes, an MTF curve will go to zero more than once. The modulations beyond the first zero of MTF are known as spurious resolution. Spurious resolution is an artifact of the periodic nature of the test pattern. No useful information about a general object can be transferred to the image for spatial frequencies beyond the first zero of MTF. This is true both for analytically calculated MTF curves and for experimentally measured curves.

Figure A.12.9 is the MTF plot for the same defocused lens used for Figure A.12.4, except the amount of defocus has been increased to 1.6 waves peak-to-valley. Note that the MTF curve goes to zero twice. All modulations beyond the first zero near 20 cycles/mm are spurious. Spurious resolution can also be seen in Figure A.12.7.

It is interesting to look at photographs of a resolution test pattern (either sine wave or square wave) imaged by a lens with spurious resolution. If spurious resolution has set in, then there will be one less target stripe in the image than in the object. The Air Force three-bar test target will appear transformed into a two-bar target. In other words, there is an apparent contrast reversal, with a 180° phase shift in the PTF.

If there is more than one zero of MTF, then the situation becomes even more complicated. But these details do not matter for the lens designer. What is important is to disregard all modulations beyond the first zero of MTF. If in doubt when measuring MTF experimentally, count the target stripes to make sure that all of them are there in the image.

A.12.10 Nyquist Frequency

The terms Nyquist criterion and Nyquist frequency are often encountered when considering the limiting resolving power of solid state television detectors (CCDs) that consist of a rectilinear grid of pixels. The Nyquist criterion states that full information in an imaged scene can be extracted only when the point spread function diameter occupies two pixels or more. In particular, a single pixel is not sufficient to define system resolution.

The Nyquist frequency (or limit) is the corresponding high spatial frequency response cutoff, and it is determined by the center-to-center spacing of the pixel grid. The finite grid mesh precludes the resolution of image spatial frequencies whose periods are less than twice the grid spacing. A one-dimensional version of the same effect is present perpendicular to the scan lines of a conventional television image.

If d is the center-to-center pixel spacing, P_N the Nyquist period, and f_N the Nyquist frequency, then,

$$f_N = \frac{1}{P_N} = \frac{1}{2d}$$

where d must be given in millimeters to get cycles/mm. If the pixels are rectangular rather than square, then the Nyquist frequency is different in different directions.

The Nyquist response cutoff is explained by the sampling theorem, which applies to all discrete or digital sampling of continuous or analog signals. The sampling theorem states that a continuous signal when discretely sampled (analog to digital) can be reconstructed (digital to analog) to only a limited extent. If the continuous signal is expressed by its Fourier transform frequency components (either temporal or spatial as the case may be), then for reconstruction, the theorem requires that a given Fourier frequency must have two or more samples per cycle (equal sample spacing). In other words, for imagery, you need two or more pixels per spatial frequency cycle. Higher (finer) frequencies with less than two samples (pixels) per cycle cannot be reconstructed. The Nyquist limit is the frequency having exactly the minimum two samples (pixels) per cycle.

In the real world, however, you can never actually reach the Nyquist frequency. Noise in the system causes things to start to fail as the theoretical limit is approached. Thus, you always need somewhat more than two samples per cycle for adequate reconstruction; that is, the practical limit is always somewhat less than the Nyquist frequency. Note that the issue of signal sampling is of vital importance in many fields, including the developing technology of hi-fi digital sound reproduction.

Chapter A.13

The Merit Function

There is a joke to the effect that lens designers do not design lenses any more, they now design merit functions. Although this is a great exaggeration, there is a grain of truth. In order to optimize a lens with one of the optical software packages, it is of critical importance that the designer be able to control the optimization process to shepherd the lens in the direction he wishes. The merit function is the means of exerting this control. In this chapter, the nature and construction of the merit function are discussed.

A.13.1 The Merit Function as a Measure of Optical Performance

The process of optimizing a lens involves entering a starting lens configuration into the computer, and then allowing the program to make changes in the lens to improve performance. Through an iterative process, performance is eventually optimized.

But to do this, the program must have a well-defined measure of what constitutes good performance. Furthermore, this measure must be expressible as a *single* number with no ambiguities. If the number gets larger, that means one thing, and if the number gets smaller, that means the opposite.

In optical design, this measure of performance is called either the merit function or the error function. In virtually all programs, the merit/error function is actually a demerit function; that is, a measure of how *bad* the lens is. The merit function for a perfect lens has a value of zero, and aberrations and other errors increase the value. Thus, the term error function is more appropriate. Nevertheless, from tradition and wide usage, the term merit function is adopted here.

During optimization, the computer program attempts to reduce the value of the merit function to a small number as close to zero as possible. But in most cases, the final value never reaches zero because residual aberrations remain.

However, some specially constructed merit functions can be exactly satisfied and have an optimized value of zero, even though the lens is not perfect. In these cases, the lens is merely doing the limited things you asked it to do.

A.13.2 The Constituents of the Merit Function

A lot of things can go into the merit function, and the merit function can evolve during the design process to reflect changing priorities and unexpected problems. But the main constituents are various measures of:

1. How well first-order properties are satisfied,

2. How well any special constraints are satisfied, and

3. How well aberrations are controlled.

First-order properties include paraxial and Gaussian parameters such as focal length, focal ratio, and the locations of pupils and images. Special constraints control the exact value or allowable range of particular optical and mechanical parameters. For example, special constraints might be placed on element center and edge thicknesses, overall length, surface curvatures, glass properties, and so forth.

Aberrations, of course, refer to image sharpness and distortion. Image sharpness is commonly measured by one of three quantities:

1. Spot size measured by ray-intercept errors in the image,
2. Wavefront imperfections measured by OPD (optical path difference) errors in the exit pupil, and
3. MTF (modulation transfer function) in the image.

Spot size and OPD errors can be incorporated into the merit function of any professional lens design computer program, and most can address MTF too. Each of these measures has advantages and disadvantages, and there are situations where one or another is more appropriate. Only one of these measures is normally used at a given time. The designer may also wish to directly control specific classical aberrations individually. Distortion, which is not related to image sharpness, must be controlled in this way.

A.13.3 Optimization Operands and Damped Least-Squares

The individual components of the merit function are called optimization operands. An operand is some function of rays, wavefronts, or various constructional parameters that you wish to control. The value that you wish an operand to assume is called a target. In many cases, although not all, the value of the target is zero.

When designing a lens, the number of operands is usually larger, often much larger, than the number of independent lens variables available. In this case, there is no solution, given the basic lens configuration, that allows all operands to exactly reach their targets simultaneously to give a merit function value of zero. The best you can do is to adjust the lens variables to allow the operands to approach their targets as closely as possible.

There are several ways to achieve this compromise, but in lens design, the most common approach is one adapted from statistics called damped least-squares. A least-squares optimization adjusts the lens variables so that the sum of the squares of the (weighted) errors for all the operands in the merit function is a minimum.

The process is iterative. In each cycle, the computer evaluates the merit function and then makes changes to the lens to reduce the merit function. To maintain approximate mathematical linearity, the sizes of the changes must not be too large. To prevent overshoot and oscillation, the damping further reduces the step size as the solution is approached. After a number of cycles, the process converges on the optical configuration giving a least-squares minimum of the merit function. For the given starting lens configuration and the given merit function, the process has then gone as far as it can go.

If further improvements are necessary, then the lens designer must intervene, modify the merit function, and restart the iterations. By this procedure, the final optimal lens configuration is eventually reached. If this design is found lacking, then a different starting configuration must be selected and the optimization procedure repeated.

A.13.4 Weighting Operands and Lagrange Multipliers

Some operands in the merit function are usually more critical than others; that is, it is desirable that after optimization, their values have relatively less departure from their targets. To accomplish this, the lens designer gives these operands higher (or heavier) weights. Higher weights give these optimization operands proportionately more importance in the least-squares solution.

In addition, there may be a few operands that must be *exactly* satisfied; that is, they must have almost zero final error. This correction can be accomplished in two ways. The first way is to give these operands very heavy weight in the least-squares solution. A very heavy weight calls attention to even a small error.

The second way to correct operands exactly is with Lagrange multipliers, an option available in most lens design programs. Using a Lagrange multiplier with a given operand always forces the optimization process to correct that operand exactly (in general, of course, your lens must have the geometry and degrees of freedom to allow the solution). Lagrange multipliers can be convenient because you do not have to fuss with finding the operand weight that gives the required close solution. All too often when correcting something close in a least-squares optimization, if the weight is too small, the program ignores the operand (or gives a big error), and if the weight is too large, the program ignores everything else.

Note, however, that Lagrange multipliers optimize without regard to the consequences to the rest of the system. Thus, you must use Lagrange multipliers carefully, especially with operands whose values are initially far from their targets. If the optimization computations "blow up," stall or stagnate (go nowhere), or otherwise fail at the beginning of a series of iterations, switch to weighted operands. Later, when the lens is closer to a final solution, you can switch back to Lagrange multipliers. How you use Lagrange multipliers will often depend on the individual characteristics of your program. Feel free to experiment.

A.13.5 Weighting Fields and Wavelengths

Optimization operands are not the only things that must be weighted by the lens designer. The relative importance of the several field positions and wavelengths must also be defined by assigning weights. The wavelength weights are used in many of the program's image evaluation options, such as spot diagram, point spread function, MTF, and so forth. But more relevant for optimization, the program uses both the field and wavelength weights to help adjust the weights of the constituents of a default merit function (see the following section for more on default merit functions). The judicious selection of the field and wavelength weights is often crucial for a successful system optimization.

A.13.6 Built-in Operands and Default Merit Functions

In the early "automatic" lens design programs, there were no built-in or default merit functions. There were only a few built-in optimization operands that the designer could use to construct a custom merit function. These operands were used either singly or in combination to address specific aberrations or system properties in specific ways. For example, to control aberrations, first the lens designer defined a set of rays. Then he selected a set of operands that used these rays. Finally, for special user-defined operands, he combined the user-selected operands.

Note the terminology. A user-selected operand is built into the program for the user to select as he wishes. A user-defined operand is not built into the program and must be constructed by the user out of parts that are built into the program.

Today, in addition to an extended assortment of optimization operands that can be user-selected, most lens design programs contain one or more ready-made default merit functions. These default merit functions are omnibus merit functions that address general image quality. Typically, a default merit function consists of a long set of operands whose purpose is to shrink spot sizes or reduce OPDs during optimization. If the designer wishes to include a default merit function as part of his total merit function, he instructs the program to construct the required operands and append them to his specially selected or defined operands. Special operands are still needed to control focal length, distortion, system constraints, particular aberrations, and so forth.

The specific construction of a default merit function is determined by which options the lens designer chooses and by the relative weights he assigns to the fields and wavelengths. For example, shrinking spot size is often truly the default, and an option controls whether the program switches to minimizing OPDs. Similarly, placing a greater weight on a certain field or wavelength increases its importance, and a weight of zero effectively turns it off.

Note that when an option is changed or when any of the field or wavelength weights are changed, the default merit function must be reconstructed. Depending on your program, this reconstruction is either automatic or must be initiated by the user.

A bigger difference is that, in some lens design programs such as ZEMAX, the constituents of the default merit function are visible and explicit. In other programs, such as CODE V, many of the details are hidden and implicit. In all cases, consult your program's user's manual for operating instructions.

In the next three sections, optimizing with merit functions based on spot size, OPD, and MTF is discussed. This is followed by an extended discussion of user-selected and user-defined optimization operands.

A.13.7 Optimizing with RMS Spot Size

For an object point radiating an array of rays covering the entrance pupil of a lens, the collective result of the ray piercing points on the image surface is a spot diagram, a measure of the geometrical point spread function (PSF). If compact, purely geometrical PSFs are of prime interest, then a merit function based on minimizing spot sizes is a very good approach to optical optimization.

The most common measure of the size of a spot is RMS spot size. For each ray in a spot, the radial distance of the ray from the chief ray is calculated and squared. For all rays in a spot, these squared distances are summed and then divided by the number of rays to get mean-squared spot size. Finally, the square root is taken to get root-mean-squared (RMS) spot size, with units of length such as millimeters. For highly asymmetric spots, some programs allow the reference point in a spot to be the image centroid instead of the chief ray.

If the system is polychromatic, then several monochromatic spots are combined, suitably weighted and normalized, in the merit function. The RMS spot size of a polychromatic spot is calculated relative to the chief ray for the reference wavelength (or relative to the polychromatic centroid). Similarly, if the system has more than one object point in the field of view and/or multiple configurations (such as zoom positions), then the various spots are again combined, suitably weighted and normalized. In most lens design computer programs, these details are handled automatically.

At the beginning of the optimization process, it is advisable to use rays and shrink spot size first. This is because the spot size optimization algorithm in the computer program is more robust than the OPD and MTF algorithms. In other words, when aberrations are huge as they will be initially, the spot size algorithm is less likely to fail. After this spot size optimization, the result will be small wavefront slope errors, and these are always accompanied by well-controlled (but not the smallest) wavefront OPD errors. If the smallest OPD errors are desired, then the lens can be reoptimized with an OPD merit function as a second step.

A.13.8 Optimizing with OPD Errors

Make no mistake, there is nothing wrong with achieving tiny spot sizes. And for systems with large aberrations where diffraction is negligible, spot size is an accurate measure of image spread. But the image quality of most optical systems is judged today, not by geometrical spot size, but by the physical measures of diffraction PSF or diffraction MTF. This is especially true if aberrations are small relative to diffraction and spot size loses its physical significance.

Recall that diffraction PSF and MTF are determined by the pupil function (Equation A.12.3), and the pupil function is determined by wavefront OPD errors. Thus, optimizing with OPDs in the exit pupil is a more fundamental approach than optimizing with transverse ray-intercept errors on the image surface.

An OPD figure of merit is constructed by a process very similar to the process used to make a spot size figure of merit. The same array of real rays is traced through the system. But now for each ray, the OPD error is calculated instead of the ray-intercept error. The OPD error is the optical path length along the ray between the actual wavefront in the exit pupil and the ideal (aberration-free) wavefront in the exit pupil whose center of curvature is at the image. The OPD figure of merit is the RMS average of the OPDs for all rays emerging from the exit pupil. For multiple wavelengths, multiple object points, and multiple configurations, OPDs are combined, suitably weighted and normalized, to give the final OPD figure of merit.

Recall that even though OPD errors have been made small, it is still possible to have undesirably large wavefront slope errors. This phonograph groove effect is caused by small amplitude but high spatial frequency OPD errors. To prevent this situation, some optical design programs have an optimization option that allows wavefront slope errors to be combined with OPD errors. The amount that slope errors are factored into the OPD merit function can be adjusted from zero to some maximum contribution. In CODE V, this option is controlled by the WVB command. The phonograph groove effect is especially likely if you are designing with high-order polynomial aspheric surfaces. Thus, use the lowest possible polynomial orders that still do the job. The optical shop will also appreciate this.

If the lens is to be used stopped down much of the time, then the central part of the wavefront in the exit pupil is arguably more important than the outer part. Thus, during optimization, you may wish to weight the center of the pupil more heavily than the outer zones. This idea applies to both OPD and spot size merit functions.

Emphasizing the center of the exit pupil during optimization also has the effect of producing PSFs with concentrated, tight cores and diffuse, extended wings. This combination yields images with lower contrast (lower low-frequency MTFs) but higher limiting resolution (higher high-frequency MTFs). For photographic lenses, this trade-off is usually preferred. However, for video lenses for TV detectors whose limiting (Nyquist) cutoff frequency is well below the resolving power of film, good high-frequency MTF values are useless and irrelevant. Thus, photographic and video lenses are usually not interchangeable.

Many camera lenses are focused wide open and are then stopped down to take the picture. When stopping down these lenses, it is very undesirable to have a noticeable focus shift caused by excessive zonal spherical aberration. Thus, in the centers of the fields of these lenses where focusing is done, controlling zonal spherical aberration with transverse rays may be preferable to reducing OPD errors. However, in the outer parts of the field, reducing OPDs may be more effective in raising MTF levels. Such a hybrid merit function may seem unusual, but it works. The reason is probably that the on-axis aberrations are so much simpler than the off-axis aberrations. The only on-axis options available to the lens designer are usually the pupil zone for correcting longitudinal color, the pupil zone for correcting zonal spherical, and occasionally defocus.

Again, do not use OPDs for initial optimization runs. The algorithm all too often will fail or find poor solutions. Shrink spot size first and then reoptimize with OPDs.

A.13.9 Optimizing with Modulation Transfer Function

Many optical design computer programs now have the option of including diffraction MTF itself in the merit function. If MTF is the final image quality criterion, then there may be advantages in optimizing MTF directly. During optimization, MTF operands are defined and target values are set. Departures from these target MTF values are the errors in the merit function.

In practice, it is usually hard to control a direct MTF optimization. Optimizing with MTF is not a matter of simply asking the program to give the highest MTF values. The spatial frequencies to be used and the corresponding target MTF values must both be determined before optimization, and this selection process may be difficult and uncertain. An unfortunate guess leads to a poor solution.

The more reliable and theoretically justified way to maximize MTFs is to minimize OPDs. The OPD approach has an additional advantage: the MTF computations in some programs take an exceedingly long time, even with a fast computer. Thus, optimizing with MTF should be reserved for unusual situations with non-standard requirements.

Finally, optimizing with MTF should never be considered until the lens has already been thoroughly optimized geometrically by first shrinking spots and then reducing OPDs. An MTF optimization is strictly a final touch-up procedure.

A.13.10 Optimizing with User-Selected and User-Defined Operands

Although it is usually preferable to optimize with either spot size or OPD (or occasionally MTF), there are times when doing an old-fashioned optimization using only user-selected and user-defined operands has advantages. For one thing, this approach is very direct and affords great control. The analogy is a rifle rather than a shotgun. For another thing, this approach uses very few rays and is thus very computationally efficient. This efficiency was especially important in previous years when computer time was expensive. But even today, there are lengthy, computationally intensive design jobs where speed is still important, such as global optimization. In these cases, highly efficient merit functions consisting entirely of user-selected and user-defined operands may give better or quicker results.

But the most common use of user-selected and user-defined optimization operands is in supplementing a default merit function to allow precise aberration and system control. Perhaps the most valuable aberration-control operands address longitudinal color, spherical aberration, coma, and distortion. System-control operands address first-order properties and special constraints. A special constraint might restrict the sign of the curvature of a lens surface to only positive or only negative values. Or the relative sizes of the airspaces in a Cooke Triplet might be controlled to select one of two possible solutions. Maybe the joke about designing merit functions rather than lenses is no joke after all.

A.13.11 Examples of User-Selected and User-Defined Optimization Operands

Listing A.13.1 is an example list of user-selected and user-defined optimization operands that the lens designer would enter into the computer by hand in order

Listing A.13.1

Merit Function Listing
Title: TESSAR, 52MM, f/3.5, 45.2DEG

Merit Function Value: 1.58615479E-003

Num	Type	Int1	Int2	Hx	Hy	Px	Py	Target	Weight	Value	% Cont
1	EFFL		2					5.20000E+001	1	5.20005E+001	0.001
2	BLNK										
3	BLNK										
4	RGLA	2	8					5.00000E-002	0	5.00000E-002	0.000
5	BLNK										
6	BLNK										
7	MXCG	2	8					1.50000E+001	0	1.50000E+001	0.000
8	MNCG	2	8					1.00000E+000	1	1.00000E+000	0.000
9	MNEG	2	8					1.00000E+000	1	1.00000E+000	0.000
10	BLNK										
11	MXCA	2	8					1.50000E+001	0	1.50000E+001	0.000
12	MNCA	2	8					2.00000E-001	1	2.00000E-001	0.000
13	MNEA	2	8					1.00000E-001	1	1.00000E-001	0.000
14	BLNK										
15	BLNK										
16	TTHI	2	8					0.00000E+000	0	2.01724E+001	0.000
17	OPLT	16						2.50000E+001	0	2.50000E+001	0.000
18	BLNK										
19	BLNK										
20	BLNK										
21	AXCL							0.00000E+000	0	1.15163E-001	0.000
22	REAY	11	1	0.0000	0.0000	0.0000	0.8000	0.00000E+000	0	1.82077E-003	0.000
23	REAY	11	3	0.0000	0.0000	0.0000	0.8000	0.00000E+000	0	1.73808E-003	0.000
24	DIFF	22	23					0.00000E+000	1000	8.26943E-005	0.043
25	BLNK										
26	BLNK										
27	LACL							0.00000E+000	0	2.28387E-002	0.000
28	REAY	11	1	0.0000	1.0000	0.0000	0.0000	0.00000E+000	0	2.16446E+001	0.000
29	REAY	11	3	0.0000	1.0000	0.0000	0.0000	0.00000E+000	0	2.16443E+001	0.000
30	DIFF	28	29					0.00000E+000	1000	2.76058E-004	0.482
31	BLNK										
32	BLNK										
33	SPHA	0	2					0.00000E+000	0	3.02768E+000	0.000
34	REAY	10	2	0.0000	0.0000	0.0000	0.9000	0.00000E+000	1000	1.22810E-005	0.001
35	BLNK										
36	BLNK										
37	COMA	0	2					0.00000E+000	0	-6.14952E+000	0.000
38	TRAY		2	0.0000	0.7000	0.0000	0.5000	0.00000E+000	0	1.80950E-002	0.000
39	TRAY		2	0.0000	0.7000	0.0000	-0.5000	0.00000E+000	0	-1.31926E-002	0.000
40	TRAY		2	0.0000	0.7000	0.0000	0.7000	0.00000E+000	0	4.01039E-002	0.000
41	TRAY		2	0.0000	0.7000	0.0000	-0.7000	0.00000E+000	0	-3.45457E-002	0.000
42	TRAY		2	0.0000	1.0000	0.0000	0.3000	0.00000E+000	0	-5.61701E-004	0.000
43	TRAY		2	0.0000	1.0000	0.0000	-0.3000	0.00000E+000	0	7.11241E-003	0.000
44	TRAY		2	0.0000	1.0000	0.0000	0.5000	0.00000E+000	0	1.46606E-002	0.000
45	TRAY		2	0.0000	1.0000	0.0000	-0.5000	0.00000E+000	0	1.07596E-003	0.000
46	SUMM	38	39					0.00000E+000	10	4.90238E-003	1.520
47	SUMM	40	41					0.00000E+000	10	5.55824E-003	1.954
48	SUMM	42	43					0.00000E+000	10	6.55071E-003	2.714
49	SUMM	44	45					0.00000E+000	10	1.57366E-002	15.661
50	BLNK										
51	TRAY		2	0.0000	0.7000	0.7000	0.0000	0.00000E+000	60	1.26409E-003	0.606
52	TRAY		2	0.0000	0.7000	1.0000	0.0000	0.00000E+000	60	1.24217E-003	0.585
53	TRAY		2	0.0000	1.0000	0.5000	0.0000	0.00000E+000	60	-5.99592E-003	13.642
54	TRAY		2	0.0000	1.0000	0.7000	0.0000	0.00000E+000	60	-1.25932E-002	60.176
55	BLNK										
56	BLNK										
57	ASTI	0	2					0.00000E+000	0	-1.57204E+001	0.000
58	TRAY		2	0.0000	0.9500	0.0000	0.1000	0.00000E+000	0	3.73989E-004	0.000
59	TRAY		2	0.0000	0.9500	0.0000	-0.1000	0.00000E+000	0	-8.26761E-006	0.000
60	TRAX		2	0.0000	0.9500	0.1000	0.0000	0.00000E+000	0	-8.31593E-005	0.000
61	TRAX		2	0.0000	0.9500	-0.1000	0.0000	0.00000E+000	0	8.31593E-005	0.000
62	DIFF	58	59					0.00000E+000	0	3.82257E-004	0.000
63	DIFF	60	61					0.00000E+000	0	-1.66319E-004	0.000
64	DIFF	62	63					0.00000E+000	1000	5.48575E-004	1.903
65	BLNK										
66	BLNK										
67	FCUR	0	2					0.00000E+000	0	2.26576E+001	0.000
68	PETZ		2					0.00000E+000	0	-1.91898E+002	0.000
69	TRAY		2	0.0000	0.8000	0.0000	0.1000	0.00000E+000	0	2.25906E-003	0.000
70	TRAY		2	0.0000	0.8000	0.0000	-0.1000	0.00000E+000	0	-2.08064E-003	0.000
71	TRAX		2	0.0000	0.8000	0.1000	0.0000	0.00000E+000	0	-2.33737E-003	0.000
72	TRAX		2	0.0000	0.8000	-0.1000	0.0000	0.00000E+000	0	2.33737E-003	0.000
73	DIFF	69	70					0.00000E+000	0	4.33970E-003	0.000
74	DIFF	71	72					0.00000E+000	0	-4.67473E-003	0.000
75	SUMM	73	74					0.00000E+000	1000	-3.35032E-004	0.710
76	BLNK										
77	BLNK										
78	DIMX	0	2					7.50000E-001	0	7.50000E-001	0.000
79	DIST	0	2					0.00000E+000	1000	-1.22345E-005	0.001
80	REAY	11	2	0.0000	0.8000	0.0000	0.0000	0.00000E+000	0	1.69580E+001	0.000
81	PARY	11	2	0.0000	0.8000	0.0000	0.0000	0.00000E+000	0	1.69763E+001	0.000
82	DIFF	80	81					0.00000E+000	0	-1.82880E-002	0.000
83	DIVI	82	81					0.00000E+000	0	-1.07727E-003	0.000
84	CONS							1.00000E+002	0	1.00000E+002	0.000
85	PROD	83	84					0.00000E+000	0	-1.07727E-001	0.000
86	BLNK										
87	BLNK										
88	DMFS										

to be used. Contained here are many of the operands that the author has found to be especially useful at one time or another. However, the list is by no means definitive and should be viewed as only suggestions. The format is that of the ZEMAX program, but the concepts are applicable to any professional lens design program.

Note that you can make (and evolve) an all-purpose collection of your favorite optimization operands such as these before doing a design. Later when doing the design, you can use the collection like an optical smorgasbord. You can select whichever operands you wish for the current application and leave the rest turned off or deleted.

The lens in these operand examples is nearly identical to the Tessar camera lens shown in Figure A.6.4. Surface 1 is a dummy plane surface in front of the lens to control the lengths of rays drawn in front of the lens on layouts. Surface 2 is the first lens surface, surface 6 is the stop surface, and surface 9 is the last lens surface. Surface 10 is a dummy plane surface at the paraxial focus, and surface 11 is the actual image surface. Recall the use of both the paraxial and actual image surfaces from Chapter A.7 and Figure A.7.2.1. In the present case, the thickness between surfaces 10 and 11 is zero to keep the actual image surface at the paraxial focus.

Three wavelengths, numbered 1, 2, and 3 in order of increasing size, are assumed to be specified elsewhere in the program. The middle wavelength, number 2, is made the reference wavelength that is used in controlling first-order properties and monochromatic aberrations. The extreme wavelengths, numbers 1 and 3, are used for controlling longitudinal and lateral color. Entrance pupil diameter and field of view have also been specified elsewhere in the program, and thus object height and pupil height are entered in these operands as values normalized to ± 1.0.

In the ZEMAX format, Num is the sequence number of the operand. Type is the four letter code for the type of operand. The type codes are intended to be descriptive, such as DIFF for difference and OPLT for operand less than. Int1 (Integer 1) is usually, but not always, the lens surface number (with zero indicating a summation over all lens surfaces). Int2 (Integer 2) is usually the wavelength number. Hx and Hy are the x- and y-direction normalized object heights. Px and Py are the x- and y-direction normalized pupil heights. Int2, Hx, Hy, Px, and Py serve to define the ray used in this operand. Target is the ideal value of the operand; the damped least-squares computations try to drive the value of each operand toward its target value. Weight is the relative emphasis given to the operand in the computations. A weight of zero instructs the program to calculate the operand for later use or for reference, but to not include it in the least-squares solution. Value is the current numerical value of the operand. % Cont is the percent contribution of the operand to the merit function; here you can quickly see which operands are most difficult to control. Note that not all of these operand parameters are used for all operands. Each of the operands in Listing A.13.1 is described below, with more details in the ZEMAX user's manual. Incidentally, BLNK is merely a blank to visually separate different groups of related operands.

In operand 1, EFFL controls paraxial effective focal length. The wavelength is number 2, the target is 52 mm, and the weight is 1.

In operand 4, RGLA restricts the index, dispersion, and partial dispersion of fictitious variable glasses to be close to those of real available glasses. In the present lens, no variable glasses are specified and RGLA is not needed, so operand 4 is weighted zero (not selected from the smorgasbord).

In operands 7 through 13, the maximum and minimum values of center and edge thicknesses for glass and air spaces are controlled for surfaces 2 through 8. These controls prevent unrealistic lens configurations, such as excessively thick elements or negative edge thicknesses.

In operand 16, TTHI sums the center thicknesses associated with surfaces 2 through 8, and thus the operand gives the overall vertex length from surface 2 to

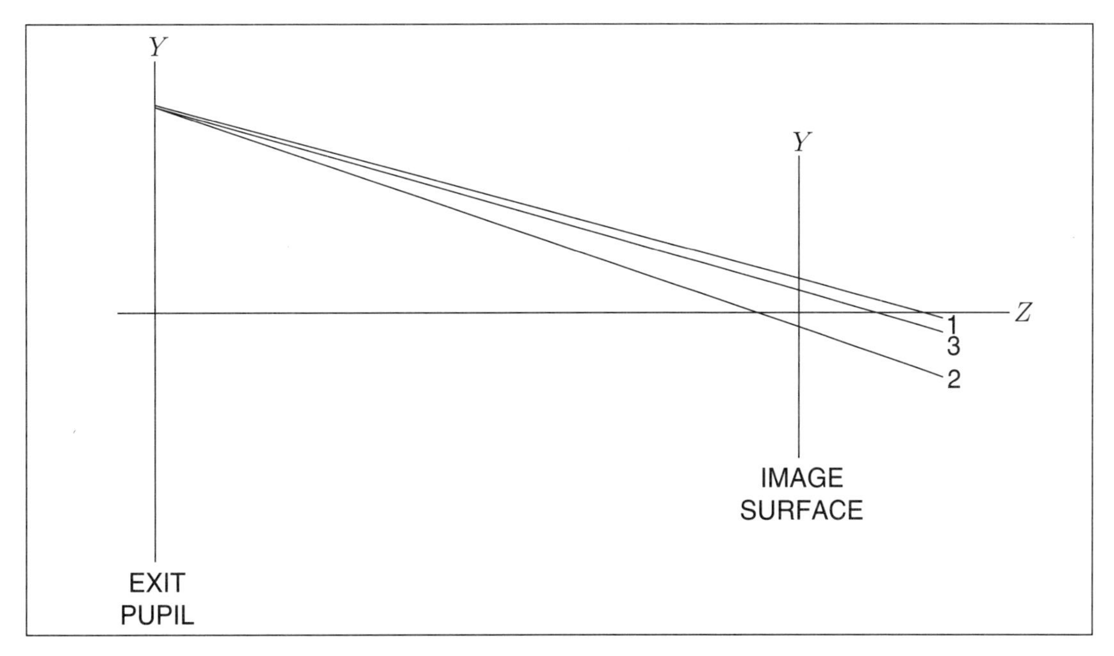

Figure A.13.1. Longitudinal Color. *To correct longitudinal color, the heights of the on-axis marginal rays for the blue (ray 1) and red (ray 3) wavelengths are made equal on the image surface.*

surface 9 (not 8). Operand 16 is weighted zero because it is not used directly but is calculated for use by operand 17. In operand 17, OPLT specifies that operand 16 is to be less than 25 mm. By this method, operands 16 and 17 together prevent the lens from becoming excessively long. The zero weight for operand 17 means that this operand is not needed for the moment and is calculated for reference.

A.13.12 Longitudinal Color

In operand 21, AXCL controls longitudinal (or axial) color, measured in linear units such as millimeters. For an object on the optical axis, AXCL calculates the difference between the first-order paraxial BFLs for wavelengths 1 and 3. To correct paraxial longitudinal color, the value of this operand is made zero during optimization. But if you wish to control longitudinal color for any other pupil zone, most programs do not supply the required operand and you must create your own.

To do this, real rays for wavelengths 1 and 3 are traced from the center of the object field, through the selected entrance pupil zone, and then to the image. Figure A.13.1 is a schematic meridional layout of image space. Rays 1 and 3 are shown emerging from the exit pupil and proceeding to and beyond the image surface. The ray labels correspond to wavelengths. Zonal ray 2 is also shown, but only for reference.

Because of longitudinal color, each of these three rays crosses the axis at a different distance from the lens. To correct primary longitudinal color, the BFLs of rays 1 and 3 are made equal. Any remaining difference between the BFLs of either rays 1 or 3 and ray 2 is secondary longitudinal color. Controlling secondary color is not addressed here.

Most lens design programs do not have optimization operands to directly control longitudinal ray properties in image space. Instead, their optimization operands control transverse ray-intercept errors on the image surface. To circumvent this limitation, note on Figure A.13.1 that if the image ray-intercept heights of rays 1 and 3 are made equal relative to the optical axis, then the BFLs of these two rays are also made equal. By this means, longitudinal color can be indirectly controlled by using transverse ray errors.

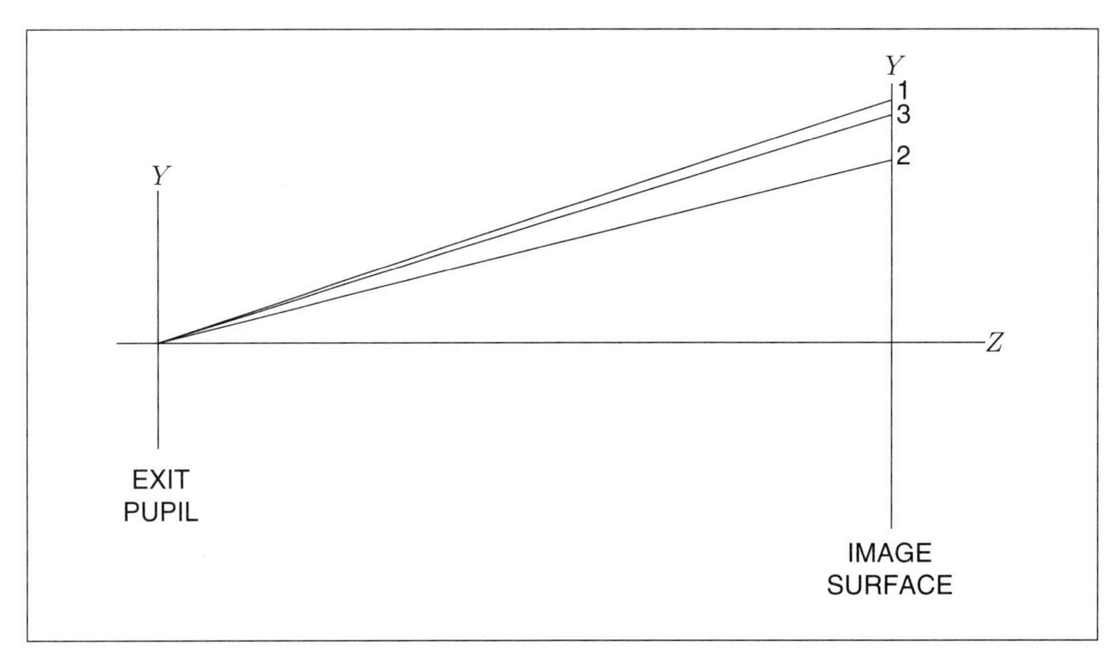

Figure A.13.2. Lateral Color. *To correct lateral color, the heights of the off-axis chief rays for the blue (ray 1) and red (ray 3) wavelengths are made equal on the image surface.*

Actually, the ray geometry in Figure A.13.1 is not quite correct because the location of the exit pupil and the ray heights on the exit pupil are slightly wavelength dependent. But for purposes of controlling longitudinal color with transverse operands, these effects are small and can be neglected.

Operands 22, 23, and 24 in Listing A.13.1 are the user-selected operands to be combined to create a user-defined operand (or group of operands) to control longitudinal color. Operands 22 and 23 are calculated for use by operand 24 and use the operand REAY. REAY calculates the y-direction intercept height of a real trigonometric ray on a lens surface relative to the optical axis. Int1 specifies surface 11, the image surface. Int2 specifies the wavelength; operand 22 uses wavelength 1, and operand 23 uses wavelength 3. The normalized (x, y) object coordinates are (0,0), the center of the field. The normalized (x, y) pupil coordinates are (0,.8); this specifies the desired pupil zone. Operand 24 takes the difference between the values of operands 22 and 23, and a target of zero is specified. A heavy weight is specified for operand 24 to force the optimization process to correct, not merely minimize, this aberration.

A.13.13 Lateral Color

In operand 27, LACL controls lateral color, measured in linear units such as millimeters. Two first-order Gaussian chief rays, one for wavelength 1 and one for wavelength 3, are traced by LACL from the edge of the object field, through the center of the entrance pupil, and finally to the Gaussian image plane tangent to the actual image surface. The height difference between these two colored chief rays on the Gaussian image plane is lateral color as measured by LACL. But if you have, for example, a curved image surface, then you may wish to create a custom lateral color operand that uses real rays.

A custom lateral color operand operates very similarly to LACL except that two real chromatic chief rays are traced and the actual image surface is used. Figure A.13.2 is a schematic layout of the resulting image space, with the ray labels corresponding to wavelengths. Ray 2, the chief ray for the reference wavelength, is also shown.

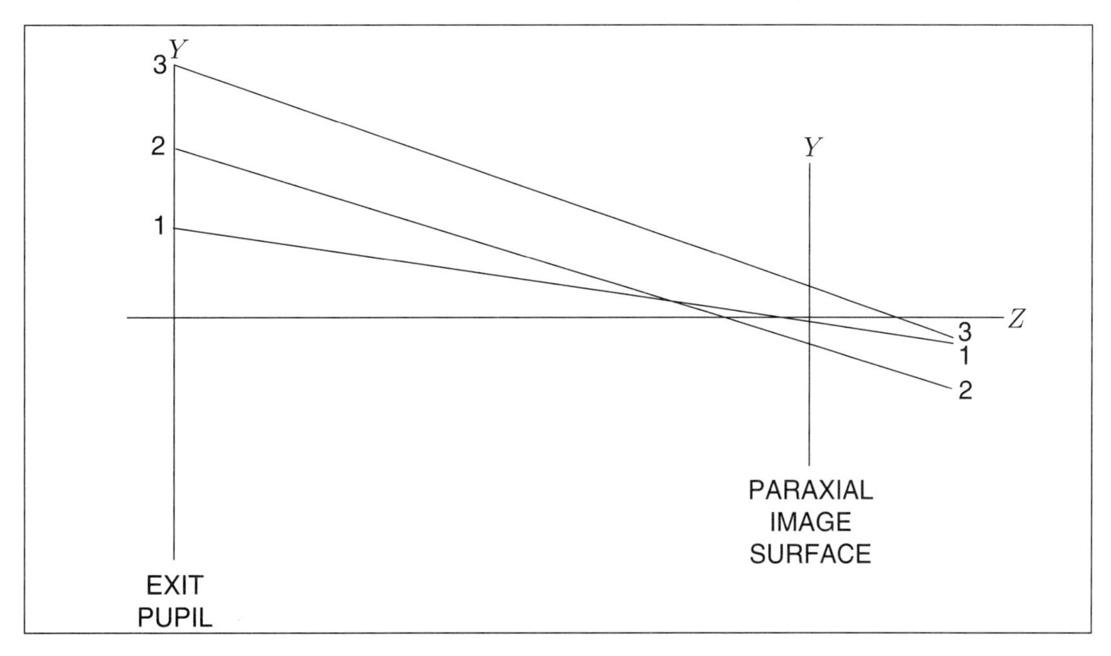

Figure A.13.3. Spherical Aberration. *To correct on-axis spherical aberration for a particular pupil zone and wavelength (such as with ray 2), the height of the ray is made zero on the paraxial image surface.*

Because of lateral color, each of these three rays intercepts the image surface at a different height from the optical axis. To correct primary lateral color, the image heights of rays 1 and 3 are made equal. Any remaining difference between the heights of either rays 1 or 3 and ray 2 is secondary lateral color. As with secondary longitudinal color, secondary lateral color in not addressed here.

Operands 28, 29, and 30 in Listing A.13.1 are the optimization operands needed to control lateral color. These are the same as operands 22, 23, and 24, except that the normalized (x, y) object coordinates are (0,1), and the normalized pupil coordinates are (0,0). As with operand 24, operand 30 is given a target of zero and a heavy weight.

A.13.14 Spherical Aberration

In operand 33, SPHA controls OPD third-order spherical aberration of the wavefront, measured in wavelengths in the exit pupil. The Int1 value of zero specifies that the aberration is summed over the entire system. Int2 specifies wavelength 2. However, if your lens has higher orders of spherical, then controlling just the third order is usually inadequate or awkward. Reducing trigonometric spot size or OPDs are ways of handling higher orders. But there are times when you may wish to make the sum of third and higher orders of spherical exactly zero at one particular pupil zone.

Figure A.13.3 is a schematic layout of image space that illustrates zonal spherical aberration. The paraxial image surface is used, and the lens has only undercorrected third-order and overcorrected fifth-order spherical aberration (seventh and higher orders are insignificant). The signs of the aberration orders are such that for inner pupil zones the rays focus too close, and for outer pupil zones the rays focus too far. Because spherical is a monochromatic aberration, all the rays shown in Figure A.13.3 use wavelength 2, and the numbers associated with the rays are now only labels for identification.

To correct spherical, a real ray is traced from the center of the object field, through the selected entrance pupil zone, and then to the image. During optimization, the height of this ray on the paraxial image surface is made equal to zero.

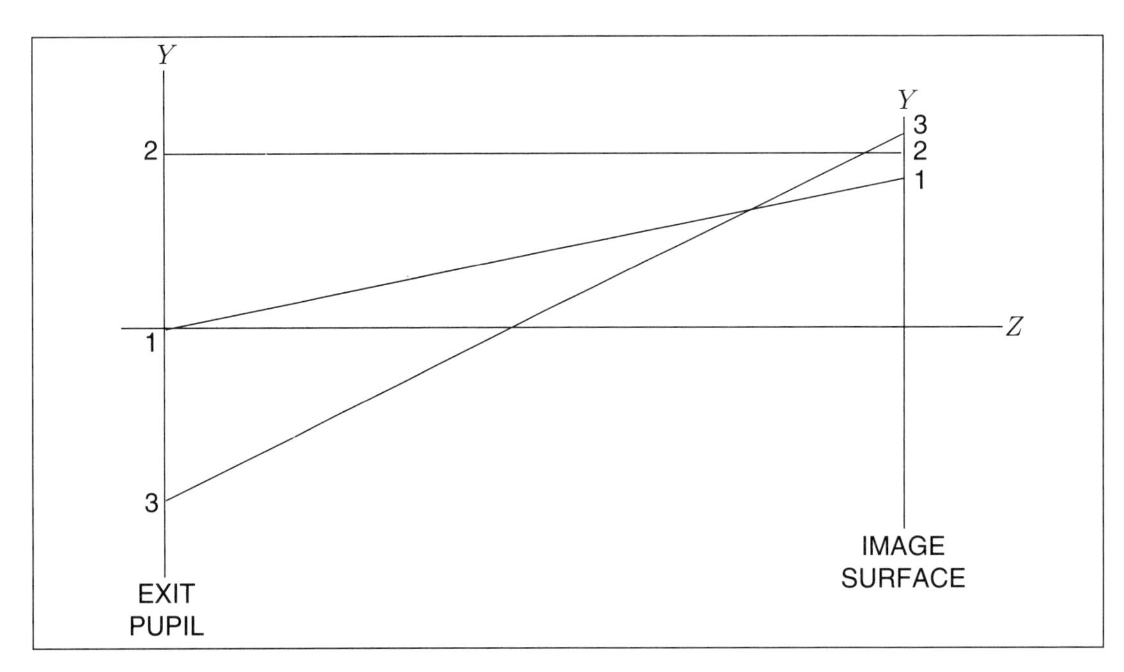

Figure A.13.4. Tangential Coma. *To correct tangential coma for a particular off-axis object height, pupil zone, and wavelength, the average image height of the two tangential rays passing through the top and bottom of the zone (rays 2 and 3) is made equal to the image height of the chief ray (ray 1).*

This can be done with just one REAY operand, operand 34 in Listing A.13.1. Int1 specifies surface 10, the paraxial image surface, not surface 11, the actual image surface. Int2 specifies wavelength 2. The (x, y) object height is (0,0), and the pupil zone height is (0,.9). A ray height target of zero is specified, and again a heavy weight is used.

Note that to isolate spherical aberration and prevent defocus from entering, the paraxial image surface must be used. This is another reason why it is good practice to include both the paraxial and actual image surfaces in the optical construction, even if the thickness between them is kept zero. Of course, if there is no paraxial defocusing, then it really does not matter which image surface you use.

After optimizing a lens in this way, look at the ray fan plots to see whether you selected the best pupil zone. If a different zone appears more promising, change the pupil height in operand 34 and reoptimize.

Incidentally, in lenses with significant seventh-order spherical, this special spherical operand can also be used to give a three term cancellation of the third, fifth, and seventh orders. However, the great majority of lenses have relatively little seventh-order spherical, and this possibility is rarely encountered. Furthermore, when seventh-order spherical is significant, it is usually more effective to shrink on-axis spot size and not use the special operand.

A.13.15 Tangential Coma

In operand 37, COMA controls OPD third-order tangential coma, measured in wavelengths in the exit pupil. The Int1 and Int2 values are the same as with SPHA above. But if you wish to control higher orders of coma with real rays, then a user-defined coma operand is needed.

Figure A.13.4 is a schematic layout of image space that shows three meridional tangential rays for a lens with coma (plus a small amount of defocus for generality). Like the rays in Figure A.13.3, all three rays in Figure A.13.4 use wavelength 2, and the numbers associated with the rays are only labels for identification. Note how rays 2 and 3, the upper and lower zonal rays, are not symmetrically arrayed about ray 1, the chief ray. To restore symmetry and correct tangential coma, an operand is created that during optimization makes the *average* height on the image surface of rays 2 and 3 equal to the height of ray 1. Equivalently, on the image surface, the chief ray is made to fall midway between the two zonal rays.

Note that when defocus is added to coma, the image intercept heights of rays 2 and 3 change in opposite directions, and the average height of the two rays is nearly unchanged. Thus, this tangential coma operand is insensitive to defocus, a property that can be very useful.

Operands 38 through 49 in Listing A.13.1 are the operands to control tangential coma for four different combinations of object and pupil height. Operands 38 through 45 trace eight rays in four pairs. These operands use the optimization operand TRAY. TRAY is similar to REAY, except that y-direction ray-intercept heights are measured relative to the chief ray, not the optical axis. In other words, these are differential heights, and the chief ray does not have to be traced separately. Int1 is not used for TRAY because the (actual) image surface is assumed. As before, Int2 specifies wavelength 2. The first two ray pairs are for an (x, y) object height of (0,.7) and for pupil heights of (0,±.5) and (0,±.7). The second two ray pairs are for an object height of (0,1) and for pupil heights of (0,±.3) and (0,±.5).

Operands 46 through 49 use the values of operands 38 through 45 to calculate coma. The differential ray heights for each pair of rays are summed. Then, in theory, the sum would be divided by two to get the average. However, the division by two is omitted here to simplify the list of operands. It is easier to compensate for the twice too large aberration values by reducing their weights by a factor of two during optimization.

Note that the pupil heights do not extend to the pupil edges. This was done deliberately because these operands were designed for use when optimizing camera lenses that have mechanical vignetting. When mechanical vignetting is present, off-axis exit pupils are clipped and reshaped by carefully chosen undersized clear apertures on some of the lens elements. By this means, some of the more severely aberrated rays need not be controlled but are simply removed. The optimization process can then concentrate on the remaining rays for a better overall solution. If a lens has no mechanical vignetting (or uses vignetting factors), then pupil heights of (0,±.7) and (0,±1) are used for both field points.

In the present example, it is desired to achieve a compromise design solution rather than correct coma to zero for just one field and pupil zone. Thus, all of the final coma operands (numbers 46 through 49) are included in the least-squares computations by giving each a nonzero weight. These weights must now be much less than the weights on the previous operands that were to be corrected exactly. Thus, a weight of 10 was (somewhat arbitrarily) chosen.

If your program has the feature that lets you see the percent contribution that each operand gives to the merit function, then you can adjust the various weights until you get reasonable contributions. Another way to adjust weights is to see if the aberration values themselves are reasonable after an optimization run. If you find that you are too far from your targets and not getting closer, increase the weights. Of course, there is always a limit beyond which the capabilities of your basic lens configuration cannot go.

A.13.16 Sagittal Coma

Although third-order tangential coma is always exactly three times larger than third-order sagittal coma, there are situations where controlling sagittal coma is advantageous. The right-left (bilateral) symmetry of axially centered lenses means that the right and left sagittal rays are also symmetrical. Thus, sagittal coma can be controlled with an operand using only two rays, one of the sagittal rays plus the chief ray, rather than three tangential rays as with tangential coma. In addition,

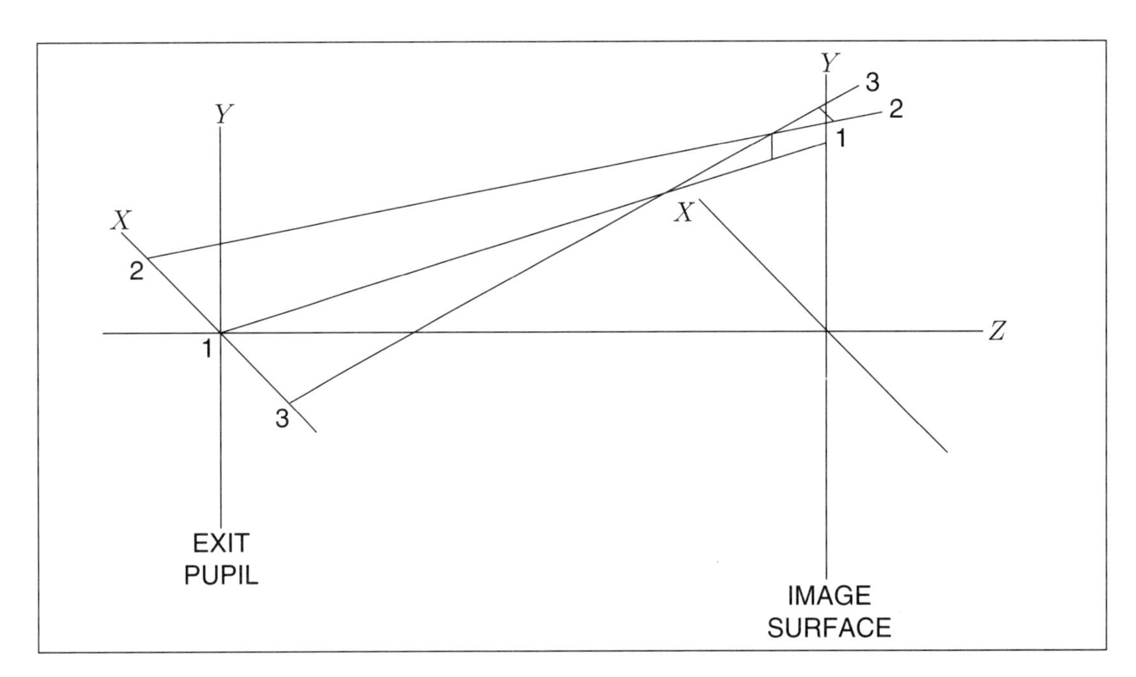

Figure A.13.5. Sagittal Coma. *To correct sagittal coma for a particular off-axis object height, pupil zone, and wavelength, the y-direction image height of a sagittal skew ray passing through either the right or left side of the pupil (either ray 2 or 3) is made equal to the image height of the chief ray (ray 1). Note the right-left (bilateral) symmetry that allows only one of the sagittal rays to be used.*

there are times when higher-order off-axis aberrations of all sorts are present, and the ray-based tangential and sagittal coma operands are not redundant.

Figure A.13.5 is a schematic layout of image space that illustrates sagittal coma. Note that whereas all of the previous layouts in this chapter have been two-dimensional meridional layouts, Figure A.13.5 is a three-dimensional representation. Three rays are shown. Ray 1 is the chief ray. Recall that the chief ray always lies in the intersection of the orthogonal tangential and sagittal planes. Rays 2 and 3 are the symmetrical sagittal rays. Again, all three rays use wavelength 2.

Rays 2 and 3 are defined to lie in the sagittal plane in the space between the object and the lens. However, because of coma, these two rays no longer lie precisely in the sagittal plane after passing through the lens. Instead, their trajectories are inclined slightly above (or below) the sagittal plane, and they focus together directly above (or below) the chief ray. This focus point actually lies in the tangential plane. In other words, when in focus, these two rays each have a nonzero y-direction image error, but a zero x-direction image error. In Figure A.13.5, the location of this focus is indicated by the short vertical or y-direction line connecting the focus with the chief ray.

If the image surface does not coincide with this sagittal focus, then defocus is added to coma. But because sagittal rays are arrayed in the x-direction relative to the chief ray, defocus introduces mainly x-direction ray displacements to sagittal rays. Thus, adding defocus to coma causes very little change in the y-direction image ray-intercept errors of sagittal rays. As with tangential coma, sagittal coma is also insensitive to defocus. In Figure A.13.5, the nonzero x-direction image ray errors caused by defocus are represented by the short x-direction line in the image surface connecting rays 2 and 3.

Operands 51 through 54 in Listing A.13.1 are the optimization operands to control sagittal coma for four different combinations of object and pupil height. Each is a TRAY operand that calculates the y-direction differential image error of a ray relative to the chief ray. Thus, these operands give sagittal coma directly without the need of combining two or more separate operands. For an object height

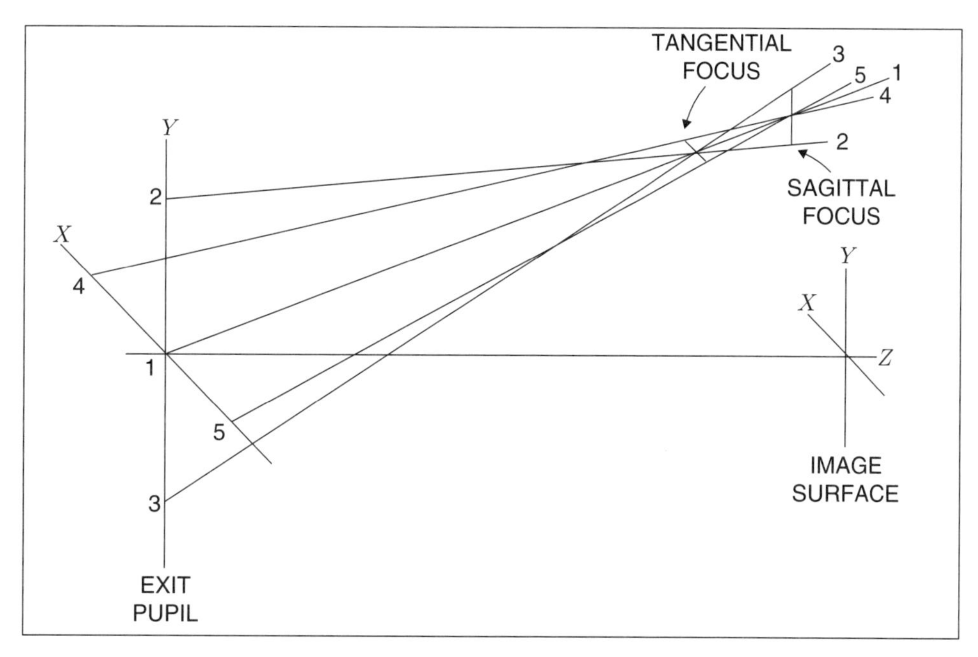

Figure A.13.6. Astigmatism and Field Curvature. *To correct astigmatism or medial field curvature, the y-direction and x-direction blur widths on the image surface are compared. Signs are important. See text for details. Ray 1 is the chief ray. Rays 2 and 3 are the top and bottom tangential rays. Rays 4 and 5 are the right and left sagittal rays. The heights of the tangential and sagittal rays in the pupil are really very small.*

of (0,.7), (x, y) pupil heights of (.7,0) and (1,0) are used. For an object height of (0,1), the pupil heights are (.5,0) and (.7,0).

For a lens with deliberate mechanical vignetting, there is more clipping of the off-axis exit pupil in the tangential (meridional) y-direction than in the sagittal x-direction. Thus, the vignetted pupil has an elongated aspect ratio that is shorter in the y-direction and longer in the x-direction. However, some clipping is usually present in both directions, especially for objects near the field edge. Accordingly, the pupil heights for the sagittal coma operands at the field edge (numbers 53 and 54) have been reduced to allow for mechanical vignetting. Note that this reduction is less than the reduction required for the corresponding tangential coma operands (numbers 42 through 45). For the 70% field, the amount of sagittal vignetting is so slight that no reduction is needed (numbers 51 and 52). If a lens has no mechanical vignetting (or uses vignetting factors), then pupil heights of (.7,0) and (1,0) are used for both field points.

To make the sagittal and tangential coma contributions comparable in the optimization computations, the sagittal operands are weighted six times as much as the tangential operands. A factor of three is included because transverse third-order sagittal coma is only a third as large as transverse third-order tangential coma. In the present case, a further factor of two is included because the tangential coma operands, as defined in the previous section, are a factor of two too large.

A.13.17 Astigmatism

Unlike asymmetrical coma, astigmatism is an off-axis aberration that is symmetrical about the chief ray. The effect of astigmatism is that the off-axis BFL is different for rays in the tangential and sagittal planes. One consequence is that if astigmatism is present, then a flat image surface can be formed by, at most, only one of these sets of rays. Thus, astigmatism is closely linked with field curvature, and together the two aberrations are often alternatively referred to as tangential and sagittal field curvature. Nevertheless, for the present, astigmatism will be separated from field curvature, with field curvature discussed in the next section.

In operand 57, ASTI controls OPD third-order astigmatism, measured in wavelengths in the exit pupil. The Int1 and Int2 values are the same as with SPHA and COMA above. But, once again, the lens designer may prefer to use real rays.

Figure A.13.6 is a schematic three-dimensional layout of image space that illustrates a lens with astigmatism and field curvature. Five rays are shown. Ray 1 is the chief ray. The remaining rays all pass through the same pupil zone, with rays 2 and 3 in the tangential plane, and rays 4 and 5 in the sagittal plane. All five rays use wavelength 2, and the numbers associated with the rays are again labels for identification. Only the central part of the image surface is indicated to avoid confusing the diagram. Also, to clarify the diagram without altering the concept, the rays are shown passing through an outer pupil zone, although astigmatism is really only defined for quasi-paraxial or parabasal rays very close to the chief ray. The tangential and sagittal foci are indicated; the medial focus, not indicated, is located halfway in between.

Operands 58 through 64 in Listing A.13.1 are the optimization operands to control astigmatism. Operands 58 and 59 use TRAY to get the y-direction image errors relative to the chief ray for tangential rays 2 and 3. The (x, y) object height is (0,.95), and the pupil heights are (0,$\pm$.1). Operands 60 and 61 use TRAX (not TRAY) to get the x-direction image errors relative to the chief ray for sagittal rays 4 and 5. The object height is again (0,.95), but now the pupil heights are ($\pm$.1,0). Thus, rays 2 through 5 are all in the same circular pupil zone, and the geometry is symmetrical for this symmetrical aberration. Note that although the pupil zone heights are small (10%), they are not infinitesimal.

Operand 62 takes the difference between the two tangential ray errors, which have opposite signs, to get the full width of the tangential image spread. Similarly, operand 63 takes the difference between the two sagittal ray errors to get the full width of the sagittal image spread.

Three image regions must be differentiated. If the image surface is closer to the lens than the nearest focus, in this case the tangential focus, then the values of operands 62 and 63 are both positive numbers. If there is no astigmatism, then the image blur is round and the two values are equal. But if there is astigmatism, then the image blur is elliptical and the two values are unequal. Operand 64, therefore, takes the difference between the values of operands 62 and 63, and an optimization target of zero is specified to give round blurs and zero astigmatism.

If the image surface is farther from the lens than the farthest focus, in this case the sagittal focus, then the values of operands 62 and 63 are both negative numbers. When subtracted, operand 64 again has a value of zero when there is no astigmatism.

If the image surface is located between the two astigmatic foci, then the signs of the values of operands 62 and 63 are opposite. When the two terms are subtracted, the values add. If there is astigmatism, then the value of operand 64 becomes either a large positive or large negative number. But once again, an optimization target of zero for operand 64 controls astigmatism.

Note that in the above discussion, the full widths of the astigmatic blurs were calculated using four zonal rays plus the chief ray. For quasi-paraxial rays very close to the chief ray, the half widths would have served just as well, and these would have required tracing only two zonal rays plus the chief ray. However, for real rays at some practical finite distance from the chief ray, any image asymmetry from coma will cause the absolute values of the image ray-intercept errors for the top and bottom tangential rays to be different. For example, note the different magnitudes of the values of operands 58 and 59. Thus, using a compromise astigmatism operand with four zonal rays plus the chief ray is more reliable.

A.13.18 Field Curvature

In operand 67, FCUR controls OPD third-order field curvature, measured in wavelengths in the exit pupil. The Int1 and Int2 values are the same as for the other third-order aberrations above. In operand 68, PETZ calculates the radius of curvature of the Petzval surface for wavelength 2. But you may wish to flatten the field (or otherwise control field curvature) by using trigonometric rays to directly place the astigmatic medial focus on the image surface. This can be done with a procedure very similar to that just used to control astigmatism. Refer again to Figure A.13.6.

Operands 69 through 75 in Listing A.13.1 are the optimization operands to control field curvature. Operands 69 through 72 are identical to operands 58 through 61 except that a different object height, (0,.8), is used. Operands 73 and 74 are identical to operands 62 and 63. Operand 75 now *adds* the values of operands 73 and 74, instead of subtracting as in operand 64.

The effect of summing in operand 75 reverses the way the errors combine in the three image distance regions. If the image surface is closer than the nearest focus or farther than the farthest focus, then the values of operands 73 and 74 are the same sign, and adding gives either a large positive or negative number. Only if the image surface is located in the region between the two foci are the values of operands 73 and 74 of opposite signs so that adding produces a small number. Furthermore, only at the medial focus are the tangential and sagittal blur widths the same size. Thus, to place the image surface at the medial focus (or vice versa), an optimization target of zero is set for operand 75.

Note that this procedure is tied to astigmatism and again uses quasi-paraxial rays close to the chief ray. But these rays do not sample the outer pupil zones. If there are significant amounts of other off-axis aberrations, such as fifth-order oblique spherical aberration, then controlling field curvature with operand 75 will not give a good total solution. Nevertheless, this optimization operand may be useful for preliminary work and for special circumstances.

A.13.19 Distortion

Although field curvature is concerned with the longitudinal location of off-axis image points, its effect is most often seen as out-of-focus images. Thus, field curvature is another of the aberrations that affect image sharpness. Distortion is also concerned with the location of off-axis image points, but only the transverse location. Thus, except for chromatic variations, distortion has nothing to do with image sharpness.

In operand 78, DIMX controls the maximum allowed absolute value of distortion in percent for the system as a whole. Int1 is the field number (zero selects the edge of the field) and Int2 is the wavelength number. A target of 0.75 (which means $\pm 0.75\%$) is specified, but the zero weight indicates that the operand is not presently used.

In operand 79, DIST controls the actual value of distortion at the edge of the field. When Int1 is zero, distortion is given in percent for the system as a whole. A target of zero is specified here, and the heavy weight indicates that the operand is to be corrected exactly during optimization.

In practice, these two built-in operands are sufficient to handle most distortion issues. However, to allow distortion to be exactly controlled at any field angle, a custom distortion optimization operand is constructed with operands 80 through 85 in Listing A.13.1. In operand 80, REAY, as before, calculates the trigonometric y-direction intercept height, relative to the optical axis, of a real ray on a surface. Int1

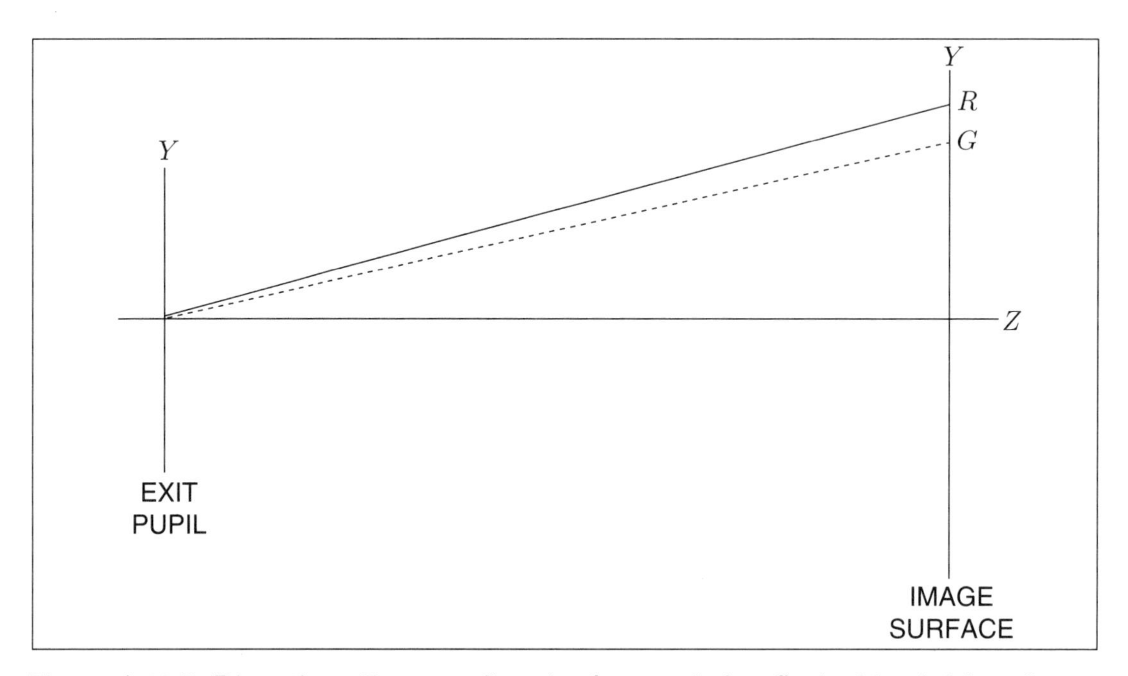

Figure A.13.7. Distortion. *To correct distortion for a particular off-axis object height and wavelength, the heights of the real trigonometric chief ray (ray R) and the first-order Gaussian chief ray (ray G) are made equal on the image surface.*

specifies surface 11, the image surface. Int2 specifies wavelength 2, the reference wavelength. An (x, y) object height of (0,.8) is selected, and a pupil height of (0,0) specifies that the ray is the chief ray.

In operand 81, PARY calculates the first-order y-direction intercept height, relative to the optical axis, of a Gaussian ray on a plane surface tangent to one of the lens surfaces. Note that this calculation uses the Gaussian approximation for finite distances from the axis, not the paraxial limiting case for infinitesimal distances from the axis. Int1 specifies that the plane surface is tangent to surface 11, the image surface. Of course, if the image surface is flat, then the tangent surface is identical to the image surface. Int2 and the object and pupil heights are the same as in operand 80; that is, the ray is the Gaussian chief ray corresponding to the trigonometric chief ray used in operand 80.

Figure A.13.7 is a schematic layout of image space that illustrates these two rays. Ray R is the real trigonometric chief ray, and ray G is the Gaussian first-order chief ray.

If you have a flat image surface and you take the height difference on this surface between the real ray and the Gaussian ray, then you are really measuring the error in the Gaussian approximation. But recall that the first-order Gaussian image properties on a flat image surface are the ideal, aberration-free image properties. Aberrations are third-, fifth-, and higher-order departures from this first-order ideal. Thus, during optimization, you try to minimize the differences between the real and Gaussian optical properties. In particular, to correct distortion during optimization, the image heights of rays R and G are made equal.

Operand 82 calculates the difference between the image ray heights of rays R and G. This value is in linear measure such as millimeters. To convert to percent distortion, the linear distortion height error is divided (in operand 83) by the ideal Gaussian height, and this dimensionless ratio is multiplied (in operand 85) by 100 (a constant entered in operand 84). A nonzero weight for operand 85 would activate this user-defined distortion operand during optimization. Note that if the object height had been (0,1), then the value of operand 85 would have been equal to the distortion value given by operand 79.

Note that in operand 82, distortion is measured by taking the height difference between the real ray on the actual image surface and the Gaussian ray on the tangent plane. Here is where the problem of distortion on a curved image surface becomes apparent. If the image surface is curved, then this height difference refers to heights on two *separate* surfaces, and the meaning and applicability of the resulting distortion measure becomes unclear. It is the old map-maker's problem: you cannot project a curved surface onto a flat plane (or vice versa) without distortion. Therefore, optical distortion is only properly applied to the mapping of a plane object surface onto a plane image surface, and is not *defined* for curved object and image surfaces. In special applications, such as scanners and fisheye lenses, other projection ideals are used.

A.13.20 Using Both Special Aberration Operands and Vignetting Factors

There are problems when the off-axis user-defined operands are used with a lens that also uses vignetting factors. More about vignetting factors will be said in Chapter B.4. But briefly, as implemented in ZEMAX (and perhaps other programs as well), vignetting factors do two things. First, off-axis pupils are compressed (or occasionally stretched) to match the shapes of off-axis beams. Thus, off-axis pupil zones become elliptical rather than circular. Second, the centers of off-axis pupils are decentered to match the locations of off-axis beams. Thus, the middle ray in the beam no longer passes exactly through the center of the stop; that is, the middle ray is no longer the true chief ray.

These consequences of using vignetting factors are not an error. They are a requirement if the Gaussian quadrature optimization option is to be usable with lenses having mechanical vignetting.

All of the off-axis user-defined operands given above make use of the chief ray. But now Px and Py equalling zero does not give the chief ray. Thus, the off-axis aberration operands have questionable validity.

Worse, the elliptical pupil zones mean that equal Px and Py pupil heights no longer give equal distances from the beam center. This asymmetry causes the astigmatism and field curvature operands to give the wrong solution during optimization.

There are two partial exceptions. When you enter vignetting factors by hand, you can specify zero pupil decenters, thereby making the middle ray the chief ray. When you optimize with vignetting factors set by the computer, you can include an operand in the merit function that forces the middle ray to pass exactly through the center of the stop. But in both exceptions, the astigmatism and field curvature operands must still be avoided.

In general, the off-axis user-defined aberration operands must be used with great caution or not at all when vignetting factors are also being used. When using these special operands, it may be better to handle vignetting in a different way. Note that the special on-axis operands are not affected by vignetting factors.

A.13.21 The DMFS Operand

In the list of special optimization operands in Listing A.13.1, the last operand, operand 88, is the DMFS operand. DMFS stands for default merit function start. In ZEMAX, when appending a default merit function to a series of special operands, it is always wise to make the last of your special operands the DMFS operand. This is a safety precaution to prevent the possibility of the computer accidently overwriting and erasing your special operands when it splices on the default merit function operands.

A.13.22 Solves

There is another way of handling some of the lens variables during optimization, and this is by using solves. Unlike optimization operands, solves are not part of the merit function. Instead, solves are part of the lens prescription itself. The purpose of solves is to adjust lens parameters to satisfy constraints. The three most frequently used types of solves are:

1. Angle solves,
2. Height solves, and
3. Pickup solves.

Angle solves are used to adjust the curvature of a surface to control the slope (angle) of a first-order ray either incident on or leaving the surface. One application is controlling the slope of the marginal ray to achieve a desired focal ratio. A second application is making the slope of the chief ray zero to create a lens with a telecentric image (the chief ray is incident on the image surface parallel to the axis).

Height solves are used to adjust the thickness between two successive lens surfaces to control the height of a first-order ray on the second surface. One application is locating the second surface where the marginal ray height is zero, thereby placing the surface at a paraxial image. Similarly, if the chief ray height is zero, then the surface is at the stop or an image of the stop (pupil). Note that the ray height need not be zero, as when the height of the chief ray is used to control image scale.

Also, note that depending on your program, some angle and height solves may have the option of using real rays, not just first-order rays.

Pickup solves specify that a given lens parameter is picked up from or is a function of another similar lens parameter elsewhere (usually further forward) in the lens. For example, you can specify that the curvature on a surface is equal, or equal with opposite sign, to the curvature on another surface. Similarly, the thickness between two successive surfaces can be made some multiple of another thickness. And pickups are often used to specify that two lens elements be made of the same type of glass.

Some programs, for example CODE V, also incorporate a second, more general form of pickup solve that might be termed a linkage. In a linkage, a lens parameter is not simply picked up. Instead, the parameter is coupled to another parameter so that the two vary together during optimization according to a functional relationship.

Because solves are part of the lens prescription, solves are continuously active, not just during optimization, but at other times too. Thus, solves may inadvertently change the lens at times and in ways that you do not intend. For example, solves are calculated using the reference wavelength, and if you change that, then all parameters controlled by solves also change. Therefore, after optimization, you may wish to delete all solves to freeze the variables they control. Be especially careful if you use solves during a tolerance analysis.

All designers use pickup or linkage solves. But the use of angle and height solves is somewhat a matter of the individual style of the lens designer and the software package he is using. Some designers like solves, whereas others choose to put most of these controls and constraints in the merit function. Often, the specific lens in question, or the stage within the optimization process, lends itself to using or not using angle and height solves. Note that solves are very efficient during optimization, but in these days of fast personal computers, the time savings may be of little significance. Feel free to do as you prefer with solves.

Chapter A.14

Finding a Starting Design

In the optimization process, the computer software can take a given optical configuration and make improvements. But the computer cannot change the basic form. For example, if you start with a triplet lens, there is no way that the computer, on its own, can convert this into a six-element Double-Gauss lens or an all-reflecting Cassegrain telescope. Thus, it is of the greatest importance that the lens designer select his lens starting point carefully.

In this chapter, selecting a promising initial optical configuration is discussed. The problem has two parts. First, you must determine what you need to do. Second, you must figure out ways to do it.

A.14.1 Determining System Requirements

When a lens designer is asked to design an optical system, he (or she) must determine from the customer what the complete system is to be; that is, the nature of the optics and how the optics will fit together with everything else. Second, he must determine the required level of optical performance. Third, he must determine how the optics will be used in practice, including the user interface. Fourth, he must ascertain the allowable limits on size, weight, cost, and schedule. Fifth, he must explore the relevant issues that lie beyond his control. These may include the specifics of the object, the photosensitive detector system, the operating environment, any data transmission link, and so forth. Sixth, he must ask whether there are any ground rules or other restrictions, such as secrecy or the required use or avoidance of a patented approach. These are general suggestions, and the list is neither exhaustive nor universally applicable.

While obtaining this overview of the project, an interdisciplinary systems approach is essential. Not only does a good engineer need this broad perspective to help him to do the job effectively, but often a perceptive designer can help the customer to better formulate his requirements and expand his options.

Following this overview, the designer must determine, with the customer (and/or management), the more detailed optical system properties, specifications, and considerations. These include:

1. The first-order system properties,
2. The imaging requirements, and
3. Any and all special considerations.

The first-order properties are central to the definition of any optical system. Thus, these are usually among the first details that the customer discusses with the lens designer. First-order properties include: entrance pupil diameter (EPD), focal length (EFL), focal ratio (f/number), magnification, field of view (FOV), image and pupil locations and orientations, back focus clearance, wavelength coverage, scanning, zooming, and so forth.

Imaging requirements concern the level of detail in the object that is to be recorded or conveyed. Image quality can be expressed in a variety of ways, and it can be given either directly or referred back to the object. Direct image measures include: geometrical spot size, geometrical wavefront error, limiting resolution, contrast transfer at a specific spatial frequency, general MTF, Strehl ratio, encircled energy, energy on a pixel, and so forth. For many applications, MTF is preferred. Imaging requirements also specify a flat or curved field and the allowable amount of distortion. In controlling image quality, the lens designer is primarily concerned with aberrations and diffraction, although atmospheric seeing, image motion, and stray light may be important too.

All other optical properties are related to special considerations. Special considerations include unusual optical features, such as those in a spectrograph, a photometer, an interferometer, a coronagraph, and so forth. Special considerations also include restrictions arising from the nature of the system and how it will be used, such as those in a minimum-loss astronomical telescope, an x-ray telescope, a thermal infrared system, a complex space-based system, a battlefield instrument, and so forth. And special considerations routinely include, either directly or indirectly, the practical issues of manufacturing and testing. The list of special considerations is often long and diverse.

Sometimes the lens designer is given a basic optical configuration that he is only allowed to modify and reoptimize. More often, the lens designer can select or derive his own optical configuration. But in either case, in the end it is not always possible to achieve the specified required performance level. Therefore, it is always wise for the lens designer to determine early-on how rigid the customer's requirements are, and whether any of these are negotiable if problems arise and relief is needed.

A.14.2 Determining the Number of Effective Independent System Variables

The lens designer has only certain types of system variables available to accomplish his optical goals. The number of these variables is also unavoidably finite. These limitations usually influence his choice of a lens starting point. System variables are also called system degrees of freedom.

It is often necessary to actually count the number of variables that a lens configuration has available for change during optimization. Of this total possible number, an estimate must then be made of how many of these variables are both *independent* and *effective* in controlling optical properties and aberrations. This subset is often much smaller than the total.

By far the majority of optical designs are collections of centered spherical surfaces. For these systems, the only variables are:

1. Surface radii/curvatures,
2. Intersurface spacings/thicknesses,
3. Glass types (indices and dispersions), and
4. Stop position.

Mirrors are considered a special case of glass type, and in the parlance of optical designers, even an all-reflecting system can still be called a "lens."

Unconventional systems usually have additional variables arising from their use of aspheres, gradient index materials, holographic surfaces, tilts and decenters, prisms and diffraction gratings, and so forth. Aspheres (any surface that is not a sphere is an asphere) are still uncommon and are used mainly in special expensive lenses, in reflecting telescopes, and in injection-molded plastic optics. Other special lens features are even less common.

Neglecting stop position, a spherical mirror has only one variable: its surface curvature. If in addition, the mirror has a general conic surface figure (ellipses, hyperbolas, etc.), then there are two variables: curvature and conic constant.

Again neglecting stop position, a singlet lens or element with spherical surfaces has five variables: two surface curvatures, one glass thickness, and the index and dispersion of the glass. The two curvatures can alternatively be viewed as one lens power and one lens shape or bending.

If a singlet lens or element is thin, then glass thickness is no longer an independent variable. Thickness is adjusted merely to give the minimum practical edge or center thickness. In addition, even in a thick lens, glass thickness may not be very effective in controlling aberrations or system parameters. Such an ineffective variable is called a weak variable. However, there are many cases where glass thickness is indeed an effective variable.

The glass type in a singlet lens or element may or may not be an independent variable. Even if variable, there are always restrictions on glass selection. The limited range of available indices and dispersions severely restricts freedom of glass choice (refer again to Figure A.10.2). And in a compound lens, system requirements often dictate that the glass of a given element be either a crown or a flint material. Sometimes the exact index and dispersion of an element are weak variables and do not matter much. In other elements, the general magnitude or specific value of index or dispersion may be of critical importance. Throughout the practice of lens design, glass selection is often the most subtle issue of all.

Two airspaced singlets have five variables each, plus two more for the airspace and stop position, for a total of twelve variables. But for example, if the two elements are to comprise an airspaced achromatic telescope doublet, then the dispersions and thus glass choices are no longer independent. The positive element must be a crown and the negative element must be a flint (for an overall positive system). Other aspects of glass choice are weak variables, and the glasses are usually chosen for convenience (avoid making the dispersion difference too small). Similarly, the glass thicknesses are also weak and are usually chosen to be conveniently thin. The airspace may or may not be a major variable. Finally, for a telescope objective lens, the stop is at the front surface.

Thus, of the twelve possible variables in an airspaced achromatic doublet, only four remain both independent and effective. They are either the four curvatures (with the airspace fixed), or three curvatures and the airspace (with a fourth curvature a variable but not independent). See Chapter B.1 for more on achromatic doublets.

If two singlets are to be cemented together, then two independent variables are lost. Two of the curvatures, one on each lens, must be made equal to allow contact at the cemented interface. And of course, the airspace becomes zero. However, for a cemented achromatic doublet, the dispersion difference at the cemented interface now becomes a major independent variable and restores the total number to four again. The index of the crown element must also be lower than the index of the flint element to allow spherical aberration to be corrected in a cemented achromatic telescope doublet.

And finally, if the stop is not at the lenses, then the designer is given another independent and very effective variable for controlling aberrations off-axis.

A.14.3 Controlling Optical Properties

In general, it takes one effective independent variable (degree of freedom) to control one optical property. But not all variables are independent or sufficiently effective, as was just discussed. Knowing which variables are effective and in-

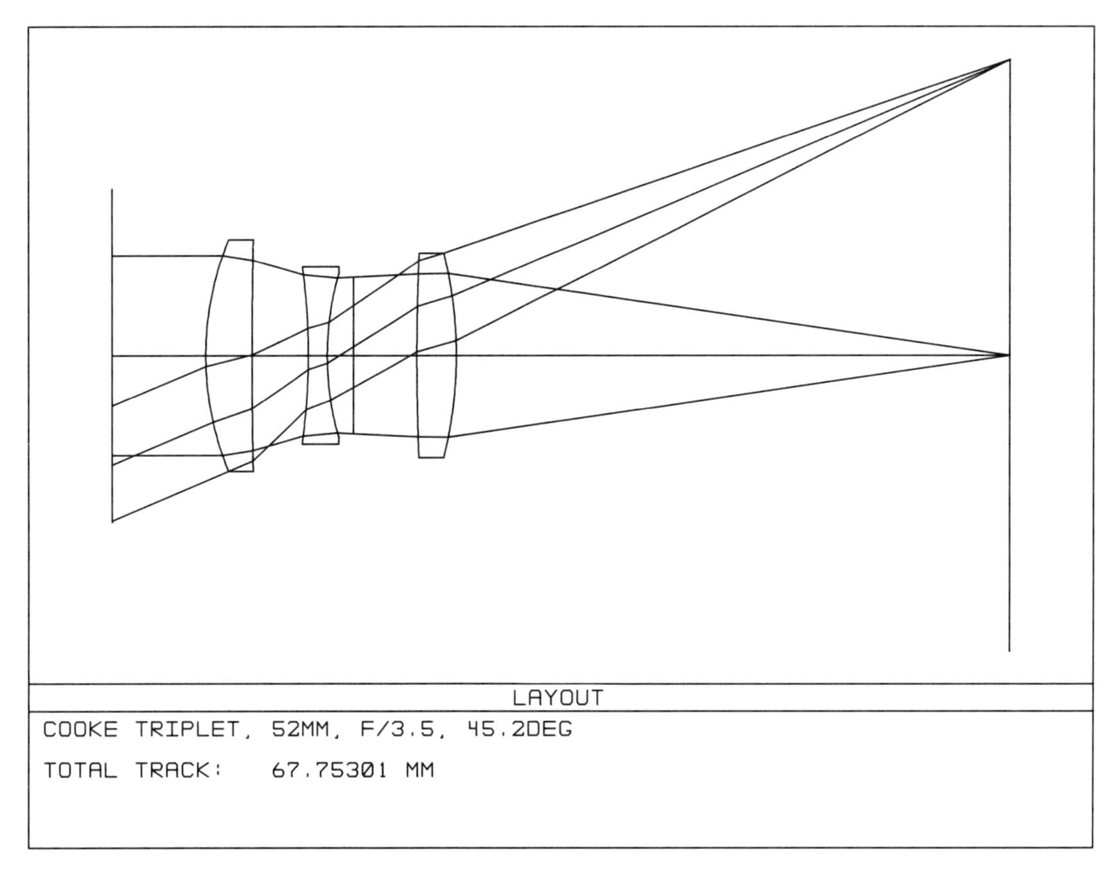

Figure A.14.1.

dependent is very much a subject where there is no substitute for experience. This experience may come either from your own work or from that of others.

In both the airspaced and cemented achromatic telescope doublets above, there are only four effective independent variables. Thus, it is possible to control only four optical properties, and no more. Usually, the lens designer will choose to control longitudinal color, spherical aberration, coma, and focal length (controlling focal length requires a degree of freedom). Because the stop is at the thin lenses, the chief ray passes through the middle with almost no deviation, and thus lateral color and distortion are automatically small. But astigmatism and field curvature are not controlled and may be large. Thus, although doublets are quite satisfactory over small fields, they cannot sharply image a wide field of view. In other words, you cannot use a telescope doublet as a camera lens covering 50°. This example illustrates why the lens designer must choose his lens starting point carefully and always be vigilant against lens forms with insufficient independent variables to do the prescribed job.

Now consider the Cooke Triplet lens shown in Figure A.14.1. Assume that the glasses have been chosen to allow color correction and to help reduce the Petzval sum to make it easier to flatten the field (two high-index positive crown elements and one negative flint element with an equal or somewhat lower index). Also, assume that the element glass thicknesses are relatively thin and have only a minor effect on the design. Finally, assume that the stop position has been chosen to be close to the central element for reasons of symmetry, thus making it easier to control lateral color, coma, and distortion. The result is eight effective independent variables: six surface curvatures (or three powers and three bendings) and two large interelement airspaces.

This is a very interesting case. Eight effective independent variables or degrees of freedom is exactly the minimum number to allow simultaneous control of focal length plus all seven of the basic aberrations (longitudinal and lateral color, spherical aberration, coma, astigmatism, field curvature, and distortion). Thus, a Cooke Triplet can be perfectly corrected to within the third-order approximation. However, higher-order aberrations and chromatic variations of the monochromatic aberrations severely compromise this rosy picture. Nevertheless, if some third-order aberrations are carefully left in to balance higher-order aberrations, the result is an excellent and very popular lens design form. See Chapter B.3 for more on Cooke Triplets.

For still more control over aberrations, or to allow additional features such as zooming, an even greater number of effective independent lens variables is needed. The consequence is more lens elements and greater complexity. Over the years, lens designers have discovered numerous solutions in this never-ending search. Some solutions have been more successful than others. Some became popular and were later abandoned, or abandoned and still later resurrected. The lens designer must therefore become familiar with what has been done before and what is being done now. That way, he learns what has been tried, what works, what to avoid, what might be worth trying again, and which of his ideas are new.

A.14.4 Following the Literature

There are several ways to learn the state of the art. Aside from taking a university degree in optics, perhaps the best way is to read. Books, published papers, manufacturer's literature, and optical patents are a vast source of information that can suggest lens starting points. Even an experienced professional lens designer must continue to read the literature to expand his understanding and keep up to date.

Books containing lens designs are not numerous, but there are a few. Some of these are included in the list of references at the back of this book. Note that an unexpectedly useful source of information on lenses is books on camera collecting.

In the United States, the two most prominent professional optical associations are the Optical Society of America (OSA) and SPIE—The International Society for Optical Engineering. Both of these societies publish books, proceedings, and journals containing a huge amount of information on all aspects of optics, including much on lens design. There are also articles on optical designs, techniques, and software in periodicals such as: *Photonics Spectra*, *Laser Focus World*, *Popular Photography*, and *Sky & Telescope*. Interestingly, it is often the ads in these periodicals that really reveal the state of the art.

Most camera and lens manufacturers publish promotional literature describing their optical products. Often, lens cross-sectional layouts are included. It is amazing how much information you can get just by looking at a lens layout. You may not be told the glass types or the exact radii and thicknesses, but the basic idea is there. More than once the author has laid a ruler on a published layout, and guessed at the glasses, to get a lens starting point.

For more detailed descriptions of successful lens designs, published patents can be very informative. Here, one hopes, is a complete discussion of an innovative idea that can inspire further departures and improvements. Often, complete optical prescriptions are included. An excellent source of patent information on lenses is the recently introduced LensVIEW, a library of tens of thousands of patented lens prescriptions that is available commercially on CD-ROM. However, there is a potential problem in using patents, aside from possible patent infringement. All too often, some of the published details are purposely wrong to prevent the design from being directly and illegally copied. But you can still get ideas and starting points.

A.14.5 Attending Meetings

Another way to learn the state of the art and get ideas for lens starting points is by talking with other lens designers. Brain-storming sessions with colleagues can be most beneficial and even fun. Outside the work place, one of the best places for talk is at meetings. OSA, SPIE, and other groups hold frequent professional meetings, both large and small. Although the formal presentations are of interest, some of the most valuable time you can spend at these gatherings is in the lobby and at social events. Here, in an informal setting, you can learn the latest news and exchange ideas directly with friends. At the larger conferences and conventions, major displays of commercial hardware and software are included, and these can be of very great interest indeed.

A different type of meeting that may be educational for the lens designer is camera shows, especially the camera collectors' shows. The author often goes to camera shows and enjoys looking at all the new and old lenses on display. Before long, many of these lenses become like old friends, and the new acquaintances are even more interesting.

In addition to looking at lenses, pick them up and look *into* them. Often you can see constructional details, such as: lens mounts, diaphragm designs, anti-reflection coatings, and vignetting (including intentional entrance pupil growth off-axis to partially counteract vignetting). The reflections off the various internal lens surfaces can be especially revealing. Two nominally identical lenses from the same manufacturer may have different sets of internal reflections, revealing that during production the design was recomputed. The differences may be slight or radical. In this case, the newer lens with the higher serial number often has an advantage.

One nasty surprise you may get when looking into a lens is finding that some of the interior surfaces are fogged up with contaminants. This fogging may be caused by lubricants outgassing from the iris diaphragm mechanism and recondensing onto the glass. Some manufacturers' lenses are more prone to fogging than others', as you will learn if you look.

A visit to a collectors' camera show can also be a lesson in how far we have progressed technically in the past several decades. Although lenses made prior to about 1960 are often masterpieces of vintage optical and mechanical skill, new materials and computer design techniques introduced in subsequent years have often made these earlier attempts obsolete.

Still another place to learn about optical designs, and in addition optical performance, is at a telescope star party or at one of the amateur telescope makers' conferences. Here, you have the chance both to look at and through a wide range of telescopes, both conventional and unusual. At the conferences, prizes are awarded for the best examples of homemade telescopes and astro-cameras, some of which may be radical designs known only to a few optical enthusiasts. It is noteworthy that many professional lens designers (and astronomers) attend these amateur gatherings. They know how much there is to be learned there, especially from people with expertise different from their own. Where else are you invited to examine both the instruments and their images, and to cross-examine the makers?

If you go to a star party, try to look through a wide range of telescopes. In each case (if possible), ask the owner to point his telescope at a bright star. Then examine the star's image under high magnification with a short focal length eyepiece.[1] After looking at the in-focus image, run the eyepiece a short distance back and forth through focus. If the out-of-focus images on the two sides of best focus look different, then the telescope may have spherical aberration (this is a very sensitive test). If the telescope has refractive elements, look for chromatic aberration. If the telescope is misaligned (out of collimation), you may see coma in the center of the field. With the eyepiece in focus, look at the image degradations caused by atmospheric seeing.

[1] An excellent reference is Harold Richard Suiter, *Star Testing Astronomical Telescopes*.

Note how much seeing can change from night to night, or even from hour to hour. Note too that seeing looks different in large and small telescopes and is a strong function of the observing site. And always, look at the diffraction pattern if the seeing will allow. If the telescope has a central obscuration supported by vanes (a spider), look for diffraction spikes emanating from the star's image. Before long, you will get a visceral feeling of what good and bad images look like (be diplomatic if the telescope has problems). It is one thing to read about images in a book, and quite another thing to see for yourself.

Chapter A.15

Optimization Techniques

Lens design can be a very creative activity. From little more than a set of requirements and some ideas, the lens designer generates the design of an optical system that does the job. The concept may be a modification of a well-known approach, or it may be a wholly new departure. Also, the concept may be as simple as a singlet lens, or it may include numerous, and sometimes moving, elements and other components, such as: mirrors, beam splitters, filters, polarizers, prisms, diffraction gratings, image intensifiers, the human eye, and so forth. Often, the designer must interact and negotiate with the final user to jointly converge on a solution that is both functional and practical; that is, it works on paper, you can build it in the real world, and you can do all this within the inevitable cost and time constraints.

After selecting a promising optical form to try, the next step is the heart of the design process: optimization. Here the designer must derive the sought-after functional and practical solution. Now is when experience and technique really pay off. Optimization is also where much of the creativity and satisfaction happens (there is also much satisfaction in seeing the system work as you intended after it is built). In this chapter, optimization techniques are discussed.

Note that none of what follows is gospel. These are only suggestions that the author has found to work more often than not.

A.15.1 Local Minima and Global Optimization

It is important for the lens designer to understand the basic processes that are going on inside the computer during an optimization run. This insight will allow him to plan his approach and know what to expect.

The lens is defined by a number of parameters, some of which are fixed (or frozen) and some of which are variables. The merit function is defined by a number of operands and their ideal values (or targets). During optimization, the computer changes the lens variables in a series of iterations to make the operands approach their targets as closely as possible. The most common mathematical procedure by which this is done is damped least-squares, although there are other approaches. In any case, the basic goals are the same.

Perhaps the best way to explain what happens during optimization is to simplify the situation and to assume, for the moment, a lens having only two independent variables. This is not unrealistic; a thin singlet of a given glass with the stop at the lens has only two independent variables, one power and one bending. A specific merit function is also assumed.

Merit functions are defined to be real (not complex), greater than or equal to zero, and single valued. For any set of values for the lens variables, there is only one corresponding value of the merit function. For the lens with two variables, if the value of the merit function is plotted as a function of all possible values of the variables, then the plot requires a three-dimensional space. Furthermore, the values of the merit function all fall on a single, generally curved surface located above the plane of the two variables. This surface is actually a two-dimensional object that exists in *three*-dimensional space; that is, if you are on the surface, then it takes only two parameters, the two independent variables, to tell you where you are located.

From the nature of a merit function, the lowest point on the merit function surface gives the best optical solution. Thus, the goal of the optimization process is to find this lowest point. Of course, if the merit function is changed, then the location of the lowest point is also changed.

To help visualize this situation and the processes involved, consider the analogous case of finding the lowest point on the surface of the Earth using lens design methods. Like the merit function surface, the surface of the Earth is also a two-dimensional object in three-dimensional space. Your position on the Earth's surface is uniquely determined if you know the values of only two parameters, your latitude and longitude (digging, diving, flying, etc. are not allowed).

Now, imagine that you are a blindfolded hiker, placed somewhere on Earth (perhaps in a range of mountains), given only a cane, and told to find the location of the lowest elevation you can. You feel around with your cane to find the greatest downward slope or gradient, and then you take a few steps in that direction. After that, you stop, feel around again to redetermine the greatest slope, and take a few more steps in the revised direction.

So downward you grope. You continue this process until eventually you find yourself in something ranging from a narrow hole to a broad basin (in this fanciful analogy, the ocean can be regarded as an extended plain). In any event, now when you feel around, all directions slope upward, and you have gone as far down as you can by this method. You are at an elevation minimum.

Now the question is whether you are in a minor depression or at the shore of the Dead Sea. If you are at the Dead Sea, you are at the lowest place on Earth, a global elevation minimum. This is the best you can do anywhere. But the more likely situation is that you are stuck in some local elevation minimum far higher than the global minimum. However, with only your cane to feel around, you cannot tell this.

Lens design optimization works in nearly the same way. You start with some initial lens (your initial location). You have n variables that, together with the merit function, yield an n dimensional surface in an $n+1$ dimensional space. The optimization routine analyzes the merit function to find the greatest local multi-dimensional downward slope. Then it makes small changes to the lens variables to reduce the merit function value. Then it repeats and does another analysis, makes more small changes, and so forth. Thus, by a series of iterations, the computer searches for a minimum of the merit function surface.

Eventually the computer stops when it reaches a local minimum. Actually, you may find in practice that the computer stops before reaching the very bottom of the local minimum toward which it is heading. This premature stop may not matter; you will probably be close enough. But if you are unsure, restart the optimizations once or twice and see what happens.

As with the blindfolded hiker, there is no guarantee that the local minimum you have found is also the global minimum. Now you see why it is so important to select a good starting lens configuration. Where you end up often depends on where you start. And you can also see why it is so important to define a good merit function free of extraneous local minima.

The search for techniques to actually find the global minimum is an active area of research. More than one approach is being tried, and the last word will not be written for a long time. In the meantime, the approach outlined here has proven to be very effective in deriving excellent optical designs and will be useful for many years to come.

A.15.2 Entering the Starting Design

When designing a lens, the first step that involves the computer is entering the starting design. Sometimes you may start with only a basic concept. At other times you may start with a previously optimized lens that you wish to modify. But in either case, you need actual values for radii, thicknesses, glasses, and so forth that you can enter into the lens design program.

If your starting design is a previously optimized lens obtained from a book, a patent, or wherever, simply enter these known numbers. Then use the program's scaling feature to scale all linear dimensions to give the required new focal length. Although it is not absolutely required, this scaling simplifies the subsequent optimization.

If you do not have a previously optimized lens, you can derive a set of starting lens parameter values by simply making them up. This rough initial design does not have to be very good because it only serves to select the basic optical form. Fortunately these days, you rarely if ever have to go through the old, lengthy, and difficult third-order predesign procedures.[1] With modern lens design software, the computer can usually take an extremely tentative design and whip it into shape.

Because you cannot count on having a previously optimized starting design, using a rough starting design is the more versatile approach. This is especially true if you wish to try something new. For the sake of versatility, the discussion that follows assumes a rough design.

A.15.3 How to Derive a Rough Starting Design

To derive a rough starting design, the first thing to do is make a sketch of the lens you have in mind, with the important features clearly visible. This sketch is also useful for sequentially numbering the lens surfaces. Make a note of all first-order properties and any special constraints, such as a limit on overall length or a required large back focus clearance. These special constraints are called boundary conditions because they define the limits or bounds of the acceptable solutions (thus, some possible solutions may be out of bounds).

Then enter a preliminary prescription into the computer. Go ahead and make guesses at the values of the various parameters, and then look at the layout. If you guessed wrong and, for example, the lens has a negative focal length when you wanted a positive focal length, or if the lens is too fast and has rays missing surfaces or has total internal reflections, then tinker by hand with the prescription in the computer and do another layout. Before long, you will have a lens that the optimization routine can accept and fix. Often the most important consideration at this earliest time is merely to get light rays through to the focal surface.

A.15.4 Optimizing in Stages

Usually your initial guess is far from a good solution. With such large system errors, it is not advisable to immediately try a full optimization run where all possible variables are allowed to be completely free. Too many independent variables too soon can hopelessly confuse the optimization routine, which may then go off on a tangent, stall or stagnate, blow up, or otherwise fail. For example, suppose you wish to design a Cooke Triplet. If you are not careful, you may end up with a closely spaced doublet out front and a singlet field flattener near focus. This latter configuration may produce good images, but it is not what you intended.

The way to avoid this problem is to optimize in stages. What follows describes a three-stage approach using early, intermediate, and final optimizations. This

[1] For a description of these procedures, an excellent text is Rudolf Kingslake, *Lens Design Fundamentals*, 1978.

method is most applicable to systems with substantial fields, such as camera lenses, where all aberrations must be controlled. For systems with narrower fields and more restricted requirements, such as many telescopes and microscopes, the optimization procedures can be correspondingly simplified. In general, different lenses require different optimization strategies. The lens designer should always feel free to experiment and improvise.

A.15.5 Early Optimizations

The purpose of the early optimizations is to shepherd your initial guess roughly toward the desired solution. You correct the first-order properties and boundary conditions and (optionally) begin to control aberrations.

To make this job easier for the computer, try to simplify the lens configuration, and also try to reduce the number of independent variables. Of course, you only use these restrictions initially. In subsequent stages of optimization, you remove the simplifications and free-up the variables.

The easiest way to reduce the number of variables is to just freeze some of them at their input values. At this early stage of optimization, glass types are nearly always frozen. Also, long-radius surfaces are often entered as planes and frozen. Another excellent tactic is to use pickup solves to make some parameters functions of other parameters; that is, a pickup allows a parameter to be a variable, but not an independent variable. For example, if you know that the final design will be nearly symmetrical about the stop, use pickups to make and keep all or parts of the lens exactly symmetrical at first.

If the lens is complicated, such as a zoom lens, then there may be a great many first-order conditions to satisfy. A possible approach here is to initially enter only a series of singlets (the simplest lens configuration), ignore all considerations of image quality, and optimize to correct only the first-order properties and boundary conditions. Once the system works to first-order, then you can add more elements and complexity to control aberrations.

During the early optimization runs, you may wish to include most or all of the first-order properties and boundary conditions in the least-squares part of the merit function. This is in contrast to insisting that these parameters be immediately corrected exactly with Lagrange multipliers. If the operands controlling first-order properties and boundary conditions are initially far from their targets, then using Lagrange multipliers is like hitting the lens with a sledge hammer rather than nudging it in the right direction. This harsh approach is a common reason for early optimization runs to stall or blow up. Note that the optimization routines in different software packages may respond differently here.

At this time, focal length should especially be included in the merit function, as opposed to using an angle solve on the rear lens surface to control marginal ray slope (focal ratio). Avoiding the solve prevents the last surface from possibly assuming unrealistic values initially, and it also allows you to use a curvature pickup on the rear surface if you wish. If you like to use solves, you can switch later.

If you attempt to control aberrations during these early optimizations, use a merit function that is robust and unlikely to fail. These robust merit functions are usually simple and based on shrinking spot size. If the system has refractive elements, then shrink polychromatic spot size. Never use OPDs, and avoid most special aberration operands. A frequently included exception is an operand to control longitudinal color.

After these early optimizations, do not be surprised if aberrations remain large. Remember, at this time you may be imposing restrictions on the system that preclude a good solution. All you are trying to do now is rough-in the lens. Often, the best criterion for success is whether the resulting layout looks like your sketch or like a published drawing. You know you have the wrong solution if the layout looks wrong in a major way (do not worry about small details).

A.15.6 Intermediate Optimizations

The purpose of the intermediate optimizations is to shepherd the lens to very near the desired solution. This stage is where most aberrations are controlled and where the essential (but not final) form of the optimized lens is determined. Thus, the intermediate optimizations are crucial.

To allow aberration control, the temporary restrictions placed on the lens during the early optimizations are removed. If there is to be deliberate mechanical vignetting and it has not previously been included, it is added now. And to concentrate on the important aberrations, some of the less important aberrations are not considered.

Recall that aberrations can be classified as chromatic and monochromatic. During the intermediate optimizations, these two types of aberrations are controlled separately. The basic chromatic aberrations are controlled using only the two extreme wavelengths. The basic monochromatic aberrations are controlled using only one central wavelength. But the secondary chromatic aberrations and the variations of the monochromatic aberrations with wavelength are not considered. The idea is to simplify the merit function and reduce the number of distracting local minima.

Transverse ray errors on the image surface continue to be the best measure of aberrations. As before, the reason is reliability; the resulting merit function is more robust and less likely to fail. Typically, the intermediate merit function uses individual optimization operands to control:

1. Longitudinal color (on-axis, one pupil zone),
2. Lateral color (chief rays, edge of field),
3. Spherical aberration (on-axis, one pupil zone), and
4. Distortion (chief ray, edge of field).

Note that numbers 1 and 2 together control the chromatic aberrations, and numbers 1 and 3 together control the on-axis aberrations. To control the more complicated off-axis aberrations, a default merit function is appended that shrinks off-axis monochromatic spot sizes. Of course, first-order properties and boundary conditions must continue to be controlled.

The individual operands are among the user-selected and user-defined operands described in Chapter A.13 and listed in Listing A.13.1. For special image control, add further individual operands; the most common example is a coma operand to reduce image spot asymmetry.

Actually, this intermediate approach is not as approximate as it may sound. On-axis zonal spherical is controlled by the pupil zone selected for number 3 above. On-axis spherochromatism is controlled by the pupil zone selected for number 1 above. Higher-order off-axis monochromatic aberrations (such as oblique spherical) are included in the spot shrink optimization. And the neglected off-axis chromatic effects are usually intrinsic to a lens configuration and are hard to change. Thus the solution derived from controlling just the basic chromatic and monochromatic aberrations is often very close to the best you can do.

As usual throughout the lens design process, during the intermediate optimizations you should monitor your progress. Do frequent layouts to check for unexpected detours giving unrealistic configurations. Do frequent ray fan plots to monitor image quality. Occasional spot diagrams, as well as listings of surface aberration contributions, may also be useful.

A.15.7 Locating the Image Surface

It has been said that a lens should be optimized and used on the paraxial image plane because that is where the good image is located. Although there are exceptions, this rule is usually true. Placing the image surface at the paraxial focus of the central (or reference) wavelength is especially valuable in avoiding poor solutions.

Recall the example in Figure A.7.7.2 where defocus plus the increasing orders of spherical aberration cause the ray-intercept ray fan plot to oscillate. The departure of the image surface from the paraxial focus produces the nonzero slope of the plot at the origin (the pupil center). At a zone some distance out in the pupil, the slope reverses because third-order spherical aberration overpowers defocus and becomes dominant there. Further out in the pupil, the slope reverses again because fifth-order spherical overpowers the combination of defocus and third-order spherical. Finally in this example, the slope reverses still again near the edge of the pupil where seventh- and higher-order spherical take over. In general, the various orders of spherical aberration become dominant one after another in order as you increase pupil zone height, and this produces the oscillations.

For most lenses, pronounced oscillations such as these are not desirable. It is usually better to have a smoother, flatter curve. One way to achieve this smoothness is to not allow paraxial defocusing. A lens similar to that of Figure A.7.7.2, but without defocusing, is shown in Figure A.7.6.2. Here the slope of the ray fan plot is zero at the origin.

Keeping the image surface at the paraxial focus can be very effective during optimization, especially during the early and intermediate stages. If this is not done and defocus is allowed, then the optimization routine may find a poor local minimum of the merit function. A large amount of defocus and large amounts of the individual orders of spherical aberration may all balance each other to yield large-amplitude zonal oscillations of the ray fan plot.

More generally, other aberrations can also get involved in the balance. If a large amount of defocus is allowed, then on-axis the computer may try a heroic balancing act with defocus, spherical aberration of various orders, and longitudinal color. The situation gets even more complicated off-axis. Here the aberration mix can include sizable amounts of defocus (inside paraxial focus), third-order spherical aberration (undercorrected), field curvature (inward curving), and fifth-order oblique spherical aberration (overcorrected). Except in unusual cases, this is not what you want and the attempt is doomed to failure.

However, if there is no paraxial defocusing during optimization, then oscillating and other extreme solutions are discouraged. It is more likely that the computer will find a local minimum giving a smoother, flatter ray fan plot and smaller contributions by the individual constituent aberrations. There will still be an aberration balance, but you will not be balancing such large opposing quantities.

Once such a solution has been found, then it may be all right to fine-tune it by allowing a *small* amount of defocus to be introduced during the final stage of optimization.

There are two additional advantages in using the paraxial focus. The first relates to a shift in best focus when the lens is stopped down. For lenses such as camera lenses that are most often focused wide open but used stopped down, or that are simply used at a variety of lens openings, a significant focus shift with changing aperture is bad.

Recall that when a lens is stopped down, aberrations that depend on the lens aperture (everything except lateral color and distortion) are suppressed. On-axis, the actual trigonometric rays approach the limiting-case paraxial rays. Thus, the paraxial focal plane is where the stopped-down image actually does fall.

It follows that, to avoid focus shift, all lens openings must form their best images on or near the paraxial focal plane. Ripples in the ray fan plot are a good indicator that focus shift is likely. Once again, designing with no paraxial defocusing to get as smooth and flat a ray fan plot as possible is advised.

The second additional advantage in using the paraxial focus arises from MTF considerations. If the lens has spherical aberration, then the PSF at the paraxial focus has a concentrated core surrounded by extended diffuse wings. If focus is shifted, the core is lost but the total blur is less. Recall the examples in Chapter A.7. A tight image core tends to raise the MTF at higher spatial frequencies and gives an image with more resolution of fine details. This is important in, for example, photographic lenses. However, if the image detector is a CCD for a television camera, then there is a sharp detector resolution cutoff (the Nyquist limit) beyond which any optical resolution is wasted. In this case, a small focal shift to raise the lower frequency MTFs (contrast) is beneficial. In a similar way, if there is significant secondary longitudinal color, it may also be beneficial to refocus for the best chromatic compromise. As always, the lens designer must make decisions based on the application.

A.15.8 Final Optimizations

When you have completed the intermediate optimizations, you should have a lens that is quite close to the final design. The purpose of the final optimizations is to make fine adjustments and polish the design. It is also the time to attempt, if you wish, an OPD optimization.

Most of the aberrations that were previously ignored are chromatic aberrations. Thus, the first part of a final optimization involves shrinking polychromatic spot sizes. Two approaches are possible.

One approach is to continue to control the on-axis aberrations with individual optimization operands and to shrink polychromatic spots only off-axis. Using individual operands on-axis is often still effective because the on-axis aberrations are relatively simple.

The second approach is to shrink polychromatic spots for all field angles, including the on-axis field. Here, a minimum of assumptions is made about what constitutes best imagery. Small spots are all that matter. This approach might be termed natural or realistic (as opposed to theoretical). By suitably selecting and weighting the fields and wavelengths, the solution can be adjusted. With either approach, when shrinking polychromatic spots, use three or preferably more wavelengths spread over the spectral waveband.

The second (and optional) part of a final optimization is to reoptimize the lens to minimize wavefront OPDs in the exit pupil rather than transverse ray errors on the image surface. Today, the performance of most lenses is specified, not by spot size, but by diffraction modulation transfer function (MTF). MTF is directly related to the pupil function, and the pupil function in turn is directly related to wavefront OPD errors in the exit pupil. Thus, in the final optimization stage, switching to a merit function based on polychromatic OPDs may be beneficial.

A final optimization with OPDs may even be beneficial for systems whose aberrations are large relative to the effects of diffraction; that is, where there are many waves of OPD errors. This suggestion may seem counterintuitive because geometrical effects predominate here and spot size is such a good measure of geometrical image spread. Nevertheless, minimizing OPDs can still be effective in boosting MTFs.

Actually, the recommendation here is that you *try* using a merit function based on OPDs. For reasons that may not be clear, an OPD optimization of some optical forms does *not* give improved image quality, even if your criterion of good imagery is MTF. In fact, the results may be unquestionably worse. Because of this uncertainty, if you reoptimize the spot solution with OPDs, save both solutions and select the better one at the end.

Note that there are some systems whose requirements preclude an OPD solution. Note too that when optimizing with OPDs, the natural or realistic approach is usually better. Thus, minimize polychromatic OPDs for all object points, including the on-axis point.

During the final optimizations, it may be appropriate to allow a small amount of paraxial defocusing. Also, the final optimizations are where you must be sure that any deliberate mechanical vignetting is realistically handled. And as usual, distortion as well as first-order and boundary conditions must continue to be separately controlled.

Now is also when you must take an especially hard and critical look at your lens and its performance. Is this the lens that you (and the customer) really want? Will it really do the job? To answer these questions, you must now monitor and document your results in greater detail. In addition to the usual and still vital layout, ray-intercept ray fan plot, and spot diagram, you may wish to add: an OPD ray fan plot, a field curvature plot, a distortion plot, a vignetting plot, a listing of surface aberration contributions, a diffraction PSF analysis, an encircled energy analysis, an MTF analysis, and, of course, a listing of the lens prescription and first-order properties.

Once again it must be stressed that there are no absolute rules in this business. Sometimes one optimization method works best; another time a different approach is more effective. An optical configuration or technique that never worked before may be just the ticket today.

A.15.9 Potential Problem Areas and Suggestions

At all stages of the optimization process, you must monitor your work to screen out poor or impractical solutions and ineffective approaches. Here are some of the common pitfalls and possible ways out.

Try to avoid lens designs that have large amounts of positive and negative power balancing each other. These solutions give large surface aberration contributions. Cancelling out these large contributions requires a precarious balancing act and tight (expensive) manufacturing tolerances. Ideally, you want the lowest possible element powers.

Similarly, unless you really need them, avoid highly curved surfaces and surfaces with grazing rays. Again, these surfaces give large aberration contributions, especially higher-order aberration contributions, and they require tight manufacturing tolerances.

Watch for individual lens elements that stand out as very strong or very weak relative to the other elements. Ideally, you want each element in a lens to do about the same amount of optical "work." If an element is too strong, consider splitting it into two elements. If an element is too weak, it may be unnecessary; consider eliminating it from the design.

Look for overly thin elements that are hard to make and mount, and, of course, look for negative edge thicknesses. Figure A.15.1 is the layout of an extremely thin meniscus lens that also has impractically sharp edges. Figure A.15.2 is the layout of a lens with negative edge thickness. Also look for overly thick elements that are heavy and expensive. Figure A.15.3 is the layout of an excessively thick lens. These are three examples of lenses to be avoided.

Try to avoid glass types that have undesirable practical problems, such as: low internal transmission; excessive inhomogeneity, striae, or bubbles; poor resistance to weathering or staining; low hardness; high thermal expansion or thermal index variation; poor availability from the manufacturer; and so forth. And if you are designing a large lens, find out whether your glass types are available in large-enough pieces.

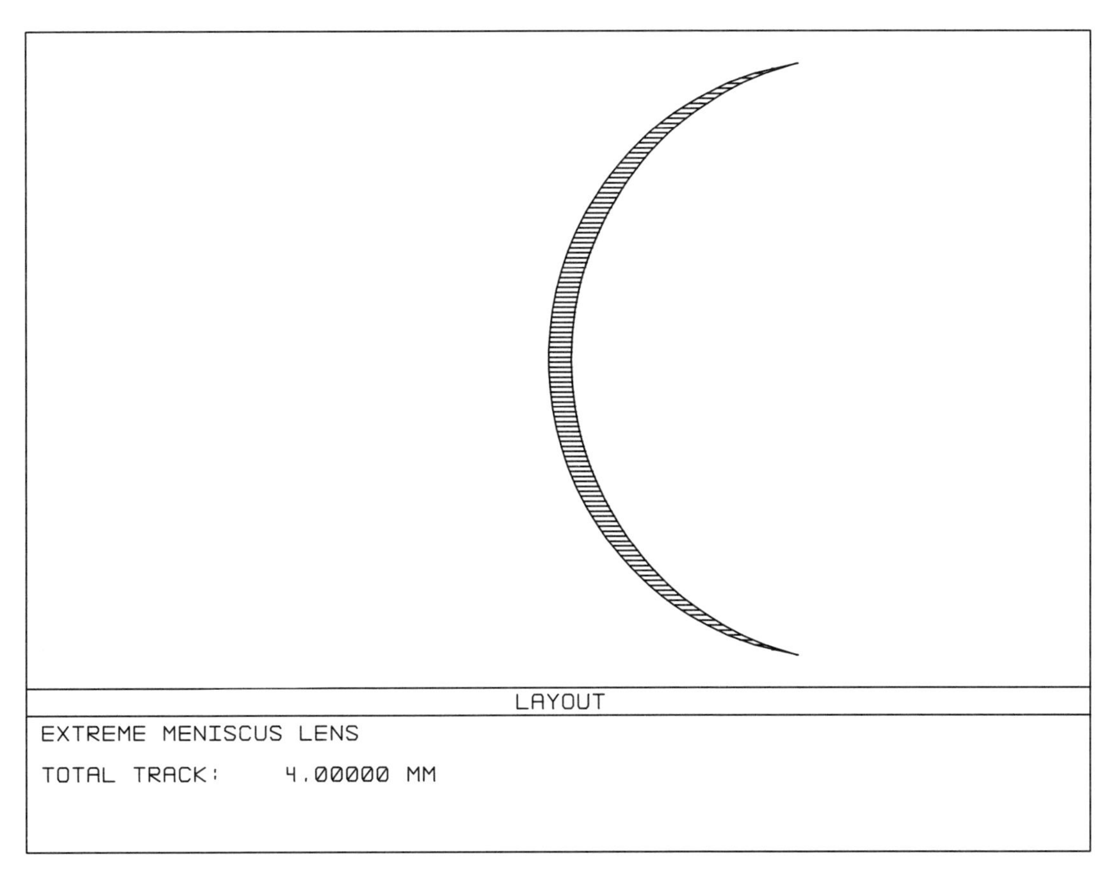

Figure A.15.1.

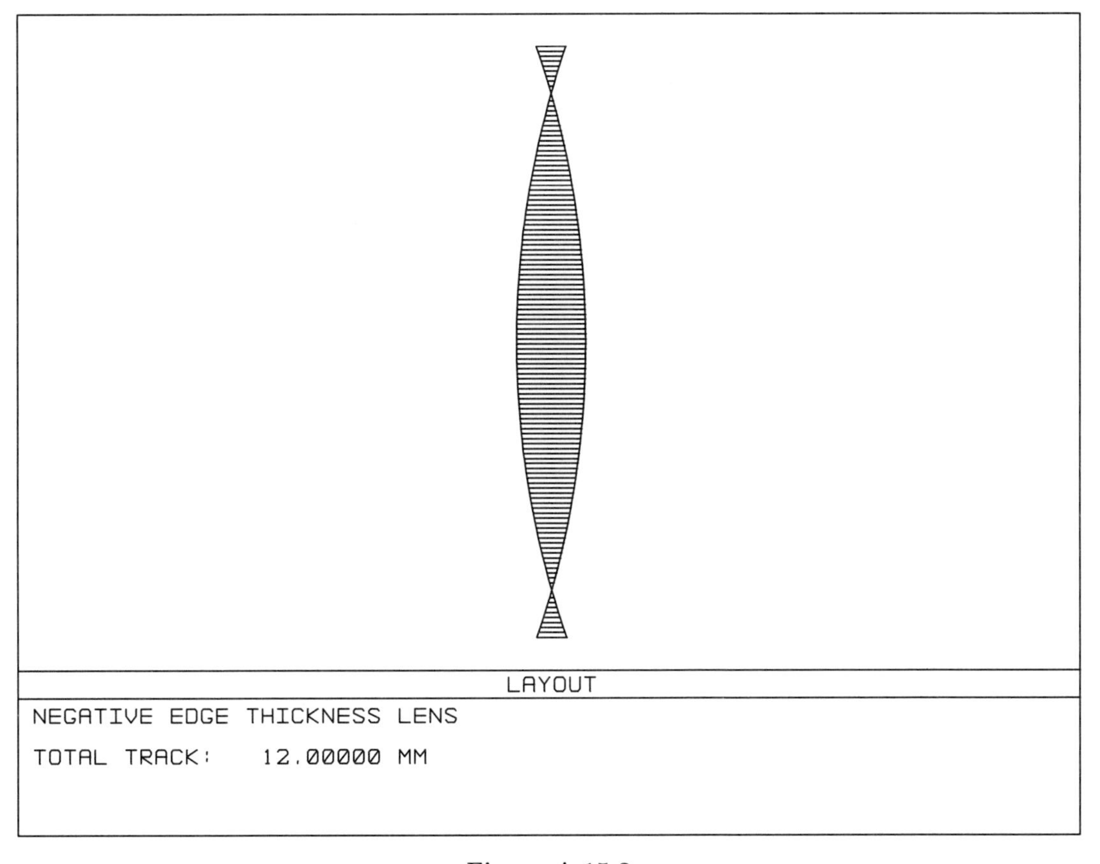

Figure A.15.2.

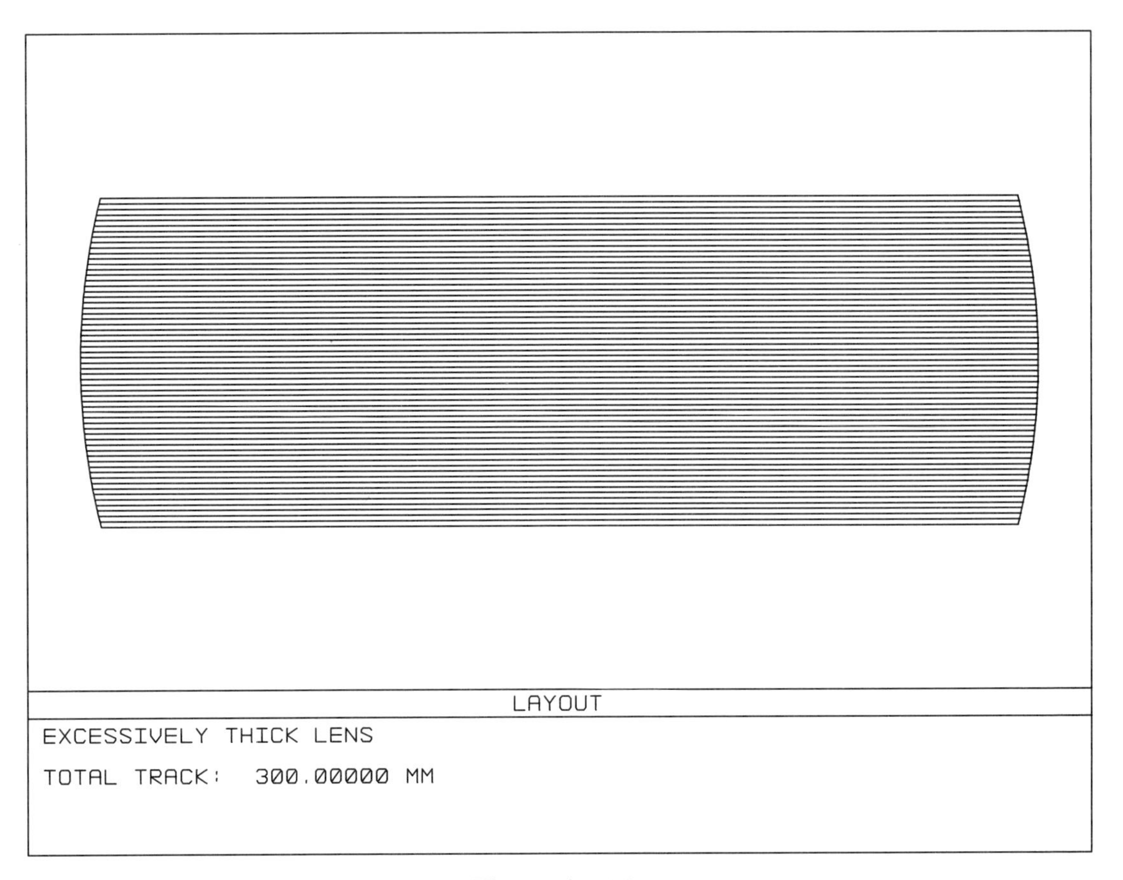

Figure A.15.3.

If you are using any aspheric surfaces (including conics), ask whether they are really necessary. Aspherics can be very effective in controlling aberrations, but they are much more expensive to make than spherical surfaces. Perhaps there are other, cheaper ways to achieve the same effect. If possible, stay with centered spherical surfaces.

If you are using polynomial aspheric surfaces, be sure to use enough rays in the entrance pupil when optimizing. If too few rays are used, then the optimization routine, which only controls the rays it is given, may produce a solution having small errors for the given rays but large errors for the uncontrolled rays in between. To check for this problem, look at the ray fan plots to verify that the curves do not oscillate wildly.

Also, if you are using higher-order polynomials, calculate the contributions of these terms to the surface figure. If these terms only change the surface sags by a small fraction of the wavelength of light, then these terms are negligible in the real world. You will do just as well with a design using only lower-order terms. The fourth- and sixth-order terms are usually enough.

Watch for designs having large amounts of zonal spherical aberration. Zonal spherical reduces image sharpness, but even worse, it causes a shift in best focus as the lens is stopped down. This focus shift is highly undesirable or fatal in many types of lenses, especially camera lenses where viewing and focusing are done wide open and the picture is taken at a smaller opening.

During optimization, watch for variables that are ineffective or only weakly effective. Such variables may cause the optimization process to do cycle after cycle, each cycle making a significant change to the lens but only a tiny reduction in the merit function. In this case, you are in a very broad minimum where all solutions have about equal performance. If this happens, stop the optimization, freeze the ineffective variables at reasonable values, and try again. You can often spot the ineffective variables by their tendency to go to extremes without much effect. One perennial suspect is glass thickness, which is often, but not always, a weakly effective variable.

At some point during a series of optimizations, you may add a variable or a group of related variables, as when you split an element. If you find that the new variables do not change much the next time you optimize, then you may be stuck in a local minimum from the previous optimization. To get out, try perturbing the lens by hand or adding a temporary constraint, and then reoptimizing.

Sometimes an optimization run fails to even get started. The most common reasons for an optimization to stall at the outset are (1) an inadequate set of lens variables and (2) a merit function with conflicting and/or unrealistic constraints. Double-check that you have not made a blunder. Are all of your constraints really necessary? Are your constraints really doing what you intended? Do you have enough variables of the right kind to do the job? Are you asking the impossible?

In summary, choose your starting design carefully. Develop an optimization approach that is systematic, rational, and flexible. Shepherd your design in the direction you intend. Be vigilant against problems and do continuous sanity checks. Discard poor solutions when you get them. But also be on the lookout, especially with complex lenses, for excellent unexpected solutions that the computer may find. Remember, the computer has no knowledge of the classical lens design forms. Look for solutions that both work on paper and can be built in practice.

Chapter A.16

Fabrication Errors and Tolerancing

Nothing can be made perfectly, and even the most precise lenses are no exception. Numerous types of fabrication errors always arise to cause a manufactured lens to depart from its nominal design. Unfortunately, if the nominal design is optimized, then these errors can only degrade optical performance. Therefore, consideration of fabrication errors must be part of the design process.

Optical fabrication is nearly always a cooperative effort by experts in several different fields. In addition to the lens designer, these people include: opticians (experts on optical fabrication, testing, and alignment), optomechanical engineers, and machinists or instrument makers. Furthermore, project administrators, businessmen, marketing researchers, and bankers may also have to be consulted.

What the people in the optical and machine shops need to know is how closely the various lens parameters must be held during manufacture to ensure that final system performance meets requirements. The types and amounts of these errors are called tolerances. What the people in the front office need to know is how much holding these tolerances will cost in time and money. To determine a practical set of tolerances, the lens designer does an optical tolerance analysis. This chapter discusses optical fabrication errors and the procedures for doing a tolerance analysis. A tolerancing example is given in Chapter B.7.

A.16.1 Types of Fabrication Errors

There are several distinct types of optical fabrication errors. The major types are outlined below.

Surface curvature error. Surface curvature error is an error in the vertex radius of curvature. There are many ways to measure curvature error, but for a small spherical surface, a common way is to place the unknown surface on a test plate of known and opposite curvature, such as a convex lens surface on a concave test plate. Ideally, the two pieces should form a perfect contact. Any error in the contact produces Newton's rings under monochromatic light, and a tolerance can be placed on the number of these interference fringes. This is equivalent to placing a tolerance on surface sag error at the edge of the clear aperture.

Surface figure error. One type of surface figure error is having the wrong basic surface shape, such as an error in the conic constant. A second type of figure error is having a series of zonal departures that resemble concentric ripples centered about the surface vertex. A third type of figure error is irregularity, such as a cylindrical error (or astigmatism) on a surface that should be rotationally symmetric. Whether the ideal surface is a sphere or an asphere, tolerances can be placed on errors in the conic constant, polynomial coefficients, sag errors, fringe number, fringe shape, dial-indicator variation, and so forth.

Index and dispersion error. The index and dispersion of a specific glass batch or melt will, in general, depart slightly from the nominal values. For a large production run, a tolerance can be placed on index and dispersion. For a small production run, the design of the lens can be reoptimized for the batch of glass being used.

Index inhomogeneity error. The index throughout a piece of glass will, in general, show variations from point to point. A small axial linear index gradient (variations fore and aft) can be focused out, as can also a small radial quadratic index gradient (variations radially outward from the axis). However, other index variations cannot be focused out and are a problem. A tolerance can be placed on non-refocusable index inhomogeneity.

Element or airspace center thickness error. A tolerance can be placed on the maximum allowable errors on axial glass thicknesses and airspaces. Note that a center thickness error shifts all subsequent surfaces by the amount of the error.

Surface axial displacement error. A single surface can be axially displaced or despaced relative to the rest of the lens without shifting all subsequent surfaces. An example is a despaced mirror surface. These shifts can also be given tolerances.

Surface transverse decentration error. A single surface can be decentered at right angles to the axis, and tolerances can be placed on these errors.

Surface tilt error. A single surface can be tilted or tipped. A general surface tilt can be decomposed into a tilt about the surface vertex plus a decenter and despace. Tolerances can be placed on surface tilts. Note that for a lens element with spherical surfaces, a tilt of one of the surfaces is nearly equivalent to a decentration of that same surface. Both processes produce element wedge. The only difference between the effect of a tilt and a decenter is a tiny change in lens element center thickness. Beware of counting the same error twice when investigating wedge.

Group tilt, decenter, and despace errors. Groups of surfaces forming one or more elements can be tilted, decentered, or despaced as a unit. A general group tilt can be decomposed into a tilt about a convenient lens surface vertex (usually the front or rear) plus a decenter and despace. Tolerances can be placed on group tilts, decenters, and despaces.

These are some of the possible ways that a real lens can be perturbed from its nominal configuration during manufacture. The above list is not exhaustive, and other errors are possible.

Note, if the system is large or will be used under arduous conditions (such as on a battlefield or in space after a rocket launch), then the values of some of the tolerances in your analysis may have to be increased to account for anticipated future mechanical drifts and decollimation.

A.16.2 Compensators

In many lenses, one or more system parameters can be adjusted during the final assembly of finished parts to partially compensate for fabrication errors. These special variables are called compensators. Compensators can be used to both correct for image sharpness degradation and for boresight (aiming) errors.

The most common compensator is focus shift. A simple refocus can make a huge improvement. Less often, internal airspaces between elements, element tilts and decenters, and image plane tilt and decenter are also used. In principle, any accessible parameter can be used as a compensator. Note that there are times, such as with large production runs of inexpensive lenses, when no compensators are allowed.

A.16.3 Measures of Performance during Tolerancing

When doing a tolerance analysis, there are several possible criteria that can be used to measure the degradation of optical performance caused by fabrication errors. Purely geometrical criteria include: (1) growth of RMS spot size and (2) increase in RMS wavefront OPD errors. Criteria with diffraction include: (3) reduction of

Strehl ratio, (4) reduction of diffraction MTF at one or more spatial frequencies, and (5) increase in diameter of diffraction PSF encircled energy for a specific percentage. In addition, (6) the increase in the value of a special tolerance merit function can be used. In this latter case, it is usually best to use a merit function containing only weighted image errors and not include first-order errors or boundary condition violations.

To make the analysis more realistic, use a polychromatic criterion with at least three wavelengths (unless the system is monochromatic or all-reflecting). Also, use two or more field angles if the system has a finite field of view.

A given lens design program may have only one or two of the above criteria from which to choose. As time progresses and software improves, more criteria will probably become available.

A.16.4 Error Budget

The central question during a tolerance analysis is how much a lens as manufactured can depart from its nominal design and still be acceptable. Although the details depend on the performance criterion selected, this allowable difference between perfect and actual yields the margin for error or the total error budget. Once a total error budget has been determined, this now fixed amount of overall error must be parceled out among all of the individual sources of error. To determine how to do this, a sensitivity analysis is done.

A.16.5 Sensitivity Analysis

To do a sensitivity analysis of an optical design, the first step is to compile a list of all possible fabrication errors or tolerance types. This is often a difficult and subtle task. It is unfortunately no joke that the general rule is Murphy's law: "If something can go wrong, it will go wrong."

The second step is to assign to each fabrication error a likely error amount or tolerance value. Here it is very important to consult with the people in the optical and machine shops to determine what they consider reasonable tolerances. Also, just in case, find out the tightest (smallest) tolerances that they can hold if they really must. Finally, investigate different ways of manufacturing various components and subassemblies. Different manufacturing methods may have important relative advantages and disadvantages.

The third step is to enter the fabrication errors and their tolerances, plus any compensators, into the computer and then run the sensitivity analysis routine. Here, the effects of both positive and negative values of each error are calculated. To predict the probable effect of all these errors together, the individual error values are combined in a root-sum-square (RSS) statistical analysis. However, an RSS analysis assumes that the errors are random, and not all errors are random; some can be systematic. For errors with a systematic tendency, the individual errors must be added directly in a special analysis separate from the usual RSS sensitivity analysis. A common example of a systematic error is a turned-down edge.

When evaluating a sensitivity analysis, look first at the overall result to see whether you remained within your total error budget. In all likelihood, you will have exceeded your budget (perhaps substantially), and thus you will have to tighten the tolerances on some of the potential errors. Next, look at the individual contributions, and, in particular, look for the big contributors. These are the errors that are most profitable to reduce. These are also the errors that can disqualify a candidate design if they cannot be controlled.

In addition, look for the small contributors. These are the errors that make little or no difference. For these errors, do not specify an excessively large tolerance just because it does not matter. Put in a reasonable tolerance and let the resultant effect on performance be small. Alternatively, you can simply delete these errors from the analysis as being negligible, and concentrate on the problem errors.

A.16.6 Iterating to Find the Final Tolerances

The above process is now repeated several times to converge on the best compromise set of tolerances. You may wish to consult again with the other experts on the project team; this is the time when the cooperative nature of a tolerance analysis is most important. If all goes well, the final tolerances are all within realizable limits and the lens can be successfully built. Otherwise, either the customer will have to relax his requirements to give relief, or the basic lens configuration will have to be changed and the whole design process repeated, or the project will have to be abandoned (this can happen).

A.16.7 Reoptimization for Known Fabrication Errors

After a particular lens component (including the raw glass) has been fabricated, its actual as-made properties can be measured. With the measured values substituted into the lens prescription and frozen, the lens can be reoptimized to allow the remaining free variables to change and compensate for the known errors.

The reoptimization for glass index errors is called a melt fit. At the beginning of optical fabrication, one of the first steps is buying the glass. When the glass arrives from the manufacturer, each piece is accompanied by a melt sheet. These sheets give actual measured indices for your particular glass melts. At this time you should substitute these indices into your lens prescription and reoptimize the design. For most applications, regular melt sheets are usually satisfactory. However, for critical applications, such as when secondary color is being controlled, the designer must request that the glass manufacturer supply precision melt sheets giving the index measured at more wavelengths and to a greater number of decimals.

As fabrication continues and the lens elements are made, actual curvatures and glass center thicknesses can be measured. When these measured values are substituted into the lens prescription and frozen, the remaining variables can be adjusted slightly in a further reoptimization. These reoptimizations are called curvature and thickness fits.

Note that the tolerance values for the measured parameters now become the errors of measurement, not the much larger statistically expected errors of fabrication. Note too that the above compensations and reoptimizations are only practical for very small production runs of expensive lenses. For larger production runs of more conventional lenses, the statistically expected errors may be unavoidable. Here, there may be no choice but to specify tight tolerances for the critical parameters.

A.16.8 Test Plate Fit

There is one further procedure that is often used if test plates are to be used to check curvatures and if the additional cost of making custom test plates is to be avoided. This is called a test plate fit. The designer requests from the optical shop a list of all their currently available test plates, each with its measured radius of curvature. The designer then tries to slightly modify the design of his lens to make each of its surface curvatures fit one of the available test plates. One surface at a time is fitted, and the lens is reoptimized after each fit. The strongest or most sensitive surfaces are fitted first. Fitting both surfaces of a given element is postponed for as long as possible. The longer an element has one variable curvature, the longer the power of the element remains a variable during the reoptimizations.

If the designer is lucky, no new test plates will have to be made and little performance will have to be sacrificed. If performance begins to degrade, one or two new test plates may have to be made. Alternatively, another optical shop, which has accumulated a different set of test plates over the years, might be considered. If the production run is large, however, the cost of making all new test plates is negligible, and the economy of a test plate fit is unnecessary.

A.16.9 Recent Advances in Fabrication/Testing, Optomechanics, and Active/Adaptive Optics

In recent decades, optical shop techniques have made major advances. It is now possible to produce optical components with unprecedented properties. In particular, aspheric and eccentric (off-axis) surfaces can be made today that would have been out of the question a short time ago. The new fabrication techniques include: computer-controlled polishing, deformable laps, stress/flex polishing, ionic polishing, and diamond turning. The new testing techniques mainly include applications of interferometers, such as computer-generated holograms. Of course, computer analysis of more conventional test data is now universal. In a related development, computers plus interferometers now allow the systematic alignment of general tilted/decentered systems.

Optomechanical design and construction have also made major advances. Of special importance are the new techniques of finite element analysis. Again, computers have made the difference. Using these techniques, the optomechanical engineer can now model complex structures and accurately predict flexures caused by loads. The result is much better control over alignment, and thus reduced tilt, decenter, and despace error tolerances.

Finally, the new techniques of active/adaptive optics are having a major impact on certain types of optical systems. There is a new generation of very large telescopes under construction that relies on active (slow time response) optics to control both system alignment (collimation) and the exact shape of the primary mirror. Without this computer control, these new, high-performance systems would not satisfy the required tolerances.

Adaptive (fast time response) optics are even more revolutionary. In a big telescope, a small auxiliary deformable mirror is computer-controlled to remove the blurring effects of atmospheric turbulence in real time. To do this seeing compensation, the errors in the incoming wavefront are sensed, and the required surface shape on the deformable mirror needed to cancel out these errors is determined. To follow the rapidly changing atmosphere, the figure on the deformable mirror must be updated approximately every millisecond. The result is a gain in angular resolution of more than ten times. By this means, large ground-based telescopes have, for the first time, been able to realize nearly their full diffraction-limited resolution capability.

This completes the discussion outlining the basic optical concepts and techniques used in computer-aided lens design. In the following chapters, these ideas will be applied to several actual design examples.

Part B

Design Examples

Chapter B.1

Achromatic and Apochromatic Doublets

The first achromatic doublet lens was made in 1729 for use as a telescope objective lens. After more than two and a half centuries, achromatic doublets remain one of the most important and widespread optical forms. In addition to telescopes (including binoculars), they are found in microscopes, magnifiers, eyepieces, camera lenses, collimators, and many other types of optical instruments.[1]

Telescopes with lenses are called refractors, as opposed to reflectors using mirrors. The largest refracting telescope ever made was the Great Paris Refractor, built for the 1900 Paris Universal Exhibition (World's Fair). Its doublet lens had a clear diameter of 1250 mm (49.2 inches). However, when the fair closed, the Paris telescope was disassembled and never used again.[2] The next largest refractor, the 40-inch Yerkes telescope (1016 mm), completed in 1897, is still in service.

Incidentally, the term objective for a lens refers to the lens nearer the object. The older term is object glass. This is in contrast to the lens nearer the eye, the eyepiece.

In the present chapter, the process is described for designing airspaced telescope doublets with either achromatic or apochromatic correction. Four examples are included. The first two are classical achromats working at $f/5$ and $f/15$. The second two are apochromats, both working at $f/15$ but with different glasses. As throughout this book, many of the optimization details are specific to the ZEMAX program, although the concepts are generally applicable.

B.1.1 Achromatization

If two singlet lenses have the same first-order properties, but one is made of crown glass and the other is made of flint glass, then the flint lens will have a larger chromatic shift in BFL than the crown lens. It then follows that two lenses, one crown and one flint, can be made whose focal lengths are of opposite signs and unequal magnitudes, but whose chromatic focal shifts are opposite and equal. When put together, longitudinal color cancels out, but the combination has a finite focal length. This type of compound lens is called an achromatic doublet or achromat. The word achromatic means no-color.

Actually, the color cancellation is not perfect. With ordinary glasses, only two wavelengths can be corrected to have exactly the same BFL. Fortunately, the lens designer can choose the two wavelengths by adjusting the powers of the two lens elements during optimization. Other wavelengths retain small BFL errors. These remaining focal shifts are called secondary longitudinal chromatic aberration, secondary color, or secondary spectrum. Lens designers say that primary color is corrected and secondary color remains. Normally, the magnitudes of secondary color focal shifts are very much smaller than primary color focal shifts.

[1]For a discussion of the development of achromats and other forms of telescopes, see Henry C. King, *The History of the Telescope*.

[2]See "The Great Paris Telescope Fiasco," in Joseph Ashbrook, *The Astronomical Scrapbook*, pp. 179–183.

Secondary color (both longitudinal and lateral) is caused by using crown and flint glasses having different partial dispersions; that is, by combining short and long glasses. This mismatch in the shapes of the dispersion curves causes the imperfect color error cancellation. Thus, controlling secondary color is a glass selection problem.

Conventional achromatic doublets have both elements made of ordinary glass types. In the visible wavelength range, all of these achromats have about the same amount of secondary color. It makes no difference whether they are cemented or airspaced. Controlling secondary color requires that one or both of the lens elements be made of unusual glass types. These unusual glasses, unfortunately, are few in number, expensive, difficult to work in the shop, susceptible to weathering, and have other undesirable properties. Many of these "glasses" are optical crystals. Nevertheless, they are indispensable. In the infrared, there are many fewer transmitting optical materials than in the visible, but some of these have unusual properties that allow amazing color correction.

If relative to ordinary glasses, the selected unusual glasses only reduce secondary color, then the resulting lens is called an achromat with reduced secondary color. But if by using very special glasses, three wavelengths are made to have the same BFL, then the lens is called an apochromat (pronounced with the accent on the first syllable, not the second). In an apochromat, secondary color is said to be corrected, and the remaining chromatic BFL errors are called tertiary color or tertiary spectrum. Tertiary color is usually much smaller than secondary color. In extreme cases, four wavelengths can be made to have the same BFL. These lenses are called super-achromats (not super-apochromats), and they have quaternary color.

Note that all of what has been said about the achromatization of a doublet lens applies equally well to more complex lenses with multiple elements of many glass types.

B.1.2 *F*/5 Achromatic Doublet with BK7 and F2 Glasses

The first design example is an airspaced $f/5$ achromatic doublet for use as a telescope objective lens. The object is at infinity (this object distance is assumed for telescopes and camera lenses unless otherwise stated). Clear aperture or entrance pupil diameter (EPD) is 150 mm (about 6 inches). The glass disks are slightly oversized to allow mounting the lenses in a cell. Focal length (EFL) is 750 mm, giving a focal ratio of $f/5$. Field of view (FOV) is $\pm 0.50°$ (twice the angular size of the full moon), with image quality emphasized in the center of the field. All lens surfaces are purely spherical. The glasses, taken from the Schott catalog, are BK7 crown and F2 flint, both of which are ordinary glasses. Glass thicknesses are arbitrary and are adjusted during optimization for minimum practical center and edge thicknesses. The airspace between elements is also arbitrary, and 2 mm has been adopted. This airspace is small, but large enough to allow a slight later respacing to compensate for minor measured fabrication errors. The image surface is flat and is located at the paraxial focus of the central wavelength.

Three wavelengths are used during optimization: 0.4861, 0.5461, and 0.6563 μm. The central wavelength, 0.5461 μm, is the reference wavelength and is used for calculating first-order properties, solves, and monochromatic aberrations. It is also the wavelength of the mercury green line (mercury emission lamps are commonly available in optical shops for testing lenses during fabrication). The two extreme wavelengths, 0.4861 and 0.6563 μm, are used for calculating the chromatic aberration. These are also the Fraunhofer F and C lines, the classical wavelengths used for achromatizing for photopic visual response.

In choosing the basic lens form, the common crown-in-front Fraunhofer configuration is selected. This choice is in preference to the less common flint-in-front Steinheil configuration, which has optical properties very simi-

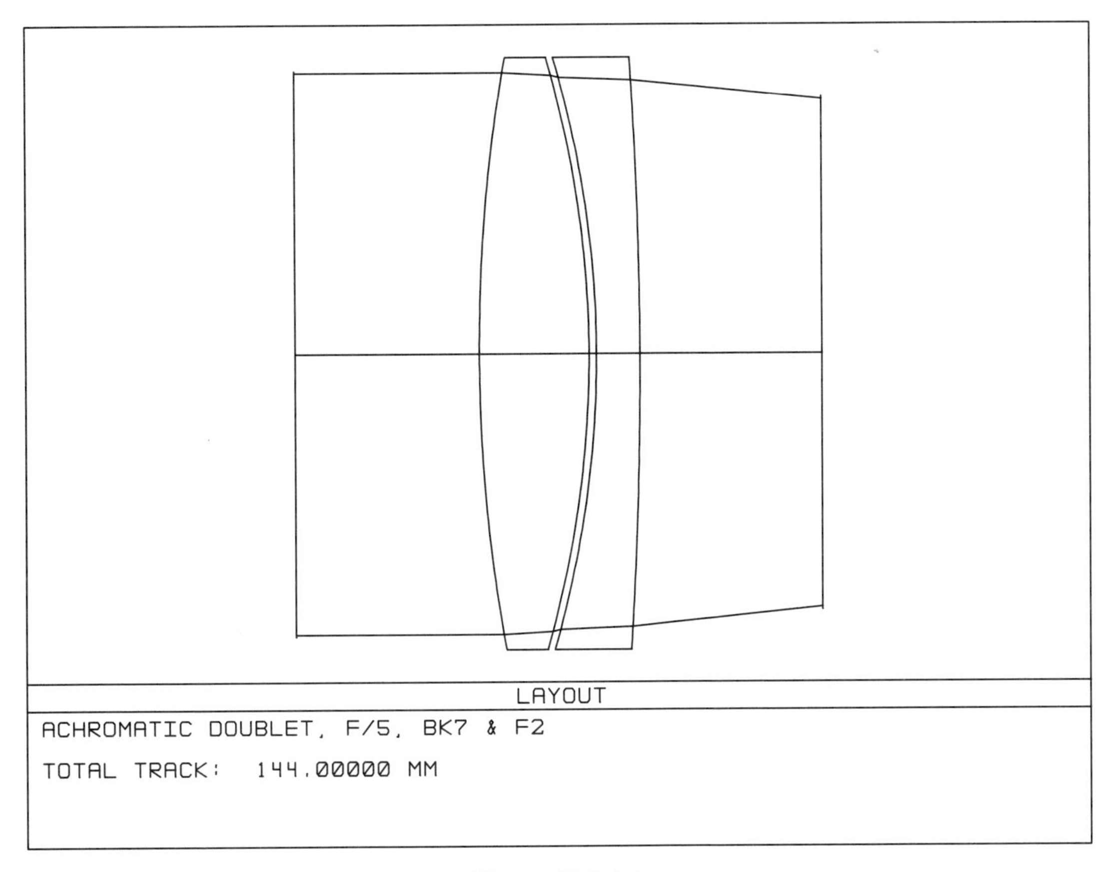

Figure B.1.1.1.

lar to the Fraunhofer form. This choice is also in preference to the alternative crown-in-front solution, the rarely used double-meniscus Gauss configuration, which has quite different optical properties.

From published drawings, you know roughly what a Fraunhofer airspaced achromat looks like. The doublet consists of (1) a front double-convex positive crown element and (2) a rear nearly plano-concave negative flint element. The two surfaces at the airspace have nearly equal, but nevertheless different, curvatures. Figure B.1.1.1 shows a layout of the final lens.

In any achromatic lens, it is of vital importance to select glasses with significantly different V_d dispersions. Dispersion difference directly affects the powers of the individual elements. If the dispersion values are too close to each other, then in order for achromatization to work, the powers of the individual elements must be large; that is, a large amount of positive power must balance a large amount of negative power. With a greater dispersion difference, the individual element powers are less. Lower powers mean weaker surface curvatures, and weaker curvatures mean less unpleasant higher-order aberrations. BK7 and F2, the classical glass pair for larger airspaced telescope doublets, have dispersions of 64.2 and 36.4, respectively, giving a dispersion difference of 27.8. This is normally sufficient.

In addition to the actual lens surfaces, three dummy surfaces have been placed in the lens prescription. Dummy surfaces have the same material, in this case air, on both sides, and they are usually flat or plane. No refraction or reflection takes place at a dummy surface, and thus they have no effect on the lens itself. They are inserted as required to model the lens in the computer or for other special reasons.

When doing a layout, it looks better to include short lengths of rays in front of the first actual lens surface. A plane dummy surface a short distance in front of the lens facilitates this drawing (with CODE V, these extra lengths are drawn automatically). Similarly, if a layout is to be drawn of just the lens (without the focal surface), dummy surfaces can be placed on both sides of the lens. Thus in the present case, dummy surfaces have been placed 50 mm in front of and following the lens. Note their use in Figure B.1.1.1.

The third dummy plane surface follows the second and is located at the paraxial focus of the lens. This placement can be accomplished in more than one way, but here a height solve is used on the thickness separating the two surfaces. The height solve adjusts the thickness so that the height of the first-order (Gaussian) marginal ray on the third dummy surface is zero, thereby placing the surface at the focus. The actual image surface follows the dummy paraxial image plane.

If the image surface is to remain at the paraxial focus, leave the thickness following the third dummy surface zero. However, to allow a small defocus from the paraxial focus, this thickness can be made an independent variable. In the present lens, defocus is kept zero. In this case, the third dummy surface could actually be eliminated, and the height solve placed on the thickness between the second dummy surface and the focal surface. However, the third dummy surface has been added here for the sake of generality and for good technique. In a later example in this chapter, defocus from the paraxial focus will be used.

Before proceeding to optimization, make a sketch of the anticipated lens. Sequentially number the surfaces, including all dummy surfaces, from surface 1 to the image surface (the object surface is surface 0). This sketch will help you to keep the geometry straight in your mind.

You are now ready to begin work at the computer. As discussed in Chapter A.15, optimization is done in stages. The first stage is a rough early optimization.

Without a previously optimized lens to enter, you must derive your own starting design. The specifications given above supply many of the required lens parameters, but the curvatures and glass thicknesses are unknown. When designing with a computer, perhaps the best way to get starting values for these unknowns is to make educated guesses. Fortunately, these guesses need not be very good because the optimization routines in modern lens design programs can fix even poor guesses (with the designer's guidance).

In a Fraunhofer doublet, any reasonable glass thicknesses can be initially chosen. To get initial curvatures, you can use two simplifying assumptions; these also shepherd the design toward the Fraunhofer configuration. First, using pickup solves, make and keep both surfaces of the crown element, and the front surface on the flint element, all the same curvature with appropriate signs. Second, make and keep the rear of the flint element flat. The result is a lens that looks about right with only one independent variable or degree of freedom. This is the single curvature value, and it must be used to correct EFL to the specified 750 mm.

Now, with your guesses, enter the starting lens configuration into the computer in the manner prescribed by the program's manual. Immediately look at the layout and the first-order data. You will probably find that focal length is quite a bit off. You may have other problems as well, such as rays missing surfaces, total internal reflection, or inappropriate glass thicknesses. If so, revise your set of guesses and try again. After a bit of tinkering, you will converge on a reasonable starting design that is close enough.

Next, create a merit function controlling only focal length; that is, select the EFFL operand and set its target to 750 mm. Then use the optimization routine to change the one independent curvature variable to get exactly this required focal length. Note that here you cannot use an angle solve on the rear lens surface to control focal ratio (and thus focal length) because the curvature of the rear surface is not yet a free variable. After optimization, do another layout and check the first-order properties to verify that the optimization did what you intended.

Note that in general, it is good practice to save your results after each optimization stage. This is insurance. If something later goes wrong, or you just wish to try something different, you do not have to go back to the beginning and start over.

For the intermediate optimization, remove the pickups and allow all four lens curvatures to be independent variables. These four degrees of freedom can also be viewed as two lens element powers and two lens element bendings. Recall that element power is the reciprocal of element focal length, and bending a lens element means changing its shape without changing its power.

In an airspaced doublet, these four curvature variables are the only effective degrees of freedom available. Unfortunately, the interelement airspace is not an effective variable because varying it duplicates the effect of varying one of the curvatures (more will be said about this later). Glass thicknesses, if kept reasonably thin, are also not effective in controlling aberrations. And glass selection has very little effect on final performance, provided that a large dispersion difference is maintained and that only common glasses are allowed. Thus, thicknesses and glasses are usually chosen for convenience or to facilitate fabrication. During the optimization computations, thicknesses and glasses are frozen (made fixed or nonvariable). Two exceptions, not considered here, are a cemented doublet (requiring perfect internal contact and careful glass selection) and an airspaced doublet with edge contact (requiring no spacer).

Recall from earlier chapters that four effective degrees of freedom allow you to control four optical properties and no more. This number is far short of the eight required to simultaneously address all of the basic seven aberrations plus focal length (not to mention higher-order aberrations). Consequently, this system limitation severely restricts the way that an achromatic doublet is optimized; that is, the optimization must be very selective and not aim for perfect imagery.

To have at least some region of good imagery, the center of the field must be as well corrected as possible. Thus, the two on-axis aberrations, longitudinal color and spherical aberration, must be controlled. And focal length must continue to be corrected. Thus, there is only one degree of freedom left for the off-axis aberrations. Because coma is the most serious off-axis aberration, it is the one chosen to be controlled. Other aberrations must remain uncontrolled. However, if the lens elements are thin and close together, and if the stop is at the lenses, then lateral color and distortion are both automatically small (the chief ray passes close to the centers of the lenses and is thus nearly undeviated). Secondary longitudinal color, zonal spherical aberration, spherochromatism, astigmatism, and field curvature all remain potentially large.

The achromatic doublet is one of those cases where correcting first-order chromatic and third-order Seidel aberrations moves the design directly and reliably to the right approximate solution. The computer will go to the Fraunhofer solution, rather than the alternative Gauss solution, because you shepherded the lens toward the Fraunhofer solution in the early optimization. The required very simple intermediate merit function contains only four operands. They are: EFFL to continue to correct paraxial focal length, AXCL to control paraxial longitudinal color, SPHA to control third-order spherical aberration, and COMA to control third-order coma. These operands are all user-selected from the operands built into the program.

Note that you have four effective independent lens variables and four optimization operands. Thus in this case, it is possible to correct the values of all four operands exactly. During optimization, the merit function converges to a small value (not quite zero in practice). Although the lens is not perfect, it is perfectly satisfying a restricted merit function. A least-squares best-fit compromise is not required, although the damped least-squares optimization routine in the computer program is still used to do the calculations.

In the optimization, the four operands can all be weighted equally. Alternatively, the three aberration operands can be given equal ordinary weights, and a Lagrange multiplier can be used to hold focal length at 750 mm from the early optimization. Another alternative is to use Lagrange multipliers to correct all four operands exactly.

How well each of these options works for you may depend on your particular computer program. Some optical design programs handle Lagrange multipliers better than others when the values of the operands are initially far from their targets. If, when using Lagrange multipliers, your program stalls or blows up, switch back to weighted operands.

As yet another alternative, EFL can now be corrected by an angle solve on the rear lens surface. As used here, the angle solve adjusts the curvature of the surface to make the angle (slope) of the first-order marginal ray leaving the surface equal to some specified value. Thus, the angle solve controls the half-angle of the on-axis cone of light incident on the image surface. This angle controls focal ratio, which in turn controls focal length for a given EPD. In the present case, the desired focal ratio is $f/5$. Thus, the slope of the marginal ray must be $^1/_{10}$ (equal to the entrance pupil semi-diameter divided by the desired EFL), with a negative sign if the ray passes through the top of the lens and is sloping down.

During intermediate optimization, the design should converge very quickly to a solution. After optimization, look at the values of the merit function operands to check that they really are near their targets. Then look at the layout as a sanity check. While looking at the layout, check that the glass thicknesses that you previously chose are still reasonable. If you wish to revise these thicknesses, make the changes by hand and then reoptimize. Always reoptimize the lens after making any changes by hand.

Next, look at the ray-intercept ray fan plots for the on-axis image, shown in Figure B.1.1.2. Because this is an on-axis image, the tangential and sagittal plots are identical and redundant, and each of the plots is symmetrical (identical when rotated about the origin by 180°). The curves for the three wavelengths are distinguished by different dashed lines.

Look at the curve for the central wavelength. First, note the zero slope at the origin, indicating correct paraxial focus. Second, note the upward on the right and downward on the left curvature, indicating spherical aberration of the overcorrected type. This spherical must be fifth-order because third-order has just been corrected to zero.

Then look at the two curves for the extreme wavelengths. First, note the general inclination, indicating defocus caused by secondary color. Second, note that the nonzero slopes at the origin are equal, indicating equal paraxial defocus. This equal paraxial defocus is just what the paraxial longitudinal color optimization operand is supposed to accomplish. Third, note the different curve shapes, indicating spherochromatism.

This intermediate design solution is now very close to the final solution. All that remains is to make some last adjustments during the final optimization while continuing to hold focal length at 750 mm.

To reduce the effect of fifth-order spherical aberration, deliberately reintroduce a small amount of third-order spherical to balance the fifth-order. One way to achieve this balance is to use real trigonometric rays and cause the on-axis RMS monochromatic spot size for the central wavelength to be minimized during optimization. But shrinking spot size requires many rays to be traced.

There is a second approach that accomplishes the same thing while requiring only one ray to be traced. Select the central wavelength ray from the on-axis object point that passes through the 0.9 pupil zone. During optimization, make the intercept height of this ray on the paraxial image surface zero. When only third- and fifth-order spherical are present (higher-orders are negligible and there is no paraxial defocusing), this approach yields nearly the same solution as a full RMS spot shrink optimization. The method is described in Chapter A.13 in the section on spherical aberration and is adopted here.

Whether shrinking spot size or using the single ray method, continue using

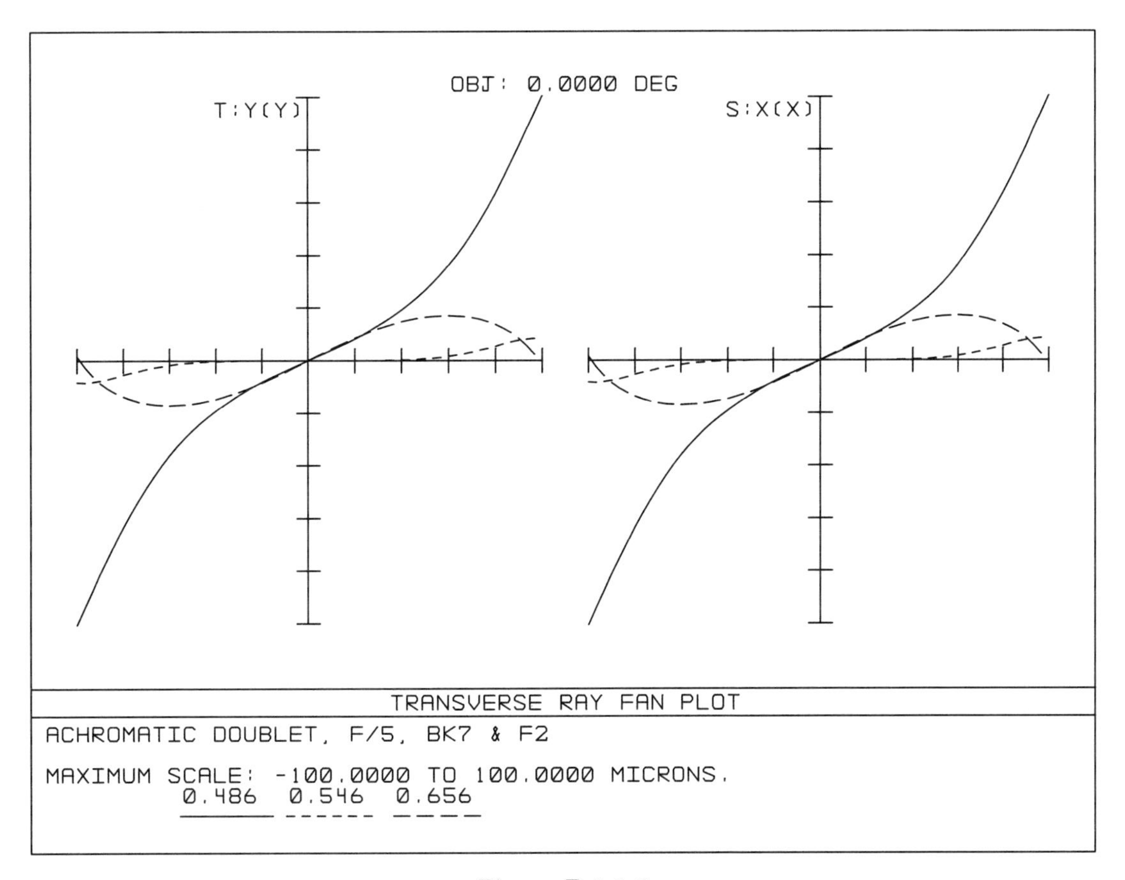

Figure B.1.1.2.

the height solve to keep the image surface at the paraxial focus for the central wavelength.

To balance secondary color and spherochromatism, longitudinal color is now corrected, not paraxially for the pupil center, but for an outer pupil zone using a special user-defined operand. The choice of zone height is a matter of preference, but the 0.8 zone is usually effective. The method is described in Chapter A.13 in the section on longitudinal color.

To correct coma, either tangential or sagittal coma can be corrected with real rays for the edge of the field and with rays passing through the edge of the pupil. Again, the method is described in Chapter A.13.

The resultant final merit function is given in Listing B.1.1. To speed convergence, focal length is weighted less than the aberrations. During optimization, all four weighted operands are exactly corrected, and the value of the merit function again converges to a small value (very close to zero this time).

Incidentally, this final merit function is so effective that the intermediate optimization step can usually be skipped. The lens derived from the early optimization can be reoptimized directly with the final merit function for the same final result.

After final optimization, again do a layout. It should look like Figure B.1.1.1. Then do a ray fan plot for the on-axis image. This should look like Figure B.1.1.3. Look at the curve for the central wavelength and examine the zonal spherical. At the origin the slope of the curve is zero, indicating that the paraxial rays are in focus. For intermediate zones, a small amount of undercorrected third-order spherical is evident. For outer zones, overcorrected fifth-order spherical predominates. There is one zero crossing and it is at the 0.9 pupil zone, as specified.

Next look at the curves for the two extreme wavelengths. First, note that both curves are inclined, indicating defocus caused by secondary color. The positive slopes indicate that their images focus beyond the image surface. Second, note that

Listing B.1.1

Merit Function Listing
Title: ACHROMATIC DOUBLET, F/5, BK7 & F2

Merit Function Value: 7.30821798E-012

Num	Type	Int1	Int2	Hx	Hy	Px	Py	Target	Weight	Value	% Cont
1	EFFL		2					7.50000E+002	1	7.50000E+002	0.012
2	BLNK										
3	BLNK										
4	AXCL							0.00000E+000	0	6.07767E-001	0.000
5	REAY	8	1	0.0000	0.0000	0.0000	0.8000	0.00000E+000	0	3.15882E-002	0.000
6	REAY	8	3	0.0000	0.0000	0.0000	0.8000	0.00000E+000	0	3.15882E-002	0.000
7	DIFF	5	6					0.00000E+000	10	1.01608E-011	62.354
8	BLNK										
9	BLNK										
10	SPHA	0	2					0.00000E+000	0	1.09552E+000	0.000
11	REAY	7	2	0.0000	0.0000	0.0000	0.9000	0.00000E+000	10	5.52092E-012	18.409
12	BLNK										
13	BLNK										
14	COMA	0	2					0.00000E+000	0	-2.60245E-002	0.000
15	TRAY		2	0.0000	1.0000	0.0000	1.0000	0.00000E+000	0	-7.72058E-003	0.000
16	TRAY		2	0.0000	1.0000	0.0000	-1.0000	0.00000E+000	0	7.72058E-003	0.000
17	SUMM	15	16					0.00000E+000	10	-5.64171E-012	19.224
18	BLNK										
19	BLNK										
20	DMFS										

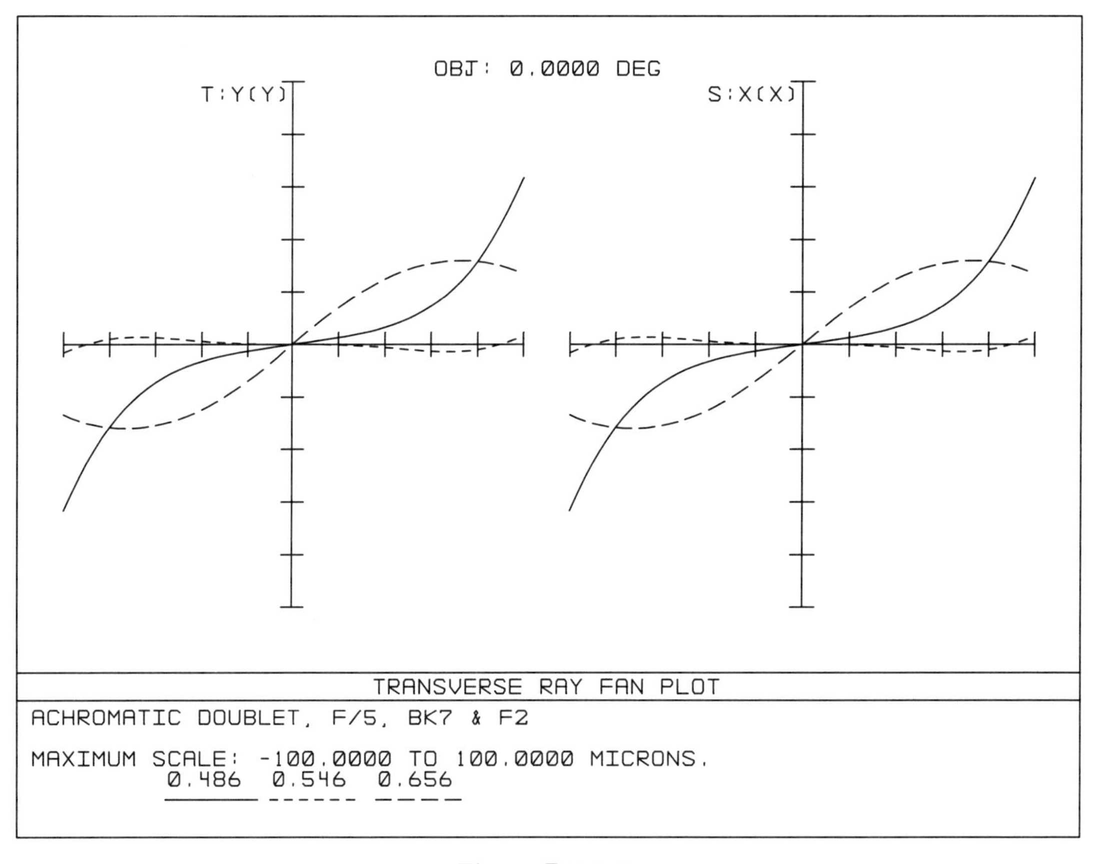

Figure B.1.1.3.

the shapes of the two curves are different, indicating spherochromatism. Third, note that the two curves cross at the 0.8 zone, as required by the special optimization operand. This gives a good balance of secondary color and spherochromatism across the pupil.

For an $f/5$ achromatic doublet with ordinary glasses, this level of aberration control is all that can be expected. The biggest on-axis image errors are secondary color and spherochromatism. Actually, the situation looks even worse if you include the entire spectral sensitivity range of the eye, panchromatic films, and many electronic image detectors. To show this, the same lens is used, the solves are removed (to prevent inadvertent refocusing), and the wavelengths are changed to 0.40,

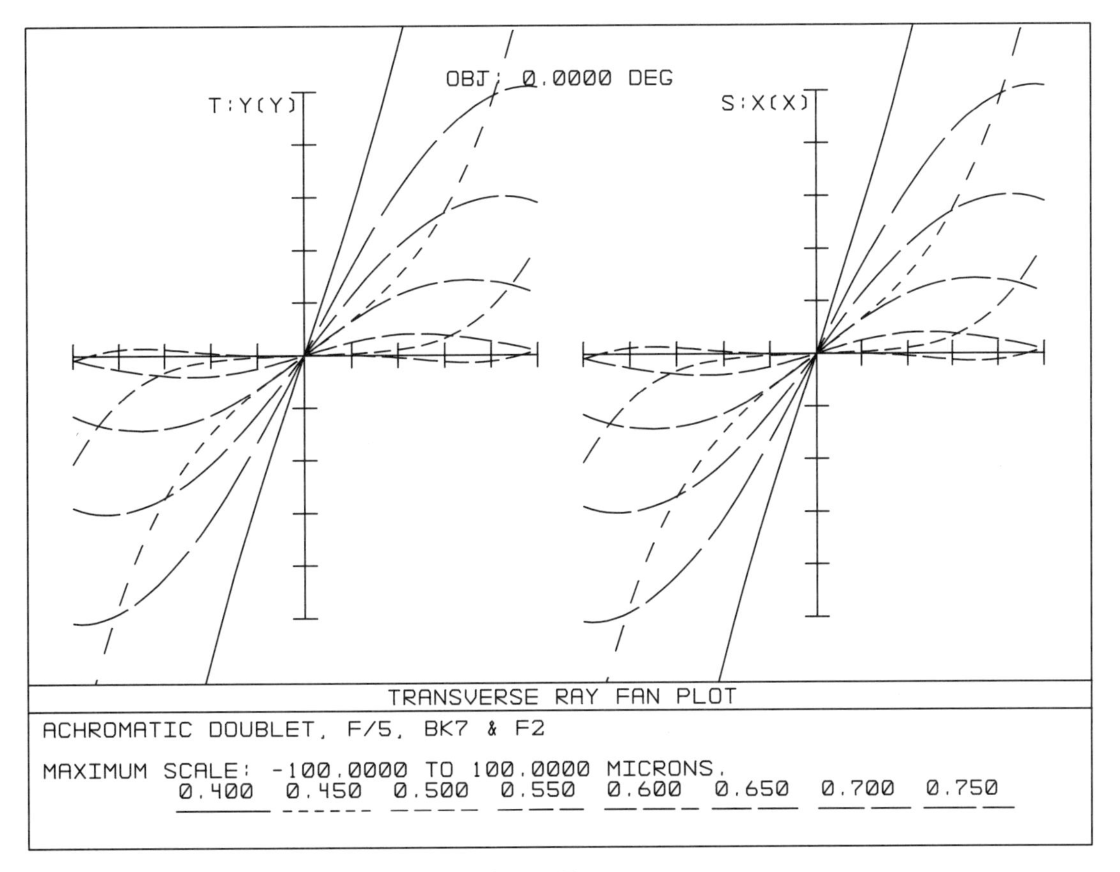

Figure B.1.1.4.

0.45, 0.50, 0.55, 0.60, 0.65, 0.70, and 0.75 μm. Figure B.1.1.4 is the resulting ray fan plot on the same scale as Figure B.1.1.3. Clearly, this lens has a lot of color!

Figure B.1.1.5 is a matrix spot diagram. The same eight wavelengths are used, and again a large amount of color flare is seen. This figure also shows off-axis performance. For a field of $\pm 0.50°$, off-axis performance is nearly identical to on-axis performance; that is, the off-axis aberrations are small compared to the on-axis aberrations (which are also present off-axis). Of course, eventually at some field angle, astigmatism and field curvature will overwhelm even these large chromatic aberrations.

Figure B.1.1.6 is the on-axis OPD ray fan plot. Note the scale: ± 25 waves. Clearly, except within a small wavelength range, this lens is nowhere near being diffraction limited. Nevertheless, lenses of this type are commonly made and widely used with success. For many applications, diffraction-limited performance is not required.

Except in very special circumstances, do not attempt to reduce the large chromatic flare by a reoptimization that tries to shrink polychromatic spot size. The computer will do a heroic balancing of secondary color, spherical aberration, zonal spherical, spherochromatism, and defocus in an effort doomed to failure. All you will get is poor images at all wavelengths. There is simply no way to reduce or mitigate the inevitable residual chromatic aberrations in a conventional achromat of a given f/number. For the same reason, do not attempt a polychromatic OPD reoptimization. The procedure outlined above gives the classical design solution for an achromatic doublet. Given the fundamental system limitations, it remains the best solution.

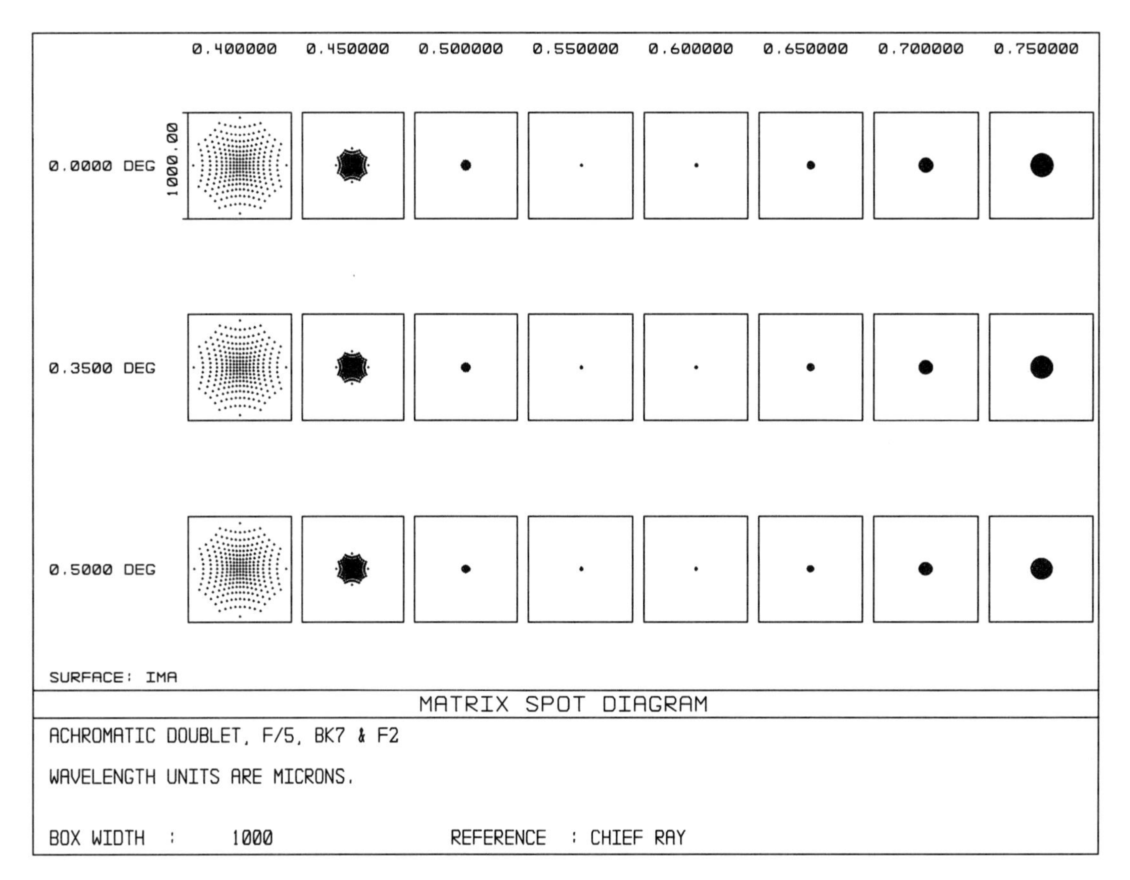

Figure B.1.1.5.

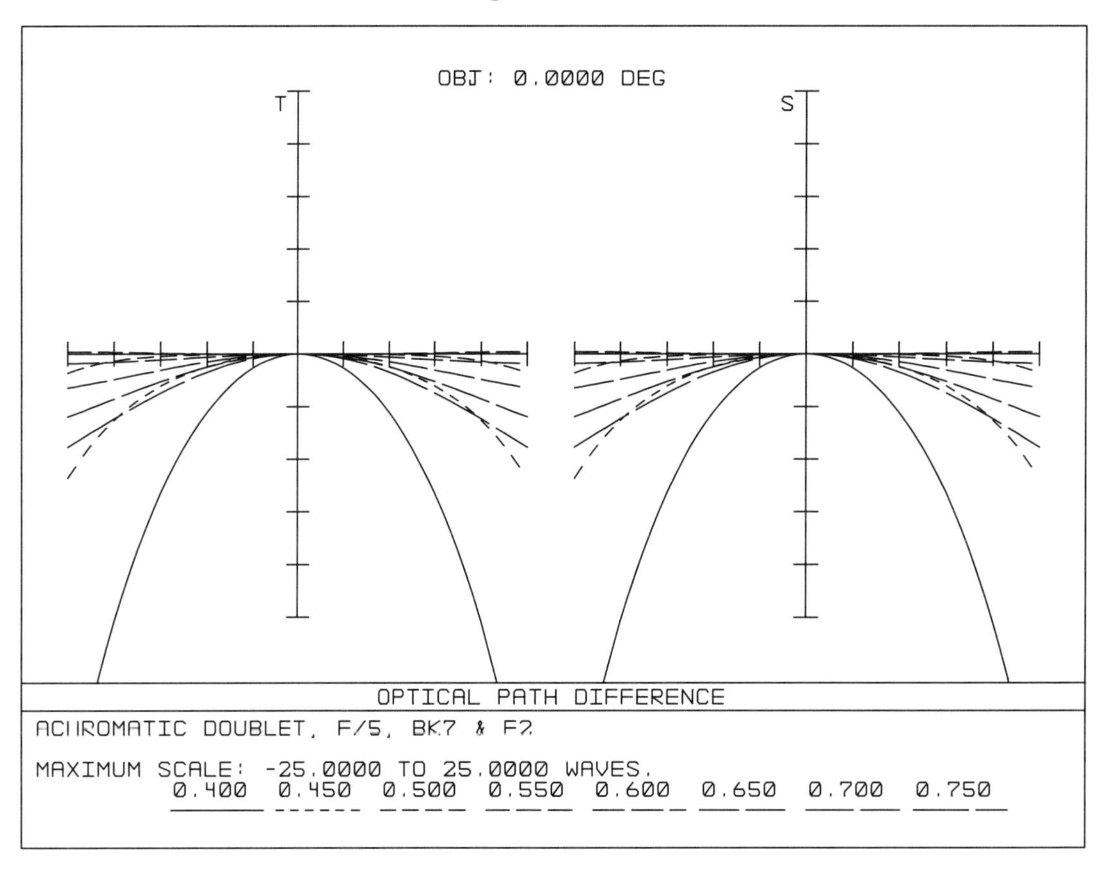

Figure B.1.1.6.

B.1.3 *F*/15 Achromatic Doublet with BK7 and F2 Glasses

If better chromatic performance is required from an achromatic doublet made with ordinary glasses, the only way is to reduce lens speed. Telescope doublets with a focal ratio of about $f/15$ have been made since Fraunhofer adopted this speed in the early nineteenth century, and they are still a standard astronomical instrument.

For an $f/15$ lens, all the system properties specified for the $f/5$ lens can be retained except focal length, which now becomes 2250 mm. The design procedure for the $f/15$ lens is also the same as for the $f/5$ lens. The final merit function is identical to Listing B.1.1. except that the EFFL target is 2250. The layout of the resulting lens is shown in Figure B.1.2.1.

Note that the internal curvatures for the lens in Figure B.1.2.1 are very close to being equal, but they are not equal. During fabrication, there may be an advantage in having these two curvatures exactly equal. This can be accomplished by a variation of the optimization procedure just described. Recall that originally the two internal curvatures were each independent variables and the airspace thickness was fixed. In the modified procedure, only the first internal curvature is an independent variable, the second is made equal by using a pickup, and the airspace is now an independent variable.

Thus, in both lens versions there are four effective independent degrees of freedom when the outside curvatures are included. Both lenses are optimized with exactly the same merit function. And when optimized, the performance of both lenses is virtually the same.

Figure B.1.2.2 is the layout of the revised $f/15$ doublet. The internal airspace has been increased from 2 mm to about 7.5 mm (still a practical value), and the four curvatures have been slightly altered.

The ray fan plot for the on-axis image is shown in Figure B.1.2.3. Note that the central wavelength is so well corrected that its curve lies on top of the horizontal axis and cannot be seen separately. Note also that spherochromatism is now small and that secondary color is by far the predominant on-axis aberration. There is still some spherochromatism, however, and the two curves for the extreme wavelengths are made to cross at the 0.8 pupil zone.

Figure B.1.2.4 is a set of ray fan plots for all three object points, located at 0°, 0.35°, and 0.50° off-axis. Note that the three plots are nearly identical, and thus image quality is nearly constant over the field of view. However, look closely at the 0.50° plot. The central wavelength curve has emerged from the horizontal axis due to a small amount of field curvature defocusing. Also note that the tangential and sagittal curves for the central wavelength show a small difference in defocusing (slope) caused by uncorrected astigmatism. Finally, note that there is no image asymmetry caused by coma. Eliminating image asymmetry is very important if the lens is to be used for astrometry or positional astronomy. The positions of astronomical images are measured to determine comet and asteroid orbits, stellar parallaxes, stellar proper motions, binary star orbits, and so forth. Asymmetric comatic images can introduce huge systematic errors.

Figure B.1.2.5 is a ray fan plot similar to Figure B.1.2.3, except that eight wavelengths are displayed (as was done in Figure B.1.1.4). Note that the scales in Figures B.1.1.3, B.1.1.4, B.1.2.3, and B.1.2.5 are all the same. Interestingly, the linear color errors of the $f/5$ and $f/15$ lenses are about the same.

Figure B.1.2.6 is a matrix spot diagram for the $f/15$ lens. When compared to the matrix spot diagram in Figure B.1.1.5 for the $f/5$ lens, the two lenses again are shown to have similar linear spot sizes.

However, the $f/15$ lens has three times the focal length and thus three times the image scale of the $f/5$ lens. Therefore, the same approximate linear spot size (on the image surface) for the two lenses translates into an angular spot size (projected onto the sky) for the $f/15$ lens that is only a third as large. This reduction is a significant advantage and is the reason why refracting telescopes are usually long

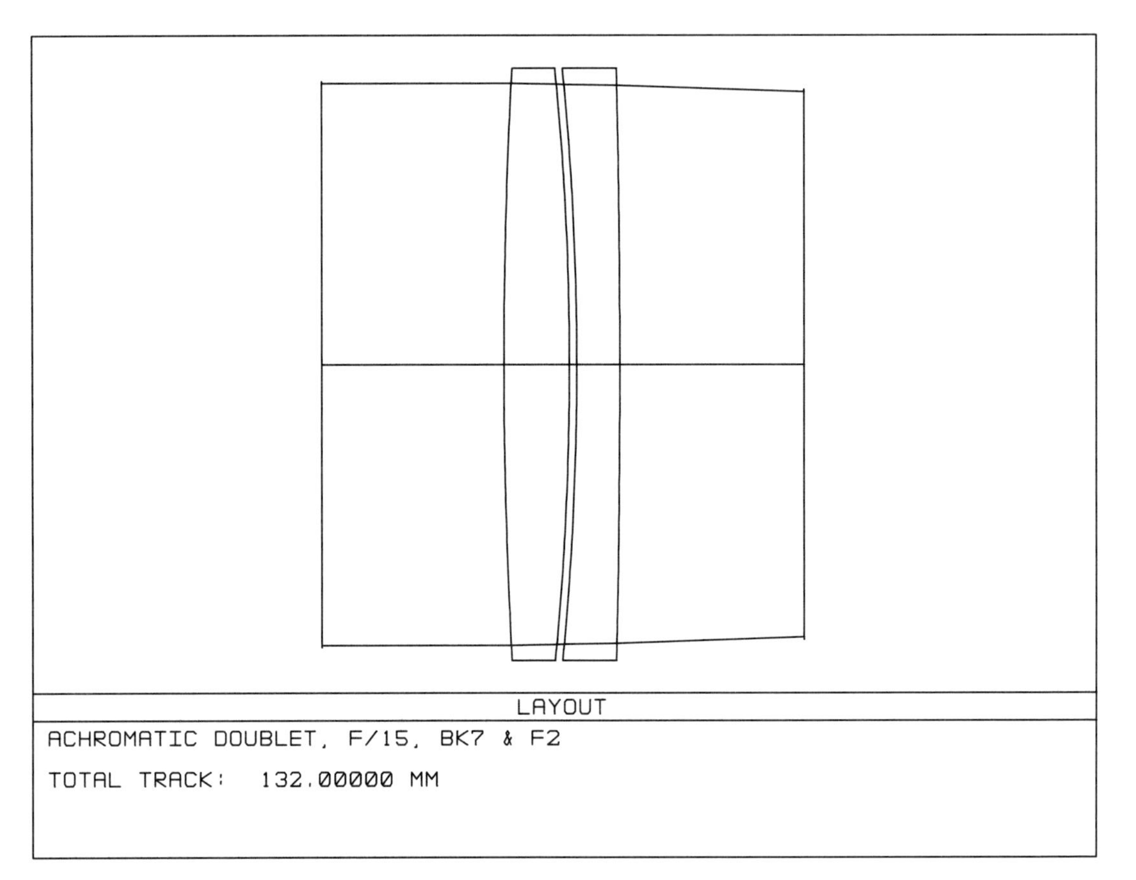

Figure B.1.2.1.

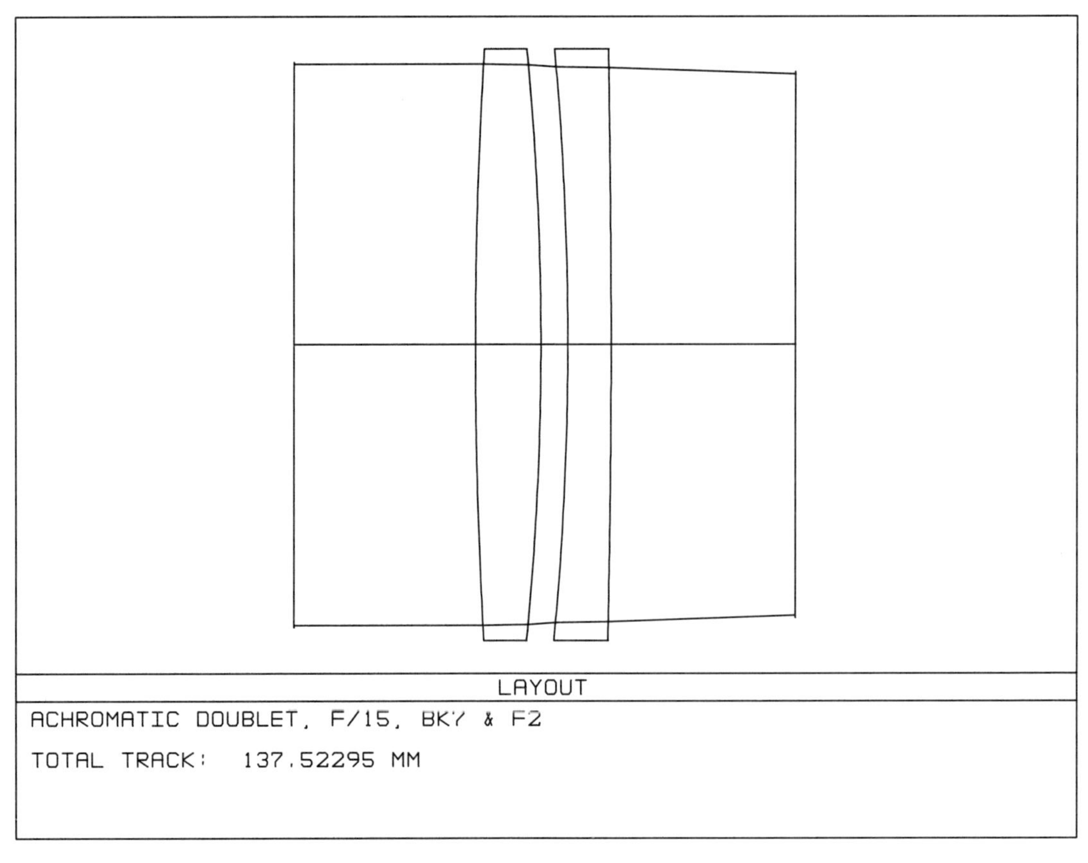

Figure B.1.2.2.

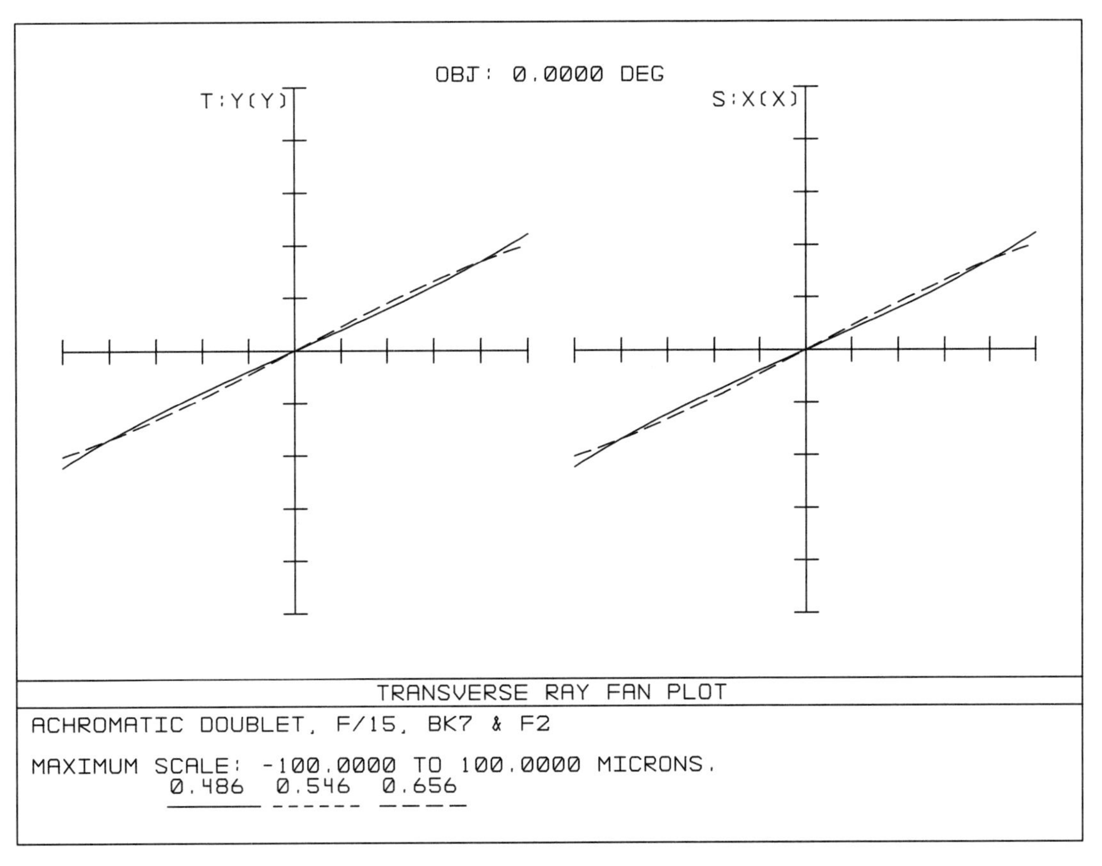

Figure B.1.2.3.

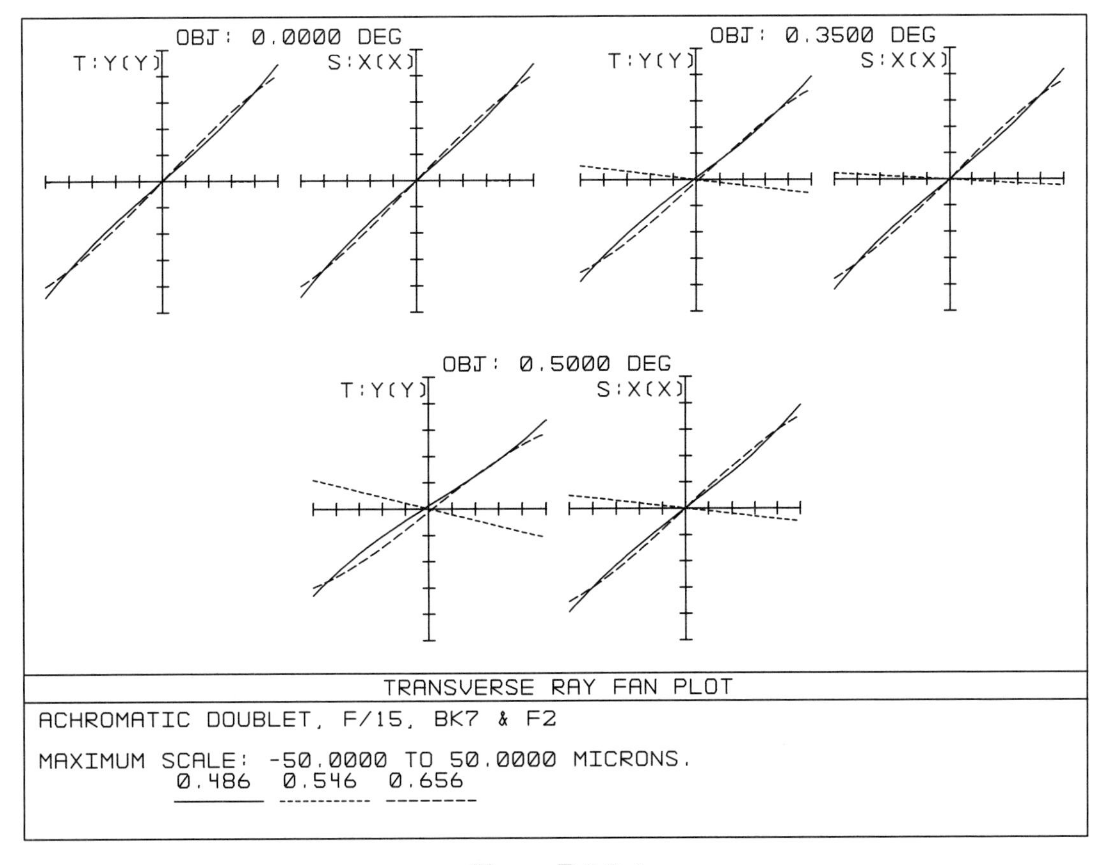

Figure B.1.2.4.

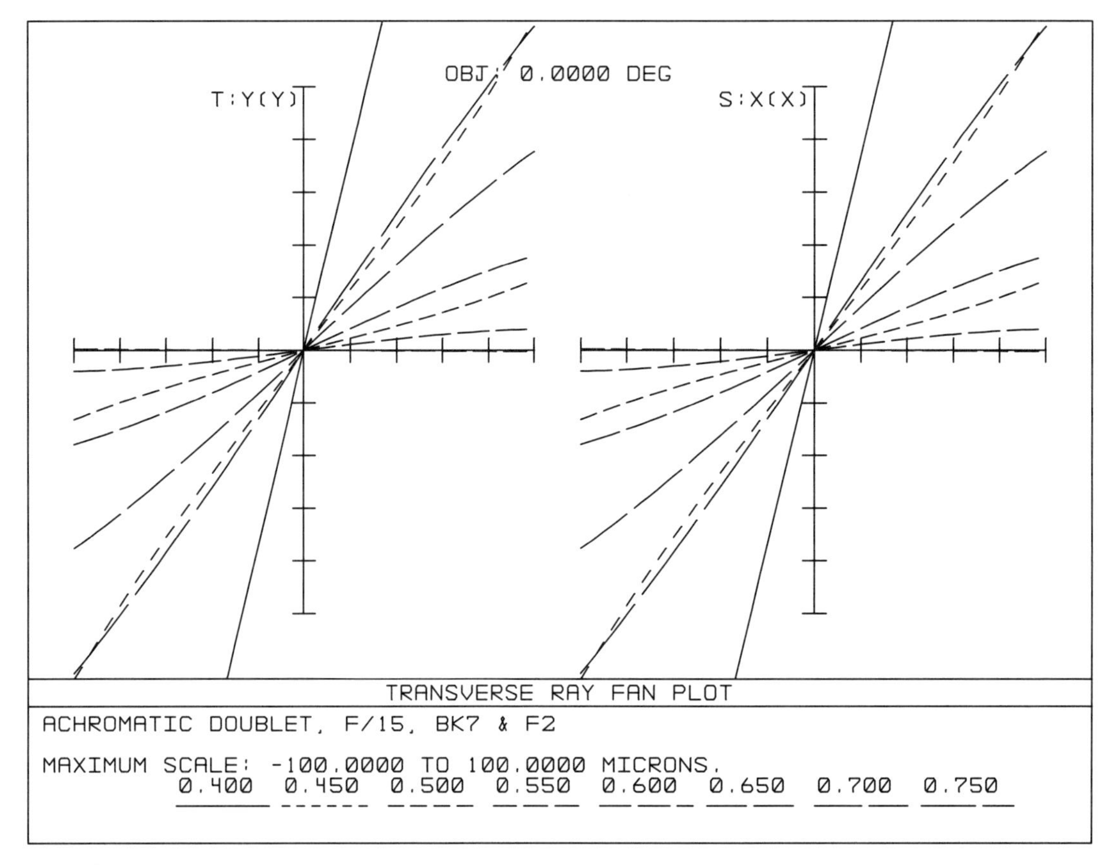

Figure B.1.2.5.

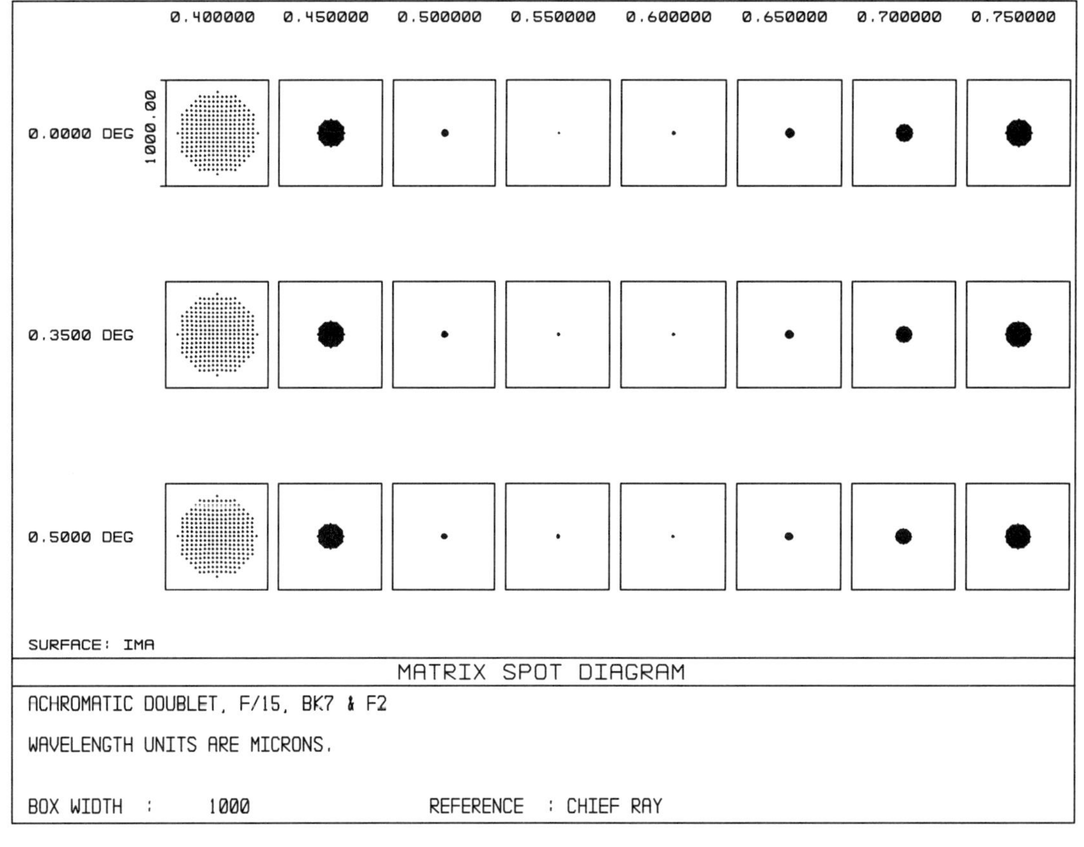

Figure B.1.2.6.

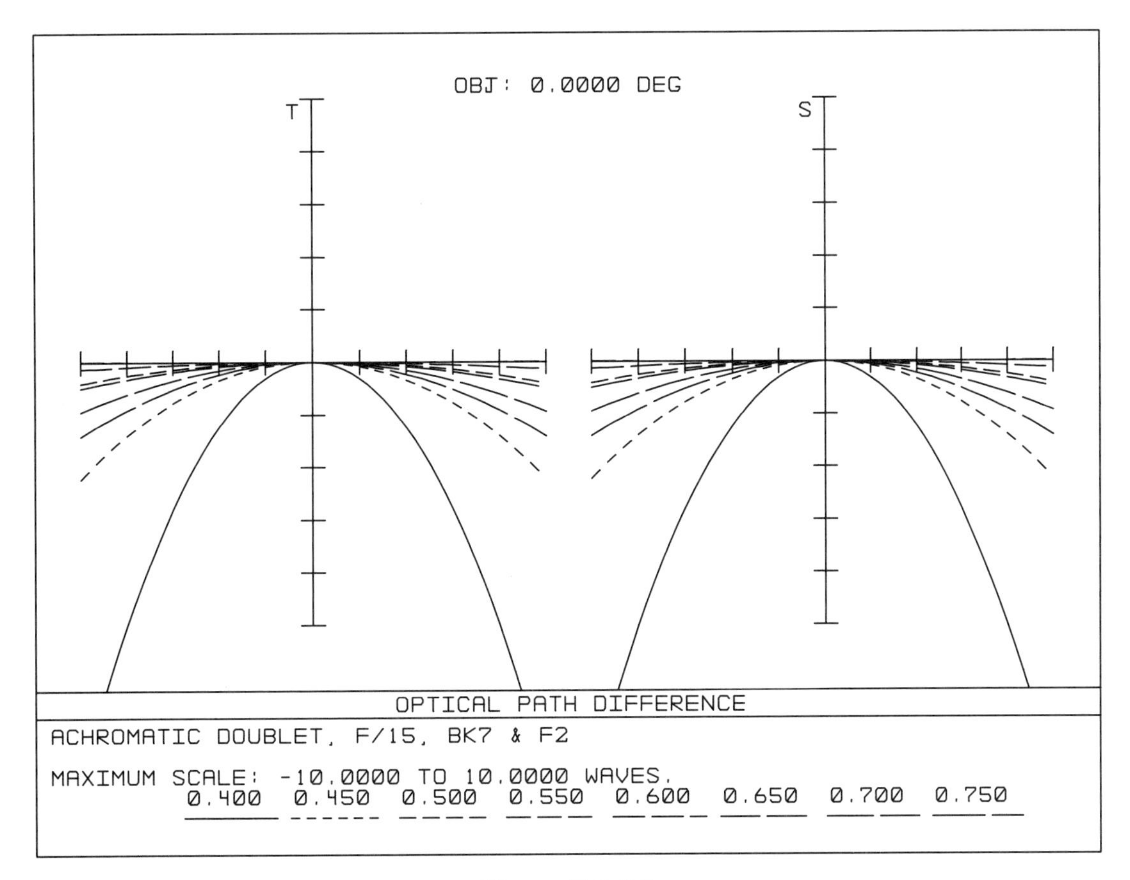

Figure B.1.2.7.

and slow.

Figure B.1.2.7 is the OPD ray fan plot. Note that compared to Figure B.1.1.6 for the $f/5$ lens, the scale for the $f/15$ lens has been reduced from ± 25 waves to ± 10 waves. The geometric linear spot sizes may be similar, but the linear Airy disk diameter for the $f/15$ lens is three times larger than for the $f/5$ lens. Thus, the aberrations in waves (relative to diffraction effects) are less.

In addition to easier fabrication, there is another advantage in using the doublet with equal internal curvatures and a larger airspace. In any lens, the unavoidable small surface reflections cause every pair of surfaces to form a ghost image of every object point. Ghosts are undesirable (especially if you are looking at bright stars on a dark background), and you want them as far out of focus as possible to dilute their effect. In a Fraunhofer doublet, the smallest and most intense ghosts are formed by reflections between the two internal lens surfaces, as illustrated in Figure B.1.2.8. In the lens with the 2 mm airspace, the ghosts are 1.4 mm in diameter. In the lens with the larger 7.5 mm airspace, the ghosts are 34 mm in diameter, as shown in Figure B.1.2.9. Clearly, the lens with the larger airspace is preferable.

The optical prescription for the $f/15$ doublet is given in Listing B.1.2. From this information, you could build the lens. However, before you did so, you might wish to replace the nominal indices with the actual indices supplied on the melt sheets for your glasses, and then reoptimize. You might also wish to do a tolerance analysis to guide the optical and machine shops in their work. And after the lens elements have been made and measured, you might wish to reoptimize the airspace between the elements to compensate for any departures of the actual lens surface curvatures and glass thicknesses from their nominal values.

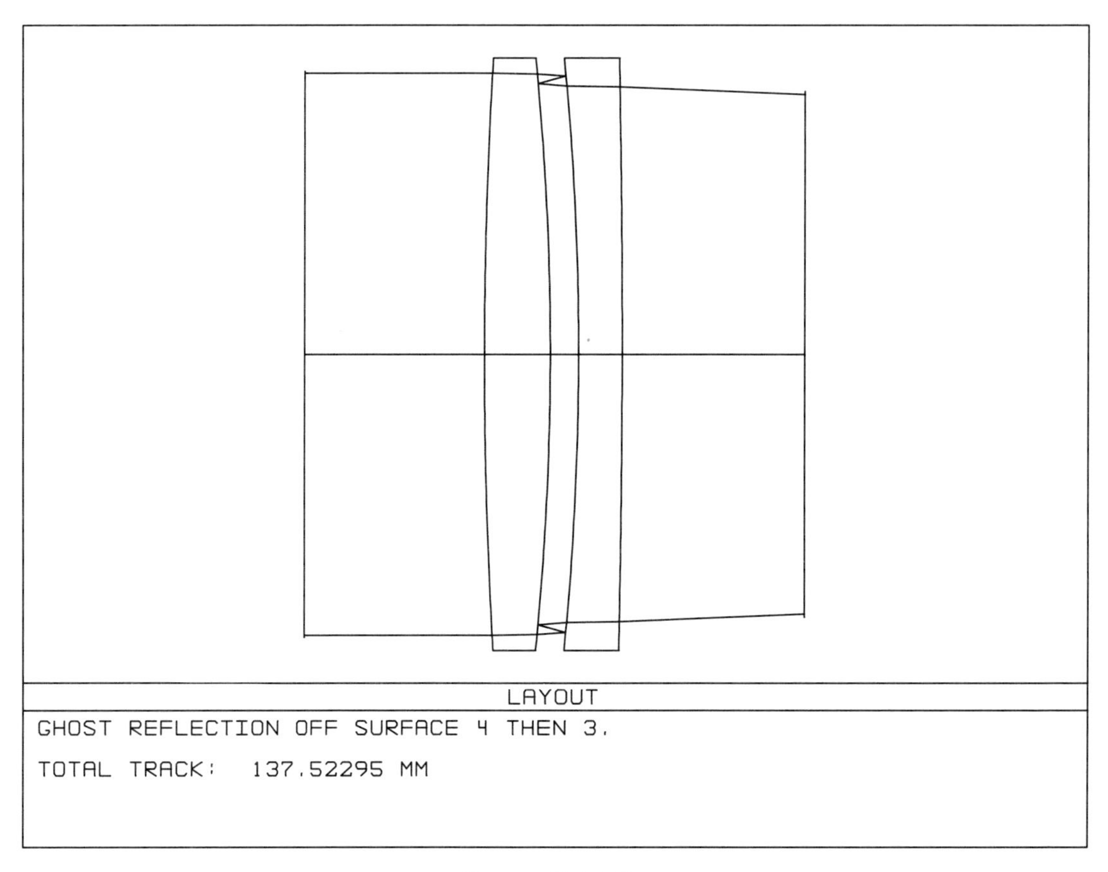

Figure B.1.2.8.

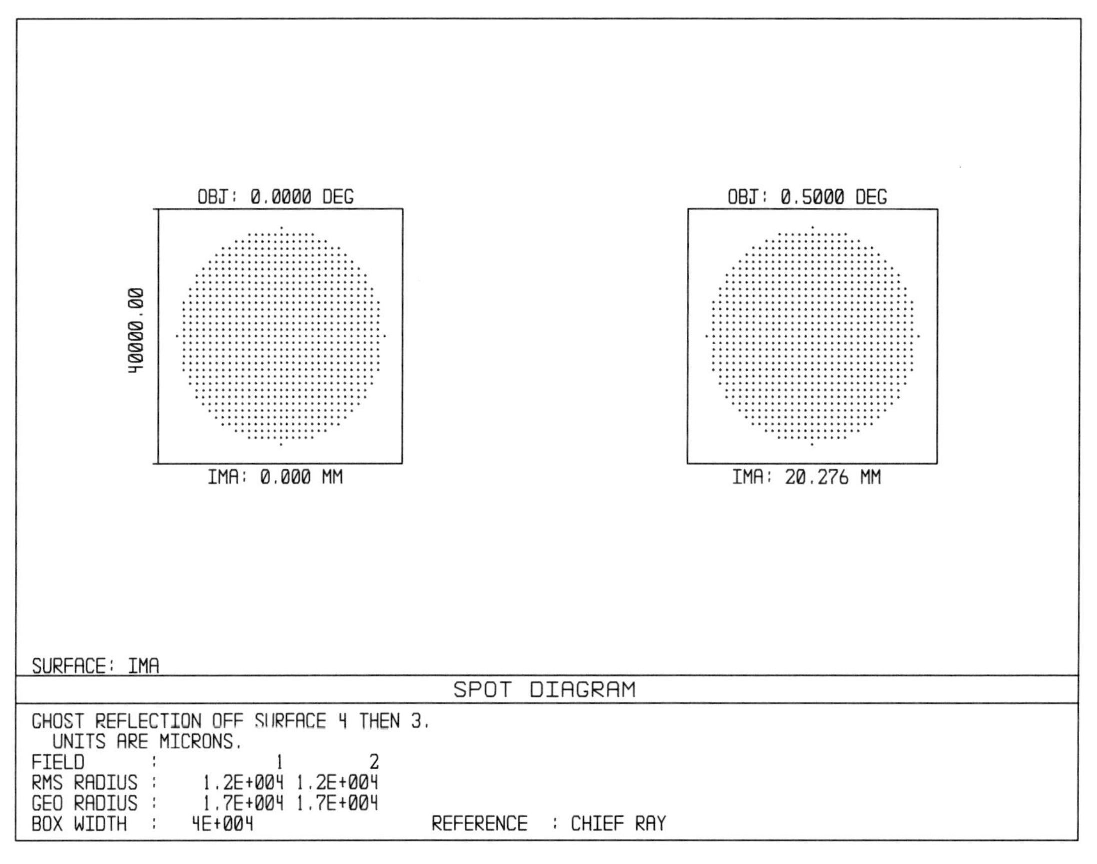

Figure B.1.2.9.

Listing B.1.2

```
System/Prescription Data
Title:  ACHROMATIC DOUBLET, F/15, BK7 & F2

GENERAL LENS DATA:

Surfaces             :              8
Stop                 :              2
System Aperture      : Entrance Pupil Diameter
Ray aiming           :  Off
Apodization          : Uniform, factor =      0.000000
Eff.  Focal Len.     :           2250 (in air)
Eff.  Focal Len.     :           2250 (in image space)
Total Track          :        2304.81
Image Space F/#      :             15
Para.  Wrkng F/#     :             15
Working F/#          :             15
Obj.  Space N.A.     :       7.5e-009
Stop Radius          :             75
Parax.  Ima.  Hgt.   :        19.6355
Parax.  Mag.         :              0
Entr.  Pup.  Dia.    :            150
Entr.  Pup.  Pos.    :             50
Exit Pupil Dia.      :        149.598
Exit Pupil Pos.      :       -2243.97
Field Type           :  Angle in degrees
Maximum Field        :            0.5
Primary Wave         :       0.546100
Lens Units           :  Millimeters
Angular Mag.         :        1.00269

Fields           : 3
Field Type:  Angle in degrees
#         X-Value          Y-Value           Weight
1        0.000000         0.000000         1.000000
2        0.000000         0.350000         0.000000
3        0.000000         0.000000         0.000000

Vignetting Factors
#            VDX             VDY             VCX             VCY
1       0.000000        0.000000        0.000000        0.000000
2       0.000000        0.000000        0.000000        0.000000
3       0.000000        0.000000        0.000000        0.000000

Wavelengths      : 3
Units:  Microns
#          Value           Weight
1       0.486100        0.000000
2       0.546100        1.000000
3       0.656300        0.000000

SURFACE DATA SUMMARY:

Surf        Type          Radius       Thickness        Glass       Diameter         Conic
 OBJ    STANDARD        Infinity        Infinity                           0             0
   1    STANDARD        Infinity              50                    150.8727             0
 STO    STANDARD        1304.015              18          BK7            158             0
   3    STANDARD       -807.6215        7.522953                         158             0
   4    STANDARD       -807.6215              12           F2            158             0
   5    STANDARD       -3607.759              50                         158             0
   6    STANDARD        Infinity        2167.284                    145.9355             0
   7    STANDARD        Infinity               0                    39.33832             0
 IMA    STANDARD        Infinity               0                    39.33832             0

SURFACE DATA DETAIL:

Surface OBJ     :  STANDARD
Surface 1       :  STANDARD
Surface STO     :  STANDARD
Surface 3       :  STANDARD
Surface 4       :  STANDARD
Surface 5       :  STANDARD
Surface 6       :  STANDARD
Surface 7       :  STANDARD
Surface IMA     :  STANDARD

SOLVE AND VARIABLE DATA:

Curvature of   2     : Variable
Semi Diam   2        : Fixed
Curvature of   3     : Variable
Thickness of   3     : Variable
Semi Diam   3        : Fixed
Curvature of   4     : Solve, pick up value from 3, scaled by 1.00000
Semi Diam   4        : Fixed
Curvature of   5     : Variable
Semi Diam   5        : Fixed
Thickness of   6     : Solve, marginal ray height = 0.00000
```

```
INDEX OF REFRACTION DATA:

Surf     Glass      0.486100       0.546100       0.656300
  0              1.00000000     1.00000000     1.00000000
  1              1.00000000     1.00000000     1.00000000
  2        BK7   1.52237866     1.51872064     1.51432149
  3              1.00000000     1.00000000     1.00000000
  4         F2   1.63208677     1.62407751     1.61503000
  5              1.00000000     1.00000000     1.00000000
  6              1.00000000     1.00000000     1.00000000
  7              1.00000000     1.00000000     1.00000000
  8              1.00000000     1.00000000     1.00000000

ELEMENT VOLUME DATA:
Units are cubic cm.
Values are only accurate for plane and spherical surfaces.
Element surf    2 to    3 volume :      291.507621
Element surf    4 to    5 volume :      264.739077
```

B.1.4 Telescope Exit Pupils

Contrary to a widespread misunderstanding, for visual observing through a telescope with an eyepiece, a slow telescope with a large f/number (even $f/15$) is no disadvantage (even for deep sky work). The reason concerns how light passes through the telescope's exit pupil and into the eye's pupil.

The exit pupil of a telescope is the image of the objective lens or mirror formed by the eyepiece. Except for a Galilean telescope, this exit pupil is a real image, and it is focused directly on the pupil of the eye.

The diameter of the telescope's exit pupil determines the area of the eye's pupil that is illuminated. Thus, it is the diameter of the exit pupil that determines the working f/number of the eye. It is the eye's f/number, *not* the telescope's, that determines the brightness of the image on the retina, where it counts.

The diameter of the eye's pupil depends on the ambient lighting and the observer's age. Under bright conditions, or when looking at the moon through a telescope, the eye's pupil closes down to about 2.5 mm. For youthful, fully dark adapted eyes, the pupil can open up to as much as 8 mm. But for an adult viewing faint objects through a telescope, 6 mm is a more likely maximum pupil diameter.

The size of the eye's pupil sets a practical upper limit on the size of the telescope's exit pupil. If the exit pupil is made larger, then the eye effectively stops down the telescope. The perceived image gets no brighter once the eye's pupil is fully illuminated.

An oversized exit pupil still works fine (but to no advantage) if the telescope has no central obscuration (as with a refractor). But there is a problem when using a reflecting telescope that does have a central obscuration. Here the exit pupil has a dark spot in the middle caused by the shadow of the secondary mirror. For example, a 6 mm exit pupil and a 33% central obscuration (a typical value) yields no light in the central 2 mm of your eye. If the exit pupil is made still larger, then the dark spot fills even more of your eye and you will have a hard time seeing anything at all. Thus, for a reflector, approximately 6 mm is an absolute upper limit to exit pupil size.

For all telescopes, there is also a practical lower limit on exit pupil size. If the exit pupil is made less than about 0.5 mm, diffraction effects become quite visible and objectionable. The image may get larger, but you will not see more. This effect is called empty magnification and is to be avoided.

Thus, for any telescope, the exit pupil diameters available to an observer should range roughly from 6 mm to 0.5 mm, with 2 or 3 mm giving the highest visual acuity.

It follows that selecting eyepieces should be done on the basis of exit pupil diameter, not magnifying power. For a given telescope f/number, it is the focal length of the eyepiece that determines the diameter of the exit pupil. You can get any desired exit pupil diameter by selecting the proper eyepiece focal length.

Exit pupil diameter is simply the focal length of the eyepiece divided by the f/number of the telescope. For a 30 mm eyepiece and $f/15$, exit pupil is 2 mm.

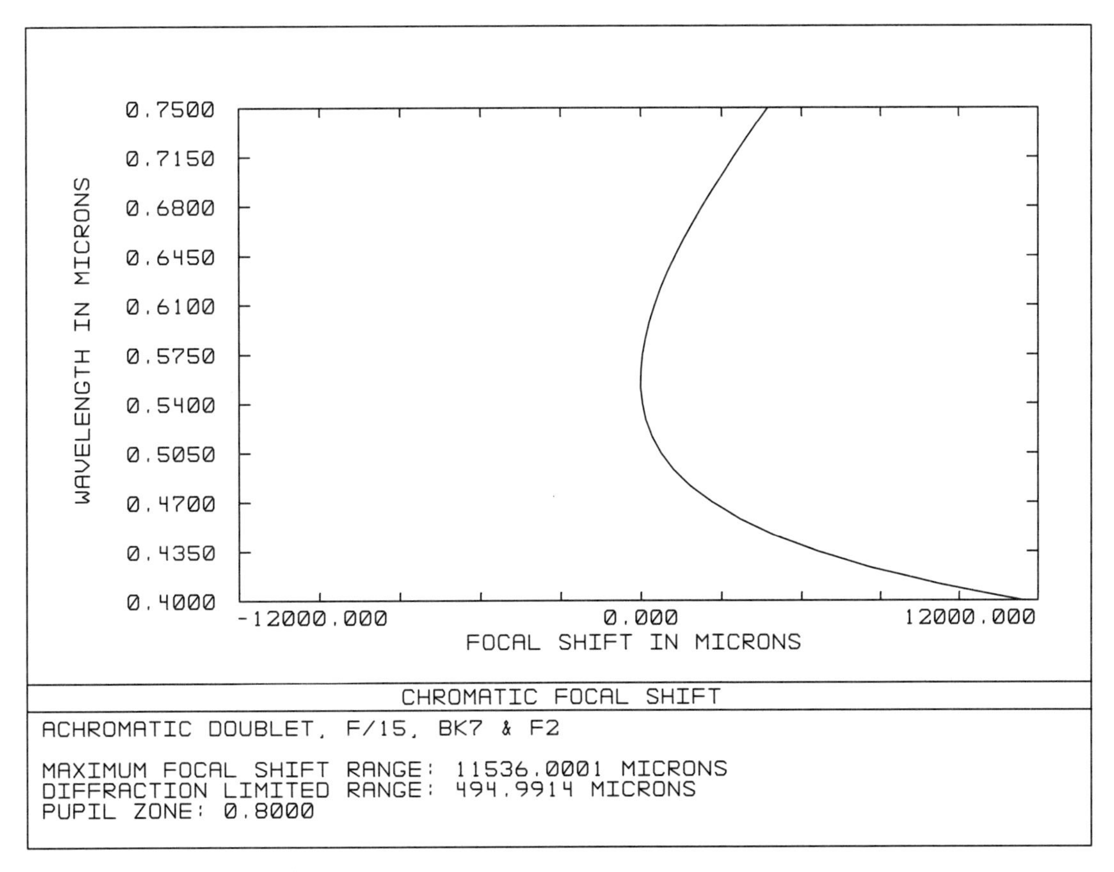

Figure B.1.3.1.

Eyepieces with focal lengths of 75 mm are readily available, and at $f/15$ these give a 5 mm exit pupil. Similarly, a 7.5 mm eyepiece gives 0.5 mm at $f/15$, and 1.5 mm at $f/5$.

The magnifying power of a telescope is simply the entrance pupil diameter divided by the exit pupil diameter (a consequence of the Lagrange invariant). For most telescopes, the entrance pupil diameter is the diameter of the objective lens or mirror. The 6 mm limit converts to $1/6$ power per millimeter of aperture diameter (4 power per inch). The 0.5 mm limit converts to 2 power per millimeter (50 power per inch).

Incidentally, there are two additional advantages in using a slow telescope. First, eyepieces are not stressed and yield their best images. Second, medium focal length eyepieces with comfortable eye relief give high powers. Of course, for direct photography, telescope f/number does matter, and fast telescopes do have an advantage there.

B.1.5 Color Curves for an Achromat

Before proceeding to apochromats, it is of interest to first look at the color curve for an ordinary achromat.[3] Figure B.1.3.1 is the chromatic defocus plot (BFL versus wavelength) for the previous $f/15$ achromat with equal internal curvatures. The 0.8 pupil zone, the same as used for color correction, is specified. Zero defocus on the horizontal plot axis indicates the location of the image surface. In this lens, the image surface is at the paraxial focus of the reference wavelength (0.5461 μm), as required by the height solve. If the image surface were moved to refocus, then the curve would shift accordingly.

[3]For another discussion of the color correction of achromats, see H. Dennis Taylor, *The Adjustment and Testing of Telescope Objectives*.

Note five things on Figure B.1.3.1. First, the color curve is U-shaped, which is characteristic of ordinary achromats. Second, intermediate wavelengths focus closer to the lens than the extreme wavelengths, which is also characteristic. Third, the slope of the color curve is zero at the closest focusing wavelength. Fourth, the two wavelengths for which longitudinal color is corrected have the same BFL. And fifth, both the slope of the curve and the amount of chromatic defocus are greater in the blue than in the red, which is again typical.

Because the color curve is U-shaped, it is clear that no matter how the image surface is allowed to refocus, the color curve can never have more than two zero crossings. Two zero crossings mean that the color correction type is achromatic, two wavelengths are in exact focus, and the remaining chromatic defocus errors are secondary color.

Near the closest focusing wavelength where the rate of change of chromatic defocus is least, an achromat approaches a lens without longitudinal color. Thus, when designing an achromat, the closest focusing wavelength is usually made to fall at the wavelength of greatest interest. This wavelength for high-light-level photopic visual response is 0.555 μm, the wavelength of the eye's greatest sensitivity.

In Figure B.1.3.1, the closest focusing wavelength is indeed near 0.555 μm. This is the direct result of choosing the Fraunhofer F (0.4861 μm) and C (0.6563 μm) wavelengths for correcting the longitudinal color. Thus, for visual instruments, these two wavelengths have become the classical pair for standard color correction.

Note that in general, a different pair of wavelengths for achromatizing gives a different closest focusing wavelength. The color in many old lenses was corrected using the Fraunhofer H (0.3968 μm) and F (0.4861 μm) wavelengths, thereby placing the closest focusing wavelength near 0.434 μm to match the old photographic blue response.

Finally, because the slopes and departures of the color curve are greater in the blue than in the red, it is clear that an achromat's color errors are harder to control in the blue.

B.1.6 Glass Selection and Color Curves for an Apochromat

The $f/15$ achromat is an improvement over the $f/5$ achromat, but secondary color remains large. This is a consequence of using two glasses whose dispersion curves (index versus wavelength) have a significantly different shape. The only way to reduce secondary color in a doublet (or any lens) is to use glasses with similarly shaped dispersion curves (with similar partial dispersions) If the glasses are carefully chosen, the result is an apochromatic doublet with three colors brought to a common focus.

Refer again to the plots of partial dispersion versus total dispersion[4] shown in Figures A.10.3.1 and A.10.3.2. Note again that the ordinary glasses cluster about the normal glass line. This nearly linear dependence has the consequence that, independent of n_d index, two ordinary glasses with moderately different V_d dispersions will have different partials. Conversely, two ordinary glasses with equal V_d dispersions will have similar partials and similarly shaped dispersion curves, even though their n_d indices may be very different. Thus, total dispersion and partial dispersion of ordinary glasses are closely linked and are not independent variables.

[4] Recall that total dispersion is conventionally measured by the Abbe number V_d, also called ν_d. The Abbe number is reciprocal dispersion, and is given by:

$$V_d = \frac{n_d - 1}{n_F - n_C}.$$

Partial dispersion for any two wavelengths relative to the standard F and C spectral lines is given by:

$$P_{1,2} = \frac{n_1 - n_2}{n_F - n_C}$$

where the first wavelength is shorter than the second, thus making the partial dispersion positive.

This linkage presents a fundamental problem for the design of apochromats. As with achromats, the elements of an apochromat must contain glasses with significantly different dispersions to prevent excessive positive and negative element powers. But controlling secondary color requires that the partial dispersions be similar. These are conflicting requirements if ordinary glasses are used.

Fortunately, there are a few unusual glasses that do fall off the normal glass line, although typically not by much. These are the abnormal-dispersion glasses. The most common way to reduce secondary color is to combine one of these abnormal-dispersion glasses with a matching ordinary glass. Less often, you might find two matching abnormal-dispersion glasses on opposite sides of the normal glass line, thus giving a larger V_d dispersion difference. In either case, the idea is the same. Only doublets with one unusual glass will be considered here.

When designing an apochromatic doublet with one unusual glass, the designer selects the unusual glass by hand. The nature of the final system depends on this initial choice; the rest of the lens is tailored accordingly. After selecting the unusual glass, the matching ordinary glass must be found. There is more than one way to do this. What follows an approach that often works with lenses corrected for visible wavelengths or for the very near ultraviolet or infrared.

This method uses the partial dispersion plots. Choose the plot corresponding to the wavelength region of interest. On the plot, find the unusual glass. Candidate matching glasses have nearly the same partial dispersion; that is, they are plotted near a horizontal line passing through the point for the unusual glass. In a variation of this method, locate the point on the normal glass line having the same partial dispersion (the same vertical height) as the unusual glass. Read off the corresponding V_d dispersion. Then on the standard index-dispersion glass map in Figure A.10.2, the ordinary glasses that are good candidates for the matching glass will lie in a vertical band of varying n_d indices that is centered on this value of V_d.

These procedures only get you a set of promising glass pairs. But apochromatic color correction depends strongly on the exact shapes of the glass dispersion curves. Thus, the only way to determine how well a given pair actually works is to try it out. If more than one glass pair works, then the ordinary glass that combines the best optical solution with the best practical properties can be selected.

To evaluate a candidate apochromatic glass pair, substitute the glasses into the lens prescription in the computer, and then optimize using the same procedure described for an achromat. The merit function in Listing B.1.1 works very well, except that the wavelength numbers must be changed to match the different wavelength set.

Note that one of the optimization operands controls longitudinal color for a particular pupil zone. If this operand is corrected to zero for the two extreme wavelengths, then these two wavelengths will have the same zonal BFL. If your choice of glass gives an apochromat, then you will find that an intermediate wavelength between the extremes also has the same BFL; that is, three wavelengths focus together.

Note that the lens designer can only select the glasses and the two extreme wavelengths. But he cannot arbitrarily select the third or intermediate wavelength. The third wavelength is determined by the dispersion properties of the glasses, and you must take what you get. For present purposes, an apochromat is defined as a lens with any intermediate wavelength having the same BFL as the extreme wavelengths. Ideally, of course, you would prefer the third wavelength to be close to the halfway point.

Two apochromatic doublet examples will be considered. The first apochromat will use Schott KzFSN4 (n_d of 1.613, V_d of 44.3) as the unusual glass. KzFSN4 is one of the special short flints and is relatively practical to work in the optical shop. As a flint, KzFSN4 requires a matching crown glass. The second apochromat example will use crystal calcium fluoride, or fluorite (n_d of 1.434, V_d of 95.0), as the unusual "glass." Fluorite is a long crown and is famous for its ability to control color. Fluorite has a very low index and very low dispersion, and thus it is located in the lower left corner of the glass map. Fluorite requires a matching flint glass, but relative to fluorite, all glasses are flints regardless of their names.

Both apochromats are to be corrected for visual use; that is, with color correction centered in the middle visible or yellow-green wavelength region and extending from the blue to the red. Unfortunately, as far as the author knows, glass manufacturers only supply partial dispersion plots for short and long wavelengths. For some reason, no one supplies partial dispersion plots for the central yellow-green. The solution is to use two plots, one in the blue and one in the red, and take an average. For convenience, Figures A.10.3.1 and A.10.3.2, the two partial dispersion plots supplied by Schott, are adopted. These are used with Figure A.10.2, the Schott index-dispersion standard glass map.

To select the matching glass for the KzFSN4 doublet, first look for KzFSN4 on Figure A.10.3.1, the g-F blue partial dispersion plot for 0.4358 and 0.4861 μm. For the same partial dispersion, the normal glass line is at V_d equal to about 49.5. Then look for KzFSN4 on Figure A.10.3.2, the C-s red partial dispersion plot for 0.6563 and 0.8521 μm. Here, for the same partial dispersion, the normal glass line is at V_d equal to about 52. Taking the average, the matching glass for a yellow-green lens should have a V_d value of about 51.

From the glass map in Figure A.10.2, note that the V_d value of Schott SSK3 is 51.2; this should be a good candidate glass to try. If you design a doublet with SSK3 and KzFSN4 using the achromat optimization procedure (merit function like Listing B.1.1, achromatization wavelengths of 0.40 and 0.75 μm, pupil zone of 0.8), you get the color curve shown in Figure B.1.3.2. Note that the curve is S-shaped, not U-shaped. An S-shaped color curve is characteristic of an apochromat. If you refocus for the achromatization wavelengths, you get three zero crossings within the waveband. Three zero crossings mean that the color correction type is apochromatic, three wavelengths are in exact focus, and the remaining chromatic defocus errors are tertiary color.

Actually, in this case, the author had some help with the glass selection. Years ago, a colleague with an interest in apochromats related that SSK3 is a good match for KzFSN4.[5] But you cannot count on prior knowledge. Furthermore, SSK3 is not the only glass that will work.

If you continue trying other combinations, you find that several glasses with different n_d indices, but all with V_d values close to 51, also give three zero crossings with KzFSN4. For Schott glasses, these lie in a band from BaLF8 (n_d equal to 1.554) to SSKN5 (n_d equal to 1.658). If Ohara glasses are used, the band can be extended to cover from NSL32 (n_d equal to 1.526) to LAL58 (n_d equal to 1.694).

This illustrates the concept of abnormal dispersion. KzFSN4 has a V_d value of 44.3, but its dispersion curve shape does not match ordinary glasses with a V_d value of about 44. Rather, KzFSN4 has a dispersion curve more like ordinary glasses with a V_d value of about 51. The shape of the dispersion curve for KzFSN4 is significantly outside its V_d class and is thus abnormal.

It is of additional interest that KzFSN4 does not yield three zero crossings when paired with lanthanum glasses with a V_d value of about 51 and a n_d greater than about 1.70. Refer again to Figures A.10.3.1 and A.10.3.2, and note that the extreme lanthanum glasses are themselves developing abnormal dispersion and are no longer ordinary glasses. These lanthanum glasses depart from the normal glass line in the same direction, to the right, as the short flints. Thus, a high-index lanthanum glass with a V_d value of about 51 no longer has the right partial dispersion to

[5]The author thanks Richard A. Buchroeder for suggesting this apochromatic glass pair.

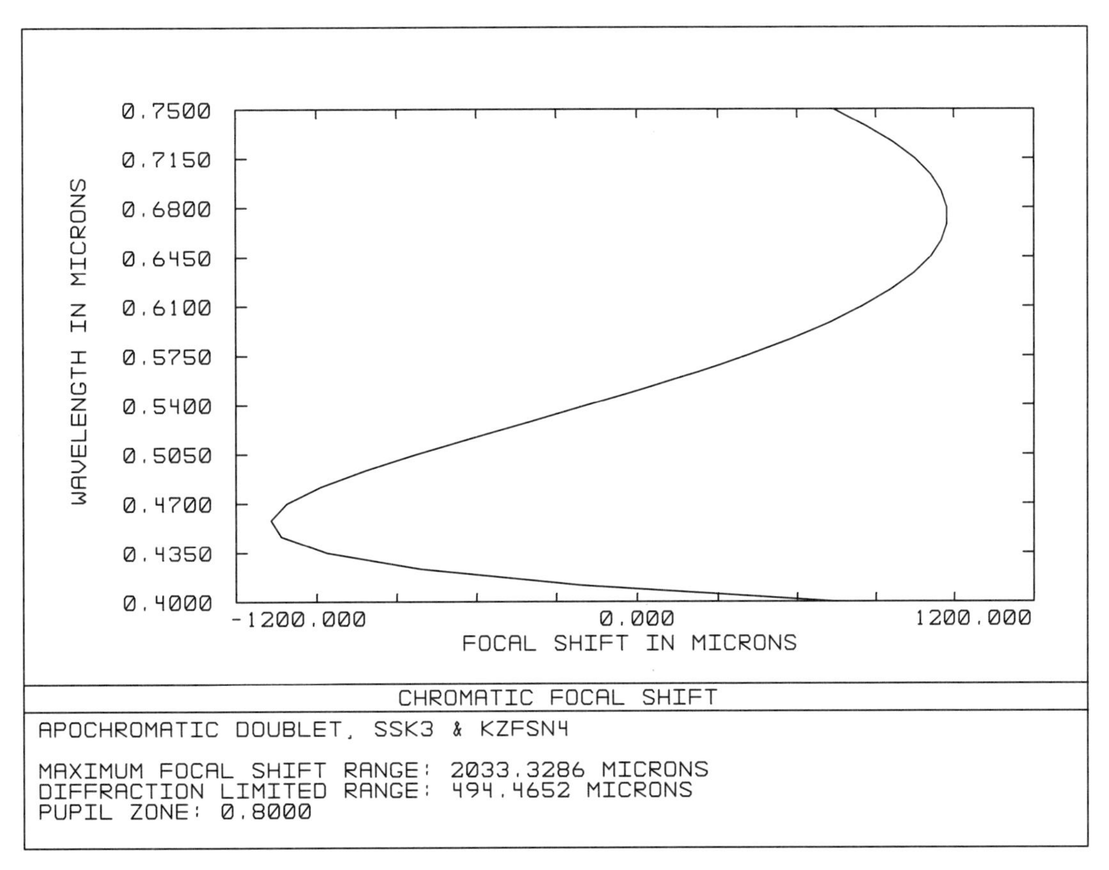

Figure B.1.3.2.

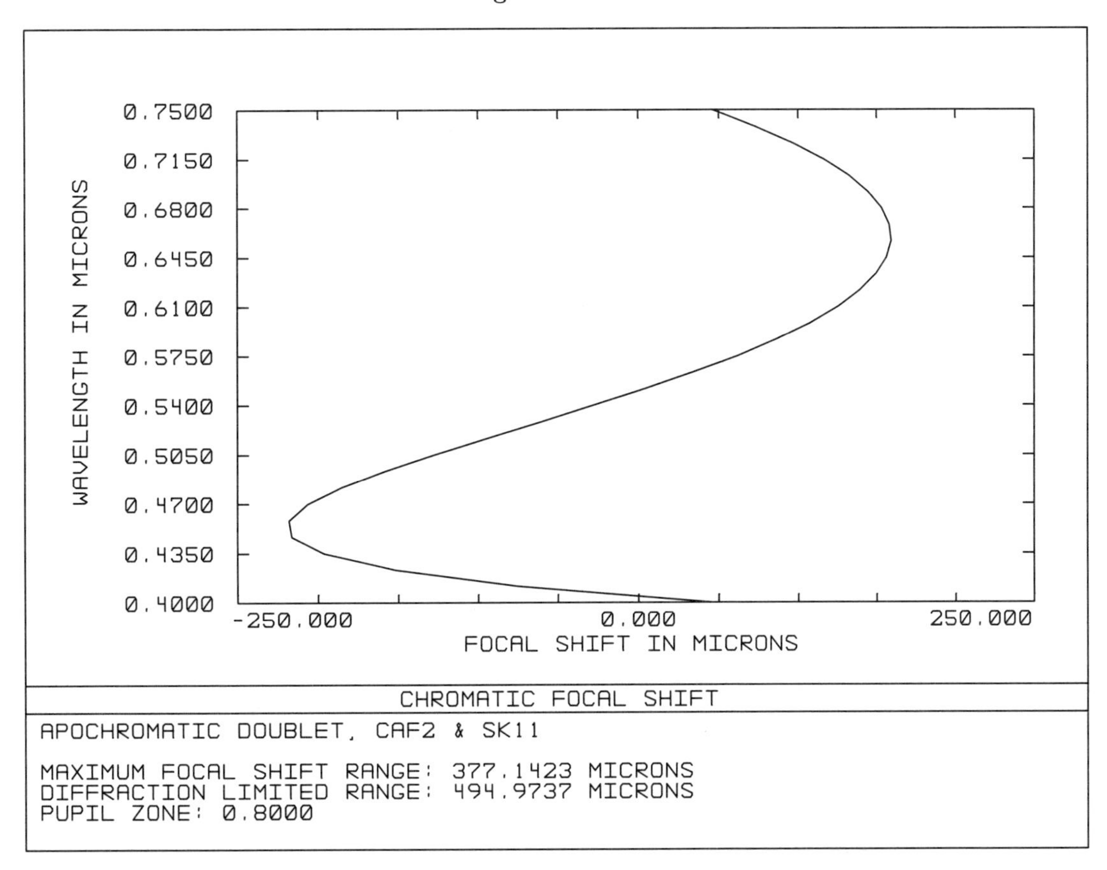

Figure B.1.3.3.

match KzFSN4. Even worse, extreme lanthanum glasses that do have the right partial dispersion no longer have a sufficiently large difference in V_d to make a good doublet with KzFSN4.

For the fluorite doublet, no matching glass was known ahead of time. Fluorite does not appear on Figures A.10.3.1 and A.10.3.2. If you wish, you can calculate the dispersions and partials, add them to the plots, and repeat the process outlined above. However, for illustrative purposes, a different approach will be used this time.

A quick but systematic search of the glass map is made. And it is quick, thanks to computers and modern software. Various glasses are selected to be paired with fluorite. For each selection, a doublet lens is optimized as just described and the color curve is examined. The results are very interesting. Incidentally, this method also works in the ultraviolet and infrared.

If fluorite is paired with glasses having a V_d value less (more flinty) than the low 60s, an ordinary U-shaped color curve similar to Figure B.1.3.1 results. As with two ordinary glasses, you are combining a relatively short crown with a relatively long flint. But if fluorite is paired with glasses having a V_d value greater (less flinty) than the low 60s, a reversed U-shaped color curves results, with the intermediate wavelengths focusing farther from the lens, not closer. Now you are combining a relatively long crown with a relatively short flint. This is the reverse of the usual situation and yields a reversed color curve.

When the search is continued, many good matches for fluorite are found with V_d values in a band between about 60 and 61 extending from Schott K7 (n_d equal to 1.511) to LaK21 (n_d equal to 1.641). Thus, the shape of the dispersion curve for fluorite, whose V_d value is 95.0, most closely matches the dispersion curves for ordinary glasses with V_d values between 60 and 61. Note that the resulting fluorite doublet has a much larger difference in V_d (35) than the KzFSN4 doublet (only 7, a factor of 5 less).

In the end, Schott SK11 was selected as the match for fluorite. The choice was a bit arbitrary, and was based partly on practical considerations such as: weathering, staining, homogeneity, thermal properties, hardness, cost, availability, and so forth. Figure B.1.3.3 is the resulting chromatic defocus plot. Again, note the S-shaped curve and three zero crossings when refocused. This color curve clearly reveals apochromatic correction. Also note that the range of the chromatic defocus errors (the amount of tertiary color) for the fluorite doublet is only about a fifth as much as the range shown in Figure B.1.3.2 for the KzFSN4 doublet. This reduced tertiary color indicates a much better match of the specific dispersion curve shapes for the fluorite doublet.

B.1.7 $F/15$ Apochromatic Doublet with SSK3 and KzFSN4 Glasses

For the doublet with SSK3 and KzFSN4, the Fraunhofer crown-in-front configuration is again selected. To allow a direct comparison with the $f/15$ achromat, the EPD of the apochromat remains 150 mm and the focal ratio remains $f/15$. With two exceptions, all other system specifications also remain the same. One exception is the airspace, which is once again 2 mm and fixed. Consequently, all four curvatures are now made independent variables. The other exception, of course, is the handling of the wavelengths. Instead of designing with three wavelengths, the apochromat is designed with eight wavelengths (any sizable number will do). These eight wavelengths are the same eight used in evaluating the achromats. They are: 0.40, 0.45, 0.50, 0.55, 0.60, 0.65, 0.70, and 0.75 μm. The reference wavelength is 0.55 μm.

When designing an apochromat, the first two optimization stages are exactly the same as when designing an achromat. In particular, it is still very effective in the second or intermediate stage, where the basic lens configuration is determined,

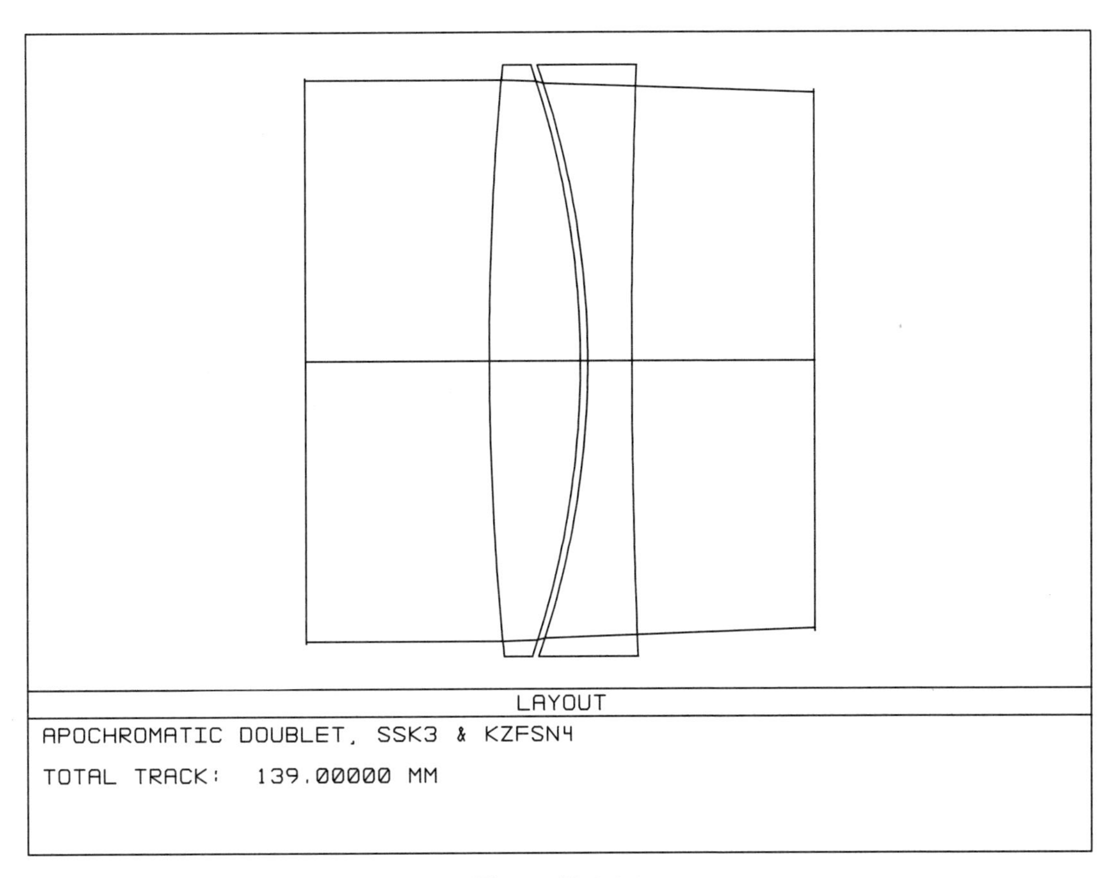

Figure B.1.4.1.

to use a simple merit function that specifically controls longitudinal color, spherical aberration, coma, and focal length (it is still a doublet and you still have only four effective independent variables). You can use either a merit function containing paraxial and third-order operands, or a merit function based on real rays (like Listing B.1.1).

Always look at the layouts before and after optimization. If your initial guesses for glass thicknesses were a bit off, readjust them and reoptimize. The layout of the final apochromat is shown in Figure B.1.4.1.

Final optimization is again for fine tuning. But the fine tuning of an apochromat is very different from the fine tuning of an achromat. Recall that with an achromat, there was just too much secondary color to be balanced out with spherical aberration, zonal spherical, spherochromatism, and defocus. Now the situation is different, and the polychromatic optimization approach that was rejected for an achromat is adopted. With an apochromat, the chromatic defocus errors are so much smaller that balancing together tertiary color, spherical aberration, zonal spherical, spherochromatism, and perhaps defocus now becomes effective. For the KzFSN4 doublet, a small amount of defocus will be used.

No prior image assumptions are made because the situation is too complex. Thus, in the final optimization stage, all on-axis aberrations are balanced together by using a default merit function that shrinks on-axis spot size for all eight equally weighted wavelengths simultaneously. This polychromatic trigonometric approach automatically achieves a best balance. Coma and EFL are still controlled separately. Except for the values of the operands, the complete resulting merit function is the same as that shown later in Listing B.1.5.

Figure B.1.4.2 is the ray-intercept ray fan plot for the on-axis image in the optimized design. Compare these curves with those in Figures B.1.1.4 and B.1.2.5. Not only are the same eight wavelengths used, but the scale for all three plots is the same: ±100 μm. Clearly, the apochromat has a very different type of color correction.

Figure B.1.4.3 is the ray fan plot for three specially chosen wavelengths: 0.423, 0.56, and 0.75 μm. Before the wavelengths were changed, the height solve locating the paraxial focal plane was removed to prevent the lens from inadvertently refocusing. Note that the ray fan curves for all three wavelengths cross near the 0.9 pupil zone and that this crossing point is on the horizontal axis. Not only do the three colors focus together, but they are all in focus on the focal plane. However, note also the large amounts of zonal spherical and spherochromatism. These are common problems for apochromats with their typically small V_d dispersion differences and consequently large element powers and surface curvatures. Large curvatures cause large surface aberration contributions and large amounts of higher-order aberrations.

Another way to understand the concept of correcting secondary color is by comparing Figure B.1.4.3 with Figure B.1.1.3. In Figure B.1.1.3, the two intersecting curves for the extreme wavelengths are inclined upward indicating secondary color defocus. But if the two inclined curves could somehow be rotated clockwise to refocus without also rotating (defocusing) the curve for the central wavelength, then all three wavelengths would be in focus together, and all three curves would lie together and look similar to the three curves in Figure B.1.4.3. This is what it means to correct secondary color.

Figure B.1.4.4 is the matrix spot diagram. Note that the spots for 0.50° off-axis are only slightly different from the on-axis spots. Compare this plot to Figure B.1.2.6 for the $f/15$ achromat. Again, the color correction of the apochromat is seen to be very different. Note that over the whole wavelength range the apochromat is better, but at no wavelength does the apochromat achieve the small spots obtained by the achromat for wavelengths near 0.55 μm. Apochromatic correction is no guaranteed panacea.

Figure B.1.4.5 is the OPD ray fan plot. Note the scale: ±2.0 waves. Most of the curves exceed by a large amount the Rayleigh quarter-wave rule for diffraction-limited performance. Figure B.1.4.6 shows the polychromatic MTF plot. Again it is demonstrated that this lens is not diffraction limited.

As is generally the case, it is not clear ahead of time whether a further final optimization using wavefront OPDs will be beneficial or detrimental. The only way to know for sure is to try it and compare the performance of the two versions. When the KzFSN4 apochromat is reoptimized with OPDs, the OPDs are slightly reduced, but the MTF curves are virtually unchanged.

This lens is intended for use in an astronomical telescope where compact and well-defined star images are of prime importance and where MTF is of secondary importance. Because an OPD reoptimization has no MTF benefit and can actually degrade the star images, the minimum spot size optimization is retained.

Recall that the performance of an ordinary achromat was improved by slowing down the lens from $f/5$ to $f/15$. Perhaps the same approach can help the KzFSN4 apochromat too. Thus, it might be worthwhile for the interested reader to recompute this lens using a focal ratio of $f/20$.

B.1.8 *F*/15 Apochromatic Doublet with Crystal Fluorite and SK11 Glass

If exceedingly good color correction is required, lens designers have known for over a century that fluorite gives excellent results. But fluorite has a practical problem. In earlier times, cost and availability were problems, but this is no longer the case. Lens blanks of artificially grown crystal fluorite are now routinely produced in diameters over 200 mm and at a cost comparable to expensive glasses. Although not immune, fluorite also resists weathering caused by atmospheric water vapor.

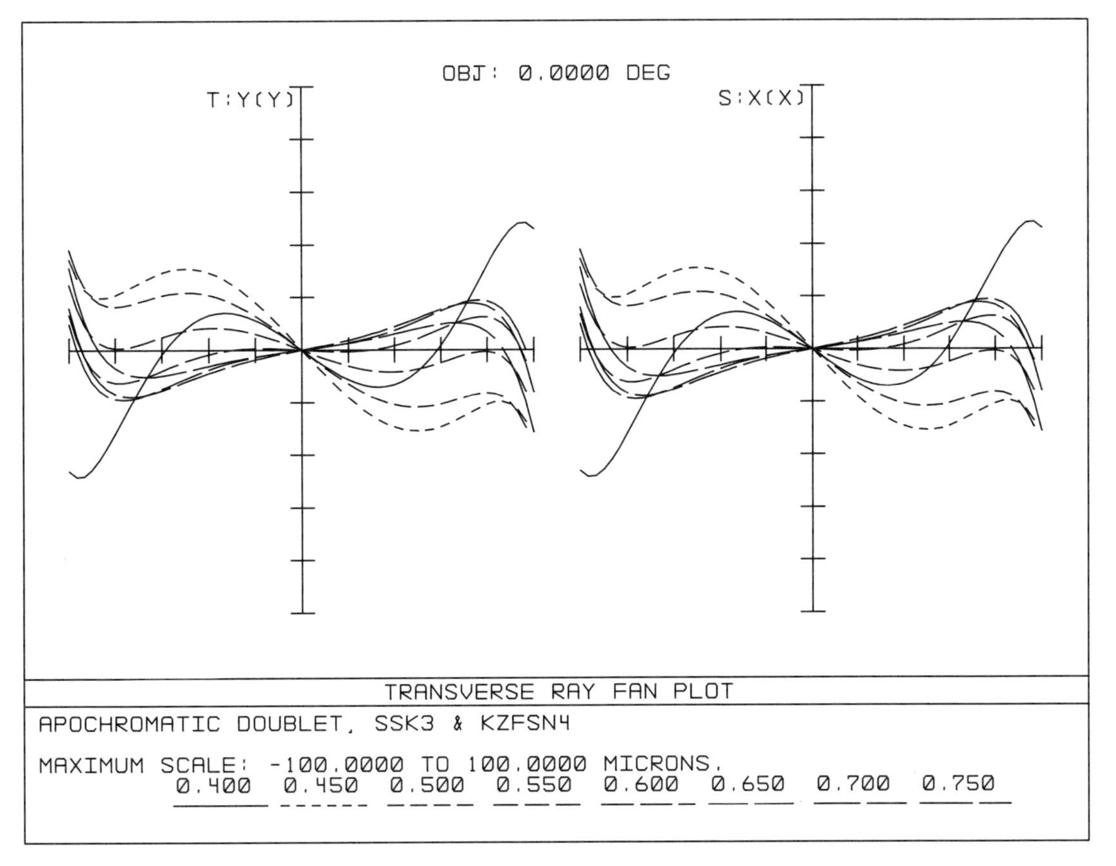

Figure B.1.4.2.

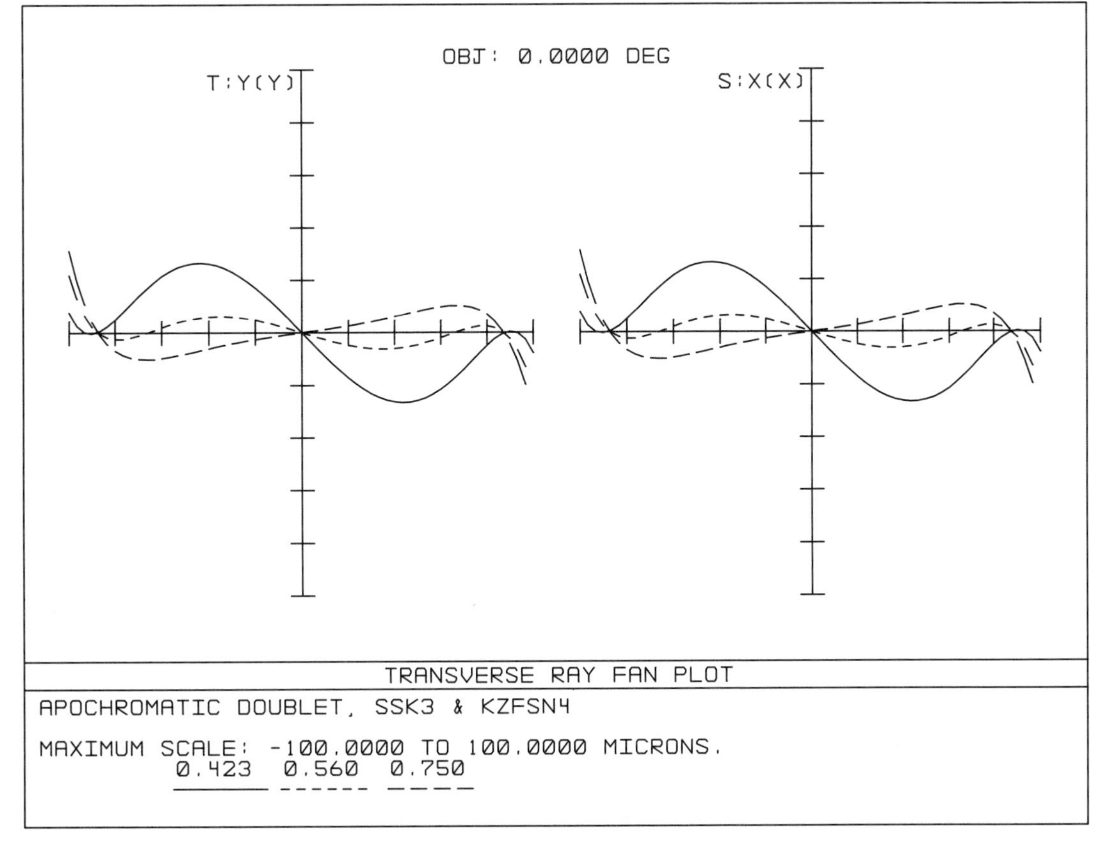

Figure B.1.4.3.

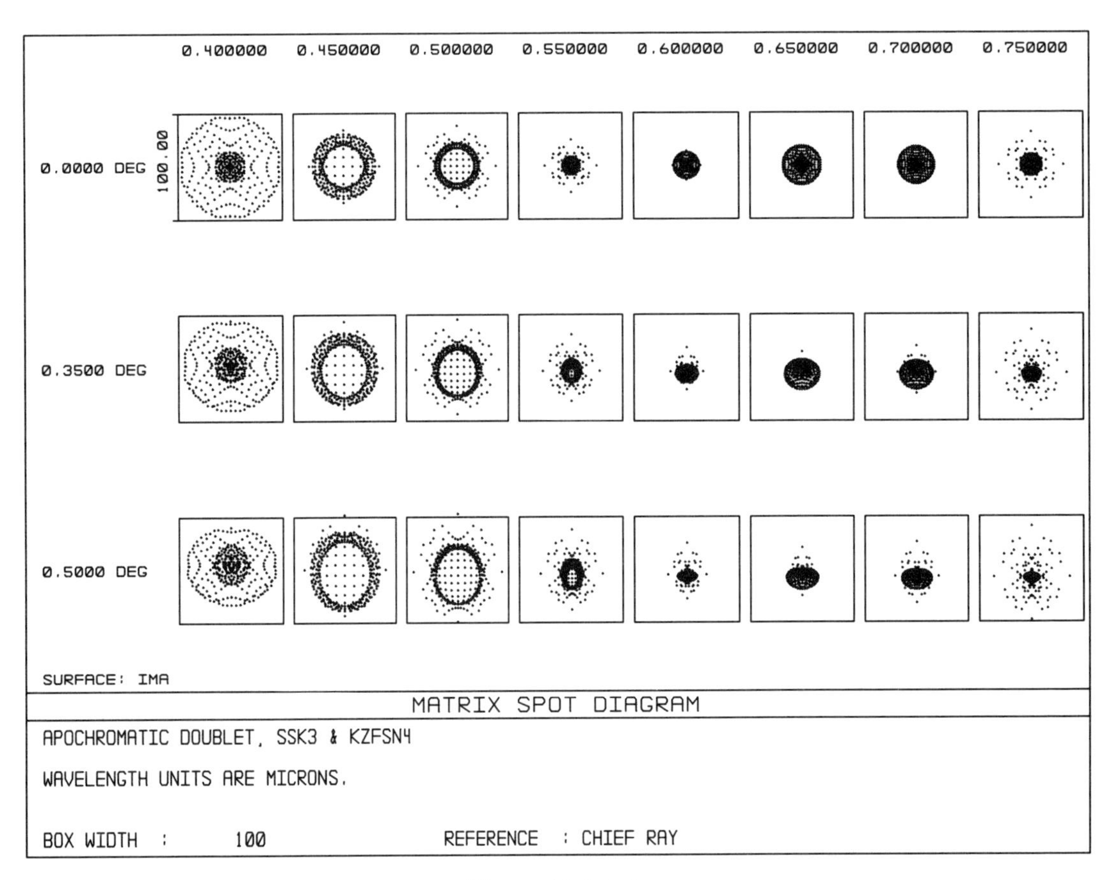

Figure B.1.4.4.

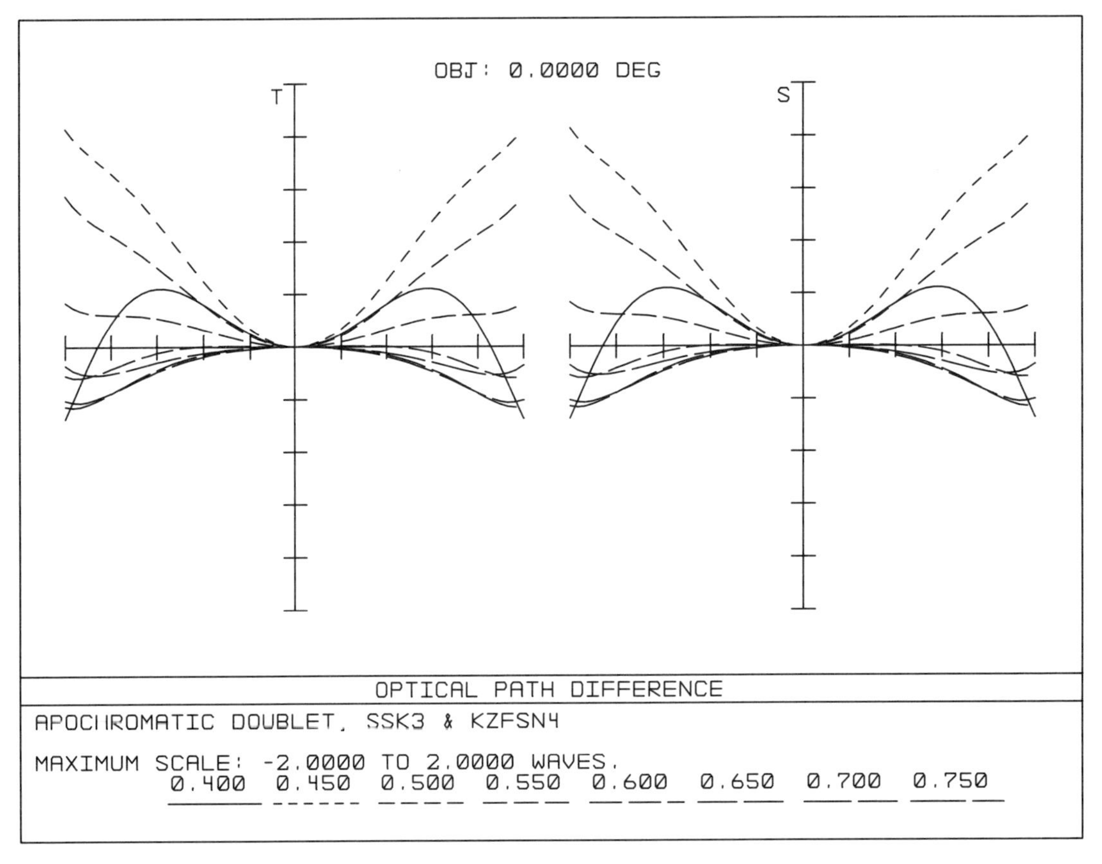

Figure B.1.4.5.

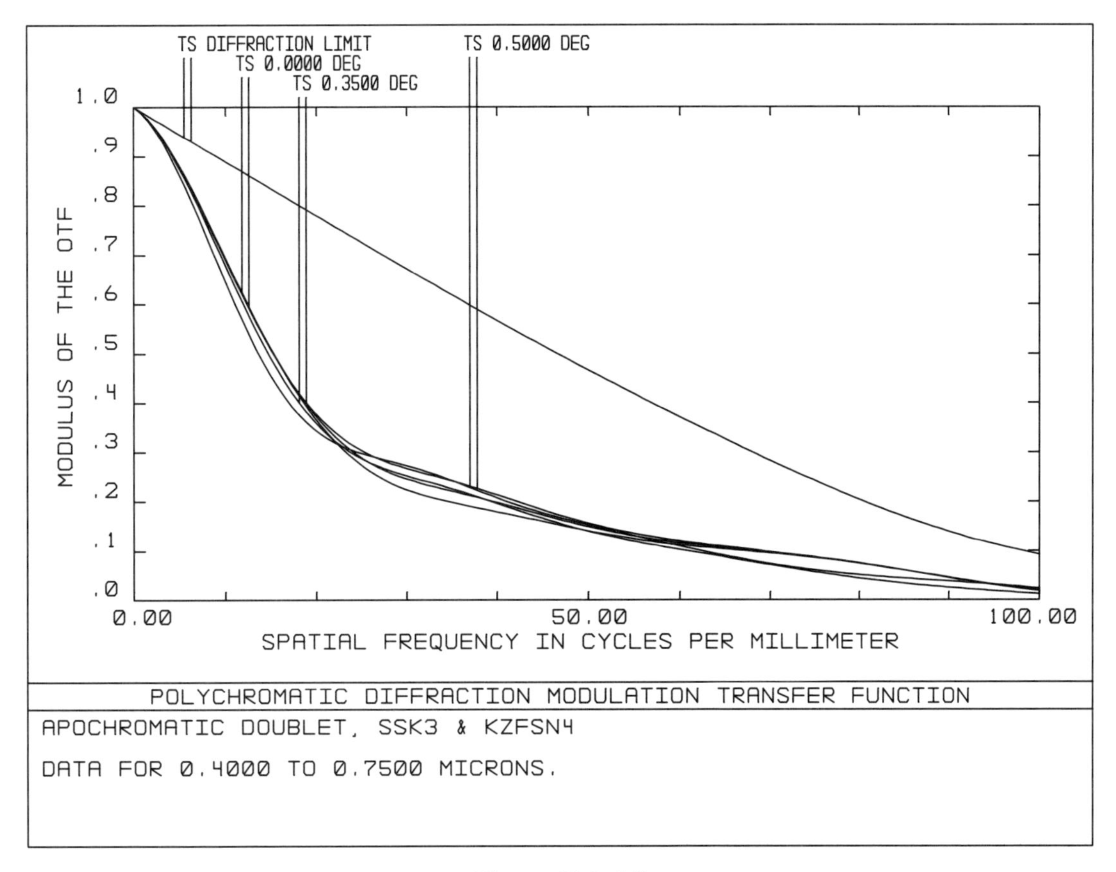

Figure B.1.4.6.

And fluorite can be worked in the optical shop with nearly standard techniques. The problem with fluorite is that it is sensitive to thermal shock. A rapid change in temperature that causes a large thermal gradient in the crystal material can fracture a fluorite lens.

This thermal shock problem is not fatal, however. Microscope objectives with fluorite have long been made, and they never fracture from thermal shock because the fluorite elements are small and are protected inside the lens structure. Recently, telescopes and camera lenses with fluorite have also become commercially available and are now used with success in real environments. The secret of success once again is to protect the fluorite elements. Thus, in a telescope doublet with fluorite, the sensitive fluorite crown element should not be the front element exposed to the weather. Rather than the Fraunhofer design, the flint-in-front Steinheil configuration should be adopted.

For the present design example, however, this change will not be made. The crown-in-front Fraunhofer configuration will be retained for purposes of comparison with the previous three lenses. But recall that the Fraunhofer and Steinheil configurations have almost identical image properties. Thus, although you would not want to build this lens, its calculated imagery illustrates the performance of a fluorite doublet.

The design process for the fluorite apochromat is identical to that just outlined for the KzFSN4 apochromat. The system specifications also remain the same, except this time the image surface is kept at the paraxial focus of the reference wavelength (defocusing is not necessary).

The layout of the final lens is shown in Figure B.1.5.1. Note that the powers of the individual elements of the fluorite doublet are much lower than those in the KzFSN4 doublet, although both lenses are $f/15$ overall. This reduction of element powers is caused by the much greater difference in dispersion between the

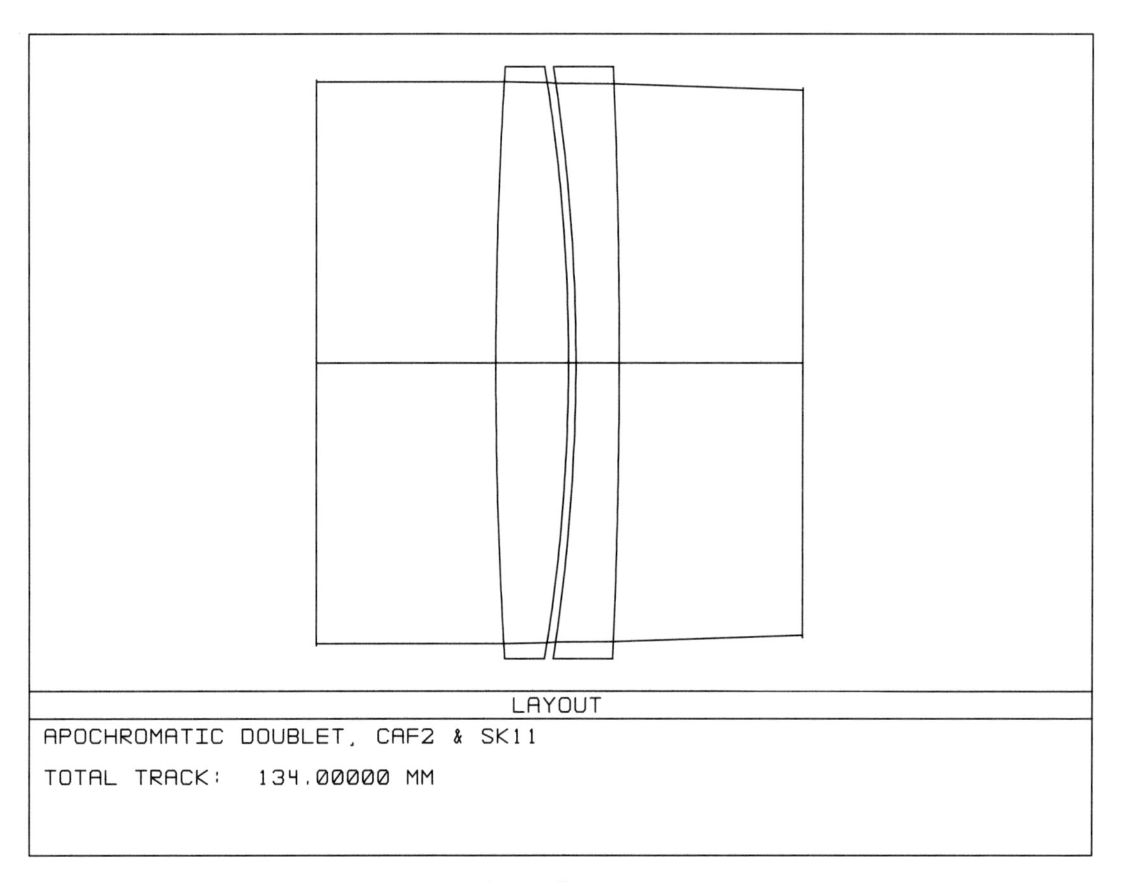

Figure B.1.5.1.

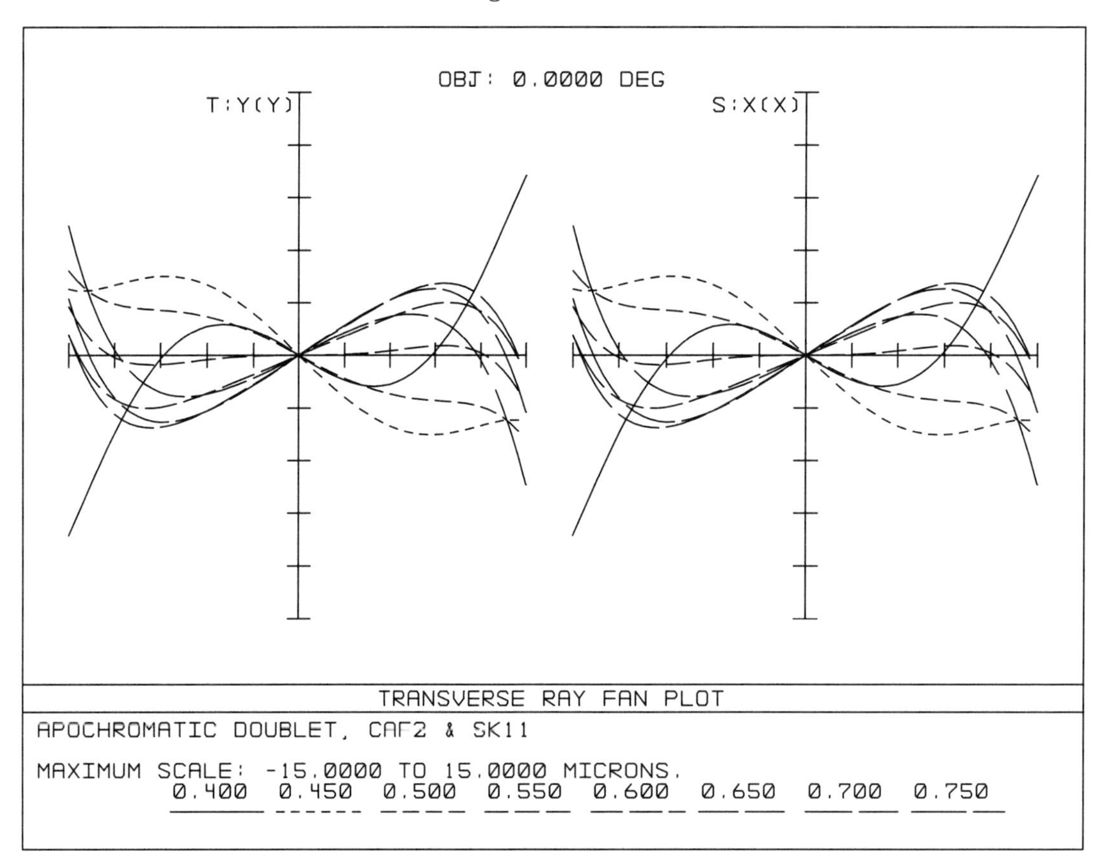

Figure B.1.5.2.

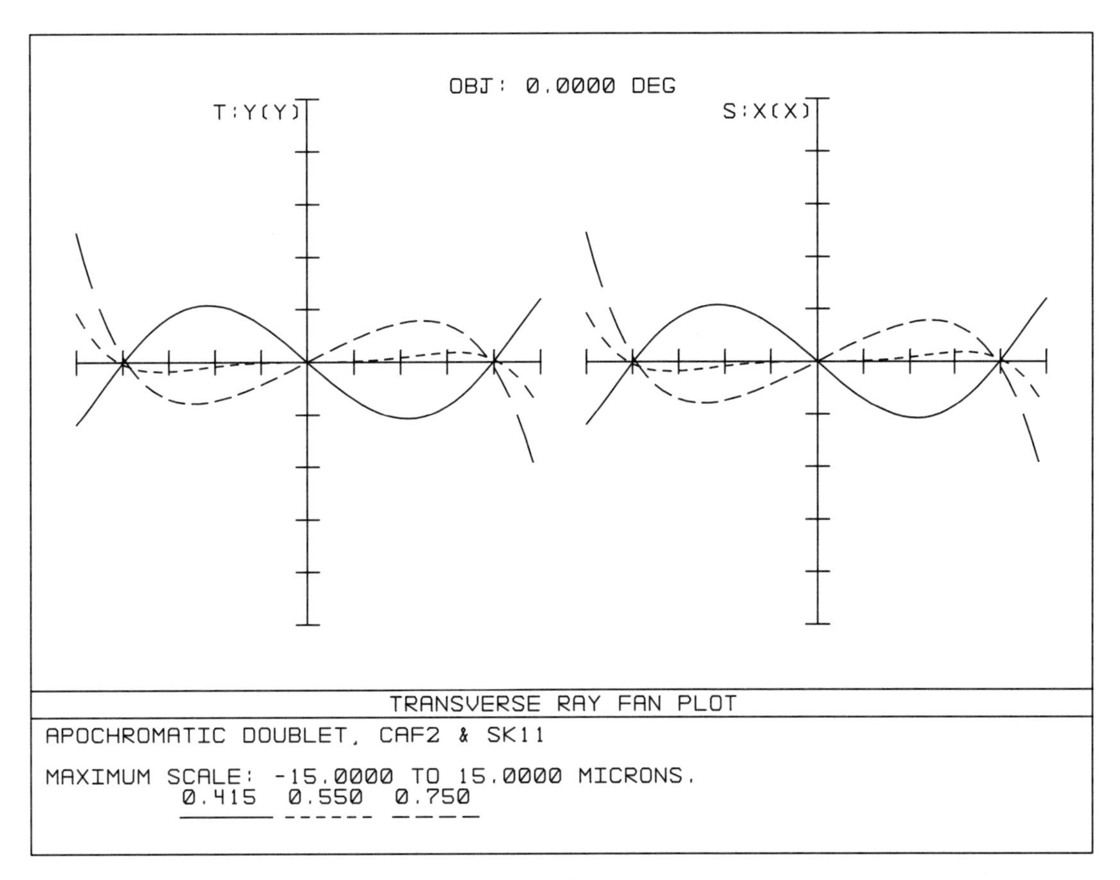

Figure B.1.5.3.

crown and flint glasses for the fluorite doublet.

Figure B.1.5.2 is the on-axis ray fan plot for the lens as optimized by shrinking polychromatic spot sizes. Note that the scale has been changed from ± 100 μm for the previous ray fan plots to only ± 15 μm now. The most significant on-axis aberrations in the fluorite apochromat are tertiary color and spherochromatism. Zonal spherical for the reference wavelength is relatively small, thanks to the low surface curvatures.

Figure B.1.5.3 is a ray fan plot similar to Figure B.1.4.3. The three wavelengths are: 0.415, 0.55, and 0.75 μm. Note that the curves for all three wavelengths cross together near the 0.8 pupil zone, and that this crossing point is in focus on the horizontal axis. Once again, secondary color is corrected.

Figure B.1.5.4 is the matrix spot diagram. Compare with Figure B.1.4.4. Note that the scales are different by a factor of five and that the spots for the fluorite doublet are much smaller. The circles drawn on Figure B.1.5.4 indicate the diameters of the Airy disks for the various wavelengths. This lens is nearly diffraction limited. Because on-axis performance is so good, at 0.50° off-axis the astigmatism and field curvature are now more noticeable but not excessive.

Figure B.1.5.5 is the on-axis polychromatic spot diagram. A concentrated spot core is surrounded by a small blue-violet flare. A circle is drawn to indicate the size of the Airy disk for the reference wavelength of 0.55 μm. The Airy disk is clearly much larger than the core of the spot. Thus, diffraction is a major factor in determining the true size of the PSF, and the real light does not go where the geometrical spot indicates. The small spot now only indicates a well-corrected image.

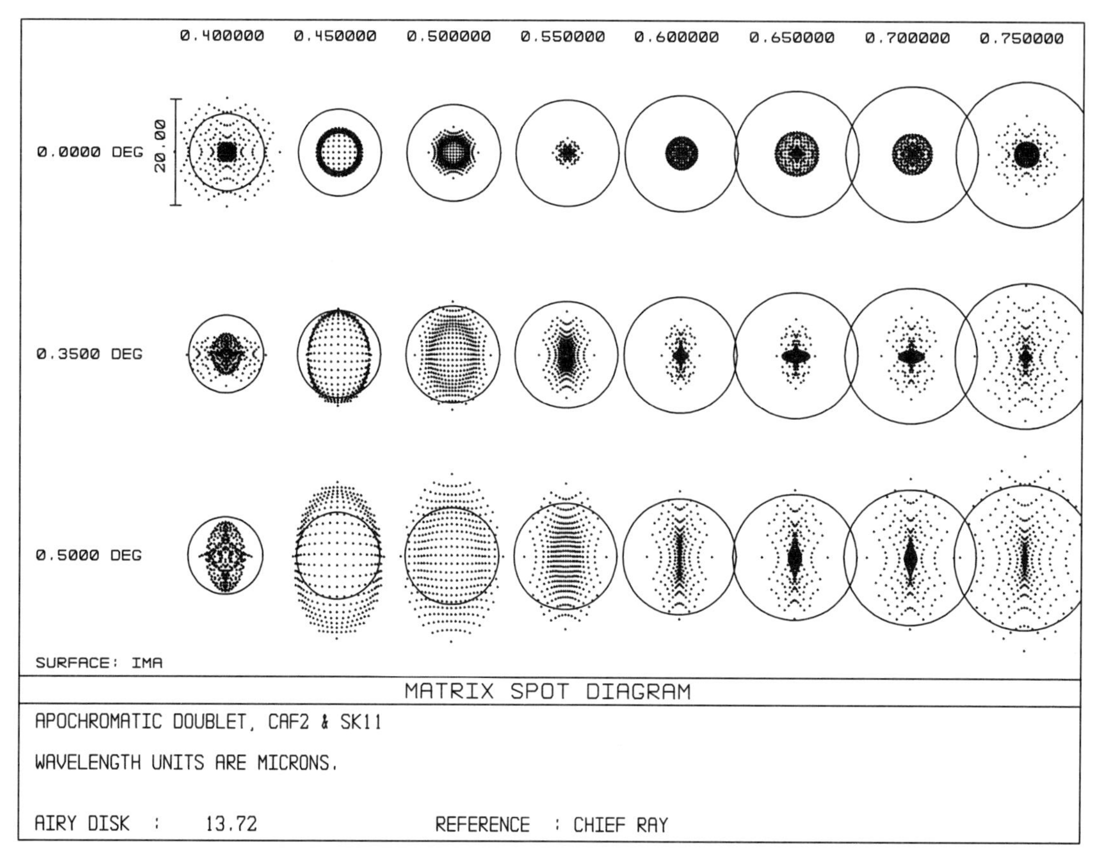

Figure B.1.5.4.

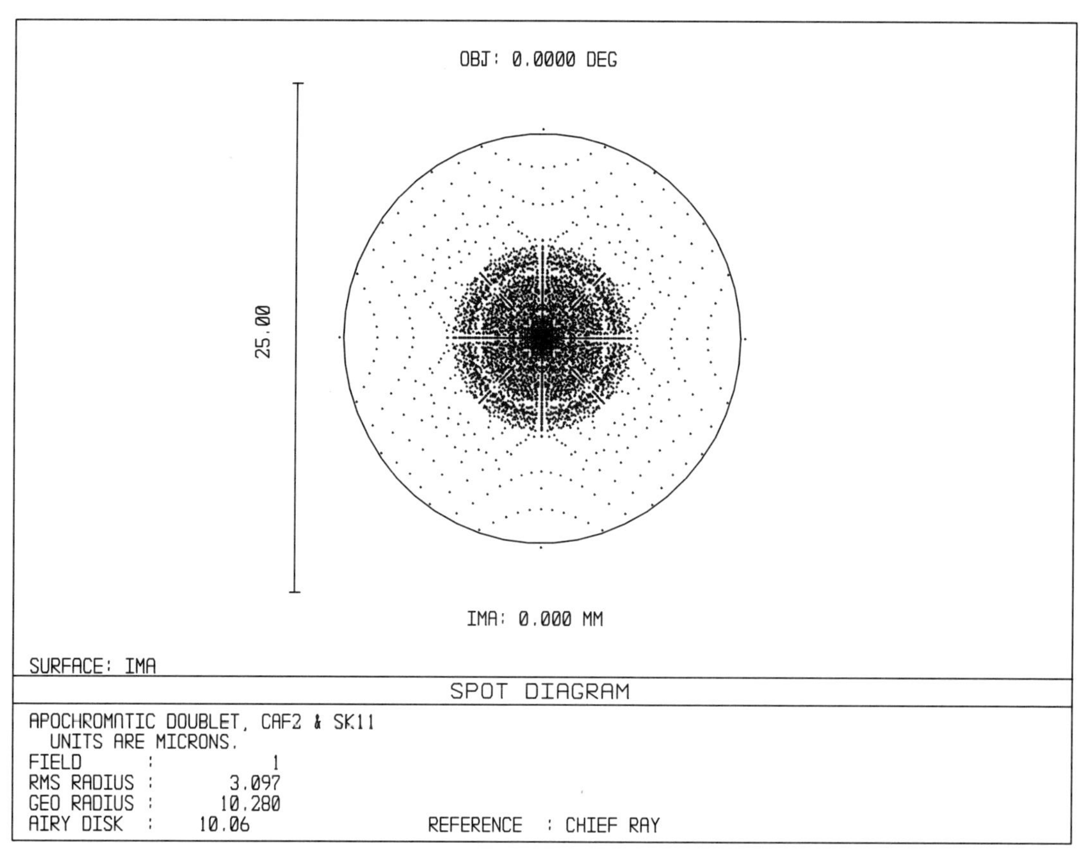

Figure B.1.5.5.

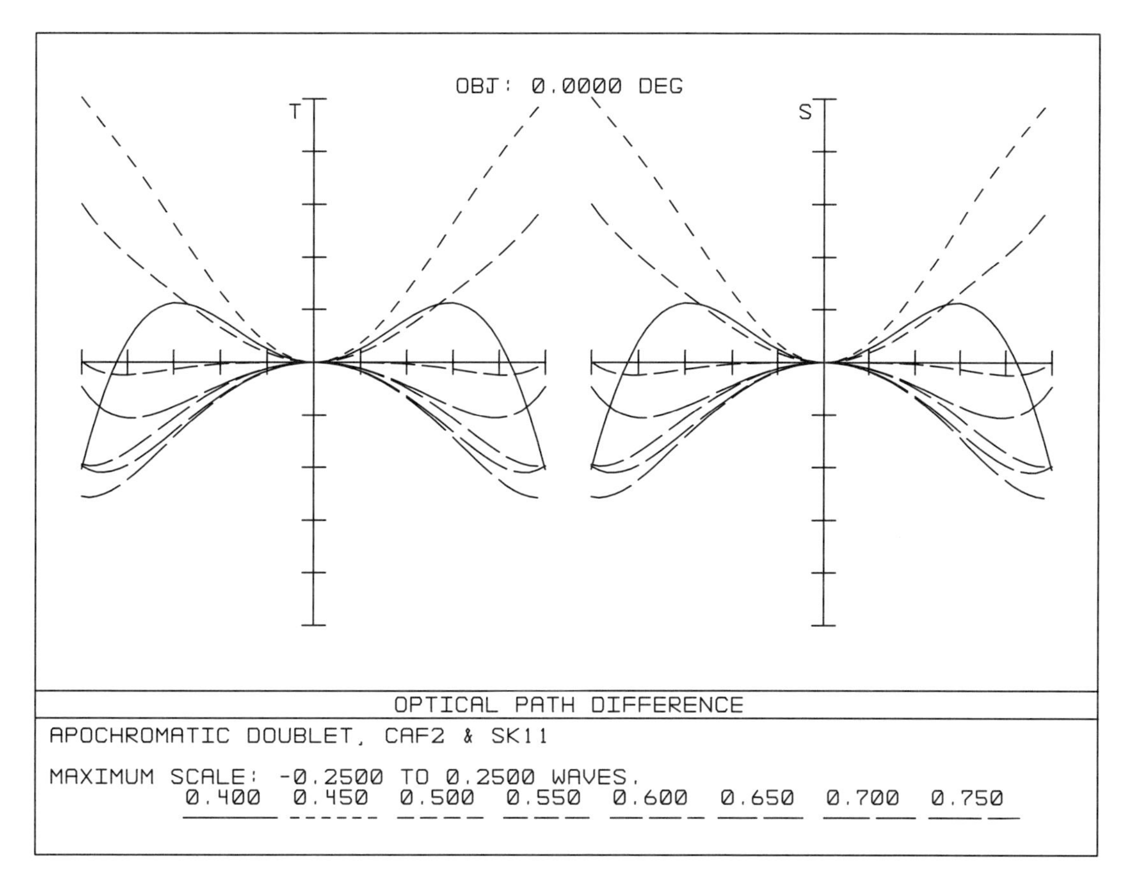

Figure B.1.5.6.

The diameter of the Airy disk is about 20 μm. For a lens with a focal length of 2250 mm, a 20 μm image corresponds to 1.8 arc-seconds projected onto the sky. This image blur is about the same as the blur caused by ordinary atmospheric seeing (turbulence).

Figure B.1.5.6 is the OPD ray fan plot. Note the scale: ± 0.25 waves. The overall peak-to-valley wavefront errors are nearly within a quarter wave; that is, the fluorite doublet is very close to satisfying the Rayleigh quarter-wave rule for diffraction-limited performance over the entire wavelength range.

To possibly make performance even closer to the diffraction limit, an OPD re-optimization is performed. As with the KzFSN4 apochromat, the OPDs for the fluorite apochromat are slightly reduced but the MTFs are unchanged. Thus once again and for the same reasons, the minimum spot size optimization is retained.

Figure B.1.5.7 shows the on-axis polychromatic diffraction PSF. Figure B.1.5.8 is the polychromatic diffraction encircled energy plot. Figure B.1.5.9 is the polychromatic diffraction MTF plot. In the latter two plots, the theoretical diffraction-limited curve is included for reference. In all three plots, the eight wavelengths are weighted equally. Listing B.1.5 is the merit function used during final optimization.

In Figure B.1.5.9, the difference between the actual on-axis MTF curve and the perfect diffraction-limited MTF curve is now very small. These tiny MTF errors suggest that diffraction, atmospheric seeing, and fabrication errors, but not aberration residuals, will determine the actual optical performance. For a small refracting telescope, this is as good as it gets. And that is very good indeed.

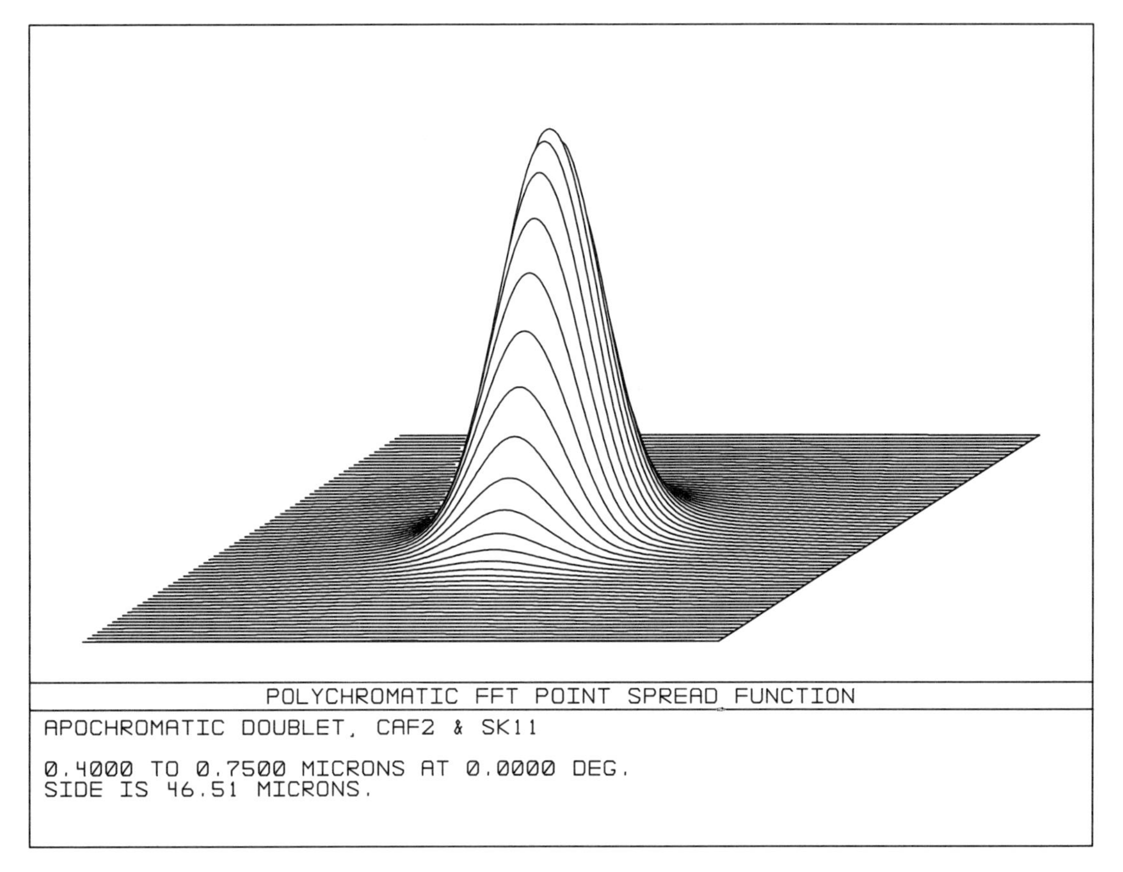

Figure B.1.5.7.

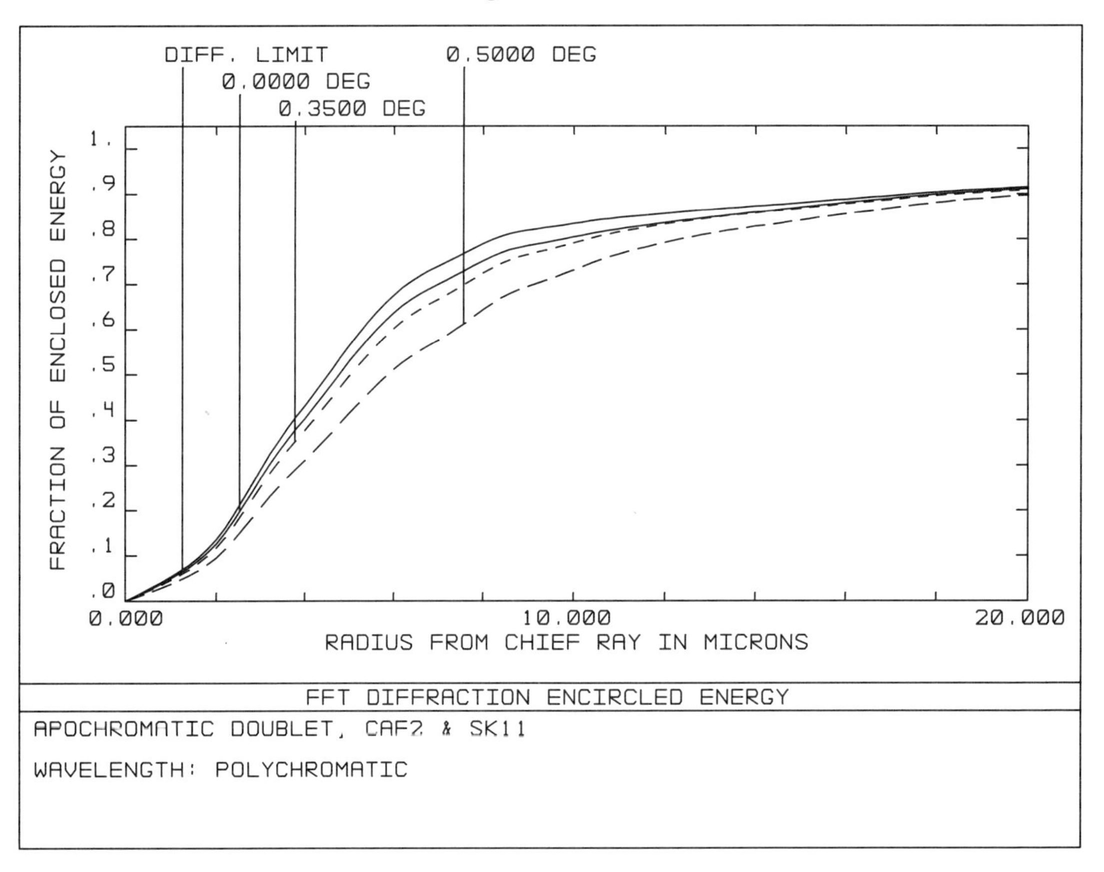

Figure B.1.5.8.

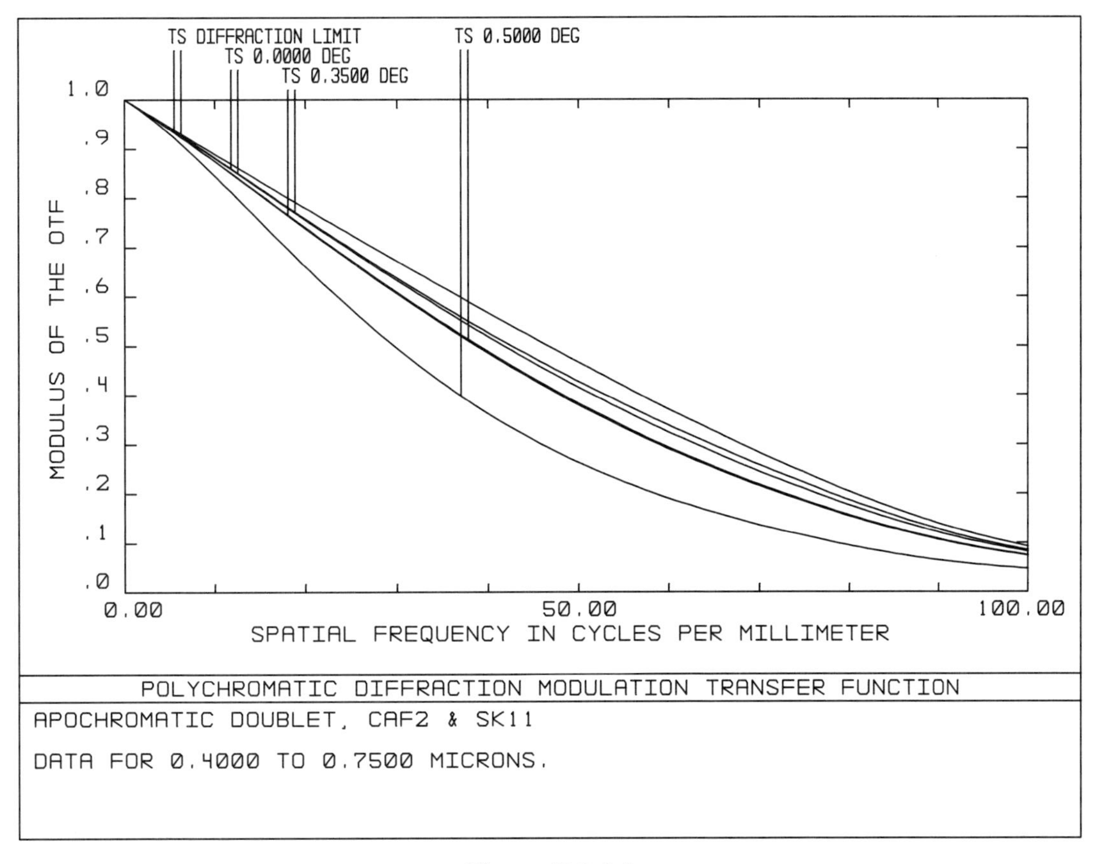

Figure B.1.5.9.

Listing B.1.5

Merit Function Listing

File : C:LENS143.ZMX
Title: APOCHROMATIC DOUBLET, CAF2 % SK11

Merit Function Value: 7.13960149E-005

Num	Type	Int1	Int2	Hx	Hy	Px	Py	Target	Weight	Value	% Cont
1	EFFL		4					2.25000E+003	1000	2.25000E+003	0.000
2	BLNK										
3	BLNK										
4	COMA	0	4					0.00000E+000	0	1.58007E-002	0.000
5	TRAY		4	0.0000	1.0000	0.0000	1.0000	0.00000E+000	0	-1.37561E-002	0.000
6	TRAY		4	0.0000	1.0000	0.0000	-1.0000	0.00000E+000	0	1.37561E-002	0.000
7	SUMM	5	6					0.00000E+000	1000	-8.82021E-009	0.000
8	BLNK										
9	BLNK										
10	DMFS										
11	TRAR		1	0.0000	0.0000	0.3357	0.0000	0.00000E+000	0.036361	1.77246E-003	1.120
12	TRAR		1	0.0000	0.0000	0.7071	0.0000	0.00000E+000	0.058178	2.20952E-003	2.784
13	TRAR		1	0.0000	0.0000	0.9420	0.0000	0.00000E+000	0.036361	8.53265E-003	25.954
14	TRAR		2	0.0000	0.0000	0.3357	0.0000	0.00000E+000	0.036361	3.61735E-003	4.665
15	TRAR		2	0.0000	0.0000	0.7071	0.0000	0.00000E+000	0.058178	4.31572E-003	10.623
16	TRAR		2	0.0000	0.0000	0.9420	0.0000	0.00000E+000	0.036361	3.68311E-003	4.836
17	TRAR		3	0.0000	0.0000	0.3357	0.0000	0.00000E+000	0.036361	1.98274E-003	1.401
18	TRAR		3	0.0000	0.0000	0.7071	0.0000	0.00000E+000	0.058178	2.69927E-003	4.156
19	TRAR		3	0.0000	0.0000	0.9420	0.0000	0.00000E+000	0.036361	3.96774E-003	5.612
20	TRAR		4	0.0000	0.0000	0.3357	0.0000	0.00000E+000	0.036361	1.63577E-004	0.010
21	TRAR		4	0.0000	0.0000	0.7071	0.0000	0.00000E+000	0.058178	4.83934E-004	0.134
22	TRAR		4	0.0000	0.0000	0.9420	0.0000	0.00000E+000	0.036361	1.51266E-003	0.816
23	TRAR		5	0.0000	0.0000	0.3357	0.0000	0.00000E+000	0.036361	1.82948E-003	1.193
24	TRAR		5	0.0000	0.0000	0.7071	0.0000	0.00000E+000	0.058178	2.96738E-003	5.022
25	TRAR		5	0.0000	0.0000	0.9420	0.0000	0.00000E+000	0.036361	4.26728E-004	0.065
26	TRAR		6	0.0000	0.0000	0.3357	0.0000	0.00000E+000	0.036361	2.71084E-003	2.620
27	TRAR		6	0.0000	0.0000	0.7071	0.0000	0.00000E+000	0.058178	4.00652E-003	9.156
28	TRAR		6	0.0000	0.0000	0.9420	0.0000	0.00000E+000	0.036361	7.19755E-004	0.185
29	TRAR		7	0.0000	0.0000	0.3357	0.0000	0.00000E+000	0.036361	2.75222E-003	2.700
30	TRAR		7	0.0000	0.0000	0.7071	0.0000	0.00000E+000	0.058178	3.41982E-003	6.670
31	TRAR		7	0.0000	0.0000	0.9420	0.0000	0.00000E+000	0.036361	9.61840E-004	0.330
32	TRAR		8	0.0000	0.0000	0.3357	0.0000	0.00000E+000	0.036361	1.98549E-003	1.405
33	TRAR		8	0.0000	0.0000	0.7071	0.0000	0.00000E+000	0.058178	1.23220E-003	0.866
34	TRAR		8	0.0000	0.0000	0.9420	0.0000	0.00000E+000	0.036361	4.64117E-003	7.679

Chapter B.2

The Wollaston Landscape Lens

Lenses for most telescopes and microscopes sharply image only a narrow field of view. This restricted coverage was adequate until about 1800 when the camera obscura became popular. A camera obscura (dark chamber) is a closed box or room with a small opening to the outside. Light admitted through the opening is made to project an image of the outside scene onto a screen for viewing.

If the opening is simply a pinhole, then a fairly sharp, wide-field image on a flat screen is produced. However, this image is quite faint. If the pinhole is enlarged to admit more light, then image sharpness suffers. If the pinhole is replaced by a larger opening containing an ordinary singlet or doublet lens (the only types available in 1800), then a brighter sharp image is formed in the center of the field, but image sharpness degrades severely away from the center.

The first lens to image a substantial field of view was developed by W.H. Wollaston in 1812. Wollaston found that a singlet meniscus lens with a displaced aperture stop sharply focuses a wide field on a flat image surface. Understandably, Wollaston's lens became known as the Landscape lens.

When the daguerreotype, the first practical photographic process, was invented in 1839, the Wollaston Landscape lens was immediately pressed into service. However, longitudinal chromatic aberration was soon found to be a problem. In those days, the way to focus a camera was by shifting the lens fore-and-aft while visually inspecting the image on a ground glass screen (the way we still focus a view camera). The difficulty was that the best focus found visually departed significantly from the best focus for daguerreotypes, which were only sensitive to blue-violet light. Consequently, an achromatic doublet version of the Landscape lens was introduced. This was the beginning of the development of the photographic camera lens, a process that continues unabated to this day.

The daguerreotype and its successor, the collodion wet-plate process, were difficult and cumbersome to use. They were definitely the province of experts. The invention of the dry gelatin emulsion in the 1870s began to change that. But the revolution came with the invention of roll film and the introduction by Eastman in 1888 of the Kodak box camera. The box camera was fundamentally new, and it popularized photography. The first Kodak came preloaded with film for 100 pictures, and after the roll was exposed, both camera and film were returned to the factory. The developed film and prints, and the camera loaded with fresh film, were returned to the user. Eastman's advertising slogan was "You press the button, we do the rest."

Box cameras must be kept inexpensive, and that includes the lens. Here, the challenge for the lens designer is not to produce a fine lens, but to produce a simple lens that takes good pictures. To minimize costs, box camera lenses are usually restricted to having no more than one or two elements. This limit does not allow a high degree of aberration control. Thus, to reduce the effects of uncorrected aberrations and maintain image quality, box camera lenses are slow.

A slow lens, however, has advantages for a box camera beyond reducing the effects of aberrations. With a small entrance pupil diameter, a slow lens has great depth of field. Similarly, with a narrow cone of light, a slow lens has great depth of focus. These large field and focus latitudes allow a slow lens to be used fixed-focus without a focusing mechanism. Having no focusing mechanism is both a cost savings and a major simplification for the user. Furthermore, with the focus fixed, the difference between visual focus and photographic focus no longer matters. The lens can be set at the factory for best photographic focus. Thus, a fixed-focus box camera lens need not be achromatic.

Clearly, the lens that satisfies these requirements is Wollaston's original design. As a box camera lens, the singlet Landscape lens has proven to be just the right combination of adequate performance and low cost. Over the past century, many millions of Landscape lenses have been fitted to box cameras, and the lens is still in production worldwide.

Because of its important position in the history of photography, and to show what a properly designed simple lens can do, the design procedure and imaging properties of the Wollaston Landscape lens are investigated in this chapter. For comparison, the wide-angle imaging properties of a singlet with the stop at the lens are presented too.

B.2.1 The Singlet Lens with the Stop at the Lens

For a singlet lens with the stop at the lens, there are only three independent degrees of freedom that are effective in controlling aberrations and focal length. These variables are the two lens curvatures, or alternatively one power and one bending, plus paraxial defocus. Lens thickness is an ineffective variable, and a thin lens is assumed. Also, the index and dispersion of the glass have only a limited effect. To minimize longitudinal color, a low-dispersion crown glass is preferred.

With only three degrees of freedom, only three optical properties can be controlled. The power of the lens is always used to correct EFL to the desired value. Lens bending is nearly always used to control spherical aberration. However, no bending of a singlet lens with spherical surfaces can correct spherical aberration to zero; the best you can do is to minimize spherical. Defocus is used to reduce the effects of the remaining spherical. With the stop at the lens, lateral color and distortion are automatically nearly zero. And by a fortunate circumstance, the same bending that minimizes spherical aberration also makes coma nearly zero.

Thus, a singlet lens with the stop at the lens has been optimized to minimize RMS spot size monochromatically on axis. Object distance is infinity. EPD is 5 mm. The glass is Schott BK7 borosilicate crown. EFL is corrected to 80 mm, yielding $f/16$. The diameter of the image format is also 80 mm, yielding a total angular field of view (FOV) of 53°, or ±26.5°. The image surface is flat. Three field angles and five wavelengths are used in the evaluation. The fields are: 0°, 18.7°, and 26.5° (or 0, 70%, and 100% of the half-field angle). The wavelengths, chosen to give roughly equal increments in refractive index, are: 0.40, 0.44, 0.49, 0.56, and 0.68 μm. The reference wavelength is the central wavelength, 0.49 μm. Figure B.2.1.1 is a layout of the optimized lens. Note that the off-axis meridional ray bundles indicate a large amount of tangential field curvature.

In Figure B.2.1.2, the left side is a plot of tangential and sagittal field curvature for the reference wavelength using quasi-paraxial rays very close to the chief ray. Note that the two types of field curvature are large and different. The large difference indicates a large amount of astigmatism. On the right side of Figure B.2.1.2 is a percent distortion plot showing that distortion is small.

Figure B.2.1.3 is the polychromatic ray fan plot for the three field angles. Note the large scale: ±1000 μm. On-axis, there is very little spherical aberration and some longitudinal color. Off-axis, the curves are nearly straight and steeply inclined, indicating defocus caused by field curvature. The difference in slope between the

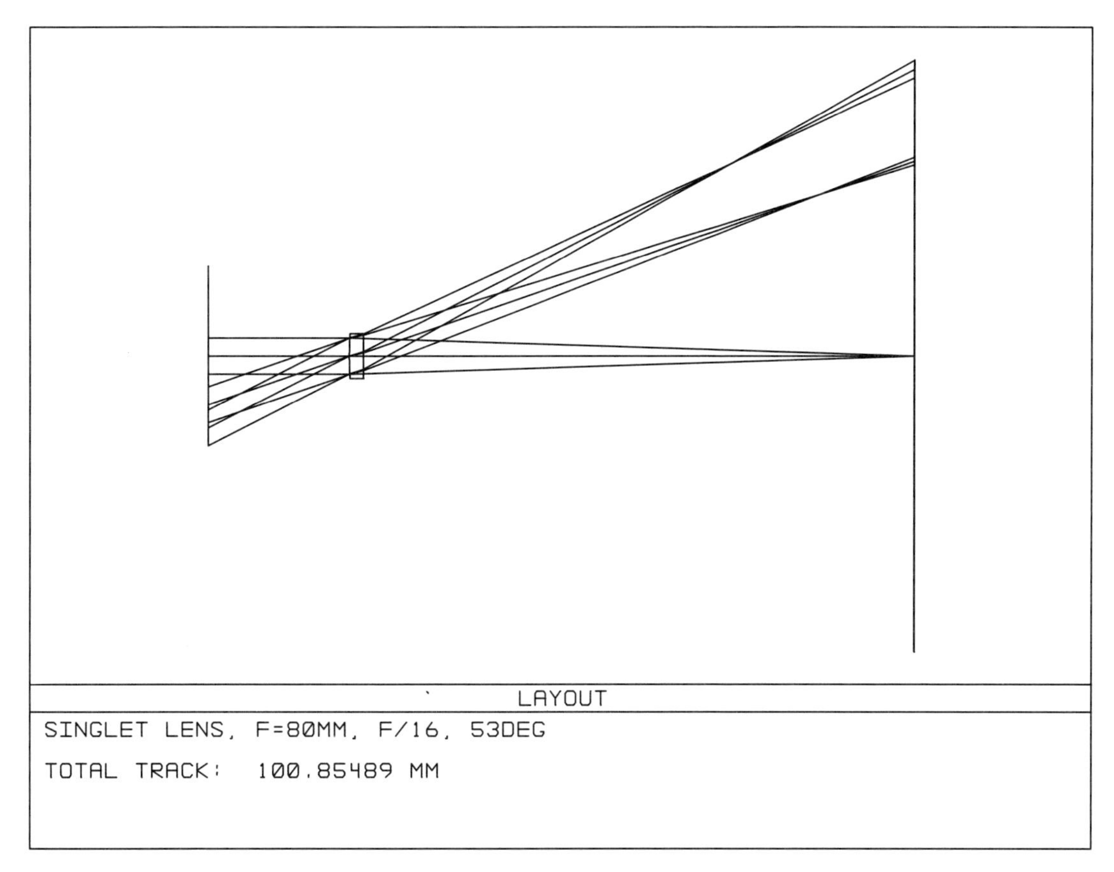

Figure B.2.1.1.

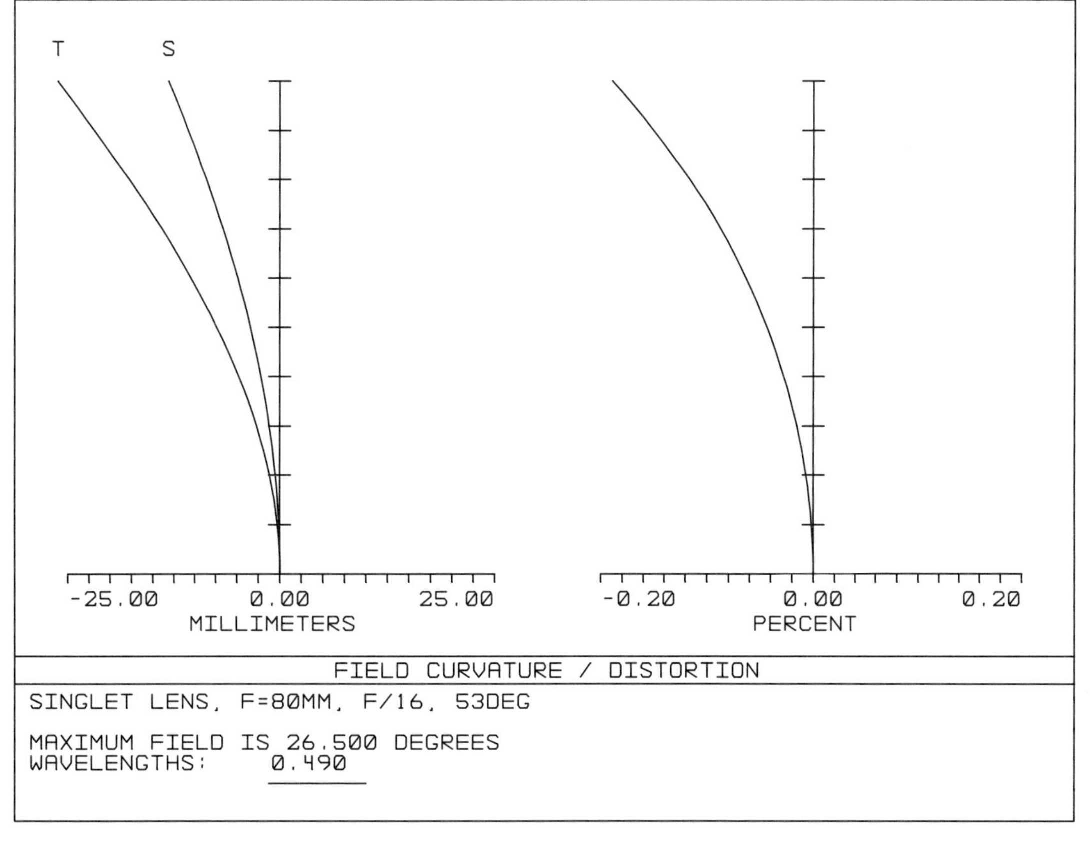

Figure B.2.1.2.

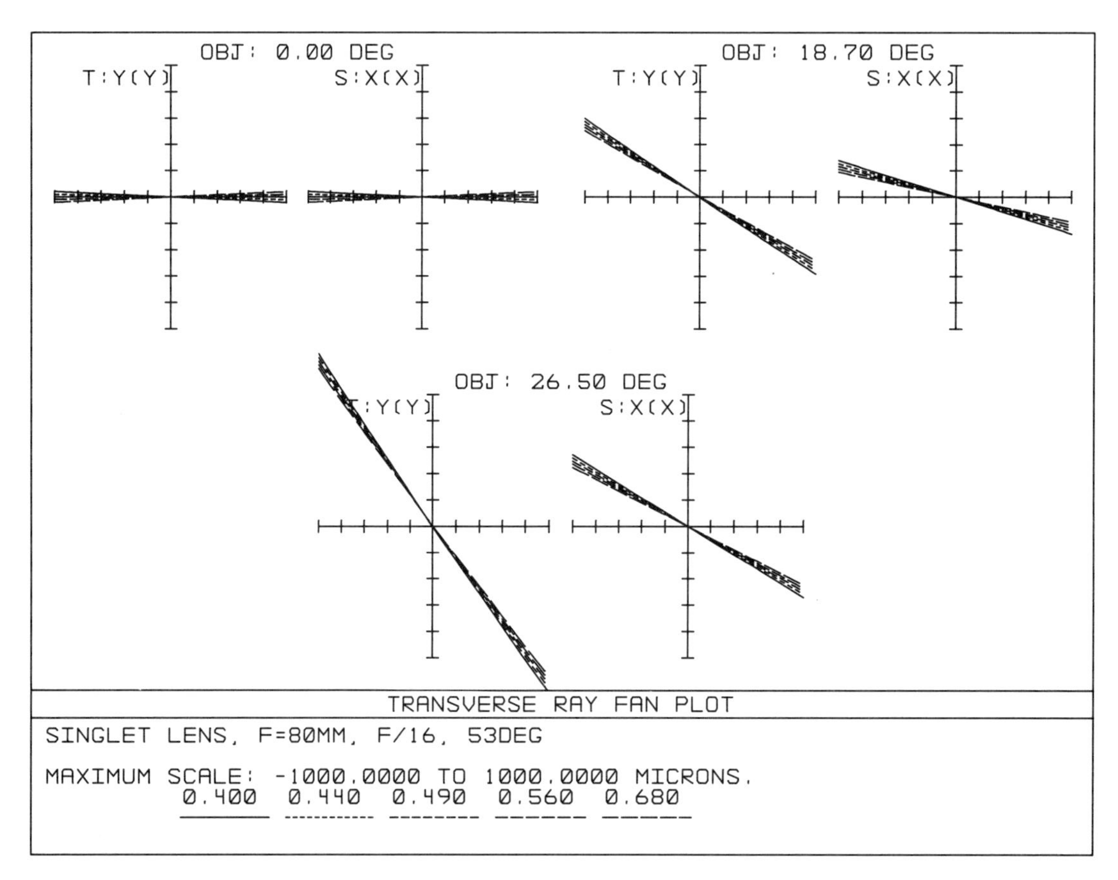

Figure B.2.1.3.

tangential and sagittal curves indicates astigmatism. Note also that the tangential curves are not U-shaped; a U-shaped curve would indicate coma. There is a relatively small amount of chromatic variation in the off-axis curves, and this is related to the longitudinal color.

Figure B.2.1.4 is the polychromatic spot diagram plot. Note again the large scale and the effects of field curvature, astigmatism, and chromatic variation of focus.

B.2.2 The Landscape Lens Optimized Polychromatically

Clearly, the simple singlet with the stop at the lens does not produce good images if the off-axis field angle is more than a few degrees. It is amazing what adding just one more effective lens variable or degree of freedom can do. To make a Landscape lens, the stop is shifted away from the lens, and the lens is bent and refocused to minimize aberrations over the entire field of view.

There are two different versions of the Landscape lens. One has the stop in front of the lens, and the other has the stop behind the lens. In both, distortion is a problem. However, the stop-in-front configuration has negative barrel distortion, whereas the stop-in-back configuration has positive pincushion distortion. Pincushion distortion is more subjectively disturbing because the picture looks like it is flying apart. Thus, the stop-in-front version is preferred and is used here.

In a Landscape lens, no attempt is made to control spherical aberration, other than by restricting lens speed. In fact, the new lens bending is very far from the minimum-spherical shape. Thus, Landscape lenses are typically about $f/11$ or slower. At these speeds, chromatic aberration is a bigger problem than spherical. To minimize color, it is important to use a low-dispersion crown glass, such as Schott BK7 or its equivalent. BK7 is also selected because it is inexpensive.

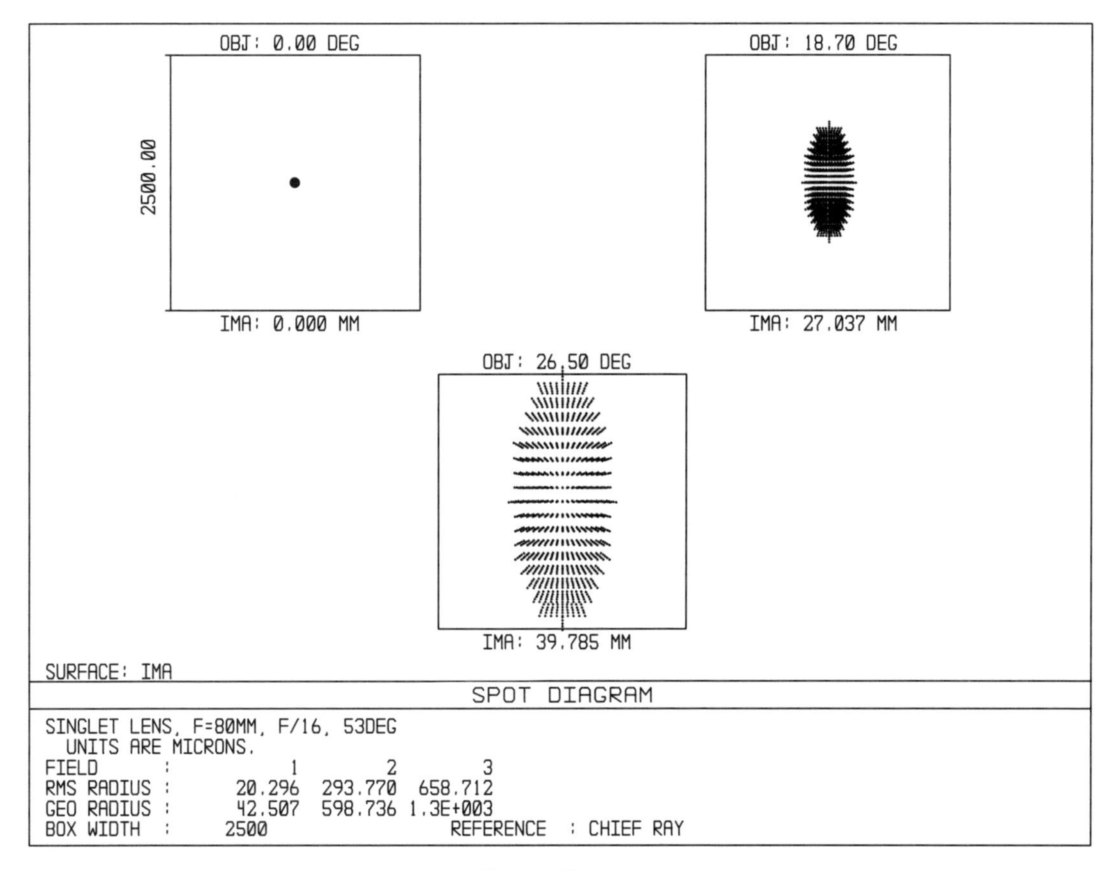

Figure B.2.1.4.

Three examples of a Landscape lens are given below. In all of these, the basic lens parameters are the same as for the above singlet. EFL, EPD, f/number, FOV, wavelengths, glass type, and so forth are all unchanged. Only stop position, lens bending, and defocus are changed to control aberrations. Glass thickness is an ineffective variable and is adjusted to give thin but practical lens edges. Lens diameter is increased as needed to pass off-axis rays.

Note that in all of the examples in this chapter, EFL is made the same as the linear image diameter, that is, made equal to the image diagonal for a rectangular film format. In particular, 80 mm is the diagonal of a square image format on 120 or 620 roll film. The resulting FOV of $\pm 26.5°$ gives photographs whose perspective appears normal to most people. Incidentally, do not confuse the term image diameter when referring to the size of the film or detector format, with image diameter when referring to the size of the PSF.

As a starting point for optimizing a Landscape lens, you can use the previously optimized singlet. By hand, place the stop a short distance in front of the lens to select the stop-in-front solution. Alternatively, you can just tinker a bit until you have a configuration that looks about right based on published drawings.

In this first version of a Landscape lens, no prior assumptions are made about how the aberrations are to be balanced. The optimization is all done in just one stage. A polychromatic optimization minimizes RMS spot size across the field of view for all colors simultaneously while also correcting focal length. With this approach, the only control you have is how you weight the fields and wavelengths when constructing the default merit function. In the present example, the five wavelengths are equally weighted at 1 1 1 1 1 (from short to long). The three fields are weighted 4 2 1 (from center to edge).

Figure B.2.2.1 is a layout of the polychromatically optimized Landscape lens. The lens is now a meniscus, concave toward the stop. Note that oblique off-axis

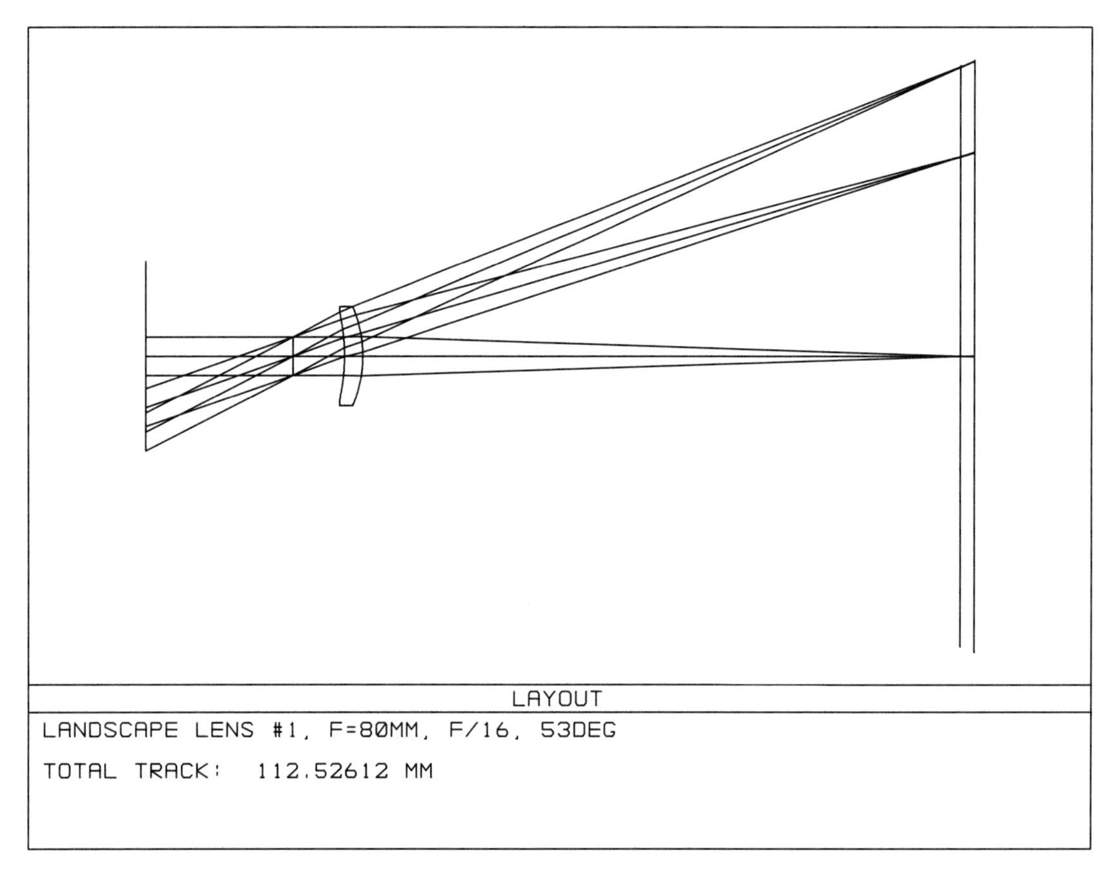

Figure B.2.2.1.

light rays pass through eccentric parts of the lens. This eccentric passage of off-axis rays is the cause of the greatly improved off-axis image quality of a Landscape lens. It is also clear that stop shift alone has no effect on on-axis imagery because the same on-axis rays are transmitted regardless of stop position.

Compare Figures B.2.2.1 and B.2.1.1. Note immediately in Figure B.2.2.1 that the field has been flattened. Also note that two surfaces are visible near the image. The surface on the far right is a dummy surface representing the paraxial focal plane; it was placed there using a paraxial marginal ray height solve. The more leftward surface is the actual image plane. Relative to the singlet with the stop at the lens, the actual image plane of a Landscape lens is considerably shifted from the paraxial focus. This shift is necessary to balance the much larger spherical aberration.

In Figure B.2.2.2, the left side is the plot of tangential and sagittal field curvature for the Landscape lens. Compare Figures B.2.2.2 and B.2.1.2, taking into account their very different scales. For the Landscape lens, note that the total amounts of both tangential and sagittal field curvature are much less. Also note that the position of the tangential image surface is now to the right of the sagittal image surface. A large amount of overcorrected astigmatism has been introduced to roughly flatten the medial image surface.

On the right side of Figure B.2.2.2 is the percent distortion plot for the Landscape lens. Distortion is no longer negligible. At the edge of the field, distortion is now −2.8% barrel. This type and amount of distortion is quite tolerable in a box camera lens.

Figure B.2.2.3 is the ray fan plot. On-axis, note the balance of third-order spherical aberration and defocus (and that higher orders of spherical are negligible).

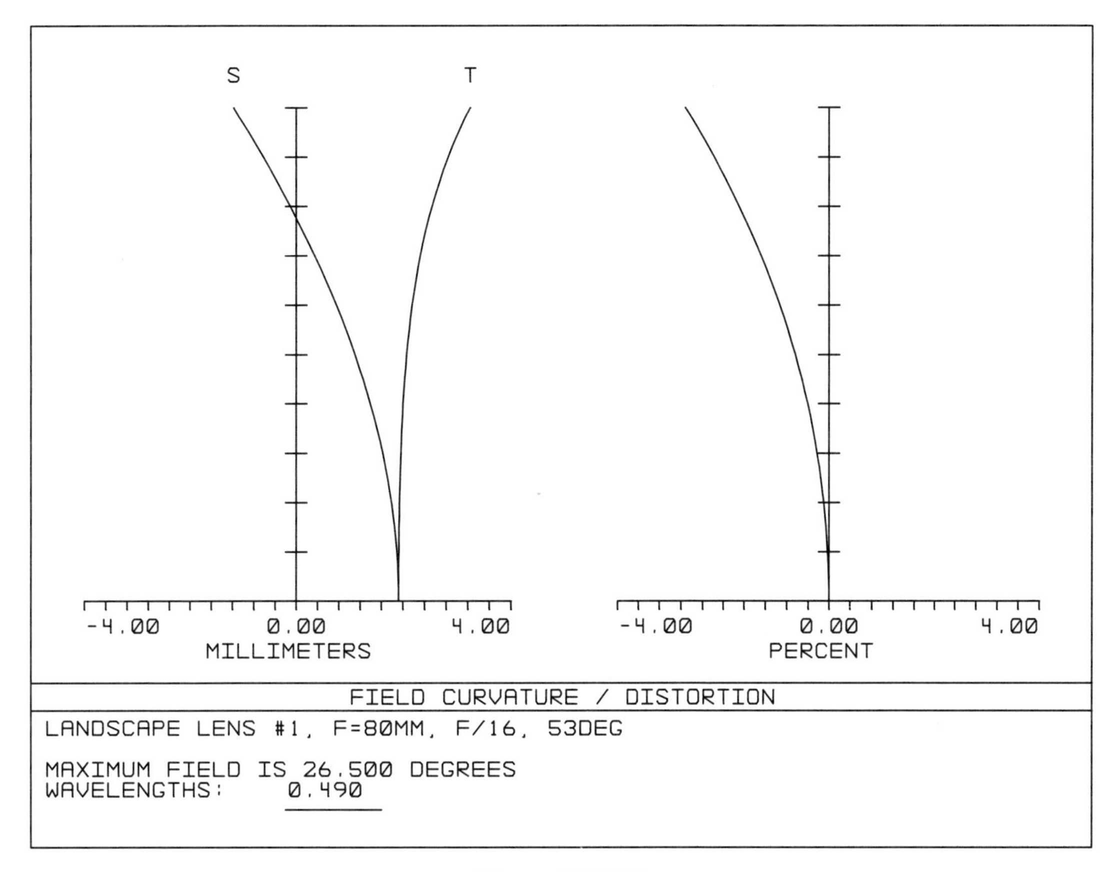

Figure B.2.2.2.

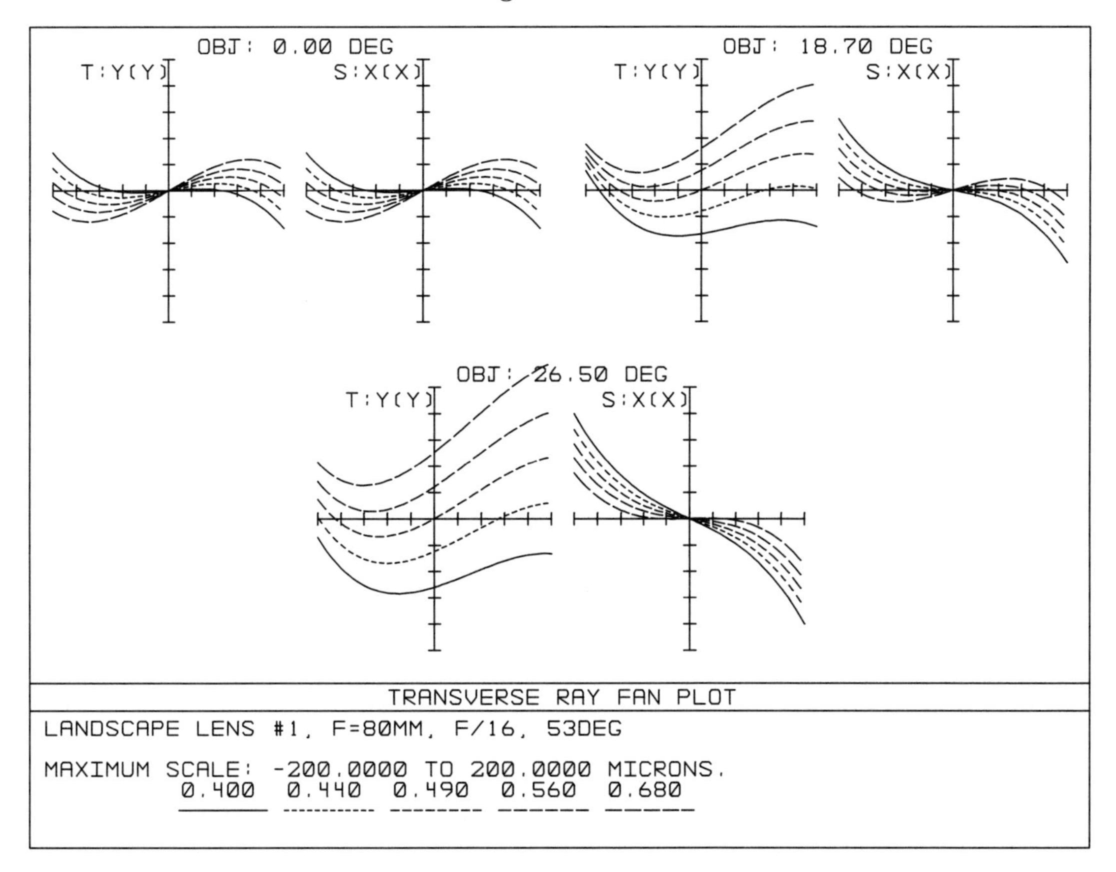

Figure B.2.2.3.

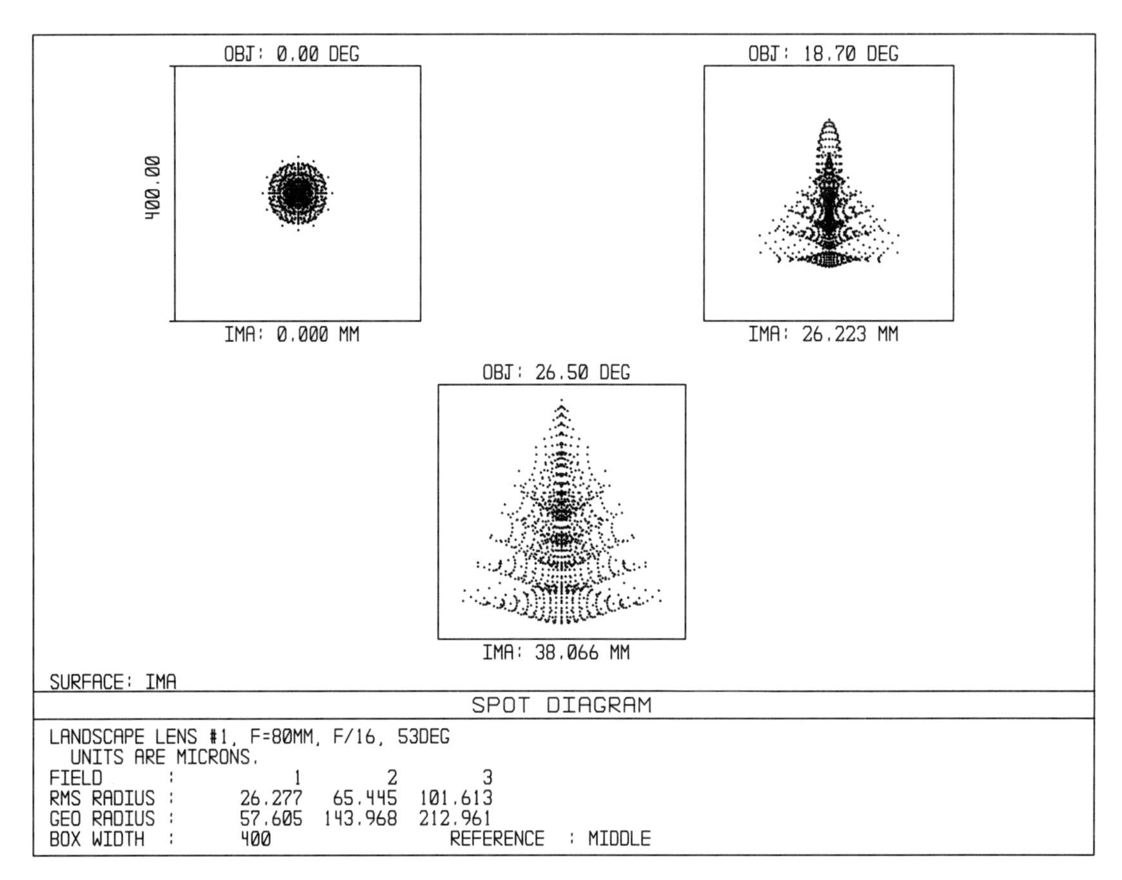

Figure B.2.2.4.

Also note that longitudinal color is more important (at $f/16$) than spherical, even with the lens bent so far away from the minimum-spherical shape. At the edge of the field, note that the tangential curves are U-shaped, indicating coma, and vertically spaced, indicating lateral color. The sagittal curves are inclined, indicating defocus. The general slopes of the tangential and sagittal curves are different, indicating astigmatism. Note too the off-axis effects of longitudinal color and spherical.

The presence of lateral color can be explained if you look at Figure B.2.2.1 and note that an off-axis beam passing through an eccentric part of the lens is effectively passing through a wedged prism with curved faces. The lens disperses the light into a small spectrum that is seen as lateral color. The presence of coma can be excused as a valiant attempt by the computer to shrink spot size by partially cancelling lateral color with coma.

Figure B.2.2.4 is the spot diagram plot. For each field angle, five individual spots, one for each wavelength, are superimposed. Off-axis, each of these component spots is shifted by lateral color and asymmetrically flared by coma.

If polychromatic spot size is the only criterion of image quality, then this Landscape lens is the best design. But coma is usually objectionable in photographs. If given the option, most photographers would rather have a bit more lateral color and be rid of coma. Fortunately with the Landscape lens, this is easy to do.

B.2.3 The Landscape Lens with No Coma and Flat Tangential Field

Most Landscape lenses are not optimized for minimum polychromatic RMS spot size across the field. There is another way that gives less spherical, no coma, less astigmatism, and a flat tangential field. This approach yields the classical Landscape lens solution.

When designing a Landscape lens, the situation is a bit reminiscent of designing a telescope achromatic doublet. You do not have enough independent lens variables to control all the basic aberrations. Therefore, you must choose very carefully which aberrations to control and let the rest remain uncontrolled.

In a singlet Landscape lens, it is impossible to correct longitudinal color. Thus, a polychromatic optimization is hopeless on-axis. With the stop separated from the lens, it is also impossible to correct lateral color. Thus, a polychromatic optimization is also hopeless off-axis. A better approach is to do a monochromatic optimization using the central wavelength, and take whatever chromatic aberrations you get.

A Landscape lens of a given glass has four effective variables or degrees of freedom: one power, one bending, stop shift, and paraxial defocus. Power is used to correct EFL, and defocus is used to balance spherical. Bending and stop shift together are used to correct coma to zero and flatten the tangential field (tangential image surface). The tangential field, rather than the medial field, is used because less overcorrected astigmatism is required, and the images are (relatively) pleasing.

In general, a ray can fall on the image surface at any azimuth or position angle relative to the chief ray. In other words, ray errors can have both tangential (y-direction) and sagittal (x-direction) components. To flatten the tangential image surface, only the tangential components of the ray errors are considered during optimization; sagittal ray errors are ignored. Controlling the tangential ray errors shrinks the spots in the y-direction. Not controlling the sagittal ray errors allows the spots to broaden in the x-direction. The result, of course, is tangential astigmatism with elongated PSF images tangent to circles centered on the field center.

One approach to flattening the tangential field is to fill the pupil with a two-dimensional array of rays from each object point. For a monochromatic solution, only rays for the central wavelength are used; the five wavelengths are weighted 0 0 1 0 0. When weighting the fields, 4 2 1 again works well. During optimization, a tangential-only solution is obtained by placing finite weights on the tangential ray errors and zero weights on the sagittal ray errors.

A second approach is to use only meridional rays and no skew rays; that is, meridional ray *fans* from each object point. Each fan causes a one-dimensional line array of rays to be incident across the pupil. In the days before computers when tracing rays (especially skew rays) was difficult, these meridional rays were often the only rays routinely used during the design of a lens. In the case of the Landscape lens, the nature of the aberrations allows this method to be quite effective. Except for the one-dimensional ray pattern in the pupil, the optimization procedure is the same as for the previous two-dimensional method.

The second approach has been adopted here. The complete tangential-only meridional merit function is given in Listing B.2.3. The operands are so simple and so few that they can be easily entered by hand (no default merit function is used). Remember, only ray errors in the y-direction are considered; in ZEMAX, use the TRAY (not the radial TRAR) operand. Note that focal length is held with a Lagrange multiplier (signaled by a weight of -1). Note too that to get the 4 2 1 field weighting, the on-axis operands had to be weighted double (8) because there are only half as many of them (5 rather than 10).

The layout of the resulting revised Landscape lens is shown in Figure B.2.3.1. Note that relative to the lens in Figure B.2.2.1, the stop is farther from the lens, the lens is larger in diameter, the curvatures on the lens are less, and the paraxial defocusing is less.

On the left side of Figure B.2.3.2 is the plot of tangential and sagittal field curvature. Note that the tangential image surface is nearly flat. Compare Figures B.2.3.2 and B.2.2.2, which have the same scales. Note that astigmatism in Figure B.2.3.2 is somewhat less, although still substantial.

On the right side of Figure B.2.3.2 is the percent distortion plot. Distortion at

Listing B.2.3

Merit Function Listing

File : C:LENS222.ZMX
Title: LANDSCAPE LENS #2, F=80MM, f/16, 53DEG

Merit Function Value: 7.08382829E-003

Num	Type	Int1	Int2	Hx	Hy	Px	Py	Target	Weight	Value	% Cont
1	EFFL		3					8.00000E+001	-1	8.00000E+001	0.000
2	BLNK										
3	BLNK										
4	TRAY		3	0.0000	0.0000	0.0000	1.0000	0.00000E+000	8	-7.51772E-003	12.871
5	TRAY		3	0.0000	0.0000	0.0000	0.8000	0.00000E+000	8	4.22171E-003	4.059
6	TRAY		3	0.0000	0.0000	0.0000	0.6000	0.00000E+000	8	9.08859E-003	B.2.813
7	TRAY		3	0.0000	0.0000	0.0000	0.4000	0.00000E+000	8	8.86204E-003	17.886
8	TRAY		3	0.0000	0.0000	0.0000	0.2000	0.00000E+000	8	5.26852E-003	6.322
9	BLNK										
10	TRAY		3	0.0000	0.7000	0.0000	1.0000	0.00000E+000	2	-1.09616E-002	6.841
11	TRAY		3	0.0000	0.7000	0.0000	0.8000	0.00000E+000	2	1.51235E-004	0.001
12	TRAY		3	0.0000	0.7000	0.0000	0.6000	0.00000E+000	2	5.29349E-003	1.595
13	TRAY		3	0.0000	0.7000	0.0000	0.4000	0.00000E+000	2	5.97852E-003	2.035
14	TRAY		3	0.0000	0.7000	0.0000	0.2000	0.00000E+000	2	3.71083E-003	0.784
15	TRAY		3	0.0000	0.7000	0.0000	-0.2000	0.00000E+000	2	-3.62511E-003	0.748
16	TRAY		3	0.0000	0.7000	0.0000	-0.4000	0.00000E+000	2	-5.60165E-003	1.787
17	TRAY		3	0.0000	0.7000	0.0000	-0.6000	0.00000E+000	2	-4.31715E-003	1.061
18	TRAY		3	0.0000	0.7000	0.0000	-0.8000	0.00000E+000	2	1.90730E-003	0.207
19	TRAY		3	0.0000	0.7000	0.0000	-1.0000	0.00000E+000	2	1.48355E-002	12.531
20	BLNK										
21	TRAY		3	0.0000	1.0000	0.0000	1.0000	0.00000E+000	1	-1.51948E-002	6.573
22	TRAY		3	0.0000	1.0000	0.0000	0.8000	0.00000E+000	1	-3.37620E-003	0.325
23	TRAY		3	0.0000	1.0000	0.0000	0.6000	0.00000E+000	1	2.71380E-003	0.210
24	TRAY		3	0.0000	1.0000	0.0000	0.4000	0.00000E+000	1	4.41089E-003	0.554
25	TRAY		3	0.0000	1.0000	0.0000	0.2000	0.00000E+000	1	3.05463E-003	0.266
26	TRAY		3	0.0000	1.0000	0.0000	-0.2000	0.00000E+000	1	-3.37067E-003	0.323
27	TRAY		3	0.0000	1.0000	0.0000	-0.4000	0.00000E+000	1	-5.63506E-003	0.904
28	TRAY		3	0.0000	1.0000	0.0000	-0.6000	0.00000E+000	1	-5.31735E-003	0.805
29	TRAY		3	0.0000	1.0000	0.0000	-0.8000	0.00000E+000	1	-8.73280E-004	0.022
30	TRAY		3	0.0000	1.0000	0.0000	-1.0000	0.00000E+000	1	9.32626E-003	2.476
31	BLNK										
32	BLNK										
33	DMFS										

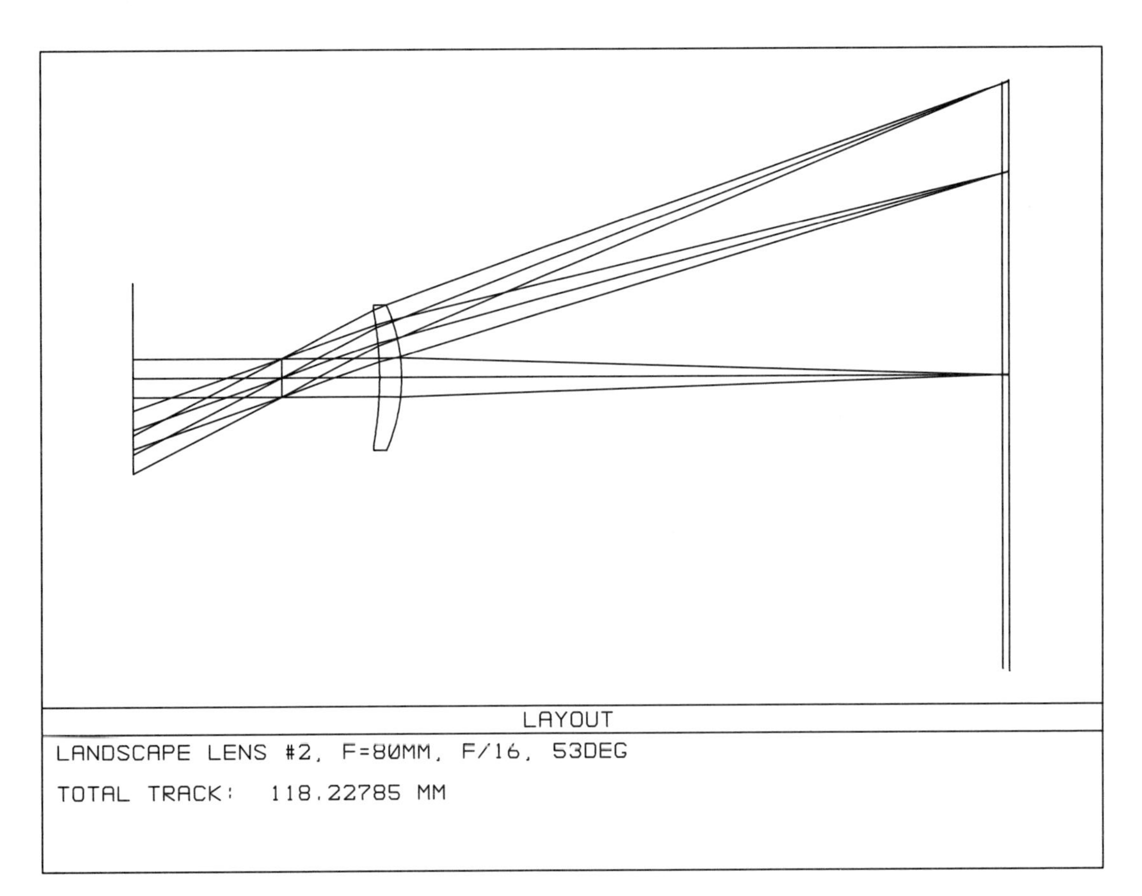

Figure B.2.3.1.

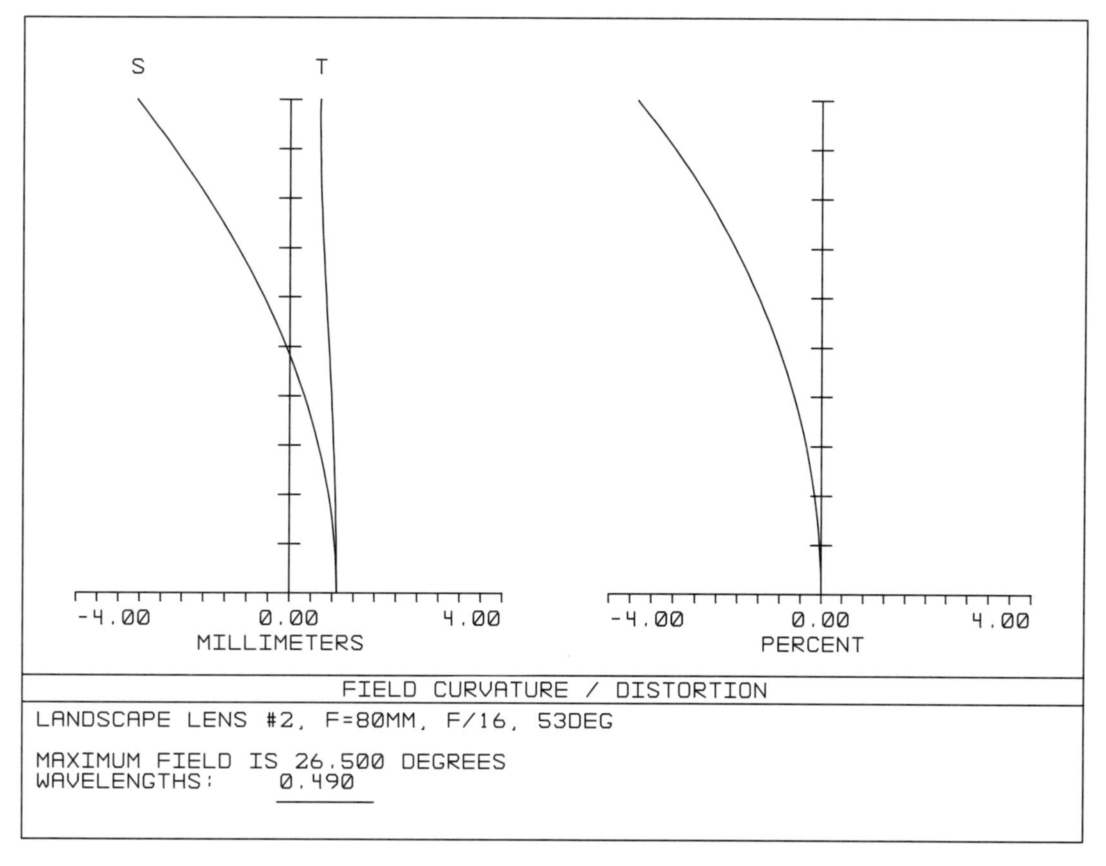

Figure B.2.3.2.

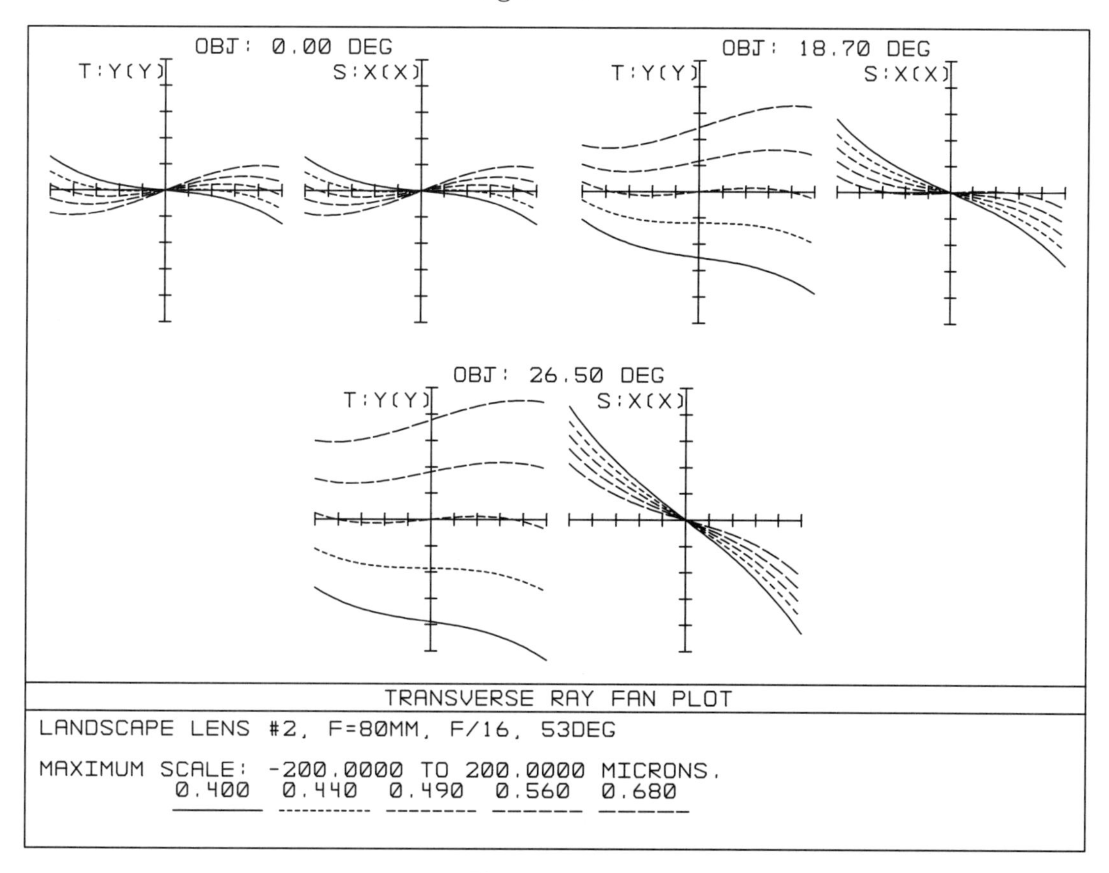

Figure B.2.3.3.

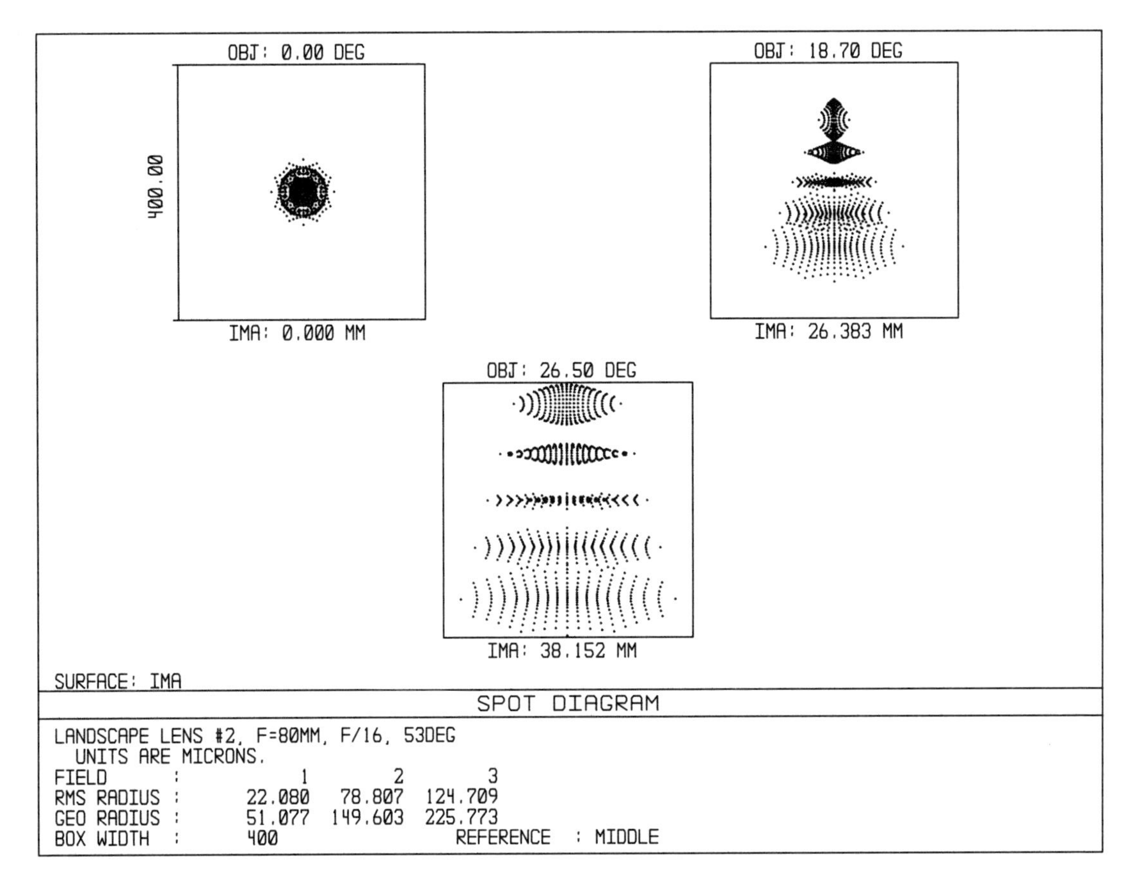

Figure B.2.3.4.

the field edge is now −3.6% barrel, an increase over the −2.8% distortion for the previous Landscape lens. Although this increased distortion is less desirable, it is still acceptable, except perhaps for architectural subjects with their straight lines.

Figure B.2.3.3 is the ray fan plot. Note that for the central wavelength, the tangential ray errors for all fields are small and nearly identical, and that coma is absent. Compare Figures B.2.3.3 and B.2.2.3, which have the same scales. On-axis, note that spherical is less for the revised lens. Off-axis, note that lateral color is significantly more, but sagittal ray errors are only slightly more.

Figure B.2.3.4 is the spot diagram plot. Compare Figures B.2.3.4 and B.2.2.4. Note that for the revised lens, the on-axis polychromatic spot is slightly smaller, and the off-axis polychromatic spots are somewhat larger with a different appearance.

B.2.4 The Landscape Lens with Mechanical Vignetting

The off-axis performance of a Landscape lens can be somewhat improved if a moderate amount of deliberate mechanical vignetting is allowed. Recall that mechanical vignetting is ray clipping and is not to be confused with cosine-fourth vignetting (see Chapter B.3 for a longer discussion of vignetting). Figure B.2.4.1 is the layout of a vignetted version of the lens in Figure B.2.3.1. The lens was not redesigned. Only its diameter was reduced to give about 40% vignetting (60% transmission) at the edge of the field.

Figure B.2.4.2 is the mechanical vignetting plot, confirming the 40% throughput drop at the edge of the field. Note also that vignetting does not begin until field angles exceed 75% of maximum field. A 40% drop may seem like a lot, but actually the latitude (exposure tolerance) of the film will make

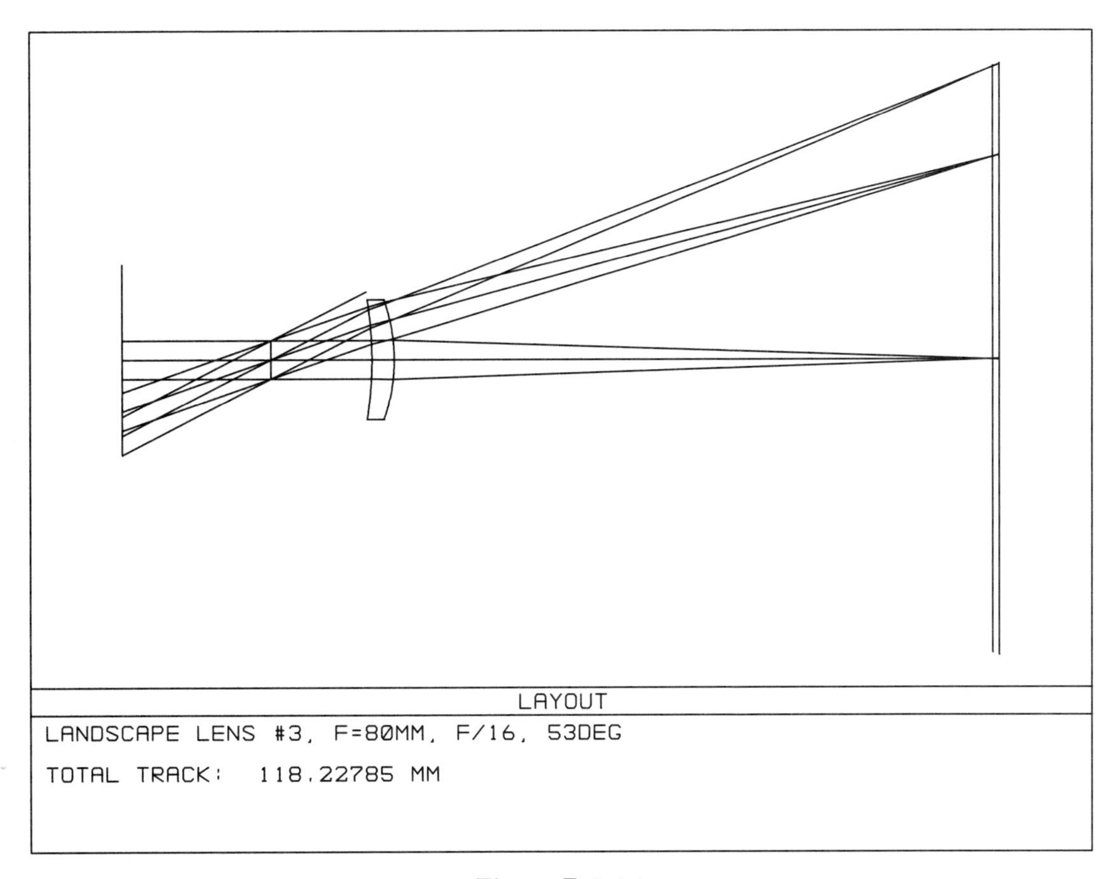

Figure B.2.4.1.

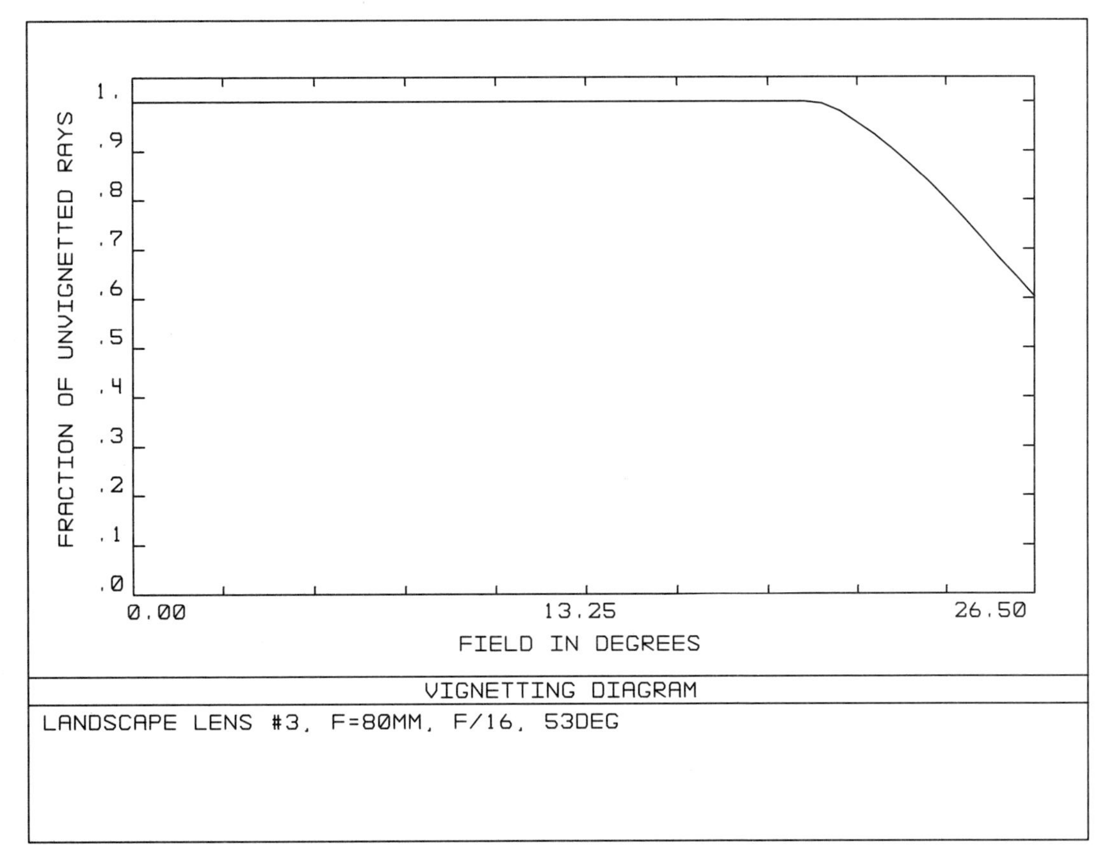

Figure B.2.4.2.

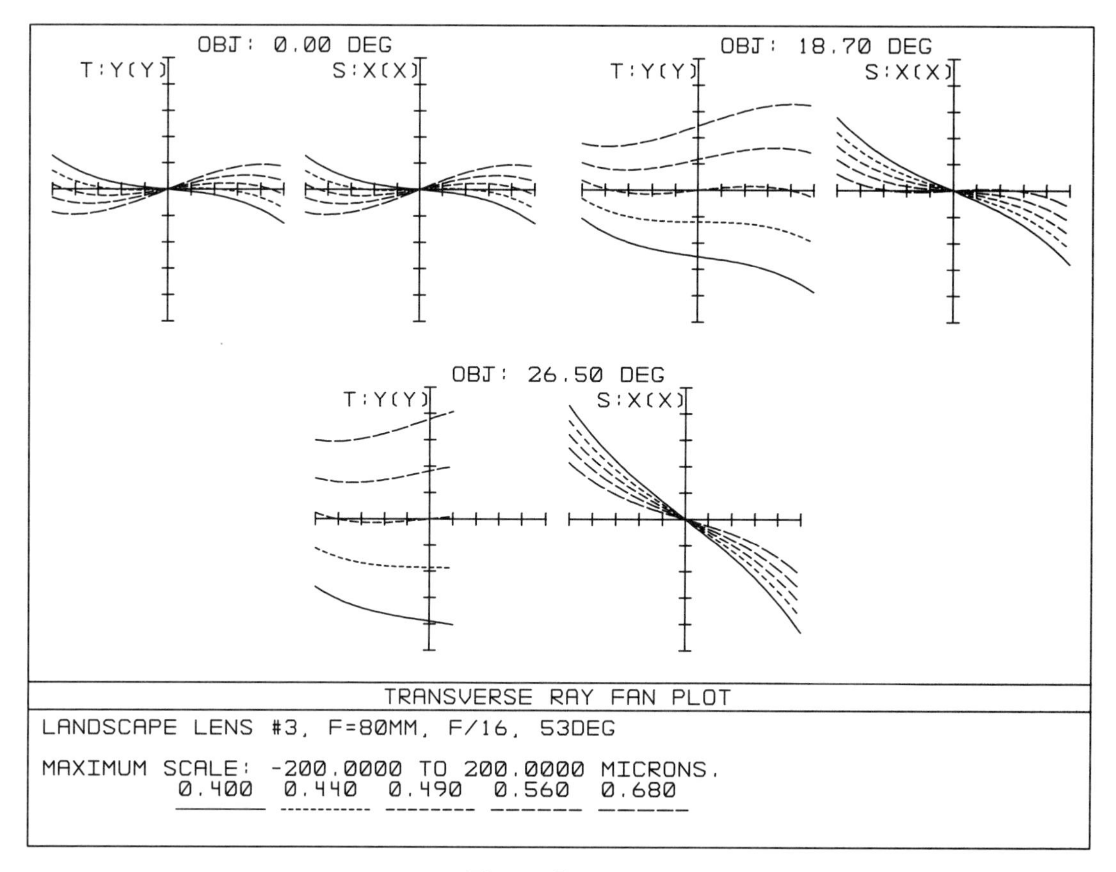

Figure B.2.4.3.

this loss hardly noticeable. Furthermore, the reduction is only in the corners of the picture.

Figure B.2.4.3 is the vignetted ray fan plot; compare this with Figure B.2.3.3. Note in Figure B.2.3.3 that the tangential ray fan curves for the field edge are more spread out for positive pupil heights (the upper part of the pupil, which is plotted on the right side of the graph). These parts of the curves are just the parts that are clipped off by the vignetting; that is, the vignetting selectively removes the most aberrated rays.

Figure B.2.4.4 is the spot diagram plot; compare this with Figure B.2.3.4. Note that there is no change for the first two field angles. Only the field edge is changed in Figure B.2.4.4.

In Figure B.2.4.4, note the sizes of the spots. On-axis, polychromatic spot size is about 100 μm across. At the edge of the field, polychromatic spot size is about 400 μm across. Diffraction, however, gives an Airy disk diameter of only about 20 μm. Thus, this lens is nowhere near the diffraction limit and geometrical spots are very good representations of actual PSFs.

These image sizes may seem huge to photographers accustomed to expensive, highly corrected, multi-element lenses. But for a box camera lens taking black and white snapshots that will not be highly enlarged, this level of image correction is perfectly satisfactory. Unfortunately, the lateral color in a Landscape lens is more noticeable with color films and with modern panchromatic black and white films (the old films were sensitive to a narrower band of wavelengths). Thus, box cameras and Landscape lenses are less popular today than in the past.

Listing B.2.4. is the optical prescription and other data for the vignetted Landscape lens.

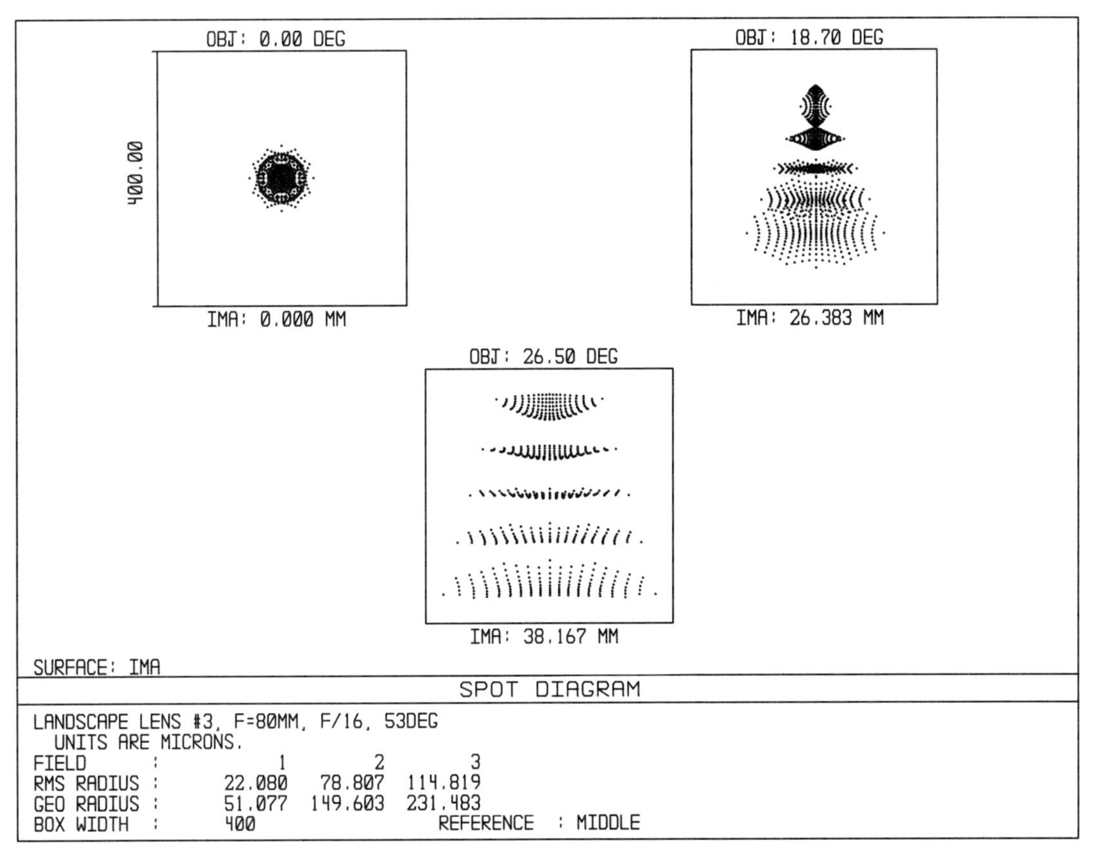

Figure B.2.4.4.

Listing B.2.4

```
System/Prescription Data

File :  C:LENS223.ZMX
Title:  LANDSCAPE LENS #3, F=80MM, f/16, 53DEG

GENERAL LENS DATA:

Surfaces          :            6
Stop              :            2
System Aperture   :Entrance Pupil Diameter
Ray aiming        : Off
Apodization       :Uniform, factor =      0.000000
Eff. Focal Len.  :           80 (in air)
Eff. Focal Len.  :           80 (in image space)
Total Track       :      118.228
Image Space f/#  :           16
Para. Wrkng f/#  :           16
Working f/#       :      15.8242
Obj. Space N.A.  :     2.5e-010
Stop Radius       :          2.5
Parax. Ima. Hgt.:      39.8865
Parax. Mag.       :            0
Entr. Pup. Dia.  :            5
Entr. Pup. Pos.  :           20
Exit Pupil Dia.  :      6.36133
Exit Pupil Pos.  :     -100.894
Field Type        : Angle in degrees
Maximum Field     :         26.5
Primary Wave      :     0.490000
Lens Units        : Millimeters
Angular Mag.      :     0.785999

Fields            : 3
Field Type:  Angle in degrees
#          X-Value           Y-Value          Weight
1         0.000000          0.000000        4.000000
2         0.000000         18.700000        2.000000
3         0.000000         26.500000        1.000000

Vignetting Factors
#            VDX               VDY               VCX               VCY
1        0.000000          0.000000          0.000000          0.000000
2        0.000000          0.000000          0.000000          0.000000
3        0.000000          0.000000          0.000000          0.000000
```

Wavelengths : 5
Units: Microns

#	Value	Weight
1	0.400000	1.000000
2	0.440000	1.000000
3	0.490000	1.000000
4	0.560000	1.000000
5	0.680000	1.000000

SURFACE DATA SUMMARY:

Surf	Type	Radius	Thickness	Glass	Diameter	Conic
OBJ	STANDARD	Infinity	Infinity		0	0
1	STANDARD	Infinity	20		24.94326	0
STO	STANDARD	Infinity	13.4926		5	0
3	STANDARD	-47.44194	3	BK7	15.6	0
4	STANDARD	-22.69406	81.73526		15.6	0
5	STANDARD	Infinity	-0.8875388		77.30986	0
IMA	STANDARD	Infinity	0		76.69215	0

SURFACE DATA DETAIL:

```
Surface OBJ      : STANDARD
Surface   1      : STANDARD
Surface STO      : STANDARD
Surface   3      : STANDARD
 Aperture        : Circular Aperture
 Minimum Radius  :            0
 Maximum Radius  :          7.8
Surface   4      : STANDARD
 Aperture        : Circular Aperture
 Minimum Radius  :            0
 Maximum Radius  :          7.8
Surface   5      : STANDARD
Surface IMA      : STANDARD
```

SOLVE AND VARIABLE DATA:

```
Thickness of   2    : Variable
Curvature of   3    : Variable
Semi Diam    3      : Fixed
Curvature of   4    : Variable
Thickness of   4    : Solve, marginal ray height = 0.00000
Semi Diam    4      : Fixed
Thickness of   5    : Variable
```

INDEX OF REFRACTION DATA:

Surf	Glass	0.400000	0.440000	0.490000	0.560000	0.680000
0		1.00000000	1.00000000	1.00000000	1.00000000	1.00000000
1		1.00000000	1.00000000	1.00000000	1.00000000	1.00000000
2		1.00000000	1.00000000	1.00000000	1.00000000	1.00000000
3	BK7	1.53084854	1.52626886	1.52209982	1.51803195	1.51361483
4		1.00000000	1.00000000	1.00000000	1.00000000	1.00000000
5		1.00000000	1.00000000	1.00000000	1.00000000	1.00000000
6		1.00000000	1.00000000	1.00000000	1.00000000	1.00000000

ELEMENT VOLUME DATA:

```
Units are cubic cm.
Values are only accurate for plane and spherical surfaces.
Element surf   3 to   4 volume :        0.504218
```

The reader may have noticed that the optimization procedures used here for the Landscape lenses are highly abbreviated versions of the general procedure outlined in Chapter A.15. Optimization is all done in one stage and only with transverse ray errors. No mention is made of OPDs or MTFs. But for lenses with so few variables and such restricted capabilities, the present approach is all that is necessary.

Chapter B.3

The Cooke Triplet and Tessar Lenses

Following the introduction of photography in 1839, the development of camera lenses progressed in an evolutionary way with occasional revolutions. Almost immediately, the singlet meniscus Landscape lens was achromatized, thus becoming a cemented doublet meniscus. Soon afterward, pairs of either singlet or doublet menisci were combined symmetrically about a stop. The most successful of the double-doublets was the Rapid Rectilinear lens, introduced in 1866. This use of symmetry was then further extended to pairs of menisci, with each meniscus consisting of three, four, or even five elements cemented together. The most successful of these lenses was the Dagor, a double-triplet (six elements), introduced in 1892. The Dagor and some of its variants are still in use today.

The first revolution or major innovation occurred in 1840 by Joseph Petzval with his introduction of mathematically designed lenses. Petzval's most famous lens is the Petzval Portrait lens. The Petzval lens is not symmetrical and has major aberrations off-axis (most noticeably, field curvature). Nevertheless, in 1840 it was an immediate success, not least because at $f/3.6$ it was 20 times faster (transmitted 20 times more light) than the $f/16$ Landscape lens, its only competition at the time. Petzval type lenses were widely used until well into the twentieth century, and they are sometimes still encountered, especially in projectors.

The second revolution, which began in 1886, was the development of new types of optical glass, especially the high-index crowns. These glasses made possible the first anastigmatic lenses; that is, lenses with astigmatism corrected on a flat field.

The third revolution was the invention by H. Dennis Taylor in 1893 of the Cooke Triplet lens (Cooke was Taylor's employer). A Cooke Triplet consists of two positive singlet elements and one negative singlet element, all of which can be thin. Two sizable airspaces separate the three elements. The negative element is located in the middle about halfway between the positive elements, thus maintaining a large amount of symmetry. By this approach, Taylor found that he could accomplish with only three elements what others required six, eight, or ten elements to do when using pairs of cemented menisci. After more than a century, the Cooke Triplet remains one of the most popular camera lens forms.

The design and performance of a Cooke Triplet is the main subject of this chapter. The construction of a Cooke Triplet is illustrated in Figure B.3.1.1. Also included here for comparison is a Tessar lens, which can be considered a derivative and extension of the Cooke Triplet (although Paul Rudolph designed the first Tessar in 1902 as a modification of an earlier anastigmat, the Protar).

The development of camera lenses has continued unabated to this day, with a huge number of evolutionary advances and several more revolutions (most notably, anti-reflection coatings and computer-aided optimization). The reader is encouraged to become familiar with optical history. It is a fascinating story.[1]

[1] Optical history is discussed in many of the references listed in the bibliography. Three particularly good sources are: Rudolf Kingslake, *A History of the Photographic Lens*; Henry C. King, *The History of the Telescope*; and Joseph Ashbrook, *The Astronomical Scrapbook*.

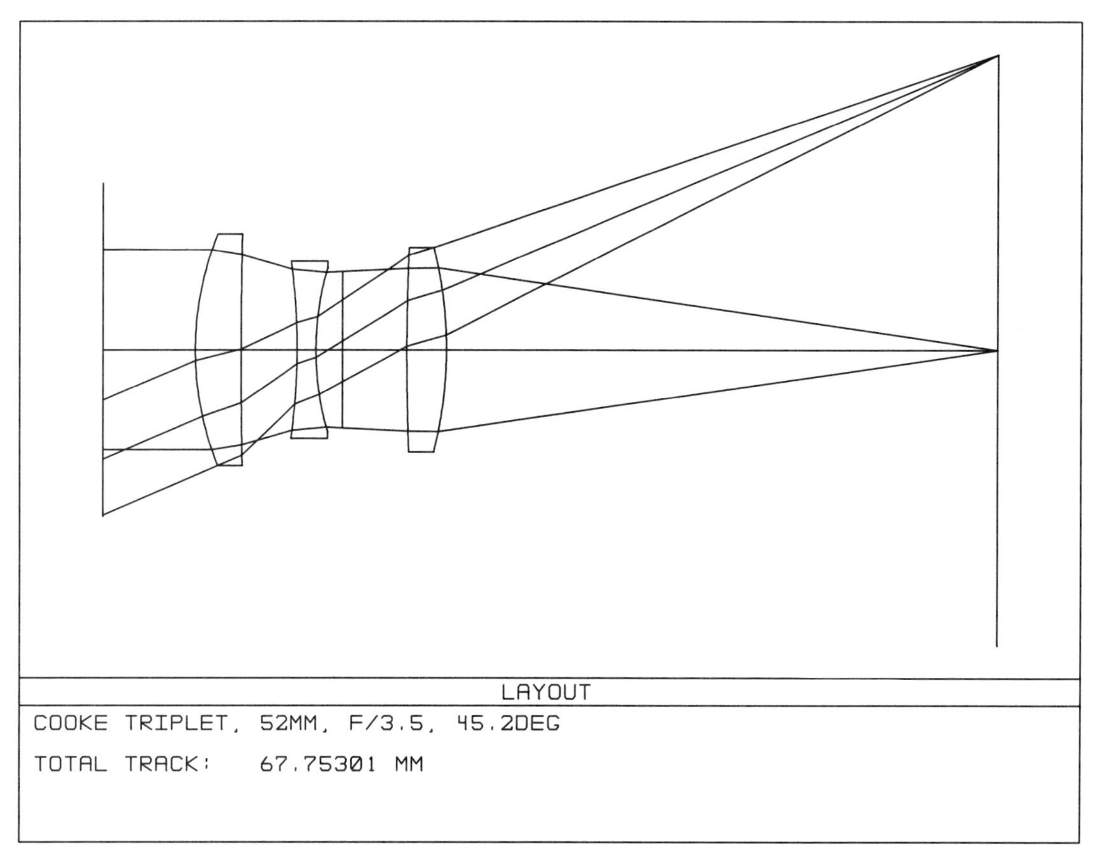

Figure B.3.1.1.

B.3.1 Lens Specifications

The lenses in this chapter are normal or standard lenses for a 35 mm type still (not movie) camera with the object at infinity. Note that 35 mm refers to the physical width of the film, not to the image size, which is only 24 mm wide (and 36 mm long). This image size was chosen for the first Leica camera in 1924 and has been the standard ever since. The remaining film area is occupied by two sets of sprocket holes, a vestige from its origin as a motion picture film (where the image is only 18 mm long).

As mentioned in the previous chapter, a normal lens gives normal-perspective photographs and has a focal length roughly equal to the image format diameter or diagonal. At 24x36 mm, the diagonal of the 35 mm format is 43.3 mm. However, the lens on the first Leica had a focal length of 50 mm, and again this value has become standard. Actually, most normal lenses for 35 mm cameras are deliberately designed to have a focal length even a bit longer still. Thus, although the nominal focal length engraved on the lens barrel is 50 mm, the true focal length is often closer to 52 mm. Thus, 52.0 mm is adopted as the focal length for the lenses in this chapter. This focal length is longer than the format diagonal, but it is close enough.

Given this focal length and image format, the diagonal field of view (from corner to corner) is 45.2°, or ±22.6°. Four field positions are used during optimization and evaluation: 0°, 9°, 15.8°, and 22.6° (or 0, 40%, 70%, and 100% of half-field). Object 1 is in the field center, object 4 is at the distance of the format corner, object 3 is at the distance of one side of an equivalent square format, and object 2 is slightly greater than halfway between objects 1 and 3. Of course, for an ordinary camera lens, the field (image surface) is flat.

For a photographic Cooke Triplet covering this angular view, it is usually not practical to increase speed beyond about $f/3.5$. Triplets with focal lengths of 52 mm and speeds of $f/3.5$ are widely manufactured and used with success. Thus, $f/3.5$ is adopted here.

During optimization, no special emphasis is given to the center of the pupil to the detriment of the pupil edges. To increase performance when stopped down, the lens is optimized and used on the paraxial focal plane. At the edge of the field, a reasonable amount of mechanical vignetting is allowed. Distortion is corrected to zero at the edge of the field.

Five wavelengths are used during optimization and evaluation. Because most films are sensitive to wavelengths between about 0.40 μm and 0.70 μm, the five wavelengths are: 0.45, 0.50, 0.55, 0.60, and 0.65 μm. For panchromatic film sensitivity, these wavelengths are weighted equally. The reference wavelength for calculating first-order properties and solves is the central wavelength, 0.55 μm.

And finally, the performance criterion for the optimized lens is diffraction MTF; that is, the form of MTF with both diffraction and aberrations included.

B.3.2 Degrees of Freedom

The Cooke Triplet is a very interesting optical configuration.[2] Refer again to Figure B.3.1.1. There are exactly eight effective independent variables or degrees of freedom available for the control of optical properties. These major variables are six lens surface curvatures and two interelement airspaces. The six curvatures can also be viewed as three lens powers and three lens bendings.

Recall that there are seven basic or primary aberrations (first-order longitudinal and lateral color, and the five monochromatic third-order Seidel aberrations). Thus, the Cooke Triplet has just enough effective independent variables to correct all first- and third-order aberrations plus focal length.

Although the airspaces in a Cooke Triplet are effective variables, the glass thicknesses are unfortunately only weak, ineffective variables that somewhat duplicate the airspaces. The glass center thicknesses are usually arbitrarily chosen for ease of fabrication.

The three glass choices can also be viewed as variables (or index and dispersion variable pairs). But the range of available glasses is limited, and the requirements of achromatization further restrict the glasses. In practice, the role of glass selection is to determine which of a multitude of possible optical solutions you get.

Stop shift is not a degree of freedom. A Cooke Triplet is nearly symmetrical about the middle element, which makes aberration control much easier. To retain as much symmetry as possible, the stop is either at the middle element or just to one side. A slightly separated stop allows the stop to be a variable iris diaphragm for changing the $f/$number. Locating the iris to the rear of the middle element, rather than to the front, is more common, although both work. In the present example, the stop is 2 mm behind the middle element.

With only eight effective variables, there are no variables available for controlling the higher-order monochromatic aberrations that will inevitably be present. Although it is possible in a Cooke Triplet to correct all seven first- and third-order aberrations exactly to zero, in practice this is never done. Controlled amounts of third-order aberrations are always deliberately left in to balance the fifth- and higher-orders. However, there is a limit to how well this cancellation works. Thus, Cooke Triplets are usually restricted to applications requiring only moderate speed and field coverage; that is, where there are only moderate amounts of fifth- and

[2]For those interested in the more analytical aspects of designing a Cooke Triplet lens, see the excellent discussions in: Warren J. Smith, *Modern Optical Engineering*, second edition, pp. 384–390; Warren J. Smith, *Modern Lens Design*, pp. 123–146; and Rudolf Kingslake, *Lens Design Fundamentals*, pp. 286–295. These three books are also highly recommended for their discussions of many other topics in optics.

higher-order aberrations. For higher performance, a lens configuration with a larger number of effective degrees of freedom is needed.

B.3.3 Glass Selection

A Cooke Triplet is an achromat. Thus, as discussed in Chapter B.1, each of the two positive elements must be made of a crown type glass (lower dispersion or higher Abbe number), and the negative element must be made of a flint type glass (higher dispersion or lower Abbe number).

For practical reasons, and with no loss of performance, both positive crown elements are usually made of the same glass type, and this will be done here.

The sizes of the airspaces in a Cooke Triplet are a strong function of the dispersion difference between the crown and flint glasses. A dispersion difference that is too small causes the lens elements to be jammed up against each other, and there are also large aberrations. Conversely, a dispersion difference that is too large causes the system to be excessively stretched out, and again there are large aberrations. One value of dispersion difference produces airspaces of the right size to yield a good optical solution. Fortunately, the required dispersion difference can be satisfied by the range of actual available glasses on the glass map.

The difference in n_d index of refraction between the crown and flint glasses also enters into the optical solution. To help reduce the Petzval sum to flatten the field, the positive elements should be made of a higher-index crown glass, and the negative element should be made of a somewhat lower-index flint glass.

To make this an exercise in determining the capability of the Cooke Triplet design form, all the more or less normal glass types will be allowed. This includes all the expensive high-index lanthanum glasses. Glass cost should not be a driving issue here because the lens elements are quite small and require only small amounts of glass. Optical and machine shop fabrication costs should be much more important than glass cost. However, this is not true for all types of lenses; the relative importance of glass cost gets greater as element sizes get larger. Note that no attempt is made here to reduce secondary color, and thus the abnormal-dispersion glasses are excluded.

Experience has shown that high-index crown glasses generally give better image quality. Not only is the Petzval sum reduced, but high indices yield lower surface curvatures that in turn reduce the higher-order aberrations. Because there are relatively few high-index crowns on the glass map, the usual procedure when selecting glasses for a Cooke Triplet is to first select the crown glass type, and to then select the matching flint glass type from among the many possibilities.

Refer back to the glass map in Figure A.10.2. Note that the boundary of the highest-index crowns is a nearly straight line on the upper left side extending from (in the Schott catalog) about SK16 (n_d of 1.620, V_d of 60.3) to about LaSFN31 (n_d of 1.881, V_d of 41.0). Although the top of this line extends well into the nominally flint glasses, the glasses along the top can function as crown glasses because they are more crowny than the very flinty dense SF glasses with which they would be paired.

The candidate crown glasses are therefore located along the line bounding the upper left of the populated region of the glass map. An excellent crown glass of very high index is Schott LaFN21 (n_d of 1.788, V_d of 47.5). Although not the most extreme high-index crown, LaFN21 is widely used and practical. Thus, LaFN21 is adopted here.

Given the crown glass selection, the basic optical design requires a matching flint having a certain dispersion difference and a low index. However, there is a limit to how low the flint index can be. This limit is the arc bordering the right and bottom of the populated region of the glass map (again see Figure A.10.2). This arc is called the old glass line. Thus, it is from the glasses along the old glass line that the flint glass of a Cooke Triplet is usually selected.

For early optimizations, make a guess at the flint glass type. Fortunately, your guess does not have to be very good because it will soon be revised. Accordingly, Schott SF15 (n_d of 1.699, V_d of 30.1), a glass on the old glass line with an index somewhat lower than LaFN21, is provisionally adopted.

For intermediate optimizations, the same crown glass selection, LaFN21, is retained, but a more optimum flint glass match must now be made. One way to find the best flint is to let flint glass type be a variable during optimization and let the computer make the selection. However, because some lens design programs may have difficulty handling variable glasses, a manual approach may be more effective, and also more revealing to the lens designer.

Using the manual approach, select several likely flint glass candidates from along the old glass line to combine with LaFN21. For each combination, optimize the Cooke Triplet with the intermediate merit function. Flint glasses down and to the left on the glass line will yield lenses with smaller airspaces; flints up and to the right will yield larger airspaces. Adjust the apertures to give roughly the required amount of vignetting. For each of the combinations, tabulate the value of the merit function and look at the layout and ray fan plot. The best glass pair will show a practical layout and have the lowest value of the merit function.

In the present Cooke Triplet example with the two crown elements made of Schott LaFN21 (n_d of 1.788, V_d of 47.5), the matching glass for the flint element is found to be, not SF15 (n_d of 1.699, V_d of 30.1) as was first guessed, but Schott SF53 (n_d of 1.728, V_d of 28.7). Both LaFN21 and SF53 are among Schott's preferred glass types.

B.3.4 Flattening the Field

When choosing glasses, it is important to consider the Petzval sum. Refer again to the discussion of Petzval sum in Chapter A.9. Recall that the Petzval sum gives the curvature of the Petzval surface, and that (when only considering astigmatism and field curvature) the Petzval surface is the best image surface when astigmatism is zero.

Thus, for a lens with little astigmatism, the Petzval sum must be made small to flatten the field. However, the sum need not be exactly zero because leaving in a small amount of Petzval curvature allows field-dependent defocus to be added to the off-axis aberration cancelling mix. The important thing is that the Petzval sum must be controllable during optimization.

For thin lens elements, the contribution to the Petzval sum by a given element goes directly as element power and inversely as element index of refraction. In a multi-element lens with an overall positive focal length, the elements with positive power necessarily predominate. Thus the Petzval sum will naturally tend to be sizable and negative, thereby giving an inward curving field (image surface concave to the light).

There are in general two ways to reduce the Petzval sum to flatten the field. They are: (1) glass selection and (2) axial separation of positive and negative optical powers. In practice, both ways are usually used together. As was mentioned in the previous section, the first way, glass selection, requires that the two positive elements of a Cooke Triplet be made of higher-index glass to give decreased negative Petzval contributions, and the negative element be made of lower-index glass to give an increased positive contribution. The second way, separation of powers, requires that the positive and negative elements be separated in a manner that causes the power of the negative element to be increased relative to the powers of the positive elements. The relatively larger negative power then gives a relatively larger positive Petzval contribution that reduces the Petzval sum.

To visualize how the second method works to flatten the field, examine the path of the upper marginal ray (for the on-axis object) as it passes through the optimized Cooke Triplet in Figure B.3.1.1. Note that the height of the ray is less on the middle negative element than on the front positive element, and consequently the ray slope is negative (downward) in the intervening front airspace. If the front airspace is made larger, and assuming the ray slope is roughly unchanged, then the height of the marginal ray on the middle element is reduced. But this reduced ray height requires that the middle element be given greater power (stronger curvatures) to bend the marginal ray by the angle necessary to give the required positive (upward) ray slope in the rear airspace. This use of airspaces is how the relative power of the negative element of a Cooke Triplet is increased to reduce the Petzval sum. The effect is greater as the airspaces are increased and the lens is stretched out.

Note that this reasoning applies to other lens types too. It even applies to lenses having thick meniscus elements. With a thick meniscus, there is again an axial separation of positive and negative powers. Now, however, the separation is between two surfaces on the same element, rather than between two different elements. Instead of an airspace, it is now a glass-space. Flattening the field with thick meniscus elements is called the thick-meniscus principle.

B.3.5 Vignetting

Like the vast majority of camera lenses, this Cooke Triplet example is to have mechanical vignetting of the off-axis pupils. Recall that mechanical vignetting is caused by undersized clear apertures on surfaces other than the stop surface, and these apertures selectively clip off-axis beams. Mechanical vignetting is useful for two reasons. First, the smaller lens elements reduce size, weight, and cost. Second and more fundamental, vignetting allows a better optical solution.

Most camera lenses of at least moderate speed and field coverage suffer from secondary and higher-order aberrations, both chromatic and monochromatic. Three prominent examples of secondary aberrations are secondary longitudinal color, secondary lateral color, and spherochromatism (chromatic variation of spherical aberration). Higher-order aberrations include the fifth-order, seventh-order, etc. counterparts of the third-order Seidel monochromatic aberrations. There are also aberrations that have no third-order form and begin with fifth- or higher-order, such as oblique spherical aberration, a fifth-order aberration. These secondary and higher-order aberrations are very resistant to control during optimization because they are usually a function of the basic optical configuration, not its specific implementation. For example, the Dagor lens just inherently has lots of on-axis zonal spherical, the result of lots of third-order spherical imperfectly balancing lots of fifth-order spherical. Similarly, the Double-Gauss lens has lots of off-axis oblique spherical.

For a lens with resistant on-axis aberrations, overall maximum system speed must be restricted. However, for resistant off-axis aberrations, system speed may need to be reduced only off-axis. To do this, you effectively stop down the lens only off-axis by using mechanical vignetting. In other words, to suppress resistant off-axis aberrations, it is often more effective to use undersized apertures to simply vignette away some of the worst offending rays rather than try to control them through optimization.

This may sound crude, but actually it is elegant when properly done. Mechanical vignetting allows you to concentrate the system's always limited degrees of freedom on reducing the remaining aberrations. The result is a much sharper lens with only a mild falloff in image illumination (irradiance), mostly in the corners. Compare the layouts in Figures B.3.1.1 and B.3.1.2. The first lens has vignetting (for simplification, the second and third fields have been omitted). The second lens has no vignetting (with all four fields drawn).

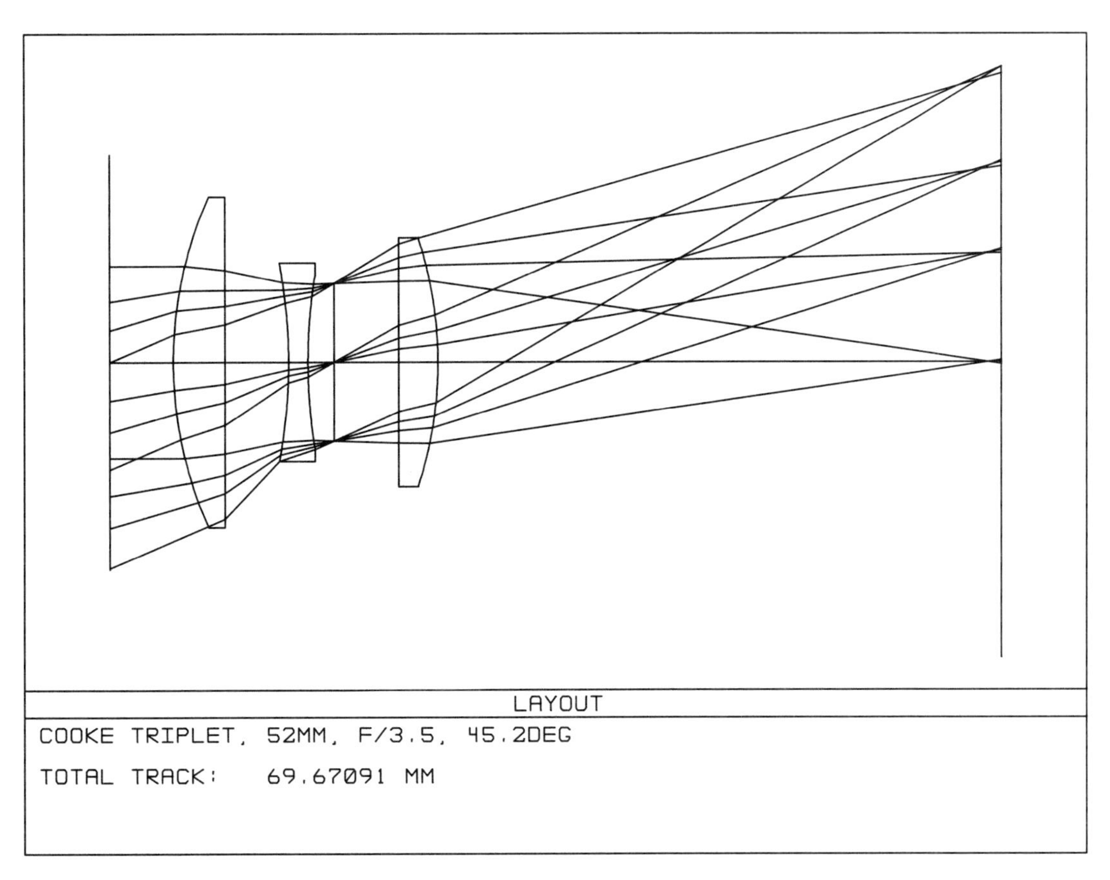

Figure B.3.1.2.

Conceptually, there are two different ways to handle mechanical vignetting when designing a lens. Both are available in ZEMAX. For this reason, and because ZEMAX happens to be the program the author uses, some of the details here apply specifically to ZEMAX, although the concepts are generally applicable.

The first way to handle vignetting uses real or hard apertures on lens surfaces to block and delete the vignetted portion of the off-axis rays. The second way uses vignetting factors or vignetting coefficients to reshape the off-axis beams to match the restricted vignetted pupil, thereby allowing all of the off-axis rays to pass through the lens to the image.

Each of these two methods has its advantages and disadvantages. The use of hard apertures is more realistic and can accommodate unusually shaped pupils and obscurations. But the use of hard apertures requires an optimization method that is less efficient and takes more computer time. The use of vignetting factors involves approximations to the actual situation and may introduce significant errors. But if vignetting factors are used (and they can be for most systems), then the computations are much faster, and finding the solution during optimization is more direct.

To illustrate both ways of handling vignetting, the present chapter uses hard apertures, and the following chapter uses vignetting factors.

In the present chapter, to construct a default merit function that includes vignetting, the entrance pupil is illuminated by simple grids of rays (the rectangular array option in ZEMAX). When projected onto the stop surface, the rectangular grids are actually square. Rays that are blocked by hard vignetting apertures are deleted from the ray sets. The same is true for central obscurations, although none are present here. Only the rays that reach the image surface are included in the construction of the operands in the default merit function. This is a very general and physically realistic approach that is applicable to any optical system.

For many lens configurations, including the Cooke Triplet, only hard apertures on the front and rear surfaces need be considered when adjusting vignetting. These are the two surfaces most distant from the stop, and they are ideally placed for defining beam clipping. All other surfaces (except the stop) are made large enough to not clip rays that can pass through the two defining surfaces.

Exceptions to this approach are lenses having surfaces located large distances from the stop; that is, where system axial length on one or both sides of the stop is considerably longer than the entrance pupil diameter. Three prominent examples are true telephoto lenses, retrofocus type wide-angle lenses, and zoom lenses. Here, the transverse location (footprint) of off-axis beams on surfaces far from the stop can shift by much more than the beam diameter. For gradual mechanical vignetting in these lenses, the defining apertures must be closer to the stop.

When selecting vignetting apertures, try to clip similar amounts off the top and bottom of the extreme off-axis beam. This maintains as much symmetry as possible. However, this rule is neither precise nor rigid, and it can be bent if one side of the pupil has worse aberrations than the other side. Use layouts and ray fan plots to determine the relative top and bottom clipping. Use the geometrical throughput option in your program to precisely calculate the fraction of unvignetted rays passing through the stop surface as a function of off-axis distance or field angle. Pay special attention to relative throughput at the edge of the field, the worst case. Of course, the defining apertures must not clip the on-axis beam; the on-axis beam must be wholly defined by the stop aperture.

For the Cooke Triplet and many other lenses, a good approach is to make the defining front and rear apertures just a little bigger than the on-axis beam diameter, as illustrated by the lens in Figure B.3.1.1. With these apertures and proper airspaces, mechanical vignetting begins about a third of the way out toward the edge of the field and increases smoothly with field angle.

For most camera lenses, a relative geometrical transmission by the vignetted off-axis pupil of about 50% or slightly less at the edge of the field is scarcely noticeable with most films and other image detectors. In fact, this amount of vignetting is quite conservative; many excellent camera lenses have much more light falloff when used wide open. Therefore, when the Cooke Triplet is wide open at $f/3.5$, relative throughput of about 50% at the edge of the field is adopted here for the allowed amount of vignetting.

Note the convention: 60% vignetting means 40% throughput. Note too that as a lens is stopped down, the apparent mechanical vignetting decreases and image illumination (irradiance) becomes more uniform.

B.3.6 Starting Design and Early Optimizations

The optimization procedure outlined in Chapter A.15 has been adopted for designing this Cooke Triplet example.

When deriving a rough starting design, you first select the starting glasses (as described above). The next thing is to make guesses at the initial values of the system parameters; that is, the six curvatures, the two airspaces, and the three glass thicknesses. To automatically reduce the three transverse aberrations (lateral color, coma, and distortion), try to make the system as symmetrical about the stop as possible. Of course, perfect symmetry is impossible because the object is at infinity and the stop is not exactly at the middle element.

Initially, make the curvatures on the outer surfaces of the two positive elements equal with opposite signs. Make the curvatures on the inner surfaces of the two positive elements equal with opposite signs. And make the two curvatures on the middle negative element equal with opposite signs. The easy way to create this symmetry and maintain it during early optimizations is to use three curvature pickup solves to make the last three surface curvatures equal to the first three surface curvatures in reverse order and with opposite signs.

Because the stop is located in the rear airspace, better symmetry can be initially achieved by making the rear airspace (glass-to-glass) a bit larger than the front airspace. The easy way to do this is to use a thickness pickup solve to make the space between the stop and rear element equal to the space between the front and middle elements.

The three elements should be thin but realistic and easy to fabricate. Make the two positive elements thick enough to avoid tiny or negative edge thicknesses. Make the negative element thick enough to avoid a delicate center. The usual procedure is to manually select the glass center thicknesses and then to freeze or fix them; that is, they are kept constant during optimization. If a glass thickness becomes inappropriate as the design evolves, change it by hand and reoptimize. In addition, remember that this lens is physically quite small. Practical edges and centers may look deceptively thick when compared to the element diameters. A life-size layout can be very valuable in giving the designer a more intuitive awareness of the true sizes involved. In fact, at some time during the design of any lens, a life-size layout should be made.

To locate the image surface at the paraxial focus of the reference wavelength, proceed as in earlier chapters. Use a marginal ray paraxial height solve on the thickness following the last lens surface to determine the paraxial focal distance (paraxial BFL). Place a dummy plane surface at this distance to represent the paraxial focal plane. The actual image surface is a plane that immediately follows the paraxial focal plane. The two surfaces, at least initially, are given a zero separation. This general procedure is good technique and retains the option of adding paraxial defocus to the aberration balance during optimization. However, this option is not always used. In particular, for both examples in this chapter, the image surface is kept at the paraxial focus; that is, no paraxial defocus is used.

Finally, because only hard apertures are to be allowed for controlling vignetting in this example, it is recommended that deliberate mechanical vignetting not be used at all in this early optimization stage of the Cooke Triplet. Vignetting can be introduced in the intermediate optimization stage. Note that this approach works for this $f/3.5$ lens, but may not work as well for a faster lens, such as the $f/2$ Double-Gauss lens in the next chapter.

Based on the above suggestions, the layout of one possible starting lens is illustrated in Figure B.3.1.2. This lens was derived by selecting the glasses and then tinkering with the curvatures and thicknesses until the layout looked about right (similar to published drawings). To simplify the guesswork, make the inner surfaces of the positive elements initially flat. Fortunately, the exact starting configuration is not too critical.

Note that because of the pickups, this starting lens has only four *independent* degrees of freedom: the first three curvatures and the first airspace. These are enough for now, but not enough to address all the basic aberrations.

The merit function for the early optimizations contains operands to correct focal length to 52 mm, prevent the airspaces from becoming too large or too small, correct paraxial longitudinal color for the two extreme wavelengths, and, using a default merit function, shrink polychromatic spots across the field. Correcting paraxial longitudinal color controls element powers and at this stage is very effective in shepherding the design in the right direction. A distortion operand is not included, but symmetry should keep distortion in check.

Listing B.3.1.1

Merit Function Listing

File : C:LENS311.ZMX
Title: COOKE TRIPLET, 52MM, F/3.5, 45.2DEG

Merit Function Value: 3.63793459E-002

Num	Type	Int1	Int2	Hx	Hy	Px	Py	Target	Weight	Value	% Cont
1	EFFL		3					5.20000E+001	1	5.20003E+001	0.000
2	BLNK										
3	BLNK										
4	MXCA	2	7					1.00000E+001	1	1.00000E+001	0.000
5	MNCA	2	7					1.00000E-001	1	1.00000E-001	0.000
6	MNEA	2	7					1.00000E-001	1	1.00000E-001	0.000
7	BLNK										
8	BLNK										
9	AXCL							0.00000E+000	10	-6.82493E-004	0.023
10	BLNK										
11	BLNK										
12	DMFS										
13	TRAR		1	0.0000	0.0000	0.4597	0.0000	0.00000E+000	0.07854	1.03524E-003	0.000
14	TRAR		1	0.0000	0.0000	0.8881	0.0000	0.00000E+000	0.07854	1.28006E-002	0.063
15	TRAR		3	0.0000	0.0000	0.4597	0.0000	0.00000E+000	0.07854	6.76146E-003	0.018
16	TRAR		3	0.0000	0.0000	0.8881	0.0000	0.00000E+000	0.07854	3.43827E-002	0.455
17	TRAR		5	0.0000	0.0000	0.4597	0.0000	0.00000E+000	0.07854	3.58840E-003	0.005
18	TRAR		5	0.0000	0.0000	0.8881	0.0000	0.00000E+000	0.07854	3.32670E-002	0.426
19	TRAR		1	0.0000	0.6991	0.2298	0.3981	0.00000E+000	0.02618	8.50000E-002	0.927
20	TRAR		1	0.0000	0.6991	0.4440	0.7691	0.00000E+000	0.02618	2.08535E-001	5.581
21	TRAR		1	0.0000	0.6991	0.4597	0.0000	0.00000E+000	0.02618	5.42000E-002	0.377
22	TRAR		1	0.0000	0.6991	0.8881	0.0000	0.00000E+000	0.02618	1.05953E-001	1.441
23	TRAR		1	0.0000	0.6991	0.2298	-0.3981	0.00000E+000	0.02618	2.04915E-002	0.054
24	TRAR		1	0.0000	0.6991	0.4440	-0.7691	0.00000E+000	0.02618	7.26274E-002	0.677
25	TRAR		3	0.0000	0.6991	0.2298	0.3981	0.00000E+000	0.02618	7.37596E-002	0.698
26	TRAR		3	0.0000	0.6991	0.4440	0.7691	0.00000E+000	0.02618	2.10582E-001	5.691
27	TRAR		3	0.0000	0.6991	0.4597	0.0000	0.00000E+000	0.02618	5.32878E-002	0.364
28	TRAR		3	0.0000	0.6991	0.8881	0.0000	0.00000E+000	0.02618	1.18142E-001	1.791
29	TRAR		3	0.0000	0.6991	0.2298	-0.3981	0.00000E+000	0.02618	2.04063E-002	0.053
30	TRAR		3	0.0000	0.6991	0.4440	-0.7691	0.00000E+000	0.02618	3.12326E-002	0.125
31	TRAR		5	0.0000	0.6991	0.2298	0.3981	0.00000E+000	0.02618	6.28696E-002	0.507
32	TRAR		5	0.0000	0.6991	0.4440	0.7691	0.00000E+000	0.02618	2.00040E-001	5.136
33	TRAR		5	0.0000	0.6991	0.4597	0.0000	0.00000E+000	0.02618	4.95950E-002	0.316
34	TRAR		5	0.0000	0.6991	0.8881	0.0000	0.00000E+000	0.02618	1.15423E-001	1.710
35	TRAR		5	0.0000	0.6991	0.2298	-0.3981	0.00000E+000	0.02618	2.41034E-002	0.075
36	TRAR		5	0.0000	0.6991	0.4440	-0.7691	0.00000E+000	0.02618	2.43297E-002	0.076
37	TRAR		1	0.0000	1.0000	0.2298	0.3981	0.00000E+000	0.02618	1.26323E-001	2.048
38	TRAR		1	0.0000	1.0000	0.4440	0.7691	0.00000E+000	0.02618	3.24527E-001	13.516
39	TRAR		1	0.0000	1.0000	0.4597	0.0000	0.00000E+000	0.02618	8.62986E-002	0.956
40	TRAR		1	0.0000	1.0000	0.8881	0.0000	0.00000E+000	0.02618	1.68672E-001	3.651
41	TRAR		1	0.0000	1.0000	0.2298	-0.3981	0.00000E+000	0.02618	6.70936E-002	0.578
42	TRAR		1	0.0000	1.0000	0.4440	-0.7691	0.00000E+000	0.02618	2.56002E-001	8.411
43	TRAR		3	0.0000	1.0000	0.2298	0.3981	0.00000E+000	0.02618	1.05619E-001	1.432
44	TRAR		3	0.0000	1.0000	0.4440	0.7691	0.00000E+000	0.02618	3.16643E-001	12.867
45	TRAR		3	0.0000	1.0000	0.4597	0.0000	0.00000E+000	0.02618	7.94419E-002	0.810
46	TRAR		3	0.0000	1.0000	0.8881	0.0000	0.00000E+000	0.02618	1.69222E-001	3.675
47	TRAR		3	0.0000	1.0000	0.2298	-0.3981	0.00000E+000	0.02618	3.34562E-002	0.144
48	TRAR		3	0.0000	1.0000	0.4440	-0.7691	0.00000E+000	0.02618	1.90654E-001	4.665
49	TRAR		5	0.0000	1.0000	0.2298	0.3981	0.00000E+000	0.02618	9.05470E-002	1.052
50	TRAR		5	0.0000	1.0000	0.4440	0.7691	0.00000E+000	0.02618	3.00899E-001	11.620
51	TRAR		5	0.0000	1.0000	0.4597	0.0000	0.00000E+000	0.02618	7.45595E-002	0.713
52	TRAR		5	0.0000	1.0000	0.8881	0.0000	0.00000E+000	0.02618	1.63310E-001	3.423
53	TRAR		5	0.0000	1.0000	0.2298	-0.3981	0.00000E+000	0.02618	2.42043E-002	0.075
54	TRAR		5	0.0000	1.0000	0.4440	-0.7691	0.00000E+000	0.02618	1.71502E-001	3.775

The early merit function is given in Listing B.3.1.1. To use the simplest allowable default merit function, some fields and wavelengths are weighted zero; this deletes the corresponding operands during the default merit function construction. Thus, the four fields are weighted 1 0 1 1 (in order from center to edge), and the five wavelengths are weighted 1 0 1 0 1 (in order from short to long). Because no beam clipping is used here to produce vignetting, a ray array based on the Gaussian quadrature algorithm is allowed and adopted (see Chapter B.4 for more on Gaussian quadrature).

After optimizing, look at the layout. This can be very revealing about how your glass choices affect the design. If your choice of matching flint glass gives a dispersion difference that is too large or too small, then the airspaces will be correspondingly too large or too small. If so, change the flint glass and reoptimize. The layout is your guide at this early optimization stage.

B.3.7 Intermediate Optimizations

You now have a good early design that is suitable for intermediate optimization. There are many ways to control aberrations. These methods are functions of both the preferences of the designer and the features in his software. What follows is an approach that the author finds to be effective for many types of optical systems.

The details are specific to the present Cooke Triplet, but the ideas are generally applicable.

As described in Chapter A.15, intermediate optimization is a combination monochromatic-polychromatic procedure. The monochromatic aberrations for a central wavelength and the chromatic aberrations for the extreme (or two widely spaced) wavelengths are controlled or corrected. For the intermediate optimizations of a Cooke Triplet, the pickups are removed and the six curvatures and two airspaces are all made independent variables. The height solve is retained to keep the image surface at the paraxial focus.

Now during the intermediate optimizations is also the time to add deliberate mechanical vignetting to the Cooke Triplet. The techniques required are discussed in an earlier section.

When constructing the intermediate merit function, select or define five special optimization operands such as those described in Chapter A.13 and Listing A.13.1. The first operand continues to correct focal length to 52 mm (here and subsequently, you can alternatively use an angle solve on the rear lens surface to control focal ratio). The second operand corrects longitudinal color to zero for the 0.8 pupil zone. The third operand corrects lateral color to zero at the edge of the field. The fourth operand corrects spherical aberration to zero on the paraxial focal plane and for the 0.9 pupil zone. The fifth operand corrects distortion to zero at the edge of the field. To correct these operands close to their targets, use relatively heavy weights or Lagrange multipliers.

Note in Listing A.13.1 that wavelengths are specified by the identifying numbers in the Int2 column and that the lens there uses only three wavelengths. However, the present lens uses five wavelengths. Thus, the special chromatic operands must now use wavelengths one and five. And the special monochromatic operands must now use wavelength three, the reference wavelength.

The only aberrations that remain to be controlled are all off-axis monochromatic aberrations. They are: coma, astigmatism, field curvature, and any number of higher-order monochromatic aberrations. These aberrations plus spherical (which is also present off-axis) interact in a complicated way. To control these aberrations during optimization, the most general and fail-safe approach is to shrink spot sizes by appending the appropriate default merit function to the special operands. Because the on-axis field is being corrected separately with special operands, turn off the axial field when constructing the default merit function and shrink only the off-axis spots. To do this, weight the on-axis field zero and weight the remaining fields equally (at least for now). The four field weights become 0 1 1 1. Similarly, to shrink only monochromatic spots for the reference wavelength, weight the five wavelengths 0 0 1 0 0.

The complete intermediate merit function for the Cooke Triplet is given in Listing B.3.1.2. To shorten the listing, the fields have actually been weighted 0 0 1 1.

Note that there is a variation to this approach that could have been used. Instead of controlling spherical aberration with a special operand, you can shrink the monochromatic on-axis spot too. To do this, omit the special spherical operand (to avoid controlling the same aberration twice), and construct the default merit function with field weights such as 5 1 1 1. The relatively heavy weight on-axis is required because spot size there is relatively small (no off-axis aberrations). The heavy weight ensures that the damped least-squares routine pays enough attention.

Listing B.3.1.2

Merit Function Listing

File : C:LENS312B.ZMX
Title: COOKE TRIPLET, 52MM, F/3.5, 45.2DEG

Merit Function Value: 1.92724453E-004

Num	Type	Int1	Int2	Hx	Hy	Px	Py	Target	Weight	Value	% Cont
1	EFFL		3					5.20000E+001	100	5.20000E+001	0.000
2	BLNK										
3	BLNK										
4	TTHI	3	3					0.00000E+000	0	4.73834E+000	0.000
5	TTHI	5	6					0.00000E+000	0	6.45093E+000	0.000
6	DIFF	5	4					0.00000E+000	0	1.71259E+000	0.000
7	OPGT	6						1.50000E+000	100	1.50000E+000	0.000
8	BLNK										
9	BLNK										
10	AXCL							0.00000E+000	0	1.12374E-001	0.000
11	REAY	10	1	0.0000	0.0000	0.0000	0.8000	0.00000E+000	0	2.36923E-003	0.000
12	REAY	10	5	0.0000	0.0000	0.0000	0.8000	0.00000E+000	0	2.37042E-003	0.000
13	DIFF	11	12					0.00000E+000	10000	-1.19209E-006	0.001
14	BLNK										
15	BLNK										
16	LACL							0.00000E+000	0	1.39704E-002	0.000
17	REAY	10	1	0.0000	1.0000	0.0000	0.0000	0.00000E+000	0	2.16434E+001	0.000
18	REAY	10	5	0.0000	1.0000	0.0000	0.0000	0.00000E+000	0	2.16434E+001	0.000
19	DIFF	17	18					0.00000E+000	10000	-1.81699E-007	0.000
20	BLNK										
21	BLNK										
22	SPHA	0	3					0.00000E+000	0	2.64913E+000	0.000
23	REAY	9	3	0.0000	0.0000	0.0000	0.9000	0.00000E+000	10000	1.01362E-006	0.001
24	BLNK										
25	BLNK										
26	DIST	0	3					0.00000E+000	1000	5.36819E-008	0.000
27	BLNK										
28	BLNK										
29	DMFS										
30	TRAR		3	0.0000	0.6991	0.1000	-0.5000	0.00000E+000	0.1	6.12221E-003	0.323
31	TRAR		3	0.0000	0.6991	0.1000	-0.3000	0.00000E+000	0.1	4.34227E-003	0.163
32	TRAR		3	0.0000	0.6991	0.1000	-0.1000	0.00000E+000	0.1	4.43055E-003	0.169
33	TRAR		3	0.0000	0.6991	0.1000	0.1000	0.00000E+000	0.1	4.43263E-003	0.170
34	TRAR		3	0.0000	0.6991	0.1000	0.3000	0.00000E+000	0.1	4.66716E-003	0.188
35	TRAR		3	0.0000	0.6991	0.1000	0.5000	0.00000E+000	0.1	6.55600E-003	0.371
36	TRAR		3	0.0000	0.6991	0.1000	0.7000	0.00000E+000	0.1	1.42184E-002	1.744
37	TRAR		3	0.0000	0.6991	0.3000	-0.5000	0.00000E+000	0.1	1.26632E-002	1.384
38	TRAR		3	0.0000	0.6991	0.3000	-0.3000	0.00000E+000	0.1	1.28531E-002	1.425
39	TRAR		3	0.0000	0.6991	0.3000	-0.1000	0.00000E+000	0.1	1.34846E-002	1.569
40	TRAR		3	0.0000	0.6991	0.3000	0.1000	0.00000E+000	0.1	1.35945E-002	1.595
41	TRAR		3	0.0000	0.6991	0.3000	0.3000	0.00000E+000	0.1	1.35749E-002	1.590
42	TRAR		3	0.0000	0.6991	0.3000	0.5000	0.00000E+000	0.1	1.40286E-002	1.698
43	TRAR		3	0.0000	0.6991	0.3000	0.7000	0.00000E+000	0.1	1.91872E-002	3.176
44	TRAR		3	0.0000	0.6991	0.5000	-0.5000	0.00000E+000	0.1	1.90906E-002	3.144
45	TRAR		3	0.0000	0.6991	0.5000	-0.3000	0.00000E+000	0.1	2.01296E-002	3.496
46	TRAR		3	0.0000	0.6991	0.5000	-0.1000	0.00000E+000	0.1	2.20930E-002	4.211
47	TRAR		3	0.0000	0.6991	0.5000	0.1000	0.00000E+000	0.1	2.26538E-002	4.428
48	TRAR		3	0.0000	0.6991	0.5000	0.3000	0.00000E+000	0.1	2.22624E-002	4.276
49	TRAR		3	0.0000	0.6991	0.5000	0.5000	0.00000E+000	0.1	2.14883E-002	3.984
50	TRAR		3	0.0000	0.6991	0.5000	0.7000	0.00000E+000	0.1	2.62421E-002	5.942
51	TRAR		3	0.0000	0.6991	0.7000	-0.3000	0.00000E+000	0.1	2.24606E-002	4.353
52	TRAR		3	0.0000	0.6991	0.7000	-0.1000	0.00000E+000	0.1	2.62597E-002	5.950
53	TRAR		3	0.0000	0.6991	0.7000	0.1000	0.00000E+000	0.1	2.77273E-002	6.633
54	TRAR		3	0.0000	0.6991	0.7000	0.3000	0.00000E+000	0.1	2.65562E-002	6.085
55	TRAR		3	0.0000	0.6991	0.7000	0.5000	0.00000E+000	0.1	2.47190E-002	5.272
56	TRAR		3	0.0000	0.6991	0.9000	-0.1000	0.00000E+000	0.1	1.64515E-002	2.335
57	TRAR		3	0.0000	0.6991	0.9000	0.1000	0.00000E+000	0.1	1.77613E-002	2.722
58	TRAR		3	0.0000	0.6991	0.9000	0.3000	0.00000E+000	0.1	1.72447E-002	2.566
59	TRAR		3	0.0000	1.0000	0.1000	-0.3000	0.00000E+000	0.1	8.95928E-003	0.693
60	TRAR		3	0.0000	1.0000	0.1000	-0.1000	0.00000E+000	0.1	3.66083E-003	0.116
61	TRAR		3	0.0000	1.0000	0.1000	0.1000	0.00000E+000	0.1	3.79969E-003	0.125
62	TRAR		3	0.0000	1.0000	0.1000	0.3000	0.00000E+000	0.1	5.54200E-003	0.265
63	TRAR		3	0.0000	1.0000	0.1000	0.5000	0.00000E+000	0.1	2.92387E-003	0.074
64	TRAR		3	0.0000	1.0000	0.3000	-0.3000	0.00000E+000	0.1	4.43901E-003	0.170
65	TRAR		3	0.0000	1.0000	0.3000	-0.1000	0.00000E+000	0.1	3.89385E-003	0.131
66	TRAR		3	0.0000	1.0000	0.3000	0.1000	0.00000E+000	0.1	7.13892E-003	0.440
67	TRAR		3	0.0000	1.0000	0.3000	0.3000	0.00000E+000	0.1	7.82365E-003	0.528
68	TRAR		3	0.0000	1.0000	0.3000	0.5000	0.00000E+000	0.1	6.33420E-003	0.346
69	TRAR		3	0.0000	1.0000	0.5000	-0.1000	0.00000E+000	0.1	5.19208E-003	0.233
70	TRAR		3	0.0000	1.0000	0.5000	0.1000	0.00000E+000	0.1	9.23102E-003	0.735
71	TRAR		3	0.0000	1.0000	0.5000	0.3000	0.00000E+000	0.1	8.34579E-003	0.601
72	TRAR		3	0.0000	1.0000	0.5000	0.5000	0.00000E+000	0.1	9.90926E-003	0.847
73	TRAR		3	0.0000	1.0000	0.7000	-0.1000	0.00000E+000	0.1	1.65419E-002	2.361
74	TRAR		3	0.0000	1.0000	0.7000	0.1000	0.00000E+000	0.1	9.28791E-003	0.744
75	TRAR		3	0.0000	1.0000	0.7000	0.3000	0.00000E+000	0.1	9.97684E-004	0.009
76	TRAR		3	0.0000	1.0000	0.9000	0.1000	0.00000E+000	0.1	3.50863E-002	10.621

Note that different optical design programs may handle weights differently. Therefore, it is risky to recommend specific weights for controlling the relative emphasis of various field positions and wavelengths during optimization. This is especially true for nonuniform weights. Thus, the weights offered here are given with a caveat. The best advice to the lens designer is to experiment with weights until you get results that you like.

In the design of a Cooke Triplet, you will often find two possible solutions; that is, two different local minima of the merit function. The first solution has the larger airspace between the front and middle elements; the second solution has the larger airspace between the middle and rear elements. As mentioned earlier, better symmetry about the stop is achieved if the rear airspace is the larger one. This more symmetrical configuration causes the off-axis beams to be more centered about their chief rays, which is better when stopping down the lens. To shepherd the design in this direction, a constraint is added to the intermediate merit function that keeps the rear airspace (glass-to-glass) somewhat larger than the front airspace.

After optimizing with the airspace constraint, look at the merit function to see if it was invoked. It may have been unused. If it was used, try removing it (perhaps by setting its weight to zero) and optimize again. The lens will then find its own best airspaces. However, do not be surprised if the negative element ends up nearly in the middle between the two positive elements, or if you occasionally have to leave in the constraint.

After each optimization run, the system parameters (curvatures and airspaces) will have changed to a greater or lesser extent, thereby also changing the size and shape of the vignetted off-axis pupils. To accommodate these changes, rebuild the default merit function to update the set of unvignetted rays, and then reoptimize. Iterate as needed. In addition, the lens designer is also often required to manually readjust the diameters of the hard apertures to maintain the desired amount of vignetting. After any such change, again rebuild the default merit function and reoptimize.

The last step in the intermediate optimizations is finding the best flint glass type. To do this, repeat the intermediate optimization with several likely glass combinations. The process is described in an earlier section. The result for the present Cooke Triplet is that Schott SF53 flint glass (n_d of 1.728, V_d of 28.7) is the best match for Schott LaFN21 crown glass (n_d of 1.788, V_d of 47.5).

B.3.8 Final Optimizations Using Spot Size

The intermediate solution for the Cooke Triplet is actually quite close to the final solution. The final solution is a refinement, which is normally accomplished in two stages. The first stage shrinks polychromatic spots on the image surface. The second stage minimizes polychromatic OPD errors in the exit pupil. The designer then compares the two solutions and chooses the better one for the given application. If MTF is the image criterion, then the OPD solution is usually, but not always, preferable.

There are two ways to do the final spot optimization. The first way uses a merit function that shrinks polychromatic spot sizes for all field positions and all wavelengths while continuing to correct distortion and focal length with special operands. Note that longitudinal and lateral color are not individually corrected; the polychromatic spot optimization includes the chromatic aberrations. Thus, when constructing the default merit function, weight the four fields equally; that is, 1 1 1 1, at least initially. Also weight the five wavelengths equally; that is, 1 1 1 1 1. As an alternative, you might do as just described plus still continue to exactly correct longitudinal color with a special operand to ensure a specific color curve.

The second way is very similar except that the on-axis aberrations (longitudinal color and spherical aberration) are corrected with special operands, and the default merit function shrinks polychromatic spots only off-axis. Again, distortion and focal length are specially corrected. When constructing the default merit function, the on-axis spot is turned off by weighting this field position zero; that is, field weights are 0 1 1 1 (at least initially). The polychromatic wavelength weights are again 1 1 1 1 1. In the present Cooke Triplet example, this second method has been adopted for constructing the merit function.

Listing B.3.1.3 gives the complete final merit function using spot size. An operand has also been included to control the relative airspaces, although it is not

Listing B.3.1.3

Merit Function Listing

File : C:LENS313B.ZMX
Title: COOKE TRIPLET, 52MM, F/3.5, 45.2DEG

Merit Function Value: 5.16722085E-004

Num	Type	Int1	Int2	Hx	Hy	Px	Py	Target	Weight	Value	% Cont
1	EFFL		3					5.20000E+001	1000	5.20000E+001	0.000
2	BLNK										
3	BLNK										
4	TTHI	3	3					0.00000E+000	0	4.32187E+000	0.000
5	TTHI	5	6					0.00000E+000	0	6.85434E+000	0.000
6	DIFF	5	4					0.00000E+000	0	2.53247E+000	0.000
7	OPGT	6						1.50000E+000	1000	1.50000E+000	0.000
8	BLNK										
9	BLNK										
10	AXCL							0.00000E+000	0	1.11785E-001	0.000
11	REAY	10	1	0.0000	0.0000	0.0000	0.8000	0.00000E+000	0	2.26244E-003	0.000
12	REAY	10	5	0.0000	0.0000	0.0000	0.8000	0.00000E+000	0	2.33925E-003	0.000
13	DIFF	11	12					0.00000E+000	10000	-7.68052E-005	0.959
14	BLNK										
15	BLNK										
16	SPHA	0	3					0.00000E+000	0	2.69377E+000	0.000
17	REAY	9	3	0.0000	0.0000	0.0000	0.9000	0.00000E+000	10000	1.10586E-004	1.989
18	BLNK										
19	BLNK										
20	DIST	0	3					0.00000E+000	1000	2.01569E-006	0.000
21	BLNK										
22	BLNK										
23	DMFS										
24	TRAR		1	0.0000	0.3982	0.2500	-0.7500	0.00000E+000	0.5	1.13189E-002	1.042
25	TRAR		1	0.0000	0.3982	0.2500	-0.2500	0.00000E+000	0.5	6.13682E-003	0.306
26	TRAR		1	0.0000	0.3982	0.2500	0.2500	0.00000E+000	0.5	6.79766E-003	0.376
27	TRAR		1	0.0000	0.3982	0.2500	0.7500	0.00000E+000	0.5	4.29339E-003	0.150
28	TRAR		1	0.0000	0.3982	0.7500	-0.2500	0.00000E+000	0.5	8.61819E-003	0.604
29	TRAR		1	0.0000	0.3982	0.7500	0.2500	0.00000E+000	0.5	1.09141E-002	0.969
30	TRAR		2	0.0000	0.3982	0.2500	-0.7500	0.00000E+000	0.5	4.39938E-003	0.157
31	TRAR		2	0.0000	0.3982	0.2500	-0.2500	0.00000E+000	0.5	7.75161E-003	0.489
32	TRAR		2	0.0000	0.3982	0.2500	0.2500	0.00000E+000	0.5	7.13903E-003	0.414
33	TRAR		2	0.0000	0.3982	0.2500	0.7500	0.00000E+000	0.5	8.18535E-003	0.545
34	TRAR		2	0.0000	0.3982	0.7500	-0.2500	0.00000E+000	0.5	1.46952E-002	1.756
35	TRAR		2	0.0000	0.3982	0.7500	0.2500	0.00000E+000	0.5	1.75552E-002	2.506
36	TRAR		3	0.0000	0.3982	0.2500	-0.7500	0.00000E+000	0.5	4.13661E-003	0.139
37	TRAR		3	0.0000	0.3982	0.2500	-0.2500	0.00000E+000	0.5	6.83198E-003	0.380
38	TRAR		3	0.0000	0.3982	0.2500	0.2500	0.00000E+000	0.5	5.77825E-003	0.272
39	TRAR		3	0.0000	0.3982	0.2500	0.7500	0.00000E+000	0.5	7.22030E-003	0.424
40	TRAR		3	0.0000	0.3982	0.7500	-0.2500	0.00000E+000	0.5	1.48809E-002	1.801
41	TRAR		3	0.0000	0.3982	0.7500	0.2500	0.00000E+000	0.5	1.74078E-002	2.464
42	TRAR		4	0.0000	0.3982	0.2500	-0.7500	0.00000E+000	0.5	5.52951E-003	0.249
43	TRAR		4	0.0000	0.3982	0.2500	-0.2500	0.00000E+000	0.5	4.88180E-003	0.194
44	TRAR		4	0.0000	0.3982	0.2500	0.2500	0.00000E+000	0.5	4.11031E-003	0.137
45	TRAR		4	0.0000	0.3982	0.2500	0.7500	0.00000E+000	0.5	5.03739E-003	0.206
46	TRAR		4	0.0000	0.3982	0.7500	-0.2500	0.00000E+000	0.5	1.21386E-002	1.198
47	TRAR		4	0.0000	0.3982	0.7500	0.2500	0.00000E+000	0.5	1.44850E-002	1.706
48	TRAR		5	0.0000	0.3982	0.2500	-0.7500	0.00000E+000	0.5	9.21534E-003	0.691
49	TRAR		5	0.0000	0.3982	0.2500	-0.2500	0.00000E+000	0.5	3.01616E-003	0.074
50	TRAR		5	0.0000	0.3982	0.2500	0.2500	0.00000E+000	0.5	3.22798E-003	0.085
51	TRAR		5	0.0000	0.3982	0.2500	0.7500	0.00000E+000	0.5	5.54136E-003	0.250
52	TRAR		5	0.0000	0.3982	0.7500	-0.2500	0.00000E+000	0.5	8.23579E-003	0.552
53	TRAR		5	0.0000	0.3982	0.7500	0.2500	0.00000E+000	0.5	1.04533E-002	0.889
54	TRAR		1	0.0000	0.6991	0.2500	-0.2500	0.00000E+000	0.5	1.18736E-002	1.146
55	TRAR		1	0.0000	0.6991	0.2500	0.2500	0.00000E+000	0.5	1.22854E-002	1.227
56	TRAR		1	0.0000	0.6991	0.2500	0.7500	0.00000E+000	0.5	1.48341E-002	1.789
57	TRAR		1	0.0000	0.6991	0.7500	-0.2500	0.00000E+000	0.5	1.72910E-002	2.431
58	TRAR		1	0.0000	0.6991	0.7500	0.2500	0.00000E+000	0.5	2.11634E-002	3.642
59	TRAR		2	0.0000	0.6991	0.2500	-0.2500	0.00000E+000	0.5	1.34572E-002	1.473
60	TRAR		2	0.0000	0.6991	0.2500	0.2500	0.00000E+000	0.5	1.23649E-002	1.243
61	TRAR		2	0.0000	0.6991	0.2500	0.7500	0.00000E+000	0.5	1.53093E-002	1.906
62	TRAR		2	0.0000	0.6991	0.7500	-0.2500	0.00000E+000	0.5	2.26614E-002	4.176
63	TRAR		2	0.0000	0.6991	0.7500	0.2500	0.00000E+000	0.5	2.82597E-002	6.494
64	TRAR		3	0.0000	0.6991	0.2500	-0.2500	0.00000E+000	0.5	1.24155E-002	1.253
65	TRAR		3	0.0000	0.6991	0.2500	0.2500	0.00000E+000	0.5	1.12590E-002	1.031
66	TRAR		3	0.0000	0.6991	0.2500	0.7500	0.00000E+000	0.5	1.80932E-002	2.662
67	TRAR		3	0.0000	0.6991	0.7500	-0.2500	0.00000E+000	0.5	2.30272E-002	4.312
68	TRAR		3	0.0000	0.6991	0.7500	0.2500	0.00000E+000	0.5	2.85360E-002	6.622
69	TRAR		4	0.0000	0.6991	0.2500	-0.2500	0.00000E+000	0.5	1.05241E-002	0.901
70	TRAR		4	0.0000	0.6991	0.2500	0.2500	0.00000E+000	0.5	9.94036E-003	0.803
71	TRAR		4	0.0000	0.6991	0.2500	0.7500	0.00000E+000	0.5	2.20493E-002	3.953
72	TRAR		4	0.0000	0.6991	0.7500	-0.2500	0.00000E+000	0.5	2.09049E-002	3.554
73	TRAR		4	0.0000	0.6991	0.7500	0.2500	0.00000E+000	0.5	2.60342E-002	5.512
74	TRAR		5	0.0000	0.6991	0.2500	-0.2500	0.00000E+000	0.5	8.88009E-003	0.641
75	TRAR		5	0.0000	0.6991	0.2500	0.2500	0.00000E+000	0.5	8.85322E-003	0.637
76	TRAR		5	0.0000	0.6991	0.2500	0.7500	0.00000E+000	0.5	2.67318E-002	5.811
77	TRAR		5	0.0000	0.6991	0.7500	-0.2500	0.00000E+000	0.5	1.80876E-002	2.660
78	TRAR		5	0.0000	0.6991	0.7500	0.2500	0.00000E+000	0.5	2.24098E-002	4.084
79	TRAR		1	0.0000	1.0000	0.2500	-0.2500	0.00000E+000	0.1	3.54277E-002	2.041
80	TRAR		1	0.0000	1.0000	0.2500	0.2500	0.00000E+000	0.1	1.64360E-002	0.439
81	TRAR		1	0.0000	1.0000	0.7500	0.2500	0.00000E+000	0.1	1.29293E-002	0.272
82	TRAR		2	0.0000	1.0000	0.2500	-0.2500	0.00000E+000	0.1	2.87876E-002	1.348
83	TRAR		2	0.0000	1.0000	0.2500	0.2500	0.00000E+000	0.1	1.25153E-002	0.255
84	TRAR		2	0.0000	1.0000	0.7500	0.2500	0.00000E+000	0.1	4.61541E-003	0.035
85	TRAR		3	0.0000	1.0000	0.2500	-0.2500	0.00000E+000	0.1	2.09629E-002	0.715
86	TRAR		3	0.0000	1.0000	0.2500	0.2500	0.00000E+000	0.1	9.28730E-003	0.140
87	TRAR		3	0.0000	1.0000	0.7500	0.2500	0.00000E+000	0.1	4.84668E-003	0.038
88	TRAR		4	0.0000	1.0000	0.2500	-0.2500	0.00000E+000	0.1	1.32269E-002	0.285
89	TRAR		4	0.0000	1.0000	0.2500	0.2500	0.00000E+000	0.1	6.43729E-003	0.067
90	TRAR		4	0.0000	1.0000	0.7500	0.2500	0.00000E+000	0.1	8.04721E-003	0.105
91	TRAR		5	0.0000	1.0000	0.2500	-0.2500	0.00000E+000	0.1	5.90391E-003	0.057
92	TRAR		5	0.0000	1.0000	0.2500	0.2500	0.00000E+000	0.1	3.94329E-003	0.025
93	TRAR		5	0.0000	1.0000	0.7500	0.2500	0.00000E+000	0.1	1.22290E-002	0.243

invoked during optimization. Field weights have been changed to 0 5 5 1 (see below). To make the figure of manageable size for printing, the default merit function is shown using 4x4 rectangular ray arrays; in reality, 10x10 or 20x20 ray arrays are used.

After optimizing, the size and shape of the vignetted off-axis pupil will have again changed. Again, rebuild the default merit function to match the new pupil and reoptimize.

As in the intermediate optimization stage, during the final optimizations, look at the layout, transverse ray fan plot, and spot diagram to monitor your solution. Repeat these tests frequently as your design continues to evolve. In addition, you may wish to look at the vignetting plot, the field curvature plot, the distortion plot, and the OPD ray fan plot. And, especially toward the end of these optimizations, the MTF plot is invaluable because MTF is your ultimate performance criterion. Except right at the end, you may wish to save computing time by using the geometrical MTF plot instead of the diffraction MTF plot; this shortcut is allowed because you are far from the diffraction limit.

By now in your design of a Cooke Triplet, it should have become apparent that the 100% field position tends to give better imagery than the 70% field position. This is not what you normally want. In a camera lens, the intermediate fields are normally more important than the edge of the field. Of course, there is less vignetting at intermediate fields, thereby allowing some of the more highly aberrated rays to get through. But a look at the tangential-sagittal field curvature plot, such as the left side of Figure B.3.1.5, reveals another problem. In a Cooke Triplet, there is a large difference between the tangential and sagittal curves at intermediate fields, but the curves cross near the field edge. Thus, at intermediate field zones you get a lot of astigmatism, but astigmatism goes to zero near the field edge.

This astigmatic behavior is the result of the interaction of relatively large amounts of third- and fifth-order astigmatism (seventh is negligible). The sum of the third- and fifth-orders is zero at the field angle where the curves cross. Unfortunately, at other field angles the sum may not be small. Third-order predominates inside the crossing, and fifth-order predominates outside the crossing. The problem is analogous to zonal spherical aberration, except that the effect is in the field, not the pupil. Significant residual field-zonal astigmatism is part of the inherent nature of a Cooke Triplet.

The use of high-index glasses generally reduces this astigmatism and is highly recommended. The only other thing you can do is tell the computer to change its priorities and emphasize the intermediate fields at the expense of the field edge; that is, you increase the relative weights on the two intermediate fields during optimization. This moves the field angle for zero astigmatism inward. Unfortunately, astigmatism increases very rapidly for field angles outside the tangential-sagittal crossing. You must therefore be very careful to avoid overdoing this brute force tactic. Also, the computer has other aberrations to worry about besides astigmatism, which is why the crossing initially fell so far out. A reasonable solution is obtained if the default part of the merit function is rebuilt with the field weights changed from 0 1 1 1 to 0 5 5 1 (while continuing to separately correct the on-axis field with special operands). The intermediate fields get better while the edge of the field gets worse.

This illustrates a recurrent situation in lens design. A given lens configuration has only a certain ability to control aberrations. If you take a previously optimized lens and reoptimize it to make one thing better, something else must get worse. There is an adage that says that the process of reoptimizing a lens is like squeezing on a water balloon. If you squeeze in one place, it pops out somewhere else.

Figure B.3.1.1 shows the layout of the Cooke Triplet as optimized by correcting focal length, distortion, and special on-axis operands; by shrinking polychromatic off-axis spots; and by emphasizing the intermediate fields.

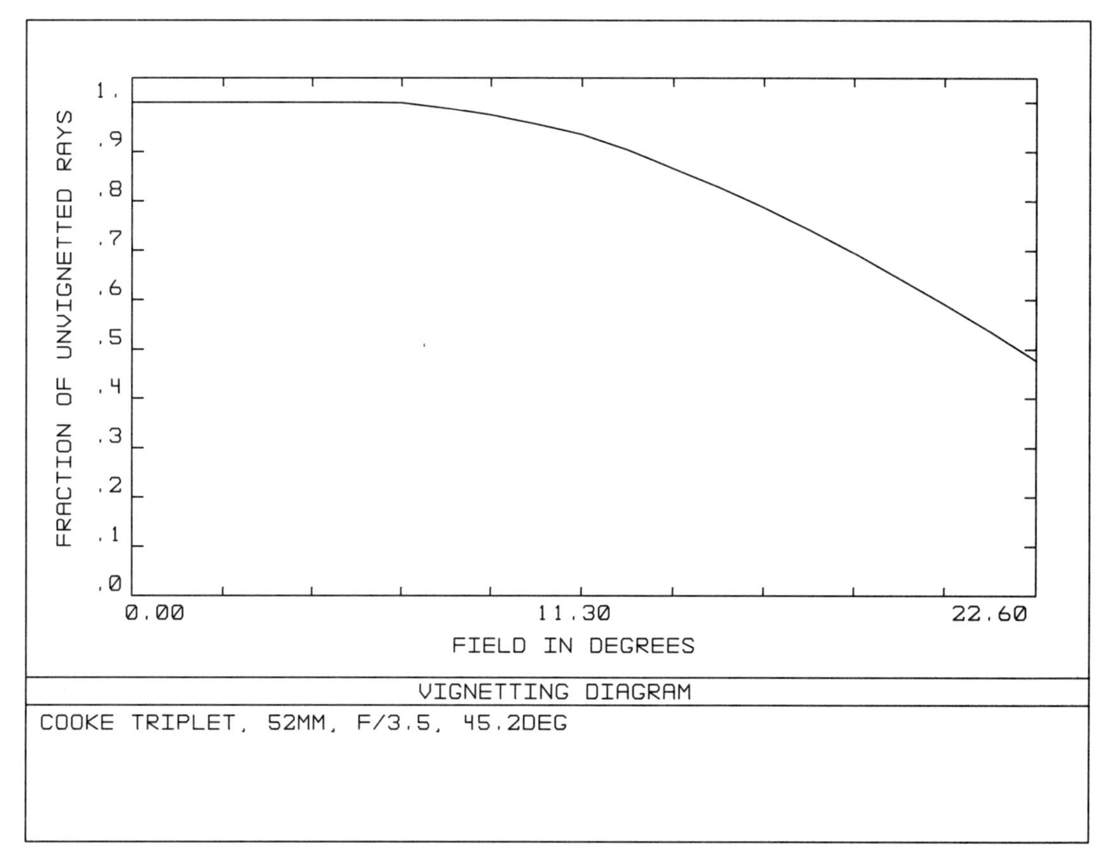

Figure B.3.1.3.

For increased clarity, the beams for only two fields have been drawn, the on-axis field and the extreme off-axis field. Note the mechanical vignetting of the off-axis beam by the front and rear outside lens surfaces. Note too that what appears to be the off-axis chief ray does not go through the center of the stop. Actually, the ray in question is not the chief ray, but is merely the middle ray of the ray bundle. The true chief ray always goes exactly through the center of the stop, by definition.

Figure B.3.1.3 is the vignetting diagram showing the fraction of unvignetted rays as a function of field angle. This plot confirms that vignetting is well-behaved and is about 50% at the edge of the field.

Figure B.3.1.4 is the transverse ray fan plot. Note the well-behaved on-axis curves. For the three off-axis fields, note how the slopes of the tangential and sagittal curves are different at the origin, indicating astigmatism. Note too the oblique spherical aberration in the off-axis fields. Finally, note that the chromatic aberrations are relatively small compared to the monochromatic aberrations.

On the left side of Figure B.3.1.5 is the field curvature plot. Note the crossing of the tangential and sagittal curves near the field edge and the large astigmatic difference at intermediate field angles. On the right side of Figure B.3.1.5 is the percent distortion plot. Distortion has been corrected to zero at the field edge, but note the higher-order distortion residuals at intermediate field angles. These residuals are tiny and totally negligible for most purposes.

Figure B.3.1.6 is the polychromatic spot diagram. Note the different appearance of the spots. The spot for the 70% field is spread horizontally by tangential astigmatism, and the spot for the field edge is spread vertically by sagittal (radial) astigmatism. However, these images are not purely astigmatic because other aberrations are complicating the situation.

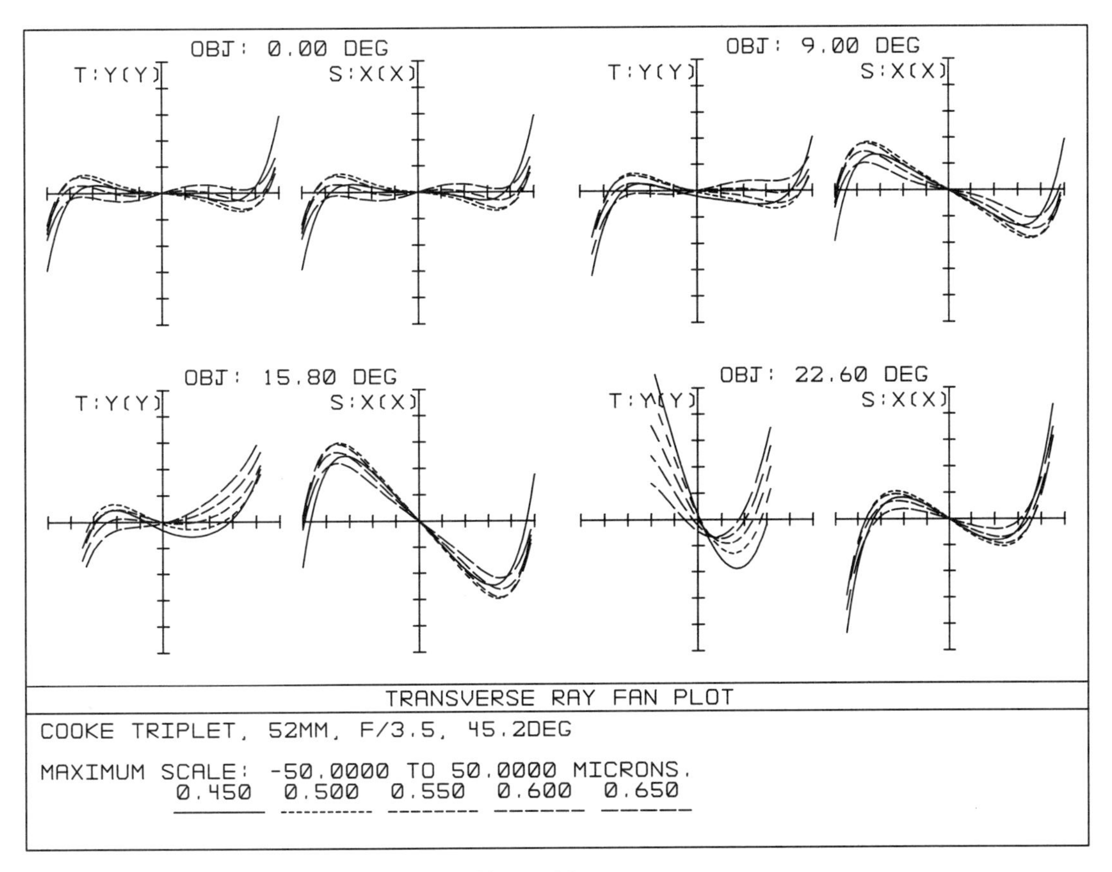

Figure B.3.1.4.

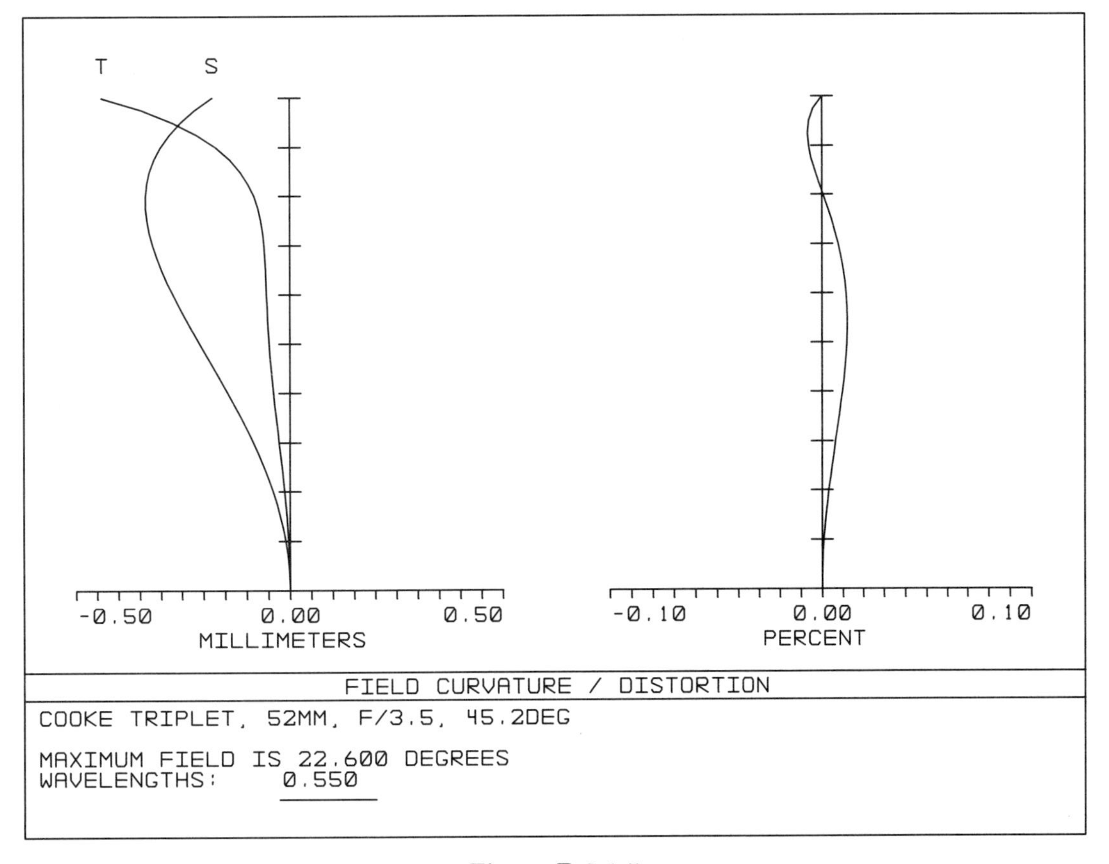

Figure B.3.1.5.

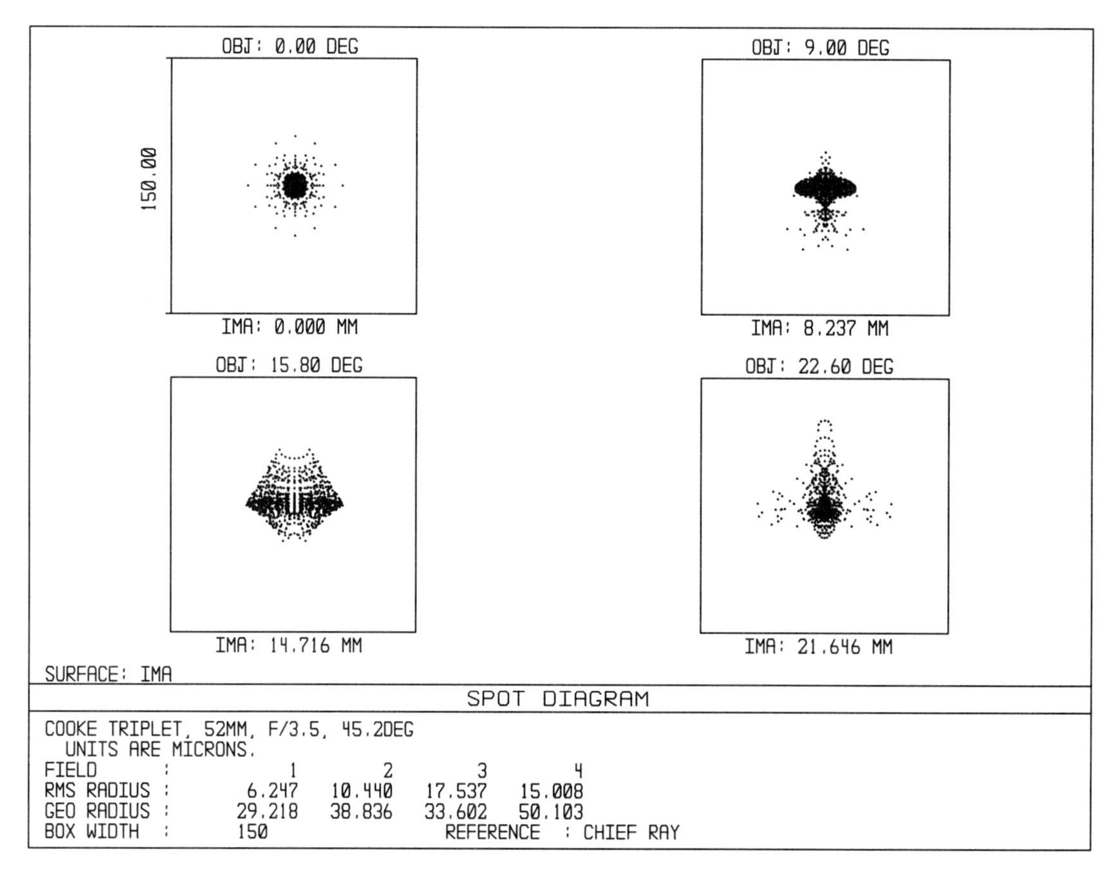

Figure B.3.1.6.

Figure B.3.1.7 is the matrix spot diagram. Note that the spots vary strongly with field but only moderately with wavelength.

Figure B.3.1.8 is the OPD ray fan plot. Note the scale: several waves. Clearly, this lens is aberration limited, not diffraction limited. Even on-axis, this lens does not satisfy the Rayleigh quarter-wave rule. But this performance level is normal for this type of lens and is no cause for concern. At $f/3.5$ and for a wavelength of 0.55 μm, the diameter of the Airy disk is only 4.7 μm and the diffraction MTF cutoff is 520 cycles/mm. These values are much finer than most films can resolve. In fact, very few camera lenses are diffraction limited when used wide open, and most of these are slow, long-focus lenses.

Figure B.3.1.9 is the diffraction MTF plot. Even with increased weight on the intermediate fields during optimization, the sagittal MTF at the 70% field is very poor. Still heavier weights are not recommended; the spot size at the edge of the field would explode without much benefit to the intermediate fields. Note the spurious resolution. This artifact of the periodic nature of a pure spatial frequency is of no help for real images; modulations beyond the first zero must be ignored when evaluating lens performance.

Listing B.3.1.4 is the optical prescription for the spot size optimized Cooke Triplet. From this listing, you could build the lens. However, for a practical camera lens intended for use with films having a resolving capability in excess of 50 cycles/mm, this design for the Cooke Triplet is probably unacceptable, at least by modern standards based on MTF.

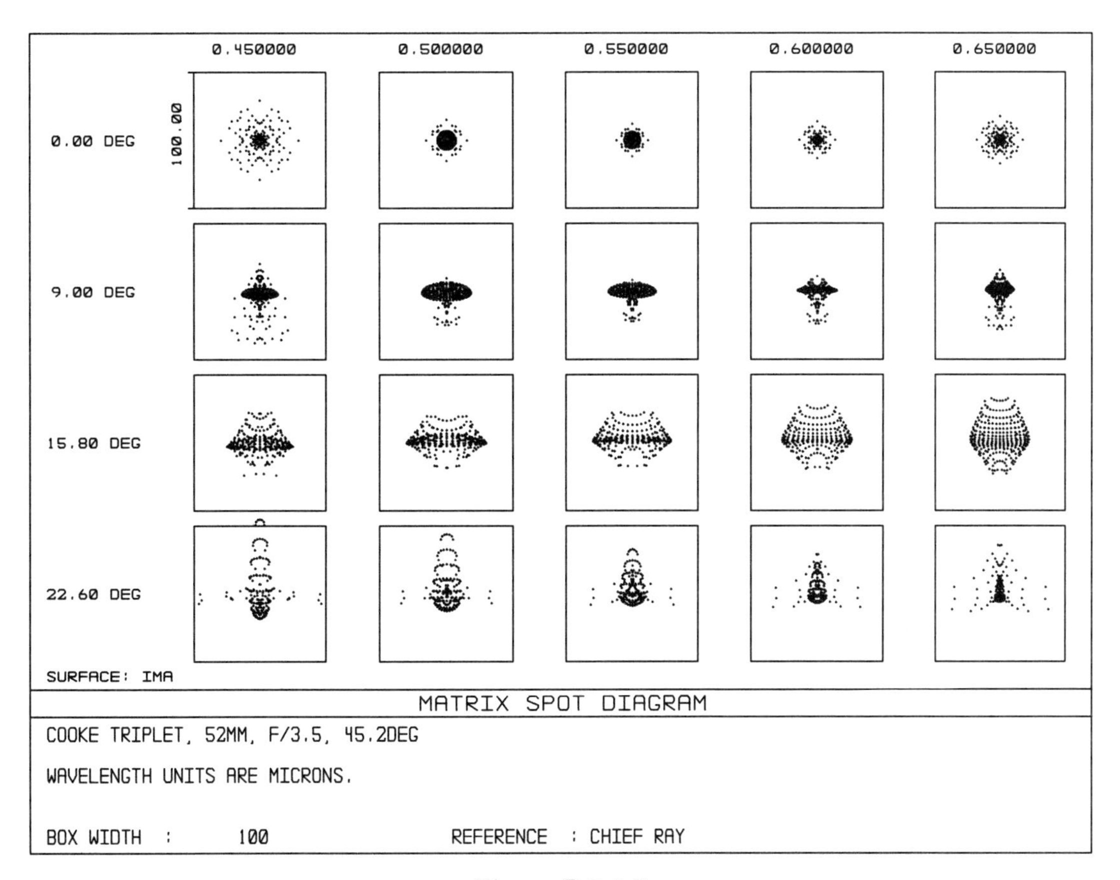

Figure B.3.1.7.

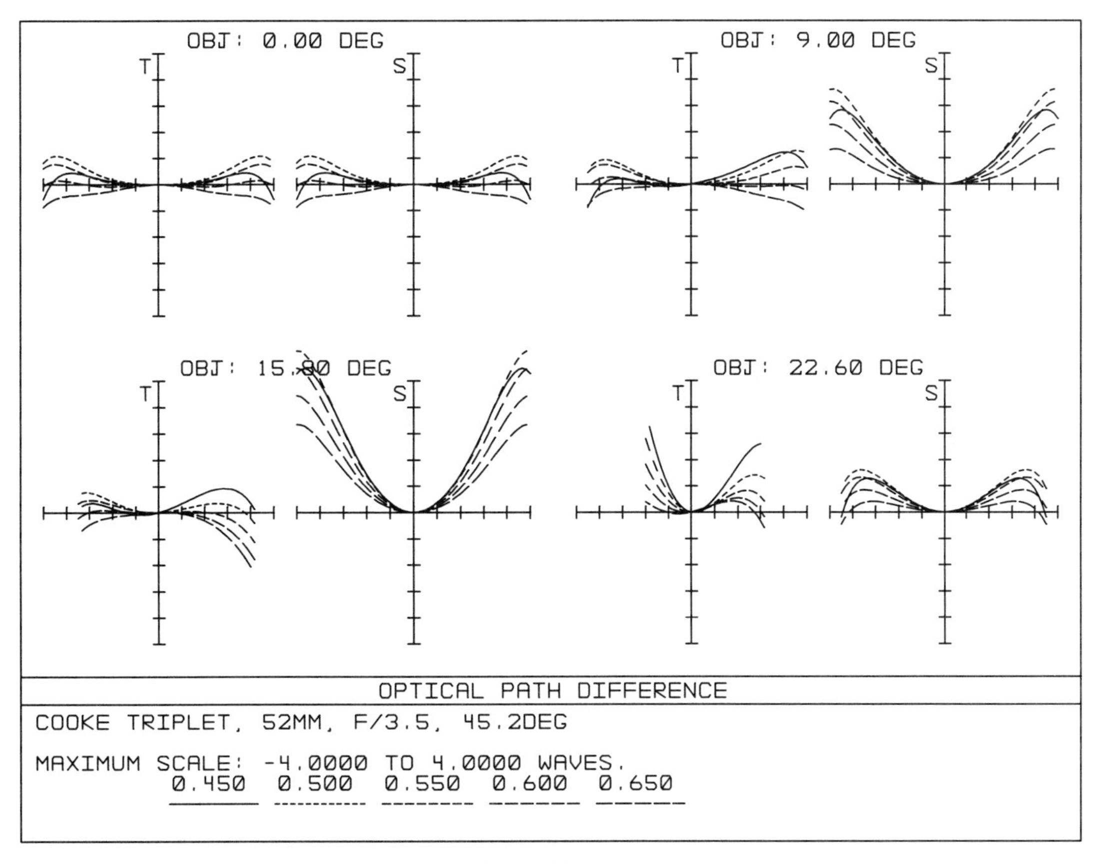

Figure B.3.1.8.

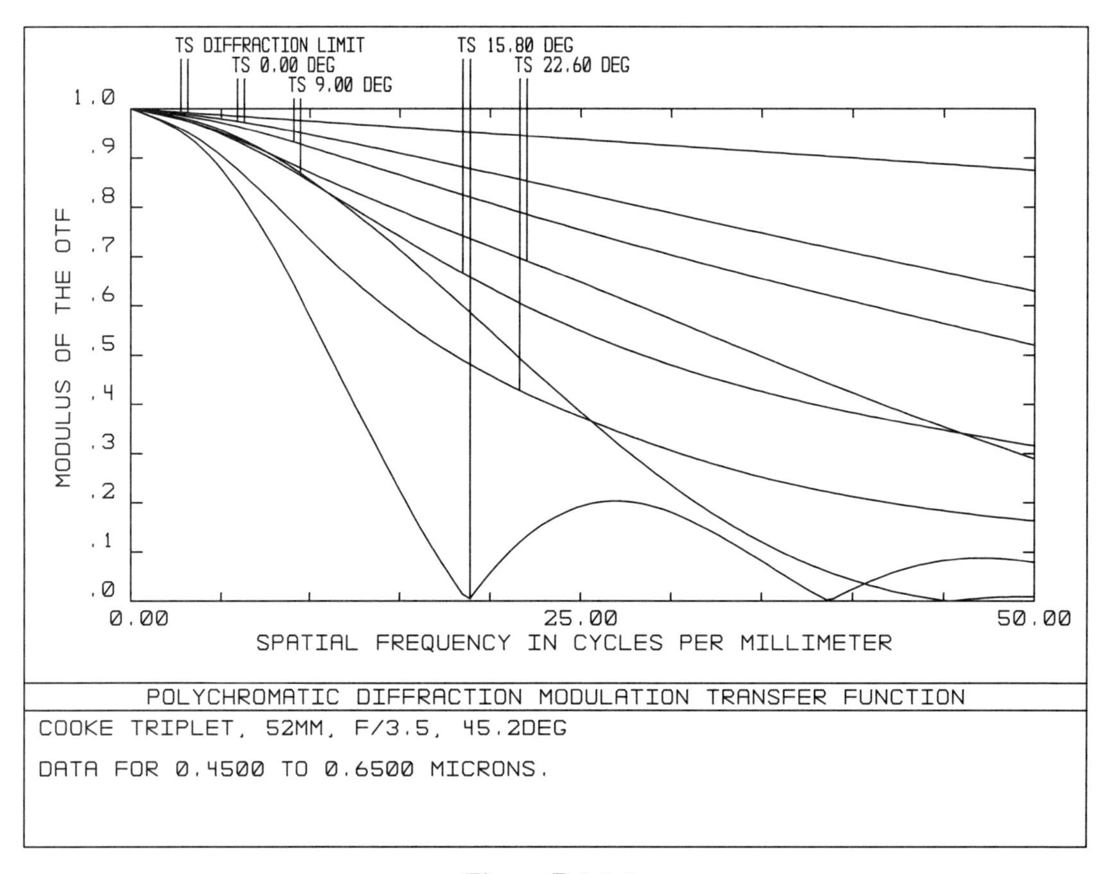

Figure B.3.1.9.

Listing B.3.1.4

```
System/Prescription Data

File : C:LENS313B.ZMX
Title: COOKE TRIPLET, 52MM, F/3.5, 45.2DEG

GENERAL LENS DATA:

Surfaces           :          10
Stop               :           6
System Aperture    :Entrance Pupil Diameter
Ray aiming         : On
 X Pupil shift     : 0
 Y Pupil shift     : 0
 Z Pupil shift     : 0
Apodization        :Uniform, factor =      0.000000
Eff. Focal Len.    :          52 (in air)
Eff. Focal Len.    :          52 (in image space)
Total Track        :      67.753
Image Space F/#    :     3.49933
Para. Wrkng F/#    :     3.49933
Working F/#        :     3.55795
Obj. Space N.A.    :   7.43e-010
Stop Radius        :     5.73158
Parax. Ima. Hgt.   :     21.6455
Parax. Mag.        :           0
Entr. Pup. Dia.    :       14.86
Entr. Pup. Pos.    :     19.9395
Exit Pupil Dia.    :     14.1746
Exit Pupil Pos.    :    -49.6015
Field Type         : Angle in degrees
Maximum Field      :        22.6
Primary Wave       :    0.550000
Lens Units         : Millimeters
Angular Mag.       :     1.04835

Fields             : 4
Field Type:  Angle in degrees
#       X-Value        Y-Value          Weight
1      0.000000       0.000000        0.000000
2      0.000000       9.000000        5.000000
3      0.000000      15.800000        5.000000
4      0.000000      22.600000        1.000000
```

```
Vignetting Factors
#           VDX           VDY           VCX           VCY
1      0.000000      0.000000      0.000000      0.000000
2      0.000000      0.000000      0.000000      0.000000
3      0.000000      0.000000      0.000000      0.000000
4      0.000000      0.000000      0.000000      0.000000

Wavelengths     : 5
Units:  Microns
#          Value           Weight
1       0.450000         1.000000
2       0.500000         1.000000
3       0.550000         1.000000
4       0.600000         1.000000
5       0.650000         1.000000

SURFACE DATA SUMMARY:

Surf      Type         Radius      Thickness        Glass      Diameter        Conic
 OBJ  STANDARD       Infinity       Infinity                          0            0
   1  STANDARD       Infinity              7                   33.36006            0
   2  STANDARD       21.74267            3.5       LAFN21            17            0
   3  STANDARD       400.8834       4.321867                         17            0
   4  STANDARD      -43.54159            1.5         SF53            13            0
   5  STANDARD       20.87459              2                         12            0
 STO  STANDARD       Infinity       4.854339                   11.46317            0
   7  STANDARD       165.5048              3       LAFN21            15            0
   8  STANDARD       -30.5771       41.57679                         15            0
   9  STANDARD       Infinity              0                   43.61234            0
 IMA  STANDARD       Infinity              0                   43.61234            0

SURFACE DATA DETAIL:

Surface OBJ      : STANDARD
Surface   1      : STANDARD
Surface   2      : STANDARD
 Aperture        : Circular Aperture
 Minimum Radius  :            0
 Maximum Radius  :          8.5
Surface   3      : STANDARD
Surface   4      : STANDARD
Surface   5      : STANDARD
Surface STO      : STANDARD
Surface   7      : STANDARD
Surface   8      : STANDARD
 Aperture        : Circular Aperture
 Minimum Radius  :            0
 Maximum Radius  :          7.5
Surface   9      : STANDARD
Surface IMA      : STANDARD

SOLVE AND VARIABLE DATA:

Curvature of   2    : Variable
Semi Diam   2       : Fixed
Curvature of   3    : Variable
Thickness of   3    : Variable
Semi Diam   3       : Fixed
Curvature of   4    : Variable
Semi Diam   4       : Fixed
Curvature of   5    : Variable
Semi Diam   5       : Fixed
Thickness of   6    : Variable
Curvature of   7    : Variable
Semi Diam   7       : Fixed
Curvature of   8    : Variable
Thickness of   8    : Solve, marginal ray height = 0.00000
Semi Diam   8       : Fixed

INDEX OF REFRACTION DATA:

Surf      Glass       0.450000     0.500000     0.550000     0.600000     0.650000
   0               1.00000000   1.00000000   1.00000000   1.00000000   1.00000000
   1               1.00000000   1.00000000   1.00000000   1.00000000   1.00000000
   2     LAFN21    1.80620521   1.79788762   1.79184804   1.78728036   1.78370773
   3               1.00000000   1.00000000   1.00000000   1.00000000   1.00000000
   4       SF53    1.75665336   1.74309602   1.73366378   1.72676633   1.72152789
   5               1.00000000   1.00000000   1.00000000   1.00000000   1.00000000
   6               1.00000000   1.00000000   1.00000000   1.00000000   1.00000000
   7     LAFN21    1.80620521   1.79788762   1.79184804   1.78728036   1.78370773
   8               1.00000000   1.00000000   1.00000000   1.00000000   1.00000000
   9               1.00000000   1.00000000   1.00000000   1.00000000   1.00000000
  10               1.00000000   1.00000000   1.00000000   1.00000000   1.00000000

ELEMENT VOLUME DATA:

Units are cubic cm.
Values are only accurate for plane and spherical surfaces.
Element surf   2 to   3 volume :       0.610995
Element surf   4 to   5 volume :       0.269441
Element surf   7 to   8 volume :       0.433018
```

B.3.9 Final Optimizations Using OPD Errors

For this Cooke Triplet example, there is not much more that you can do to improve image quality by shrinking spots. Because MTF is the criterion for image quality, a better approach may be to reduce wavefront OPD errors. Even though this lens is not near the diffraction limit, an OPD optimization may be beneficial. But bear in mind that if you optimize for minimum OPDs, then spot sizes may be small but they are no longer minimized. You cannot have it both ways. Remember the water balloon.

In the OPD optimization, all special optimization operands are removed from the merit function except those correcting or controlling distortion, focal length, and the relative airspace thicknesses. The bulk of the merit function is the default part that controls exit pupil OPDs for all fields and all wavelengths. The five wavelengths are again weighted 1 1 1 1 1. The four fields might be initially weighted 5 1 1 1. As suggested earlier, the on-axis field is more heavily weighted to ensure that the damped least-squares routine pays enough attention. The pupil is uniformly weighted with no extra emphasis on the center. No other prior assumptions are made. The goal is to do a polychromatic OPD optimization in a natural or realistic way (as it was called in Chapter A.15).

Listing B.3.2.1 is the complete final merit function for an OPD optimization. The field weights have been changed to 4 3 2 1 (see below). Again for practical reasons, 4x4 ray arrays are shown, although 10x10 or 20x20 arrays are actually used when optimizing. Note that the control on the relative airspace thicknesses is not invoked.

After optimizing with OPDs, the lens elements will have shifted by some amount relative to their positions for the minimum spot size solution. Thus, after the first OPD optimization, the vignetting will have changed and you must rebuild the default merit function and reoptimize. In addition, you may wish to fine tune the vignetting curve to give a specific throughput value at the edge of the field. This is done by slightly changing the vignetting apertures, again rebuilding the merit function, and again reoptimizing.

Next, do the usual layout, transverse ray fan plot, and spot diagram as sanity checks. Then do an OPD ray fan plot to check your OPD optimization. Finally, do the crucial MTF plot. As with a spot size optimization, an OPD optimization with uniform weights on the off-axis fields yields poorer performance for the intermediate fields than for the extreme field. The (partial) remedy once again is to increase the weights on the intermediate fields. Try different sets of weights, reoptimize for each set, and compare the MTF curves. Select the weights that give the best overall compromise. In the present case, field weights of 4 3 2 1 are found to work well.

Recall that in the present example, vignetting is handled during optimization by deleting the vignetted rays from a set of incident rays. This deletion has the subtle side effect of reducing the working weights on the fields with more vignetting and fewer transmitted rays. Fewer rays yield fewer optimization operands. Fewer operands make less of a contribution to the damped least-squares solution. This complication is another reason why the lens designer may wish to experiment with field weights.

Figure B.3.2.1 is the layout of the final OPD optimized Cooke Triplet. Note that the general configuration is very similar to the spot size optimized lens in Figure B.3.1.1, but the front airspace is relatively smaller and the surface curves are somewhat different.

Figure B.3.2.2 is the vignetting plot. Note that the fraction of unvignetted rays at the edge of the field is 0.47 or 47% (1.09 stops down from the field center). This throughput value is close enough to the 50% target.

However, note that the vignetting diagram in Figure B.3.2.2 does not include the effect of cosine-fourth vignetting. Cosine-fourth vignetting darkens the edge of the field by an additional amount beyond any mechanical vignetting. For 22.6° off-axis, cosine-fourth is 0.73 or 73% (0.46 stops down from the field center). This value assumes negligible pupil growth from pupil aberrations, a reasonable assumption for a Cooke Triplet. Thus, total estimated light falloff from both types of vignetting gives relative illumination (irradiance) at the edge of the field of 0.34 or 34% (1.55

Listing B.3.2.1

Merit Function Listing

File : C:LENS314B.ZMX
Title: COOKE TRIPLET, 52MM, F/3.5, 45.2DEG

Merit Function Value: 7.50829989E-003

Num	Type	Int1	Int2	Hx	Hy	Px	Py	Target	Weight	Value	% Cont
1	EFFL		3					5.20000E+001	1e+005	5.20001E+001	0.005
2	BLNK										
3	BLNK										
4	TTHI	3	3					0.00000E+000	0	4.01727E+000	0.000
5	TTHI	5	6					0.00000E+000	0	7.19832E+000	0.000
6	DIFF	5	4					0.00000E+000	0	3.18105E+000	0.000
7	OPGT	6						1.50000E+000	1e+005	1.50000E+000	0.000
8	BLNK										
9	BLNK										
10	DIST	0	3					0.00000E+000	1e+005	7.33768E-006	0.000
11	BLNK										
12	BLNK										
13	DMFS										
14	OPDC		1	0.0000	0.0000	0.2500	0.2500	0.00000E+000	0.8	8.28033E-003	0.000
15	OPDC		1	0.0000	0.0000	0.2500	0.7500	0.00000E+000	0.8	-1.93433E-001	0.177
16	OPDC		1	0.0000	0.0000	0.7500	0.2500	0.00000E+000	0.8	-1.93433E-001	0.177
17	OPDC		2	0.0000	0.0000	0.2500	0.2500	0.00000E+000	0.8	8.69492E-002	0.036
18	OPDC		2	0.0000	0.0000	0.2500	0.7500	0.00000E+000	0.8	4.49848E-001	0.957
19	OPDC		2	0.0000	0.0000	0.7500	0.2500	0.00000E+000	0.8	4.49848E-001	0.957
20	OPDC		3	0.0000	0.0000	0.2500	0.2500	0.00000E+000	0.8	3.10335E-002	0.005
21	OPDC		3	0.0000	0.0000	0.2500	0.7500	0.00000E+000	0.8	3.05747E-001	0.442
22	OPDC		3	0.0000	0.0000	0.7500	0.2500	0.00000E+000	0.8	3.05747E-001	0.442
23	OPDC		4	0.0000	0.0000	0.2500	0.2500	0.00000E+000	0.8	-5.79185E-002	0.016
24	OPDC		4	0.0000	0.0000	0.2500	0.7500	0.00000E+000	0.8	-6.32398E-002	0.019
25	OPDC		4	0.0000	0.0000	0.7500	0.2500	0.00000E+000	0.8	-6.32398E-002	0.019
26	OPDC		5	0.0000	0.0000	0.2500	0.2500	0.00000E+000	0.8	-1.48682E-001	0.105
27	OPDC		5	0.0000	0.0000	0.2500	0.7500	0.00000E+000	0.8	-4.74652E-001	1.066
28	OPDC		5	0.0000	0.0000	0.7500	0.2500	0.00000E+000	0.8	-4.74652E-001	1.066
29	OPDC		1	0.0000	0.3982	0.2500	-0.7500	0.00000E+000	0.3	-1.07468E+000	2.048
30	OPDC		1	0.0000	0.3982	0.2500	-0.2500	0.00000E+000	0.3	8.80501E-002	0.014
31	OPDC		1	0.0000	0.3982	0.2500	0.2500	0.00000E+000	0.3	2.50417E-001	0.111
32	OPDC		1	0.0000	0.3982	0.2500	0.7500	0.00000E+000	0.3	9.56563E-003	0.000
33	OPDC		1	0.0000	0.3982	0.7500	-0.2500	0.00000E+000	0.3	7.60901E-001	1.027
34	OPDC		1	0.0000	0.3982	0.7500	0.2500	0.00000E+000	0.3	1.18858E+000	2.506
35	OPDC		2	0.0000	0.3982	0.2500	-0.7500	0.00000E+000	0.3	-1.67402E-001	0.050
36	OPDC		2	0.0000	0.3982	0.2500	-0.2500	0.00000E+000	0.3	2.29260E-001	0.093
37	OPDC		2	0.0000	0.3982	0.2500	0.2500	0.00000E+000	0.3	2.18183E-001	0.084
38	OPDC		2	0.0000	0.3982	0.2500	0.7500	0.00000E+000	0.3	3.01055E-001	0.161
39	OPDC		2	0.0000	0.3982	0.7500	-0.2500	0.00000E+000	0.3	1.39278E+000	3.441
40	OPDC		2	0.0000	0.3982	0.7500	0.2500	0.00000E+000	0.3	1.60406E+000	4.564
41	OPDC		3	0.0000	0.3982	0.2500	-0.7500	0.00000E+000	0.3	-2.28682E-001	0.093
42	OPDC		3	0.0000	0.3982	0.2500	-0.2500	0.00000E+000	0.3	1.71230E-001	0.052
43	OPDC		3	0.0000	0.3982	0.2500	0.2500	0.00000E+000	0.3	1.31326E-001	0.031
44	OPDC		3	0.0000	0.3982	0.2500	0.7500	0.00000E+000	0.3	6.06463E-002	0.007
45	OPDC		3	0.0000	0.3982	0.7500	-0.2500	0.00000E+000	0.3	1.18771E+000	2.502
46	OPDC		3	0.0000	0.3982	0.7500	0.2500	0.00000E+000	0.3	1.33141E+000	3.144
47	OPDC		4	0.0000	0.3982	0.2500	-0.7500	0.00000E+000	0.3	-5.82113E-001	0.601
48	OPDC		4	0.0000	0.3982	0.2500	-0.2500	0.00000E+000	0.3	5.92456E-002	0.006
49	OPDC		4	0.0000	0.3982	0.2500	0.2500	0.00000E+000	0.3	4.09036E-002	0.003
50	OPDC		4	0.0000	0.3982	0.2500	0.7500	0.00000E+000	0.3	-3.06883E-001	0.167
51	OPDC		4	0.0000	0.3982	0.7500	-0.2500	0.00000E+000	0.3	7.46006E-001	0.987
52	OPDC		4	0.0000	0.3982	0.7500	0.2500	0.00000E+000	0.3	8.79151E-001	1.371
53	OPDC		5	0.0000	0.3982	0.2500	-0.7500	0.00000E+000	0.3	-1.00425E+000	1.789
54	OPDC		5	0.0000	0.3982	0.2500	-0.2500	0.00000E+000	0.3	-6.03986E-002	0.006
55	OPDC		5	0.0000	0.3982	0.2500	0.2500	0.00000E+000	0.3	-4.01545E-002	0.003
56	OPDC		5	0.0000	0.3982	0.2500	0.7500	0.00000E+000	0.3	-6.78202E-001	0.816
57	OPDC		5	0.0000	0.3982	0.7500	-0.2500	0.00000E+000	0.3	2.62630E-001	0.122
58	OPDC		5	0.0000	0.3982	0.7500	0.2500	0.00000E+000	0.3	4.08174E-001	0.296
59	OPDC		1	0.0000	0.6991	0.2500	-0.2500	0.00000E+000	0.2	2.91949E-001	0.101
60	OPDC		1	0.0000	0.6991	0.2500	0.2500	0.00000E+000	0.2	3.83947E-001	0.174
61	OPDC		1	0.0000	0.6991	0.2500	0.7500	0.00000E+000	0.2	-1.75633E+000	3.648
62	OPDC		1	0.0000	0.6991	0.7500	-0.2500	0.00000E+000	0.2	1.60934E+000	3.063
63	OPDC		1	0.0000	0.6991	0.7500	0.2500	0.00000E+000	0.2	1.95819E+000	4.534
64	OPDC		2	0.0000	0.6991	0.2500	-0.2500	0.00000E+000	0.2	4.02041E-001	0.191
65	OPDC		2	0.0000	0.6991	0.2500	0.2500	0.00000E+000	0.2	2.71910E-001	0.087
66	OPDC		2	0.0000	0.6991	0.2500	0.7500	0.00000E+000	0.2	-1.67794E+000	3.329
67	OPDC		2	0.0000	0.6991	0.7500	-0.2500	0.00000E+000	0.2	2.15371E+000	5.485
68	OPDC		2	0.0000	0.6991	0.7500	0.2500	0.00000E+000	0.2	2.24437E+000	5.956
69	OPDC		3	0.0000	0.6991	0.2500	-0.2500	0.00000E+000	0.2	2.97159E-001	0.104
70	OPDC		3	0.0000	0.6991	0.2500	0.2500	0.00000E+000	0.2	1.67794E-001	0.033
71	OPDC		3	0.0000	0.6991	0.2500	0.7500	0.00000E+000	0.2	-1.89446E+000	4.244
72	OPDC		3	0.0000	0.6991	0.7500	-0.2500	0.00000E+000	0.2	1.86086E+000	4.095
73	OPDC		3	0.0000	0.6991	0.7500	0.2500	0.00000E+000	0.2	1.91078E+000	4.317
74	OPDC		4	0.0000	0.6991	0.2500	-0.2500	0.00000E+000	0.2	1.40219E-001	0.023
75	OPDC		4	0.0000	0.6991	0.2500	0.2500	0.00000E+000	0.2	8.22733E-002	0.008
76	OPDC		4	0.0000	0.6991	0.2500	0.7500	0.00000E+000	0.2	-2.15451E+000	5.489
77	OPDC		4	0.0000	0.6991	0.7500	-0.2500	0.00000E+000	0.2	1.34077E+000	2.126
78	OPDC		4	0.0000	0.6991	0.7500	0.2500	0.00000E+000	0.2	1.42478E+000	2.400
79	OPDC		5	0.0000	0.6991	0.2500	-0.2500	0.00000E+000	0.2	-1.87010E-002	0.000
80	OPDC		5	0.0000	0.6991	0.2500	0.2500	0.00000E+000	0.2	1.39299E-002	0.000
81	OPDC		5	0.0000	0.6991	0.2500	0.7500	0.00000E+000	0.2	-2.39172E+000	6.764
82	OPDC		5	0.0000	0.6991	0.7500	-0.2500	0.00000E+000	0.2	7.89234E-001	0.737
83	OPDC		5	0.0000	0.6991	0.7500	0.2500	0.00000E+000	0.2	9.32214E-001	1.028
84	OPDC		1	0.0000	1.0000	0.2500	-0.2500	0.00000E+000	0.1	1.73150E+000	1.773
85	OPDC		1	0.0000	1.0000	0.2500	0.2500	0.00000E+000	0.1	3.86223E-001	0.088
86	OPDC		1	0.0000	1.0000	0.7500	0.2500	0.00000E+000	0.1	-2.10541E+000	2.621
87	OPDC		2	0.0000	1.0000	0.2500	-0.2500	0.00000E+000	0.1	1.24723E+000	0.920
88	OPDC		2	0.0000	1.0000	0.2500	0.2500	0.00000E+000	0.1	3.24755E-001	0.062
89	OPDC		2	0.0000	1.0000	0.7500	0.2500	0.00000E+000	0.1	-1.31885E+000	1.028
90	OPDC		3	0.0000	1.0000	0.2500	-0.2500	0.00000E+000	0.1	7.64997E-001	0.346
91	OPDC		3	0.0000	1.0000	0.2500	0.2500	0.00000E+000	0.1	2.79604E-001	0.046
92	OPDC		3	0.0000	1.0000	0.7500	0.2500	0.00000E+000	0.1	-1.23002E+000	0.895
93	OPDC		4	0.0000	1.0000	0.2500	-0.2500	0.00000E+000	0.1	3.57826E-001	0.076
94	OPDC		4	0.0000	1.0000	0.2500	0.2500	0.00000E+000	0.1	2.45048E-001	0.036
95	OPDC		4	0.0000	1.0000	0.7500	0.2500	0.00000E+000	0.1	-1.36724E+000	1.105
96	OPDC		5	0.0000	1.0000	0.2500	-0.2500	0.00000E+000	0.1	2.66086E-002	0.000
97	OPDC		5	0.0000	1.0000	0.2500	0.2500	0.00000E+000	0.1	2.17879E-001	0.028
98	OPDC		5	0.0000	1.0000	0.7500	0.2500	0.00000E+000	0.1	-1.57081E+000	1.459

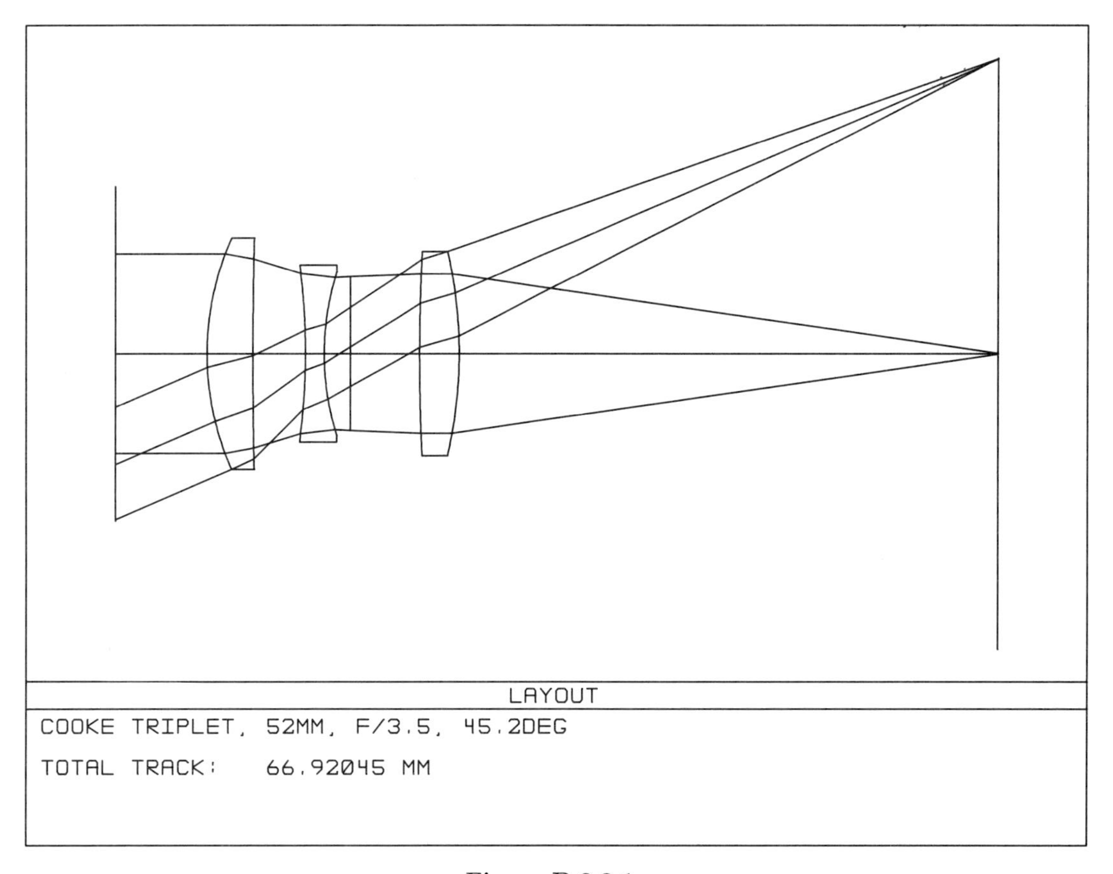

Figure B.3.2.1.

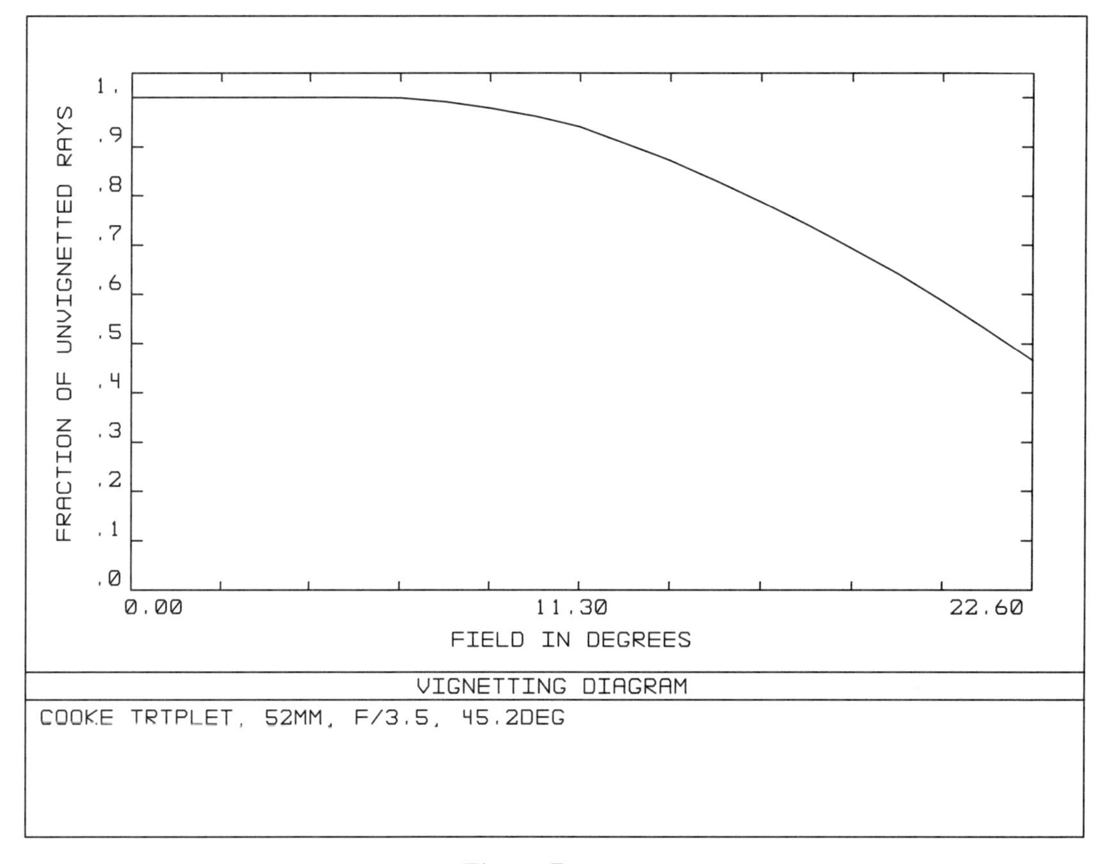

Figure B.3.2.2.

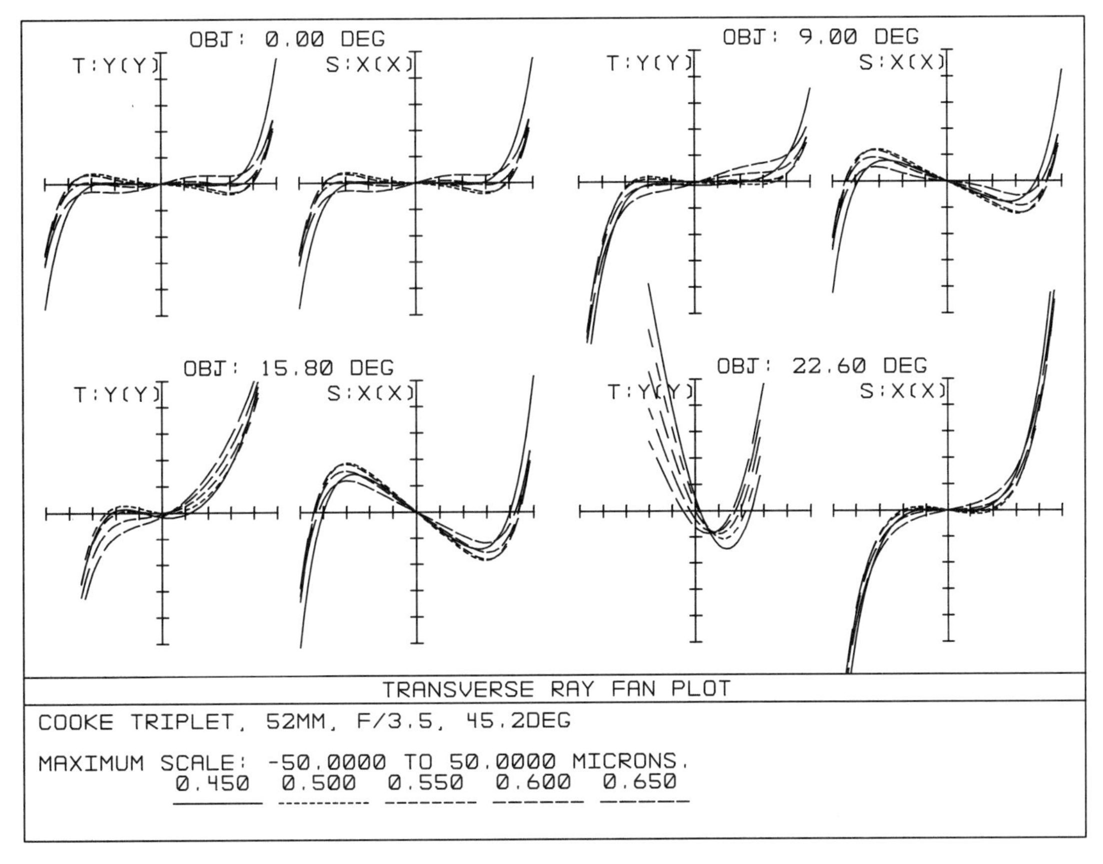

Figure B.3.2.3.

stops down from the field center). Again, this amount of light falloff will scarcely be noticed in practice.

Figure B.3.2.3 is the transverse ray fan plot; compare it to Figure B.3.1.4. Figure B.3.2.4 gives the field curvature and percent distortion plots; compare these with the plots in Figure B.3.1.5. Figure B.3.2.5 is the polychromatic spot diagram; compare this with Figure B.3.1.6. Clearly, Figures B.3.2.3 and B.3.2.5 show that the OPD optimized lens is worse when evaluated by transverse ray errors.

Figure B.3.2.6 is the OPD ray fan plot; compare it to Figure B.3.1.8. Now the OPD optimized lens is better. Note that there is still astigmatism, as revealed by the different curvatures at the origins of the tangential and sagittal off-axis plots. Note too that the astigmatism changes sign between the 70% field and the 100% field; this agrees with the field curvature plot where the curves cross near the 90% field.

Figure B.3.2.7 is the diffraction MTF plot; compare this with Figure B.3.1.9. Again the OPD optimized lens is better. Furthermore, the lens as optimized with OPD now gives a level of MTF performance that is acceptable for a normal, moderate speed 35 mm camera lens.

Camera lenses are often used stopped down. Figure B.3.2.8 is the MTF plot for the OPD optimized $f/3.5$ Cooke Triplet when stopped down to $f/8$. At this commonly used opening, image quality is quite good. However, the effects of the field-zonal astigmatism are still noticeable. Refer again to the field curvature plot in Figure B.3.2.4. At 70% field, the sagittal image surface is quite far from the actual image surface, and thus the sagittal MTF is degraded. At 100% field, past the tangential-sagittal curve crossing, the tangential image surface is now quite far from the actual image surface, and thus now the tangential MTF suffers.

Finally, Listing B.3.2.2 gives the optical prescription for the OPD optimized Cooke Triplet.

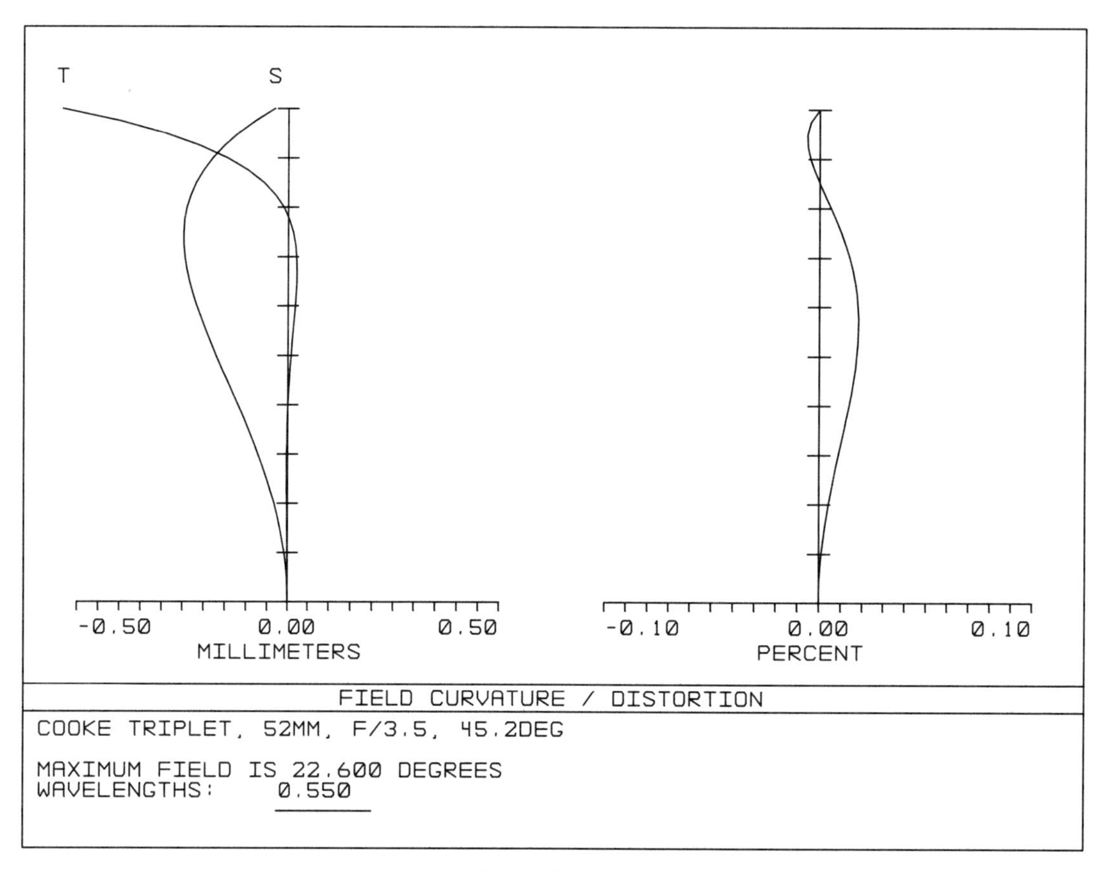

Figure B.3.2.4.

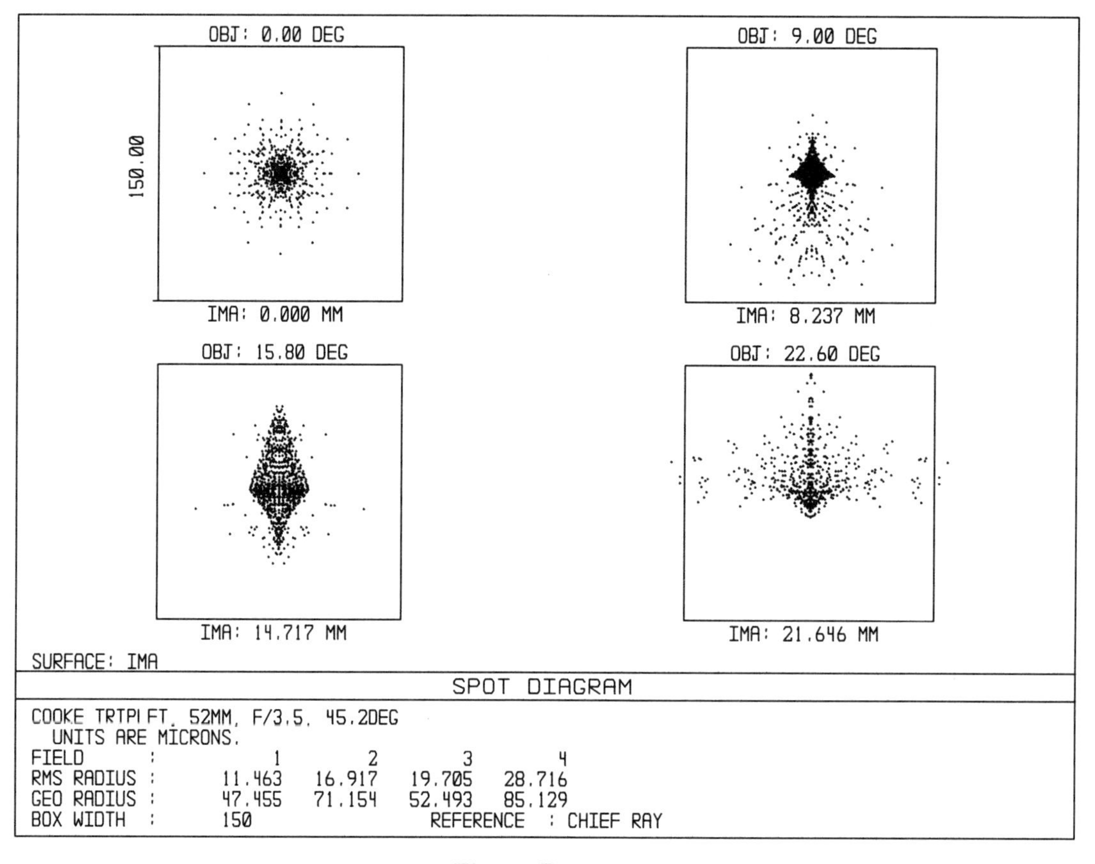

Figure B.3.2.5.

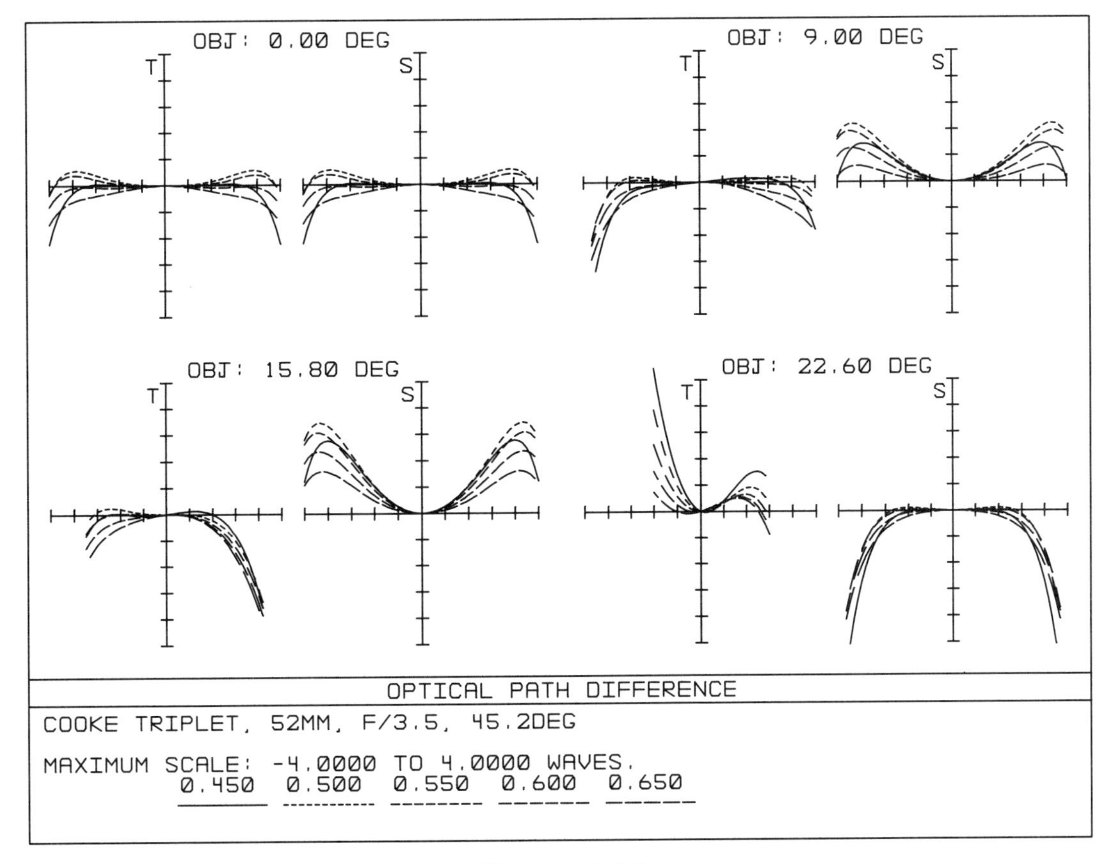

Figure B.3.2.6.

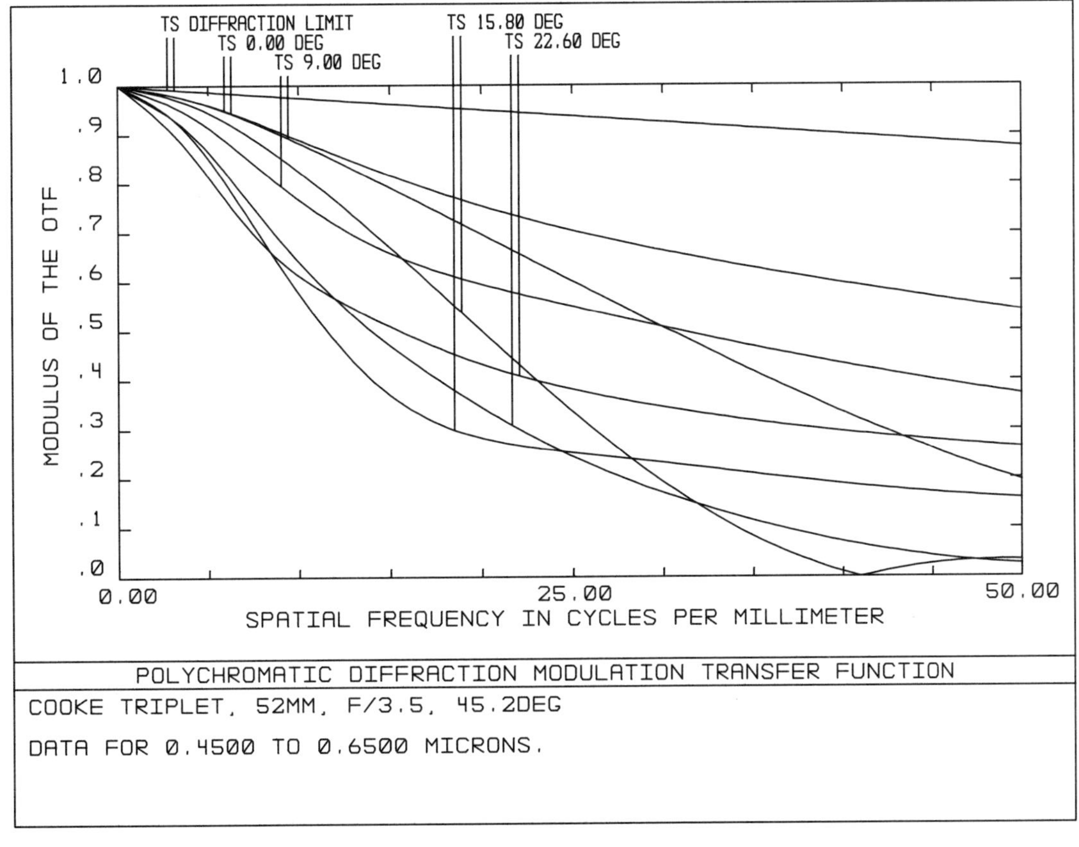

Figure B.3.2.7. Lens wide open.

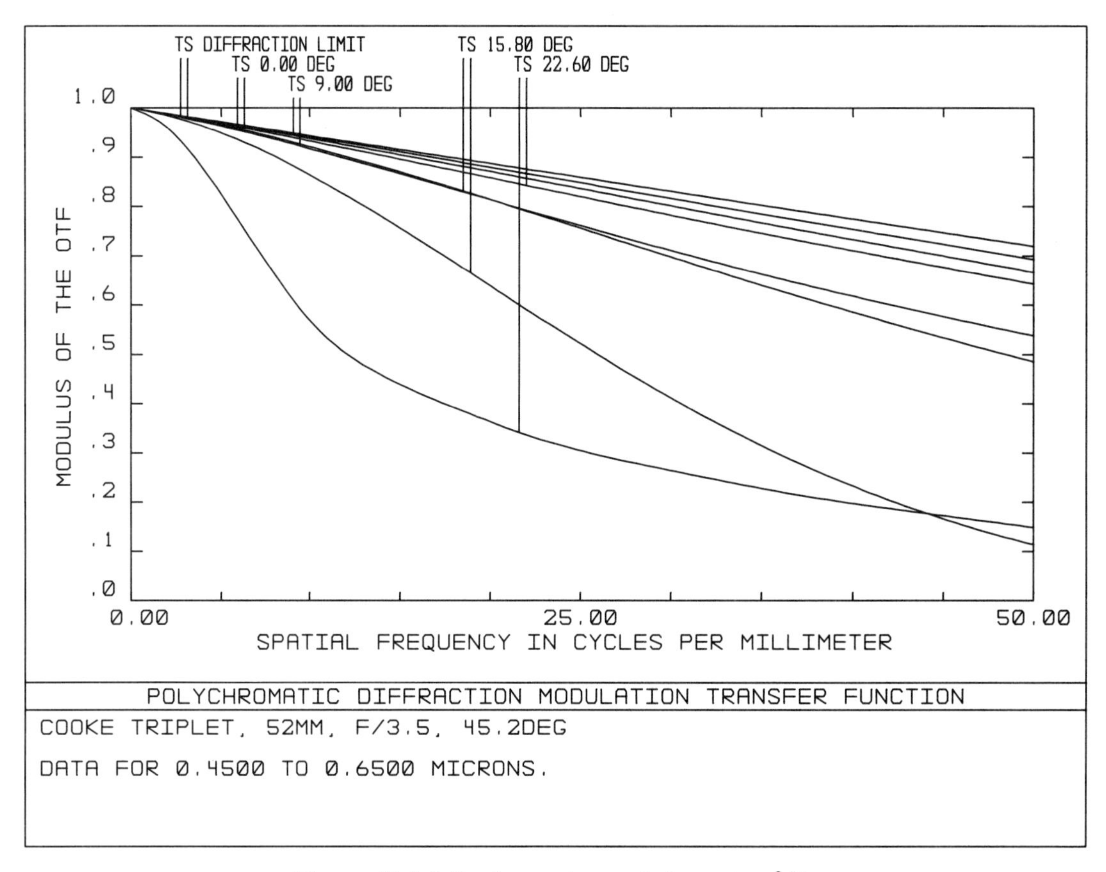

Figure B.3.2.8. Lens stopped down to *f*/8.

Listing B.3.2.2

```
System/Prescription Data

File : C:LENS314B.ZMX
Title: COOKE TRIPLET, 52MM, F/3.5, 45.2DEG

GENERAL LENS DATA:

Surfaces          :           10
Stop              :            6
System Aperture   :Entrance Pupil Diameter
Ray aiming        : On
 X Pupil shift    : 0
 Y Pupil shift    : 0
 Z Pupil shift    : 0
Apodization       :Uniform, factor =      0.000000
Eff. Focal Len.  :      52.0001 (in air)
Eff. Focal Len.  :      52.0001 (in image space)
Total Track       :      66.9205
Image Space F/#  :      3.49933
Para. Wrkng F/#  :      3.49933
Working F/#       :      3.56044
Obj. Space N.A.  :    7.43e-010
Stop Radius       :      5.65521
Parax. Ima. Hgt.:      21.6455
Parax. Mag.       :            0
Entr. Pup. Dia.  :        14.86
Entr. Pup. Pos.  :      19.7086
Exit Pupil Dia.  :      14.0556
Exit Pupil Pos.  :     -49.1852
Field Type        : Angle in degrees
Maximum Field     :         22.6
Primary Wave      :     0.550000
Lens Units        : Millimeters
Angular Mag.      :      1.05723

Fields            : 4
Field Type:  Angle in degrees
#         X-Value           Y-Value           Weight
1        0.000000          0.000000         4.000000
2        0.000000          9.000000         3.000000
3        0.000000         15.800000         2.000000
4        0.000000         22.600000         1.000000
```

Vignetting Factors

#	VDX	VDY	VCX	VCY
1	0.000000	0.000000	0.000000	0.000000
2	0.000000	0.000000	0.000000	0.000000
3	0.000000	0.000000	0.000000	0.000000
4	0.000000	0.000000	0.000000	0.000000

Wavelengths : 5
Units: Microns

#	Value	Weight
1	0.450000	1.000000
2	0.500000	1.000000
3	0.550000	1.000000
4	0.600000	1.000000
5	0.650000	1.000000

SURFACE DATA SUMMARY:

Surf	Type	Radius	Thickness	Glass	Diameter	Conic
OBJ	STANDARD	Infinity	Infinity		0	0
1	STANDARD	Infinity	7		33.76985	0
2	STANDARD	20.21384	3.5	LAFN21	17	0
3	STANDARD	357.8886	4.01727		17	0
4	STANDARD	-46.76389	1.5	SF53	13	0
5	STANDARD	18.91477	2		12	0
STO	STANDARD	Infinity	5.198322		11.31043	0
7	STANDARD	142.092	3	LAFN21	15	0
8	STANDARD	-32.48564	40.70486		15	0
9	STANDARD	Infinity	0		43.7664	0
IMA	STANDARD	Infinity	0		43.7664	0

SURFACE DATA DETAIL:

```
Surface OBJ      : STANDARD
Surface   1      : STANDARD
Surface   2      : STANDARD
 Aperture        : Circular Aperture
 Minimum Radius  :            0
 Maximum Radius  :          8.5
Surface   3      : STANDARD
Surface   4      : STANDARD
Surface   5      : STANDARD
Surface STO      : STANDARD
Surface   7      : STANDARD
Surface   8      : STANDARD
 Aperture        : Circular Aperture
 Minimum Radius  :            0
 Maximum Radius  :          7.5
Surface   9      : STANDARD
Surface IMA      : STANDARD
```

SOLVE AND VARIABLE DATA:

```
Curvature of   2   : Variable
Semi Diam    2     : Fixed
Curvature of   3   : Variable
Thickness of   3   : Variable
Semi Diam    3     : Fixed
Curvature of   4   : Variable
Semi Diam    4     : Fixed
Curvature of   5   : Variable
Semi Diam    5     : Fixed
Thickness of   6   : Variable
Curvature of   7   : Variable
Semi Diam    7     : Fixed
Curvature of   8   : Variable
Thickness of   8   : Solve, marginal ray height = 0.00000
Semi Diam    8     : Fixed
```

INDEX OF REFRACTION DATA:

Surf	Glass	0.450000	0.500000	0.550000	0.600000	0.650000
0		1.00000000	1.00000000	1.00000000	1.00000000	1.00000000
1		1.00000000	1.00000000	1.00000000	1.00000000	1.00000000
2	LAFN21	1.80620521	1.79788762	1.79184804	1.78728036	1.78370773
3		1.00000000	1.00000000	1.00000000	1.00000000	1.00000000
4	SF53	1.75665336	1.74309602	1.73366378	1.72676633	1.72152789
5		1.00000000	1.00000000	1.00000000	1.00000000	1.00000000
6		1.00000000	1.00000000	1.00000000	1.00000000	1.00000000
7	LAFN21	1.80620521	1.79788762	1.79184804	1.78728036	1.78370773
8		1.00000000	1.00000000	1.00000000	1.00000000	1.00000000
9		1.00000000	1.00000000	1.00000000	1.00000000	1.00000000
10		1.00000000	1.00000000	1.00000000	1.00000000	1.00000000

ELEMENT VOLUME DATA:

```
Units are cubic cm.
Values are only accurate for plane and spherical surfaces.
Element surf   2 to   3 volume :       0.596651
Element surf   4 to   5 volume :       0.273760
Element surf   7 to   8 volume :       0.435456
```

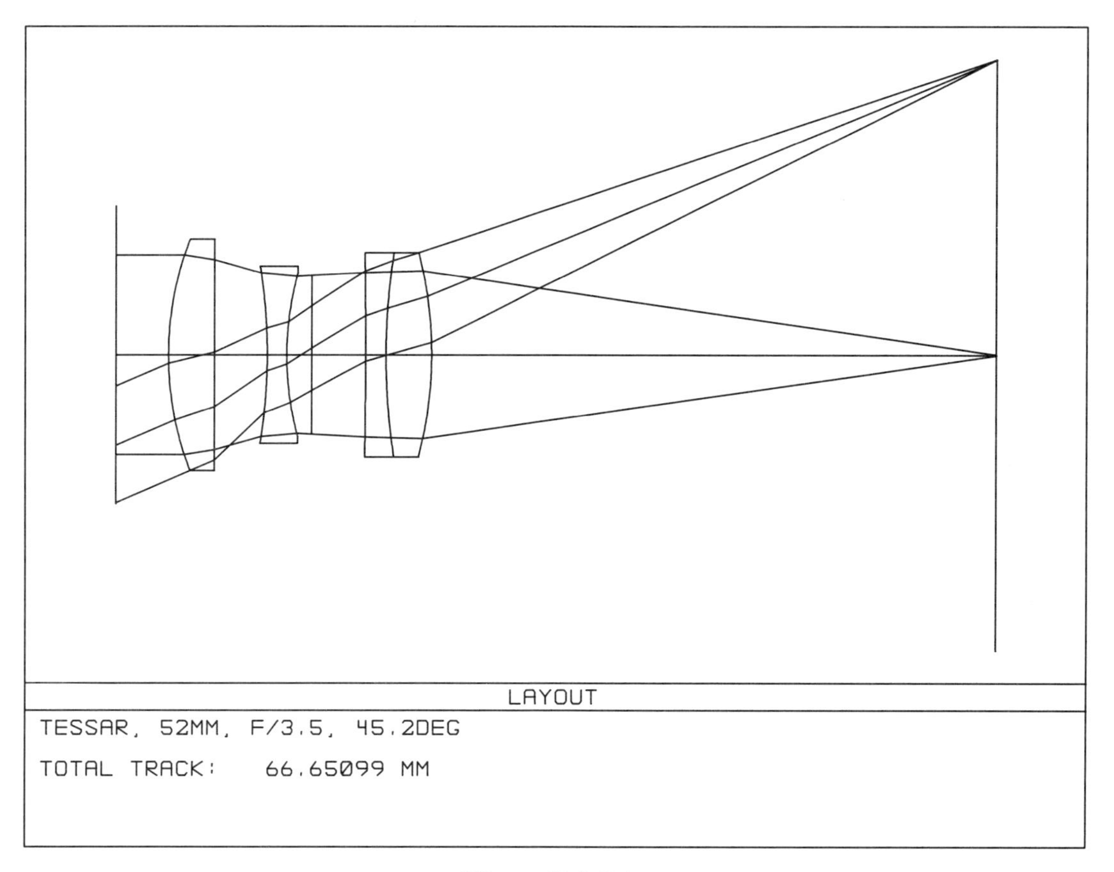

Figure B.3.3.1.

B.3.10 The Tessar Lens

If you require a bit more optical performance than the Cooke Triplet can deliver, then the four element Tessar configuration, illustrated in Figure B.3.3.1, may be the answer. Clearly, the Tessar, whose name derives from the Greek word meaning four, closely resembles the Cooke Triplet. The big difference is that the rear singlet in the Cooke Triplet has been replaced by a cemented doublet. The additional element gives the lens designer one more surface curvature to vary during optimization to control aberrations. With the Tessar, there are nine effective independent variables: seven curvatures and two airspaces. Once again, the glass thicknesses are only weak, ineffective variables and are fixed at practical values. Of course, glass selection is still very important.

The lens shown here is the classic Tessar with the outside element of the doublet having positive power. There is a Tessar derivative where the positive and negative elements of the doublet are reversed in order. There is still another derivative where the whole lens is reversed. The performance of all three versions is about the same. By far, the majority of Tessars are the classic variety.

In the present Tessar example, the basic lens specifications are the same as in the Cooke Triplet example. This includes the same amount of vignetting. The idea is to create two designs that can be directly compared. Once again, MTF is the image quality criterion.

The Tessar is not strictly symmetrical, but nevertheless the design still contains much symmetry. This symmetry can help in deriving a starting configuration. In addition, a look at published Tessar layouts will suggest some useful initial constraints that can shepherd the lens in the right direction. Make the inside surface of the front element flat. Make the two curvatures of the singlet negative element equal with opposite signs. Make the inside air-to-glass surface of the doublet flat.

Make the two curvatures of the positive element of the doublet equal with opposite signs. And make the front airspace equal to the space between the stop and the first surface of the cemented doublet. Use pickups for the coupled parameters. For glasses, make a guess based on published accounts (see below for more on glass). It is no accident that to a large extent this starting configuration looks very much like a Cooke Triplet.

With these constraints, you optimize as you did for the Cooke Triplet. Focal length is corrected to 52 mm, paraxial longitudinal color is corrected to zero, and polychromatic spot sizes are reduced. Ignore distortion for now. If the airspaces become too large or too small, change the glass for the negative singlet and reoptimize. The resulting lens configuration will be good enough for use at the start of the intermediate optimizations (with the initial constraints removed, of course).

The biggest issue throughout the design of a Tessar is glass selection. The one thing that is clear is that for best results, the positive elements should be made of high-index crown glass. The reasons are the same as those outlined for the Cooke Triplet: crown glass for achromatization, and high index to reduce the naturally negative Petzval sum and to reduce higher-order aberrations. Older Tessar designs use the barium crowns (the SK glasses in the Schott catalog). Newer Tessars use the even higher-index lanthanum crowns.

Thus, the problem becomes selecting the glasses for the two negative elements. Most published accounts suggest that relative to the crowns, both negative elements be flints and be selected from the glasses along the old glass line. The negative singlet usually has a somewhat lower index and is very flinty relative to the crowns. The cemented negative element usually has a much lower index and is only moderately more flinty relative to the crowns.

But this approach is not universal. The famous 50 mm $f/3.5$ Elmar lens for the original Leica camera is a Tessar derivative with the stop in the front airspace rather than the rear airspace. In order from front to rear, the glasses for the Elmar are: SK7, F5, BK7, and SK15.[3] Relative to SK15 (n_d of 1.623, V_d of 58.1), BK7 (n_d of 1.517, V_d of 64.2) is a crown, not a flint. In fact, it is possible to design excellent Tessars with equal dispersions for the two elements of the cemented doublet.

For the present Tessar example, the glass type adopted for both of the crowns is the same as for the crowns in the Cooke Triplet, Schott LaFN21. However, when the computer is asked to find the matching glasses for the negative elements, the results are inconclusive. To understand this behavior, a manual approach is used. Four separate Tessars were optimized with the glass for the cemented negative element fixed successively at LLF2, F5, SF2, and SF15. For each case, a matching flint for the negative singlet was found by a search along the old glass line and by repeatedly optimizing with the intermediate merit function. The remarkable result is that all four solutions give almost identical optical performance. No wonder the automatic glass selection is inconclusive.

Thus, Schott F5 (n_d of 1.603, V_d of 38.0) is arbitrarily adopted for the negative cemented element. The reason is historical; relative to LaFN21 (n_d of 1.788, V_d of 47.5), F5 has the classical position on the glass map, down and to the right. The matching flint is found to be Schott SF15 (n_d of 1.699, V_d of 30.1). Thus, the glass set for the Tessar becomes: LaFN21, SF15, F5, LaFN21. These are all preferred glasses.

The intermediate and final optimizations for a Tessar are very similar to those for a Cooke Triplet. During the final optimizations, both a spot size optimization and an OPD optimization are done. MTFs are plotted and the results compared. As with the Cooke Triplet, the OPD optimized Tessar is found to give better MTF performance.

Figure B.3.3.1 shows the layout of the final OPD optimized Tessar. Note that the inside surface of the front element and the air-to glass surface of the cemented

[3]Dierick Kossel, "Glass Compositions," *Leica Fotografie*, English edition, Feb. 1978, pp. 20–25.

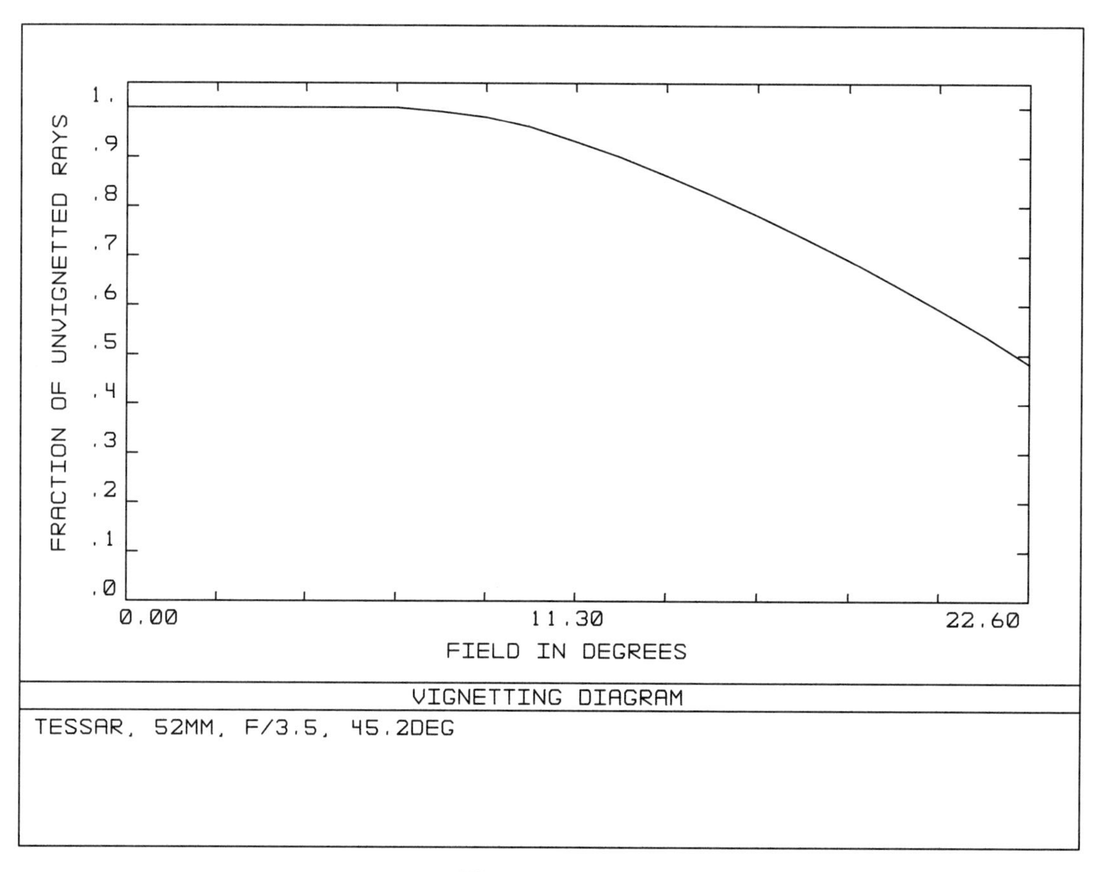

Figure B.3.3.2.

negative element are both nearly flat. If this Tessar were to be actually fabricated, one or both of these surfaces would probably be made exactly flat and the lens reoptimized with little loss of performance. Surfaces with very long radii are hard to make and are often not worth the effort.

Figure B.3.3.2 is the vignetting plot of geometrical throughput versus field angle for the Tessar. Note that the fraction of unvignetted rays at the edge of the field is 0.48 or 48% (1.06 stops down from the field center). This value is nearly identical to the corresponding value for the Cooke Triplet as shown on Figure B.3.2.2.

Figure B.3.3.3 shows the transverse ray fan plots for the Tessar. Compare these to the plots in Figure B.3.2.3 for the Cooke Triplet. Clearly, the two sets of curves are very similar. The most noticeable difference is that the off-axis tangential curves for the Tessar show somewhat smaller ray errors. Like the Triplet, the Tessar suffers from off-axis oblique spherical aberration and field-zonal astigmatism.

Figure B.3.3.4 shows the field curvature and percent distortion plots. Compare these to the plots in Figure B.3.2.4. On the left, note that the Tessar has noticeably less field-zonal astigmatism than the Cooke Triplet. Consequently, the Tessar needs less emphasis on the intermediate field positions during optimization. The field weights of 4 3 2 1 for the Triplet were changed to 5 4 3 2 for the Tessar. On the right, note that like the Cooke Triplet, distortion in the Tessar has been corrected to zero at the edge of the field by balancing third-order distortion with higher-orders. Note too that at intermediate fields there is a small amount of residual third-order distortion of little or no practical consequence.

Figure B.3.3.5 gives the polychromatic spot diagrams. Compare these to Figure B.3.2.5. As with the ray fan plots, the spots indicate that the Tessar and Triplet have very similar geometrical performance. Qualitatively, the spot shapes are nearly the same; quantitatively, the spot sizes are somewhat less for the Tessar. Thus, the

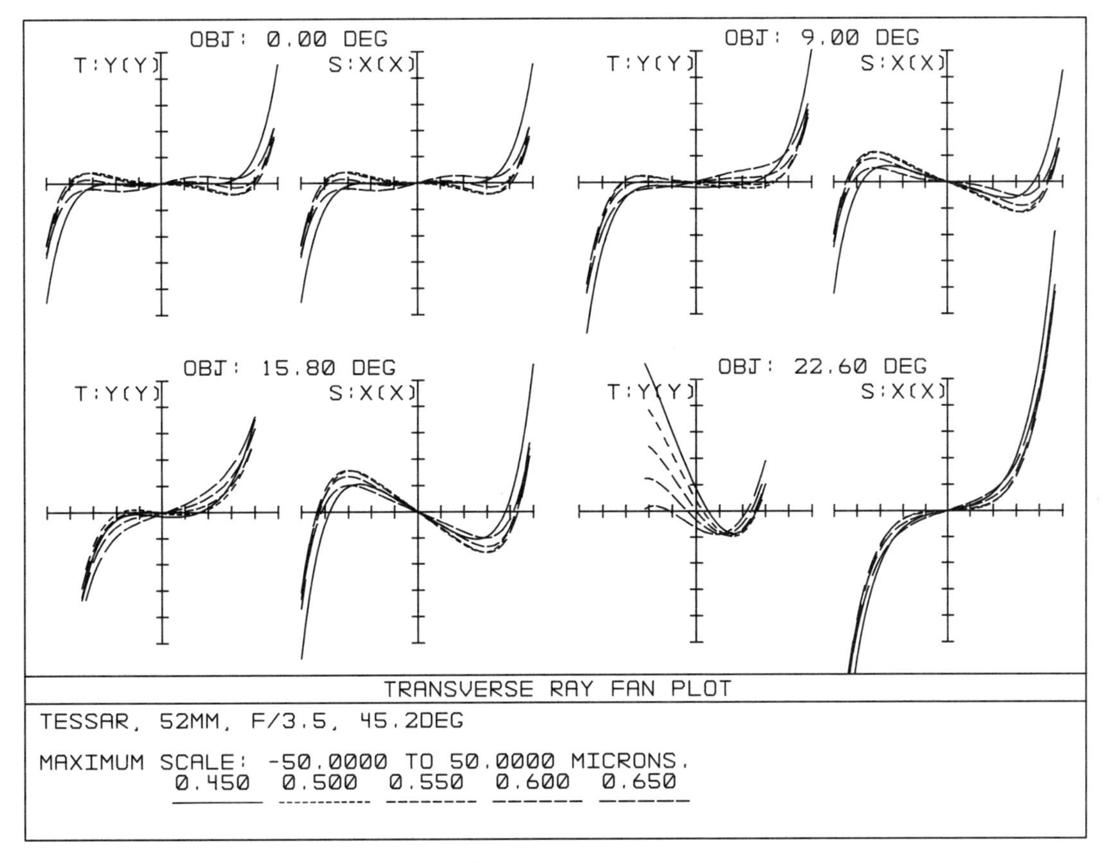

Figure B.3.3.3.

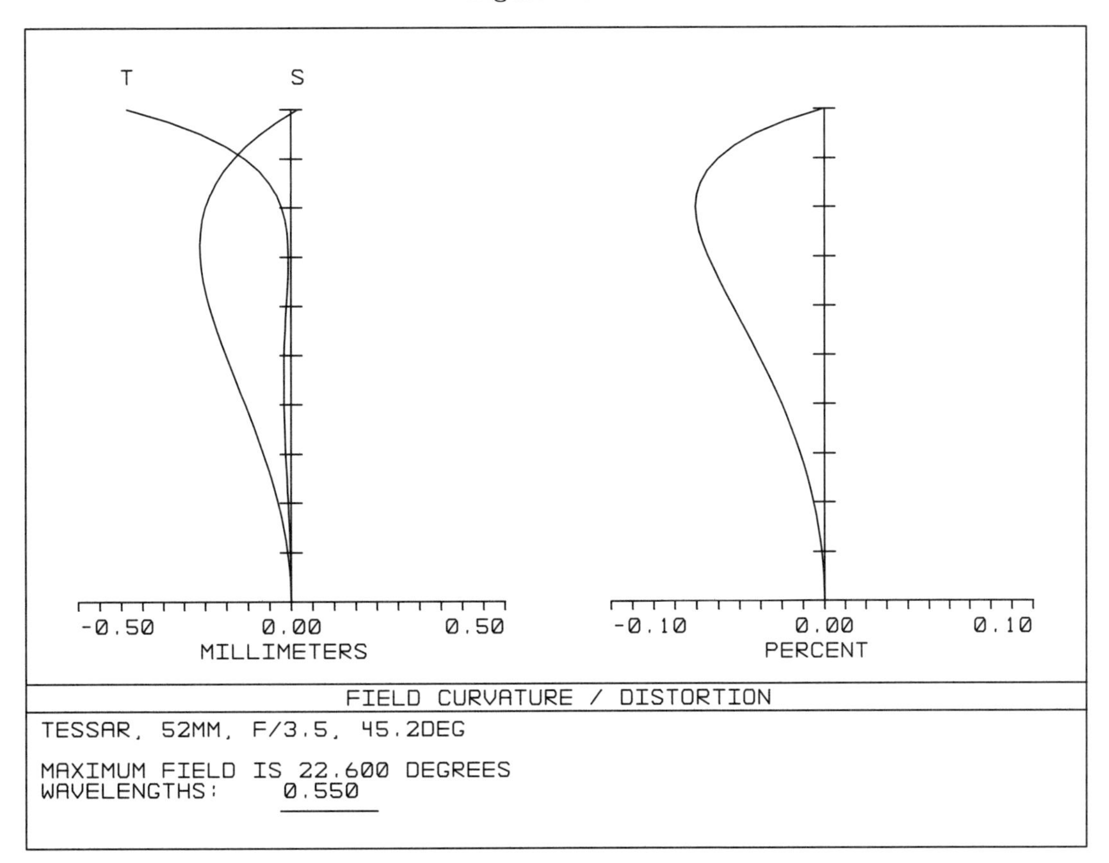

Figure B.3.3.4.

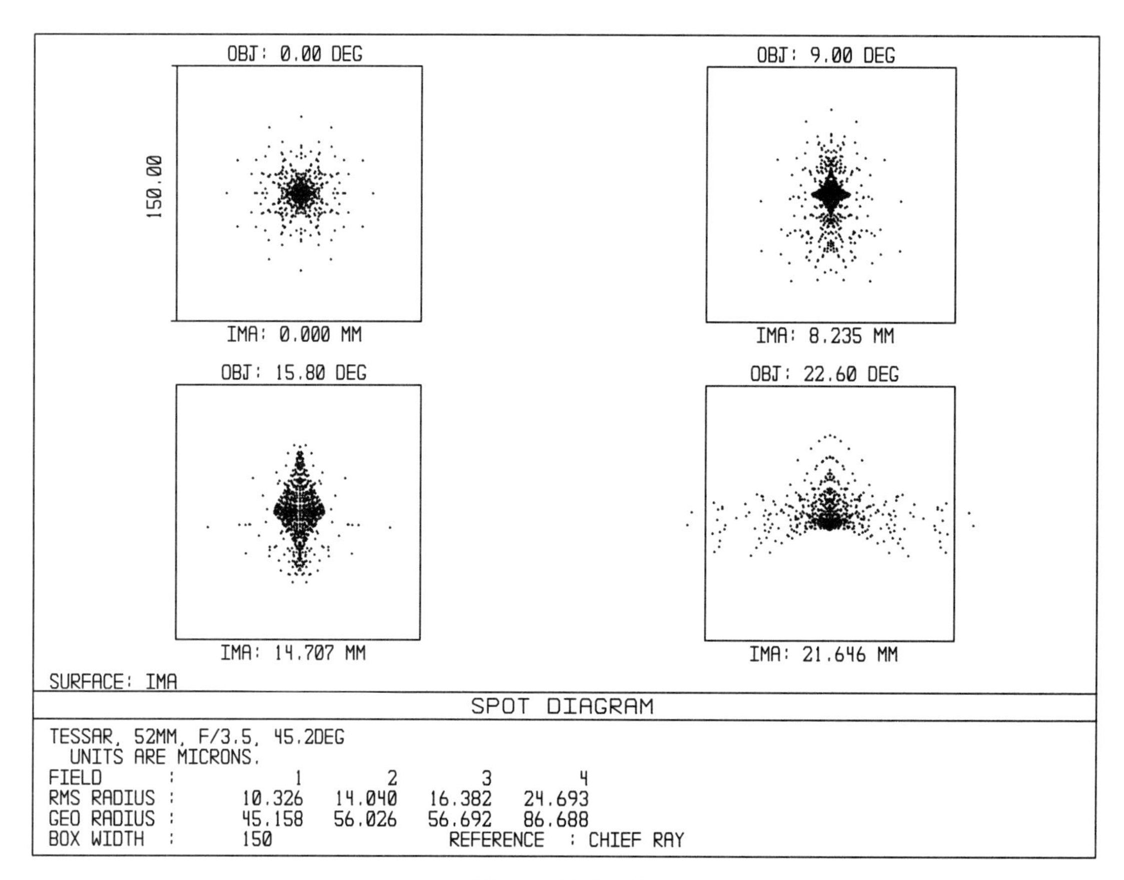

Figure B.3.3.5.

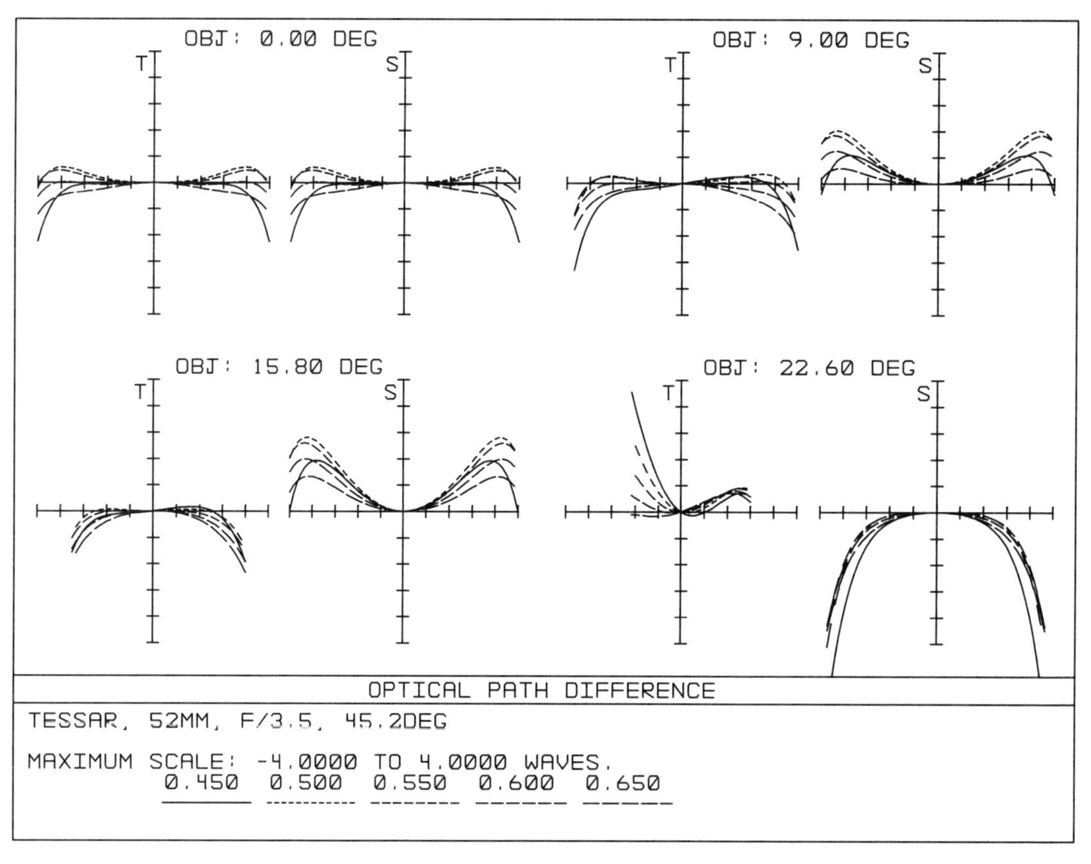

Figure B.3.3.6.

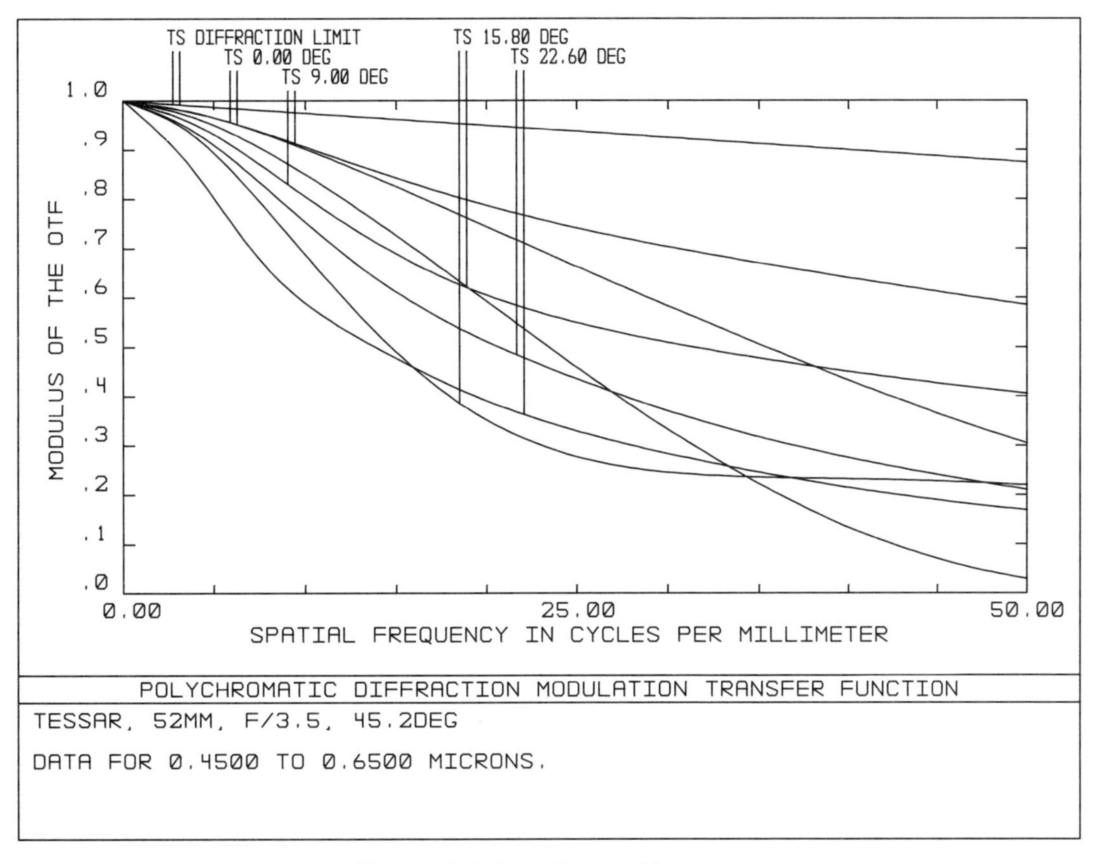

Figure B.3.3.7. Lens wide open.

Tessar is definitely the better lens, although only moderately so.

Figure B.3.3.6 gives the OPD ray fan plots. Compare these to Figure B.3.2.6. Again the Tessar is somewhat better.

Figure B.3.3.7 gives the MTF curves for the Tessar wide open at $f/3.5$. Compare these with the MTF curves in Figure B.3.2.7 for the Cooke Triplet, also wide open at $f/3.5$. Clearly, the Tessar has higher performance, but again the advantage is not great. However, this modest advantage may be the difference between success and failure in the marketplace. Accordingly, many lens makers have chosen to expend the extra effort to add a fourth element to make a Tessar rather than a Cooke Triplet.

Figure B.3.3.8 gives the MTF curves for the Tessar when stopped down to $f/8$. Compare these with the MTF curves in Figure B.3.2.8 for the Cooke Triplet, also stopped down to $f/8$. In addition, compare the MTF performance of both lenses at $f/8$ versus $f/3.5$. Clearly, both the Tessar and Triplet have much better MTF performance at $f/8$ than at $f/3.5$. But this should be no surprise to any photographer who knows that his lens gets sharper when stopped down. When both lenses are at $f/8$, the Tessar is once again somewhat better than the Cooke Triplet.

Finally, Listing B.3.3 gives the optical prescription for the OPD optimized Tessar. This Tessar lens is toleranced in Chapter B.7.

The Cooke Triplet and Tessar lens configurations are two of the most popular in the world. But their speeds are limited. For a 35 mm camera, many photographers now insist on buying a faster standard lens. Over the years, several basic approaches to a fast lens have been tried. But only one is widely used today. This configuration is known among lens designers as the Double-Gauss. In the next chapter, the Double-Gauss lens will be examined, and its performance will be compared to that of the Cooke Triplet and Tessar.

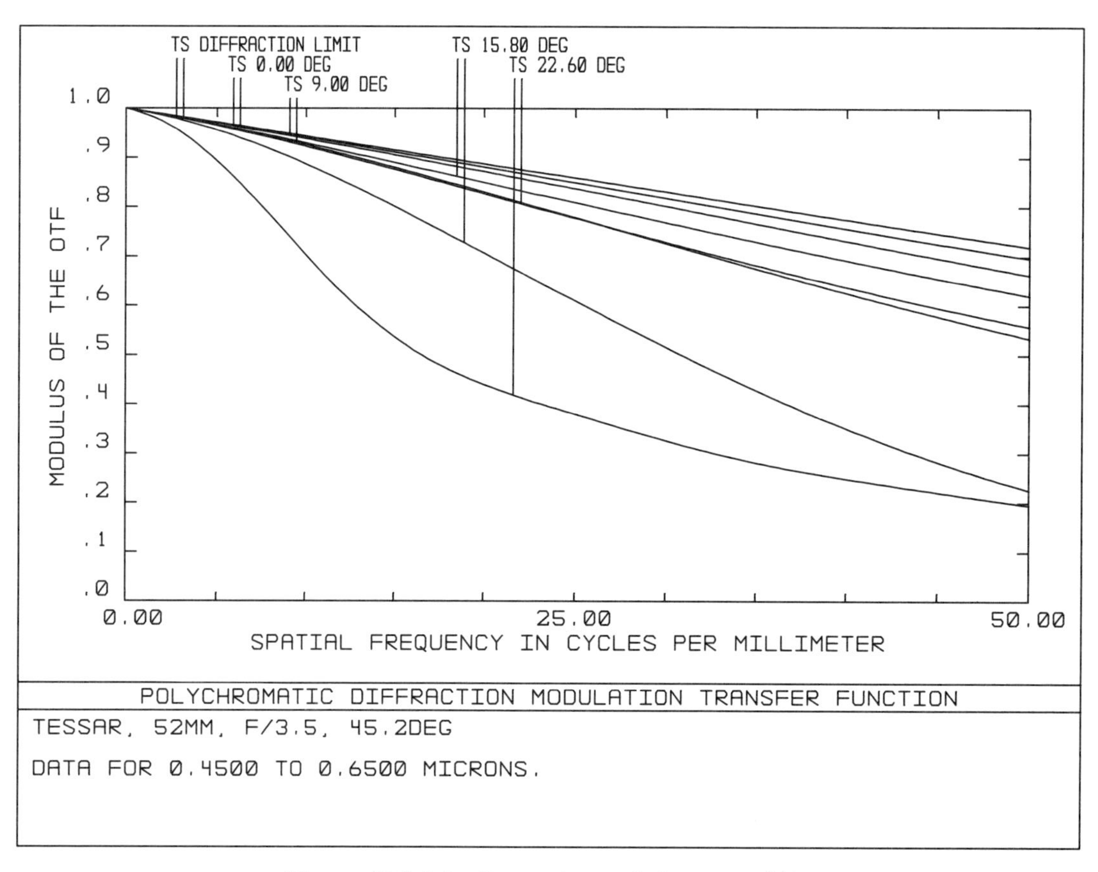

Figure B.3.3.8. Lens stopped down to *f*/8.

Listing B.3.3

```
System/Prescription Data

File : C:LENS324B.ZMX
Title: TESSAR, 52MM, F/3.5, 45.2DEG

GENERAL LENS DATA:

Surfaces          :           11
Stop              :            6
System Aperture   :Entrance Pupil Diameter
Ray aiming        : On
 X Pupil shift    : 0
 Y Pupil shift    : 0
 Z Pupil shift    : 0
Apodization       :Uniform, factor =     0.000000
Eff. Focal Len.   :       52.001 (in air)
Eff. Focal Len.   :       52.001 (in image space)
Total Track       :       66.651
Image Space F/#   :      3.49939
Para. Wrkng F/#   :      3.49939
Working F/#       :      3.56248
Obj. Space N.A.   :    7.43e-010
Stop Radius       :      5.82916
Parax. Ima. Hgt.  :      21.6459
Parax. Mag.       :            0
Entr. Pup. Dia.   :        14.86
Entr. Pup. Pos.   :      16.2583
Exit Pupil Dia.   :      14.6565
Exit Pupil Pos.   :     -51.2887
Field Type        : Angle in degrees
Maximum Field     :         22.6
Primary Wave      :     0.550000
Lens Units        : Millimeters
Angular Mag.      :      1.01389


Fields           : 4
Field Type:  Angle in degrees
#        X-Value          Y-Value          Weight
1        0.000000         0.000000         5.000000
2        0.000000         9.000000         4.000000
3        0.000000        15.800000         3.000000
4        0.000000        22.600000         2.000000
```

Vignetting Factors

#	VDX	VDY	VCX	VCY
1	0.000000	0.000000	0.000000	0.000000
2	0.000000	0.000000	0.000000	0.000000
3	0.000000	0.000000	0.000000	0.000000
4	0.000000	0.000000	0.000000	0.000000

Wavelengths : 5
Units: Microns

#	Value	Weight
1	0.450000	1.000000
2	0.500000	1.000000
3	0.550000	1.000000
4	0.600000	1.000000
5	0.650000	1.000000

SURFACE DATA SUMMARY:

Surf	Type	Radius	Thickness	Glass	Diameter	Conic
OBJ	STANDARD	Infinity	Infinity		0	0
1	STANDARD	Infinity	4		30.51105	0
2	STANDARD	22.5851	3.5	LAFN21	17	0
3	STANDARD	3174.661	4.005808		17	0
4	STANDARD	-39.77737	1.5	SF15	13	0
5	STANDARD	20.74764	2		12	0
STO	STANDARD	Infinity	4.06086		11.65832	0
7	STANDARD	-502.9552	1.5	F5	15	0
8	STANDARD	47.47455	3.5	LAFN21	15	0
9	STANDARD	-28.85977	42.58433		15	0
10	STANDARD	Infinity	0		43.63454	0
IMA	STANDARD	Infinity	0		43.63454	0

SURFACE DATA DETAIL:

Surface OBJ : STANDARD
Surface 1 : STANDARD
Surface 2 : STANDARD
Aperture : Circular Aperture
Minimum Radius : 0
Maximum Radius : 8.5
Surface 3 : STANDARD
Surface 4 : STANDARD
Surface 5 : STANDARD
Surface STO : STANDARD
Surface 7 : STANDARD
Surface 8 : STANDARD
Surface 9 : STANDARD
Aperture : Circular Aperture
Minimum Radius : 0
Maximum Radius : 7.5
Surface 10 : STANDARD
Surface IMA : STANDARD

SOLVE AND VARIABLE DATA:

Curvature of 2 : Variable
Semi Diam 2 : Fixed
Curvature of 3 : Variable
Thickness of 3 : Variable
Semi Diam 3 : Fixed
Curvature of 4 : Variable
Semi Diam 4 : Fixed
Curvature of 5 : Variable
Semi Diam 5 : Fixed
Thickness of 6 : Variable
Curvature of 7 : Variable
Semi Diam 7 : Fixed
Curvature of 8 : Variable
Semi Diam 8 : Fixed
Curvature of 9 : Variable
Thickness of 9 : Solve, marginal ray height = 0.00000
Semi Diam 9 : Fixed

INDEX OF REFRACTION DATA:

Surf	Glass	0.450000	0.500000	0.550000	0.600000	0.650000
0		1.00000000	1.00000000	1.00000000	1.00000000	1.00000000
1		1.00000000	1.00000000	1.00000000	1.00000000	1.00000000
2	LAFN21	1.80620521	1.79788762	1.79184804	1.78728036	1.78370773
3		1.00000000	1.00000000	1.00000000	1.00000000	1.00000000
4	SF15	1.72487043	1.71248942	1.70386313	1.69754461	1.69273708
5		1.00000000	1.00000000	1.00000000	1.00000000	1.00000000
6		1.00000000	1.00000000	1.00000000	1.00000000	1.00000000
7	F5	1.62084299	1.61262327	1.60678655	1.60244994	1.59911057
8	LAFN21	1.80620521	1.79788762	1.79184804	1.78728036	1.78370773
9		1.00000000	1.00000000	1.00000000	1.00000000	1.00000000
10		1.00000000	1.00000000	1.00000000	1.00000000	1.00000000
11		1.00000000	1.00000000	1.00000000	1.00000000	1.00000000

ELEMENT VOLUME DATA:

Units are cubic cm.
Values are only accurate for plane and spherical surfaces.
Element surf 2 to 3 volume : 0.609663
Element surf 4 to 5 volume : 0.272423
Element surf 7 to 8 volume : 0.322578
Element surf 8 to 9 volume : 0.478834

Chapter B.4

The Double-Gauss Lens

The Fraunhofer type lens, as described in Chapter B.1, is not the only way to make a crown-in-front achromatic doublet. In the analytical simultaneous correction of spherical aberration and coma, a quadratic equation is obtained, and this equation has two solutions. One solution yields the Fraunhofer configuration. The other solution yields a doublet consisting of two airspaced meniscus elements. This second solution was first discovered by Gauss in 1817 and has thus been named after him.

The Gauss configuration has the advantage of having less spherochromatism than the Fraunhofer configuration. However, the Gauss configuration has a little more secondary color and a lot more zonal spherical aberration. Furthermore, it is considerably more difficult to make. In practice, the extra work is rarely justified, and historically very few telescopes have been made with Gauss objective lenses.

In 1888, the idea occurred to Alvan G. Clark that two Gauss-type telescope objectives placed back-to-back symmetrically about a stop might make a good wide-angle camera lens. Thus was born the Double-Gauss lens configuration. Later, other designers developed the idea further. Wide-angle lenses based on Clark's original configuration are still in use and have names such as Topogon and Metrogon.

In 1896, Paul Rudolph, who six years later would invent the Tessar, modified the basic Double-Gauss lens. To reduce the field-zonal astigmatism and oblique spherical aberration, Rudolph thickened the negative elements and reduced the airspaces between the positive and negative elements. To better correct the color, he also added a cemented buried surface inside each negative meniscus. The glasses on each side of a cemented buried surface have similar indices but different dispersions. The effect is to simulate a glass with greater dispersion than any available. These modifications worked very well, and he called his $f/4.5$ lens the Planar.

In 1920, H.W. Lee went even further. Lee expanded on Rudolph's approach by (1) making the front and rear halves of the lens slightly unsymmetrical, (2) using flints with lower indices than the crowns, and (3) using a shorter focal length to match a smaller film format. The result was a breakthrough: a speed of $f/2$ with about a 50° field. This type of Double-Gauss lens became very popular and remains so to this day. The trade names of these lenses are too many to list. But if you have a recent 35 mm camera with a roughly 50 mm, $f/2$ to $f/1$ lens, then it is almost certain that your lens is a Double-Gauss or one of its variations.

Note that the reduced focal length is crucial in making the $f/2$ speed practical. When a lens is scaled down, the geometrical aberrations are scaled down too, but the film resolution remains the same. Thus, a lens configuration that is incapable of giving the high angular resolution required for a large format view camera may be quite suitable for a small format roll film camera or movie camera.

In this chapter, the classical Double-Gauss camera lens is investigated. The construction is illustrated in Figure B.4.1.1. For increased clarity, only the on-axis and extreme off-axis beams have been drawn.

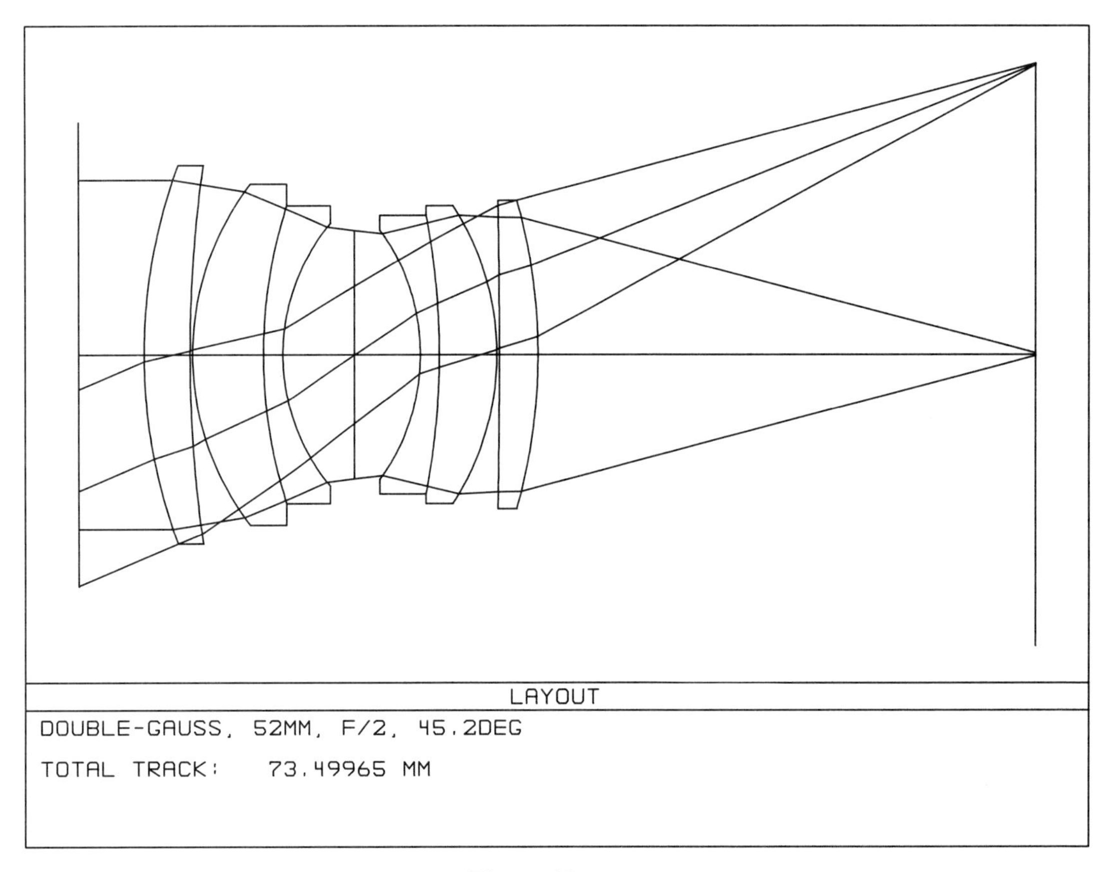

Figure B.4.1.1.

However, before proceeding, there must be a note about the treatment. It is perhaps unavoidable that much emphasis is placed on only one lens design program. To cover all or even several programs, each with its individual format and approach, would be impractical. Because the author uses the ZEMAX program and because ZEMAX has the required features, the details in what follows are based on ZEMAX.

But ZEMAX is by no means the only first rate lens design program that is currently available (or will be available in the future). In all cases, the lens designer must consult his program's user's manual for operational details. What applies to ZEMAX may not apply to, for example, CODE V or OSLO. This book is not a substitute for your user's manual.

B.4.1 Lens Specifications

The Double-Gauss lens in this chapter has specifications very similar to the Cooke Triplet and Tessar lenses in Chapter B.3; that is, it is a normal or standard lens for use on a 35 mm type still camera with the object at infinity. The most important difference is that the lens is faster.

Maximum speed is $f/2$. Focal length is 52 mm. Field coverage is 45.2°, or ±22.6°, with object points at 0°, 9°, 15.8°, and 22.6° (or 0, 40%, 70%, and 100% of half-field). The wavelengths for optimization and evaluation match the panchromatic sensitivity of films: 0.45, 0.50, 0.55, 0.60, and 0.65 μm. The reference wavelength for calculating first-order properties and solves is the central wavelength, 0.55 μm. Distortion is corrected to zero at the edge of the field. The lens is optimized and used on the paraxial focal plane (refer again to Section A.15.7), and this plane must be at least 38 mm behind the rear outside lens surface to allow clearance for a single-lens-reflex (SLR) mirror (this back focal length is realistic; it is the minimum clearance required for Nikon SLR lenses according to information supplied by the manufacturer). Roughly 50% mechanical vignetting is allowed at the edge of the field. All high-index lanthanum crown glasses are allowed. And finally, the performance criterion for image quality is diffraction MTF.

Note that the use of expensive lanthanum crown glasses may be controversial. An $f/2$ Double-Gauss lens is much larger and heavier than an $f/3.5$ Cooke Triplet or Tessar lens, and thus glass cost now becomes more significant. Pretend that the lens in this chapter is for an expensive camera, such as a Nikon or Leica reflex.

B.4.2 Multiple Configurations

Optimizing a lens wide open places all the emphasis on wide open performance. But a fast $f/2$ lens is not always used wide open. More commonly used openings are between $f/4$ and $f/8$, with $f/2$ reserved for low-light situations or when shallow depth of field is required. Thus, it makes sense when optimizing an $f/2$ (or faster) lens to also give consideration to how it performs when stopped down; that is, to give extra consideration to the more central area of the entrance pupil.

Two approaches are possible. When optimizing, more emphasis can be given to rays passing through the central pupil area by weighting these rays more heavily. In CODE V, the control command is WTA in AUTO. In ZEMAX, the apodization feature is used when constructing the default merit function.

The other approach is to set up the lens as a multiple configuration lens operating both wide open and stopped down to some smaller opening. During optimization, both configurations are optimized simultaneously. To make the damped least-squares optimizer pay attention to the anticipated smaller aberrations when stopped down, the smaller opening is given a greater weight.

Of the two approaches, the multiple configuration approach is more realistic. For this reason, and also to illustrate a multiple configuration lens, this approach is adopted here. Thus in the present example, the lens is optimized wide open at $f/2$ and also stopped down to $f/4$ (half its maximum entrance pupil diameter).

In a Double-Gauss lens, the worst off-axis aberration is oblique spherical, a fifth-order aberration. Measured by transverse ray errors on the image surface, oblique spherical varies as the third power of entrance pupil diameter (and as the second power of off-axis field angle). Measured by OPD errors in the exit pupil, oblique spherical varies as the fourth power of the entrance pupil diameter. Thus, when optimizing with spots and transverse ray errors, the $f/4$ configuration is weighted eight times (two cubed) as much as the $f/2$ configuration. Similarly, when optimizing with OPDs, the $f/4$ configuration is weighted sixteen times (two to the fourth power) as much as the $f/2$ configuration. How appropriate these weightings are can be verified by examining the percent contributions of the various operands in the resulting merit function. The two configurations should contribute roughly equally.

Note that the best-known type of multiple configuration lens is a zoom lens where focal length can be continuously varied by the user. Thus for brevity, some lens designers and design programs may use the term zoom in place of multiple configuration. Similarly, designers might say that a parameter is zoomed when they mean varied between multiple configurations. In the present example, entrance pupil diameter is zoomed although the lens is not a zoom lens. Do not be confused by this jargon.

B.4.3 Vignetting Factors

As mentioned in the previous chapter, there are conceptually two ways to handle mechanical vignetting. The first way uses simple and uniform ray grids incident on the entrance pupil plus real or hard apertures to delete rays that are blocked by vignetting. Figure B.4.1.2 illustrates this first method. The second way reshapes the incident ray bundles to make them conform with the vignetted pupils, thereby allowing all the rays to be transmitted to the image surface. Figure B.4.1.3 illustrates this second method. In the previous chapter, the first method was used; in the present chapter, the second method is used.

In the second method, the reshaping of the ray bundles is controlled by vignetting factors. As implemented in ZEMAX, for each object point there is a set of four vignetting factors. These control beam compressions in the y-direction and x-direction, and beam decenters in the y-direction and x-direction. Of course, there must be no vignetting in the center of the field, and thus only the off-axis vignetting factors are nonzero.

Before vignetting factors are applied, an incident beam is centered and circular on the stop surface. After vignetting factors are applied, a general off-axis beam is shifted (decentered) and elliptical (usually compressed but occasionally stretched) on the stop surface to best match the vignetted and/or aberrated pupil. However, the fit is usually not perfect, and significant errors may be introduced. Areas in the actual pupil can be missed by the ellipse (pupil undersampling), or areas outside the actual pupil can be included in the ellipse (pupil oversampling).

Sometimes, vignetting factors are clearly not appropriate and should not be used. Imagine ZEMAX trying to fit an ellipse to the D-shaped vignetted pupil at the edge of the field of the Landscape lens in Chapter B.2. Or imagine an optical system with a large, strangely shaped central obscuration. For systems like these, it is better to use hard apertures and delete rays. However, for most axially centered lenses, vignetting factors are indeed appropriate.

Vignetting factors can be set in two ways. In the first way, you set the surface diameters and the computer sets the corresponding vignetting factors. In the second way, you set the vignetting factors based on experience and on the desired amount of vignetting. The computer then sets the minimum surface diameters required to pass all the rays from all the object points.

Figure B.4.1.4 shows the layout of a Tessar lens with the diameters of the front and rear outside surfaces set by hand and with vignetting factors calculated to match by the computer. Note that the beams are defined by and are consistent with the specified diameters.

Figure B.4.1.5 shows the layout of a similar Tessar with specific vignetting factors set by hand. The computer has set the minimum surface diameters. Note here that the beams are not perfectly consistent with the diameters. In the real world, of course, the surface diameters do define the beam sizes. Thus, this model of the lens is physically incorrect, although the errors are not large.

You might ask why anyone would ever want to purposely model a lens incorrectly with specific vignetting factors. Setting vignetting factors by hand is very useful during the early and intermediate optimization stages to control roughly the amount of vignetting without having to fuss with surface diameters. The lens and its diameters can change, but the amount of vignetting remains the same.

If 50% vignetting is desired at the edge of the field and if object points are located at 0, 40%, 70%, and 100% of half-field, then an often not too inaccurate set of beam compression vignetting factors is 0 .1 .25 .45 in the y-direction and 0 0 0 .1 in the x-direction. All beam shifts (decenters) are left zero.

Note that if the beam shifts are kept zero, then you are designing with beams that all pass exactly through the center of the stop surface (and iris diaphragm). These centered beam paths are preferred in a lens that later will be used stopped down. The iris diaphragm will then close down on the beam centers, which usually

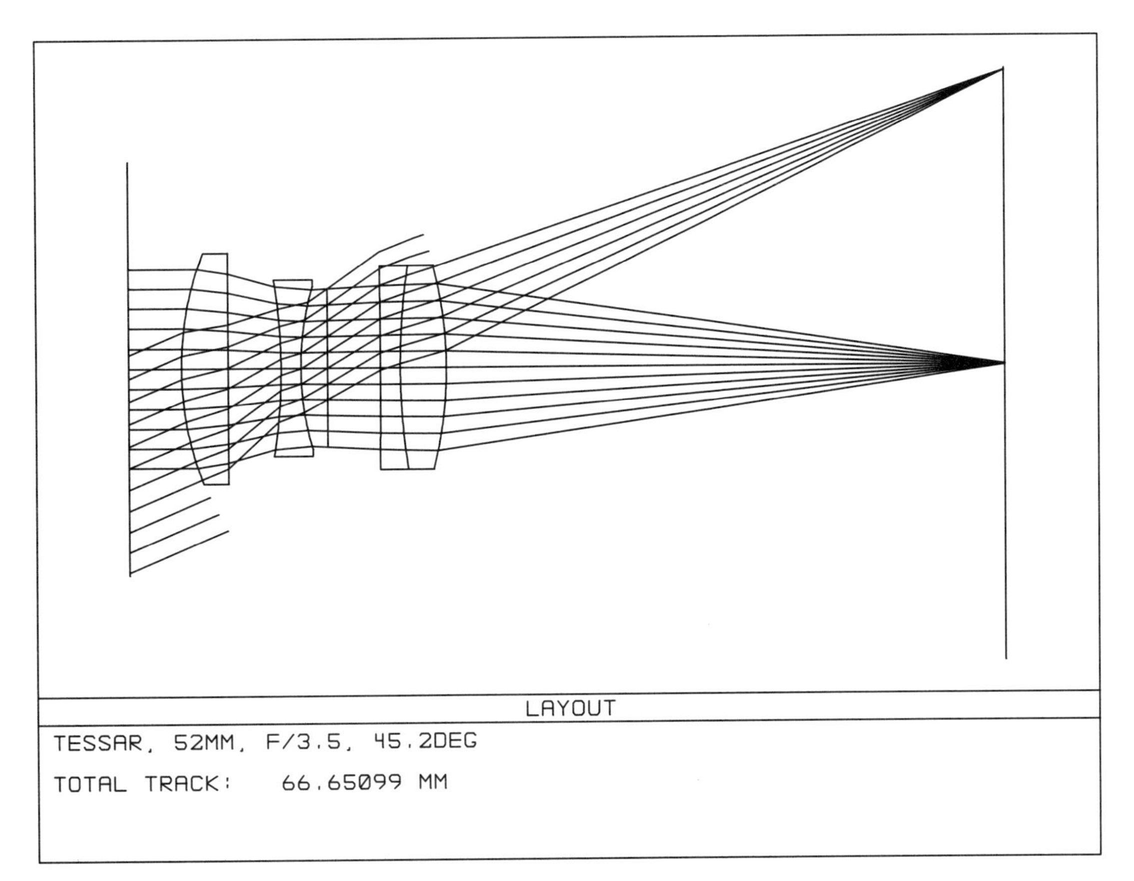

Figure B.4.1.2.

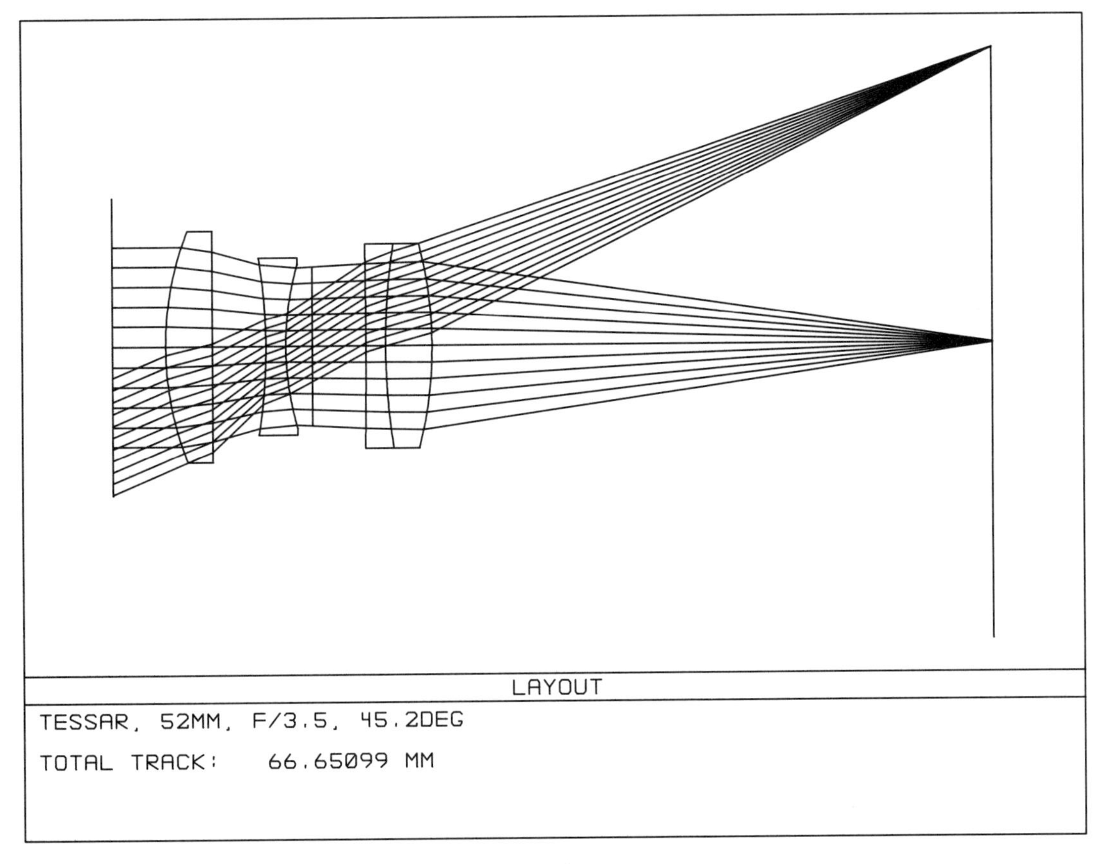

Figure B.4.1.3.

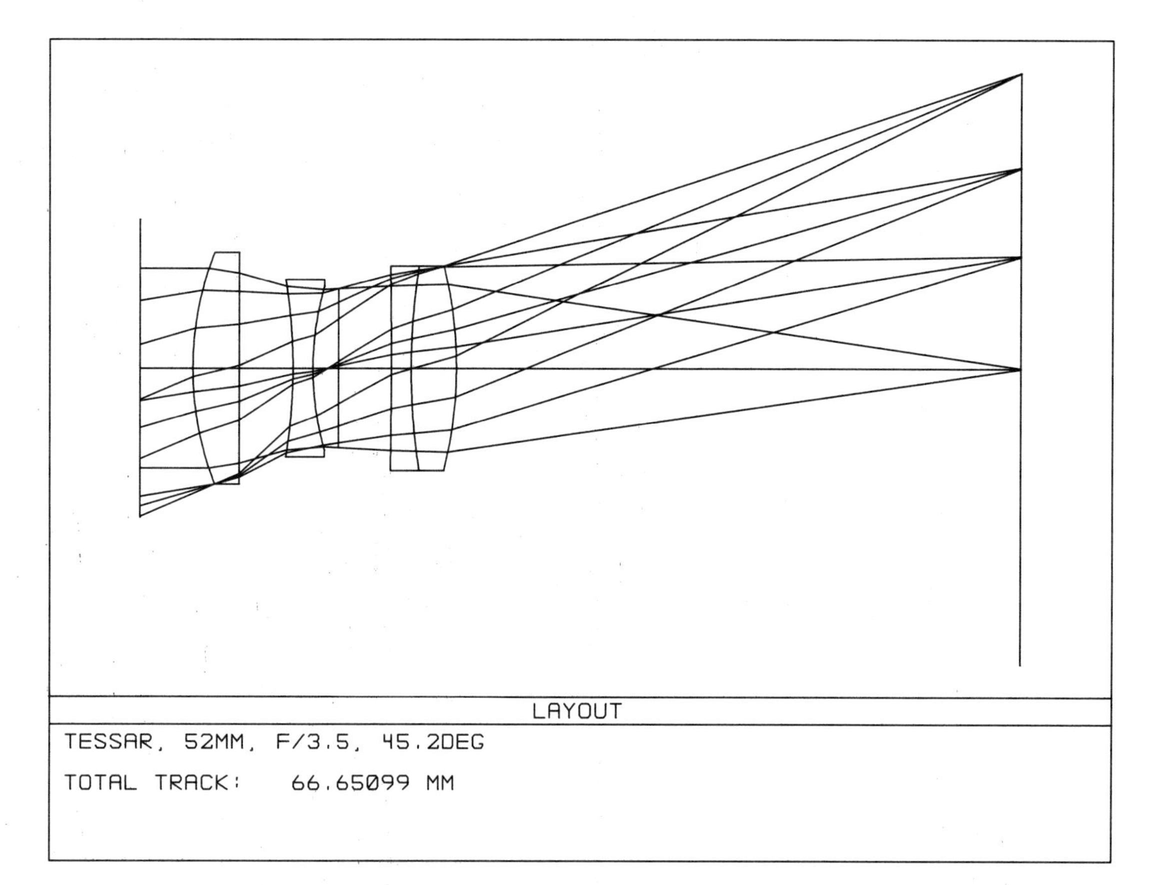

Figure B.4.1.4.

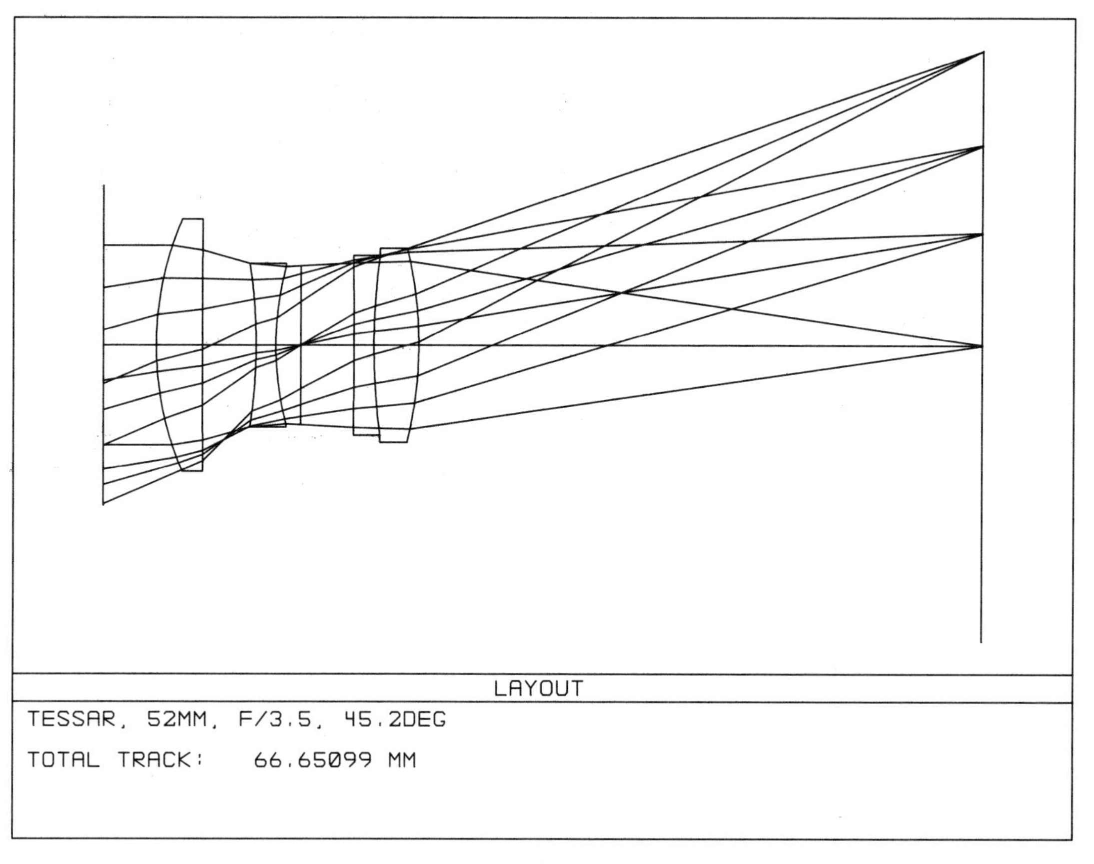

Figure B.4.1.5.

have the least aberrations. Mechanical vignetting will also disappear at the widest possible opening.

But caution! Designing with specific vignetting factors is just a temporary expedient that is appropriate only during the early and intermediate optimizations. During the final optimizations, you must return to reality. One way to do this is to have the computer set the vignetting factors to match specified surface diameters. But even better, for careful work during the final optimizations, it is preferable to not use vignetting factors at all (zero them out) and revert back to the use of even more realistic hard apertures that delete vignetted rays.

There is an additional reason for not using vignetting factors during final optimizations. As presently implemented in ZEMAX, when you have the computer set general vignetting factors that include beam shifts, what the program calls the chief ray is actually the middle ray in an off-axis beam. Unless the beam shifts are all zero, this middle ray does not pass through the center of the stop surface and is thus not the true chief ray, by definition. Refer again to the layouts in Figures B.3.1.1 and B.4.1.4, which were plotted using general vignetting factors.

This convention in ZEMAX can lead to a subtle problem when optimizing a lens. When you ask spot sizes or OPDs to be minimized relative to the chief ray, the optimization is now done relative to the middle ray. During optimization, the middle ray can move around on the stop; that is, you have a moving target. Furthermore, it is not the target you want if you want beams centered on the stop.

However, if the beam shifts are all zero, then you are guaranteed to be using the true chief ray. This happens both when you set user-specified vignetting factors and when you zero them all out. Thus during final optimizations, zeroing out the vignetting factors and using hard apertures not only guarantees accurate pupil sampling, it also guarantees that optical properties will be properly referenced to the true chief ray.

Note finally that vignetting factors are part of the lens prescription in ZEMAX and are active, not just during optimizations, but also during performance evaluations. Nearly all of the evaluation options are affected by pupil sampling errors and by the use of the middle, not chief, ray. Thus, to not be misled during evaluations, it is suggested that here too you should zero out all of the vignetting factors and use hard apertures to delete rays.

B.4.4 Gaussian Quadrature

In ZEMAX, the main advantage in using vignetting factors is that they allow the Gaussian quadrature[1] algorithm to be used during the optimization of a lens with mechanical vignetting. The advantages of a Gaussian quadrature optimization over a rectangular array optimization are computing speed and optical accuracy. Gaussian quadrature is especially effective during the crucial intermediate optimizations when the basic lens configuration is being determined.

In a rectangular array optimization, many rays are needed, and rays can be deleted from the ray set to account for mechanical vignetting. Gaussian quadrature, however, requires relatively few rays, but the ray pattern is unusual and very specific. In addition, the algorithm only works if the entire ray set is present; that is, no rays can be deleted. Thus, if the lens has mechanical vignetting, then vignetting factors must be used to get all the rays through to the image.

Gaussian quadrature requires that the rays must pass through either the entrance pupil or the stop surface at special points located at the intersections of a series of radial spokes and concentric circular rings. The spokes (or arms) are equally spaced in angle, but the rings are unequally spaced out from the center.

[1] For the original paper on using Gaussian quadrature in lens design, see G. W. Forbes, "Optical System Assessment for Design: Numerical Ray Tracing in the Gaussian Pupil," *J. Opt. Soc. Am. A*, Vol. 5, No. 11, November 1988, pp. 1943–1956.

The angular separation of the spokes, the various diameters of the rings, and the relative ray weights are all calculated to exactly target with no redundancy the several orders of aberrations in an optical system. More spokes and rings merely address higher orders of aberrations.

Gaussian quadrature works best if there are twice the number of spokes as rings. In particular, for each field position and wavelength, six spokes (60° apart) and three rings address defocus, third-order, and fifth-order aberrations. Similarly, eight spokes and four rings address defocus plus third-, fifth-, and seventh-order aberrations. Avoid using more spokes and rings than you really need. For most lenses, six or eight spokes and three or four rings are enough.

If vignetting factors are active, then the circular rings of rays are compressed into appropriate ellipses of rays that still satisfy the requirements of the Gaussian quadrature algorithm.

Incidentally, if you are using ZEMAX and the stop is located inside the lens, turn on the ray aiming feature. By an iterative trigonometric procedure, this option ensures the correct locations of the piercing points of the real rays as they pass through the physical stop surface. Ray aiming is especially important in lenses with significant pupil aberrations. All too often you are asking for subtle and unexpected errors if the ray positions are defined on the approximate first-order Gaussian entrance pupil rather than on the actual stop surface. This caution applies to both Gaussian quadrature and rectangular array optimizations and also to lens performance evaluations.

B.4.5 Starting Design and Early Optimizations

As with the Cooke Triplet and Tessar lenses, the optimization procedure outlined in Chapter A.15 is used when designing a Double-Gauss lens. Also, as usual, it is very useful to read the literature to find what has worked well for others and what the layout of a good solution looks like. Of course, if instead you are developing a new type of optical system for which there is no prior art, then you proceed based on your knowledge of good design techniques and on how any good lens must perform.

The first thing to do when designing a Double-Gauss lens is to make a guess at the initial glass set. Many published designs for Double-Gauss lenses have nearly every element made of a different type of glass. In earlier times, this was done for reasons related to the design methods then in use. Today, it is done to save money by using less expensive glass types where it matters less. But some lens designers now feel that it is better to use only two glass types, one crown and one flint. This limitation simplifies glass procurement and imposes no optical penalty (given computers and optimizing software). In the present example, this simplified glass choice procedure is adopted.

Published accounts of Double-Gauss designs report that (as you might expect) using high-index crown glass yields better image quality. Furthermore, for comparison, it might be of interest to use the same high-index crown glass type that was used in the previous Cooke Triplet and Tessar lenses. Thus, all four positive elements of the Double-Gauss lens are made of Schott LaFN21 (n_d of 1.788, V_d of 47.5).

Similarly, experience has shown that the matching flint glass type in a Double-Gauss lens should lie on the old glass line and have an index somewhat lower than the crown type. As an initial guess, the same flint glass used for the Cooke Triplet, Schott SF53 (n_d of 1.728, V_d of 28.7), is selected for the two negative elements. This choice will be verified later.

Like the Cooke Triplet and Tessar lenses, a Double-Gauss camera lens must have deliberate mechanical vignetting to achieve the best overall performance compromise. When designing the earlier, moderate-speed lenses, vignetting was not added until the intermediate optimization stage. However, for a fast lens such

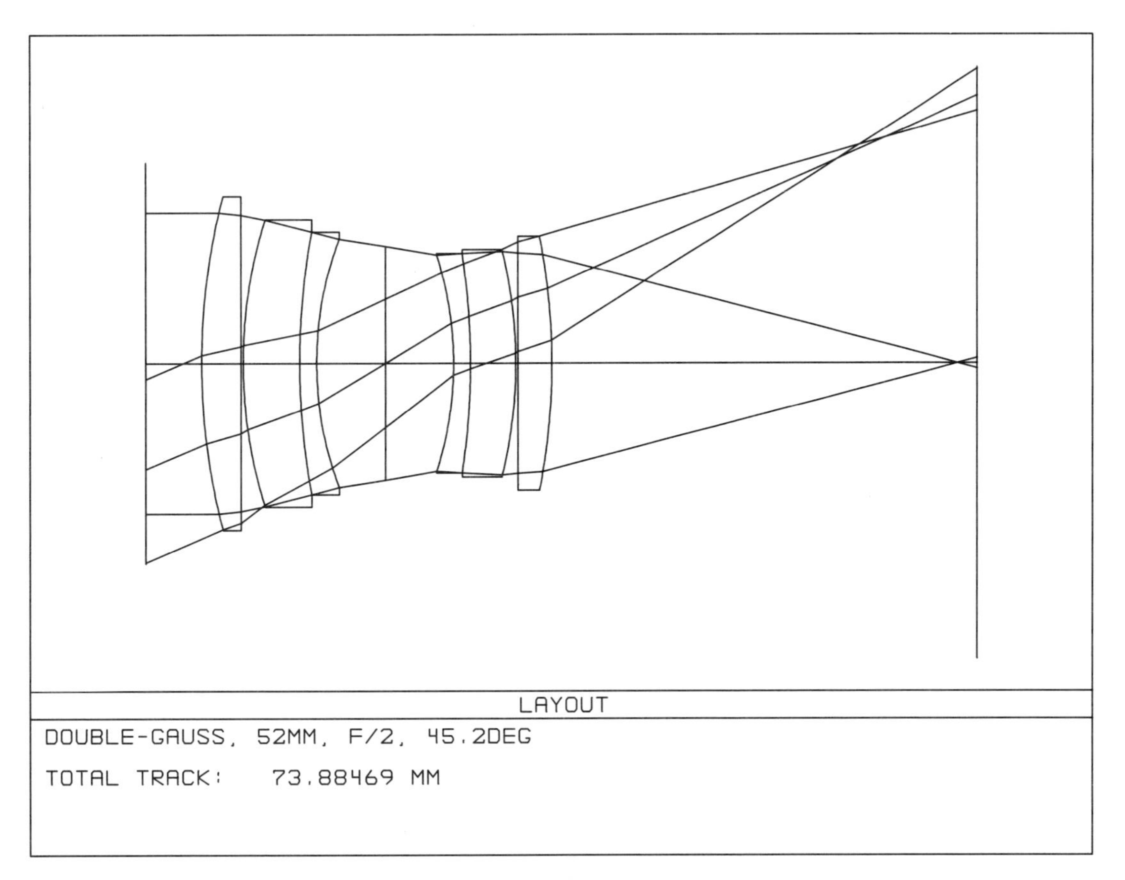

Figure B.4.2.1.

as a Double-Gauss, the geometrical configuration of a lens without vignetting may be quite different from that of a lens with vignetting. Designing an unvignetted lens in the early optimization stage may not do a good job of shepherding the design in the right direction. Thus, it is advisable in this case to include vignetting from the start.

The early lens need not be optimized at both $f/2$ and $f/4$; $f/2$ is sufficient. The exact vignetting amounts are not important at this stage, and you can make guesses. As outlined earlier, use the four specified field angles, set the y-direction compression vignetting factors to 0 .1 .25 .45, set the x-direction compression vignetting factors to 0 0 0 .1, and keep the vignetting factors controlling beam shifts (decenters) all zero. During optimization, use the Gaussian quadrature option.

A Double-Gauss lens is quasi-symmetrical; that is, relative to the stop, the rear half is nearly a mirror image of the front half (refer again to Figure B.4.1.1). To simplify and restrict the early optimizations, make and keep the system (almost) exactly symmetrical. Do this with pickups or linkages. The precise form of the starting design is not very critical. Tinker with the lens until the geometry shown in the layout has the correct basic configuration. You may wish to make some of the air and glass thicknesses slightly different on the two sides. An example of a starting layout is shown in Figure B.4.2.1.

The early merit function contains an operand to correct focal length to 52 mm. Another operand keeps the back focus greater than 38 mm. A third operand corrects paraxial longitudinal color to zero (to shepherd the design in the right direction). A fourth operand keeps the central edge airspaces greater than some small number, such as 0.5 mm. And a default merit function is appended to shrink spot sizes for all fields and wavelengths (the shotgun approach). Distortion is ignored, but symmetry keeps it small. As in earlier chapters, a height solve places a dummy plane at the paraxial focus, and the image surface immediately follows with zero separation.

Because of the pickup solves maintaining symmetry, the only independent degrees of freedom during early optimization are in the front half of the lens. The five curvatures in front of the stop are independent degrees of freedom. And the airspace between the front negative element and the stop surface is an independent degree of freedom. However, all the glass thicknesses are set at reasonable values and frozen, and the airspaces between the positive singlets and cemented menisci are set at minimum values and frozen.

The five free curvatures and one free airspace total only six effective independent variables. This total is less than the number necessary to address all seven major aberrations plus constraints. Thus, after early optimization, optical performance is still quite poor. But this does not matter. You will have accomplished your goal of pushing the design quite far toward the desired configuration. Your verification is whether the layout looks reasonable based on your knowledge of published designs.

B.4.6 Intermediate Optimizations

For intermediate optimizations, remove the pickups on the curvatures. Now all 10 surface curvatures are effective independent variables.

The center glass thicknesses of the positive elements in the two cemented menisci are also important variables because the thick-meniscus principle is used to help flatten the field of a Double-Gauss lens. However, the two glass thicknesses together really constitute only one effective degree of freedom. Thus, make the thickness of the front element a regular variable, and using a pickup make the thickness of the rear element equal to 0.8 times the thickness of the front element. The 0.8 factor gives a good physical balance. This constraint is retained throughout the design and causes no significant reduction in final optical performance.

In a similar way, the two central airspaces on each side of the stop surface also constitute only one effective independent variable. Thus, make the front airspace a regular variable, and control the rear airspace with a pickup from the front airspace. To give the maximum clearance around both sides of the iris diaphragm, the stop can be located roughly equidistant from the edges (not centers) of the surrounding concave glass surfaces by using a pickup factor of 0.9.

The 10 surface curvatures, one glass thickness, and one air thickness thus give a total of 12 effective independent variables for the correction or control of aberrations, focal length, and back focus.

The Double-Gauss lens has several minor variables that are usually set at convenient values and frozen. For the front and rear singlets, center glass thicknesses are chosen that give minimum practical edge thicknesses. For the negative elements in the cemented menisci, minimum practical center thicknesses are chosen. And for the airspaces between the singlets and menisci, minimum values are chosen. If these variables were allowed to be free during optimization, they would bump up against constraints. In particular, all of these variables want to be small, and it is easier to just freeze them at small values. Of course, as the design progresses, check for problems, such as negative edge thicknesses and so forth. If any of the frozen parameters become inappropriate, change them by hand and reoptimize.

Recall that intermediate optimization is a combination monochromatic and polychromatic procedure, and that more than one approach is possible. An approach similar to the one used to optimize the Cooke Triplet has been adopted. With the lens wide open at $f/2$, user-selected or user-defined merit function operands are used to: (1) correct longitudinal color to zero for the 0.8 pupil zone, (2) correct lateral color to zero at the edge of the field, (3) correct spherical aberration to zero on the paraxial focal plane for the 0.9 pupil zone, (4) correct distortion to zero at the edge of the field, (5) correct focal length to 52 mm, (6) control back focus to be greater than 38 mm, and (7) keep the central airspace edge thicknesses greater than some small number, such as 0.5 mm. With the lens at both $f/2$ and $f/4$, off-axis aberrations are controlled by shrinking monochromatic (central wavelength only) spot sizes with a default merit function.

When constructing the default merit function, you must weight the field positions, wavelengths, and "zoom" configurations. Because you are already controlling the on-axis aberrations with special operands, you must avoid double jeopardy and not include the on-axis image in the default merit function; thus, on-axis weight is zero. Also, you need not worry about the 40% field at this stage. Therefore, good field weights might be 0 0 2 1. The wavelength weights can serve double duty; that is, they can also be used to weight the configurations. Thus, for the $f/2$ configuration, use wavelength weights of 0 0 1 0 0, and for the $f/4$ configuration, use wavelength weights of 0 0 8 0 0.

As in the early optimizations, user-specified vignetting factors are used to maintain a constant amount of vignetting at the edge of the field without having to fuss with apertures. For $f/2$, use the same compression vignetting factors of 0 .1 .25 .45 and 0 0 0 .1; the shifts are again all zero. For $f/4$, all vignetting factors are zero (no vignetting). During optimization, again use the Gaussian quadrature option.

The prescription of the lens with two configurations as optimized with the intermediate merit function is given in Listing B.4.2.1. The intermediate merit function, again for two configurations, is given in Listing B.4.2.2.

Listing B.4.2.1

```
System/Prescription Data

File : C:LENS332B.ZMX
Title: DOUBLE-GAUSS, 52MM, F/2, 45.2DEG

GENERAL LENS DATA:

Surfaces          :          14
Stop              :           7
System Aperture   :Entrance Pupil Diameter
Ray aiming        : On
 X Pupil shift    : 0
 Y Pupil shift    : 0
 Z Pupil shift    : 0
Apodization       :Uniform, factor =     0.000000
Eff. Focal Len.   :          52 (in air)
Eff. Focal Len.   :          52 (in image space)
Total Track       :     73.3085
Image Space F/#   :           2
Para. Wrkng F/#   :           2
Working F/#       :     2.00873
Obj. Space N.A.   :    1.3e-009
Stop Radius       :     9.33465
Parax. Ima. Hgt.  :     21.6455
Parax. Mag.       :           0
Entr. Pup. Dia.   :          26
Entr. Pup. Pos.   :     23.0714
Exit Pupil Dia.   :     27.0316
Exit Pupil Pos.   :    -54.0632
Field Type        : Angle in degrees
Maximum Field     :        22.6
Primary Wave      :    0.550000
Lens Units        : Millimeters
Angular Mag.      :    0.961838

Fields            : 4
Field Type:  Angle in degrees
#          X-Value          Y-Value          Weight
1         0.000000         0.000000         0.000000
2         0.000000         9.000000         0.000000
3         0.000000        15.800000         2.000000
4         0.000000        22.600000         1.000000

Vignetting Factors
#          VDX          VDY          VCX          VCY
1       0.000000     0.000000     0.000000     0.000000
2       0.000000     0.000000     0.000000     0.100000
3       0.000000     0.000000     0.000000     0.250000
4       0.000000     0.000000     0.100000     0.450000

Wavelengths       : 5
Units:   Microns
#          Value          Weight
1        0.450000        0.000000
2        0.500000        0.000000
3        0.550000        1.000000
4        0.600000        0.000000
5        0.650000        0.000000
```

SURFACE DATA SUMMARY:

Surf	Type	Radius	Thickness	Glass	Diameter	Conic
OBJ	STANDARD	Infinity	Infinity		0	0
1	STANDARD	Infinity	5		34.47182	0
2	STANDARD	36.43697	3.5	LAFN21	28.12233	0
3	STANDARD	83.63113	0.2		26.98263	0
4	STANDARD	20.14916	4.913594	LAFN21	24.6048	0
5	STANDARD	32.77678	1.5	SF53	21.96137	0
6	STANDARD	14.45029	6.08634		19.23433	0
STO	STANDARD	Infinity	5.477706		18.6693	0
8	STANDARD	-15.2194	1.5	SF53	18.0894	0
9	STANDARD	-47.22007	3.930875	LAFN21	19.74308	0
10	STANDARD	-19.98523	0.2		21.10466	0
11	STANDARD	-296.0465	3	LAFN21	22.20774	0
12	STANDARD	-37.21029	37.99998		23.00104	0
13	STANDARD	Infinity	0		43.37011	0
IMA	STANDARD	Infinity	0		43.37011	0

SURFACE DATA DETAIL:

```
Surface OBJ     : STANDARD
Surface   1     : STANDARD
Surface   2     : STANDARD
Surface   3     : STANDARD
Surface   4     : STANDARD
Surface   5     : STANDARD
Surface   6     : STANDARD
Surface STO     : STANDARD
Surface   8     : STANDARD
Surface   9     : STANDARD
Surface  10     : STANDARD
Surface  11     : STANDARD
Surface  12     : STANDARD
Surface  13     : STANDARD
Surface IMA     : STANDARD
```

MULTI-CONFIGURATION DATA:

```
 Configuration   1:
 Aperture       :           26
 Wave wgt     1 :            0
 Wave wgt     2 :            0
 Wave wgt     3 :            1
 Wave wgt     4 :            0
 Wave wgt     5 :            0
 Field vcx    1 :            0
 Field vcx    2 :            0
 Field vcx    3 :            0
 Field vcx    4 :          0.1
 Field vcy    1 :            0
 Field vcy    2 :          0.1
 Field vcy    3 :         0.25
 Field vcy    4 :         0.45

 Configuration   2:
 Aperture       :           13
 Wave wgt     1 :            0
 Wave wgt     2 :            0
 Wave wgt     3 :            8
 Wave wgt     4 :            0
 Wave wgt     5 :            0
 Field vcx    1 :            0
 Field vcx    2 :            0
 Field vcx    3 :            0
 Field vcx    4 :            0
 Field vcy    1 :            0
 Field vcy    2 :            0
 Field vcy    3 :            0
 Field vcy    4 :            0
```

SOLVE AND VARIABLE DATA:

```
Curvature of   2   : Variable
Curvature of   3   : Variable
Curvature of   4   : Variable
Thickness of   4   : Variable
Curvature of   5   : Variable
Curvature of   6   : Variable
Thickness of   6   : Variable
Thickness of   7   : Solve, pick up value from 6, scaled by 0.90000
Curvature of   8   : Variable
Curvature of   9   : Variable
Thickness of   9   : Solve, pick up value from 4, scaled by 0.80000
Curvature of  10   : Variable
Curvature of  11   : Variable
Curvature of  12   : Variable
Thickness of  12   : Solve, marginal ray height = 0.00000
```

Listing B.4.2.2

Merit Function Listing

File : C:LENS332B.ZMX
Title: DOUBLE-GAUSS, 52MM, F/2, 45.2DEG

Merit Function Value: 2.41615334E-003

Num	Type	Int1	Int2	Hx	Hy	Px	Py	Target	Weight	Value	% Cont
1	CONF	1									
2	EFFL		3					5.20000E+001	10	5.20000E+001	0.000
3	BLNK										
4	BLNK										
5	CTGT	12						3.80000E+001	10	3.80000E+001	0.000
6	BLNK										
7	BLNK										
8	MNEA	6	7					5.00000E-001	10	5.00000E-001	0.000
9	BLNK										
10	BLNK										
11	AXCL							0.00000E+000	0	4.94419E-002	0.000
12	REAY	14	1	0.0000	0.0000	0.0000	0.8000	0.00000E+000	0	-1.35380E-002	0.000
13	REAY	14	5	0.0000	0.0000	0.0000	0.8000	0.00000E+000	0	-1.35307E-002	0.000
14	DIFF	12	13					0.00000E+000	100	-7.32077E-006	0.000
15	BLNK										
16	BLNK										
17	LACL							0.00000E+000	0	1.80346E-003	0.000
18	REAY	14	1	0.0000	1.0000	0.0000	0.0000	0.00000E+000	0	2.16452E+001	0.000
19	REAY	14	5	0.0000	1.0000	0.0000	0.0000	0.00000E+000	0	2.16452E+001	0.000
20	DIFF	18	19					0.00000E+000	100	-1.10323E-006	0.000
21	BLNK										
22	BLNK										
23	SPHA	0	3					0.00000E+000	0	2.16447E+001	0.000
24	REAY	13	3	0.0000	0.0000	0.0000	0.9000	0.00000E+000	100	-8.58960E-005	0.029
25	BLNK										
26	BLNK										
27	DIST	0	3					0.00000E+000	100	-4.94604E-006	0.000
28	BLNK										
29	BLNK										
30	DMFS										
31	CONF	1									
32	TRAR		3	0.0000	0.6991	0.1679	0.2907	0.00000E+000	0.029089	1.43927E-002	0.238
33	TRAR		3	0.0000	0.6991	0.3536	0.6124	0.00000E+000	0.046542	1.28455E-002	0.303
34	TRAR		3	0.0000	0.6991	0.4710	0.8158	0.00000E+000	0.029089	5.87906E-002	3.966
35	TRAR		3	0.0000	0.6991	0.3357	0.0000	0.00000E+000	0.029089	2.01523E-002	0.466
36	TRAR		3	0.0000	0.6991	0.7071	0.0000	0.00000E+000	0.046542	1.53064E-002	0.430
37	TRAR		3	0.0000	0.6991	0.9420	0.0000	0.00000E+000	0.029089	1.40469E-001	22.642
38	TRAR		3	0.0000	0.6991	0.1679	-0.2907	0.00000E+000	0.029089	1.61482E-002	0.299
39	TRAR		3	0.0000	0.6991	0.3536	-0.6124	0.00000E+000	0.046542	1.79143E-002	0.589
40	TRAR		3	0.0000	0.6991	0.4710	-0.8158	0.00000E+000	0.029089	5.74361E-002	3.785
41	TRAR		3	0.0000	1.0000	0.1679	0.2907	0.00000E+000	0.014544	8.40109E-003	0.040
42	TRAR		3	0.0000	1.0000	0.3536	0.6124	0.00000E+000	0.023271	3.44033E-003	0.011
43	TRAR		3	0.0000	1.0000	0.4710	0.8158	0.00000E+000	0.014544	3.94417E-002	0.893
44	TRAR		3	0.0000	1.0000	0.3357	0.0000	0.00000E+000	0.014544	8.50771E-003	0.042
45	TRAR		3	0.0000	1.0000	0.7071	0.0000	0.00000E+000	0.023271	3.92630E-002	1.415
46	TRAR		3	0.0000	1.0000	0.9420	0.0000	0.00000E+000	0.014544	2.08246E-001	24.881
47	TRAR		3	0.0000	1.0000	0.1679	-0.2907	0.00000E+000	0.014544	4.38720E-003	0.011
48	TRAR		3	0.0000	1.0000	0.3536	-0.6124	0.00000E+000	0.023271	1.53412E-002	0.216
49	TRAR		3	0.0000	1.0000	0.4710	-0.8158	0.00000E+000	0.014544	6.80250E-002	2.655
50	CONF	2									
51	TRAR		3	0.0000	0.6991	0.1679	0.2907	0.00000E+000	0.23271	8.98574E-003	0.741
52	TRAR		3	0.0000	0.6991	0.3536	0.6124	0.00000E+000	0.37234	1.67885E-002	4.140
53	TRAR		3	0.0000	0.6991	0.4710	0.8158	0.00000E+000	0.23271	1.70803E-002	2.678
54	TRAR		3	0.0000	0.6991	0.3357	0.0000	0.00000E+000	0.23271	9.57816E-003	0.842
55	TRAR		3	0.0000	0.6991	0.7071	0.0000	0.00000E+000	0.37234	2.12765E-002	6.649
56	TRAR		3	0.0000	0.6991	0.9420	0.0000	0.00000E+000	0.23271	2.76139E-002	7.000
57	TRAR		3	0.0000	0.6991	0.1679	-0.2907	0.00000E+000	0.23271	9.74043E-003	0.871
58	TRAR		3	0.0000	0.6991	0.3536	-0.6124	0.00000E+000	0.37234	2.02859E-002	6.044
59	TRAR		3	0.0000	0.6991	0.4710	-0.8158	0.00000E+000	0.23271	2.31151E-002	4.905
60	TRAR		3	0.0000	1.0000	0.1679	0.2907	0.00000E+000	0.11636	6.52024E-003	0.195
61	TRAR		3	0.0000	1.0000	0.3536	0.6124	0.00000E+000	0.18617	8.71692E-003	0.558
62	TRAR		3	0.0000	1.0000	0.4710	0.8158	0.00000E+000	0.11636	3.30150E-003	0.050
63	TRAR		3	0.0000	1.0000	0.3357	0.0000	0.00000E+000	0.11636	5.97515E-003	0.164
64	TRAR		3	0.0000	1.0000	0.7071	0.0000	0.00000E+000	0.18617	8.32741E-003	0.509
65	TRAR		3	0.0000	1.0000	0.9420	0.0000	0.00000E+000	0.11636	6.59656E-003	0.200
66	TRAR		3	0.0000	1.0000	0.1679	-0.2907	0.00000E+000	0.11636	4.04948E-003	0.075
67	TRAR		3	0.0000	1.0000	0.3536	-0.6124	0.00000E+000	0.18617	1.77251E-003	0.023
68	TRAR		3	0.0000	1.0000	0.4710	-0.8158	0.00000E+000	0.11636	1.77325E-002	1.443

Now the question of the best matching flint glass type must be addressed. Initially, Schott SF53 was selected as the match for Schott LaFN21, but is there a better match? To find out, several flint glasses (same type for both negative elements) were selected from along the old glass line. With each, the Double-Gauss lens was optimized with the intermediate merit function. The solutions were documented with layouts, ray fan plots, and by noting the values of the merit function. The surprising result is that over a wide range, flint glass type scarcely matters. For SF15, SF53, and SF55, there is very little difference. SF5 gives more of a difference, and SF2 gives a poor solution. The merit function variation is U-shaped with a broad minimum near SF53.

After considering various details, such as the appearance of the layout, the exact ray fan plots, glass cost, and glass transmission, it appears that SF53 is indeed a very good match for LaFN21 in the present classical Double-Gauss example. This glass will therefore be retained.

Note the mention of glass transmission. This consideration can be important in the selection of flint glasses. If you consult a Schott glass catalog (or any other glass manufacturer's catalog), you find that the higher-index SF type flints have significant bulk or internal absorption in the near ultraviolet, violet, and even the blue. The effect sets in for SF glasses whose indices are above roughly that of SF53 (n_d of 1.728). These glasses are often called extra dense flints because they contain such a large fraction of lead oxide and have a correspondingly high specific gravity. To the eye, extra dense flints appear pale yellow. When incorporated in sufficient thickness into a camera lens, the resulting color photographs (especially slides) have a warm cast (photographers call yellowish or reddish pictures warm and bluish pictures cool). Thus, to avoid making a warm lens, avoid SF flint glasses above SF53.

Fortunately, two of the most useful very-high-index lanthanum crown glasses, LaFN21 (n_d of 1.788) and LaSFN30 (n_d of 1.803), do not have much blue absorption.

B.4.7 Final Optimizations

As with the Cooke Triplet and Tessar, the final optimization stage for a Double-Gauss lens is polychromatic and is for fine tuning the design.

Previously, user-specified vignetting factors controlled the off-axis beams, but these beams were not quite consistent with real apertures. Now the vignetting factors must be removed, and the beams must be redefined by actual surface openings.

In ZEMAX, surface openings are handled in two separate places. One place is as a semi-diameter (the term semi-diameter is used in place of radius to avoid confusion with radius of curvature). The other place is as a hard aperture. The reason is versatility; the functions and capabilities of the two specifications are different. For example, vignetting factors are based only on semi-diameters, but only hard apertures can block and delete vignetted or centrally obscured rays. All semi-diameters are purely circular, but hard apertures can be circular, annular, rectangular, elliptical, and so forth. In the present example, all hard apertures are also purely circular. To make the two aperture descriptions agree, simply transfer the semi-diameter values to the corresponding circular hard apertures.

The procedure for converting from user-specified vignetting factors to hard apertures follows. First, examine the intermediate solution based on vignetting factors. Note the semi-diameters that the computer calculated as the minimum necessary to pass all the beams. Only the $f/2$ configuration is considered; the smaller $f/4$ beams can use the $f/2$ apertures (only the stop opening is smaller at $f/4$).

Second, select which two lens surfaces will define the vignetting (in this case, the front and rear outside surfaces). Replace the variable semi-diameters on just these two surfaces with fixed user-specified semi-diameters of roughly the computer-calculated values. At this time, also set equivalent circular hard apertures on these same two surfaces.

Third, zero out the vignetting factors and do a vignetting throughput plot. Vignetting plots use only the hard apertures and may not work properly with nonzero vignetting factors. Check to see that throughput is close to the required 50% at the edge of the field. If it is not, adjust the sizes of the two hard apertures (and do not forget to transfer the new aperture values to the semi-diameter column too).

Fourth, have the computer set all of the vignetting factors and the remaining minimum semi-diameters. These values will now be consistent with the two defining semi-diameters. Then replace all of the computer calculated semi-diameters (except on the stop) with similar (or slightly larger) user-specified values. These fixed semi-diameters will be necessary later for correct layouts. For completeness, also set the equivalent hard apertures.

Fifth, zero out the vignetting factors again. From now on, you will use the hard apertures to block and delete vignetted rays, and you will also use only the rectangular array default merit function option.

After setting apertures and semi-diameters, the next step is to enter the field and wavelength weights (and implicitly the configuration weights). Then the merit function is constructed.

In all final optimizations, focal length, back focus, and distortion are corrected or controlled with special operands. In addition, include an operand to keep overall vertex length no more than roughly the value derived in the intermediate optimizations. During the intermediate optimizations, vignetting was fixed and semi-diameters were variable. The lens was free to find its own best vertex length consistent with system requirements.

Now, however, the semi-diameters and apertures are fixed, but nothing specifically controls vignetting as such. Unless you are careful, during final optimizations, you may become the victim of vignetting creep. During these optimizations, the lens may stretch itself out to allow the fixed apertures to vignette additional off-axis rays. Eliminating these rays and their aberrations improves the merit function. But this tactic is no good because the maximum percent vignetting at the edge of the field is one of the basic system specifications. Thus, to keep vignetting within bounds, overall length must now be constrained.

Incidentally, the same problem occurs if you optimize with fixed semi-diameters plus general computer-set vignetting factors and the SVIG operand, which causes the vignetting factors to be updated between each optimization cycle.

When shrinking spots, continue to control the on-axis aberrations (longitudinal color and spherical aberration) with special operands (at $f/2$), and shrink only polychromatic off-axis spots (at both $f/2$ and $f/4$). Weight the fields 0 3 2 1, for example. Weight the wavelengths 1 1 1 1 1 for $f/2$, and 8 8 8 8 8 for $f/4$. When constructing the default merit function, choose the options that minimize RMS spot radius about the chief ray, use a 10x10 rectangular ray grid, delete all operands corresponding to vignetted rays, and use symmetry to reduce redundant computations. For a final pass, use a 20x20 ray grid. Also, do not forget to use the DMFS operand to protect your special operands from being accidently overwritten when the default merit function is appended.

When minimizing OPDs, remove the special on-axis aberration operands and do a natural or realistic optimization. Weight the fields 8 3 2 1, for example. Weight the wavelengths 1 1 1 1 1 for $f/2$, and 16 16 16 16 16 for $f/4$. Construct the default merit function in the same way as above, except choose the options that minimize RMS wavefront errors relative to the chief ray; that is, relative to the reference sphere in the exit pupil centered on the piercing point of the chief ray on the image surface.

B.4.8 Final Results

When the Double-Gauss lens is optimized first for minimum spot sizes and then for minimum OPDs, the OPD solution is found to clearly give the better set of MTF curves. The complete performance comparison is very similar to that given for the Cooke Triplet. Therefore, the results from optimizing the Double-Gauss lens using spots are omitted here, and only the results from optimizing using OPDs are given.

Figure B.4.3.1 is the layout of the final optimized classical Double-Gauss lens at $f/2$. This layout is the same as Figure B.4.1.1 except that all four of the beams are drawn. Note three things. First, the vignetting of the off-axis beams is realistically defined by the user-specified front and rear semi-diameters (to make this layout, vignetting factors were set by the computer and then removed afterwards). Second,

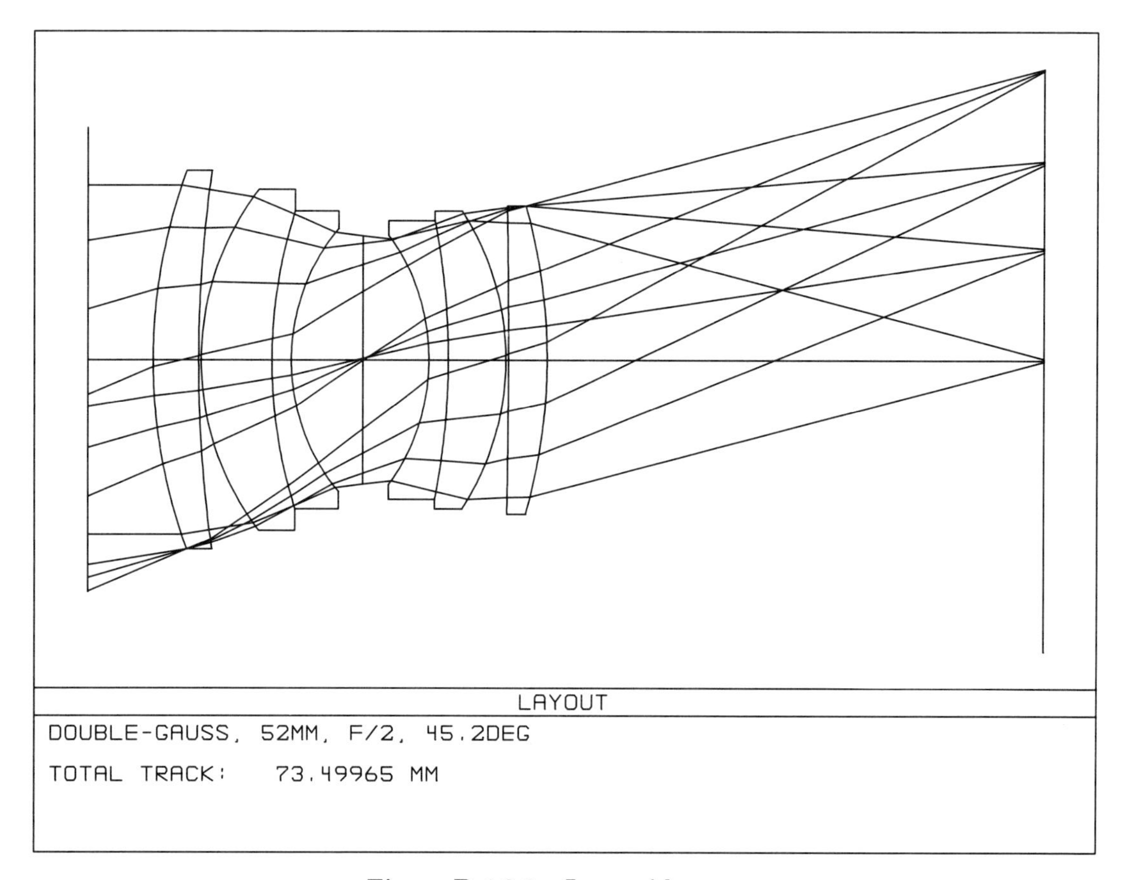

Figure B.4.3.1. Lens wide open.

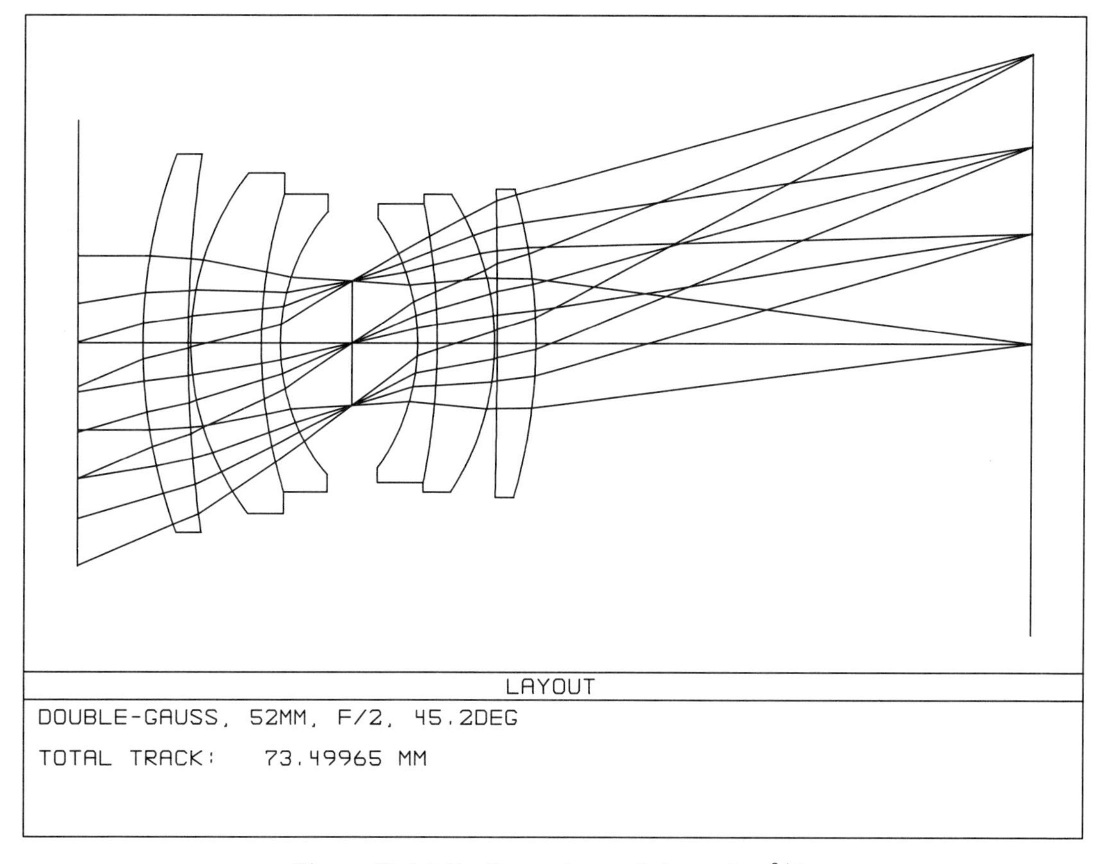

Figure B.4.3.2. Lens stopped down to $f/4$.

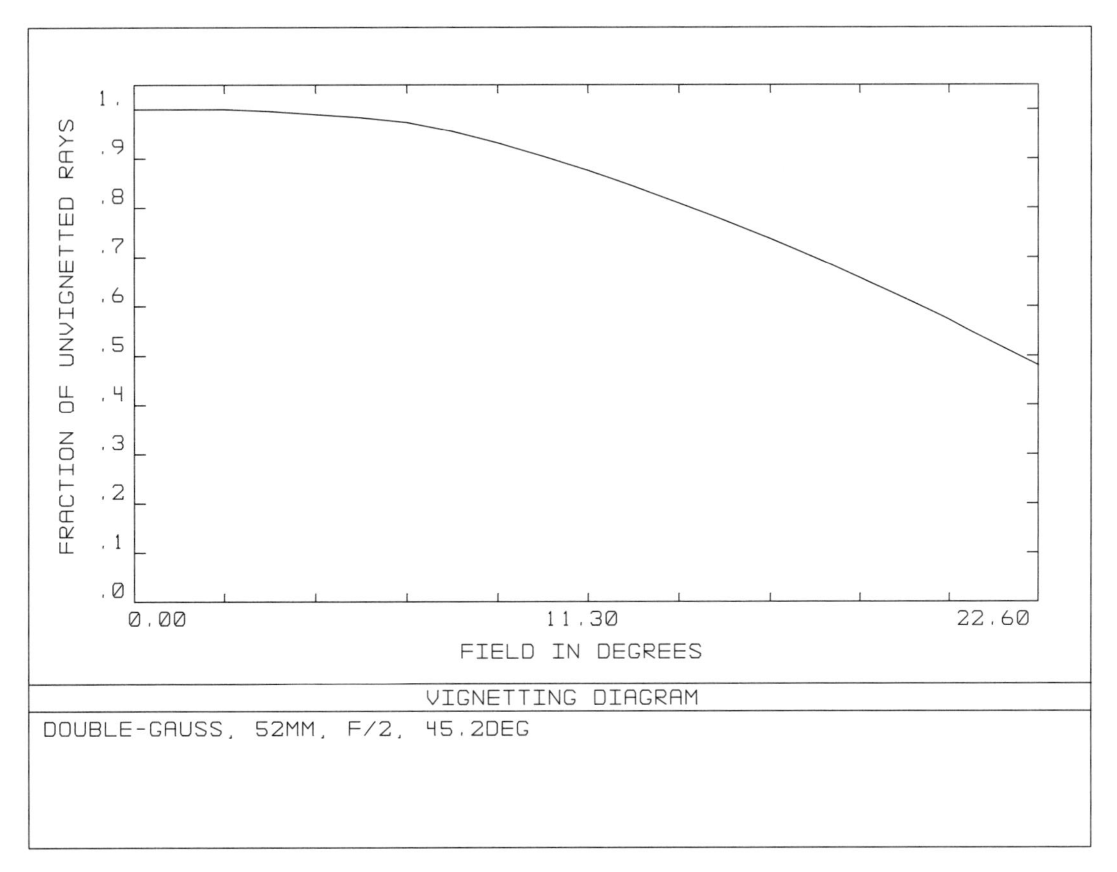

Figure B.4.3.3. Lens wide open.

the other semi-diameters, which are also user-specified, are just large enough to pass all of the beams. And third, the stop surface is about midway between the edges of the surrounding surfaces.

Figure B.4.3.2 is the layout of the same lens when stopped down to $f/4$. Note that the surface diameters (except on the stop surface) are the same as at $f/2$, and now there is no vignetting.

Figure B.4.3.3 is the $f/2$ mechanical vignetting plot of relative throughput. Note that vignetting begins a little before a third of the way out from the field center, and that throughput rolls off smoothly to about 50% at the edge of the field. Not included in this analysis is the additional effect of cosine-fourth vignetting.

Figure B.4.3.4 is an unusual type of spot diagram showing the transmitted beams at the stop. To create these spots, the rays that successfully make it all the way through the lens are reverse-traced back to the stop surface where the "image" plane is temporarily relocated. By this procedure, for each object point, the area within the stop opening where unvignetted rays pass can be mapped. At the field edge, note that the beam has a lenticular shape, which is not a perfect match to an ellipse.

Figure B.4.3.5 is the $f/2$ transverse ray fan plot (the image surface has now been repositioned to its normal location at the paraxial focus). Note that color errors are small and that the biggest source of image degradation is off-axis sagittal oblique spherical aberration (much of the tangential oblique spherical has been vignetted away).

Figure B.4.3.6 is the $f/4$ transverse ray fan plot. Note that relative to the $f/2$ plot, the scale has been enlarged by a factor of four. Clearly, this lens is much better when stopped down.

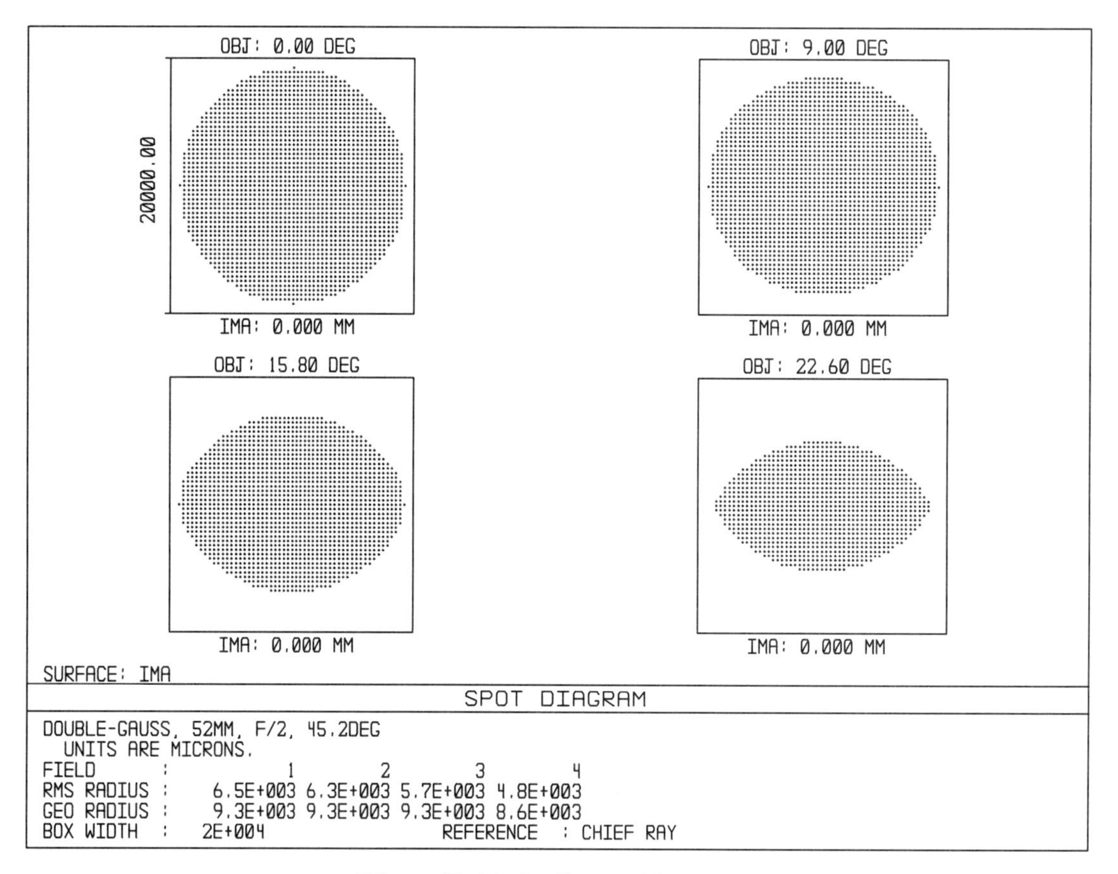

Figure B.4.3.4. Lens wide open.

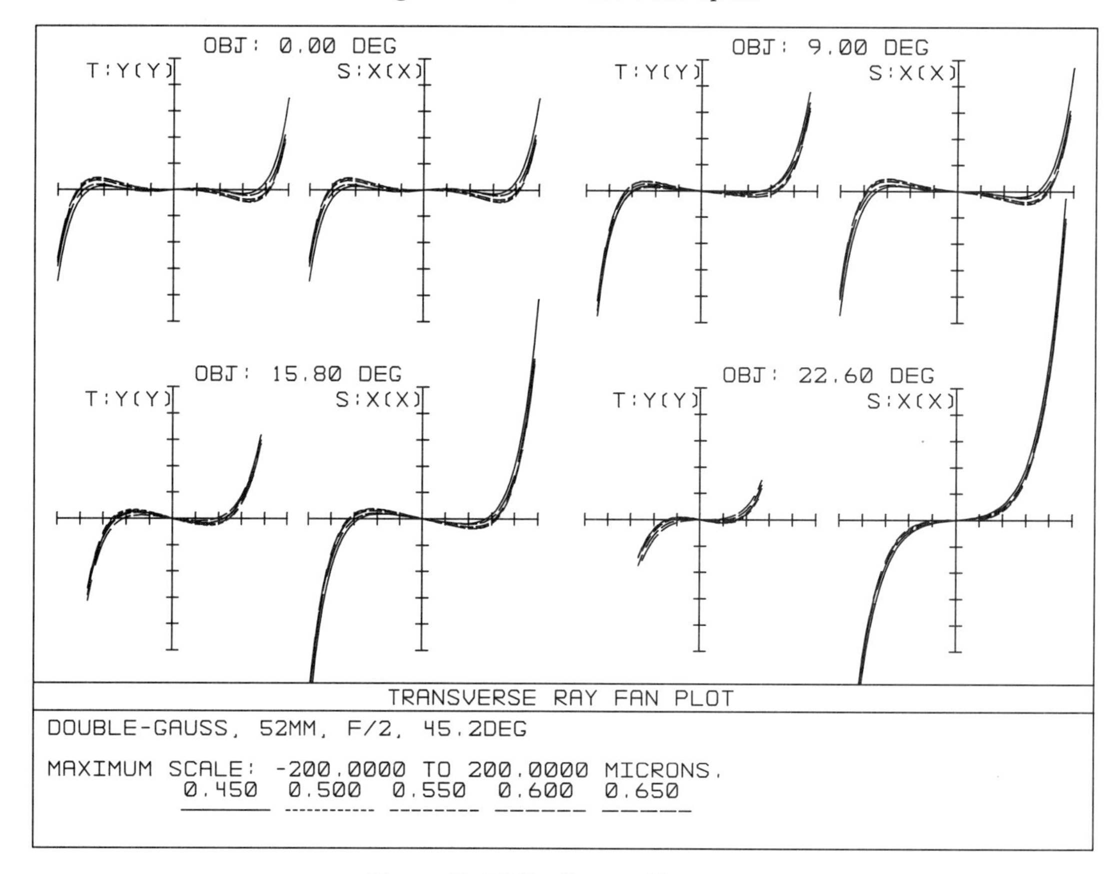

Figure B.4.3.5. Lens wide open.

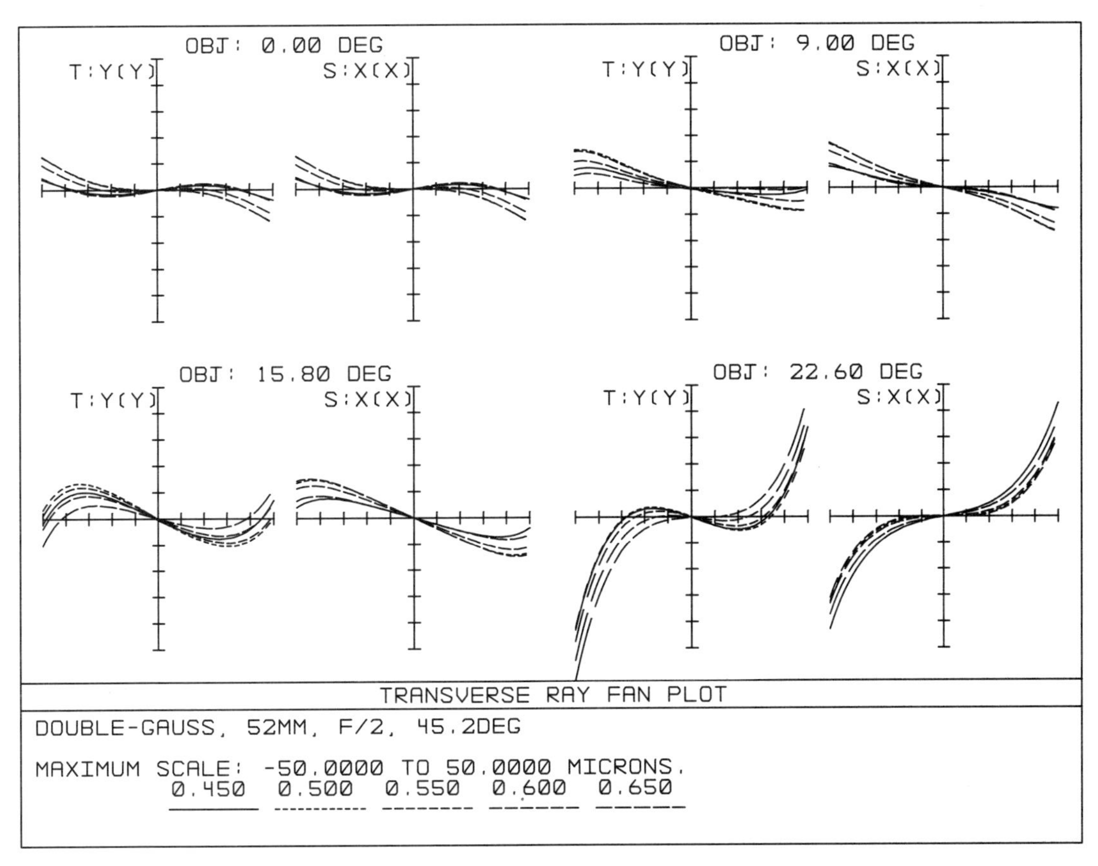

Figure B.4.3.6. Lens stopped down to $f/4$.

The left side of Figure B.4.3.7 shows the field curvature plot for rays very near the chief ray. Note that the tangential and sagittal curves are close to each other, indicating that astigmatism is small. The low astigmatism is also visible on Figures B.4.3.5 and B.4.3.6, where the slopes of the tangential and sagittal curves at the plot origins are similar. In addition, the left side of Figure B.4.3.7 reveals some field-zonal field curvature.

The right side of Figure B.4.3.7 is the percent distortion plot. Distortion is corrected to zero at the edge of the field. At intermediate field angles, tiny distortion residuals are present of little or no consequence.

Figure B.4.3.8 is the $f/2$ polychromatic spot diagram. Note that each of the images shows a tight or hard core surrounded by extended wings or flare. It is these cores that raise the MTFs. Note too that the image at the field edge appears either squashed in the y-direction or stretched in the x-direction. At first, you might think this is tangential astigmatism. But actually it is the oblique spherical as selectively vignetted by the elongated off-axis pupil. Recall that transverse oblique spherical varies as the third power of pupil width. The pupil, being wider in the x-direction, lets through much more sagittal oblique spherical than tangential oblique spherical. This effect is also visible in Figure B.4.3.5.

Figure B.4.3.9 is the $f/4$ polychromatic spot diagram. As with the two ray fan plots, the scale of the $f/4$ spot diagram has been enlarged by a factor of four relative to the $f/2$ plot. After stopping down, it is mainly the image cores that are left.

Figure B.4.3.10 is the $f/2$ OPD ray fan plot of wavefront errors in the exit pupil. Figure B.4.3.11 is the $f/4$ OPD plot (again with the scale enlarged by a factor of four). These are the errors that the OPD optimization tried to minimize, and thus these are the curves that should look good. Both on- and off-axis, the color errors appear a bit more serious here than in the transverse ray fan plots. For the 40% and 70% fields at $f/2$, note the balance between field curvature (off-axis defocus)

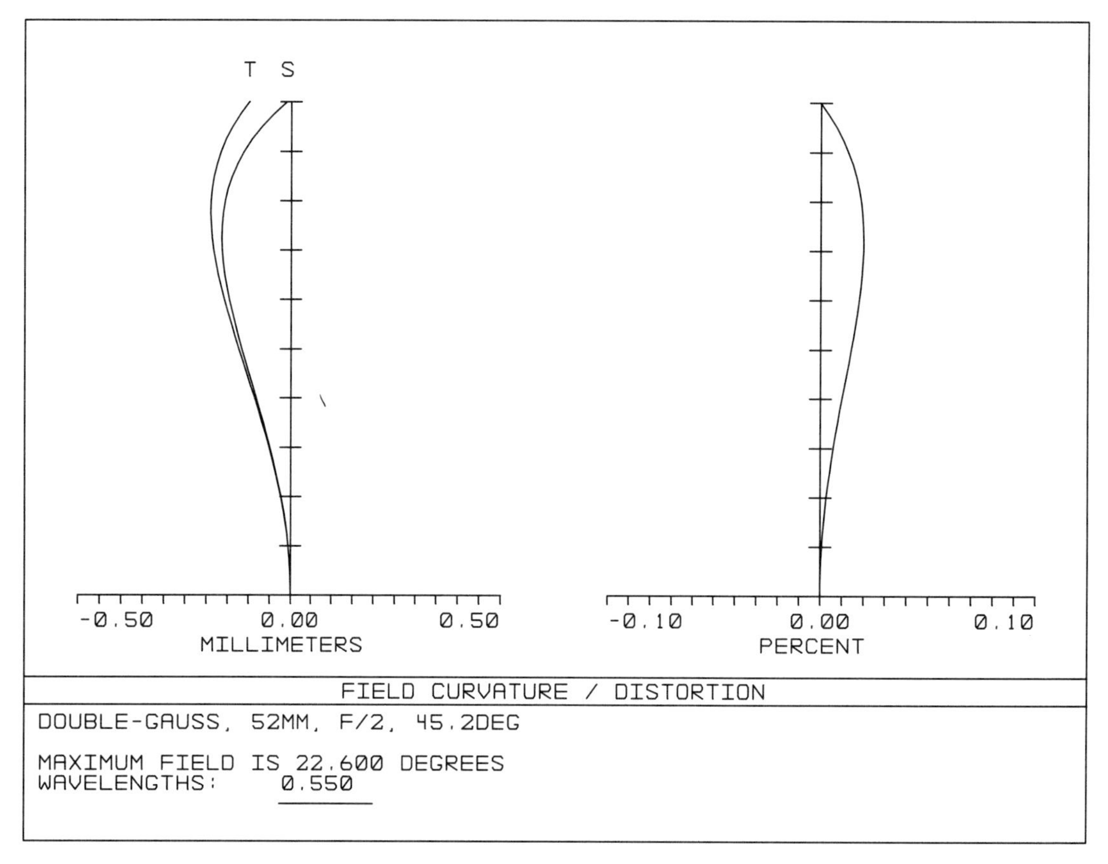

Figure B.4.3.7. Lens wide open.

and oblique spherical aberration. This balance is similar to the balance between paraxial defocus and ordinary spherical aberration that was discussed in Chapter A.7 and shown in Figures A.7.2.1 through A.7.2.3. At the edge of the field, however, oblique spherical is unbalanced. Note too the sizes of the wavefront errors; nearly all are much more than a quarter wave. Clearly, this lens is not diffraction limited, even at $f/4$. As with the Cooke Triplet, this level of performance is normal for a standard camera lens.

Figures B.4.3.12, B.4.3.13, and B.4.3.14 show the acid test, the polychromatic diffraction MTF curves for the lens at $f/2$, $f/4$, and $f/8$, respectively. First, note that none of the $f/2$ curves but many of the $f/8$ curves approach the diffraction-limited curve. Second, for each field position, the small astigmatism causes most of the tangential and sagittal MTF curves to be close to each other. Third, in nearly all cases, the MTF curves for the 40% and 70% fields are higher than the curves for the 100% field (this is as you would wish). And fourth, the $f/4$ and $f/8$ MTF curves compare favorably with the similar-speed curves in Figures B.3.2.7, B.3.2.8, B.3.3.7, and B.3.3.8 for the Cooke Triplet and Tessar. In fact, many of the Double-Gauss curves are better. This performance is both a function of the basic capability of the Double-Gauss form and of the multiple configuration optimization method (it would be interesting to reoptimize the Cooke Triplet and Tessar lenses with multiple configurations at $f/3.5$ and $f/8$).

Listing B.4.3.1 is the system prescription for the final classical Double-Gauss lens. Listing B.4.3.2 is the final merit function for an OPD optimization. To abbreviate the merit function to make it practical for printing, the second and fourth wavelengths have been weighted zero and only a 4x4 rectangular ray grid is shown. In practice, all wavelengths are weighted nonzero, and either 10x10 or 20x20 grids are used.

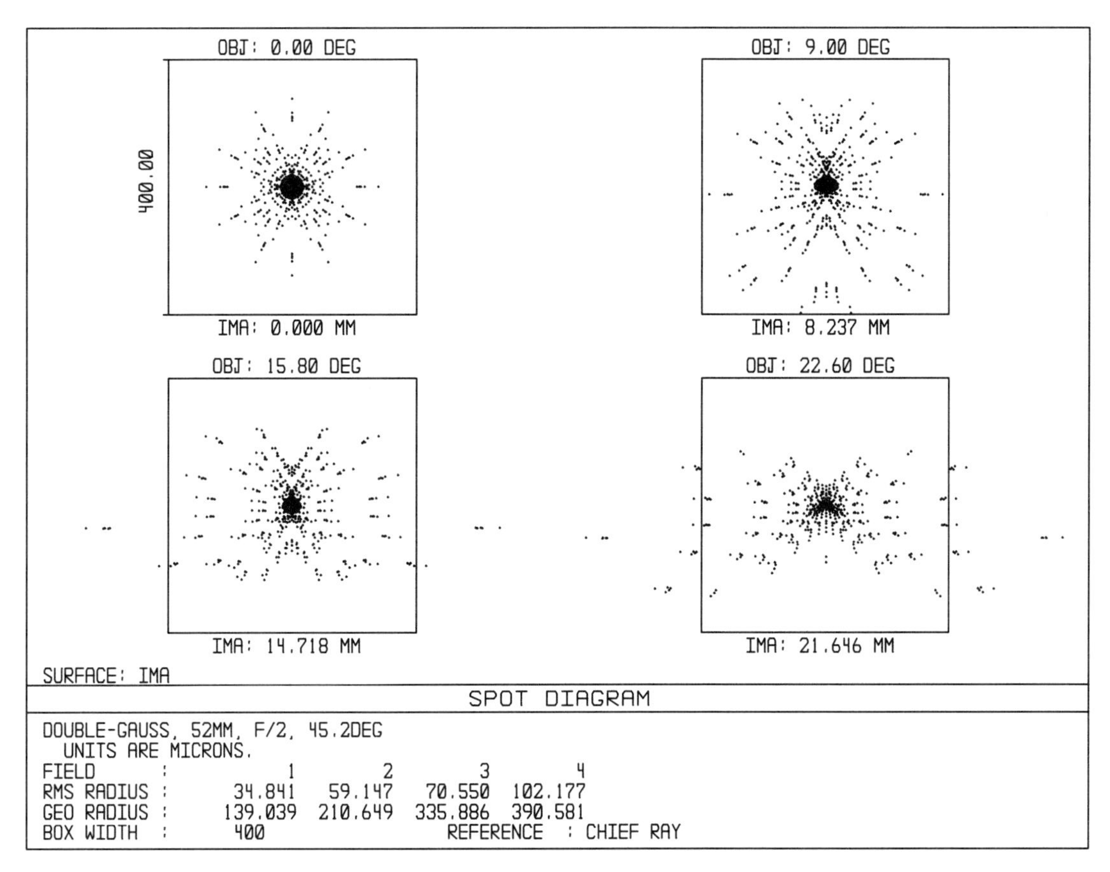

Figure B.4.3.8. Lens wide open.

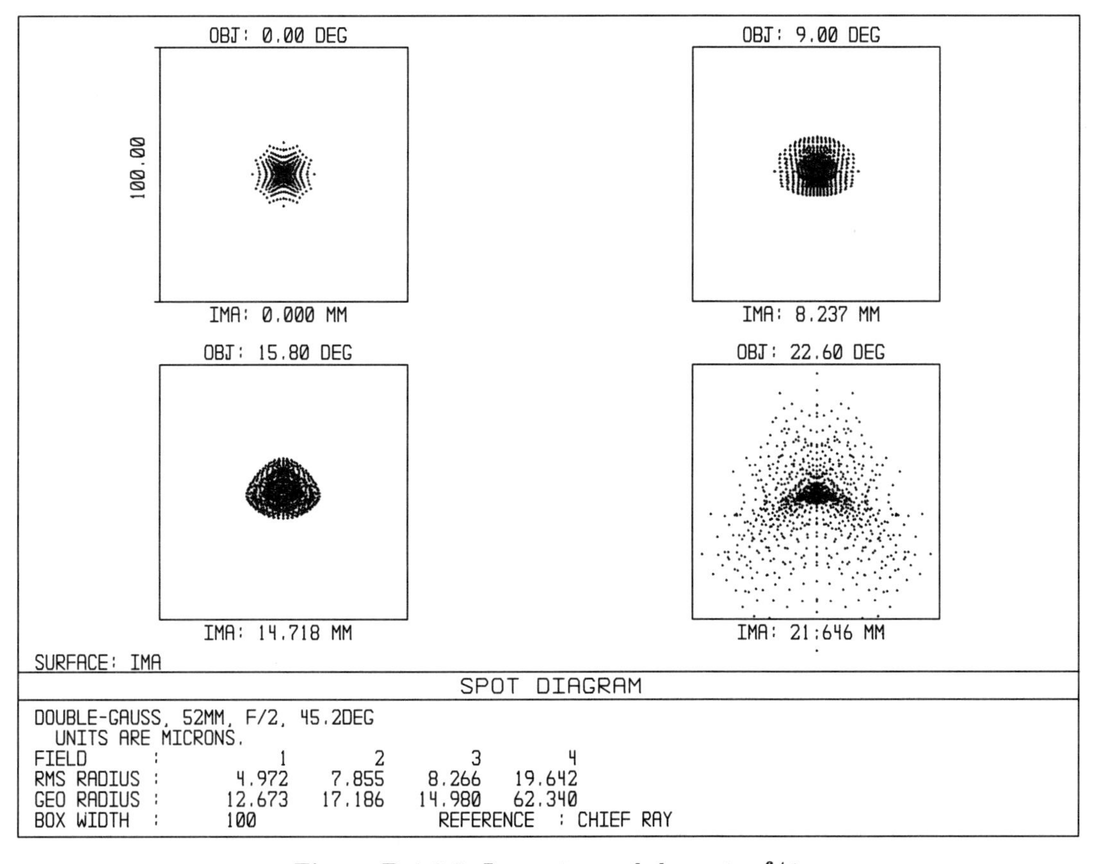

Figure B.4.3.9. Lens stopped down to $f/4$.

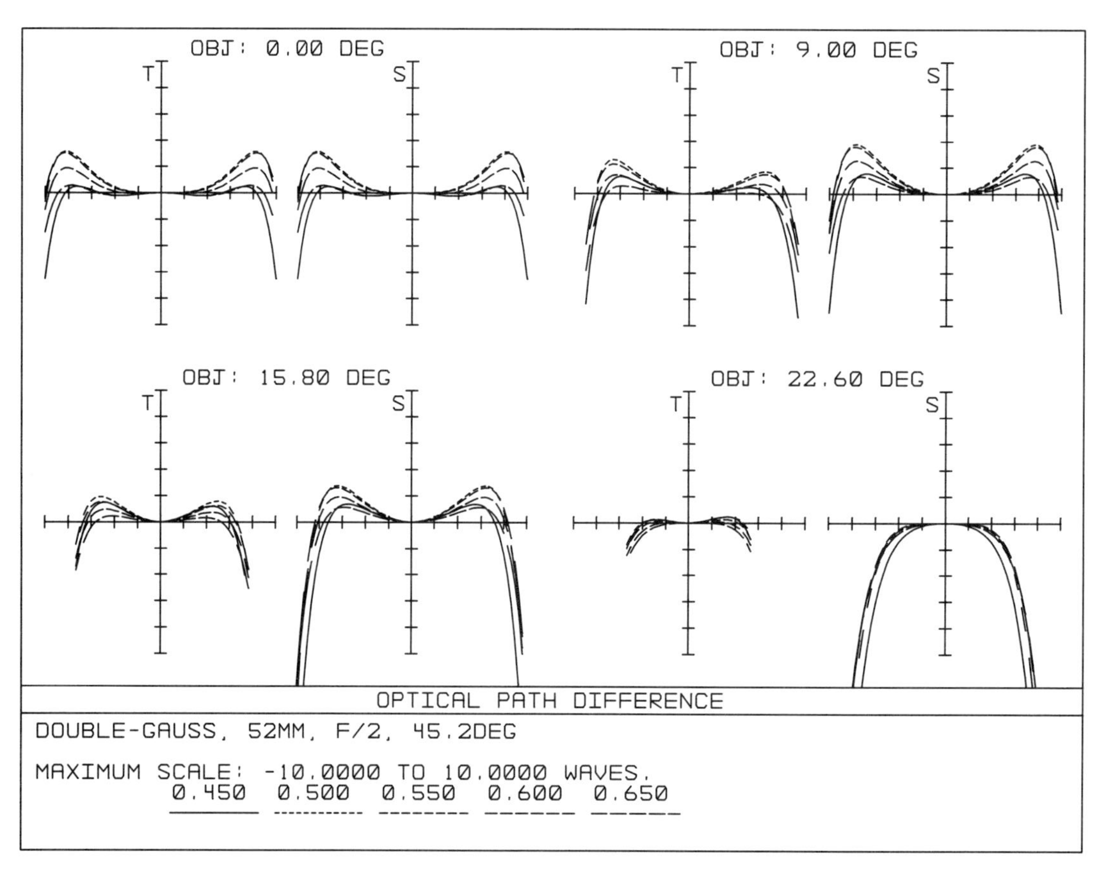

Figure B.4.3.10. Lens wide open.

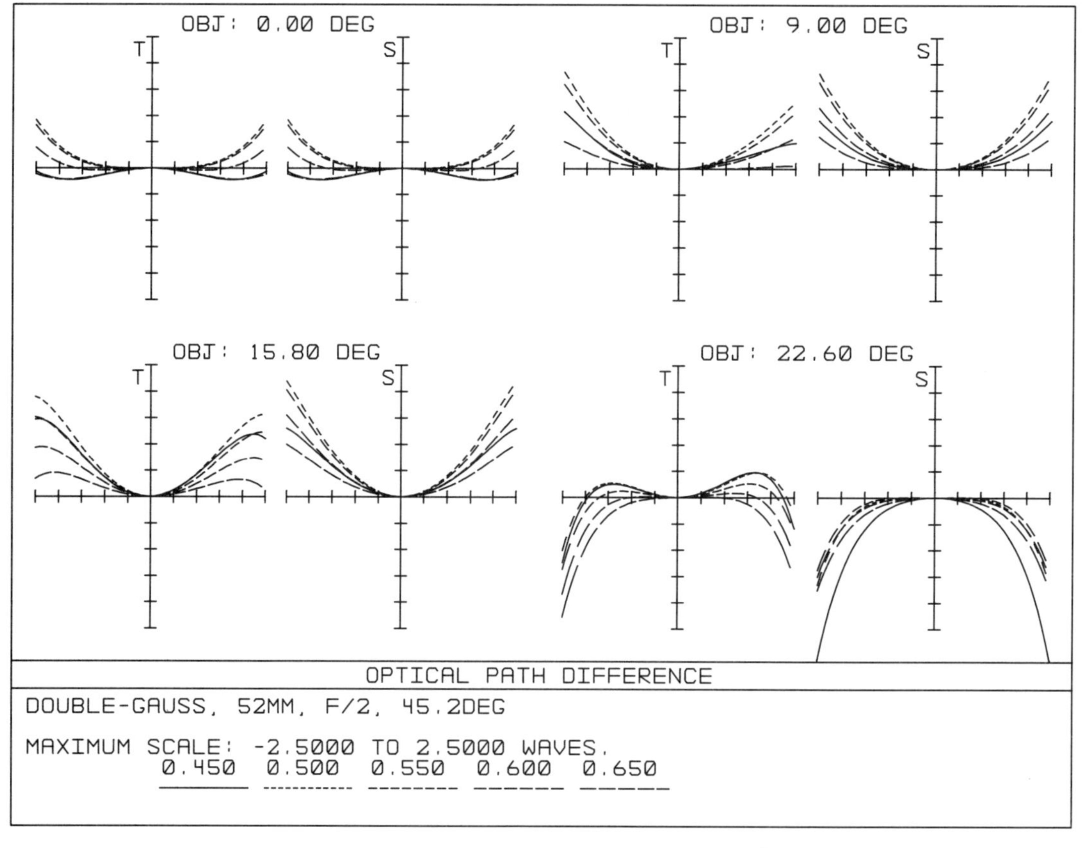

Figure B.4.3.11. Lens stopped down to $f/4$.

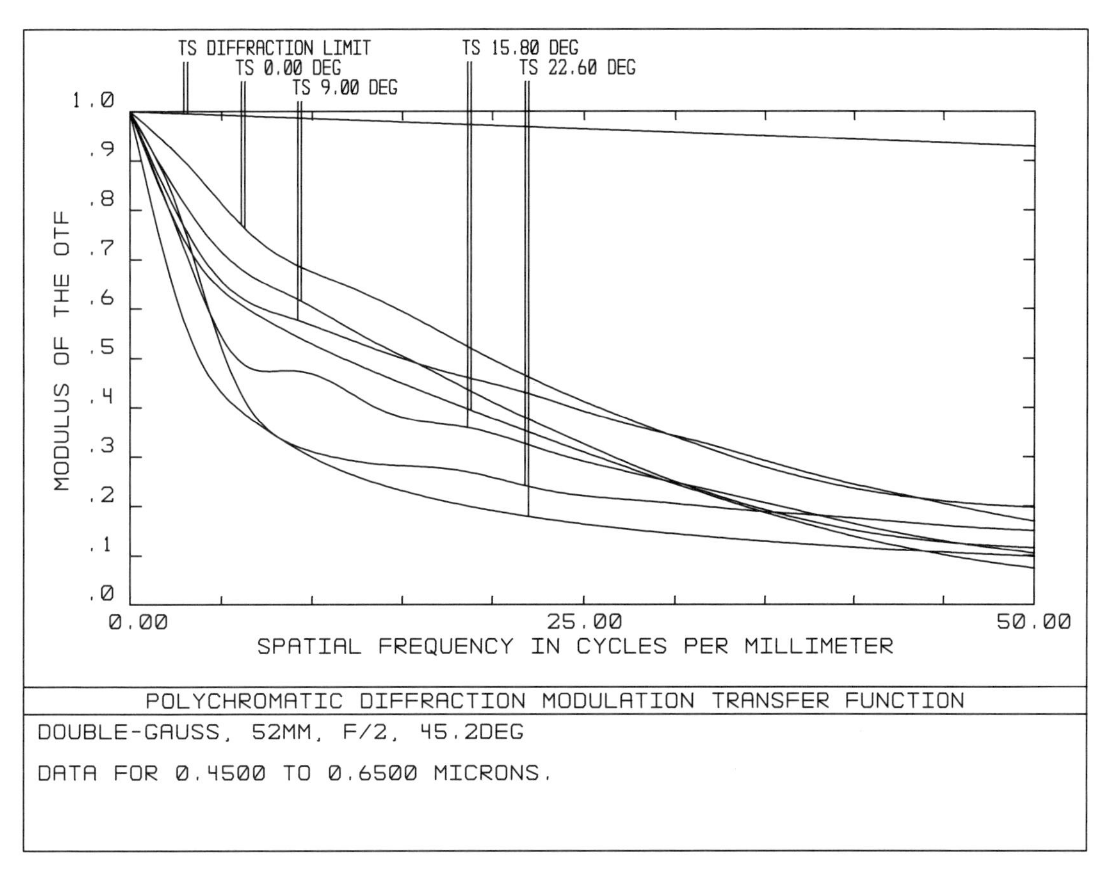

Figure B.4.3.12. Lens wide open.

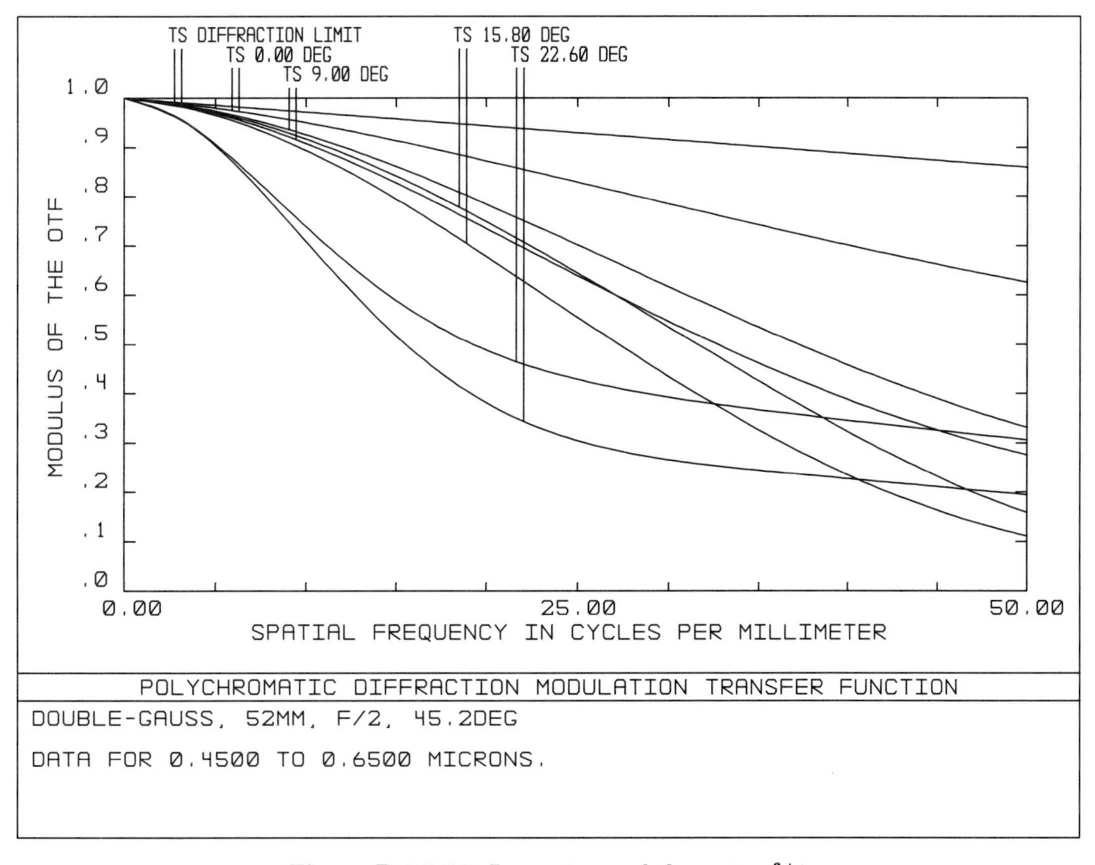

Figure B.4.3.13. Lens stopped down to $f/4$.

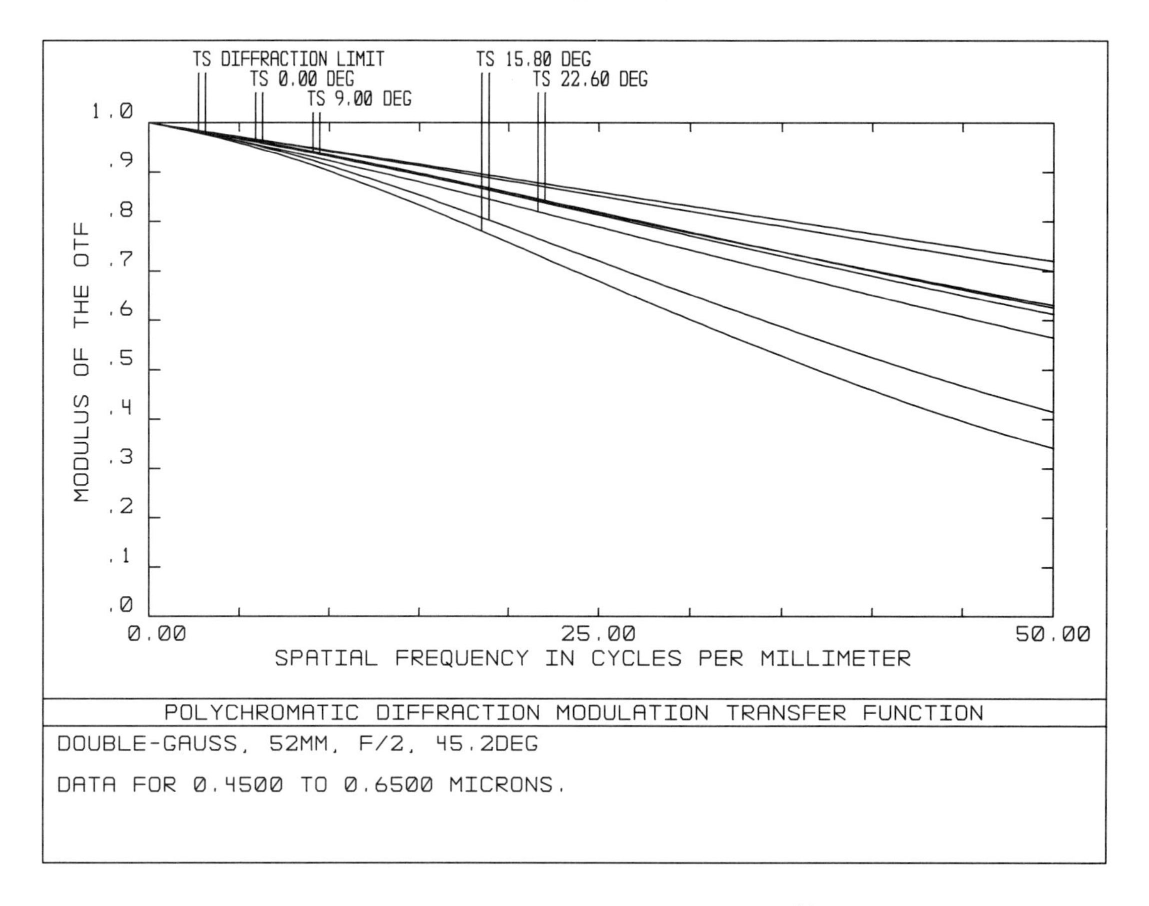

Figure B.4.3.14. Lens stopped down to $f/8$.

Listing B.4.3.1

```
System/Prescription Data

File : C:LENS334B.ZMX
Title: DOUBLE-GAUSS, 52MM, F/2, 45.2DEG

GENERAL LENS DATA:

Surfaces          :          14
Stop              :           7
System Aperture   :Entrance Pupil Diameter
Ray aiming        : On
 X Pupil shift    : 0
 Y Pupil shift    : 0
 Z Pupil shift    : 0
Apodization       :Uniform, factor =     0.000000
Eff. Focal Len.   :     52.0009 (in air)
Eff. Focal Len.   :     52.0009 (in image space)
Total Track       :     73.4997
Image Space F/#   :     2.00003
Para. Wrkng F/#   :     2.00003
Working F/#       :     2.01071
Obj. Space N.A.   :    1.3e-009
Stop Radius       :     9.26263
Parax. Ima. Hgt.  :     21.6459
Parax. Mag.       :           0
Entr. Pup. Dia.   :          26
Entr. Pup. Pos.   :     22.9614
Exit Pupil Dia.   :     26.9341
Exit Pupil Pos.   :    -53.8691
Field Type        : Angle in degrees
Maximum Field     :        22.6
Primary Wave      :    0.550000
Lens Units        : Millimeters
Angular Mag.      :    0.965319

Fields            : 4
Field Type:  Angle in degrees
#         X-Value           Y-Value            Weight
1        0.000000          0.000000          8.000000
2        0.000000          9.000000          3.000000
3        0.000000         15.800000          2.000000
4        0.000000         22.600000          1.000000
```

Vignetting Factors

#	VDX	VDY	VCX	VCY
1	0.000000	0.000000	0.000000	0.000000
2	0.000000	0.000000	0.000000	0.000000
3	0.000000	0.000000	0.000000	0.000000
4	0.000000	0.000000	0.000000	0.000000

Wavelengths : 5
Units: Microns

#	Value	Weight
1	0.450000	1.000000
2	0.500000	1.000000
3	0.550000	1.000000
4	0.600000	1.000000
5	0.650000	1.000000

SURFACE DATA SUMMARY:

Surf	Type	Radius	Thickness	Glass	Diameter	Conic
OBJ	STANDARD	Infinity	Infinity		0	0
1	STANDARD	Infinity	5		45.34406	0
2	STANDARD	39.71916	3.5	LAFN21	28.2	0
3	STANDARD	97.22563	0.2		28.2	0
4	STANDARD	20.33088	5.466869	LAFN21	25.4	0
5	STANDARD	35.06101	1.5	SF53	22.2	0
6	STANDARD	14.55009	5.66305		19.6	0
STO	STANDARD	Infinity	5.096745		18.52526	0
8	STANDARD	-15.27681	1.5	SF53	18.6	0
9	STANDARD	-52.65105	4.373495	LAFN21	20.8	0
10	STANDARD	-20.14363	0.2		22.2	0
11	STANDARD	-603.5136	3	LAFN21	23	0
12	STANDARD	-40.51005	37.99949		23	0
13	STANDARD	Infinity	0		45.65839	0
IMA	STANDARD	Infinity	0		45.65839	0

SURFACE DATA DETAIL:

```
Surface OBJ      : STANDARD
Surface   1      : STANDARD
Surface   2      : STANDARD
 Aperture        : Circular Aperture
 Minimum Radius  :           0
 Maximum Radius  :        14.1
Surface   3      : STANDARD
 Aperture        : Circular Aperture
 Minimum Radius  :           0
 Maximum Radius  :        14.1
Surface   4      : STANDARD
 Aperture        : Circular Aperture
 Minimum Radius  :           0
 Maximum Radius  :        12.7
Surface   5      : STANDARD
 Aperture        : Circular Aperture
 Minimum Radius  :           0
 Maximum Radius  :        11.1
Surface   6      : STANDARD
 Aperture        : Circular Aperture
 Minimum Radius  :           0
 Maximum Radius  :         9.8
Surface STO      : STANDARD
Surface   8      : STANDARD
 Aperture        : Circular Aperture
 Minimum Radius  :           0
 Maximum Radius  :         9.3
Surface   9      : STANDARD
 Aperture        : Circular Aperture
 Minimum Radius  :           0
 Maximum Radius  :        10.4
Surface  10      : STANDARD
 Aperture        : Circular Aperture
 Minimum Radius  :           0
 Maximum Radius  :        11.1
Surface  11      : STANDARD
 Aperture        : Circular Aperture
 Minimum Radius  :           0
 Maximum Radius  :        11.5
Surface  12      : STANDARD
 Aperture        : Circular Aperture
 Minimum Radius  :           0
 Maximum Radius  :        11.5
Surface  13      : STANDARD
Surface IMA      : STANDARD
```

MULTI-CONFIGURATION DATA:

```
 Configuration   1:
 Aperture       :            26
 Wave wgt    1 :              1
 Wave wgt    2 :              1
 Wave wgt    3 :              1
 Wave wgt    4 :              1
 Wave wgt    5 :              1

 Configuration   2:
 Aperture       :            13
 Wave wgt    1 :             16
 Wave wgt    2 :             16
 Wave wgt    3 :             16
 Wave wgt    4 :             16
 Wave wgt    5 :             16
```

```
SOLVE AND VARIABLE DATA:

Curvature of   2    : Variable
Semi Diam   2       : Fixed
Curvature of   3    : Variable
Semi Diam   3       : Fixed
Curvature of   4    : Variable
Thickness of   4    : Variable
Semi Diam   4       : Fixed
Curvature of   5    : Variable
Semi Diam   5       : Fixed
Curvature of   6    : Variable
Thickness of   6    : Variable
Semi Diam   6       : Fixed
Thickness of   7    : Solve, pick up value from 6, scaled by 0.90000
Curvature of   8    : Variable
Semi Diam   8       : Fixed
Curvature of   9    : Variable
Thickness of   9    : Solve, pick up value from 4, scaled by 0.80000
Semi Diam   9       : Fixed
Curvature of  10    : Variable
Semi Diam  10       : Fixed
Curvature of  11    : Variable
Semi Diam  11       : Fixed
Curvature of  12    : Variable
Thickness of  12    : Solve, marginal ray height = 0.00000
Semi Diam  12       : Fixed

INDEX OF REFRACTION DATA:

Surf     Glass      0.450000    0.500000    0.550000    0.600000    0.650000
  0               1.00000000  1.00000000  1.00000000  1.00000000  1.00000000
  1               1.00000000  1.00000000  1.00000000  1.00000000  1.00000000
  2     LAFN21    1.80620521  1.79788762  1.79184804  1.78728036  1.78370773
  3               1.00000000  1.00000000  1.00000000  1.00000000  1.00000000
  4     LAFN21    1.80620521  1.79788762  1.79184804  1.78728036  1.78370773
  5       SF53    1.75665336  1.74309602  1.73366378  1.72676633  1.72152789
  6               1.00000000  1.00000000  1.00000000  1.00000000  1.00000000
  7               1.00000000  1.00000000  1.00000000  1.00000000  1.00000000
  8       SF53    1.75665336  1.74309602  1.73366378  1.72676633  1.72152789
  9     LAFN21    1.80620521  1.79788762  1.79184804  1.78728036  1.78370773
 10               1.00000000  1.00000000  1.00000000  1.00000000  1.00000000
 11     LAFN21    1.80620521  1.79788762  1.79184804  1.78728036  1.78370773
 12               1.00000000  1.00000000  1.00000000  1.00000000  1.00000000
 13               1.00000000  1.00000000  1.00000000  1.00000000  1.00000000
 14               1.00000000  1.00000000  1.00000000  1.00000000  1.00000000

ELEMENT VOLUME DATA:

Units are cubic cm.
Values are only accurate for plane and spherical surfaces.
Element surf   2 to   3 volume :       1.707637
Element surf   4 to   5 volume :       2.075421
Element surf   5 to   6 volume :       0.948876
Element surf   8 to   9 volume :       0.834728
Element surf   9 to  10 volume :       1.241797
Element surf  11 to  12 volume :       0.925399
```

Listing B.4.3.2, Part 1

Merit Function Listing

File : C:LENS334B.ZMX
Title: DOUBLE-GAUSS, 52MM, F/2, 45.2DEG

Merit Function Value: 4.75699955E-003

Num	Type	Int1	Int2	Hx	Hy	Px	Py	Target	Weight	Value	% Cont
1	CONF	1									
2	EFFL		3					5.20000E+001	1e+006	5.20009E+001	0.262
3	BLNK										
4	BLNK										
5	TTHI	2	11					0.00000E+000	0	3.05002E+001	0.000
6	OPLT	5						3.05000E+001	1e+006	3.05002E+001	0.009
7	BLNK										
8	BLNK										
9	CTGT	12						3.80000E+001	1e+006	3.79995E+001	0.088
10	BLNK										
11	BLNK										
12	DIST	0	3					0.00000E+000	1e+007	-1.05070E-004	0.038
13	BLNK										
14	BLNK										
15	DMFS										
16	CONF	1									
17	OPDC		1	0.0000	0.0000	0.2500	0.2500	0.00000E+000	1.6	-2.16243E-001	0.025
18	OPDC		1	0.0000	0.0000	0.2500	0.7500	0.00000E+000	1.6	3.30992E-001	0.060
19	OPDC		1	0.0000	0.0000	0.7500	0.2500	0.00000E+000	1.6	3.30992E-001	0.060
20	OPDC		3	0.0000	0.0000	0.2500	0.2500	0.00000E+000	1.6	2.29681E-001	0.029
21	OPDC		3	0.0000	0.0000	0.2500	0.7500	0.00000E+000	1.6	2.93791E+000	4.694
22	OPDC		3	0.0000	0.0000	0.7500	0.2500	0.00000E+000	1.6	2.93791E+000	4.694
23	OPDC		5	0.0000	0.0000	0.2500	0.2500	0.00000E+000	1.6	-2.09763E-001	0.024
24	OPDC		5	0.0000	0.0000	0.2500	0.7500	0.00000E+000	1.6	5.73014E-001	0.179
25	OPDC		5	0.0000	0.0000	0.7500	0.2500	0.00000E+000	1.6	5.73014E-001	0.179
26	OPDC		1	0.0000	0.3982	0.2500	-0.7500	0.00000E+000	0.3	-1.12814E+000	0.130
27	OPDC		1	0.0000	0.3982	0.2500	-0.2500	0.00000E+000	0.3	4.49101E-001	0.021
28	OPDC		1	0.0000	0.3982	0.2500	0.2500	0.00000E+000	0.3	3.43217E-001	0.012
29	OPDC		1	0.0000	0.3982	0.2500	0.7500	0.00000E+000	0.3	-1.67187E+000	0.285
30	OPDC		1	0.0000	0.3982	0.7500	-0.2500	0.00000E+000	0.3	7.04797E-001	0.051
31	OPDC		1	0.0000	0.3982	0.7500	0.2500	0.00000E+000	0.3	7.85770E-001	0.063
32	OPDC		3	0.0000	0.3982	0.2500	-0.7500	0.00000E+000	0.3	1.23848E+000	0.156
33	OPDC		3	0.0000	0.3982	0.2500	-0.2500	0.00000E+000	0.3	7.64559E-001	0.060
34	OPDC		3	0.0000	0.3982	0.2500	0.2500	0.00000E+000	0.3	6.19319E-001	0.039
35	OPDC		3	0.0000	0.3982	0.2500	0.7500	0.00000E+000	0.3	1.02305E+000	0.107
36	OPDC		3	0.0000	0.3982	0.7500	-0.2500	0.00000E+000	0.3	3.23001E+000	1.064
37	OPDC		3	0.0000	0.3982	0.7500	0.2500	0.00000E+000	0.3	3.36867E+000	1.157
38	OPDC		5	0.0000	0.3982	0.2500	-0.7500	0.00000E+000	0.3	-9.79456E-001	0.098
39	OPDC		5	0.0000	0.3982	0.2500	-0.2500	0.00000E+000	0.3	2.40577E-001	0.006
40	OPDC		5	0.0000	0.3982	0.2500	0.2500	0.00000E+000	0.3	1.10242E-001	0.001
41	OPDC		5	0.0000	0.3982	0.2500	0.7500	0.00000E+000	0.3	-1.10720E+000	0.125
42	OPDC		5	0.0000	0.3982	0.7500	-0.2500	0.00000E+000	0.3	8.51527E-001	0.074
43	OPDC		5	0.0000	0.3982	0.7500	0.2500	0.00000E+000	0.3	9.84799E-001	0.099
44	OPDC		1	0.0000	0.6991	0.2500	-0.2500	0.00000E+000	0.2	8.84696E-001	0.053
45	OPDC		1	0.0000	0.6991	0.2500	0.2500	0.00000E+000	0.2	9.31140E-001	0.059
46	OPDC		1	0.0000	0.6991	0.7500	-0.2500	0.00000E+000	0.2	-3.49561E+000	0.831
47	OPDC		1	0.0000	0.6991	0.7500	0.2500	0.00000E+000	0.2	-1.93762E+000	0.255
48	OPDC		3	0.0000	0.6991	0.2500	-0.2500	0.00000E+000	0.2	1.04192E+000	0.074
49	OPDC		3	0.0000	0.6991	0.2500	0.2500	0.00000E+000	0.2	1.02693E+000	0.072
50	OPDC		3	0.0000	0.6991	0.7500	-0.2500	0.00000E+000	0.2	-1.50051E-001	0.002
51	OPDC		3	0.0000	0.6991	0.7500	0.2500	0.00000E+000	0.2	1.27943E+000	0.111
52	OPDC		5	0.0000	0.6991	0.2500	-0.2500	0.00000E+000	0.2	4.58045E-001	0.014
53	OPDC		5	0.0000	0.6991	0.2500	0.2500	0.00000E+000	0.2	4.55536E-001	0.014
54	OPDC		5	0.0000	0.6991	0.7500	-0.2500	0.00000E+000	0.2	-1.90510E+000	0.247
55	OPDC		5	0.0000	0.6991	0.7500	0.2500	0.00000E+000	0.2	-6.56900E-001	0.029
56	OPDC		1	0.0000	1.0000	0.2500	-0.2500	0.00000E+000	0.1	-6.50910E-001	0.014
57	OPDC		1	0.0000	1.0000	0.2500	0.2500	0.00000E+000	0.1	-1.04694E-001	0.000
58	OPDC		1	0.0000	1.0000	0.7500	-0.2500	0.00000E+000	0.1	-2.34726E+001	18.728
59	OPDC		1	0.0000	1.0000	0.7500	0.2500	0.00000E+000	0.1	-1.82381E+001	11.307
60	OPDC		3	0.0000	1.0000	0.2500	-0.2500	0.00000E+000	0.1	-3.50213E-001	0.004
61	OPDC		3	0.0000	1.0000	0.2500	0.2500	0.00000E+000	0.1	1.14294E-001	0.000
62	OPDC		3	0.0000	1.0000	0.7500	-0.2500	0.00000E+000	0.1	-1.62212E+001	8.944
63	OPDC		3	0.0000	1.0000	0.7500	0.2500	0.00000E+000	0.1	-1.16852E+001	4.641
64	OPDC		5	0.0000	1.0000	0.2500	-0.2500	0.00000E+000	0.1	-7.44091E-001	0.019
65	OPDC		5	0.0000	1.0000	0.2500	0.2500	0.00000E+000	0.1	-2.92020E-001	0.003
66	OPDC		5	0.0000	1.0000	0.7500	-0.2500	0.00000E+000	0.1	-1.52342E+001	7.889
67	OPDC		5	0.0000	1.0000	0.7500	0.2500	0.00000E+000	0.1	-1.13288E+001	4.363
68	CONF	2									
69	OPDC		1	0.0000	0.0000	0.2500	0.2500	0.00000E+000	25.6	-9.90972E-002	0.085
70	OPDC		1	0.0000	0.0000	0.2500	0.7500	0.00000E+000	25.6	-2.04659E-001	0.364
71	OPDC		1	0.0000	0.0000	0.7500	0.2500	0.00000E+000	25.6	-2.04659E-001	0.364
72	OPDC		3	0.0000	0.0000	0.2500	0.2500	0.00000E+000	25.6	1.54199E-002	0.002
73	OPDC		3	0.0000	0.0000	0.2500	0.7500	0.00000E+000	25.6	3.49635E-001	1.064
74	OPDC		3	0.0000	0.0000	0.7500	0.2500	0.00000E+000	25.6	3.49635E-001	1.064
75	OPDC		5	0.0000	0.0000	0.2500	0.2500	0.00000E+000	25.6	-8.88529E-002	0.069
76	OPDC		5	0.0000	0.0000	0.2500	0.7500	0.00000E+000	25.6	-2.07796E-001	0.376
77	OPDC		5	0.0000	0.0000	0.7500	0.2500	0.00000E+000	25.6	-2.07796E-001	0.376
78	OPDC		1	0.0000	0.3982	0.2500	-0.7500	0.00000E+000	4.8	6.49202E-001	0.688
79	OPDC		1	0.0000	0.3982	0.2500	-0.2500	0.00000E+000	4.8	6.79027E-002	0.008
80	OPDC		1	0.0000	0.3982	0.2500	0.2500	0.00000E+000	4.8	9.44227E-002	0.015

Listing B.4.3.2, Part 2

```
Merit Function Listing

File :  C:LENS334B.ZMX
Title:  DOUBLE-GAUSS, 52MM, F/2, 45.2DEG

Merit Function Value:  4.75699955E-003
```

Num	Type	Int1	Int2	Hx	Hy	Px	Py	Target	Weight	Value	% Cont
81	OPDC		1	0.0000	0.3982	0.2500	0.7500	0.00000E+000	4.8	3.99520E-001	0.260
82	OPDC		1	0.0000	0.3982	0.7500	-0.2500	0.00000E+000	4.8	5.38728E-001	0.474
83	OPDC		1	0.0000	0.3982	0.7500	0.2500	0.00000E+000	4.8	4.71035E-001	0.362
84	OPDC		3	0.0000	0.3982	0.2500	-0.7500	0.00000E+000	4.8	9.87146E-001	1.590
85	OPDC		3	0.0000	0.3982	0.2500	-0.2500	0.00000E+000	4.8	1.63340E-001	0.044
86	OPDC		3	0.0000	0.3982	0.2500	0.2500	0.00000E+000	4.8	1.43008E-001	0.033
87	OPDC		3	0.0000	0.3982	0.2500	0.7500	0.00000E+000	4.8	7.08049E-001	0.818
88	OPDC		3	0.0000	0.3982	0.7500	-0.2500	0.00000E+000	4.8	9.72086E-001	1.542
89	OPDC		3	0.0000	0.3982	0.7500	0.2500	0.00000E+000	4.8	8.91545E-001	1.297
90	OPDC		5	0.0000	0.3982	0.2500	-0.7500	0.00000E+000	4.8	3.13208E-001	0.160
91	OPDC		5	0.0000	0.3982	0.2500	-0.2500	0.00000E+000	4.8	3.77711E-002	0.002
92	OPDC		5	0.0000	0.3982	0.2500	0.2500	0.00000E+000	4.8	1.46019E-002	0.000
93	OPDC		5	0.0000	0.3982	0.2500	0.7500	0.00000E+000	4.8	6.84334E-002	0.008
94	OPDC		5	0.0000	0.3982	0.7500	-0.2500	0.00000E+000	4.8	3.28334E-001	0.176
95	OPDC		5	0.0000	0.3982	0.7500	0.2500	0.00000E+000	4.8	2.57144E-001	0.108
96	OPDC		1	0.0000	0.6991	0.2500	-0.7500	0.00000E+000	3.2	1.20723E+000	1.585
97	OPDC		1	0.0000	0.6991	0.2500	-0.2500	0.00000E+000	3.2	2.36710E-001	0.061
98	OPDC		1	0.0000	0.6991	0.2500	0.2500	0.00000E+000	3.2	2.88404E-001	0.090
99	OPDC		1	0.0000	0.6991	0.2500	0.7500	0.00000E+000	3.2	1.11659E+000	1.356
100	OPDC		1	0.0000	0.6991	0.7500	-0.2500	0.00000E+000	3.2	9.45336E-001	0.972
101	OPDC		1	0.0000	0.6991	0.7500	0.2500	0.00000E+000	3.2	1.01201E+000	1.114
102	OPDC		3	0.0000	0.6991	0.2500	-0.7500	0.00000E+000	3.2	1.22924E+000	1.644
103	OPDC		3	0.0000	0.6991	0.2500	-0.2500	0.00000E+000	3.2	2.84616E-001	0.088
104	OPDC		3	0.0000	0.6991	0.2500	0.2500	0.00000E+000	3.2	2.80462E-001	0.086
105	OPDC		3	0.0000	0.6991	0.2500	0.7500	0.00000E+000	3.2	1.09744E+000	1.310
106	OPDC		3	0.0000	0.6991	0.7500	-0.2500	0.00000E+000	3.2	1.31775E+000	1.889
107	OPDC		3	0.0000	0.6991	0.7500	0.2500	0.00000E+000	3.2	1.35195E+000	1.988
108	OPDC		5	0.0000	0.6991	0.2500	-0.7500	0.00000E+000	3.2	4.58903E-001	0.229
109	OPDC		5	0.0000	0.6991	0.2500	-0.2500	0.00000E+000	3.2	1.34132E-001	0.020
110	OPDC		5	0.0000	0.6991	0.2500	0.2500	0.00000E+000	3.2	1.32886E-001	0.019
111	OPDC		5	0.0000	0.6991	0.2500	0.7500	0.00000E+000	3.2	3.66952E-001	0.146
112	OPDC		5	0.0000	0.6991	0.7500	-0.2500	0.00000E+000	3.2	6.40661E-001	0.446
113	OPDC		5	0.0000	0.6991	0.7500	0.2500	0.00000E+000	3.2	6.75213E-001	0.496
114	OPDC		1	0.0000	1.0000	0.2500	-0.7500	0.00000E+000	1.6	-2.67283E-001	0.039
115	OPDC		1	0.0000	1.0000	0.2500	-0.2500	0.00000E+000	1.6	-2.99317E-002	0.000
116	OPDC		1	0.0000	1.0000	0.2500	0.2500	0.00000E+000	1.6	5.69553E-002	0.002
117	OPDC		1	0.0000	1.0000	0.2500	0.7500	0.00000E+000	1.6	2.38977E-001	0.031
118	OPDC		1	0.0000	1.0000	0.7500	-0.2500	0.00000E+000	1.6	-1.51447E+000	1.247
119	OPDC		1	0.0000	1.0000	0.7500	0.2500	0.00000E+000	1.6	-1.00846E+000	0.553
120	OPDC		3	0.0000	1.0000	0.2500	-0.7500	0.00000E+000	1.6	-3.48101E-001	0.066
121	OPDC		3	0.0000	1.0000	0.2500	-0.2500	0.00000E+000	1.6	2.49383E-002	0.000
122	OPDC		3	0.0000	1.0000	0.2500	0.2500	0.00000E+000	1.6	8.29407E-002	0.004
123	OPDC		3	0.0000	1.0000	0.2500	0.7500	0.00000E+000	1.6	1.19594E-001	0.008
124	OPDC		3	0.0000	1.0000	0.7500	-0.2500	0.00000E+000	1.6	-6.48180E-001	0.229
125	OPDC		3	0.0000	1.0000	0.7500	0.2500	0.00000E+000	1.6	-2.19509E-001	0.026
126	OPDC		5	0.0000	1.0000	0.2500	-0.7500	0.00000E+000	1.6	-9.74486E-001	0.516
127	OPDC		5	0.0000	1.0000	0.2500	-0.2500	0.00000E+000	1.6	-9.86033E-002	0.005
128	OPDC		5	0.0000	1.0000	0.2500	0.2500	0.00000E+000	1.6	-2.10302E-002	0.000
129	OPDC		5	0.0000	1.0000	0.2500	0.7500	0.00000E+000	1.6	-4.84900E-001	0.128
130	OPDC		5	0.0000	1.0000	0.7500	-0.2500	0.00000E+000	1.6	-9.71944E-001	0.514
131	OPDC		5	0.0000	1.0000	0.7500	0.2500	0.00000E+000	1.6	-5.80661E-001	0.183

The classical Double-Gauss lens as presented here is the original form that made the lens type famous. However, every possible modification has been tried at one time or another. Some of these variations have been successful in raising performance and are widely seen in commercial lenses. One common variation opens the cemented interface in one of the two cemented menisci into an airspace. Usually the front meniscus is opened; less often the rear meniscus is opened. However, rarely, if ever, are both menisci opened. A second common variation splits the rear singlet element into two airspaced singlet elements. Less often, the rear singlet is converted into a cemented doublet.

B.4.9 Comparison with Star Photos

As a reality check, it is of interest to see how actual Double-Gauss lenses perform in practice. Stars are point sources, and their images are point spread functions. Thus, the author took photographs of fields of stars at night using two commercial Double-Gauss lenses for 35 mm cameras. Both nominal focal lengths were 50 mm, and nominal maximum speeds were $f/1.8$ and $f/1.4$. Exposures were made with the lenses wide open and at several smaller lens openings. Note that during the required several minute time exposure, the motion of the stars caused by the earth's rotation had to be removed by attaching the camera to an equatorial mount with a clock drive turning once per day.

When the developed negatives were examined with a 10-power hand magnifier or under a low-power microscope (much is lost if you make prints), the recorded star images looked remarkably like the spots in Figures B.4.3.8 and B.4.3.9. In particular, the images taken wide open had copious amounts of sagittal oblique spherical flare. Furthermore, a long-pass yellow (minus blue) filter did little to improve the images, confirming that chromatic aberrations are minor. These results are reassuring and indicate that the Double-Gauss design presented here is typical of real lenses.

Note that in astro-photos the large asymmetrical sagittal flare around stars is very disturbing. To reduce this flare to acceptable levels, a normal Double-Gauss lens must be stopped down to about $f/4$. But stars are the most critical objects for any lens to image. All lenses seem to work much better when photographing continuous pictorial subjects.

Finally, you might think that when photographing stars, a lens optimized for minimum spot sizes would give better images than a lens optimized for minimum OPDs. You also might think this is especially true where diffraction is small and the spot is a good representation of the point spread function. In fact, you do get (somewhat) more compact star images with less flare with a spot size optimization. Thus, if well-defined star images, and not MTF, is your performance criterion, then you may prefer the spot optimization balance. In the next two chapters, two different types of astronomical telescopes will be considered. In both cases, optimization will be done by minimizing spots, not OPDs.

Chapter B.5

Cassegrain Telescopes

In this chapter, the Cassegrain reflecting telescope is investigated. First, an example of a classical Cassegrain is designed and evaluated. Then, an example of the most important Cassegrain variation, the Ritchey-Chrétien, is similarly treated. Finally, refractive field correctors, located near the focus, are added to both types of Cassegrains to widen their fields of good imagery.

The examples in this chapter are all ground-based telescopes; that is, they are not space telescopes. Also, they are hypothetical and represent no actual telescopes, real or planned. But to make things more interesting, the largest primary mirror and the largest CCD image detector likely to be available at the turn of the twenty-first century are assumed. Of course, this requires making some guesses about future developments. But if the guesses later turn out to be a little off, it will not matter. The designs will still serve to illustrate the concepts and techniques.

B.5.1 The Reflecting Telescope

In 1666, Isaac Newton performed a series of experiments that first demonstrated that white light is a mixture of light of different colors. In the process, he showed that, when passed through a prism, white light exhibits both refraction and dispersion; that is, the different colors are refracted by different amounts. He immediately realized that this chromatic differential refraction is present in lenses too and was the cause of the chromatic aberration then plaguing the singlet-lens telescopes of his time.

Newton then made a mistake. He performed a poorly conceived experiment that incorrectly suggested that the ratio of dispersion to refraction is the same for all transparent materials. Based on this misapprehension, he concluded that it would be impossible to correct chromatic aberration by combining two lens elements made of different types of glass. Newton's prestige was so great that his error was accepted until 1729, when the achromat was finally invented.

But something good came of it. Newton reasoned that because reflection is the same for all colors, a telescope made with mirrors would be free of chromatic aberration. Reflecting telescopes had previously been suggested by others, most notably by Marin Mersenne in 1636 and James Gregory in 1663. But none had been successfully built. Newton devised his own type of reflecting telescope, and in 1668 he succeeded in building a prototype. In 1671, Newton built a second example, which is still in existence. His optical configuration has thus been known ever since as a Newtonian telescope.[1]

Figure B.5.1.1 shows the layout of a Newtonian telescope. The system uses two mirrors. The light first falls on the primary mirror, which has a concave paraboloidal shape or figure. This mirror reflects and focuses the light. But if nothing more were

[1]See Henry C. King, *The History of the Telescope.*

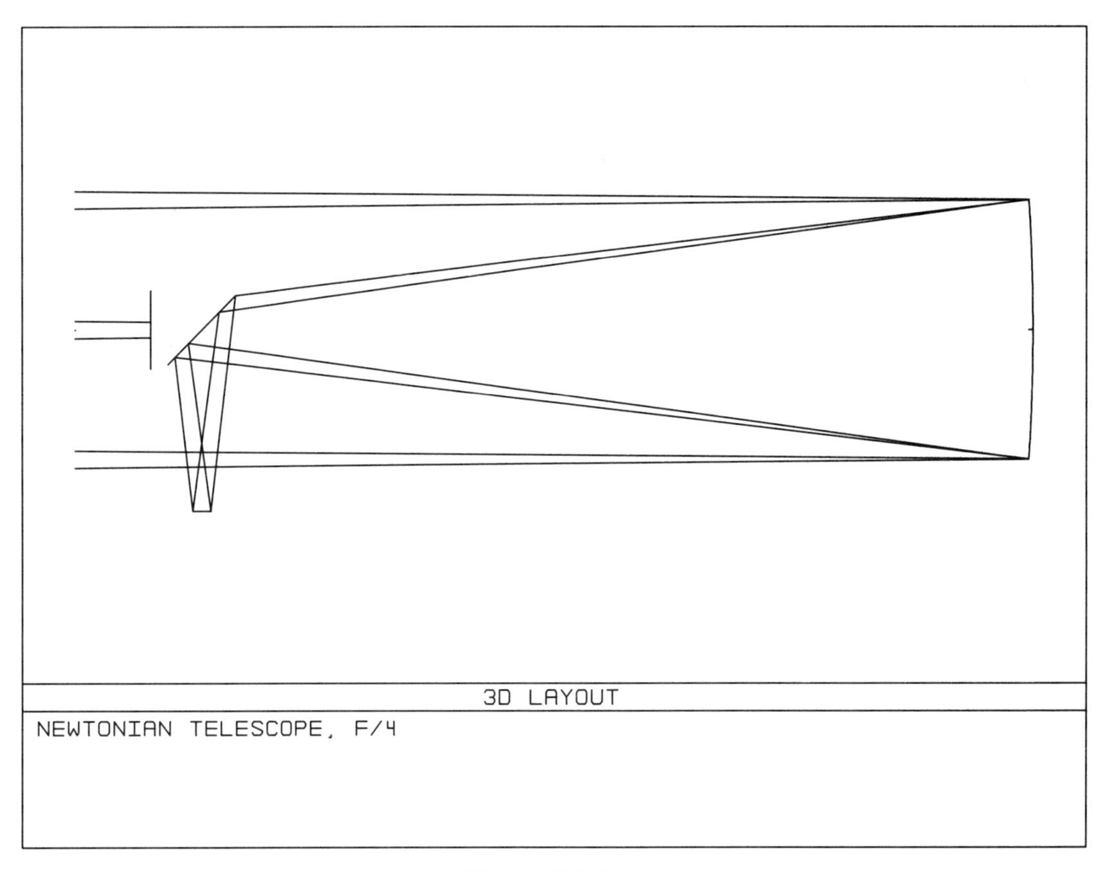

Figure B.5.1.1.

done, the light would come to a focus right in the middle of the incoming beam. This arrangement, called the prime focus, would place the image in an inconvenient location. Thus, Newton placed a secondary mirror in the converging beam a short distance before focus. This secondary is a small flat mirror tilted at 45° that reflects the light out to the side of the telescope tube for convenient viewing at the Newtonian focus.

Note that except for handedness, the optical properties at the Newtonian focus are identical to those at the prime focus and are determined only by the primary mirror. The flat secondary mirror only redirects the beam. Note too that the secondary mirror, because of its location, blocks some of the incoming light. This blockage is typical of axially centered reflecting systems; the mirrors get in each other's way. But if the central obscuration is carefully controlled, then the light loss is acceptable. Note finally that the terms primary and secondary for the mirrors refer to their order in the light path. If a third mirror were placed after the secondary, then it would be a tertiary, and so forth.

In 1672, Cassegrain in France suggested another type of reflecting telescope. His optical configuration, illustrated in Figure B.5.1.2, also uses two mirrors. The primary mirror is a concave paraboloid similar to that in a Newtonian. But unlike a Newtonian, the Cassegrain secondary mirror is an untilted convex hyperboloid that reflects the light back through a hole in the center of the primary mirror for viewing at the Cassegrain focus.

Note that the convex secondary mirror of a Cassegrain has negative (diverging) optical power and that the two mirrors are separated. Thus, the positive primary and negative secondary acting together yield an overall effective focal length that is longer than the focal length of the primary alone. Conceptually, a Cassegrain telescope is a reflecting version of a telephoto lens, which was illustrated previously in Figure A.6.3.1.

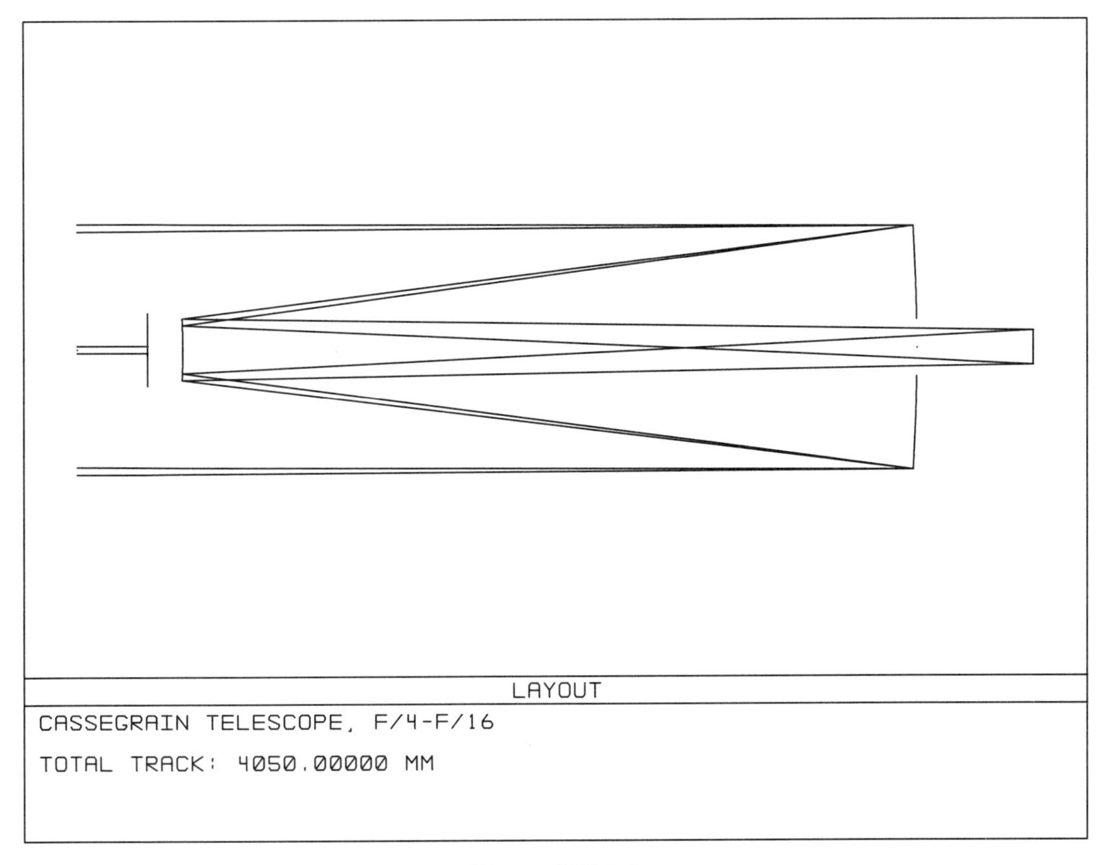

Figure B.5.1.2.

The difficulties in making good mirrors resulted in very few reflecting telescopes of any type being made in the seventeenth century. In fact, Newton left his primary mirror spherical, although he knew theoretically that it should have been a paraboloid. No one, not even Newton, could make a paraboloid in those days. And of course, a convex hyperboloid for a Cassegrain was out of the question. However, a few Cassegrains and quite a few Newtonians were made in the eighteenth century. The first proper Newtonian with a paraboloidal mirror was made by John Hadley by 1721. In the decades following 1773, William Herschel made many excellent reflectors of various designs. By the nineteenth century, the techniques of fabrication (and testing) had improved sufficiently that both Newtonian and Cassegrain telescopes could be routinely made. But the nineteenth century was the era of the great achromatic refractors, and reflectors were considered definitely second rate.

All that changed around the beginning of the twentieth century. As late as the 1860s, reflecting telescopes still used mirrors made of solid speculum metal whose optical surfaces tarnished and had to be periodically repolished and thus refigured. But by the late nineteenth century, speculum mirrors had been superseded by the much more practical silvered (later aluminized) glass mirrors that could be recoated without changing their underlying optical figure. Note that these glass mirrors are first-surface mirrors with their reflecting surface on the front side.

There were three other major changes at about the same time. First, between 1890 and 1900, photography largely replaced visual observing. The flatter but mainly blue wavelength response of the early photographic plates made the secondary chromatic aberration of a refractor a more serious problem. Also, the great refractors were slow (about $f/19$); reflectors could be made much faster ($f/5$ or less), which was a big advantage for photography. Second, after 1900, astronomy evolved into astrophysics, and the required heavy spectrographs for analyzing starlight were difficult or impossible to hang on the end of a long refractor tube. And third, there was a push for ever larger apertures to gather more light.

The practical limit on lens size had been reached in 1897 with the 40-inch (1.0-meter) diameter Yerkes refractor, but mirrors could be made much larger. A lens can only be supported around its edge, and thus a big lens will sag under its own weight. However, a mirror can be supported both around its edge and across its back. Also, big lenses are thick with lots of absorption, whereas the reflectivity of a first-surface mirror is independent of its size. Thus, by 1900, reflecting telescopes were quickly becoming much more attractive.

In 1908, the Mt. Wilson 60-inch (1.5-meter) reflector, the archetype of modern telescopes, was completed. This telescope, one of the most scientifically productive in history, is still in service and offers a Newtonian focus ($f/5$) and two different Cassegrain foci ($f/16$ and $f/30$). The $f/30$ focus is a special configuration called the coudé (elbowed) focus because the light beam is bent by a tertiary flat and sent to a big spectrograph attached solidly to the ground. Originally, the telescope was also equipped with a prime focus ($f/5$) and a third Cassegrain focus ($f/20$). The huge success of the 60-inch resulted in nearly all large observatory telescopes made after 1908 being reflectors too, either Newtonians, Cassegrains, combinations, or variations. And at the same time, the number of amateur reflecting telescopes, mostly Newtonians, has become truly astronomical.

Incidentally, the apertures of older telescopes are still often given, at least initially, in inches rather than metric measure. This convention is followed because the aperture of the telescope in inches has long ago become its *name*. For example, the 100-inch reflecting telescope on Mt. Wilson, completed in 1917, will always be the 100-inch and never the 2.5-meter. However, the 4-meter telescope on Kitt Peak, completed in 1973, never became known as the 158-inch and has always been the 4-meter.

Just as observational astronomy was revolutionized by photography in the 1890s, a similar process has been occurring since about 1985 with the advent of the CCD (charge coupled device). A CCD is a solid-state electronic television type image detector that consists of a large-scale integrated circuit on a silicon wafer. The elements of a CCD are light sensitive and are called pixels (picture elements). The array of pixels on a CCD is rectangular or square with hundreds or thousands of pixels on a side. When an optical system focuses light on a CCD, photon events (interactions) occur in the silicon, and these create charges that are read out and recorded.

For the astronomer, CCDs offer four big advantages over photography. First, CCDs have much higher quantum efficiencies. Second, the response of a CCD is linear with exposure (photography is very nonlinear). Third, a CCD can be used more than once and can thus be properly calibrated (photography is one-shot). And fourth, the output form of a CCD image is ideally suited for computer processing.

The quantum efficiency of a detector is the ratio of photon events to incident photons. The ideal, of course, is 100%. At the wavelength of peak response, the highest quantum efficiency of the photographic process is normally only about 1%, although some exceptions approach 4% (hypersensitized Kodak IIIa-J).[2] However, the peak quantum efficiencies of CCDs range from 40% to 80%. This is a huge gain. Although there are major exceptions, in most cases today it makes no sense to go to great effort and expense to build a large telescope to collect lots of light, and then to effectively throw away 99% of it with photography. In addition, the calibrated linear CCD response means that every pixel is a photometer. Thus, astronomers have embraced CCDs with enthusiasm.

The main disadvantage of CCDs is that they are relatively small. The maximum size of a CCD is limited by the size of the largest available silicon wafer. A big CCD is currently about 50 mm on a side, whereas astronomical photographic plates are

[2]For an excellent discussion of the properties of photographic emulsions, see *Scientific Imaging with Kodak Films and Plates*, Kodak Publication P-315.

available as large as 500 mm (20 inches) on a side. However, truly large slabs of electronics-grade silicon may be on the way.

B.5.2 Types of Cassegrain Telescopes

In a Cassegrain telescope, there are only five effective independent variables. Four of these are the vertex curvature (power) and conic constant (aspheric deformation) on each of the two mirrors. The fifth variable is the separation between the mirrors. The curvatures and separation are used to control first-order properties. The conics are used to control aberrations. The location of the final image is usually chosen to be somewhat behind the primary mirror.

Four different Cassegrain variations are generally recognized. These differ only in the conics and resulting aberration correction. You cannot tell one type of Cassegrain from another just by looking at it. Note that instead of conic-section aspherics, a Cassegrain telescope can be designed with polynomial aspherics, but this is normally not necessary and is almost never done because polynomial surfaces are much harder to make.

The original or classical Cassegrain configuration is one of the very few optical systems that can be designed mathematically in a closed form solution. The primary mirror is a paraboloid. As mentioned in Chapter A.7, a concave paraboloidal mirror has the property that a collimated (mutually parallel) ray bundle incident parallel to the paraboloid's axis is reflected and perfectly focused at the paraboloid's focus. All orders of spherical aberration are identically zero. The secondary mirror is a hyperboloid with one of its two foci located at the focus of the paraboloid. A convex hyperboloidal mirror has the property that a bundle of rays converging toward one focus is reflected to the other focus. In this configuration, the hyperboloid is also free of all orders of spherical aberration.

Because both the primary and secondary mirrors are free of spherical aberration, the overall system is free of spherical aberration and the on-axis image is geometrically perfect (neglecting diffraction, atmospheric seeing, fabrication and alignment errors, etc.). Furthermore, because reflection is independent of wavelength, the on-axis image is perfect both monochromatically and polychromatically. However, this perfection does not extend to off-axis images. Off-axis, a classical Cassegrain telescope suffers from a large amount of coma and, to a lesser extent, astigmatism and field curvature. Note that this astigmatism is Seidel astigmatism and not on-axis cylindrical aberration, which if present would be caused by fabrication errors.

It can be shown analytically that the amount of coma in a classical Cassegrain is exactly the same as the amount of coma in a Newtonian or prime-focus reflector of the same diameter and f/number. Astigmatism, however, is greater in a Cassegrain by a factor roughly equal to the magnification of the secondary (the ratio of the overall focal length to the focal length of the primary mirror alone). But even this amount of astigmatism remains relatively small compared to the coma.

Rigorously, the focus of the paraboloid and one of the two foci of the hyperboloid must be exactly coincident for zero spherical aberration. In practice, however, a small longitudinal separation is tolerable. Thus, the instruments at the focus of most large Cassegrains are solidly attached to the telescope tube, and focus is adjusted by moving the secondary mirror.

The classical Cassegrain does not use its degrees of freedom most effectively; that is, it is overconstrained. The primary and secondary mirrors are each independently free of spherical aberration. In practice, however, all the astronomer cares about is whether the overall system is free of spherical. And he also asks whether something might be done about the coma.

In 1910, George Ritchey, who designed and built the 60-inch at Mt. Wilson, had an idea that he thought might reduce the coma in a Cassegrain. He told his friend Henri Chrétien, a mathematician, who soon worked out the details.[3] The mirror powers and separation still control the basic layout. But now the conic constant on the primary mirror, as well as the conic constant on the secondary mirror, are independent variables. The additional degree of freedom allows both spherical aberration and coma to be simultaneously controlled to give an aplanatic system. The primary mirror becomes slightly hyperboloidal, and the secondary mirror becomes a hyperboloid with a somewhat greater eccentricity than that used in a classical Cassegrain. This type of Cassegrain telescope is now known as a Ritchey-Chrétien telescope.[4]

The advantage of a Ritchey-Chrétien is a much wider field of good imagery than is possible with a classical Cassegrain. With the coma gone, the remaining third-order aberrations are astigmatism and field curvature. These aberrations are much less serious than coma, and higher-order aberrations are even smaller.

Note that unlike a classical Cassegrain, a Ritchey-Chrétien does not correct all orders of spherical aberration to zero. Instead, a Ritchey-Chrétien allows only third-order spherical to be controlled. Thus, as is done in many lenses, third-order spherical is adjusted to cancel fifth-order spherical (the sum of third and fifth is zero) in a particular pupil zone. In other pupil zones, there is a small amount of residual zonal spherical. Similarly, only third-order coma is controlled.

Note too that in most Cassegrains of both the classical and Ritchey-Chrétien type, the power (vertex curvature) of the negative secondary is actually greater than the power of the positive primary. Thus, Newtonians and Cassegrains have Petzval sums and Petzval curvatures with opposite signs. In a Newtonian, the Petzval surface is curved convex to the light. In a Cassegrain, the Petzval surface is curved concave to the light.

Thus, in most Ritchey-Chrétien telescopes, the field is not flat. Both when designing and using the telescope, the best images can only be obtained on a curved image surface. Accordingly, when being exposed in a Ritchey-Chrétien, photographic plates and films are nearly always temporarily sucked by a vacuum platen into the required mild concave shape. Yes, you can bend large, thin glass plates if you do not do it too much. CCDs, however, are always flat and require one of several other possible solutions.

Of special interest to users of flat detectors is the flat-field Ritchey-Chrétien solution where the power of the secondary is equal and opposite to the power of the primary. In this case, the Petzval sum is zero and the Petzval surface is flat. However, the Petzval surface is the best image surface only in the absence of astigmatism, and a Ritchey-Chrétien has astigmatism. Thus, for a flat-field Ritchey-Chrétien, the mirror curvatures are actually adjusted to flatten the medial astigmatic surface, not the Petzval surface.

Unfortunately, a flat-field Ritchey-Chrétien has a relatively large central obscuration that limits its practicality. Fortunately, there are other ways to flatten the field, such as refractive field flatteners and correctors.

A big observatory reflector nearly always has more than one focal configuration. The most common are those mentioned earlier: the prime focus, Newtonian focus, Cassegrain focus, and coudé focus. A Ritchey-Chrétien is no exception.

In a Ritchey-Chrétien, the images are free of both spherical aberration and coma only at the single Cassegrain focus. At the coudé focus, which requires substituting a different convex hyperboloidal secondary mirror, the images have no spherical but they do have coma of a somewhat greater amount than in an equivalent classical Cassegrain. In either case, however, coma is not serious at the typically slow coudé focal ratio of about $f/30$. Furthermore, the coudé is used primarily for spectroscopy where a wide field is unnecessary.

[3] At that time, Ritchey and Chrétien were apparently unaware of the earlier work by Karl Schwarzschild, who in 1905 derived a different but related optical configuration.

[4] See Donald E. Osterbrock, *Pauper and Prince*, pp. 115–116.

At the prime (or Newtonian) focus, where the primary mirror is used without a convex secondary, a Ritchey-Chrétien has severe spherical aberration because the primary mirror is a hyperboloid, not a paraboloid, and there is also severe coma. The prime focus of a classical paraboloid also has severe coma but no spherical aberration. In both cases, auxiliary corrector lenses are required to control the coma (and also the astigmatism and field curvature) to get a usably wide field of view. The only difference is that a Ritchey-Chrétien corrector must also control the spherical. In practice, the two types of correctors are very similar; each takes the form of an airspaced triplet located a short distance in front of focus.

Therefore, unless you are primarily interested in a narrow field at the prime or Newtonian focus, there is no reason not to make a Ritchey-Chrétien.

Interestingly, the aplanatic Ritchey-Chrétien telescope was not accepted by astronomers for half a century. To a large extent, this delay was caused by the difficult and obscure optical techniques available in those days. It was hard to make a clear and convincing case for the design's advantages; that is, very few people could really understand how the telescope worked. However, the first director of Kitt Peak National Observatory, Aden Meinel, did understand optics (he later founded the Optical Sciences Center at the University of Arizona). Thus, the Kitt Peak 84-inch (2.1-meter) reflector, completed in 1964, was made as a Ritchey-Chrétien. The success of this telescope finally convinced astronomers, and most major observatory telescopes made since then have been of the Ritchey-Chrétien form.[5]

Incidentally, Chrétien went on to design other interesting optical systems. One of these, which he patented in 1929, later became known as CinemaScope.[6]

There are two other variations of the Cassegrain. These, however, are modifications for ease of fabrication, not for good imagery. The more common one is called a Dall-Kirkham and has a spherical secondary instead of the harder-to-make hyperboloidal secondary. Now, to control spherical aberration, the primary mirror must have a prolate ellipsoidal figure. Unfortunately, coma in a Dall-Kirkham is considerably worse than in a classical Cassegrain.

The other simplified Cassegrain has a spherical primary and an oblate ellipsoidal secondary. However, the spherical-primary configuration is almost never encountered in practice because it has even more coma than the spherical-secondary Dall-Kirkham.

B.5.3 System Specifications

In the design examples in this chapter, only the Cassegrain focus is considered. Other focal configurations are not considered, although they might be implemented in practice on a real telescope.

In the foreseeable future, the largest available monolithic mirror blank will probably have a diameter of about 8.4 meters (8400 mm or 331 inches). Three different organizations have (or are developing) the capability to make mirrors of this class: Schott, Corning, and the University of Arizona. Therefore, an 8.4-meter mirror is assumed here.

Predicting the availability of CCDs is much more difficult because the technology of making CCDs is improving very rapidly. But a few trends are clear. The largest pixel size currently available is 24 μm (0.024 mm) square; that is, the center-to-center pixel spacing is 24 μm. A big telescope requires a big pixel; otherwise, its focal length must be too short. Also, a big pixel has a greater charge-storage capacity, which gives a greater dynamic range before saturation. Because it appears unlikely that larger pixels will become available, a 24 μm pixel is assumed in these examples.

---

[5]See Donald E. Osterbrock, *Pauper and Prince*, pp. 279–287.

[6]U.S. patent 1,962,892; British patent 356,955 (1929).

Today, the largest CCD composed of 24 μm pixels is a 2048x2048 pixel square array. The total array dimensions become 49.15 mm on a side. A 4096x4096 pixel CCD is under development, but its pixels are 15 μm, giving a square array 61.44 mm on a side. In a few years, however, silicon wafers almost certainly will become available that will allow CCDs to be made with 24 μm pixels in a 4096 pixel square format. This image detector would be 98.3 mm square and have a diagonal of 139.0 mm. In the present telescope examples, such a CCD is assumed.

Next, a guess must be made at the expected effects of atmospheric seeing. Outside the atmosphere, stars are effectively point objects. But during the transit of the light through the turbulent and inhomogeneous atmosphere, these points are smeared out into small extended objects called seeing disks. For a ground-based telescope, it is highly desirable that aberrations, diffraction, and the pixel size should all be no larger than the image of the seeing disk; that is, the angular resolving power of the telescope should be limited only by seeing, even on the best of nights.

Until recently, seeing typically enlarged stars to diameters of about one arc-second or greater (3600 arc-seconds per degree). But new observatories are now being located at exotic sites with better seeing, such as Mauna Kea, the Andes, and the Canary Islands. Just as important, the design of the telescopes and their domes (enclosures) have reduced seeing effects. Thus, today it is not uncommon to have seeing disks of 0.5 arc-seconds or less in diameter. To allow for very good seeing, it is assumed that the telescope examples in this chapter must accommodate seeing as small as about 0.25 arc-seconds.

Note that these images are obtained without the use of adaptive optics to remove the effects of seeing in real time. Adaptive optics are not included here because the corrected field of view (the isoplanatic patch) is too small to fill the whole CCD area.

To reduce the size of the dome, and to facilitate the control of seeing originating inside the dome, most telescopes built today are very short and stubby. This aspect ratio is achieved by making the f/number of the primary mirror very fast. The latest fabrication techniques allow large telescope mirrors to be made with speeds of $f/1.2$ or faster. Thus, somewhat arbitrarily, $f/1.2$ is adopted here for the speed of the 8400 mm diameter primary mirror.

The overall effective focal length (EFL) at the Cassegrain focus must give an image scale that causes the seeing disk to match the pixel size. Recall the sampling theorem and the Nyquist limit, which require two or more samples per cycle. The same concept applies to recording the point spread function. For the seeing disk to be resolved (not just detected), its width must be sampled by two or more pixels. Thus, for 0.25 arc-second seeing, two adjacent pixels, when projected onto the sky, must have an angular width of about 0.25 arc-seconds.

Accordingly, the adopted Cassegrain focal length is 50400 mm, which gives an $f/6.0$ beam. Note that the secondary mirror magnifies the focal length of the $f/1.2$ primary mirror by a factor of 5. With this focal length, the side of a 24 μm square pixel corresponds to 0.098 arc-seconds in the sky. From the sampling theorem, four of these pixels in a 2x2 square matrix (48x48 μm) should cover the smallest seeing disk. The four pixels correspond to 0.196 arc-seconds on a side and 0.278 arc-seconds on the diagonal. This is a close match to 0.25 arc-second seeing.

For easy access to the image, the Cassegrain focus is usually located somewhat behind the primary mirror. The amount of this back focal clearance is determined by the thickness of the mirror and its supporting cell, and by the thickness of the attachment mechanism for the auxiliary instruments. In the present examples, the focus is chosen to be 2000 mm behind the 8400 mm diameter mirror. This clearance is not generous, but it yields the minimum central obscuration.

For a CCD with a 139.0 mm diagonal, the smallest (circular) linear field of view at the Cassegrain focus is ± 69.5 mm. For a telescope with an effective focal length of 50400 mm, ± 69.5 mm on the image corresponds to $\pm 0.079°$ in the sky.

However, in practice, some extra image area is often required for auxiliary detectors, such as off-axis guiders. Thus, in the present examples, a larger field of view of $\pm 0.10°$ (± 88.0 mm) is adopted. During design and evaluation, four field points are specified: 0°, 0.04°, 0.07°, and 0.10°. No mechanical vignetting is allowed; that is, the outside of even the furthest off-axis beam is not clipped (although there is, of course, a central obscuration).

For an all-reflecting system (composed entirely of mirrors), geometrical optical properties are independent of wavelength. Thus, when designing a Cassegrain or a Ritchey-Chrétien telescope, one wavelength is sufficient. In the present examples, a wavelength of 0.55 μm is adopted. Of course, to evaluate polychromatic diffraction performance, several wavelengths are necessary.

Several wavelengths are also necessary when designing an auxiliary refractive field corrector for one of these telescopes. In this case, the selection of wavelengths requires extra thought. When used to make astronomical observations, a CCD responds to wavelengths extending from the ultraviolet atmospheric cutoff near 0.30 μm (the atmosphere does not transmit shorter wavelengths) out to the infrared detector cutoff near 1.10 μm (the CCD does not respond to longer wavelengths). However, this full range is seldom used all at once. Instead, the astronomer nearly always employs filters to select several smaller spectral regions to observe one after another. Consequently, the lens designer must design his system to work with any possible filter passband between 0.30 and 1.10 μm, and all are equally likely and important.

Thus, for this application, it can be argued that a good set of wavelengths should have its members spaced to give equal fractional or relative changes in optical properties. For a thin prism:

$$\frac{\Delta D}{D} = \frac{\Delta n}{n-1}$$

where D and ΔD are the angular deviation and deviation increment, and n and Δn are the index and index increment. Note that the quotient on the right is wavelength-dependent dispersion as discussed in Chapter A.10.

To a first approximation, equal index increments give roughly equal fractional changes. If the wavelength range is not too great, this approximation is adequate. For all optical materials transmitting in the extended visible region of CCD sensitivity, index varies more rapidly at shorter wavelengths than at longer wavelengths, as illustrated by the dispersion curves in Figures A.10.1.1 and A.10.1.2. Thus, for a wavelength set giving equal index increments, the wavelengths are more closely spaced in the blue and more widely spaced in the red.

More accurately, equal index increments give equal absolute changes, not equal relative changes. For equal fractional changes, the variation of the $n-1$ term in the denominator of the dispersion relation must be included. Relative to the wavelength set giving equal index increments, the wavelength set giving equal fractional changes has somewhat wider spacings in the blue and narrower spacings in the red to match the varying index. Nevertheless, the wavelengths are still more closely spaced in the blue than in the red.

A set of wavelengths giving equal index increments can be read directly off a dispersion curve. Assuming fused silica (a good glass for field correctors) and the above wavelength range, a set of nine such wavelengths is: 0.32, 0.34, 0.365, 0.40, 0.44, 0.505, 0.60, 0.76, and 1.05 μm. However, because of the significant variation in the $n-1$ term, a wavelength set giving more equal fractional changes is: 0.32, 0.34, 0.37, 0.41, 0.47, 0.55, 0.70, 0.85, and 1.05 μm. Accordingly, this latter set, with equal unity weights, is adopted for use during optimization of the refractive field correctors. The middle wavelength, 0.47 μm, is adopted as the reference wavelength. During evaluation, the same wavelengths and weights are again used. This arrangement does not correspond to any commonly used filter-

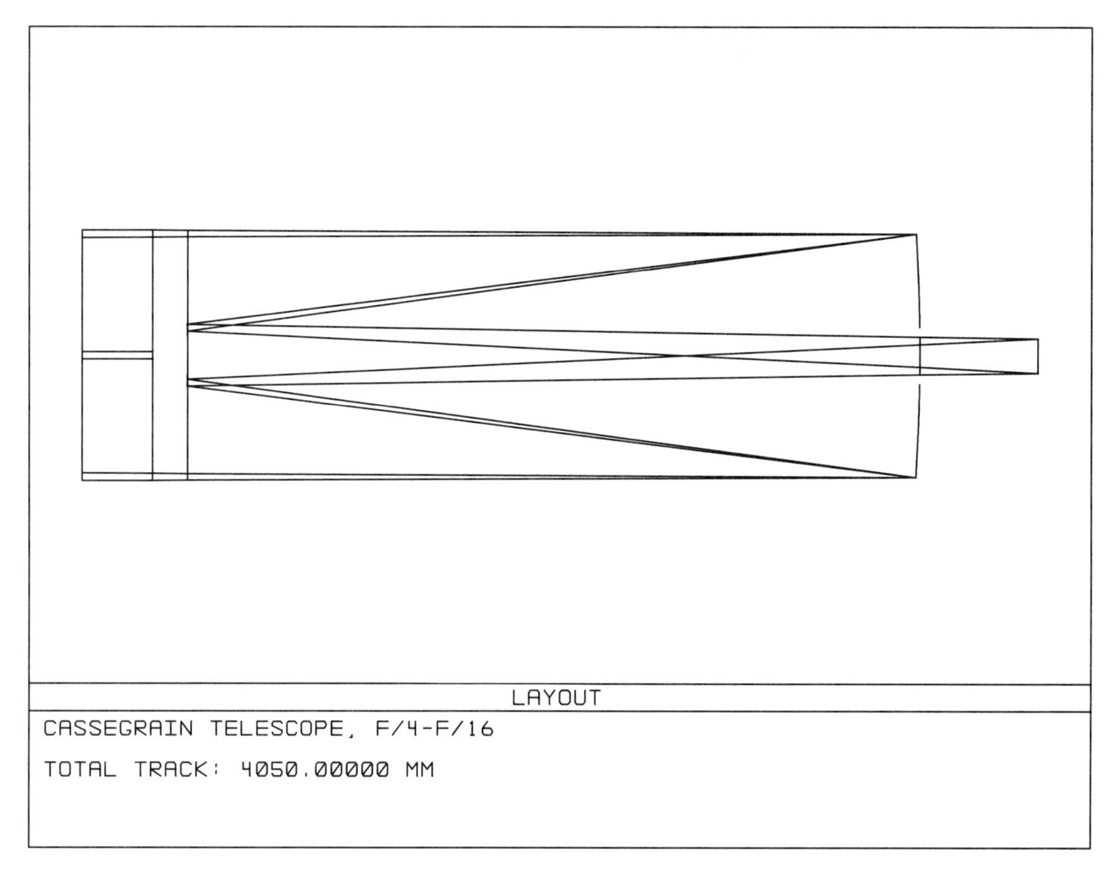

Figure B.5.1.3.

detector response curve, but it again emphasizes equal changes in optical properties. Specific filter-detector combinations must be evaluated separately.

B.5.4 The Classical Cassegrain

As usual, when entering an optical system into the computer, it is very helpful to first make a rough sketch of the configuration with the several surfaces identified. Figure B.5.1.3 is a layout of the same Cassegrain system shown in Figure B.5.1.2, except now all of the dummy surfaces are explicitly drawn. Although this layout is not of the specific 8.4-meter example, the basic arrangement is the same. Of course, this computer-generated layout is far better than your sketch needs to be.

Surface 1 is a dummy surface in front to control where the rays begin on layouts. Surface 2 is a dummy surface that contains the central obscuration. Surface 3 is a dummy surface at the same location as the secondary mirror. The thickness between surfaces 2 and 3 corresponds to the thickness of the secondary support structure. Surface 4 is the primary mirror and the stop of the system. Surface 5 is the secondary mirror. Surface 6 is a dummy surface back at the location of the primary mirror. The thicknesses following surfaces 3, 4, and 5 are all the same, with proper signs (use pickups). This value is the free thickness variable during optimization. The thickness following surface 6 is the back focal clearance. Surface 7 is a dummy surface at the paraxial focus. And surface 8 is the actual image surface.

Note that the stop is on the primary mirror, which is also the first surface with power in the system. Thus, there are no entrance pupil aberrations (the Gaussian pupil is the same as the trigonometric pupil), and the ray aiming feature can be turned off to save computing time.

After defining and numbering the surfaces, the next step is to enter the system quantities into the computer. Enter the entrance pupil diameter, fields, wavelengths, weights, and any other general system data (such as a lens title). Then fill in known or reasonable surface parameters. The glass type for both the primary and secondary mirrors is, of course, "mirror" (or whatever your program uses to specify a reflection). For the present telescope, a reasonable value for the thicknesses following both surfaces 1 and 2 is 1000 mm.

The radius of curvature of a mirror is twice its focal length. Thus, the radius of curvature of the 8400 mm $f/1.2$ primary mirror is 8400 times 1.2 times 2, or -20160 mm (the negative sign is because its center of curvature is to the left). The primary mirror is a paraboloid, and thus this radius is the radius of curvature at the mirror vertex. The conic constant of a paraboloid is -1.0 in most programs (refer again to Equation A.3.2).

None of the starting parameters for the secondary mirror are very critical, and guesses can be made. Place the secondary at any reasonable location inside the focus of the primary. Alternatively, the position of the secondary can be calculated by noting that its magnifying power is 5, and thus the distance from the secondary to the Cassegrain focus is 5 times the distance from the secondary to the prime focus. The approximate radius of curvature of the secondary can be guessed with the help of layouts. Tinker until the image at Cassegrain is roughly in focus. Alternatively, you can use Equation A.3.3. And the conic constant can be guessed; the mirror is a hyperboloid, so enter -2 or -3 as a start. There is an equation to calculate the conic constant, but the idea in this book is to let the computer handle these things.

According to the prescribed back focal clearance, the spacing must be 2000 mm exactly between the primary mirror at surface 4 (or surface 6) and the paraxial focal plane at surface 7. One way to ensure this is to make the thickness 2000 mm following surface 6 and to make this parameter fixed. An operand in the merit function is then used to focus the paraxial image onto surface 7.

Surface 8 is the actual image surface. Keep the vertex of surface 8 at the paraxial focal plane by making the thickness zero following surface 7. Initially, make surface 8 flat.

Optimization of a Cassegrain telescope departs from the general method outlined in Chapter A.15. The system is so conceptually simple, and there are so few independent variables, that the optimization can be done all at once.

The only independent variables during optimization are: (1) the spacing between the primary and secondary, (2) the vertex curvature of the secondary, (3) the conic constant on the secondary, and (4) the curvature of the image surface. The spacing between the two mirrors determines the magnification of the secondary, a first-order property. The curvature of the secondary focuses the paraxial Cassegrain image, another first-order property. The conic constant on the secondary corrects spherical aberration (all orders, in this case). And the image surface curvature selects the best off-axis image surface. Note that no variables are available to control coma or astigmatism.

Listing B.5.1.1 gives the merit function. There are three operands that are exactly corrected to their targets during optimization, either with heavy weights or Lagrange multipliers. In the ZEMAX format, EFFL corrects focal length to 50400 mm. PARY focuses the paraxial image on surface 7 by making the paraxial marginal ray height zero on this surface. REAY corrects spherical aberration by making the trigonometric ray height zero on surface 7 for the axial ray that passes through the 0.9 pupil zone (or any nonzero zone, in this case).

A flat CCD is specified. However, to find the best off-axis images, surface 8, the actual image surface, is allowed to be temporarily curved. If the performance with a curved image surface is unacceptable, then a flat image surface would be even worse. But if things look promising, the curved surface can be replaced by a flat surface and the resultant imagery evaluated. Note that if you change to a flat

Listing B.5.1.1

```
Merit Function Listing

File :  C:LENS301.ZMX
Title:  CLASSICAL CASS, 8.4 M, F/1.2-F/6.0

Merit Function Value:  5.03144279E-003
```

Num	Type	Int1	Int2	Hx	Hy	Px	Py	Target	Weight	Value	% Cont
1	EFFL		1					5.04000E+004	10	5.04000E+004	0.000
2	BLNK										
3	BLNK										
4	PARY	7	1	0.0000	0.0000	0.0000	1.0000	0.00000E+000	1000	8.93289E-006	0.000
5	REAY	7	1	0.0000	0.0000	0.0000	0.9000	0.00000E+000	1000	-5.87555E-006	0.000
6	BLNK										
7	BLNK										
8	DMFS										
9	TRAR		1	0.0000	0.4000	0.1679	0.2907	0.00000E+000	0.072722	1.72230E-002	0.042
10	TRAR		1	0.0000	0.4000	0.3536	0.6124	0.00000E+000	0.11636	7.88819E-002	1.421
11	TRAR		1	0.0000	0.4000	0.4710	0.8158	0.00000E+000	0.072722	1.41146E-001	2.844
12	TRAR		1	0.0000	0.4000	0.3357	0.0000	0.00000E+000	0.072722	7.01648E-003	0.007
13	TRAR		1	0.0000	0.4000	0.7071	0.0000	0.00000E+000	0.11636	3.06850E-002	0.215
14	TRAR		1	0.0000	0.4000	0.9420	0.0000	0.00000E+000	0.072722	5.43528E-002	0.422
15	TRAR		1	0.0000	0.4000	0.1679	-0.2907	0.00000E+000	0.072722	1.92839E-002	0.053
16	TRAR		1	0.0000	0.4000	0.3536	-0.6124	0.00000E+000	0.11636	8.30027E-002	1.574
17	TRAR		1	0.0000	0.4000	0.4710	-0.8158	0.00000E+000	0.072722	1.46334E-001	3.057
18	TRAR		1	0.0000	0.7000	0.1679	0.2907	0.00000E+000	0.072722	2.89211E-002	0.119
19	TRAR		1	0.0000	0.7000	0.3536	0.6124	0.00000E+000	0.11636	1.35505E-001	4.194
20	TRAR		1	0.0000	0.7000	0.4710	0.8158	0.00000E+000	0.072722	2.43812E-001	8.486
21	TRAR		1	0.0000	0.7000	0.3357	0.0000	0.00000E+000	0.072722	1.27410E-002	0.023
22	TRAR		1	0.0000	0.7000	0.7071	0.0000	0.00000E+000	0.11636	5.42078E-002	0.671
23	TRAR		1	0.0000	0.7000	0.9420	0.0000	0.00000E+000	0.072722	9.56629E-002	1.306
24	TRAR		1	0.0000	0.7000	0.1679	-0.2907	0.00000E+000	0.072722	3.52135E-002	0.177
25	TRAR		1	0.0000	0.7000	0.3536	-0.6124	0.00000E+000	0.11636	1.48100E-001	5.010
26	TRAR		1	0.0000	0.7000	0.4710	-0.8158	0.00000E+000	0.072722	2.59647E-001	9.624
27	TRAR		1	0.0000	1.0000	0.1679	0.2907	0.00000E+000	0.072722	3.97080E-002	0.225
28	TRAR		1	0.0000	1.0000	0.3536	0.6124	0.00000E+000	0.11636	1.90096E-001	8.254
29	TRAR		1	0.0000	1.0000	0.4710	0.8158	0.00000E+000	0.072722	3.43906E-001	16.883
30	TRAR		1	0.0000	1.0000	0.3357	0.0000	0.00000E+000	0.072722	1.91795E-002	0.053
31	TRAR		1	0.0000	1.0000	0.7071	0.0000	0.00000E+000	0.11636	7.85532E-002	1.409
32	TRAR		1	0.0000	1.0000	0.9420	0.0000	0.00000E+000	0.072722	1.37860E-001	2.713
33	TRAR		1	0.0000	1.0000	0.1679	-0.2907	0.00000E+000	0.072722	5.24652E-002	0.393
34	TRAR		1	0.0000	1.0000	0.3536	-0.6124	0.00000E+000	0.11636	2.15732E-001	10.630
35	TRAR		1	0.0000	1.0000	0.4710	-0.8158	0.00000E+000	0.072722	3.76134E-001	20.196

image surface, do not reoptimize; the best on-axis image is the same. To find the curved image surface, weight the four fields 0 1 1 1 and use a default merit function to shrink off-axis spots, as shown in Listing B.5.1.1.

Note that when optimizing a system with a central obscuration, you may choose to ignore the obscuration and optimize as though the entire pupil area transmitted. This somewhat unrealistic simplification allows you to use either a Gaussian quadrature or a rectangular array default merit function. However, for careful final work, you may wish to include the central obscuration, which precludes Gaussian quadrature. In most cases, ignoring the obscuration makes very little difference during optimization.

During evaluation, however, a central obscuration has major effects and cannot be ignored. Spot diagrams and all of the diffraction calculations are simply wrong if the obscuration is not included. Refer back to Table A.11.1 for examples of how a central obscuration can greatly modify the light distribution in a diffraction pattern. Two exceptions to including the central obscuration in evaluations are transverse and OPD ray fan plots; these plots are usually much clearer if the complete unobscured curves are drawn.

Incidentally, do not be concerned about optimizing or evaluating relative to the chief ray, which in reality is usually blocked by a central obscuration. The program can once again ignore the obscuration and trace the chief ray, which remains an excellent reference.

Figures B.5.1.4 and B.5.1.5 show two different layouts of the optimized 8.4-meter Cassegrain telescope. Figure B.5.1.4 is a conventional layout with only the central, top, and bottom pupil rays drawn. Figure B.5.1.5 is a similar layout, but now 61 rays from each object point are incident on the pupil. Note how the rays in the middle of the beams terminate on the central obscuration on surface 2. By using so many rays, the envelope of the rays can be seen. In both layouts, only the object points at the top and bottom of the field ($\pm 0.10°$) are used, thereby making the diagram clearer.

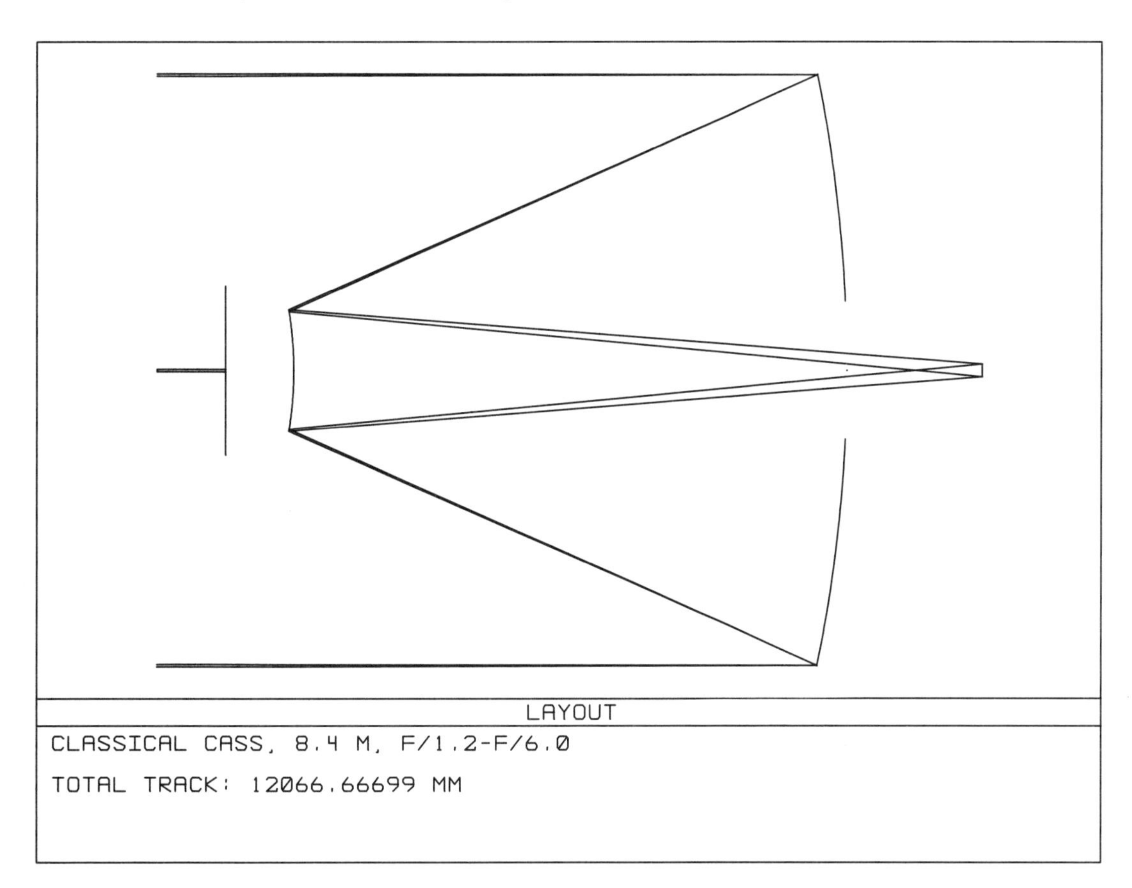

Figure B.5.1.4.

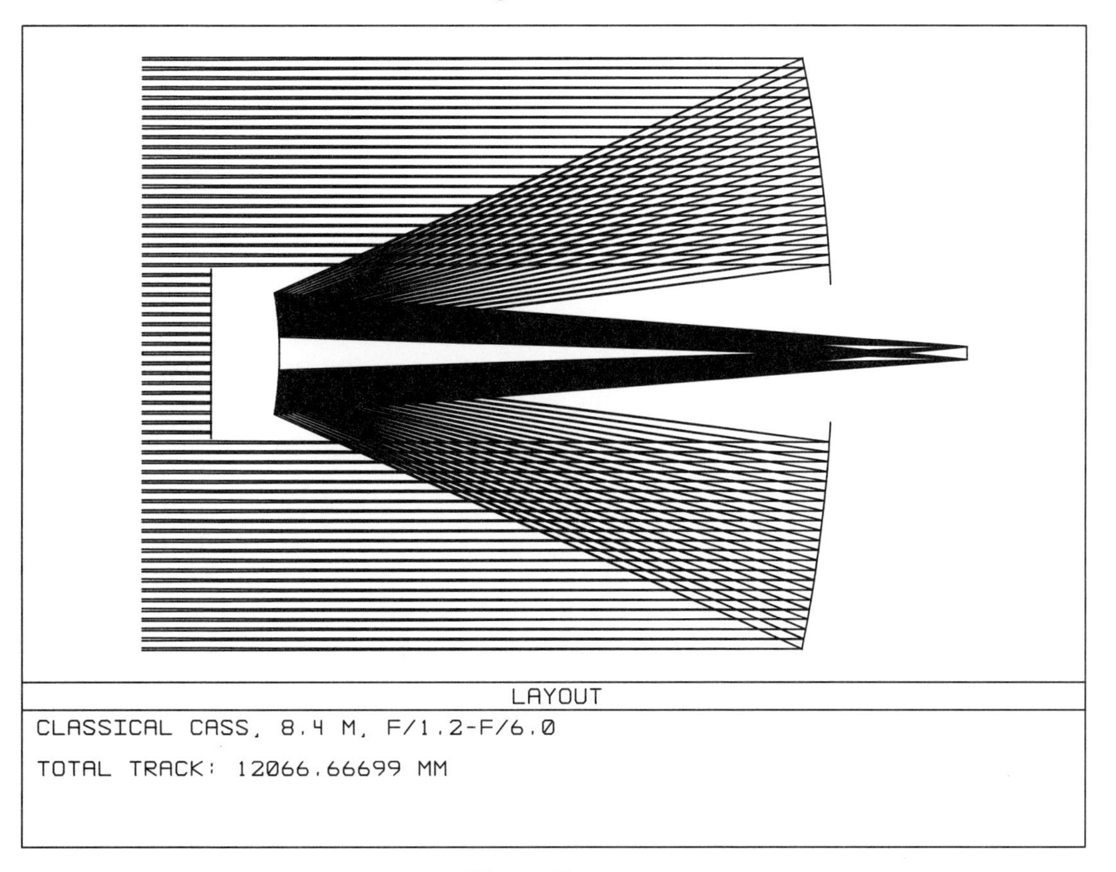

Figure B.5.1.5.

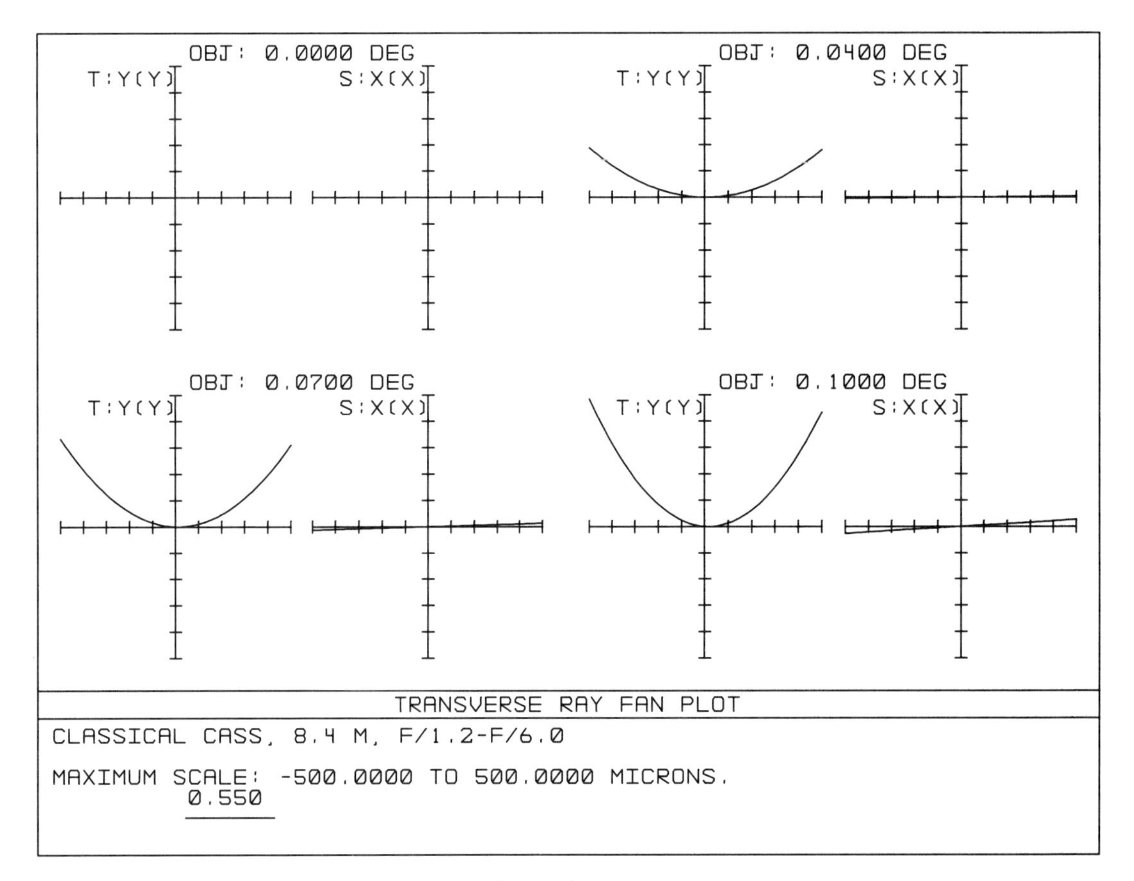

Figure B.5.1.6.

Figure B.5.1.5 is used in designing the Cassegrain baffles. The night sky is not dark, even on a moonless night far from city lights. There are still stars, airglow, zodiacal light, and aurora. Baffles are needed to prevent the CCD from viewing the sky directly rather than through the optics. These baffles consist of two hollow cylindrical tubes coaxial with the system optical axis. The secondary baffle (surrounding the secondary mirror) is the diameter of the central obscuration and extends down past the surface of the secondary mirror to the place where the ray bundles cross. The primary baffle is somewhat smaller in diameter but is much longer. It extends upward through the hole in the primary mirror to the place where the ray bundles again cross. When these baffles are properly designed, any CCD pixel should see only the secondary mirror plus the surrounding obscuration and the insides of the two baffle tubes. To suppress grazing reflections off the inside of the primary baffle, this tube may be fitted with additional annular ring baffles.

Note that the design of a baffle system must be done by hand. There is no software package that does this job, although several stray-light programs (GUERAP, APART, ASAP) are available to evaluate the performance of previously designed baffles.

Unfortunately, geometry dictates that the central obscuration of a Cassegrain telescope must be larger in diameter than the secondary mirror. In the present case, the secondary mirror for no vignetting has a clear diameter of 1722 mm. To determine the obscuration diameter, make a guess, enter this value as a hard obscuration on surface 2, and construct a layout like Figure B.5.1.5. Look at the envelope of rays and mentally add the baffles. Then with a straightedge, check what the pixels can see. Iterate until you get an obscuration slightly larger than the minimum. In the present example, the required obscuration diameter is about 2400 mm. The hole in the primary is smaller than the central obscuration and is

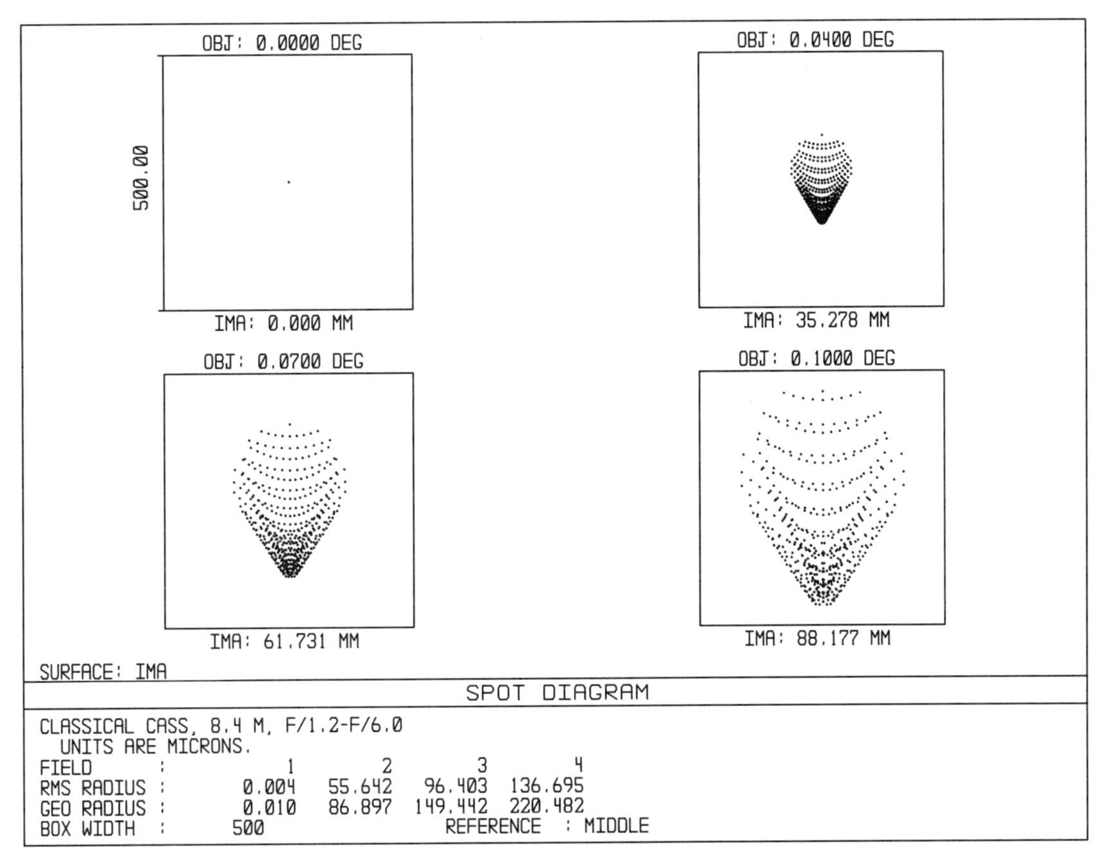

Figure B.5.1.7.

about 2000 mm in diameter.

A telescope with an 8400 mm primary mirror and a 2400 mm central obscuration has a linear obscuration ratio of 28.6% and an area obscuration ratio (light blockage) of 8.2%. These values are typical for Cassegrain telescopes working in the visible (but not in the thermal infrared). Note that to minimize the blockage, select the fastest (smallest) possible primary $f/$number, the slowest (largest) possible overall $f/$number, the smallest possible back focal clearance, and the smallest possible field. Reducing the central obscuration is one reason that the $f/$number on the primary is made $f/1.2$ here, which is very fast indeed. Of course, other considerations affect the selection of these parameters too.

Figure B.5.1.6 shows the transverse ray fan plots (with the central obscuration ignored). Note the perfect imagery on-axis and the large amount of coma off-axis. And this really is a lot of coma; the scale is ±500 μm. At the edge of the field, note that the U-shaped tangential coma curve is slightly higher on the left, and that the slightly inclined sagittal curve is lower on the left. This asymmetry is caused by a tiny amount of astigmatism.

Figure B.5.1.7 gives the spot diagrams (with the central obscuration included). Again the off-axis images are shown to be overwhelmed by coma. And this is on the best curved image surface. Note that the tips of the comatic spots are rounded; this is caused by the shadow of the central obscuration.

Listing B.5.1.2 gives the final optical prescription of the classical Cassegrain example.

Clearly, a classical Cassegrain with 500 μm spots comes nowhere near to providing the image quality to match a 24 μm pixel. The problems lie in the great size of the telescope, the fast $f/6$ Cassegrain $f/$number, and the superb (and demanding) seeing. Recall that geometrical aberrations scale directly as system size, and coma blur varies as the square of system speed. Smaller classical Cassegrains, with

Listing B.5.1.2

```
System/Prescription Data
File : C:LENS301.ZMX
Title: CLASSICAL CASS, 8.4 M, F/1.2-F/6.0

GENERAL LENS DATA:
Surfaces          :            8
Stop              :            4
System Aperture   :Entrance Pupil Diameter
Ray aiming        : Off
Apodization       :Uniform, factor =      0.000000
Eff. Focal Len.   :        50400 (in air)
Eff. Focal Len.   :        50400 (in image space)
Total Track       :      12066.7
Image Space F/#   :            6
Para. Wrkng F/#   :            6
Working F/#       :      6.01042
Obj. Space N.A.   :     4.2e-007
Stop Radius       :         4200
Parax. Ima. Hgt.  :      87.9647
Parax. Mag.       :            0
Entr. Pup. Dia.   :         8400
Entr. Pup. Pos.   :      10066.7
Exit Pupil Dia.   :      1997.48
Exit Pupil Pos.   :     -11984.9
Field Type        : Angle in degrees
Maximum Field     :          0.1
Primary Wave      :     0.550000
Lens Units        : Millimeters
Angular Mag.      :       4.2053

Fields            : 4
Field Type:  Angle in degrees
#         X-Value          Y-Value          Weight
1        0.000000         0.000000         0.000000
2        0.000000         0.040000         1.000000
3        0.000000         0.070000         1.000000
4        0.000000         0.100000         1.000000

Vignetting Factors
#          VDX              VDY              VCX              VCY
1       0.000000         0.000000         0.000000         0.000000
2       0.000000         0.000000         0.000000         0.000000
3       0.000000         0.000000         0.000000         0.000000
4       0.000000         0.000000         0.000000         0.000000

Wavelengths       : 5
Units:  Microns
#          Value           Weight
1       0.550000         1.000000

SURFACE DATA SUMMARY:

Surf      Type         Radius      Thickness          Glass        Diameter          Conic
 OBJ  STANDARD       Infinity       Infinity                              0              0
   1  STANDARD       Infinity           1000                       8435.139              0
   2  STANDARD       Infinity           1000                       8431.649              0
   3  STANDARD       Infinity       8066.667                       8428.158              0
 STO  STANDARD         -20160      -8066.667         MIRROR        8401.528             -1
   5  STANDARD      -5033.333       8066.667         MIRROR        1721.955          -2.25
   6  STANDARD       Infinity           2000                       481.3342              0
   7  STANDARD       Infinity              0                       177.2024              0
 IMA  STANDARD      -2179.441              0                       176.8765              0

SURFACE DATA DETAIL:

Surface OBJ       : STANDARD
Surface   1       : STANDARD
Surface   2       : STANDARD
 Aperture         : Circular Obscuration
 Minimum Radius   :              0
 Maximum Radius   :           1200
Surface   3       : STANDARD
Surface STO       : STANDARD
 Aperture         : Circular Aperture
 Minimum Radius   :           1000
 Maximum Radius   :           4200
Surface   5       : STANDARD
Surface   6       : STANDARD
Surface   7       : STANDARD
Surface IMA       : STANDARD

SOLVE AND VARIABLE DATA:

Thickness of   3   : Variable
Thickness of   4   : Solve, pick up value from 3, scaled by -1.00000
Curvature of   5   : Variable
Thickness of   5   : Solve, pick up value from 3, scaled by 1.00000
Conic of   5       : Variable
Curvature of   8   : Variable
```

speeds of about $f/16$, that are used under more normal (poorer) seeing, are quite satisfactory. But in the present case, something more must be done.

B.5.5 The Ritchey-Chrétien

As outlined earlier, a Ritchey-Chrétien telescope is an aplanatic Cassegrain where both spherical aberration and coma are controlled. To design a Ritchey-Chrétien, first optimize the telescope as a classical Cassegrain with the required first-order properties. Then make the conic constant on the primary mirror another free variable, and reoptimize.

The merit functions for both the classical and Ritchey-Chrétien Cassegrain versions are identical. Refer again to Listing B.5.1.1. Thus, in the present Ritchey-Chrétien example, focal length is still corrected to 50400 mm, and the paraxial image is still focused on surface 7. To control spherical aberration, the height of the axial ray that passes through the 0.9 pupil zone is still made zero on surface 7. In fact, even the way you optimize the off-axis imagery is the same. Now, however, when you shrink off-axis spots, not only do you make surface 8 the best curved image surface, but you also control the coma at the same time.

Note that $f/6$ is relatively fast for a Ritchey-Chrétien, and consequently additional higher-order off-axis aberrations are present here that would be unnoticeable at a more conventional speed of $f/8$. Thus, the best way to optimize the off-axis field of the present example is simply by shrinking spots; that is, do not use a special coma operand.

After optimizing, the remaining aberration is mainly astigmatism. The best off-axis image is near the medial focus halfway between the sagittal and tangential foci. This imagery is similar to that illustrated in Figures A.8.4.1 through A.8.6.3.

A layout of the optimized Ritchey-Chrétien would be identical to the layouts of the classical Cassegrain in Figures B.5.1.4 and B.5.1.5.

Figure B.5.2.1 shows the transverse ray fan plots. Compare these with the curves in Figure B.5.1.6. Note that the scale has been changed by a factor of 20, to ± 25 μm. On-axis, note the zonal spherical aberration. The slope of the curve is zero at the origin, indicating that the image surface is at the paraxial focus (done by the PARY optimization operand). The curve also crosses the horizontal axis at the 0.9 pupil zone (done by the REAY optimization operand). But at other pupil zones, there are small aberration residuals. Because these residuals are so small, the on-axis imagery of the Ritchey-Chrétien would be indistinguishable in practice from the geometrically perfect on-axis imagery of the classical Cassegrain.

Off-axis, the curves are roughly straight, with the tangential and sagittal curves having nearly equal and opposite slopes. This confirms astigmatism at the medial focus. But note that the off-axis curves are not perfectly straight, revealing the small amounts of higher-order aberrations.

Figure B.5.2.2 gives the spot diagrams. Note the small on-axis spot and the roughly round off-axis spots with the shadow of the central obscuration showing. Note too the effects of the higher-order off-axis aberrations. These off-axis spots are typical of not quite pure astigmatism at the best focus. Finally, note that the image spot at the edge of the field nearly fills the 72 μm square, which is three pixels on a side.

Listing B.5.2 gives the final optical prescription of the Ritchey-Chrétien example. Compare this with the prescription of the classical Cassegrain in Listing B.5.1.2. Note that most of the system parameters in the two versions are identical or nearly so (even the curvature of the image surface). Of course, the conic constants are different. It is interesting that the conic constant on the primary mirror of the Ritchey-Chrétien is only slightly hyperboloidal.

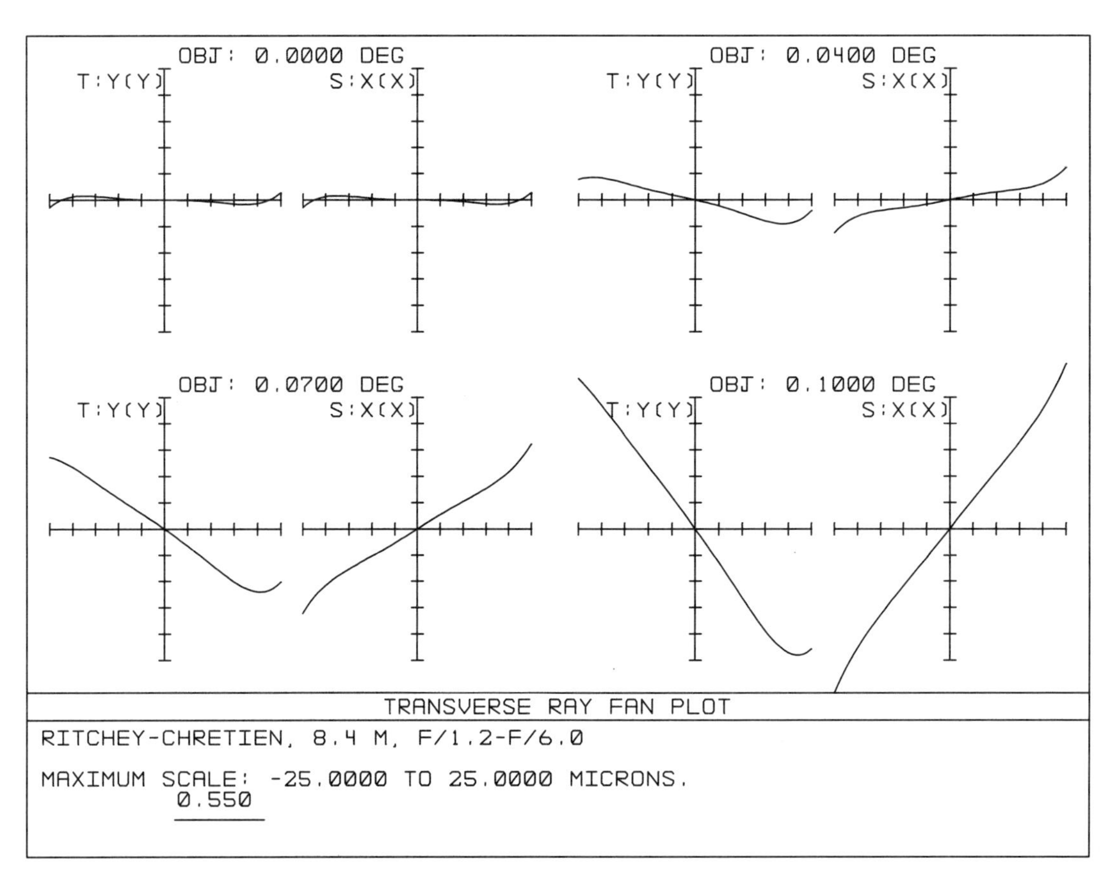

Figure B.5.2.1.

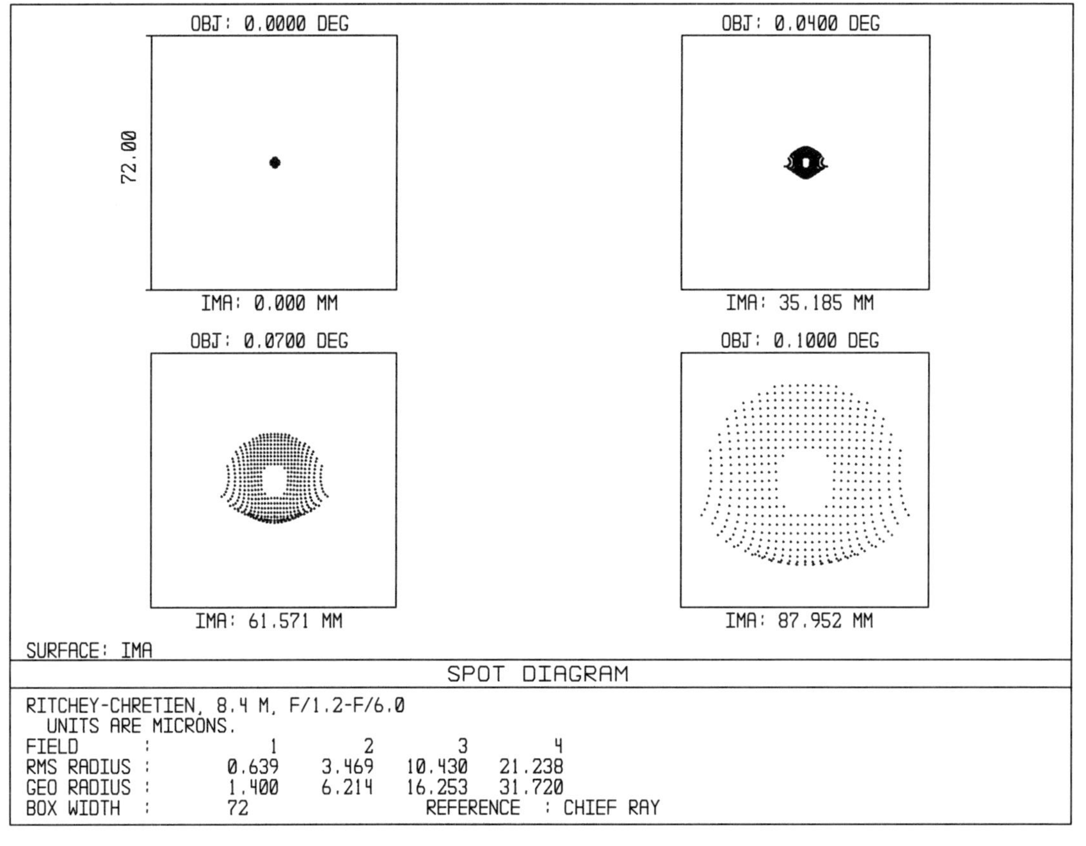

Figure B.5.2.2.

Listing B.5.2

```
System/Prescription Data

File : C:LENS302.ZMX
Title: RITCHEY-CHRETIEN, 8.4 M, F/1.2-F/6.0

GENERAL LENS DATA:

Surfaces          :          8
Stop              :          4
System Aperture   :Entrance Pupil Diameter
Ray aiming        : Off
Apodization       :Uniform, factor =      0.000000
Eff. Focal Len.   :      50400 (in air)
Eff. Focal Len.   :      50400 (in image space)
Total Track       :    12066.7
Image Space F/#   :          6
Para. Wrkng F/#   :          6
Working F/#       :    5.99999
Obj. Space N.A.   :   4.2e-007
Stop Radius       :       4200
Parax. Ima. Hgt.  :    87.9647
Parax. Mag.       :          0
Entr. Pup. Dia.   :       8400
Entr. Pup. Pos.   :    10066.7
Exit Pupil Dia.   :    1997.48
Exit Pupil Pos.   :   -11984.9
Field Type        : Angle in degrees
Maximum Field     :        0.1
Primary Wave      :   0.550000
Lens Units        : Millimeters
Angular Mag.      :     4.2053

Fields            : 4
Field Type:  Angle in degrees
#         X-Value          Y-Value          Weight
1        0.000000         0.000000        0.000000
2        0.000000         0.040000        1.000000
3        0.000000         0.070000        1.000000
4        0.000000         0.100000        1.000000

Vignetting Factors
#         VDX            VDY            VCX            VCY
1      0.000000       0.000000       0.000000       0.000000
2      0.000000       0.000000       0.000000       0.000000
3      0.000000       0.000000       0.000000       0.000000
4      0.000000       0.000000       0.000000       0.000000

Wavelengths       : 1
Units:  Microns
#           Value           Weight
1        0.550000        1.000000
SURFACE DATA SUMMARY:

Surf      Type        Radius      Thickness        Glass       Diameter         Conic
 OBJ  STANDARD      Infinity       Infinity                           0             0
   1  STANDARD      Infinity           1000                    8435.139             0
   2  STANDARD      Infinity           1000                    8431.649             0
   3  STANDARD      Infinity       8066.667                    8428.158             0
 STO  STANDARD        -20160      -8066.667       MIRROR       8401.527     -1.020205
   5  STANDARD     -5033.333       8066.667       MIRROR       1724.944     -2.447703
   6  STANDARD      Infinity           2000                    481.1789             0
   7  STANDARD      Infinity              0                    176.2966             0
 IMA  STANDARD     -2102.052              0                    175.9618             0

SURFACE DATA DETAIL:

Surface OBJ      : STANDARD
Surface   1      : STANDARD
Surface   2      : STANDARD
 Aperture        : Circular Obscuration
 Minimum Radius  :               0
 Maximum Radius  :            1200
Surface   3      : STANDARD
Surface STO      : STANDARD
 Aperture        : Circular Aperture
 Minimum Radius  :            1000
 Maximum Radius  :            4200
Surface   5      : STANDARD
Surface   6      : STANDARD
Surface   7      : STANDARD
Surface IMA      : STANDARD

SOLVE AND VARIABLE DATA:

Thickness of   3   : Variable
Thickness of   4   : Solve, pick up value from 3, scaled by -1.00000
Conic of   4       : Variable
Curvature of   5   : Variable
Thickness of   5   : Solve, pick up value from 3, scaled by 1.00000
Conic of   5       : Variable
Curvature of   8   : Variable
```

Clearly, a Ritchey-Chrétien type of Cassegrain telescope produces better off-axis images than a classical Cassegrain does. But unfortunately, even on a curved image surface, image quality is still not adequate for a 24 μm pixel. Like the classical Cassegrain, the problem lies mainly in the great size of the telescope. Again, something more must be done.

B.5.6 Refractive Field Correctors

For both the classical Cassegrain and the Ritchey-Chrétien examples above, refractive field correctors offer the possibility of reducing the excessive off-axis aberrations to acceptable levels. Field correctors also allow the image surface to be flattened to match a flat CCD without the penalty of a large central obscuration.

It was first suggested in 1873 by Piazzi Smyth that a curved image surface could be flattened by adding a singlet lens just in front of focus. Near the focus, the main effect of this field flattener is to reduce the Petzval sum. But it was not until 1913 that it was first suggested that a compound lens system located a short distance in front of the focus of a reflecting telescope could act as a more general field corrector, thereby removing coma as well as flattening the field.

At first, the idea was not pursued. Then in 1933, F.E. Ross at Mt. Wilson designed and built the first operational field corrector. Over subsequent years, Ross designed many more, including the first field correctors for the 200-inch (5-meter) telescope on Palomar Mountain. At observatories, these refractive field correctors soon became known as Ross correctors. Later, C.G. Wynne designed and built a series of similar but more advanced field correctors with higher performance, and these became known as Wynne correctors.[7]

The Ross and Wynne type field correctors belong to a larger class of afocal optical systems known as zero-power correctors. A Ross or Wynne corrector consists of two or more airspaced singlet lenses with all-spherical surfaces (no aspheres). Although the individual elements have power, the corrector as a whole has little or no net optical power when inserted into the converging beam.

Because Ross and Wynne correctors have almost no power, all of their elements can be made of the same type of glass without introducing serious chromatic aberrations. In other words, these zero-power correctors are not achromats. Note that there are other types of field correctors that do have overall power and do alter the focal length, and these must be achromatic with different glasses; they are not considered here.

Published reports by Wynne indicate that at the $f/6$ Cassegrain focus of the present examples, an airspaced doublet field corrector may be adequate to control aberrations. This would not be the case for a faster Newtonian or prime-focus corrector, where an airspaced triplet or four element configuration is necessary.

A possible design approach might be to use one of Wynne's correctors as a starting point. However, it is easier and just as effective to place two initially plane-parallel plates of glass near the focus and ask the computer to optimize their curvatures from scratch. Of course, this procedure must be done separately for the classical Cassegrain and Ritchey-Chrétien telescopes.

An excellent glass for the field corrector lenses is fused silica (or fused quartz). Throughout the extended visible wavelength region of interest and relative to other glasses, fused silica has a low index, a low dispersion (very crowny), a more uniform dispersion (a short glass), and a high bulk transmission. The low index is especially important because it reduces reflections from uncoated lens surfaces. The large

[7] For a review of the development of field correctors, see C.G. Wynne, "Field Correctors," in *Progress in Optics*, Vol. 10, Ch. 4, 1972. Also see "A New Wide-Field Triple Lens Paraboloid Field Corrector," *Mon. Not. R. Astron. Soc.*, Vol. 167, pp. 189–197, 1974. Another excellent reference is R. N. Wilson, *Reflecting Telescope Optics I: Basic Design Theory and its Historical Development*, pp. 315–379.

wavelength range here makes broadband anti-reflection coatings questionable, and the lenses may be better left uncoated. Thus, fused silica is adopted.

Because a colored filter with finite glass thickness is also required in the beam, a third plane-parallel plate is placed in the starting system. Recall that a plane-parallel glass plate in a converging (or diverging) beam introduces aberrations. Thus, although its surfaces remain flat, the filter must be included in the system during optimization. For convenience, the filter is located halfway between the two corrector lenses and is kept there with a pickup. Actually, the relative location of the filter does not matter. After optimization, the filter can be moved back and forth between the two lenses, provided that the overall separation between the lenses is not changed.

In the present examples, the filter is made, not of fused silica, but of Schott BK7, which is probably closer to filter glass. If a filter of the same thickness but different index is substituted, then the image surface can be refocused to give only a slight degradation in image quality. For no degradation, the thickness of the filter must be adjusted to match its index. In addition, if the thickness of the filter is adjusted, no refocusing is necessary, which is very desirable when you have several filters on a filter wheel.

Both the classical Cassegrain and Ritchey-Chrétien field correctors are configured the same way and optimized with nearly the same merit function. In each case, to allow the previously optimized main telescope to be usable alone without the field corrector, all variables and solves are deleted before the field corrector elements are added; that is, all of the main telescope parameters are fixed or frozen. After the field corrector is added, only its elements are allowed to vary during subsequent optimization.

The curvatures on all four corrector lens surfaces are made variables. The lens thicknesses, however, are not variables and are given minimum practical thin values. The airspaces (1) between the telescope and the front corrector lens, (2) between the two lenses, and (3) between the rear lens and the image are all allowed to vary during optimization.

As usual, both a paraxial focal plane and an actual image surface are included. The thickness between the paraxial focal plane and actual image surface is made variable during optimization. It is important that the image surface not be constrained to be at the paraxial focus. The image surface is made and kept flat during optimization.

And finally, to facilitate drawing layouts of the field corrector alone, a dummy surface is added a short distance in front of the first corrector lens.

The merit function listing for the classical Cassegrain corrector is given in Listing B.5.3.1. When building the merit function, first, do not include any constraint on system focal length. The best solution for the field corrector may have a small amount of net power. Let the computer adjust net power as it wishes. The system focal length will be close to that of the main telescope.

Second, the minimum of the merit function is very broad and flat, causing the airspaces to become large with little benefit. Include a constraint on overall length (front lens to image) to prevent the field corrector from spreading out too much. Note that the only difference between the merit functions for the two systems is that the classical Cassegrain corrector is constrained to be no longer than 375 mm, whereas the Ritchey-Chrétien corrector is constrained to be no longer than 300 mm.

Third, add a default merit function to shrink spot sizes for all four field positions and all nine wavelengths. Note that many positions and wavelengths have been omitted in Listing B.5.3.1 to save space in printing, but the idea should be clear. Weight the fields 3 3 3 2. Use equal unity weights for all wavelengths. Use no special optimization operands to control aberrations. Either the Gaussian quadrature or the rectangular array option may be selected; Gaussian quadrature is used here, thereby ignoring the central obscuration.

Listing B.5.3.1

Merit Function Listing

File : C:LENS303.ZMX
Title: CLASSICAL CASS WITH FIELD CORRECTOR

Merit Function Value: 2.80186141E-002

Num	Type	Int1	Int2	Hx	Hy	Px	Py	Target	Weight	Value	% Cont
1	TTHI	8	14					3.75000E+002	-1	3.75000E+002	0.000
2	BLNK										
3	BLNK										
4	DMFS										
5	TRAR		1	0.0000	0.0000	0.3357	0.0000	0.00000E+000	0.14544	1.76666E-002	1.325
6	TRAR		1	0.0000	0.0000	0.7071	0.0000	0.00000E+000	0.23271	1.36759E-002	1.271
7	TRAR		1	0.0000	0.0000	0.9420	0.0000	0.00000E+000	0.14544	1.44170E-002	0.883
8	TRAR		3	0.0000	0.0000	0.3357	0.0000	0.00000E+000	0.14544	1.87111E-002	1.487
9	TRAR		3	0.0000	0.0000	0.7071	0.0000	0.00000E+000	0.23271	1.65512E-002	1.861
10	TRAR		3	0.0000	0.0000	0.9420	0.0000	0.00000E+000	0.14544	9.65608E-003	0.396
11	TRAR		5	0.0000	0.0000	0.3357	0.0000	0.00000E+000	0.14544	1.96124E-002	1.633
12	TRAR		5	0.0000	0.0000	0.7071	0.0000	0.00000E+000	0.23271	1.91764E-002	2.498
13	TRAR		5	0.0000	0.0000	0.9420	0.0000	0.00000E+000	0.14544	5.15770E-003	0.113
14	TRAR		7	0.0000	0.0000	0.3357	0.0000	0.00000E+000	0.14544	2.02341E-002	1.738
15	TRAR		7	0.0000	0.0000	0.7071	0.0000	0.00000E+000	0.23271	2.11446E-002	3.037
16	TRAR		7	0.0000	0.0000	0.9420	0.0000	0.00000E+000	0.14544	1.62814E-003	0.011
17	TRAR		9	0.0000	0.0000	0.3357	0.0000	0.00000E+000	0.14544	2.05620E-002	1.795
18	TRAR		9	0.0000	0.0000	0.7071	0.0000	0.00000E+000	0.23271	2.22411E-002	3.361
19	TRAR		9	0.0000	0.0000	0.9420	0.0000	0.00000E+000	0.14544	3.92101E-004	0.001
20	TRAR		1	0.0000	1.0000	0.1679	0.2907	0.00000E+000	0.032321	4.30065E-002	1.745
21	TRAR		1	0.0000	1.0000	0.3536	0.6124	0.00000E+000	0.051713	9.27635E-003	0.130
22	TRAR		1	0.0000	1.0000	0.4710	0.8158	0.00000E+000	0.032321	1.02197E-001	9.855
23	TRAR		1	0.0000	1.0000	0.3357	0.0000	0.00000E+000	0.032321	2.79509E-002	0.737
24	TRAR		1	0.0000	1.0000	0.7071	0.0000	0.00000E+000	0.051713	2.85510E-002	1.231
25	TRAR		1	0.0000	1.0000	0.9420	0.0000	0.00000E+000	0.032321	2.86916E-002	0.777
26	TRAR		1	0.0000	1.0000	0.1679	-0.2907	0.00000E+000	0.032321	1.43939E-002	0.195
27	TRAR		1	0.0000	1.0000	0.3536	-0.6124	0.00000E+000	0.051713	5.48627E-003	0.045
28	TRAR		1	0.0000	1.0000	0.4710	-0.8158	0.00000E+000	0.032321	5.96673E-002	3.359
29	TRAR		3	0.0000	1.0000	0.1679	0.2907	0.00000E+000	0.032321	4.61321E-002	2.008
30	TRAR		3	0.0000	1.0000	0.3536	0.6124	0.00000E+000	0.051713	2.84485E-002	1.222
31	TRAR		3	0.0000	1.0000	0.4710	0.8158	0.00000E+000	0.032321	6.71226E-002	4.251
32	TRAR		3	0.0000	1.0000	0.3357	0.0000	0.00000E+000	0.032321	2.76606E-002	0.722
33	TRAR		3	0.0000	1.0000	0.7071	0.0000	0.00000E+000	0.051713	3.11420E-002	1.464
34	TRAR		3	0.0000	1.0000	0.9420	0.0000	0.00000E+000	0.032321	2.41649E-002	0.551
35	TRAR		3	0.0000	1.0000	0.1679	-0.2907	0.00000E+000	0.032321	2.35969E-002	0.525
36	TRAR		3	0.0000	1.0000	0.3536	-0.6124	0.00000E+000	0.051713	7.35614E-003	0.082
37	TRAR		3	0.0000	1.0000	0.4710	-0.8158	0.00000E+000	0.032321	4.78768E-002	2.163
38	TRAR		5	0.0000	1.0000	0.1679	0.2907	0.00000E+000	0.032321	4.70234E-002	2.086
39	TRAR		5	0.0000	1.0000	0.3536	0.6124	0.00000E+000	0.051713	4.63307E-002	3.241
40	TRAR		5	0.0000	1.0000	0.4710	0.8158	0.00000E+000	0.032321	3.26538E-002	1.006
41	TRAR		5	0.0000	1.0000	0.3357	0.0000	0.00000E+000	0.032321	2.79482E-002	0.737
42	TRAR		5	0.0000	1.0000	0.7071	0.0000	0.00000E+000	0.051713	3.36951E-002	1.714
43	TRAR		5	0.0000	1.0000	0.9420	0.0000	0.00000E+000	0.032321	2.05809E-002	0.400
44	TRAR		5	0.0000	1.0000	0.1679	-0.2907	0.00000E+000	0.032321	3.59789E-002	1.221
45	TRAR		5	0.0000	1.0000	0.3536	-0.6124	0.00000E+000	0.051713	2.23642E-002	0.755
46	TRAR		5	0.0000	1.0000	0.4710	-0.8158	0.00000E+000	0.032321	3.53280E-002	1.178
47	TRAR		7	0.0000	1.0000	0.1679	0.2907	0.00000E+000	0.032321	4.56072E-002	1.963
48	TRAR		7	0.0000	1.0000	0.3536	0.6124	0.00000E+000	0.051713	5.98571E-002	5.409
49	TRAR		7	0.0000	1.0000	0.4710	0.8158	0.00000E+000	0.032321	4.51901E-003	0.019
50	TRAR		7	0.0000	1.0000	0.3357	0.0000	0.00000E+000	0.032321	3.01654E-002	0.859
51	TRAR		7	0.0000	1.0000	0.7071	0.0000	0.00000E+000	0.051713	3.64185E-002	2.002
52	TRAR		7	0.0000	1.0000	0.9420	0.0000	0.00000E+000	0.032321	1.89179E-002	0.338
53	TRAR		7	0.0000	1.0000	0.1679	-0.2907	0.00000E+000	0.032321	4.91915E-002	2.283
54	TRAR		7	0.0000	1.0000	0.3536	-0.6124	0.00000E+000	0.051713	3.75089E-002	2.124
55	TRAR		7	0.0000	1.0000	0.4710	-0.8158	0.00000E+000	0.032321	2.63395E-002	0.655
56	TRAR		9	0.0000	1.0000	0.1679	0.2907	0.00000E+000	0.032321	4.37639E-002	1.807
57	TRAR		9	0.0000	1.0000	0.3536	0.6124	0.00000E+000	0.051713	6.70034E-002	6.778
58	TRAR		9	0.0000	1.0000	0.4710	0.8158	0.00000E+000	0.032321	1.17820E-002	0.131
59	TRAR		9	0.0000	1.0000	0.3357	0.0000	0.00000E+000	0.032321	3.30422E-002	1.030
60	TRAR		9	0.0000	1.0000	0.7071	0.0000	0.00000E+000	0.051713	3.87128E-002	2.263
61	TRAR		9	0.0000	1.0000	0.9420	0.0000	0.00000E+000	0.032321	1.91881E-002	0.347
62	TRAR		9	0.0000	1.0000	0.1679	-0.2907	0.00000E+000	0.032321	5.82728E-002	3.204
63	TRAR		9	0.0000	1.0000	0.3536	-0.6124	0.00000E+000	0.051713	4.76263E-002	3.424
64	TRAR		9	0.0000	1.0000	0.4710	-0.8158	0.00000E+000	0.032321	2.42126E-002	0.553

Note that this system will be used to image stars, and a spot size optimization gives more compact and well-defined star images than an OPD optimization. Furthermore, this is one of those cases where a spot optimization actually gives better Strehl ratios than an OPD optimization. You may wish to prove this to yourself. In any event, no results from OPD optimizations will be given here.

B.5.7 The Classical Cassegrain with Field Corrector

For the classical Cassegrain, Figure B.5.3.1 shows a layout of the optimized field corrector without the rest of the telescope. The front element has positive power, and the rear element has negative power. For any object point, examine the cone of light being focused. Because the rear element is closer to the converging beam focus, the beam footprint on the rear element is smaller than the beam footprint on the front element. Thus, to maintain an effective overall power near zero, the curvatures on the rear element must be significantly greater than the curvatures on

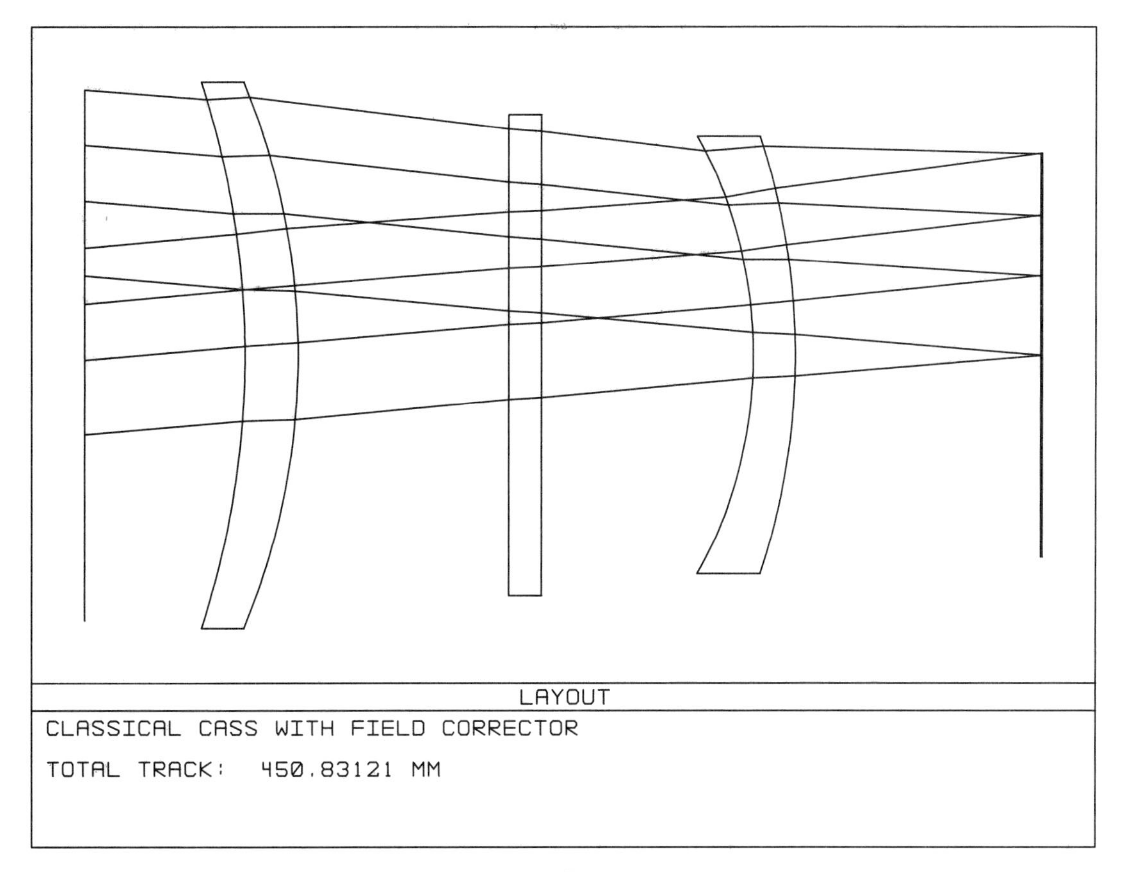

Figure B.5.3.1.

the front element. In other words, as single elements, the negative power of the rear element is greater than the positive power of the front element.

For the Cassegrain telescope alone, the Petzval sum is negative. For the field corrector alone, the greater negative power of the rear element gives a positive Petzval sum. When the telescope and field corrector are combined, the two together have a Petzval sum near zero, thereby flattening the image surface. The principle here is the same as that outlined in Chapter B.3 for flattening the field of a Cooke Triplet camera lens.

Figure B.5.3.2 shows the transverse ray fan plots for the classical Cassegrain with its optimized field corrector. On-axis, defocus balances third-order spherical, with a small amount of chromatic variation evident. Going off-axis, there is increasing oblique spherical adding to the already present ordinary spherical. This is balanced by increasing field curvature defocus adding to the already present axial defocus. Off-axis, there are also increasing chromatic variations of the monochromatic aberrations and a relatively small amount of lateral color. Note that the curves for the different wavelengths are roughly equally spaced, as expected from the chosen wavelength set. Note too the large scale: ± 100 μm.

Figure B.5.3.3 gives the spot diagrams. Again note the large scale. Although these spots are definitely smaller than the spots for the unassisted classical Cassegrain telescope in Figure B.5.1.7, they are still larger than the size of a 24 μm pixel. Once again, something more must be done.

As before, the problems lie in the great size of the telescope, its fast Cassegrain f/number, and the superb seeing. For a smaller classical Cassegrain with a slower speed used under more normal seeing, a simple doublet field corrector may be quite satisfactory. But in the present application, you are really asking a lot.

Listing B.5.3.2 gives the optical prescription of the classical Cassegrain telescope with field corrector. Note that the field corrector changes the effective focal length

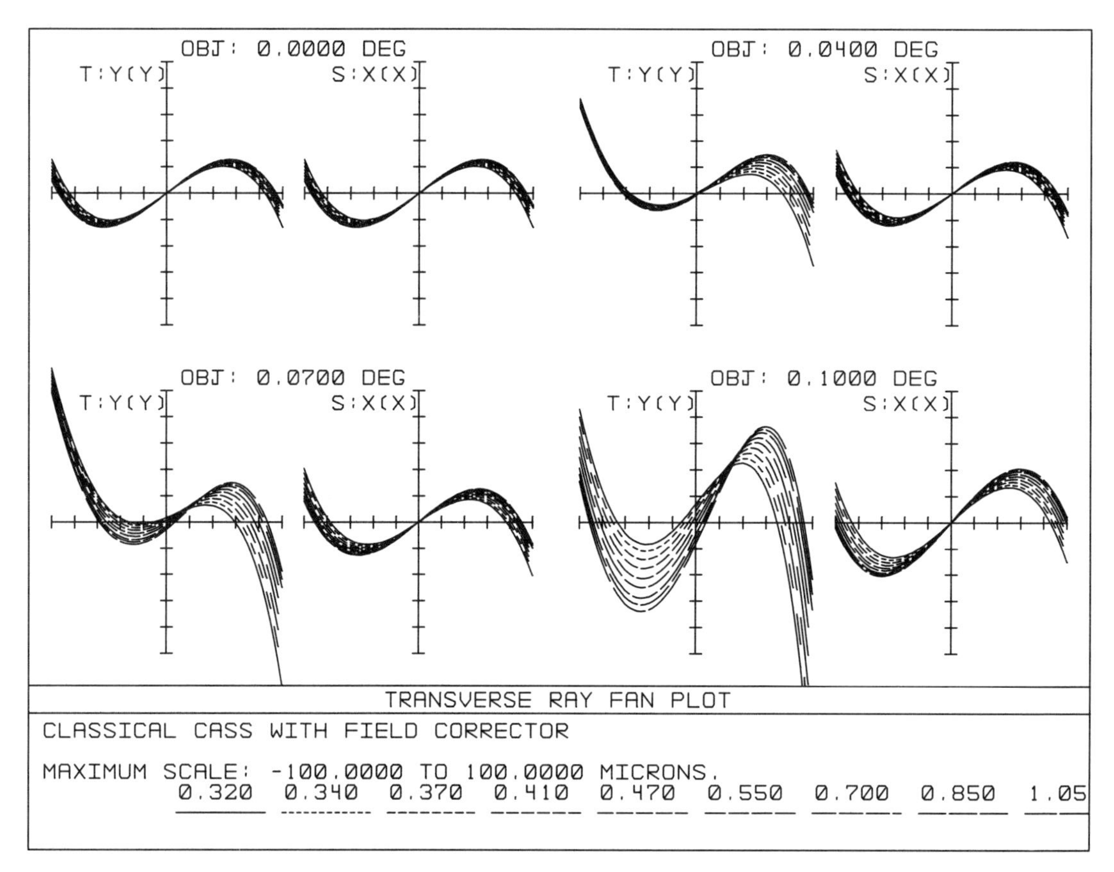

Figure B.5.3.2.

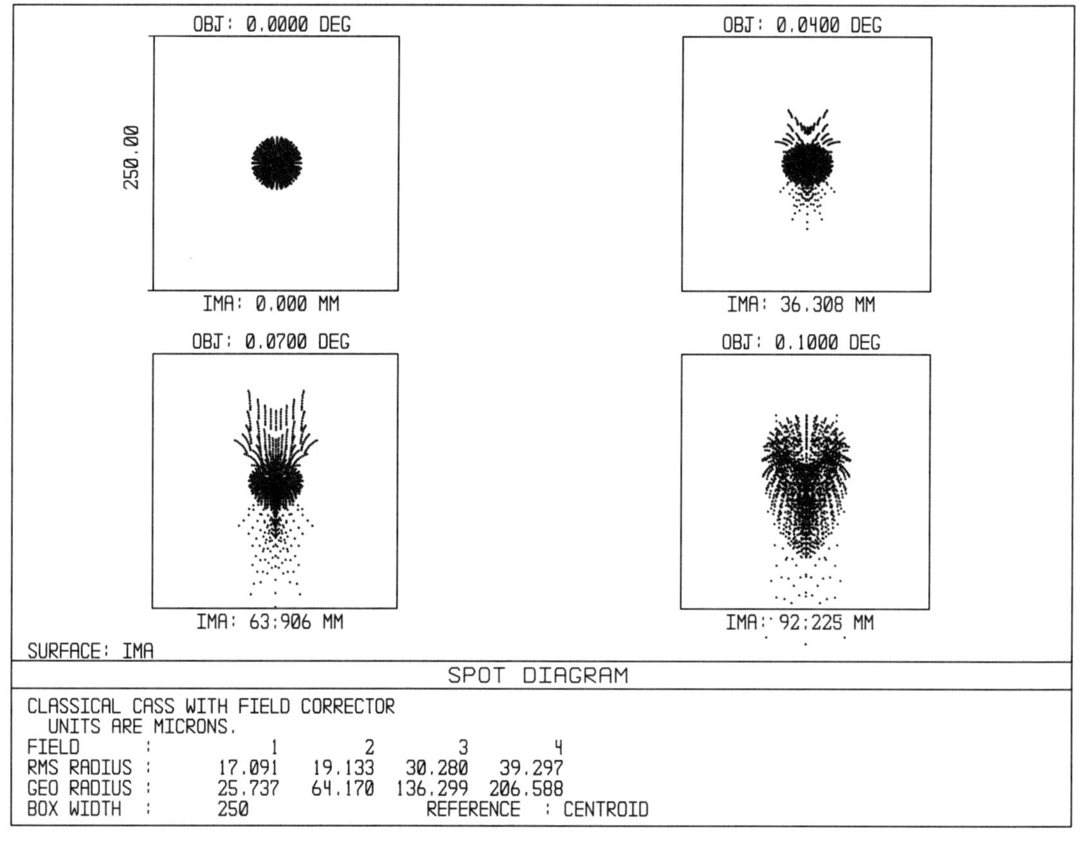

Figure B.5.3.3.

from 50400 mm to 51879 mm, a 2.9% increase. Thus, the field corrector has a small overall negative power in the converging beam. Astronomers would say that in addition to controlling aberrations, the field corrector acts as a very weak Barlow lens.

Listing B.5.3.2

```
System/Prescription Data

File : C:LENS303.ZMX
Title: CLASSICAL CASS WITH FIELD CORRECTOR

GENERAL LENS DATA:

Surfaces          :            15
Stop              :             4
System Aperture   :Entrance Pupil Diameter
Ray aiming        : Off
Apodization       :Uniform, factor =      0.000000
Eff. Focal Len.   :       51879.1 (in air)
Eff. Focal Len.   :       51879.1 (in image space)
Total Track       :       12081.4
Image Space F/#   :       6.17608
Para. Wrkng F/#   :       6.17608
Working F/#       :       6.17495
Obj. Space N.A.   :     4.2e-007
Stop Radius       :          4200
Parax. Ima. Hgt.  :       90.5462
Parax. Mag.       :             0
Entr. Pup. Dia.   :          8400
Entr. Pup. Pos.   :       10066.7
Exit Pupil Dia.   :       341.946
Exit Pupil Pos.   :      -2111.05
Field Type        : Angle in degrees
Maximum Field     :           0.1
Primary Wave      :      0.470000
Lens Units        : Millimeters
Angular Mag.      :       24.5653

Fields          : 4
Field Type:  Angle in degrees
#         X-Value          Y-Value          Weight
1        0.000000         0.000000        3.000000
2        0.000000         0.040000        3.000000
3        0.000000         0.070000        3.000000
4        0.000000         0.100000        2.000000

Vignetting Factors
#         VDX            VDY            VCX            VCY
1      0.000000       0.000000       0.000000       0.000000
2      0.000000       0.000000       0.000000       0.000000
3      0.000000       0.000000       0.000000       0.000000
4      0.000000       0.000000       0.000000       0.000000

Wavelengths      : 9
Units:  Microns
#         Value          Weight
1       0.320000       1.000000
2       0.340000       1.000000
3       0.370000       1.000000
4       0.410000       1.000000
5       0.470000       1.000000
6       0.550000       1.000000
7       0.700000       1.000000
8       0.850000       1.000000
9       1.050000       1.000000

SURFACE DATA SUMMARY:

Surf      Type        Radius      Thickness      Glass      Diameter      Conic
 OBJ  STANDARD      Infinity      Infinity                         0          0
   1  STANDARD      Infinity          1000                  8435.139          0
   2  STANDARD      Infinity          1000                  8431.649          0
   3  STANDARD      Infinity      8066.667                  8428.158          0
 STO  STANDARD        -20160     -8066.667     MIRROR       8401.528         -1
   5  STANDARD     -5033.333      8066.667     MIRROR       1721.955      -2.25
   6  STANDARD      Infinity      1563.929                  481.3342          0
   7  STANDARD      Infinity            75                  242.9646          0
   8  STANDARD      -393.596            25     SILICA            250          0
   9  STANDARD     -318.2755      99.25632                       250          0
  10  STANDARD      Infinity            15        BK7            220          0
  11  STANDARD      Infinity      99.25632                       220          0
  12  STANDARD     -201.8971            20     SILICA            200          0
  13  STANDARD      -307.306      117.3186                       200          0
  14  STANDARD      Infinity    -0.8312169                  184.8221          0
 IMA  STANDARD      Infinity             0                  184.6084          0
```

```
SURFACE DATA DETAIL:

Surface OBJ      : STANDARD
Surface   1      : STANDARD
Surface   2      : STANDARD
 Aperture        : Circular Obscuration
 Minimum Radius  :             0
 Maximum Radius  :          1200
Surface   3      : STANDARD
Surface STO      : STANDARD
 Aperture        : Circular Aperture
 Minimum Radius  :          1000
 Maximum Radius  :          4200
Surface   5      : STANDARD
Surface   6      : STANDARD
Surface   7      : STANDARD
Surface   8      : STANDARD
Surface   9      : STANDARD
Surface  10      : STANDARD
Surface  11      : STANDARD
Surface  12      : STANDARD
Surface  13      : STANDARD
Surface  14      : STANDARD
Surface IMA      : STANDARD

SOLVE AND VARIABLE DATA:

Thickness of   6    : Variable
Curvature of   8    : Variable
Semi Diam   8       : Fixed
Curvature of   9    : Variable
Thickness of   9    : Variable
Semi Diam   9       : Fixed
Semi Diam  10       : Fixed
Thickness of  11    : Solve, pick up value from 9, scaled by 1.00000
Semi Diam  11       : Fixed
Curvature of  12    : Variable
Semi Diam  12       : Fixed
Curvature of  13    : Variable
Thickness of  13    : Solve, marginal ray height = 0.00000
Semi Diam  13       : Fixed
Thickness of  14    : Variable

INDEX OF REFRACTION DATA:
```

Surf	Glass	0.320000	0.340000	0.370000	0.410000	0.470000
0		1.00000000	1.00000000	1.00000000	1.00000000	1.00000000
1		1.00000000	1.00000000	1.00000000	1.00000000	1.00000000
2		1.00000000	1.00000000	1.00000000	1.00000000	1.00000000
3		1.00000000	1.00000000	1.00000000	1.00000000	1.00000000
4	MIRROR	1.00000000	1.00000000	1.00000000	1.00000000	1.00000000
5	MIRROR	1.00000000	1.00000000	1.00000000	1.00000000	1.00000000
6		1.00000000	1.00000000	1.00000000	1.00000000	1.00000000
7		1.00000000	1.00000000	1.00000000	1.00000000	1.00000000
8	SILICA	1.48273942	1.47865136	1.47382577	1.46906629	1.46414628
9		1.00000000	1.00000000	1.00000000	1.00000000	1.00000000
10	BK7	1.54642859	1.54134356	1.53539019	1.52956877	1.52360494
11		1.00000000	1.00000000	1.00000000	1.00000000	1.00000000
12	SILICA	1.48273942	1.47865136	1.47382577	1.46906629	1.46414628
13		1.00000000	1.00000000	1.00000000	1.00000000	1.00000000
14		1.00000000	1.00000000	1.00000000	1.00000000	1.00000000
15		1.00000000	1.00000000	1.00000000	1.00000000	1.00000000

Surf	Glass	0.550000	0.700000	0.850000	1.050000
0		1.00000000	1.00000000	1.00000000	1.00000000
1		1.00000000	1.00000000	1.00000000	1.00000000
2		1.00000000	1.00000000	1.00000000	1.00000000
3		1.00000000	1.00000000	1.00000000	1.00000000
4	MIRROR	1.00000000	1.00000000	1.00000000	1.00000000
5	MIRROR	1.00000000	1.00000000	1.00000000	1.00000000
6		1.00000000	1.00000000	1.00000000	1.00000000
7		1.00000000	1.00000000	1.00000000	1.00000000
8	SILICA	1.45991088	1.45529247	1.45249829	1.44979976
9		1.00000000	1.00000000	1.00000000	1.00000000
10	BK7	1.51852239	1.51306400	1.50984013	1.50682021
11		1.00000000	1.00000000	1.00000000	1.00000000
12	SILICA	1.45991088	1.45529247	1.45249829	1.44979976
13		1.00000000	1.00000000	1.00000000	1.00000000
14		1.00000000	1.00000000	1.00000000	1.00000000
15		1.00000000	1.00000000	1.00000000	1.00000000

```
ELEMENT VOLUME DATA:

Units are cubic cm.
Values are only accurate for plane and spherical surfaces.
Element surf   8 to   9 volume :    1103.951232
Element surf  10 to  11 volume :     570.199067
Element surf  12 to  13 volume :     774.631632
```

B.5.8 The Ritchey-Chrétien with Field Corrector

Perhaps a more complex refractive field corrector would be able to produce acceptable images from the classical Cassegrain. But more lens elements would present other practical problems, such as more light loss, more ghost images and

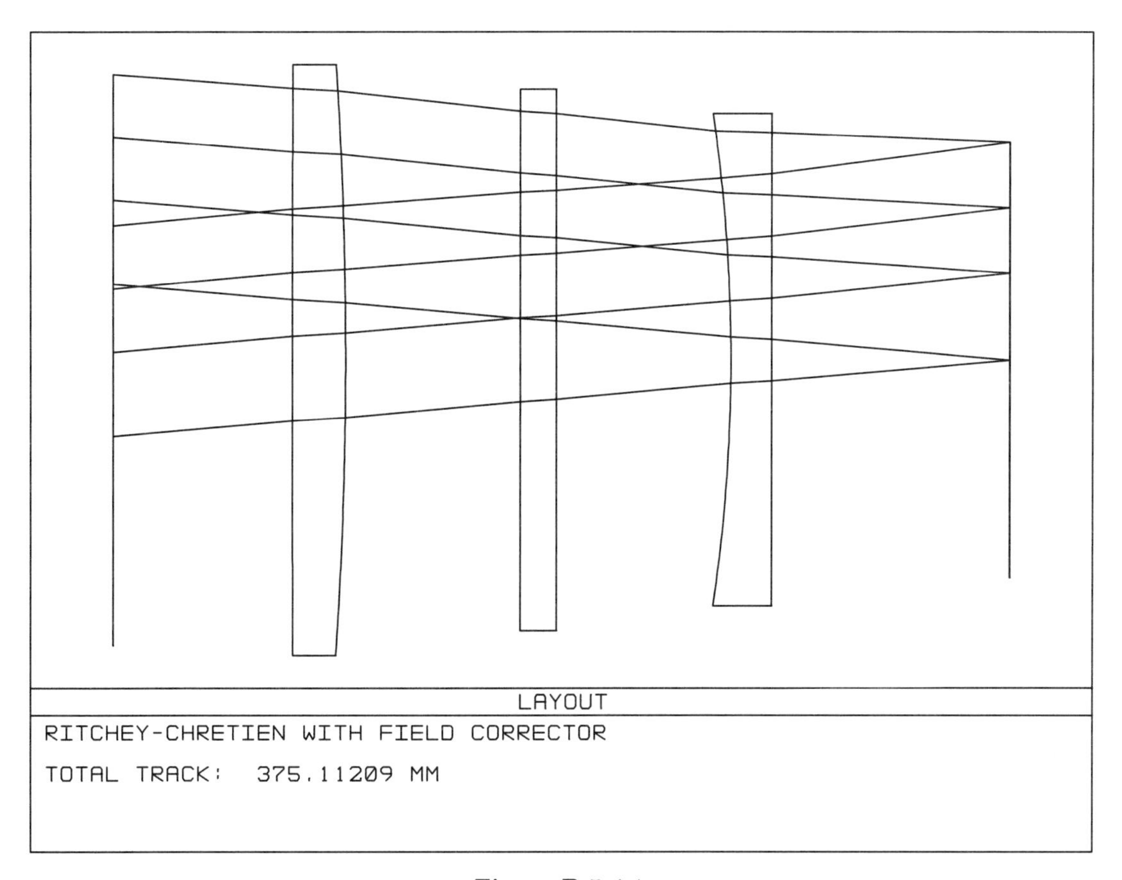

Figure B.5.4.1.

stray light, more fabrication difficulties, and so forth. Aspheric surfaces, if they proved useful, would be even harder to make. A better approach is to change the main telescope to a Ritchey-Chrétien type of Cassegrain, which is inherently coma-free. Then the field corrector does not have to be concerned with coma and need only remove astigmatism and field curvature. As mentioned earlier, the design procedure for a Ritchey-Chrétien field corrector is the same as for a classical Cassegrain field corrector.

Figure B.5.4.1 shows the layout of the optimized Ritchey-Chrétien field corrector without the rest of the telescope. Compare this with the layout of the classical Cassegrain corrector in Figure B.5.3.1. Note that the lens elements of the Ritchey-Chrétien corrector are no longer highly meniscus. Instead, each element has one surface that is nearly plano. These are very easy lenses to fabricate, especially if one or both of the nearly plano surfaces can be made exactly flat (the reoptimized system must not be too degraded). Again, the rear negative element has greater power than the front positive element to flatten the field.

Figure B.5.4.2 shows the transverse ray fan plots. Immediately note the scale: ± 10 μm. The aberrations are now well controlled. On-axis, defocus balances third-order spherical, with very little chromatic variation. Off-axis, spherical and defocus remain and are augmented by higher-order aberrations and by chromatic variations.

The most important chromatic variation is primary lateral color. Note carefully, however, that the lateral color changes sign between the 0.7 and 1.0 fields. This sign reversal is more clearly shown by the lateral color plot in Figure B.5.4.3. Lateral color is zero at the 0.83 field. Note that the residual lateral color at other field angles is primary lateral color, not secondary lateral color. The ray fan curves for the different wavelengths are all arrayed in order; secondary color requires a U-shaped progression. Eliminating primary lateral color would require more elements and

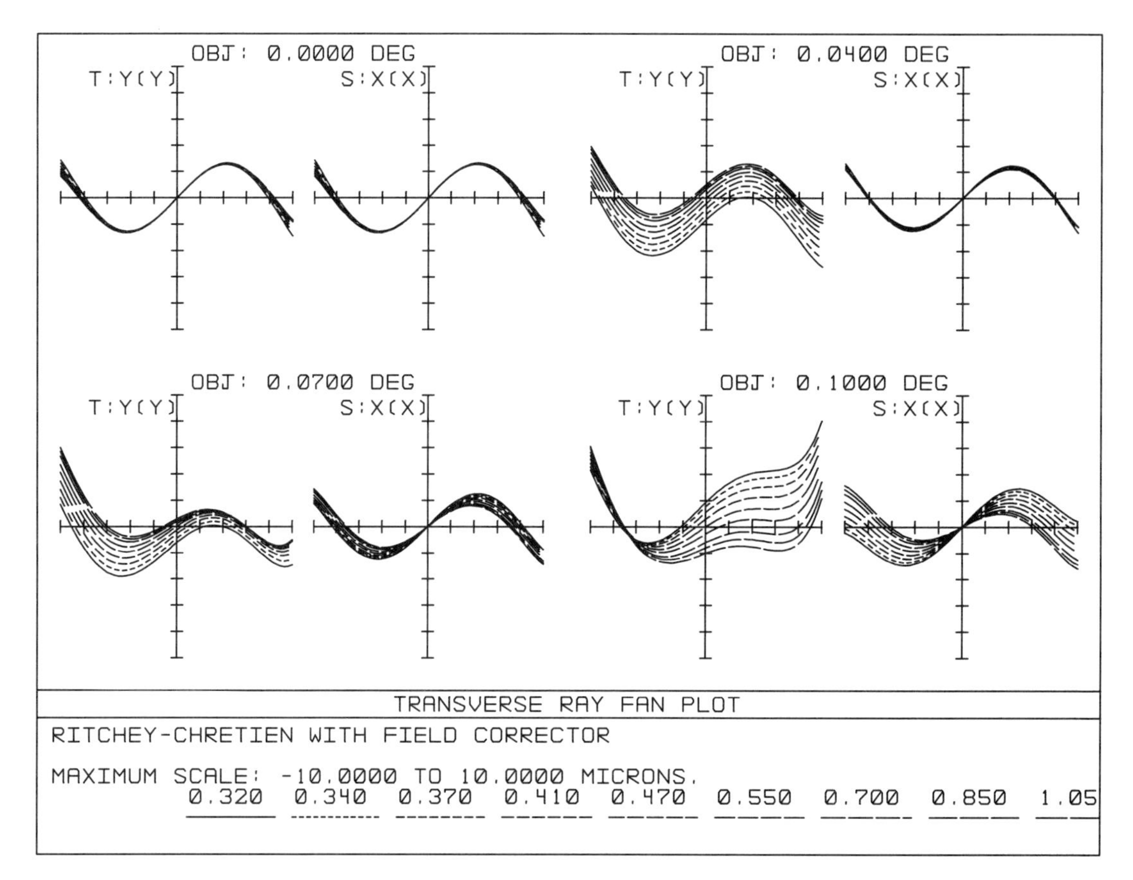

Figure B.5.4.2.

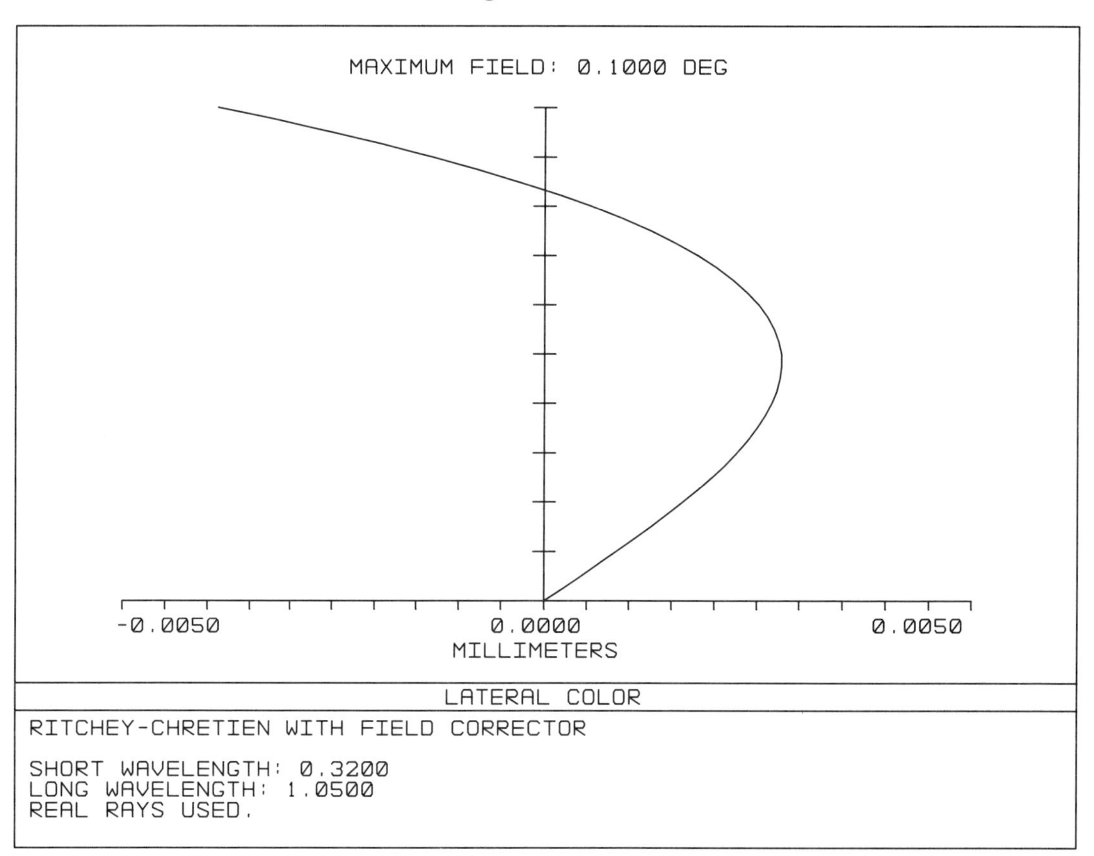

Figure B.5.4.3.

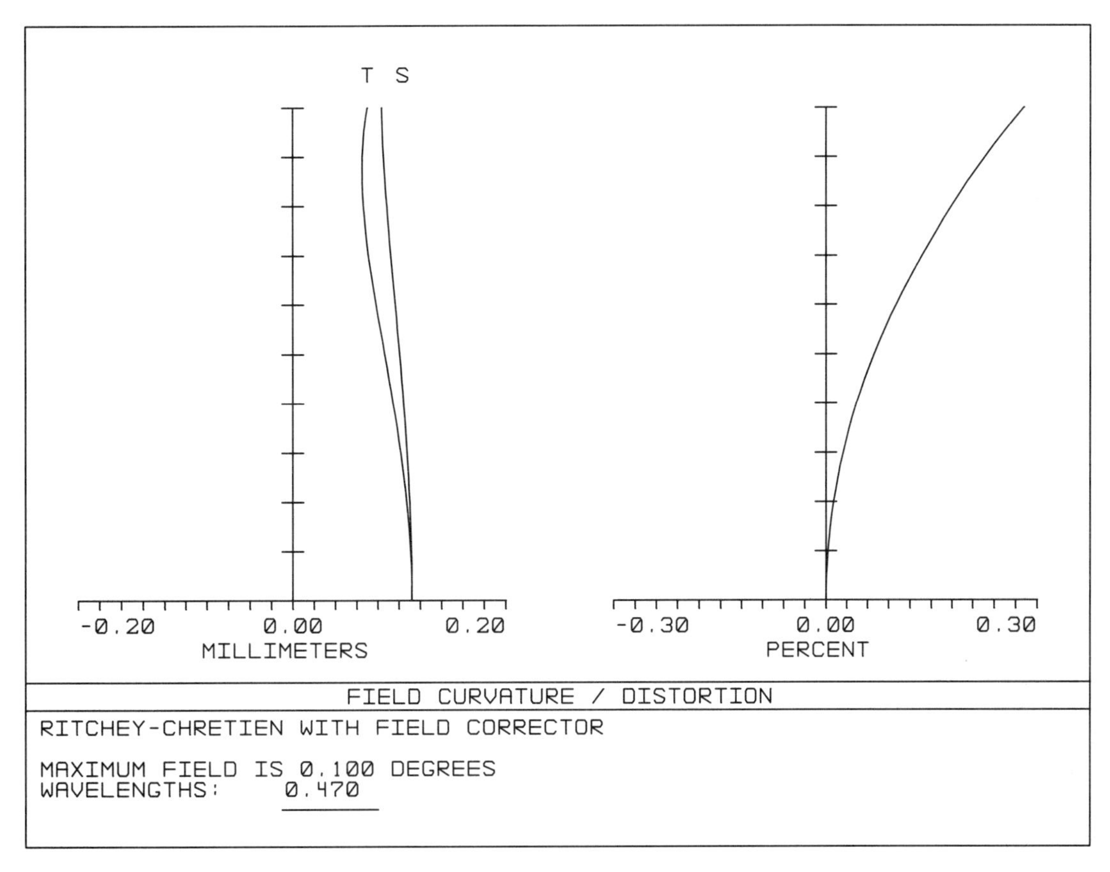

Figure B.5.4.4.

two different glass types.

The left side of Figure B.5.4.4 shows the astigmatic field curvature plot for rays very near the chief ray. The tangential and sagittal curves are almost coincident and straight, indicating very little astigmatism on a flat image surface. Both curves are displaced by the amount of the paraxial defocus. The right side of Figure B.5.4.4 shows the percent distortion plot. There is a very small amount of distortion that, if necessary, can be calibrated and removed by later image processing.

Figure B.5.4.5 is the polychromatic spot diagram. The squares are 24 μm on a side, the same size as the pixel. The spots are now considerably smaller than the pixel size and easily satisfy system requirements.

Figure B.5.4.6 is the matrix spot diagram. Now the wavelength variations are more clearly seen. For each off-axis field, note the monotonic displacements with wavelength caused by primary lateral color. For the two outer fields, note the lateral color sign change. Surrounding each spot is a circle that represents the diameter of the Airy diffraction disk for that wavelength. Amazingly, across the field and even in the ultraviolet, this 8.4-meter telescope with field corrector is nearly diffraction limited. The residual geometrical aberrations are almost lost in the diffraction, and both are completely lost in the seeing.

Figures B.5.4.7 through B.5.4.10 show three-dimensional polychromatic diffraction point spread functions. Again, image quality is almost diffraction limited and is much better than required for a 24 μm pixel. Note that in these and all other diffraction evaluations given here, no atmospheric seeing effects are included. In practice, seeing will enlarge these point spread functions beyond recognition, which is as it should be if seeing is to be the fundamental limit on resolution.

Figure B.5.4.11 is a plot of polychromatic Strehl ratio versus field angle. Not only are the Strehl ratios high, but the curve has a second peak near 0.085°, close to where the lateral color passes through zero.

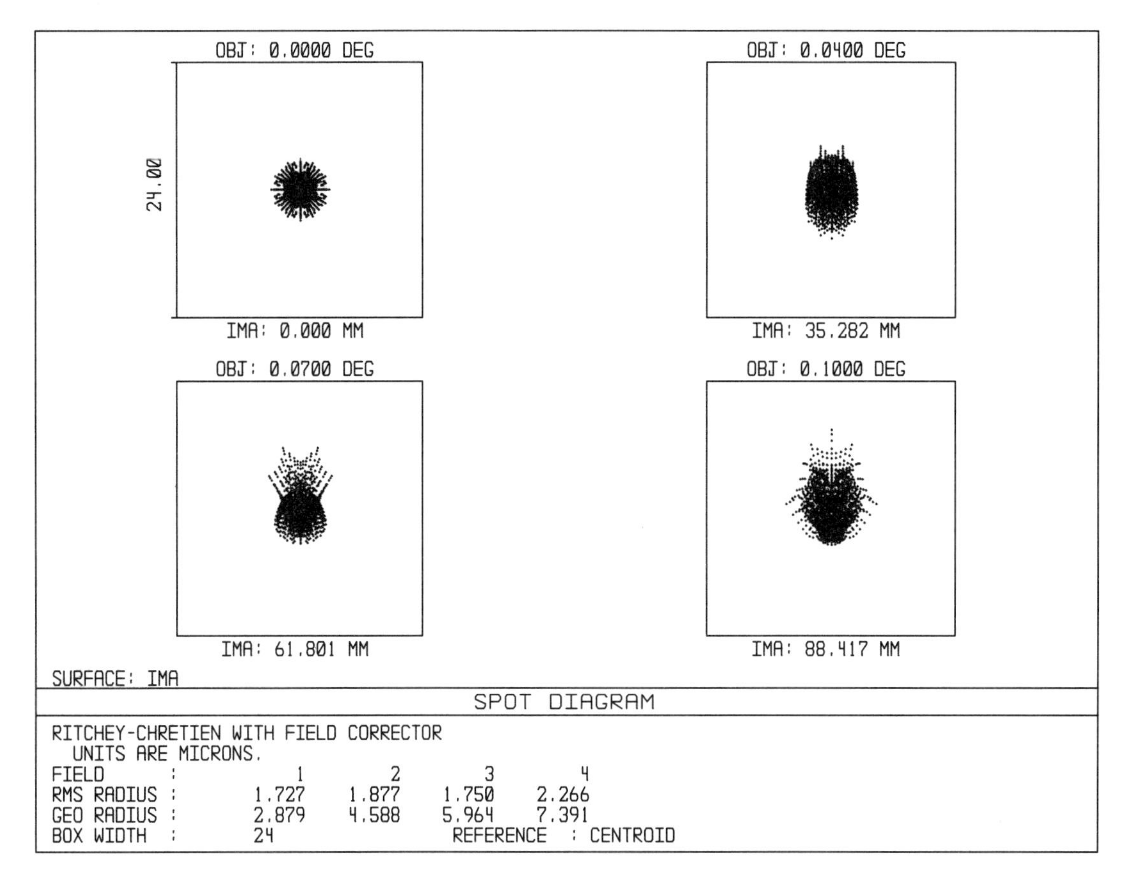

Figure B.5.4.5.

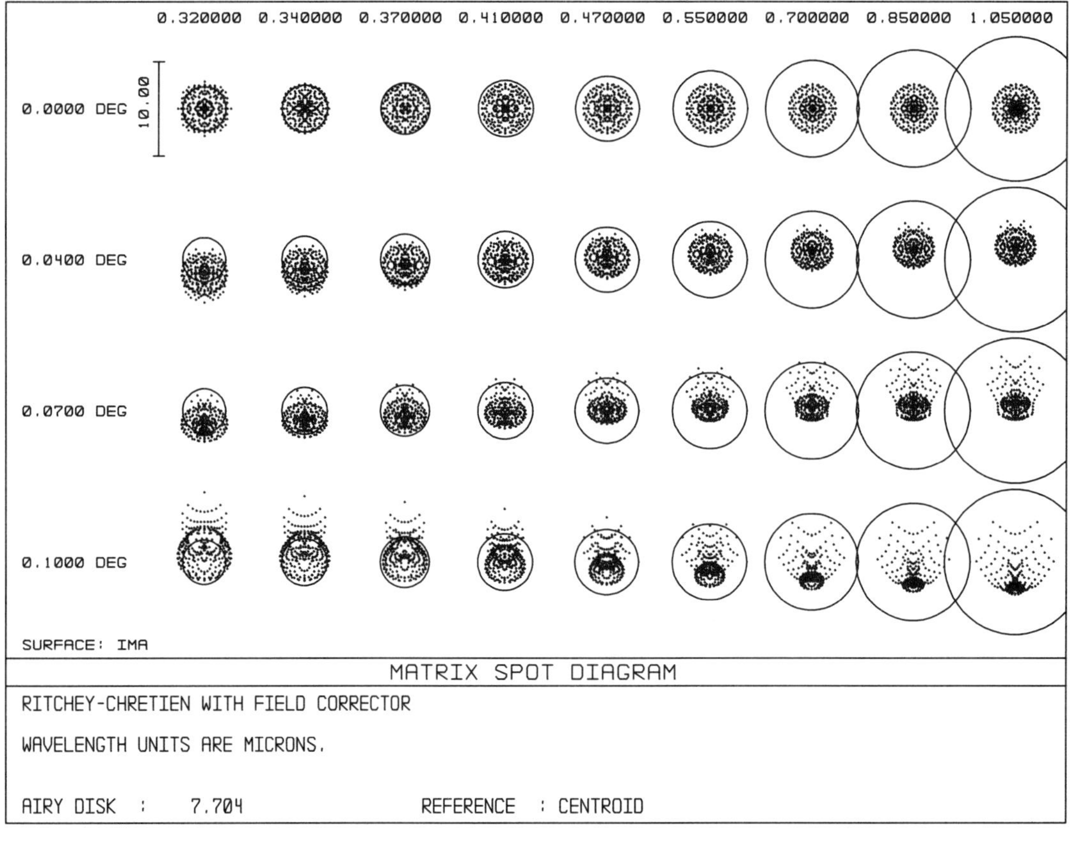

Figure B.5.4.6.

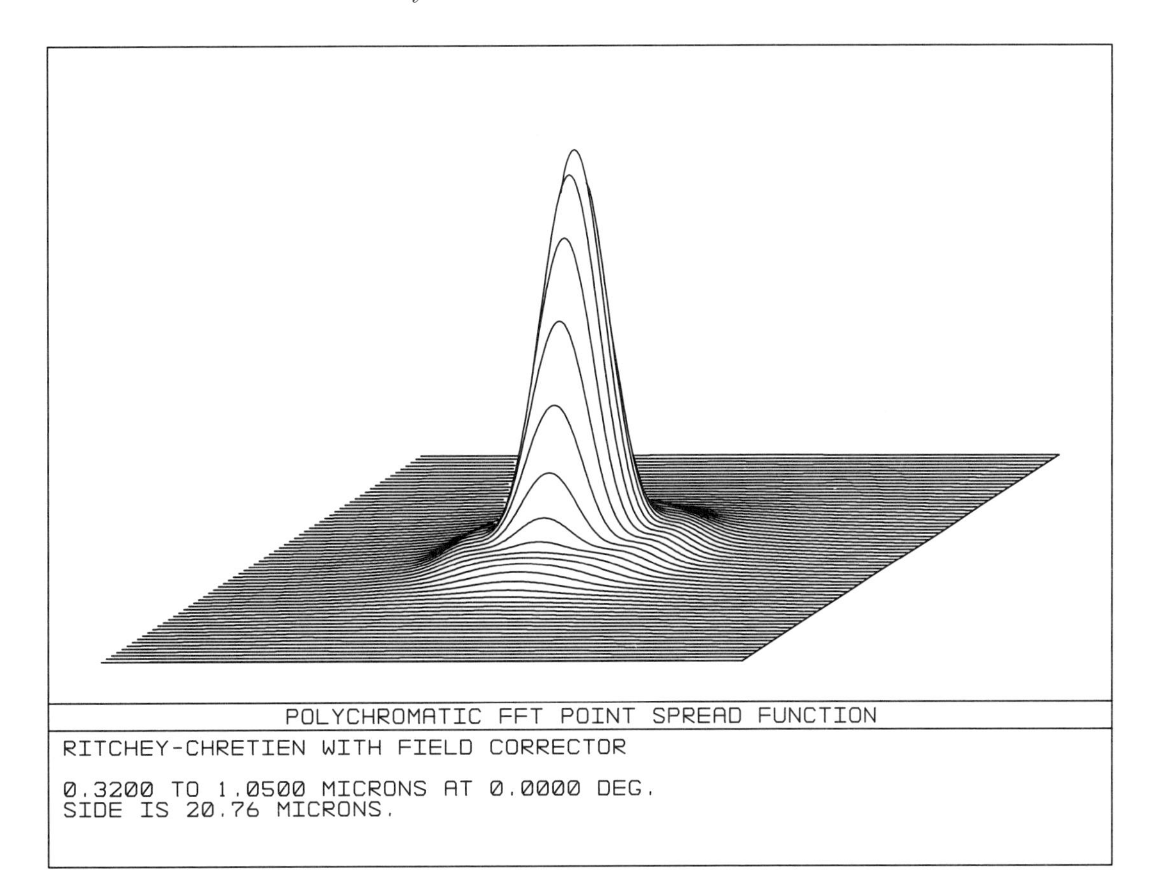

Figure B.5.4.7.

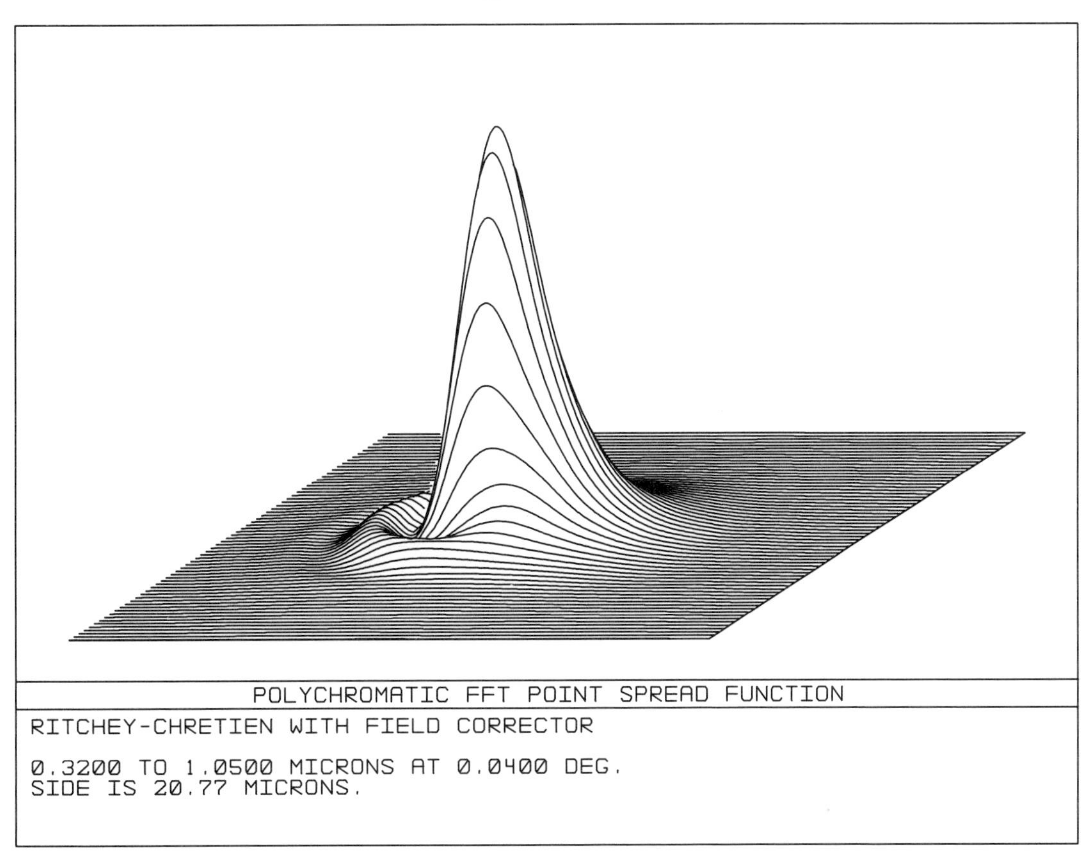

Figure B.5.4.8.

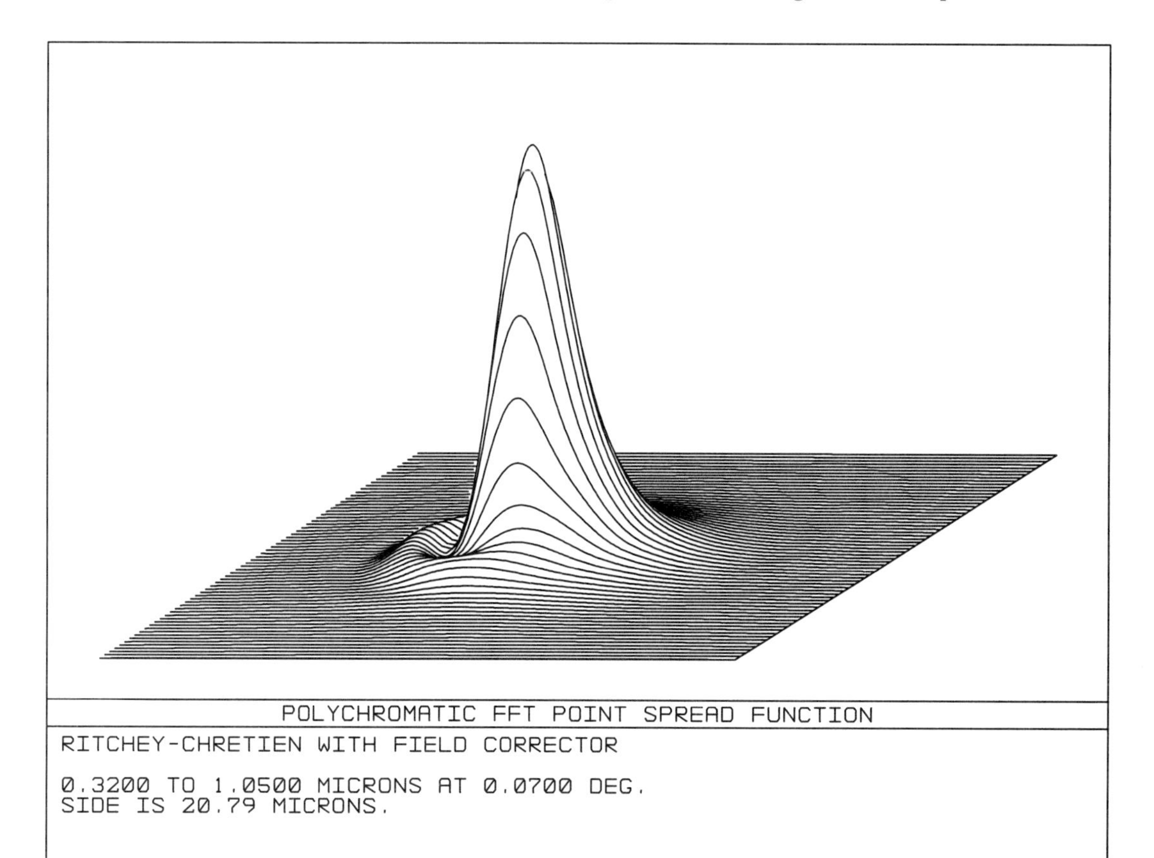

Figure B.5.4.9.

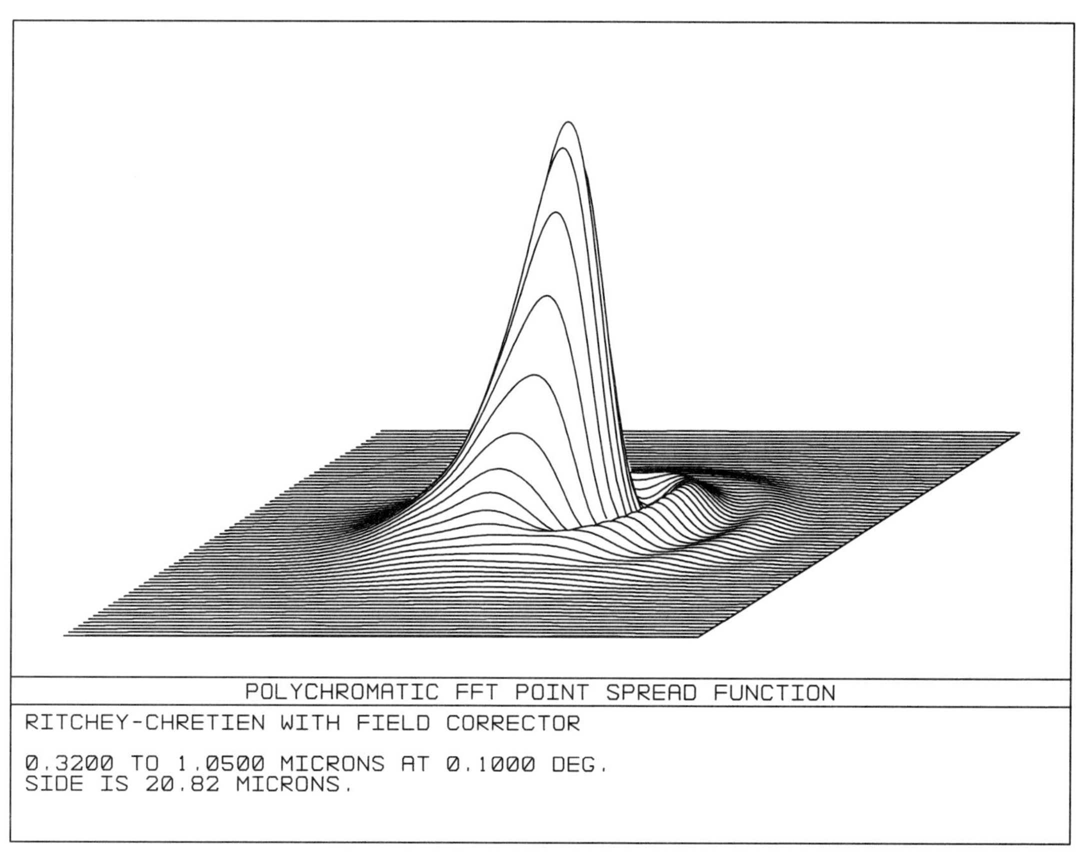

Figure B.5.4.10.

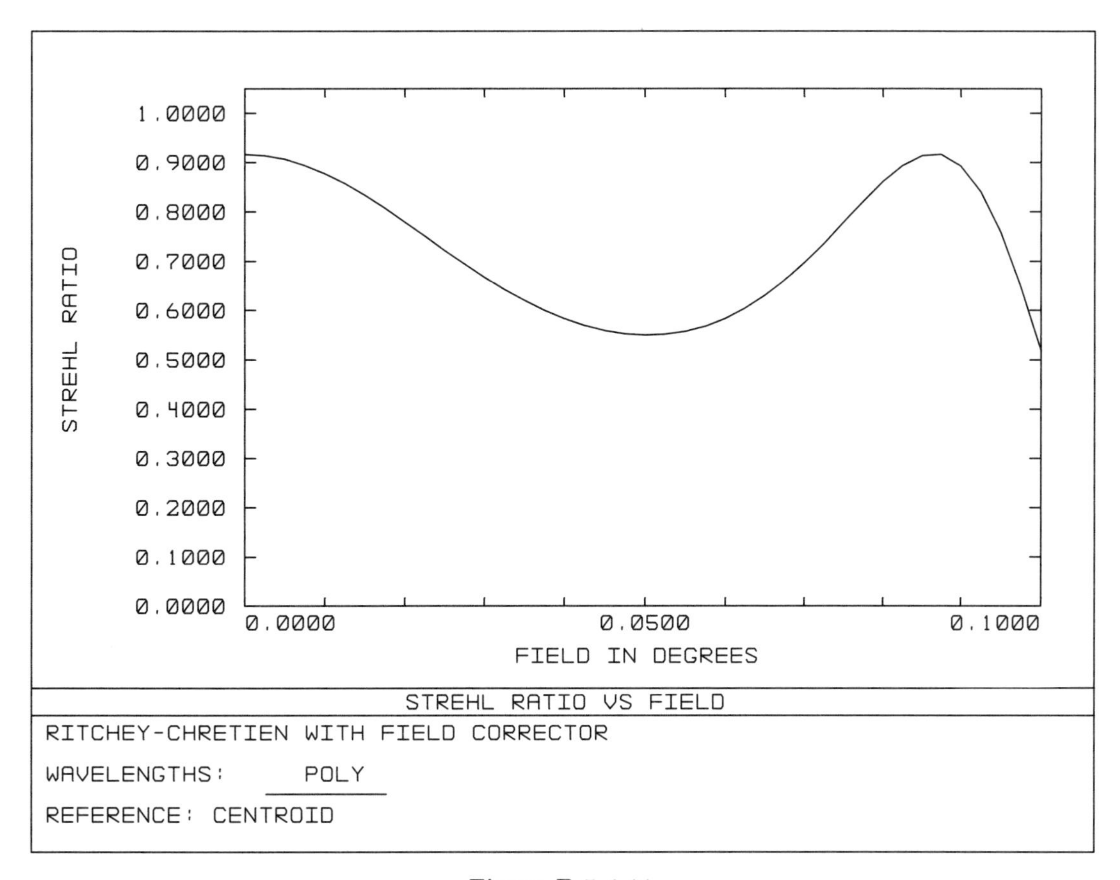

Figure B.5.4.11.

Figures B.5.4.12 and B.5.4.13 are the polychromatic diffraction encircled energy and MTF curves, respectively. Again the telescope plus field corrector is shown to be almost diffraction limited.

But before becoming too pleased and confident, a word must be said about flexure tolerances. A big telescope is very massive. The optics, hanging on the ends of a long tube, cause the ends of the tube to sag. In older telescopes, such as those at Mt. Wilson, the tube becomes slightly curved. Such a sag produces both tilt and decenter errors. Thus, the optical collimation slightly changes as the telescope is moved to different parts of the sky. For high-quality imaging with a field corrector, these alignment errors are quite noticeable.

During the 200-inch telescope project in the 1930s, it was realized that the required tube could never be made sufficiently rigid. The tube is 55 feet (16.8 meters) long, and at each end are roughly 35 ton (32 metric ton) weights (the primary mirror alone without its cell weighs 14.5 tons). Sag was unavoidable. Thus, Mark Serrurier developed what is now called the Serrurier truss. A Serrurier truss is a parallelogram tube design that allows the two ends to sag, but by equal amounts and without tilting. Thus, the optics stay much closer to alignment as the tube sags.[8]

Measurements of the actual 200-inch reveal that at no orientation of the telescope do the two ends of the tube differentially decenter by more than ± 0.01 inches (± 0.25 mm). This alignment tolerance is quite adequate for Palomar, where the primary mirror is $f/3.3$. But the sensitivity of a Cassegrain telescope to decollimation varies as the square of the speed of the primary mirror. Thus, the faster $f/1.2$ primary speed in the present design example requires an even tighter tolerance.

[8]See David O. Woodbury, *The Glass Giant of Palomar*, pp. 249–252. Also see I.S. Bowen, "The 200-Inch Hale Telescope," in Gerald P. Kuiper and Barbara M. Middlehurst, eds., *Telescopes*, pp. 1–15.

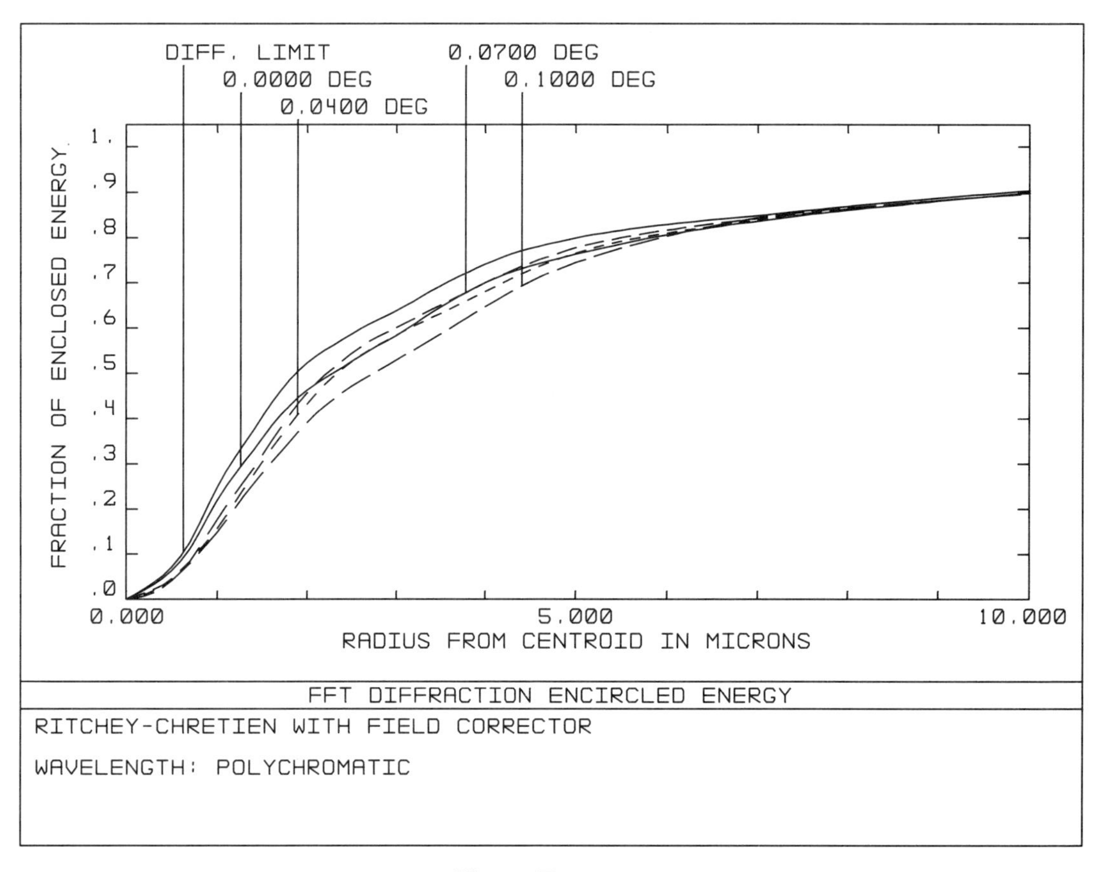

Figure B.5.4.12.

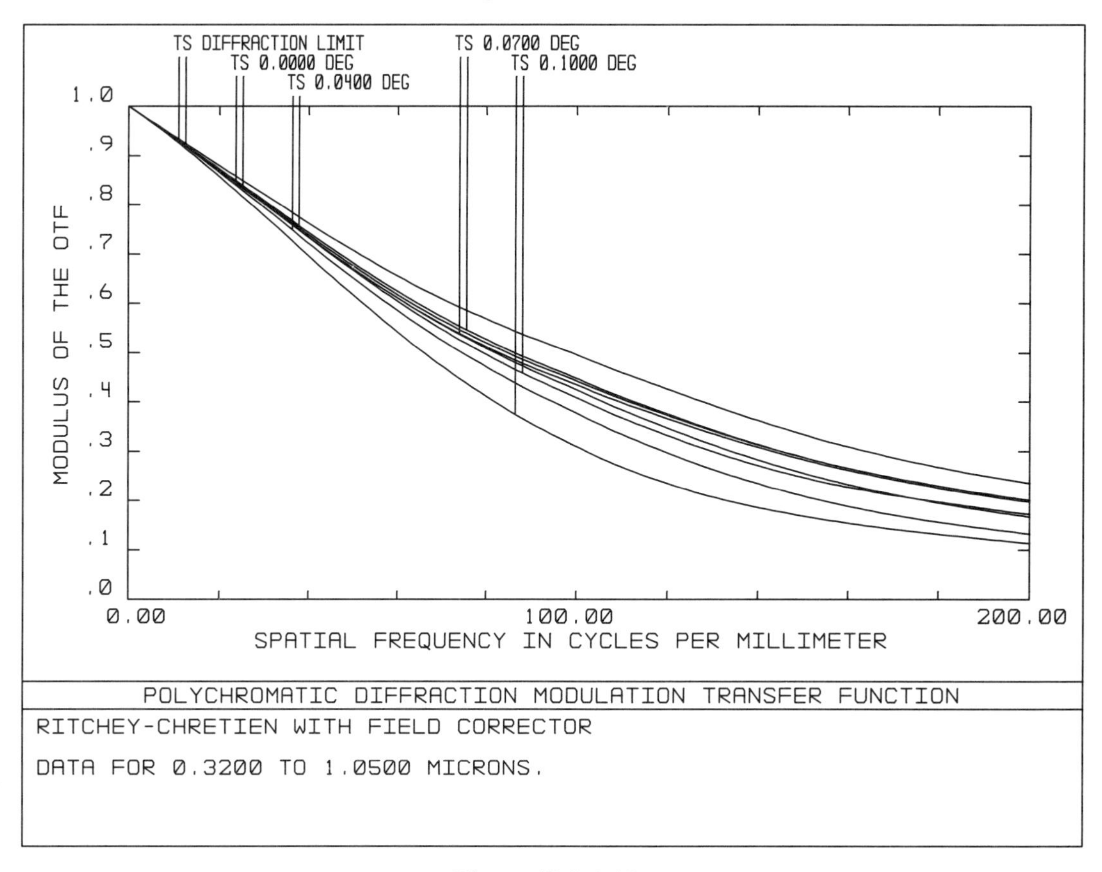

Figure B.5.4.13.

Figure B.5.4.14 is the polychromatic spot diagram for the 8.4-meter Ritchey-Chrétien telescope plus field corrector when the secondary mirror is decentered (without tilt) with respect to the primary mirror by 0.05 mm. This error is 5 times less than the Palomar error. It is assumed that the field corrector is rigidly connected to the primary mirror. Compare these spots to those for the perfectly collimated telescope in Figure B.5.4.5.

The squares in Figure B.5.4.14 are 48 μm on a side, equal to the size of the 2x2 matrix of pixels designed to match the seeing disk. Apparently, given the seeing, a decentration of ± 0.05 mm is close to the limit that can be allowed. A more precise tolerance value can only be found by doing a complete system tolerance analysis that includes the other errors that inevitably will be present. But assuming ± 0.05 mm, it is unlikely that any passive system, not even a Serrurier truss, can hold this tight tolerance. Fortunately, in recent years, active computer-controlled alignment techniques have been developed that can do the job.

There is another practical problem, one that is present in any optical system containing lenses. This problem is multiple reflections between the several lens surfaces. These unwanted reflections produce ghost images and other stray light. When imaging a field of stars, every combination of two lens surfaces produces a ghost of every star. Ideally, you want these ghosts to be so far out of focus that they are undetectable, except perhaps for the ghosts of the very brightest stars. Being out of focus also identifies the ghosts as ghosts and not additional stars. Anti-reflection coatings can help to suppress ghosts, but they are not completely effective (especially over the present very large wavelength range). Thus, a ghost image analysis was performed on the Ritchey-Chrétien telescope with field corrector to determine whether the field corrector produces any in-focus ghosts.

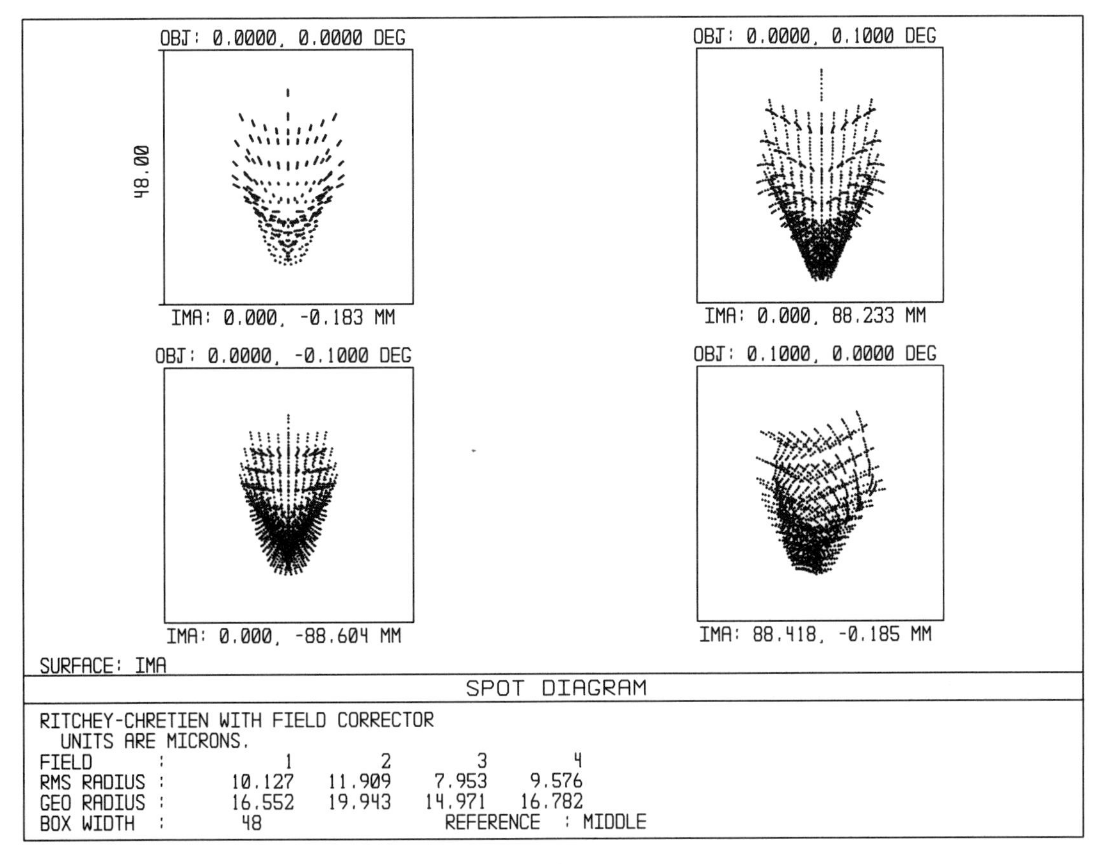

Figure B.5.4.14.

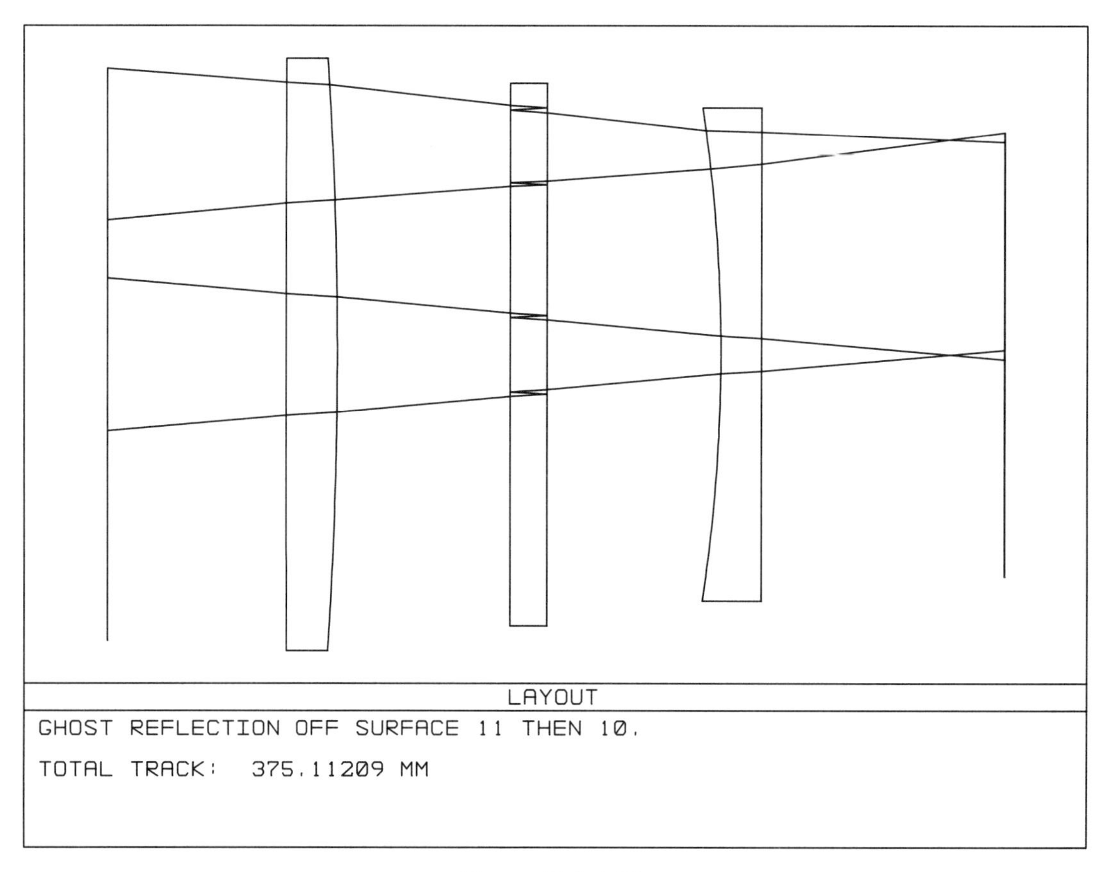

Figure B.5.4.15.

Figure B.5.4.15 is a layout showing the path of the ghost light reflecting between the two surfaces of the filter. Note that plane-parallel plates are notorious for producing in-focus ghosts. Figure B.5.4.16 is the spot diagram of the filter ghost on the CCD, both on-axis and at the edge of the field. In the present case, the filter is thick enough and the beam is fast enough that the ghost is about 4 mm in diameter anywhere across the field and is easily identifiable as a ghost.

Figure B.5.4.17 is a layout showing the path of the ghost light reflecting between the two surfaces of the rear element. The ghosts here do not remain near their originating stars. Figure B.5.4.18 shows the spot diagrams. The ghosts are increasingly aberrated with off-axis distance. Recall that the edge of the $\pm 0.10°$ field is 88 mm off-axis. Thus, the more aberrated and more concentrated ghosts are outside the field of view. Within the field of view, all ghosts remain larger than about 3 mm in diameter and are identifiable as ghosts.

All other combinations of two lens surfaces were also tested for ghosts. Even the image surface, which can also reflect light, was included in the analysis. Fortunately, the ghosts given above are the worst ones. The remaining ghosts are much more out of focus, with correspondingly lower surface brightnesses (irradiances) on the CCD.

Finally, there is one more practical problem for any ground-based astronomical telescope with very high angular resolution. This problem is atmospheric dispersion. Every star is drawn out by differential atmospheric refraction into a small spectrum. The effect increases as the star moves farther away from the zenith. Removing this chromatic spreading of the point spread function requires that additional auxiliary optics be added some distance in front of focus. These optics are not considered here.[9]

[9]For a discussion of atmospheric dispersion correctors, see R.N. Wilson, *Reflecting Telescope Optics I: Basic Design Theory and its Historical Development*, pp. 379–389.

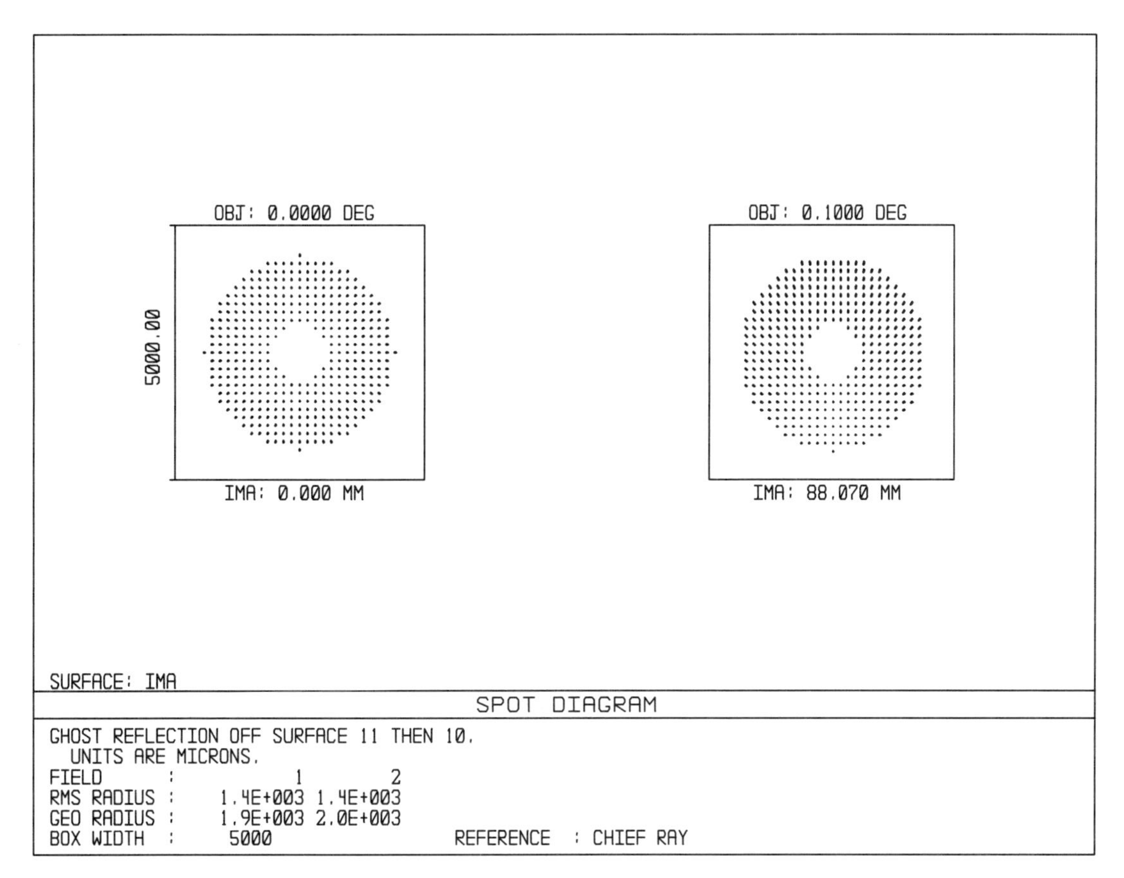

Figure B.5.4.16.

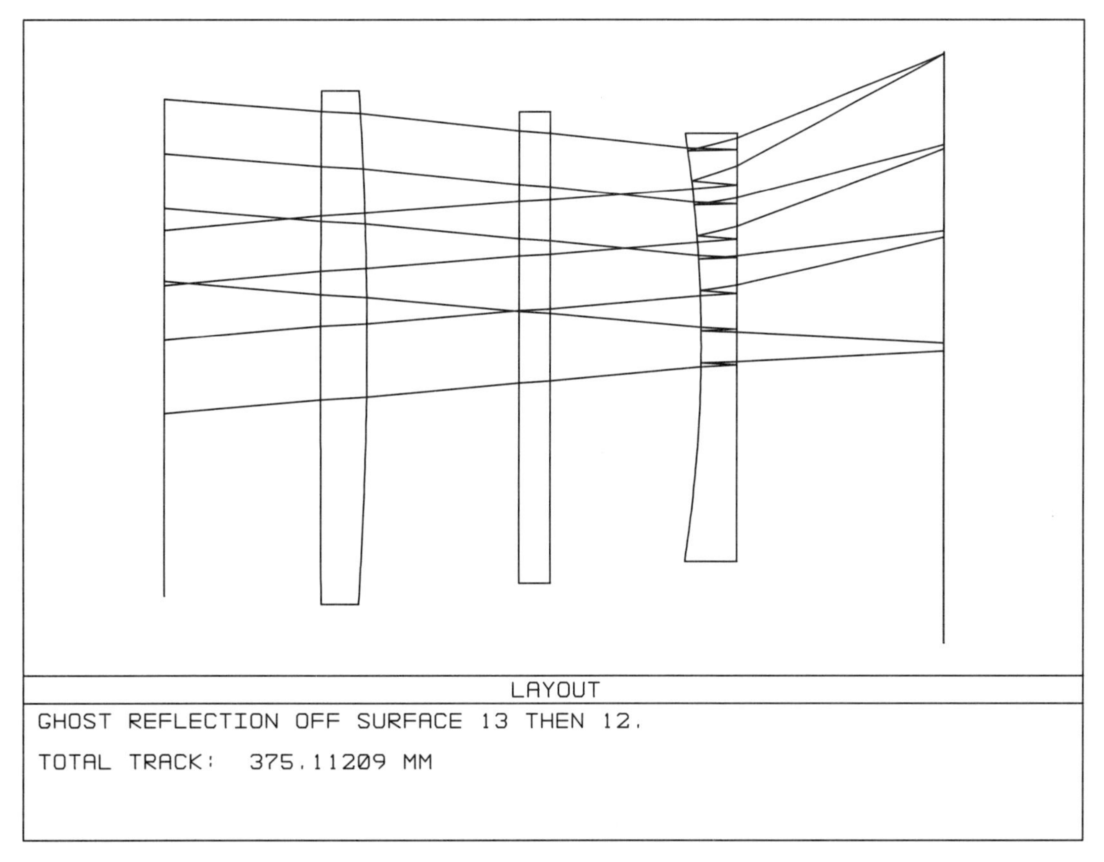

Figure B.5.4.17.

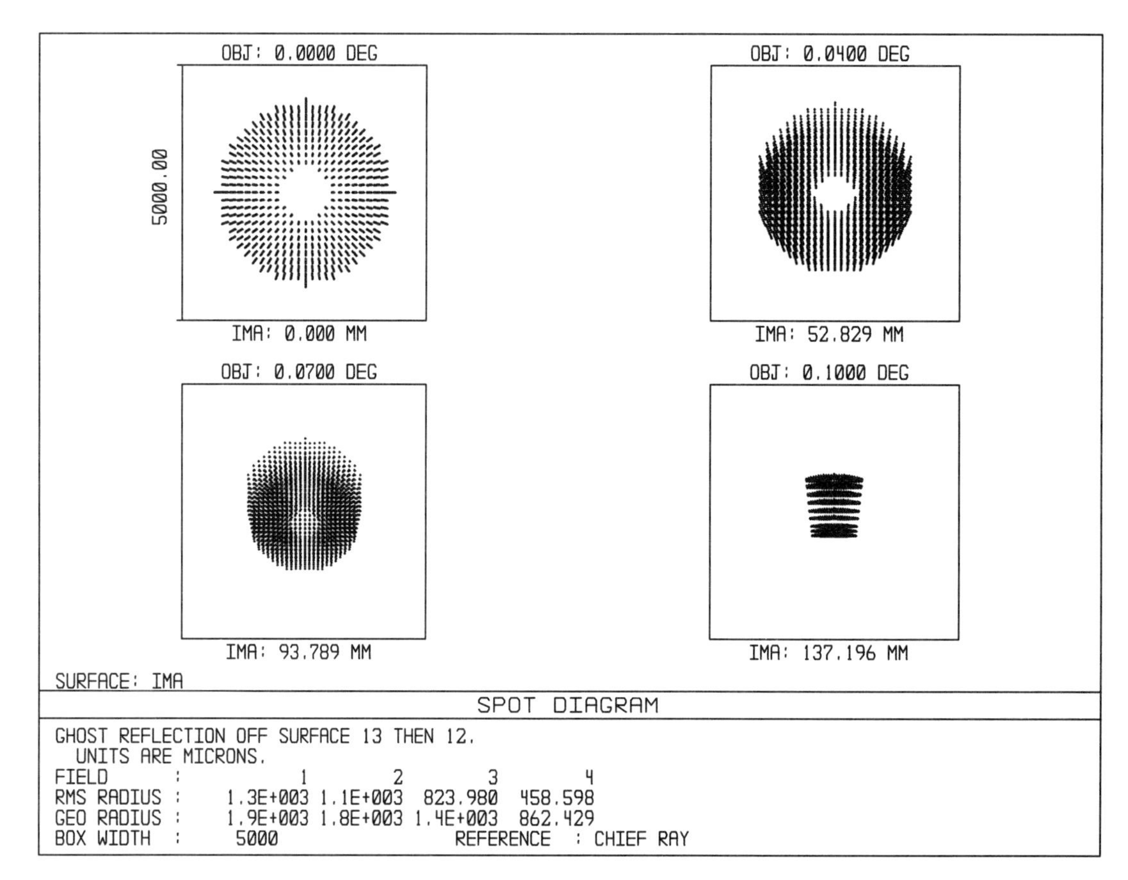

Figure B.5.4.18.

Listing B.5.4 gives the final optical prescription of the Ritchey-Chrétien telescope with field corrector. Note that the field corrector changes the effective focal length from 50400 mm to 50519 mm, only a 0.24% increase. Thus, the Ritchey-Chrétien field corrector has a much smaller overall negative power than the classical Cassegrain field corrector.

The entire field corrector is very close to the focus. Thus, the field corrector lenses are not much larger than the linear field of view. As a practical system, the field corrector lenses should probably be mechanically part of the detector module rather than part of the telescope. When the CCD is removed from the telescope to allow another instrument, such as a spectrograph, to be attached instead, the field corrector lenses would also be removed. This procedure has the advantage of avoiding the inevitable light losses in the field corrector when a wide field is not required. The rule in observational astronomy, where every photon counts, is that you always try to use the least number of optical surfaces that will do the job (otherwise you may end up with a system that will do everything but transmit light). This need for efficiency and versatility is the reason that the main telescope must be usable alone without the field corrector. It is fortunate, but no coincidence, that the aplanatic Ritchey-Chrétien configuration is preferable when used alone as well as with the field corrector.

In this chapter, afocal zero-power correctors were considered that have all-spherical surfaces and are located in the converging beam close to focus. In the next chapter, a different type of zero-power corrector will be considered that has a polynomial aspheric surface and is located in front of the rest of the optics.

Listing B.5.4

```
System/Prescription Data

File : C:LENS304.ZMX
Title: RITCHEY-CHRETIEN WITH FIELD CORRECTOR

GENERAL LENS DATA:

Surfaces          :          15
Stop              :           4
System Aperture   :Entrance Pupil Diameter
Ray aiming        : Off
Apodization       :Uniform, factor =      0.000000
Eff. Focal Len.   :     50519.1 (in air)
Eff. Focal Len.   :     50519.1 (in image space)
Total Track       :     12072.5
Image Space F/#   :     6.01418
Para. Wrkng F/#   :     6.01418
Working F/#       :     6.01368
Obj. Space N.A.   :    4.2e-007
Stop Radius       :        4200
Parax. Ima. Hgt.  :     88.1725
Parax. Mag.       :           0
Entr. Pup. Dia.   :        8400
Entr. Pup. Pos.   :     10066.7
Exit Pupil Dia.   :     344.544
Exit Pupil Pos.   :    -2072.03
Field Type        : Angle in degrees
Maximum Field     :         0.1
Primary Wave      :    0.470000
Lens Units        : Millimeters
Angular Mag.      :     24.3801

Fields            : 4
Field Type:  Angle in degrees
#          X-Value           Y-Value           Weight
1         0.000000          0.000000         3.000000
2         0.000000          0.040000         3.000000
3         0.000000          0.070000         3.000000
4         0.000000          0.100000         2.000000

Vignetting Factors
#          VDX              VDY              VCX              VCY
1       0.000000         0.000000         0.000000         0.000000
2       0.000000         0.000000         0.000000         0.000000
3       0.000000         0.000000         0.000000         0.000000
4       0.000000         0.000000         0.000000         0.000000

Wavelengths       : 9
Units:  Microns
#          Value             Weight
1         0.320000          1.000000
2         0.340000          1.000000
3         0.370000          1.000000
4         0.410000          1.000000
5         0.470000          1.000000
6         0.550000          1.000000
7         0.700000          1.000000
8         0.850000          1.000000
9         1.050000          1.000000

SURFACE DATA SUMMARY:

Surf      Type         Radius      Thickness          Glass      Diameter         Conic
 OBJ   STANDARD      Infinity       Infinity                            0             0
   1   STANDARD      Infinity           1000                     8435.139             0
   2   STANDARD      Infinity           1000                     8431.649             0
   3   STANDARD      Infinity       8066.667                     8428.158             0
 STO   STANDARD        -20160      -8066.667         MIRROR      8401.527     -1.020205
   5   STANDARD     -5033.333       8066.667         MIRROR      1724.944     -2.447703
   6   STANDARD      Infinity       1630.716                     481.1789             0
   7   STANDARD      Infinity             75                     232.0045             0
   8   STANDARD      25265.93             22         SILICA           240             0
   9   STANDARD     -1822.528       72.68064                          240             0
  10   STANDARD      Infinity             15            BK7           220             0
  11   STANDARD      Infinity       72.68064                          220             0
  12   STANDARD     -659.3149             17         SILICA           200             0
  13   STANDARD     -890946.3       100.7508                          200             0
  14   STANDARD      Infinity     -0.1120924                     176.8721             0
 IMA   STANDARD      Infinity              0                     176.8479             0
```

SURFACE DATA DETAIL:

```
Surface OBJ      : STANDARD
Surface   1      : STANDARD
Surface   2      : STANDARD
 Aperture        : Circular Obscuration
 Minimum Radius  :          0
 Maximum Radius  :       1200
Surface   3      : STANDARD
Surface STO      : STANDARD
 Aperture        : Circular Aperture
 Minimum Radius  :       1000
 Maximum Radius  :       4200
Surface   5      : STANDARD
Surface   6      : STANDARD
Surface   7      : STANDARD
Surface   8      : STANDARD
Surface   9      : STANDARD
Surface  10      : STANDARD
Surface  11      : STANDARD
Surface  12      : STANDARD
Surface  13      : STANDARD
Surface  14      : STANDARD
Surface IMA      : STANDARD
```

SOLVE AND VARIABLE DATA:

```
Thickness of   6   : Variable
Curvature of   8   : Variable
Semi Diam   8      : Fixed
Curvature of   9   : Variable
Thickness of   9   : Variable
Semi Diam   9      : Fixed
Semi Diam  10      : Fixed
Thickness of  11   : Solve, pick up value from 9, scaled by 1.00000
Semi Diam  11      : Fixed
Curvature of  12   : Variable
Semi Diam  12      : Fixed
Curvature of  13   : Variable
Thickness of  13   : Solve, marginal ray height = 0.00000
Semi Diam  13      : Fixed
Thickness of  14   : Variable
```

INDEX OF REFRACTION DATA:

Surf	Glass	0.320000	0.340000	0.370000	0.410000	0.470000
0		1.00000000	1.00000000	1.00000000	1.00000000	1.00000000
1		1.00000000	1.00000000	1.00000000	1.00000000	1.00000000
2		1.00000000	1.00000000	1.00000000	1.00000000	1.00000000
3		1.00000000	1.00000000	1.00000000	1.00000000	1.00000000
4	MIRROR	1.00000000	1.00000000	1.00000000	1.00000000	1.00000000
5	MIRROR	1.00000000	1.00000000	1.00000000	1.00000000	1.00000000
6		1.00000000	1.00000000	1.00000000	1.00000000	1.00000000
7		1.00000000	1.00000000	1.00000000	1.00000000	1.00000000
8	SILICA	1.48273942	1.47865136	1.47382577	1.46906629	1.46414628
9		1.00000000	1.00000000	1.00000000	1.00000000	1.00000000
10	BK7	1.54642859	1.54134356	1.53539019	1.52956877	1.52360494
11		1.00000000	1.00000000	1.00000000	1.00000000	1.00000000
12	SILICA	1.48273942	1.47865136	1.47382577	1.46906629	1.46414628
13		1.00000000	1.00000000	1.00000000	1.00000000	1.00000000
14		1.00000000	1.00000000	1.00000000	1.00000000	1.00000000
15		1.00000000	1.00000000	1.00000000	1.00000000	1.00000000

Surf	Glass	0.550000	0.700000	0.850000	1.050000
0		1.00000000	1.00000000	1.00000000	1.00000000
1		1.00000000	1.00000000	1.00000000	1.00000000
2		1.00000000	1.00000000	1.00000000	1.00000000
3		1.00000000	1.00000000	1.00000000	1.00000000
4	MIRROR	1.00000000	1.00000000	1.00000000	1.00000000
5	MIRROR	1.00000000	1.00000000	1.00000000	1.00000000
6		1.00000000	1.00000000	1.00000000	1.00000000
7		1.00000000	1.00000000	1.00000000	1.00000000
8	SILICA	1.45991088	1.45529247	1.45249829	1.44979976
9		1.00000000	1.00000000	1.00000000	1.00000000
10	BK7	1.51852239	1.51306400	1.50984013	1.50682021
11		1.00000000	1.00000000	1.00000000	1.00000000
12	SILICA	1.45991088	1.45529247	1.45249829	1.44979976
13		1.00000000	1.00000000	1.00000000	1.00000000
14		1.00000000	1.00000000	1.00000000	1.00000000
15		1.00000000	1.00000000	1.00000000	1.00000000

ELEMENT VOLUME DATA:

```
Units are cubic cm.
Values are only accurate for plane and spherical surfaces.
Element surf  8 to   9 volume :     899.386512
Element surf 10 to  11 volume :     570.199067
Element surf 12 to  13 volume :     653.560755
```

Chapter B.6

Schmidt Telescopes

Most astronomical telescopes are restricted by aberrations to a narrow field of view of only a few tenths of a degree. But for wide-angle astrophotography, the astronomer requires a coverage of several degrees or more. From the late nineteenth century and extending into the 1930s, astrographic lenses were built to satisfy this need. But these lenses, which were often special versions of Cooke Triplet or Tessar camera lenses, left much to be desired. To achieve the very high image quality required for astronomical work, speeds had to be kept slow, typically $f/5$ to $f/10$. Limited glass-size availability kept apertures small; 20 inches (508 mm) diameter was considered large, and the camera used to discover Pluto was only 13 inches (330 mm). Worst of all, wavelength coverage was so limited that a completely different lens was required to photograph in the blue and in the red.

Then in 1931, Bernhard Schmidt published a paper[1] describing a radically different approach to designing a wide-angle telescope. His new approach was so effective that almost immediately Schmidt telescopes supplanted the older astrographic lenses. Because of its nearly exclusive use as a photographic instrument, a Schmidt telescope is also called a Schmidt camera (or simply a Schmidt).

This chapter discusses the design of Schmidt telescopes.[2] Two variations are given: the original or classical Schmidt, and the modified achromatic Schmidt. Note that the examples presented here are hypothetical and represent no actual telescopes. But for greater interest, the designs have been inspired by the two nearly twin 48-inch (1.2-meter) Schmidts on Palomar Mountain in California and on Siding Spring Mountain in Australia.

B.6.1 The Schmidt Approach

The vast majority of optical systems are axially symmetric; that is, they consist of collections of rotationally symmetric surfaces arrayed along a common axis. All of the lenses considered in this book are axially symmetric. The problem with an axially symmetric system is that a light beam originating at an object away from the axis passes through the system differently than the axial beam does. This difference gives rise to the now familiar off-axis aberrations. When an object is too far off-axis, image quality degrades beyond what is acceptable, and this limits the field of view.

Schmidt's approach to controlling off-axis aberrations was to (nearly) eliminate the axis. He noted that when viewed from its center of curvature, every portion of the inside of a sphere appears the same; that is, there is no unique axis. Thus,

---

[1] Bernhard V. Schmidt, "Ein lichtstarkes komafreies Spiegelsystem," *Zentralzeitung für Optik und Mechanik*, Vol. 52, pp. 25–26, 1931, translated by Nicholas U. Mayall as "A Rapid Coma-Free Mirror System," *Pub. Astron. Soc. of the Pacific*, Vol. 58, pp. 285–290, 1946, also in Lang and Gingerich, eds., *A Source Book in Astronomy and Astrophysics, 1900–1975*, pp. 27–29.

[2] Also see I.S. Bowen, "Schmidt Cameras," in Gerard P. Kuiper and Barbara M. Middlehurst, eds., *Telescopes*, pp. 43–61.

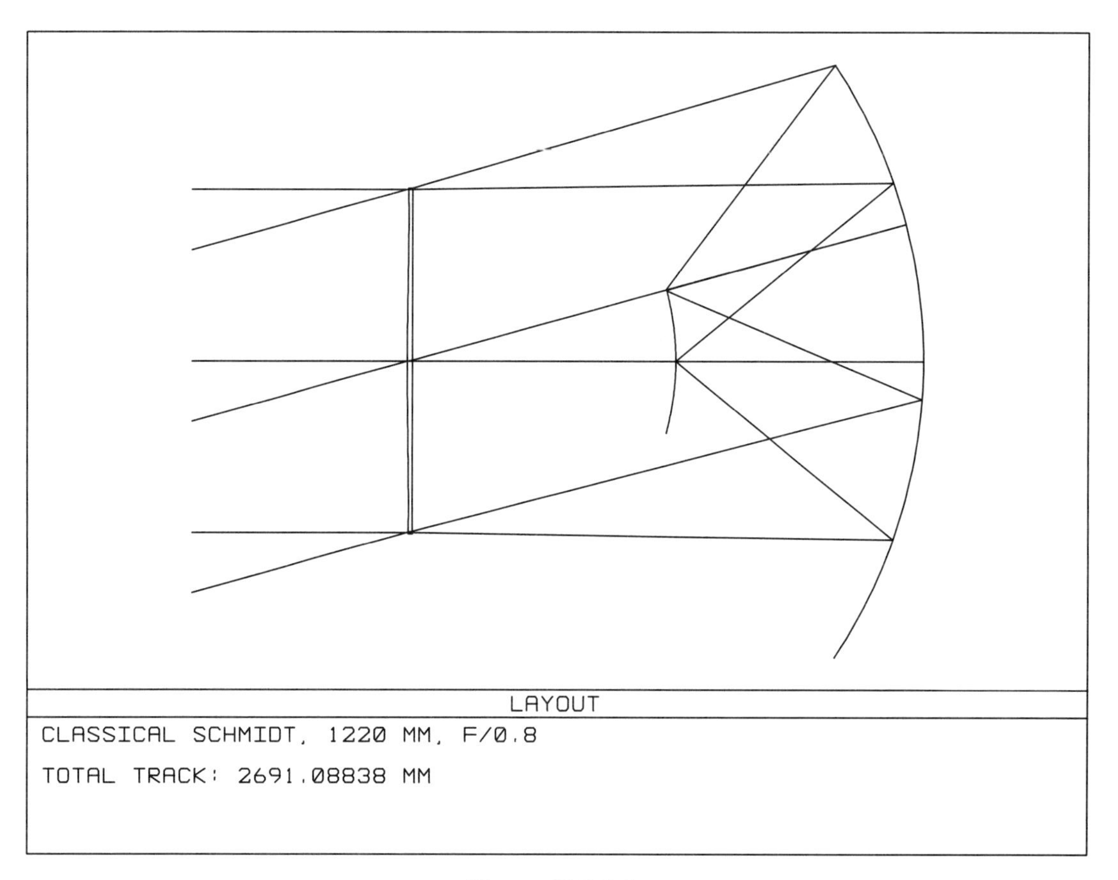

Figure B.6.1.1.

if a stop is placed at the center of curvature of a concave spherical mirror, light beams passing through the stop at different angles are all reflected and focused by the mirror in the same way. There is now point symmetry about the center of the stop (and the center of curvature of the mirror), rather than rotational symmetry about an axis. This point symmetry is the basis of the Schmidt telescope.

Figure B.6.1.1 is the layout of an $f/0.8$ Schmidt telescope that has been selected to show how different light beams pass through the system. On the left is a thin glass plate at the location of the stop. For the moment, pretend that this plate has been removed, leaving only a stop opening of the same size. On the right is a spherical mirror whose center of curvature is located in the middle of the stop opening. The diameter of the mirror is considerably larger than the diameter of the stop.

Because the focal length of a concave mirror is half its radius of curvature, light passing through the stop from distant objects is reflected by the mirror and focused halfway between the stop and mirror. This image is at the prime focus, and a detector placed there would form a central obscuration whose diameter is directly proportional to the angular field of view. For illustrative purposes, the blocking of rays by the central obscuration has been ignored in Figure B.6.1.1, thereby allowing all the incoming rays to be drawn through to the image surface.

Note that there is nothing in the above system to define an optical axis. Because there is no axis, there can be no off-axis aberrations. Of course, the stop opening becomes elliptically foreshortened for beams not normal to its plane, but this only introduces vignetting, not aberrations.

Actually, things are not quite so simple. The point symmetry of the system means that the image surface must also be curved with a center of curvature in the center of the stop. Thus, the radius of curvature of the image is half the mirror's radius of curvature. It is really more accurate to say that the system has no off-axis

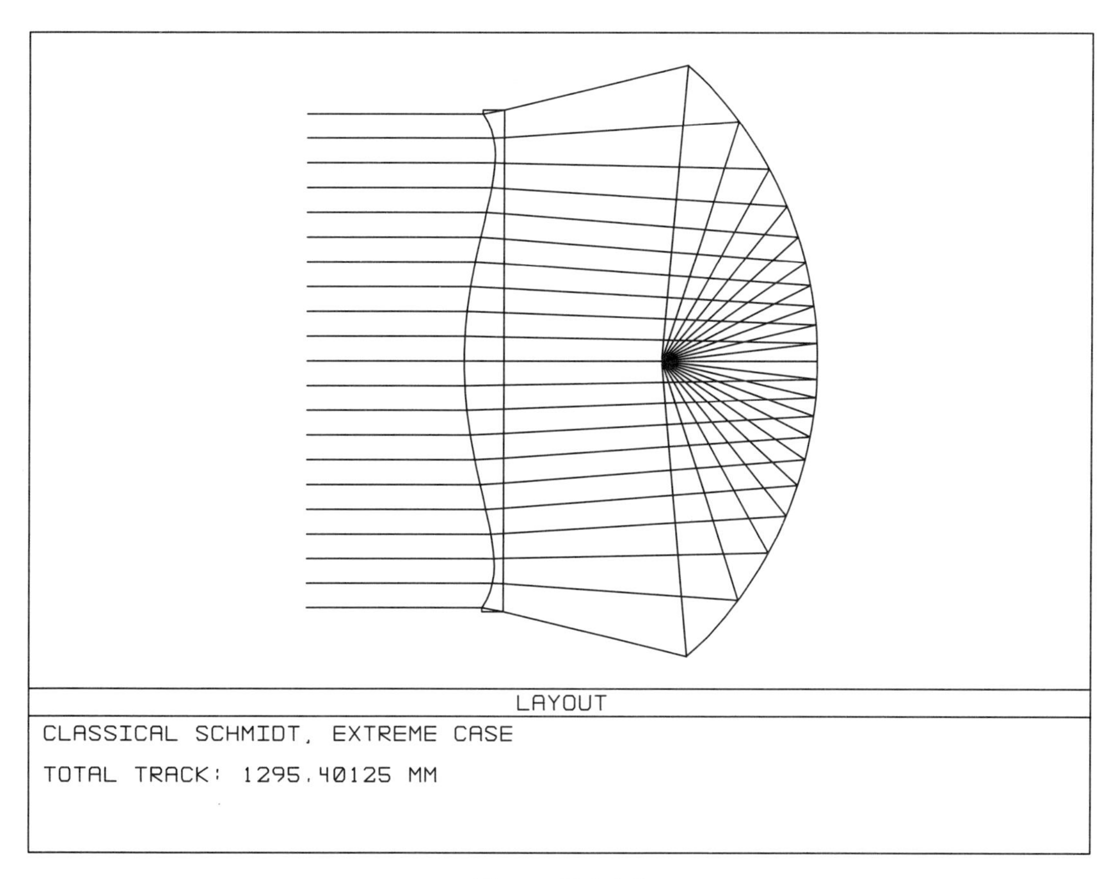

Figure B.6.1.2.

aberrations except field curvature. On a curved image surface, images are free of coma, astigmatism, and other higher-order off-axis aberrations. Recall that on a curved image surface, distortion is not defined, although you can define a mapping function.

The system just described may be free of off-axis aberrations on a curved image surface, but it is by no means free of all aberrations. The use of a spherical mirror causes the images to suffer from serious spherical aberration, an "on-axis" aberration. Like all on-axis aberrations, this aberration is equally present over the entire field.

One possible way to control the spherical aberration might be to stop down the system (recall that transverse spherical aberration varies as the third power of the entrance pupil diameter). Of course, this tactic only reduces the problem but does not solve it. Worse, for any sizable field, stopping down the system would cause the central obscuration to block most or all of the light. The geometry of a Schmidt system requires a fast enough speed to get light around the central obscuration (thus the $f/0.8$ speed in Figure B.6.1.1).

Schmidt's solution to the spherical aberration problem was to introduce a thin transparent corrector plate into the stop opening (thus the glass plate in Figure B.6.1.1). An extreme example of a Schmidt corrector plate is shown in Figure B.6.1.2. One side of the plate is flat. The other side has polynomial aspheric deformations superimposed on an underlying long-radius sphere. The aspheric side can be visualized as resembling either a set of concentric circular ripples, or a crater with a big central peak. By adjusting the polynomials, the right amount of spherical aberration is added to cancel out the spherical aberration of the spherical mirror. The underlying sphere on the corrector is not strictly necessary, but there are advantages to including it.

A Schmidt corrector plate can be designed with only a fourth-order polynomial deformation. A fourth-order deformation allows third-order spherical aberration to be controlled. However, for a better solution, both fourth- and sixth-order polynomial deformations are usually used to control both third- and fifth-order spherical. Eighth- and higher-order polynomials are only rarely used because their practical effect is usually negligible.

Note carefully that placing an aspheric corrector plate in the stop opening reintroduces an optical axis; that is, the system is no longer perfectly point symmetric. However, the curves or ripples on most correctors are very mild (much smaller than those shown in Figure B.6.1.2). Thus, the new axis is weak and the field is still wide. Even so, the field is eventually limited by the off-axis oblique spherical aberration (and to a lesser extent, astigmatism) introduced by the corrector.

Optical performance is further limited by chromatic aberrations. The corrector plate in Schmidt's original design is a singlet, which can only focus and correct the spherical aberration for one wavelength at a time. Other wavelengths suffer from varying amounts of primary longitudinal color and spherochromatism. These chromatic aberrations are the limiting aberrations near the field center and restrict the wavelength range that can be covered. Nevertheless, a Schmidt can cover a much wider wavelength range than any of the old astrocamera lenses.

For even better chromatic performance, a modified Schmidt with a two-element achromatic corrector plate can be designed. In this system, primary longitudinal color is corrected and secondary longitudinal color remains.

To be historically accurate, it must be noted that Gustav A. H. Kellner (not to be confused with Carl Kellner, the inventor of the Kellner eyepiece in 1848) was granted a patent[3] in 1910 for a system very similar to Schmidt's consisting of a spherical mirror and an aspheric corrector plate to correct the spherical aberration. But Kellner's patent only covered a narrow-angle system that was intended for use in searchlights and automobile headlights, and no explicit mention was made of off-axis aberrations or of a preferred location for the corrector plate. Schmidt's new and revolutionary contribution was to recognize the special geometry and the consequent capability for high-quality wide-angle imaging.[4]

Incidentally, optical systems containing only lenses are called dioptric. Optical systems containing only mirrors are called catoptric. And optical systems containing both lenses and mirrors with power are called catadioptric. In practice, the terms dioptric and catoptric are rarely encountered, but the term catadioptric is widely used. The Schmidt telescope is the most famous catadioptric system in optics.

B.6.2 System Specifications

For the classical Schmidt telescope example, the clear diameter of the singlet corrector plate is 1220 mm (48 inches), the system speed is $f/2.5$, and thus the effective focal length is 3050 mm. As usual for a telescope (unless otherwise stated), the object is at infinity.

Figure B.6.2.1 shows the basic layout. In addition to the essential optical components, there are three plane dummy surfaces in the optical prescription. The first is a surface in front to control how the rays are drawn in layouts. The second is a surface between the corrector and mirror that contains the central obscuration. The third is the paraxial focal plane close to the actual image surface. The order in which light encounters the various surfaces is: dummy, corrector plate, dummy with obscuration, mirror, dummy at paraxial focus, image.

The size of the image is 14x14 inches square (355.6 mm on a side and 502.9 mm across the diagonal). This image corresponds to a square field of view 6.68° on a

[3]United States patent 969,785, issued September 13, 1910.

[4]See Jenkins and White, *Fundamentals of Optics*, fourth edition, pp. 208–209. Also see Hilbert, *A Study of Concentric and Schmidt Type Catadioptric Systems,* pp. 14–15.

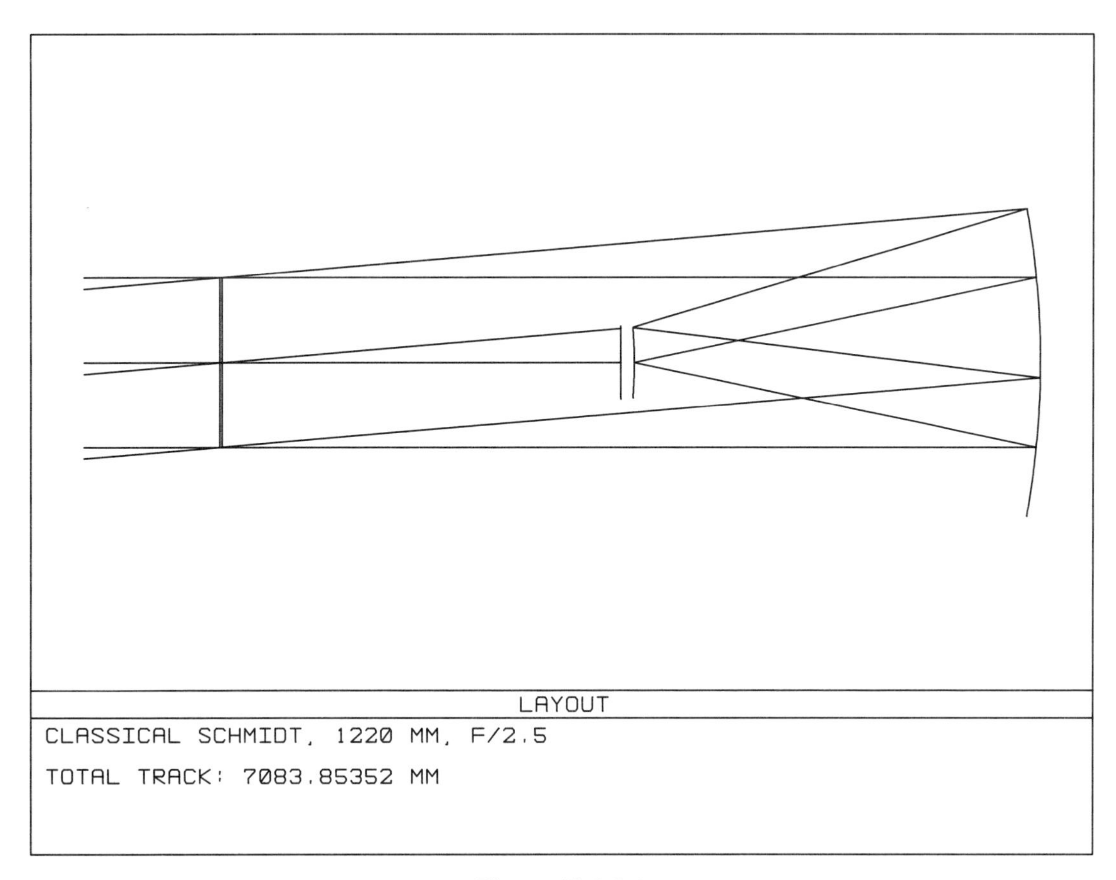

Figure B.6.2.1.

side and 9.44° (±4.72°) across the diagonal. Thus, the system must be designed to cover a ±4.72° circular field that circumscribes the square. For optimization and evaluation, the following field angles are used: 0°, 1.89°, 3.30°, and 4.72° (which correspond to 0, 40%, 70%, and 100% of the radial field).

Although the natural image surface of a Schmidt is curved, the field can be flattened by adding an appropriate convex lens just in front of focus. But a field flattener is a gross departure from the point symmetry Schmidt approach, and thus serious aberrations are introduced. The best way to handle these aberrations is simply to avoid them by retaining the curved image surface. Of course, this requires a curved image detector.

Unfortunately, all CCDs are flat, and they are also far too small for the present application. For recording images, the only practical detectors for a big Schmidt are photographic films and plates. Photographic films and plates can be easily bent into the required curved shape, and they are available in large sizes.

Note that using photography here is not an embarrassing reversion to obsolete technology. Although a photographic emulsion has a much lower quantum efficiency than a CCD, the detector area (sky coverage) and limiting resolution (pixel size) enter when considering the information collection rate. For the present photographically recorded Schmidt, the information collection rate is very high indeed. This is one of those applications where the photographic process is still supreme.

Most Schmidts use photographic film, but in the present example, thin glass photographic plates are used. Glass has the advantage of long-term dimensional stability. During exposure, the plates are temporarily bent or warped into the required convex spherical shape. In the plateholder, the plate edges are gently pressed down against a convex platen. The curve is mild enough that the plates can be bent without breaking (usually).

Because the photographic plates and their plateholder are square, the minimum central obscuration is also square. However, if a square obscuration is used, then the straight sides would produce diffraction spikes radiating from star images. These spikes would be in addition to the spikes already produced by the spider vanes supporting the plateholder. Thus, to reduce diffraction spikes, a circular baffle (opaque disk) is placed just forward of the plateholder. The baffle diameter is 520 mm, and yields a linear obscuration of 43% and an area obscuration (light blockage) of 18%. However, if you prefer more light with stronger diffraction spikes, then use the square obscuration; it blocks only about 12%.

The stop of a Schmidt telescope is at the corrector plate, and collimated light beams coming in from the various object points at different off-axis angles all cross there. After passing through the corrector plate, these beams diverge and spread out relative to each other. When the beams reach the primary mirror, they cover an area considerably greater than the area of the corrector. Thus, to avoid mechanical vignetting (which would cause a light loss in addition to the loss caused by the central obscuration), the primary mirror diameter must be larger than the corrector diameter.

In a Schmidt telescope, the diameter of the primary mirror for no vignetting is equal to the diameter of the corrector plus twice the diameter of the image. In the present example, this becomes 1220 mm plus twice 502.9 mm, or 2225.8 mm (87.6 inches). Clearly, this mirror is getting big. To save money, many Schmidts have undersized primaries and thus some vignetting. Nevertheless, no vignetting is allowed here.

As with the Cassegrain telescopes with field correctors in the previous chapter, extra thought must be given to the wavelengths when designing and evaluating the classical Schmidt telescope example. For today's observational programs, photographic plate (emulsion) sensitivities ranging from the near ultraviolet to the near infrared can be expected. However, when photographing star fields, it is now realized that, for astrophysical reasons, it is important to have none of the wavelength response curves straddle the Balmer discontinuity at 0.3646 μm. Thus, wavelengths shorter than 0.3646 μm are almost never used today for direct imaging with Schmidts, and they are not included here.

In astronomical photography, wavelength response curves are determined by combining the sensitivity curve of a photographic emulsion and the transmission curve of a filter (the system transmission may be considered as part of the filter). All emulsions are sensitive in the ultraviolet and blue. Sensitivity can be extended to include longer wavelengths by adding special dyes to the emulsion during manufacture. How far sensitivity is extended depends on which dyes are used. The most extreme infrared emulsion is hypersensitized Kodak I-Z, which is sensitive out to nearly 1.2 μm. But in all cases, there is a long-wavelength cutoff beyond which the emulsion does not respond. By selecting an emulsion with the right cutoff, the red side of the response curve can be placed at any desired wavelength. The blue side of the response curve is then determined by an appropriate absorption type long-pass filter; that is, a filter that transmits wavelengths longer than a specific wavelength and absorbs shorter wavelengths. For example, if the emulsion is sensitive out to 0.70 μm, and the filter transmits wavelengths longer than 0.60 μm, then the plate-filter response curve is between 0.60 and 0.70 μm.

Incidentally, it is easy to make long-pass absorption filters, but hard to make good short-pass and band-pass absorption filters. Short-pass absorption filters are really band-pass filters, and band-pass absorption filters often have an insidious red leak. For example, blue or green band-pass absorption filters may also transmit unintentionally in the infrared and sometimes even in the visible red. The optical engineer must be very careful here. To avoid red leaks, well-blocked thin-film interference filters must be used in place of absorption filters.

For photography with Schmidt telescopes, three plate-filter combinations are most commonly encountered today. The first uses Kodak IIIa-J to get a response curve covering the blue and green. The second uses Kodak IIIa-F to get a curve covering the orange and red. The third uses Kodak IV-N to cover the infrared out to about 0.90 μm. The required filters are assumed to be dyed gelatin absorption filters, such as Kodak Wratten filters. Because these filters are so thin, they have not been included explicitly in the optical prescription.

Thus, when designing the classical Schmidt, the extreme wavelengths are determined by the Balmer discontinuity and the cutoff of IV-N emulsions. These wavelengths are 0.375 μm and 0.90 μm.

Using the same reasoning outlined in the previous chapter, nine wavelengths are used to cover this wavelength range, and these wavelengths are chosen to give roughly equal fractional changes in optical properties. Because the wavelength range in the present chapter is not so large, equal increments in refractive index give very nearly equal fractional changes. Also, because ultraviolet transmission is not an issue, the Schmidt corrector plate can be made of Schott BK7, which is the best general purpose crown (low dispersion) glass available. Reading the wavelengths off the dispersion curve for BK7, the nine adopted wavelengths are: 0.375, 0.40, 0.425, 0.455, 0.495, 0.55, 0.62, 0.73, 0.90 μm. The central wavelength, 0.495 μm, is made the reference wavelength.

Selecting the aspheric shape of the corrector plate also takes some extra thought. There is actually a family of possible shapes, but one works better than the others. Examine carefully the shape of the corrector plate illustrated in Figure B.6.1.2. The underlying sphere causes the inner region of the corrector to be convex. The polynomials cause the outer region of the corrector to turn up and be concave. At one zone, there is a transition from convex to concave, and at this transition, the slope of the corrector surface becomes zero. This zone is called the neutral zone because a ray passing through is undeviated.

The parameter that selects which member of the corrector family you have is the location of the neutral zone. The best location is 0.866 (86.6%) of the way from the center to the edge of the clear aperture of the plate. With this neutral zone, the maximum surface slope in the inner corrector region is equal and opposite to the surface slope at the edge of the corrector. This solution gives minimum overall surface slopes across the whole corrector. Minimum surface slopes mean minimum ray angles of incidence and refraction. Minimum refraction means minimum differential refraction with wavelength, and thus minimum chromatic aberration. Because chromatic aberration is the limiting aberration in the center of the field, it is important to minimize color by selecting the 0.866 neutral zone. As an added benefit, the minimum surface slope case is also easiest for the optician to make, although no Schmidt plate is easy.

Finally, the aspheric side of the corrector plate should face the sky and not face the primary mirror. The reason concerns ghost images. More will be said about ghost images later.

B.6.3 Optimizing the Classical Schmidt

There are two basic approaches to optimizing a classical (nonachromatic) Schmidt telescope. They are the polychromatic approach and the monochromatic approach.

In the polychromatic approach, all wavelengths and fields are optimized simultaneously in a giant spot-shrink operation. When constructing the default merit function, the fields and wavelengths are usually all given equal unity weight. The resulting optical performance is quite good, but there is a disadvantage. The image for no wavelength and field is perfectly corrected to within the limits imposed by diffraction. Thus, it is difficult for the optician to do the null test (knife-edge test) during fabrication.

This lack of a simple null test is important. At least three independent tests should be done when making the optics for a telescope, especially a big, expensive telescope. An overall system null test is an excellent test. Two others might be tests of individual components and a final star test. The purpose of multiple tests is to reveal systematic errors (as opposed to random errors). It is an issue of precision versus accuracy. If there is a major systematic error in one of the tests, it takes three tests to identify which test is wrong; it becomes two against one.

Optical testing and quality control are especially relevant today following the embarrassing error that was made during the figuring of the primary mirror for the Hubble Space Telescope. For various reasons, the Hubble mirror was only tested as an individual component and only by a single method. After launch, it was found that the telescope formed images loaded with spherical aberration. A subsequent investigation revealed that the single test method had a large systematic error. The mirror's very precisely polished surface had inadvertently been given the wrong conic constant. Unfortunately, no overall test of the telescope optics had been done on the ground. This is a shame because even a crude overall test would have exposed the error in time for it to be corrected *before* launch.[5]

The monochromatic design approach does yield a perfectly diffraction-limited image that is suitable for the null test. Furthermore, you get this solution with no penalty in polychromatic optical performance. As mentioned earlier, the corrector is a singlet that can only control the spherical aberration for one wavelength at a time. Thus, a monochromatic approach is actually quite appropriate and is the approach adopted here.

During a monochromatic optimization, you shrink on-axis spot size for just one wavelength to change the shape of the corrector to control third- and fifth-order spherical aberration. Chromatic aberration is minimized by constraining the neutral zone to be at the 0.866 position. The off-axis aberrations are controlled by locating the stop (corrector plate) near the center of curvature of the mirror, and by properly curving the image surface.

The whole solution depends on the wavelength used for controlling the spherical aberration. Because spherical varies with wavelength, the chosen wavelength should have an index close to the middle of the index (refraction) range. This wavelength in the present example is 0.495 μm. Unfortunately, 0.495 μm is not close to any of the laser or emission lamp wavelengths normally available in the optical shop. As an exercise, the Schmidt telescope was optimized at the common mercury green line at 0.5461 μm; it was then refocused for the best polychromatic image. The result was unsatisfactory. Thus, the present system must be designed and tested with 0.495 μm light, perhaps isolated from a white light source with a narrow-passband interference filter.

The variables are: the vertex radius of curvature of the corrector, the fourth- and sixth-order polynomial coefficients on the corrector, the radius of curvature of the mirror, all airspaces including the space between the paraxial focal plane and the actual image surface, and the radius of curvature of the image surface. Note that the relative curvatures and thicknesses given earlier were paraxial and neglected the effect of the corrector plate. During optimization, the system must depart slightly from these simplifications.

When doing the optimization, it is best to proceed in two stages. Because of the special nature of the present system, neither of these stages corresponds to the optimization stages outlined in Chapter A.15. The first stage corrects or controls everything except the curvature of the image surface, which is only approximately controlled. The second stage exactly controls the curvature of the image surface.

Listing B.6.2.1 gives the first-stage merit function in ZEMAX format. Many of these operands have been adapted from Listing A.13.1. Operand 1 corrects focal

[5]For a definitive discussion of the Hubble error, see *The Hubble Space Telescope Optical Systems Failure Report*, NASA TM-103443, November 1990. This is commonly known as "The Allen Report."

Listing B.6.2.1

Merit Function Listing

File : C:LENS402.ZMX
Title: CLASSICAL SCHMIDT, 1220 MM, F/2.5

Merit Function Value: 1.82884720E-005

Num	Type	Int1	Int2	Hx	Hy	Px	Py	Target	Weight	Value	% Cont
1	EFFL		5					3.05000E+003	-1	3.05000E+003	0.020
2	BLNK										
3	BLNK										
4	COMA	0	5					0.00000E+000	0	2.20507E-002	0.000
5	TRAY		5	0.0000	0.7000	0.0000	0.8660	0.00000E+000	0	3.38561E-003	0.000
6	TRAY		5	0.0000	0.7000	0.0000	-0.8660	0.00000E+000	0	-3.38561E-003	0.000
7	SUMM	5	6					0.00000E+000	-1	1.97815E-010	0.000
8	BLNK										
9	BLNK										
10	TRAY		5	0.0000	0.7000	0.0000	0.8660	0.00000E+000	0	3.38561E-003	0.000
11	TRAY		5	0.0000	0.7000	0.0000	-0.8660	0.00000E+000	0	-3.38561E-003	0.000
12	TRAX		5	0.0000	0.7000	0.8660	0.0000	0.00000E+000	0	-3.38561E-003	0.000
13	TRAX		5	0.0000	0.7000	-0.8660	0.0000	0.00000E+000	0	3.38561E-003	0.000
14	DIFF	10	11					0.00000E+000	0	6.77122E-003	0.000
15	DIFF	12	13					0.00000E+000	0	-6.77122E-003	0.000
16	SUMM	14	15					0.00000E+000	-1	4.63928E-010	0.000
17	BLNK										
18	BLNK										
19	RETY	2	5	0.0000	0.0000	0.0000	0.8660	0.00000E+000	-1	-8.23551E-015	0.000
20	BLNK										
21	TTHI	4	6					1.00000E+002	-1	1.00000E+002	0.012
22	BLNK										
23	BLNK										
24	DMFS										
25	TRAR		5	0.0000	0.0000	0.2635	0.0000	0.00000E+000	0.015178	2.99084E-005	46.516
26	TRAR		5	0.0000	0.0000	0.5745	0.0000	0.00000E+000	0.028455	1.85332E-005	33.486
27	TRAR		5	0.0000	0.0000	0.8185	0.0000	0.00000E+000	0.028455	1.30079E-005	16.496
28	TRAR		5	0.0000	0.0000	0.9647	0.0000	0.00000E+000	0.015178	8.17010E-006	3.471

length to exactly 3050 mm. Although a heavy weight could have been used, a Lagrange multiplier (signaled by the weight of -1) is used here (and elsewhere) instead.

Operands 5 through 7 correct coma. A Schmidt telescope is used to image stars (plus a variety of other objects). When measuring the positions of star images on a photographic plate, it is important that the images be round and symmetrical; asymmetrical coma is the worst aberration. Thus, a user-defined coma operand must be included in the merit function. During optimization, this operand moves the stop and corrector to the best location. To avoid refraction effects, control coma through the 0.866 pupil neutral zone. To reduce the confusion of higher-order off-axis aberrations at the edge of the field, control coma at the 0.7 field. Operand 4 (third-order coma) is included for reference.

Operands 10 through 16 approximately control the image radius of curvature. The approach is to curve the image surface to find the off-axis monochromatic medial focus. Fortunately, the off-axis images need not be exactly in focus for the previous coma operand to work.

Operand 19 controls the location of the neutral zone. All rays from the distant on-axis object point are incident on the corrector plate parallel to the optical axis; that is, their slopes are zero. Select the reference wavelength ray incident on the 0.866 pupil zone. To make this zone the neutral zone, during optimization the slope of this ray is forced to remain zero right after passing through the aspheric surface.

Operand 21 controls the location of the central obscuration relative to the image surface. The overall length from the obscuration to the image surface is about the thickness of the plateholder and its supports. A thickness of 100 mm is reasonable.

The default part of the merit function controls spherical aberration. Select a default merit function that shrinks on-axis monochromatic spot size. To do this, the four field weights are 1 0 0 0 and the nine wavelength weights are 0 0 0 0 1 0 0 0 0. Either a Gaussian quadrature ray set or a rectangular array ray set can be used. In the present example, Gaussian quadrature is used, which ignores the central obscuration. Four rings are used to include seventh-order effects. During evaluation, of course, the central obscuration must not be ignored.

As usual, the location of the paraxial focal plane is controlled by a paraxial marginal ray height solve.

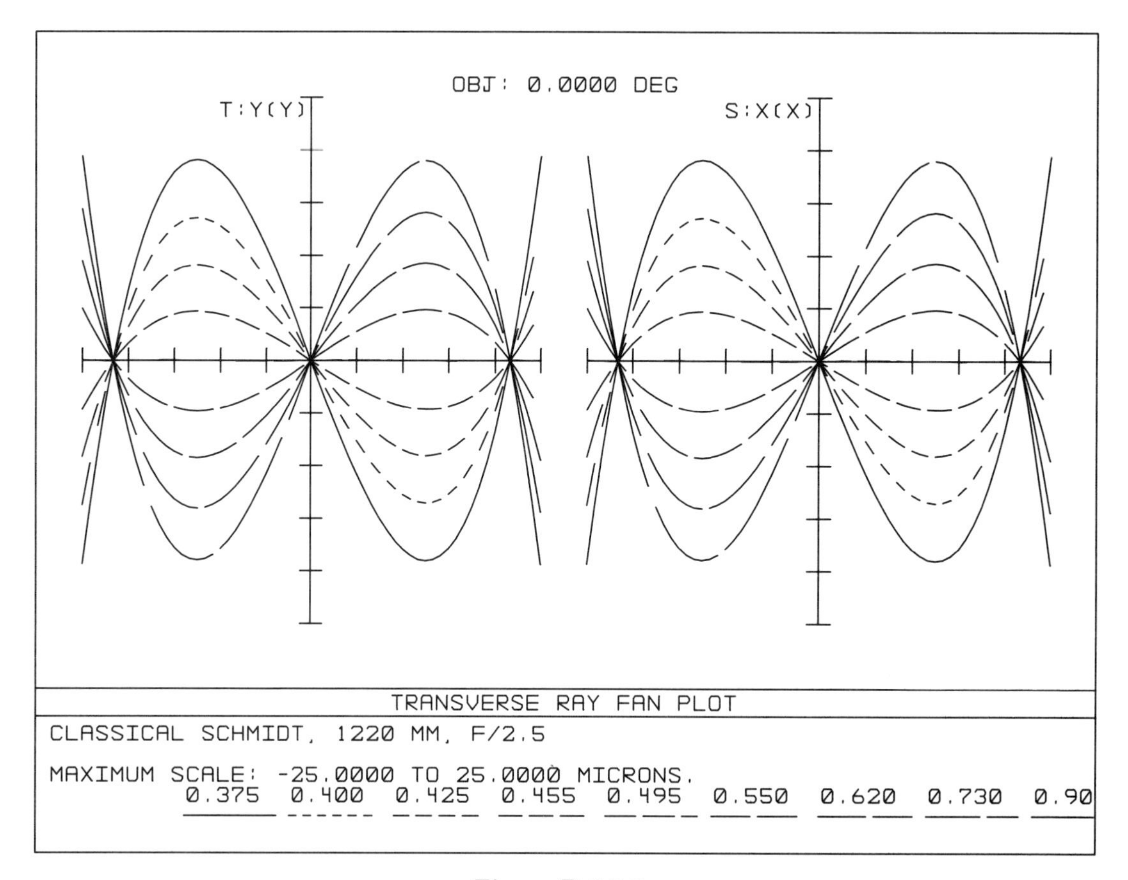

Figure B.6.2.2.

Initially during the first stage, you may wish to optimize with only the fourth-order polynomial and keep the sixth-order polynomial coefficient zero. After an initial optimization, the sixth-order coefficient can then be made a variable and the system reoptimized. This procedure can be extended to higher orders and is recommended in general for systems with polynomial surfaces to make it less likely that the optimization routine will go to a poor local minimum giving excessive zonal spherical. For the same reason, you may also wish to initially optimize with no paraxial defocus. Defocus can be added at the end of the first stage.

For the second optimization stage, remove (freeze) all variables and solves except the radius of curvature of the image surface. Use a default merit function that shrinks polychromatic spot sizes for only the off-axis images. Field weights are 0 1 1 1. Wavelength weights are all unity. No special operands are included. Optimization then gives the best curved image surface to match the remaining (and unavoidable) monochromatic and chromatic off-axis aberrations.

Up to now, nothing has been said about an OPD optimization. As has been noted, this Schmidt system will be used to image stars. In this application, compact, symmetrical, and well-defined point spread functions are paramount. Although an OPD optimization might give better MTFs, the extended PSF wings produced by an OPD optimization are undesirable in this application. Thus, no OPD optimization is attempted.

B.6.4 Evaluating the Classical Schmidt

Figure B.6.2.1 is the layout of the optimized classical Schmidt. Note the location of the central obscuration and that for off-axis beams it is no longer centered. Note too that unlike a conventional reflecting telescope, the overall tube length is twice the focal length. Thus, when compared to other types of telescopes of the same

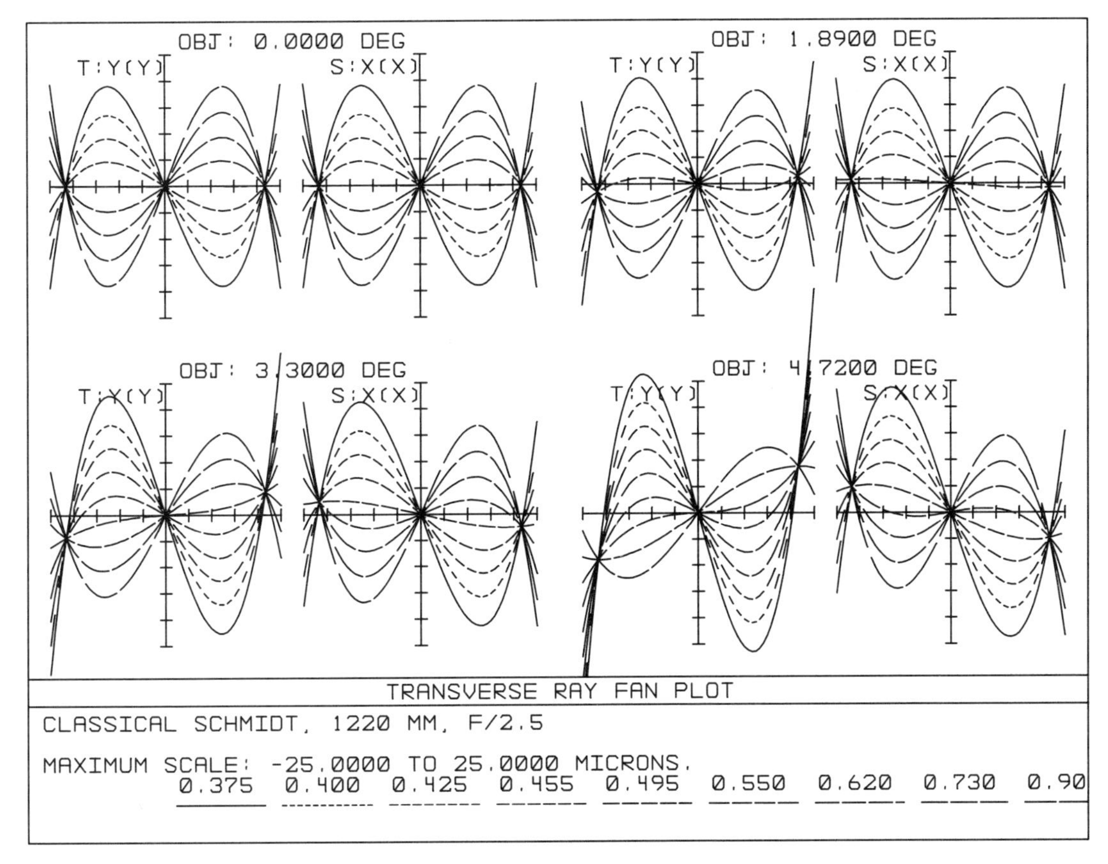

Figure B.6.2.3.

aperture, the required dome (enclosure) is relatively big and expensive.

To allow maximum scale, Figure B.6.2.2 shows the transverse ray fan plot for just the on-axis image. The curves for all nine wavelengths are shown. Note several features of these curves. First, the curve for the optimization wavelength, 0.495 μm, is indistinguishable from the horizontal axis, which indicates virtually perfect correction. Second, the curves for the extreme wavelengths, 0.375 μm and 0.90 μm, are nearly identical except for reversed signs. This symmetry is a consequence of optimizing with the middle refraction wavelength. Third, the curves for all wavelengths cross at the 0.866 pupil neutral zone. At the neutral zone, there is no refraction and thus no chromatic variation of refraction. Fourth, for any given wavelength, the maximum ray error inside the neutral zone is equal and opposite to the ray error at the edge of the pupil. This equality is a consequence of the corrector surface slopes when the neutral zone is at 0.866. Fifth, it follows that for any given wavelength, chromatically varying defocus (longitudinal color) balances chromatically varying third-order spherical aberration (spherochromatism). All across the spectrum, no refocus can improve monochromatic image quality. Sixth, the nine curves are monotonically arrayed with wavelength (from primary color) and are nearly equally spaced vertically (from the choice of wavelengths). And seventh, the scale is ± 25 μm.

Figure B.6.2.3 gives the transverse ray fan plots for all field positions. As you go off-axis, note the oblique spherical aberration and that its amount is different in the tangential and sagittal directions. Also note that the corresponding tangential and sagittal curves have different slopes at the origins, indicating astigmatism. However, the effect of the astigmatism is much less than the effect of the oblique spherical. These aberrations are caused by the altered appearance of the corrector plate to off-axis beams. To these inclined beams, the corrector appears foreshortened, and the polynomial aspheric zones appear elliptical rather than circular.

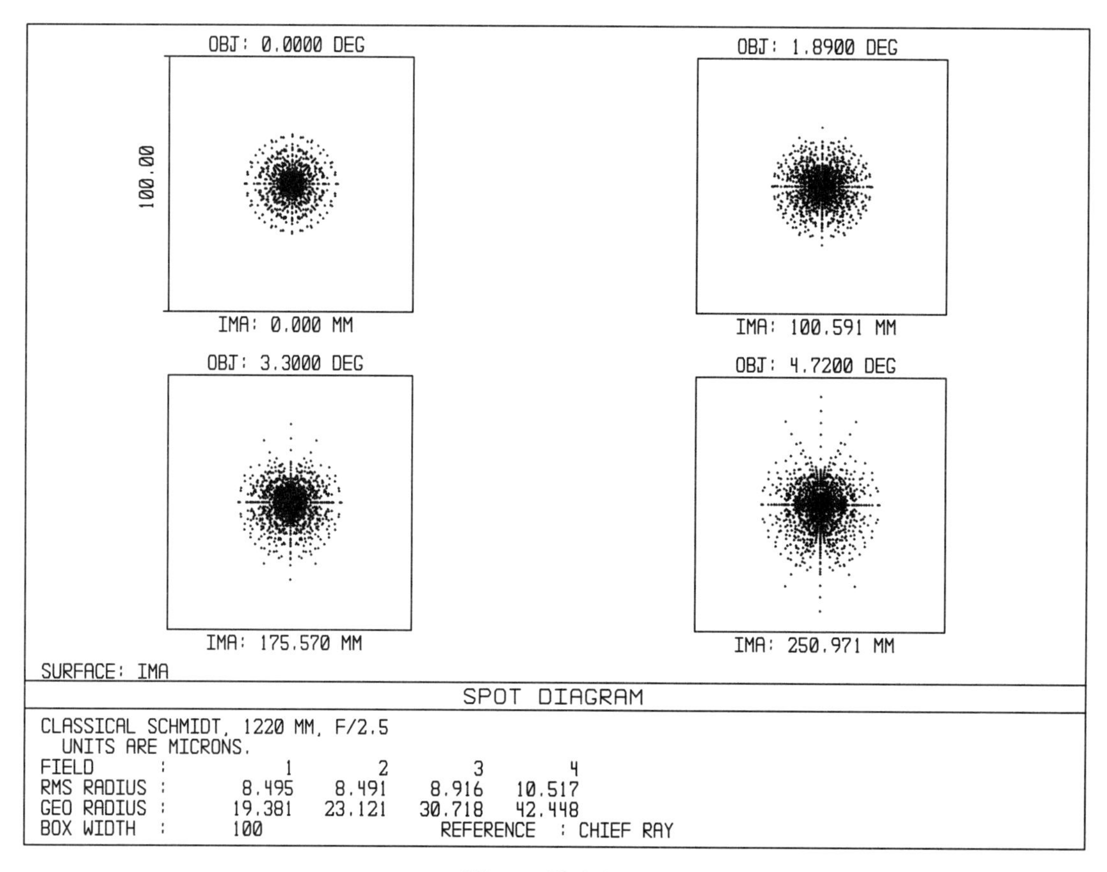

Figure B.6.2.4.

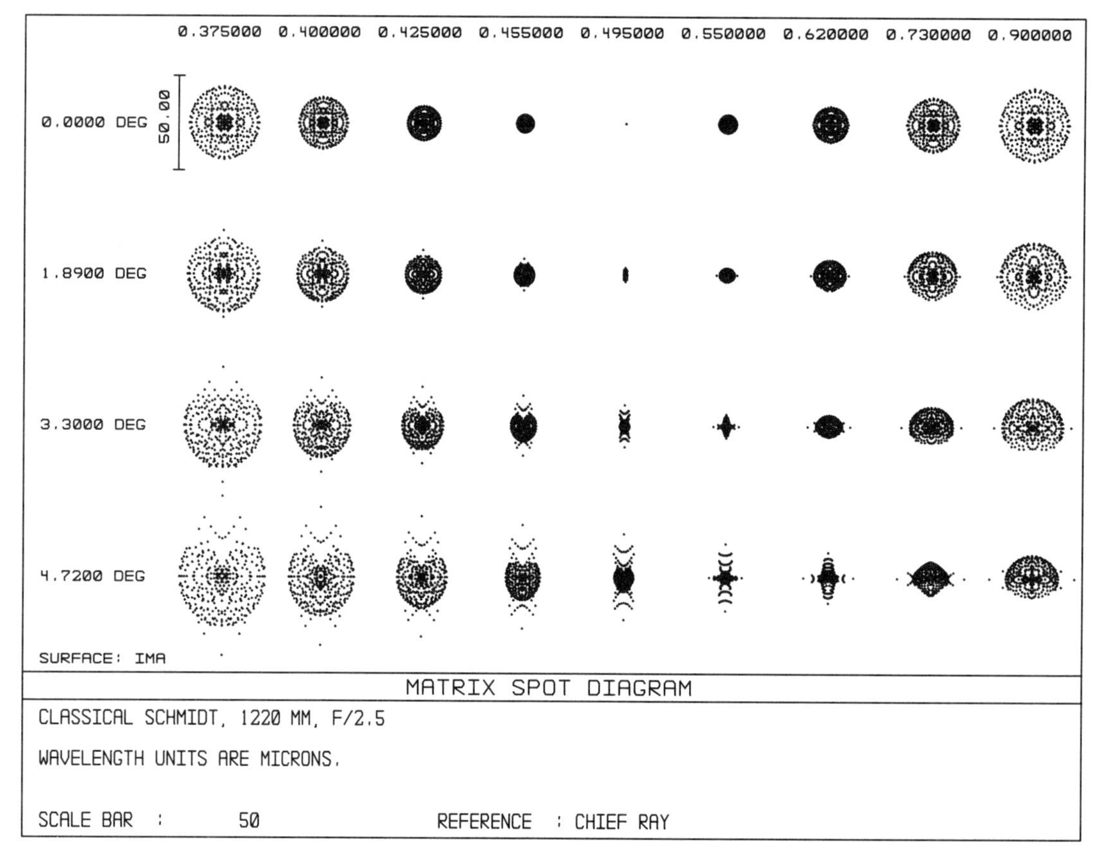

Figure B.6.2.5.

Figure B.6.2.4 shows the polychromatic spot diagrams. Note that the spots are nearly circular and are remarkably uniform across the field. The slightly asymmetrical appearance of the off-axis spots is primarily caused by the effective decentration of the central obscuration.

Figure B.6.2.5 is the matrix spot diagram with the spots for different wavelengths displayed separately. This diagram is very revealing. The on-axis correction for the central wavelength is virtually perfect. For other wavelengths and fields, the spots grow. Clearly, the biggest problem is color.

For the larger spots, spot diameter is about 40 μm. This linear diameter corresponds to an angular diameter of 2.7 arc-seconds in the sky, which is about the same size as the seeing blur under ordinary seeing. Note that the diameter of the Airy diffraction disk is only about 3.0 μm, and thus the system is not diffraction limited polychromatically.

Listing B.6.2.2 is the prescription for the classical Schmidt. Listing B.6.2.3 is the sag table for the aspheric surface of the corrector plate. The values in the sag column give the departure of the surface from a plane tangent to the vertex. The surface shape is similar to the aspheric curve illustrated in Figure B.6.1.2, although not nearly so extreme. The other columns concern the option of using the best-fit sphere during fabrication. After shaping the corrector to the best-fit sphere, the optician can remove the minimum amount of glass to produce the aspheric deformation. However, in the present instance, the radius of the best-fit sphere is so long that it may be easier to figure the surface from a plane. In either case, aspherizing requires up to about 0.11 or 0.17 mm of glass to be removed very selectively from across the corrector plate. To do this well is a true test of the optician's skill.

Listing B.6.2.2

```
System/Prescription Data

File : C:LENS403.ZMX
Title: CLASSICAL SCHMIDT, 1220 MM, F/2.5

GENERAL LENS DATA:

Surfaces          :            7
Stop              :            2
System Aperture   :Entrance Pupil Diameter
Ray aiming        : Off
Apodization       :Uniform, factor =       0.000000
Eff. Focal Len.   :         3050 (in air)
Eff. Focal Len.   :         3050 (in image space)
Total Track       :      7083.85
Image Space F/#   :          2.5
Para. Wrkng F/#   :          2.5
Working F/#       :          2.5
Obj. Space N.A.   :     6.1e-008
Stop Radius       :          610
Parax. Ima. Hgt.  :      251.828
Parax. Mag.       :            0
Entr. Pup. Dia.   :         1220
Entr. Pup. Pos.   :         1000
Exit Pupil Dia.   :      1219.97
Exit Pupil Pos.   :     -3049.91
Field Type        : Angle in degrees
Maximum Field     :         4.72
Primary Wave      :     0.495000
Lens Units        : Millimeters
Angular Mag.      :      1.00003

Fields            : 4
Field Type:  Angle in degrees
#          X-Value           Y-Value            Weight
1         0.000000          0.000000          0.000000
2         0.000000          1.890000          1.000000
3         0.000000          3.300000          1.000000
4         0.000000          4.720000          1.000000

Vignetting Factors
#            VDX              VDY              VCX              VCY
1        0.000000         0.000000         0.000000         0.000000
2        0.000000         0.000000         0.000000         0.000000
3        0.000000         0.000000         0.000000         0.000000
4        0.000000         0.000000         0.000000         0.000000
```

```
Wavelengths      : 9
Units:  Microns
#          Value          Weight
1        0.375000        1.000000
2        0.400000        1.000000
3        0.425000        1.000000
4        0.455000        1.000000
5        0.495000        1.000000
6        0.550000        1.000000
7        0.620000        1.000000
8        0.730000        1.000000
9        0.900000        1.000000

SURFACE DATA SUMMARY:

Surf     Type        Radius       Thickness         Glass       Diameter        Conic
 OBJ  STANDARD     Infinity       Infinity                             0            0
   1  STANDARD     Infinity           1000                             0            0
 STO  EVENASPH     418820.8             20            BK7       1220.025            0
   3  STANDARD     Infinity       2936.945                      1222.199            0
   4  STANDARD     Infinity       3126.909                           520            0
   5  STANDARD     -6076.91       -3026.91         MIRROR       2209.983            0
   6  STANDARD     Infinity    0.001030853                             0            0
 IMA  STANDARD    -3041.733              0                      502.0274            0

SURFACE DATA DETAIL:

Surface OBJ      : STANDARD
Surface   1      : STANDARD
Surface STO      : EVENASPH
 Coeff on r  2   :             0
 Coeff on r  4   : -2.101874e-012
 Coeff on r  6   : -8.87637e-020
 Coeff on r  8   :             0
 Coeff on r 10   :             0
 Coeff on r 12   :             0
 Coeff on r 14   :             0
 Coeff on r 16   :             0
Surface   3      : STANDARD
Surface   4      : STANDARD
 Aperture        : Circular Obscuration
 Minimum Radius  :             0
 Maximum Radius  :           260
Surface   5      : STANDARD
Surface   6      : STANDARD
Surface IMA      : STANDARD

SOLVE AND VARIABLE DATA:

Semi Diam    1      : Fixed
Semi Diam    4      : Fixed
Semi Diam    6      : Fixed
Curvature of    7   : Variable

INDEX OF REFRACTION DATA:

Surf      Glass      0.375000     0.400000     0.425000     0.455000     0.495000
   0               1.00000000   1.00000000   1.00000000   1.00000000   1.00000000
   1               1.00000000   1.00000000   1.00000000   1.00000000   1.00000000
   2        BK7    1.53454939   1.53084854   1.52782658   1.52486897   1.52175196
   3               1.00000000   1.00000000   1.00000000   1.00000000   1.00000000
   4               1.00000000   1.00000000   1.00000000   1.00000000   1.00000000
   5     MIRROR    1.00000000   1.00000000   1.00000000   1.00000000   1.00000000
   6               1.00000000   1.00000000   1.00000000   1.00000000   1.00000000
   7               1.00000000   1.00000000   1.00000000   1.00000000   1.00000000

Surf      Glass      0.550000     0.620000     0.730000     0.900000
   0               1.00000000   1.00000000   1.00000000   1.00000000
   1               1.00000000   1.00000000   1.00000000   1.00000000
   2        BK7    1.51852239   1.51553950   1.51230442   1.50899686
   3               1.00000000   1.00000000   1.00000000   1.00000000
   4               1.00000000   1.00000000   1.00000000   1.00000000
   5     MIRROR    1.00000000   1.00000000   1.00000000   1.00000000
   6               1.00000000   1.00000000   1.00000000   1.00000000
   7               1.00000000   1.00000000   1.00000000   1.00000000
```

Listing B.6.2.3

```
Listing of surface sag

File : C:LENS403.ZMX
Title: CLASSICAL SCHMIDT, 1220 MM, F/2.5

Units are Millimeters.

Semi diameter of surface 2: 610.012271.
Best Fit Sphere curvature : 0.000001.
Best Fit Sphere radius    : 850258.017476.
Best Fit Sphere residual  : 0.174328. (rms)
```

Y-coord	Sag	BFS Sag	Deviation	Remove
0.000000000	0.000000000	0.000000000	0.000000000	-0.070202417
10.000000000	0.000119362	0.000058806	-0.000060556	-0.070262973
20.000000000	0.000477195	0.000235223	-0.000241972	-0.070444389
30.000000000	0.001072743	0.000529251	-0.000543492	-0.070745908
40.000000000	0.001904744	0.000940891	-0.000963853	-0.071166270
50.000000000	0.002971432	0.001470142	-0.001501290	-0.071703707
60.000000000	0.004270537	0.002117004	-0.002153532	-0.072355949
70.000000000	0.005799281	0.002881478	-0.002917803	-0.073120220
80.000000000	0.007554384	0.003763563	-0.003790821	-0.073993237
90.000000000	0.009532057	0.004763260	-0.004768797	-0.074971213
100.000000000	0.011728005	0.005880568	-0.005847437	-0.076049854
110.000000000	0.014137428	0.007115487	-0.007021941	-0.077224357
120.000000000	0.016755016	0.008468018	-0.008286998	-0.078489414
130.000000000	0.019574951	0.009938160	-0.009636791	-0.079839208
140.000000000	0.022590908	0.011525913	-0.011064995	-0.081267411
150.000000000	0.025796049	0.013231278	-0.012564771	-0.082767188
160.000000000	0.029183028	0.015054254	-0.014128774	-0.084331191
170.000000000	0.032743986	0.016994841	-0.015749144	-0.085951561
180.000000000	0.036470551	0.019053040	-0.017417511	-0.087619927
190.000000000	0.040353839	0.021228850	-0.019124988	-0.089327405
200.000000000	0.044384449	0.023522272	-0.020862177	-0.091064593
210.000000000	0.048552466	0.025933305	-0.022619161	-0.092821577
220.000000000	0.052847456	0.028461949	-0.024385507	-0.094587923
230.000000000	0.057258468	0.031108205	-0.026150263	-0.096352680
240.000000000	0.061774030	0.033872072	-0.027901958	-0.098104375
250.000000000	0.066382149	0.036753550	-0.029628599	-0.099831016
260.000000000	0.071070310	0.039752640	-0.031317670	-0.101520087
270.000000000	0.075825472	0.042869341	-0.032956131	-0.103158547
280.000000000	0.080634068	0.046103654	-0.034530414	-0.104732831
290.000000000	0.085482005	0.049455577	-0.036026428	-0.106228844
300.000000000	0.090354659	0.052925113	-0.037429546	-0.107631963
310.000000000	0.095236875	0.056512259	-0.038724616	-0.108927032
320.000000000	0.100112965	0.060217017	-0.039895947	-0.110098364
330.000000000	0.104966705	0.064039387	-0.040927319	-0.111129735
340.000000000	0.109781336	0.067979368	-0.041801968	-0.112004385
350.000000000	0.114539557	0.072036960	-0.042502597	-0.112705013
360.000000000	0.119223527	0.076212163	-0.043011363	-0.113213780
370.000000000	0.123814861	0.080504978	-0.043309882	-0.113512299
380.000000000	0.128294629	0.084915405	-0.043379224	-0.113581641
390.000000000	0.132643351	0.089443442	-0.043199909	-0.113402326
400.000000000	0.136840999	0.094089092	-0.042751908	-0.112954324
410.000000000	0.140866990	0.098852352	-0.042014638	-0.112217055
420.000000000	0.144700186	0.103733224	-0.040966962	-0.111169379
430.000000000	0.148318890	0.108731707	-0.039587183	-0.109789599
440.000000000	0.151700846	0.113847802	-0.037853043	-0.108055460
450.000000000	0.154823232	0.119081508	-0.035741723	-0.105944140
460.000000000	0.157662661	0.124432826	-0.033229836	-0.103432252
470.000000000	0.160195179	0.129901755	-0.030293424	-0.100495841
480.000000000	0.162396255	0.135488295	-0.026907960	-0.097110377
490.000000000	0.164240788	0.141192447	-0.023048341	-0.093250758
500.000000000	0.165703095	0.147014210	-0.018688885	-0.088891302
510.000000000	0.166756915	0.152953585	-0.013803330	-0.084005747
520.000000000	0.167375400	0.159010571	-0.008364829	-0.078567245
530.000000000	0.167531115	0.165185168	-0.002345947	-0.072548363
540.000000000	0.167196036	0.171477377	0.004281341	-0.065921075
550.000000000	0.166341543	0.177887197	0.011545655	-0.058656762
560.000000000	0.164938418	0.184414629	0.019476211	-0.050726205
570.000000000	0.162956844	0.191059672	0.028102829	-0.042099588
580.000000000	0.160366397	0.197822327	0.037455930	-0.032746486
590.000000000	0.157136046	0.204702593	0.047566547	-0.022635870
600.000000000	0.153234148	0.211700471	0.058466322	-0.011736094
610.000000000	0.148628444	0.218815960	0.070187515	-0.000014901
610.012271212	0.148622347	0.218824763	0.070202417	0.000000000

B.6.5 Ghost Images

As mentioned in the previous chapter, all optical systems with lenses have problems with ghost images because transmitting surfaces also reflect a small percent of the light. This stray light must go somewhere, and all too often some of it ends up roughly focused on the image surface. A Schmidt telescope is no exception. This is true even if the corrector plate is anti-reflection coated. Unfortunately, because of its size and broad wavelength coverage, the corrector plate of a big Schmidt is often left uncoated, causing each corrector surface to reflect about 4% of the light. In addition, the photographic emulsion also reflects part of the incident light, and this can be reflected back to form more ghosts.

A Schmidt telescope produces ghost images two different ways. The first way is by a double reflection between the two surfaces of the corrector plate. Light is reflected by the rear corrector surface and then by the front corrector surface. For the present Schmidt example, Figure B.6.2.6 shows the ghost spots on the image surface produced by stars in the field center and field edge. The spots are about 10 mm in diameter and are thus easily differentiated from real stars. Each ghost is centered on its originating star.

The second way that Schmidts produce ghosts is illustrated on the layout in Figure B.6.2.7. Only one ray is drawn to avoid confusion. Light enters the telescope and is focused onto the photographic emulsion. The emulsion reflects some of the light, both specularly and diffusely. Only the specular ray is shown in Figure B.6.2.7. The system then acts as a retro-reflector, with most of the light reflected by the emulsion going back out through the corrector. But some of this light is further reflected by one or the other of the corrector surfaces. It then returns to the mirror and is refocused as a pair of ghost images.

Figure B.6.2.8 shows the ghost spots on the image surface when the light is reflected by the rear corrector surface. Similarly, Figure B.6.2.9 shows the ghost spots when the light is reflected by the front corrector surface. In the former case, the spots are about 3.2 mm in diameter; in the latter case, the spots are about 6 mm in diameter. Both spots are easily recognizable as ghosts. The location of these ghosts in the field is diametrically opposite the originating star.

Thus, every star produces three ghosts, one of the first type and two of the second. However, only the brightest stars yield ghosts that are intense enough to be recorded by the photographic process (there is an exposure threshold below which there is no response). In practice, the few ghosts that are recorded are mainly annoyances; they have little effect on the scientific value of the photographs.

It should be noted that the corrector plate of a Schmidt telescope is very thin. When the telescope is pointed upward, the plate will significantly sag under its own weight. This flexure does not change the relative thickness of the glass across the pupil, and thus the optical path lengths of transmitted rays are also unchanged. Consequently, corrector plate sag does not affect the primary images produced by a Schmidt.

This sag does, however, affect the ghost images produced by the second method; that is, one reflection from the emulsion and one reflection from the corrector. Without the sag, these ghosts are aberrated by a balance of defocus and third-order spherical aberration. With the sag, additional defocus is introduced, which is actually an advantage. The photographs in the original Palomar Sky Survey, which were taken with the Palomar 48-inch Schmidt fitted with a singlet corrector, clearly show ghosts. These ghosts are considerably larger and have much more defocus than the ghost spots in Figures B.6.2.8 and B.6.2.9. It appears that the corrector plate was sagging.

Recall that earlier it was said that the aspheric ripples on the corrector plate should be on the front surface, not the rear surface. For a moment, however, imagine that the corrector has been reversed and that the flat surface is forward. Also, consider initially only a single off-axis object point and only monochromatic light of the optimization wavelength. The ghost light path from the emulsion to the front of the corrector and back to the emulsion is the same as before. The difference is that now the spherical aberration is removed when the light passes *through* the aspheric surface and into the glass of the corrector. Immediately after reflection from the flat front of the corrector, these unaberrated rays resemble collimated rays from a distant object. Thus, the ghost image is sharply focused.

The same is nearly true polychromatically. Relative to a true star image, the ghost has three times as much color from having passed through the aspheric surface two extra times. But this amount of aberration is not sufficient to allow the ghost to be easily recognized as a ghost. Instead, the ghost looks all too much like an extra star. Furthermore, because the ghost is small, its surface brightness (power density

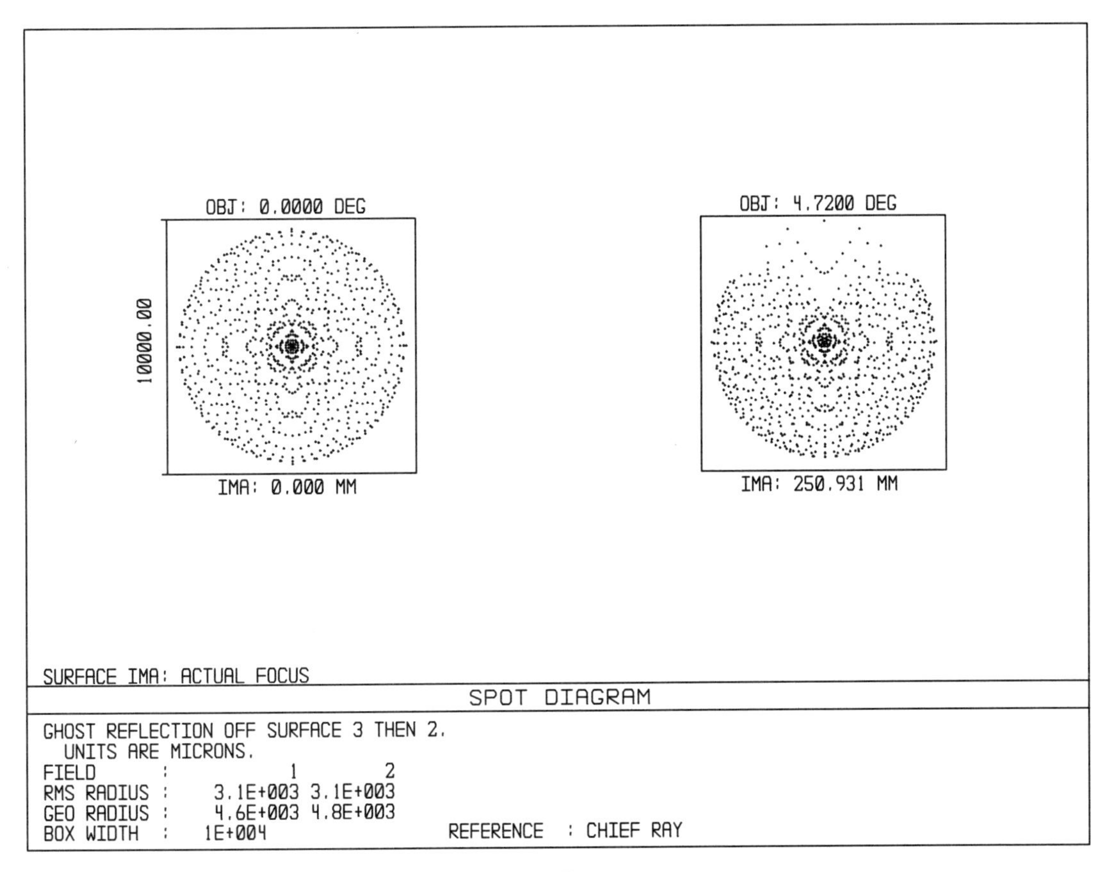

Figure B.6.2.6.

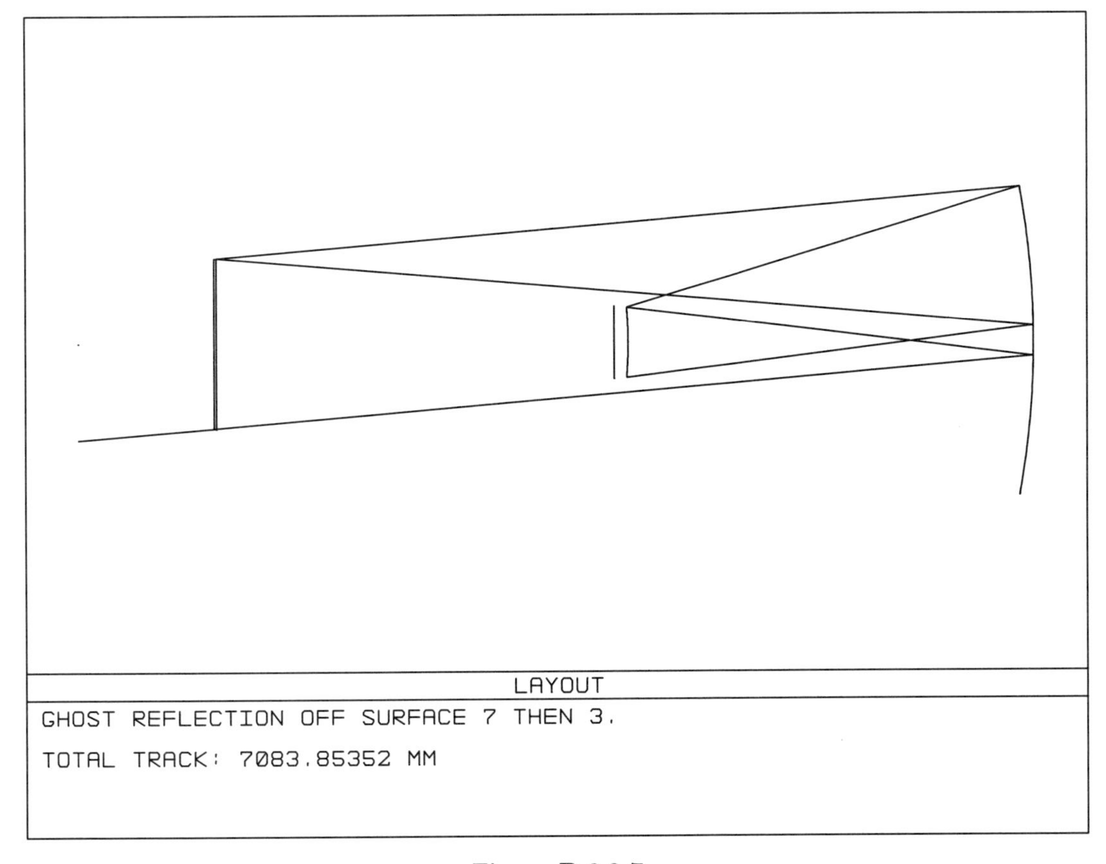

Figure B.6.2.7.

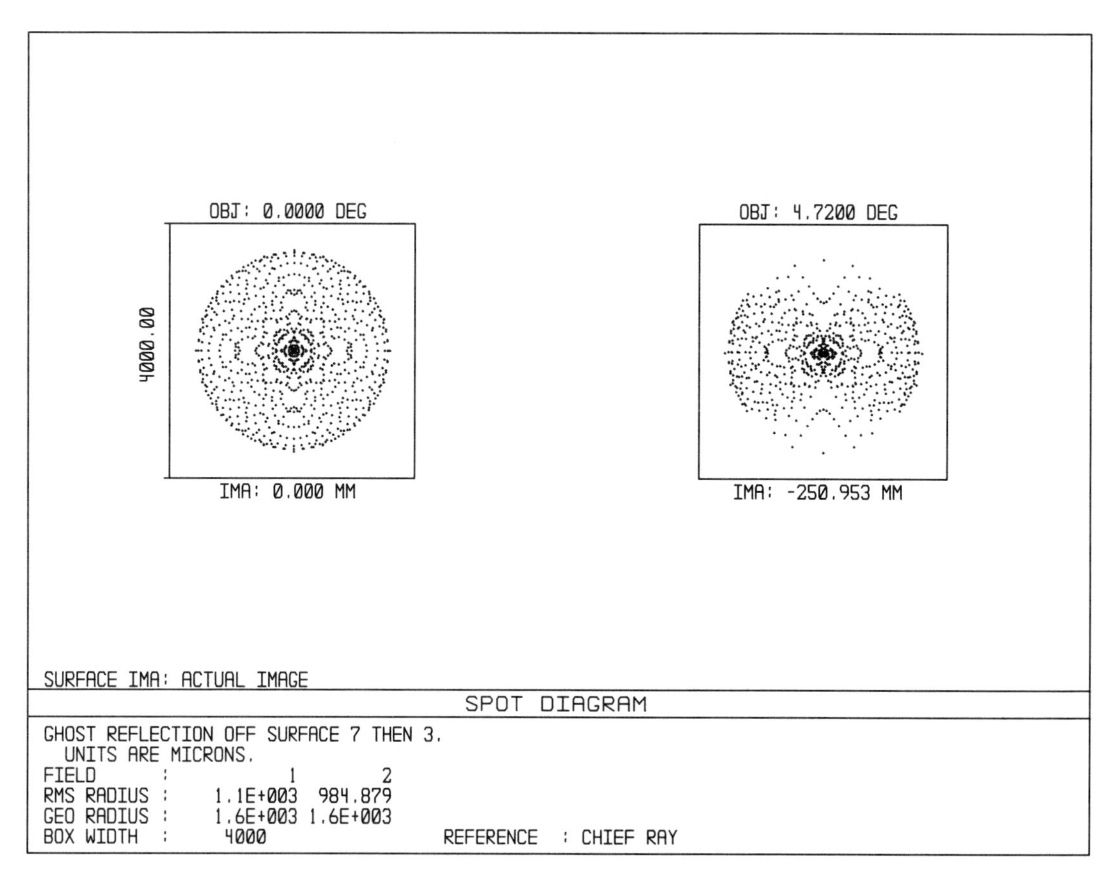

Figure B.6.2.8.

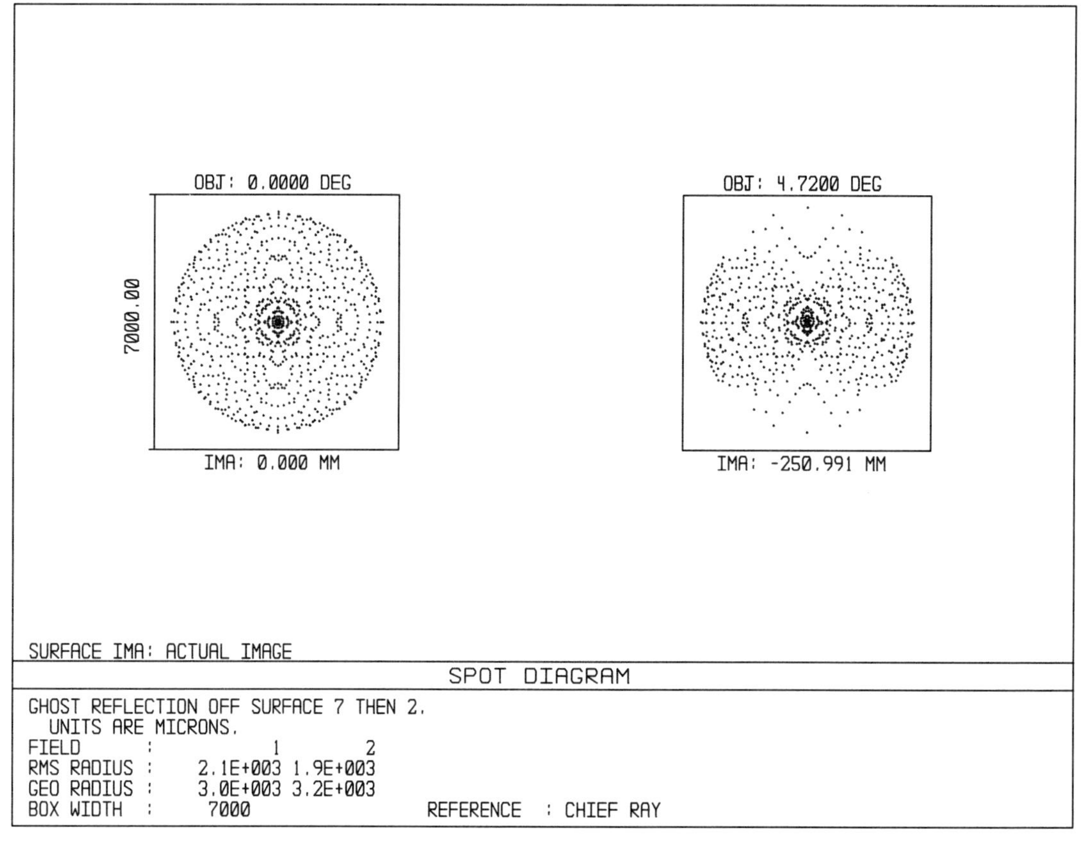

Figure B.6.2.9.

or irradiance) is relatively high, and thus its presence is more readily recorded by the photographic emulsion. Now, extend this situation to photographing a field of stars where every star produces a separate in-focus ghost. This multitude of spurious images is intolerable.

The usual way to avoid this confusion is to place the aspheric ripples on the front surface of the corrector. With this configuration, the ghost light never passes through the aspheric surface and never has the spherical aberration removed. The ghosts remain aberrated, are large with low surface brightness, and are identifiable. An alternate but rarely used approach to handling ghosts is to design and fabricate the corrector as a slight meniscus (with a deliberate sag) to throw the ghosts out of focus.

B.6.6 The Achromatic Schmidt

In 1948, when the Palomar 48-inch Schmidt first went into service, optical performance similar to that just described was quite adequate. In those days, the usual astronomical photographic emulsions were Kodak 103a-O (blue sensitive) and 103a-E (red sensitive). These emulsions had moderately coarse grain and a limiting resolution, as given by Kodak, of 80 line pairs/mm for a high-contrast target. Thus, the optical and photographic resolutions were similar, and both matched ordinary seeing. When first completed in 1973, the Siding Spring 48-inch Schmidt for the Southern Hemisphere was also fitted with a singlet corrector.

In the 1960s, a new generation of astronomical emulsions became available. These had much improved performance, including finer grain and higher resolution. Today, the most widely used examples are the previously mentioned Kodak IIIa-J, IIIa-F, and IV-N. For IIIa-J and IIIa-F, Kodak gives a limiting resolution of 200 line pairs/mm for a high-contrast target. For the latest IV-N infrared emulsion, resolution is 250 line pairs/mm.

These new emulsions presented a problem for the big Schmidts. The emulsions could now resolve much better than the optics. This optical limitation was especially noticeable on nights of good seeing (seeing blur diameter less than one arc-second). Because color is the worst aberration in a classical Schmidt, the solution was to achromatize. In 1977, an achromatic doublet corrector was installed at Siding Spring, and in 1984 a similar corrector was installed at Palomar.[6]

An example of an achromatic Schmidt follows. To facilitate comparisons with the classical Schmidt example, all wavelengths and first-order properties are kept the same. Thus for each, the system dimensions are very similar, although not exactly identical.

The glasses chosen for the achromatic corrector are Schott BK7 and LLF6. BK7 is the same crown glass used for the singlet corrector. LLF6 flint glass has a coefficient of thermal expansion that is fairly close to that of BK7. The coefficients for the more common flints, such as F2, are not as close. Having similar coefficients of thermal expansion is important because the two corrector elements are cemented together and must not break when the temperature changes. The crown element is placed forward facing the sky because BK7 has greater resistance to weathering.

As with the classical Schmidt, there is more than one way to optimize an achromatic Schmidt. An expanded monochromatic approach that includes color is preferred. This method is nearly identical to the method used earlier. Listing B.6.3.1 gives the required merit function for the first of two optimization stages.

[6] C.G. Wynne, "The Optics of the Achromatized UK Schmidt Telescope," *Q. Jl. R. Astr. Soc.*, Vol. 22, pp. 146–153, 1981.

Listing B.6.3.1

Merit Function Listing

File : C:LENS502.ZMX
Title: ACHROMATIC SCHMIDT, 1220 MM, F/2.5

Merit Function Value: 1.87008697E-005

Num	Type	Int1	Int2	Hx	Hy	Px	Py	Target	Weight	Value	% Cont
1	EFFL		5					3.05000E+003	-1	3.05000E+003	0.003
2	BLNK										
3	BLNK										
4	AXCL							0.00000E+000	0	4.00056E-003	0.000
5	REAY	8	1	0.0000	0.0000	0.0000	0.5000	0.00000E+000	0	4.08731E-003	0.000
6	REAY	8	9	0.0000	0.0000	0.0000	0.5000	0.00000E+000	0	4.08731E-003	0.000
7	DIFF	5	6					0.00000E+000	-1	2.60032E-011	0.000
8	BLNK										
9	BLNK										
10	COMA	0	5					0.00000E+000	0	7.80831E-001	0.000
11	TRAY		5	0.0000	0.7000	0.0000	0.8660	0.00000E+000	0	3.18274E-003	0.000
12	TRAY		5	0.0000	0.7000	0.0000	-0.8660	0.00000E+000	0	-3.18274E-003	0.000
13	SUMM	11	12					0.00000E+000	-1	-8.79368E-011	0.000
14	BLNK										
15	BLNK										
16	TRAY		5	0.0000	0.7000	0.0000	0.8660	0.00000E+000	0	3.18274E-003	0.000
17	TRAY		5	0.0000	0.7000	0.0000	-0.8660	0.00000E+000	0	-3.18274E-003	0.000
18	TRAX		5	0.0000	0.7000	0.8660	0.0000	0.00000E+000	0	-3.18274E-003	0.000
19	TRAX		5	0.0000	0.7000	-0.8660	0.0000	0.00000E+000	0	3.18274E-003	0.000
20	DIFF	16	17					0.00000E+000	0	6.36549E-003	0.000
21	DIFF	18	19					0.00000E+000	0	-6.36549E-003	0.000
22	SUMM	20	21					0.00000E+000	-1	-6.56311E-010	0.000
23	BLNK										
24	BLNK										
25	RETY	2	5	0.0000	0.0000	0.0000	0.8660	0.00000E+000	-1	3.73229E-010	0.000
26	RETY	4	5	0.0000	0.0000	0.0000	0.8660	0.00000E+000	-1	-9.53451E-010	0.000
27	BLNK										
28	BLNK										
29	TTHI	5	7					1.00000E+002	-1	1.00000E+002	0.003
30	BLNK										
31	BLNK										
32	DMFS										
33	TRAR		5	0.0000	0.0000	0.2166	0.0000	0.00000E+000	0.010338	1.73468E-005	10.193
34	TRAR		5	0.0000	0.0000	0.4804	0.0000	0.00000E+000	0.020884	1.37912E-005	13.015
35	TRAR		5	0.0000	0.0000	0.7071	0.0000	0.00000E+000	0.024822	1.19421E-005	11.599
36	TRAR		5	0.0000	0.0000	0.8771	0.0000	0.00000E+000	0.020884	2.58287E-005	45.651
37	TRAR		5	0.0000	0.0000	0.9763	0.0000	0.00000E+000	0.010338	2.40149E-005	19.535

Once again, a perfect null for testing the final system is provided by shrinking monochromatic spot size for a specific wavelength. The neutral zone is selected as before, except now the ray slope is made zero immediately after each of the two aspheric surfaces. The stop is at the corrector plate, and both are again located near the center of curvature of the mirror to control coma. An approximate image radius of curvature is again determined by curving the image surface to find the monochromatic medial focus. And focal length and the location of the central obscuration are controlled as before.

The main difference between Listings B.6.2.1 and B.6.3.1 is that operands 5 through 7 in Listing B.6.3.1 achromatize the corrector plate. For an achromatic Schmidt, longitudinal color is corrected for the two extreme wavelengths and for the 0.5 pupil zone.

Note that in an achromatic Schmidt, the off-axis aberrations are relatively more severe, and thus a consideration of off-axis performance is now required when optimizing. Perhaps surprisingly, this is done by the selection of the wavelength used for shrinking the on-axis monochromatic spot size. Once again, the central wavelength is best. The issues will be clearer when the ray fan plots are examined.

Note also that when optimizing an achromatic Schmidt, you will find many local minima of the merit function. There are many combinations of the four polynomial coefficients plus defocus that yield a monochromatic on-axis spot solution. But most of these solutions have excessive zonal spherical aberration and must be rejected. Perseverance and experimentation will eventually pay off. The best solution is a balance of defocus, third-, fifth-, and residual seventh-order spherical. The monochromatic transverse ray fan plot looks very similar to Figure A.7.7.2, except the scale is much smaller.

The second optimization stage for an achromatic Schmidt is identical to the second optimization stage for a classical Schmidt. All variables and solves are deleted except image surface curvature. The best image surface is then found by a polychromatic off-axis spot shrink.

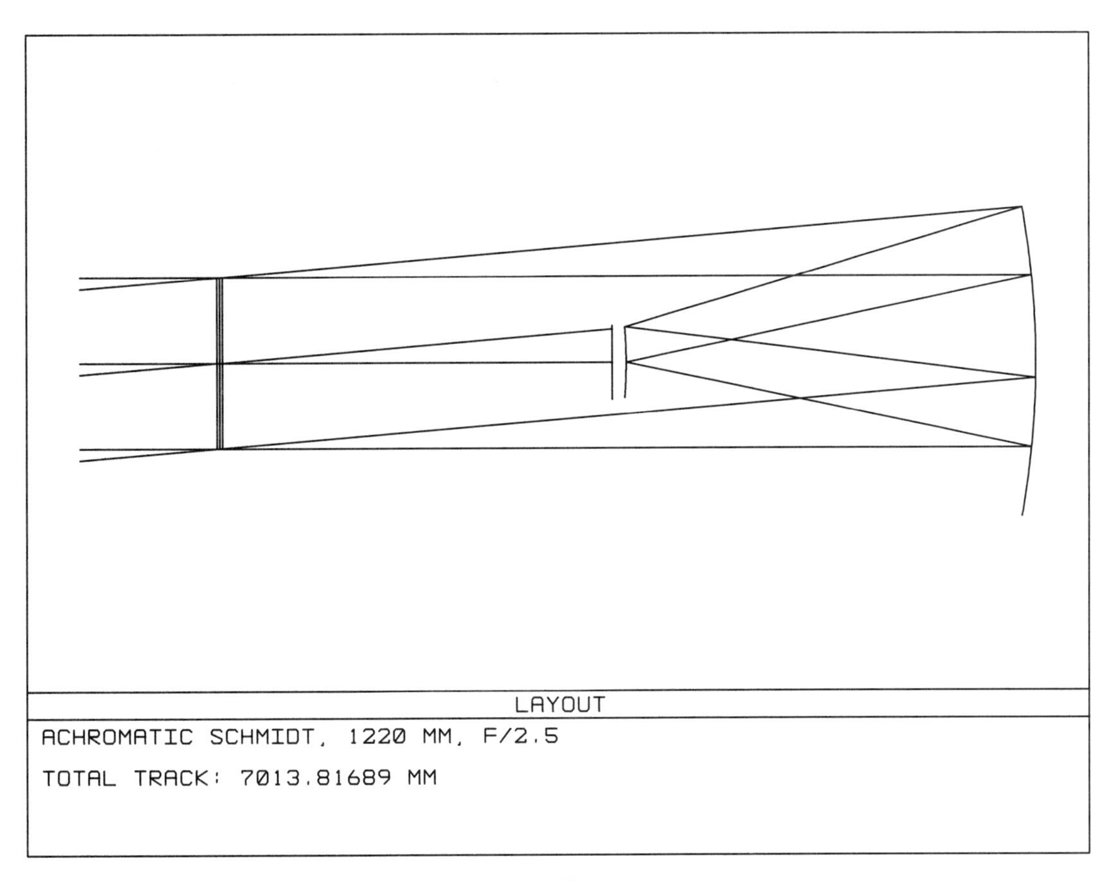

Figure B.6.3.1.

Figure B.6.3.1 is a layout of the optimized achromatic Schmidt. Except for the corrector plate, this layout is virtually the same as the layout in Figure B.6.2.1.

Figure B.6.3.2 is the transverse ray fan plot for just the on-axis image formed by the achromatic Schmidt. Compare this plot with Figure B.6.2.2 for the classical Schmidt. Note several similarities. First, the curve for the optimization wavelength, which is again 0.495 μm, is indistinguishable from the horizontal axis. This indicates virtually perfect correction and a good null. Second, the curves for all wavelengths cross at the 0.866 pupil neutral zone. Third, for any given wavelength, chromatically varying defocus (longitudinal color) balances chromatically varying third-order spherical aberration (spherochromatism).

But also note several differences. First, the curves are no longer arrayed vertically in order of wavelength; instead, the order reverses. This reversal is just what achromatization seeks to achieve. Primary color is corrected and secondary color remains (recall the U-shaped secondary color curve). Second, the curves for the two extreme wavelengths, 0.375 μm and 0.90 μm, are nearly superimposed and cross at the 0.5 zone, as required by the chromatic optimization operand. Third, the curves no longer cluster symmetrically about the horizontal axis. Instead, for a given pupil zone, all of the transverse errors have the same sign. Fourth, the scale has been changed from ± 25 μm to ± 5 μm.

Actually, a more symmetrical clustering of the on-axis curves could have been achieved if a different optimization wavelength had been selected. Perhaps the monochromatic spot could have been shrunk using either 0.400 or 0.730 μm light. But the improved on-axis performance would have been obtained at the expense of off-axis performance.

Figure B.6.3.3 is the set of ray fan plots for all four field angles. Compare these with the plots in Figure B.6.2.3 for the classical Schmidt. Note immediately that the achromatic Schmidt has much stronger off-axis aberrations, especially oblique

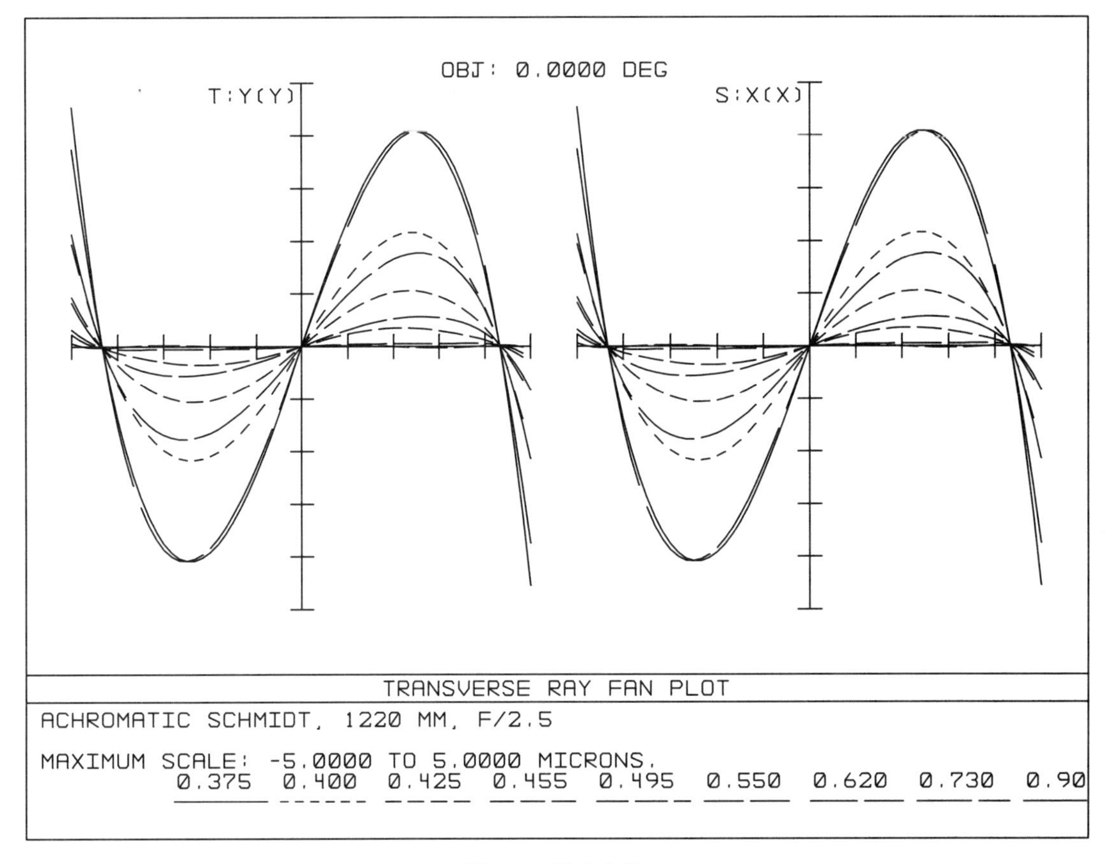

Figure B.6.3.2.

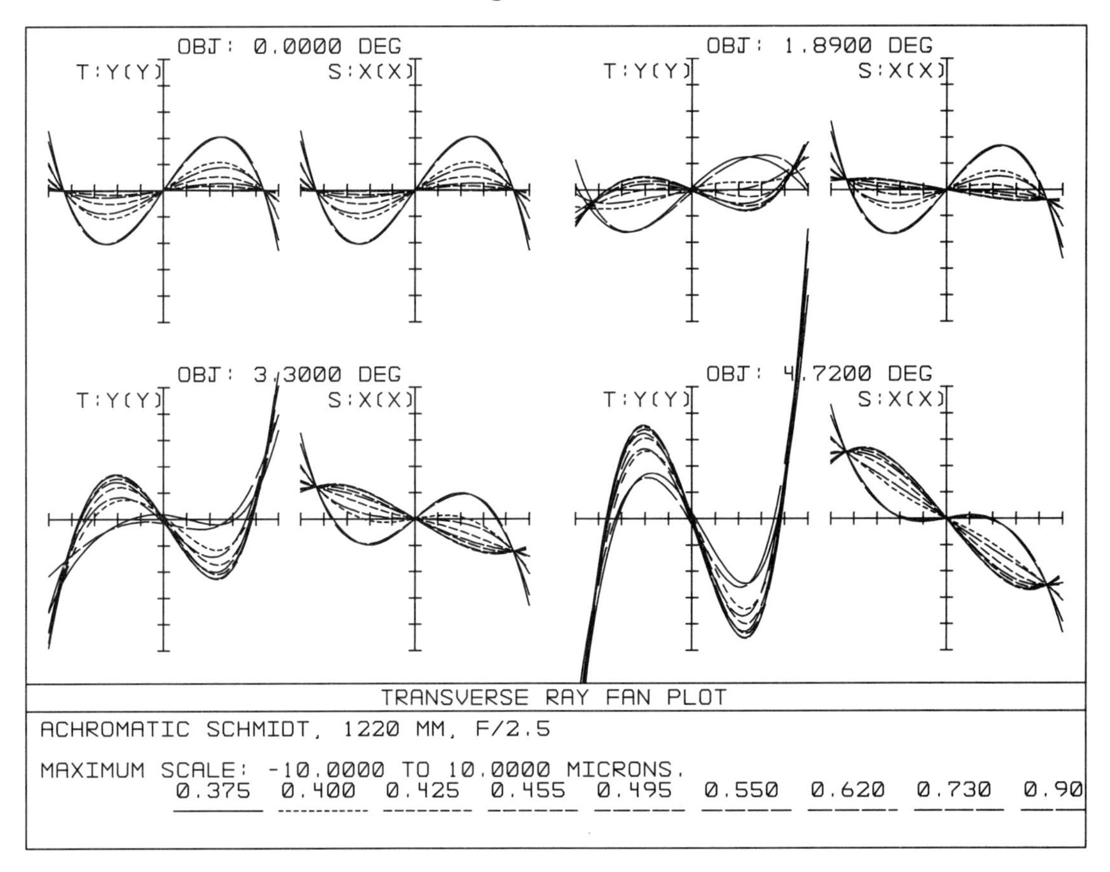

Figure B.6.3.3.

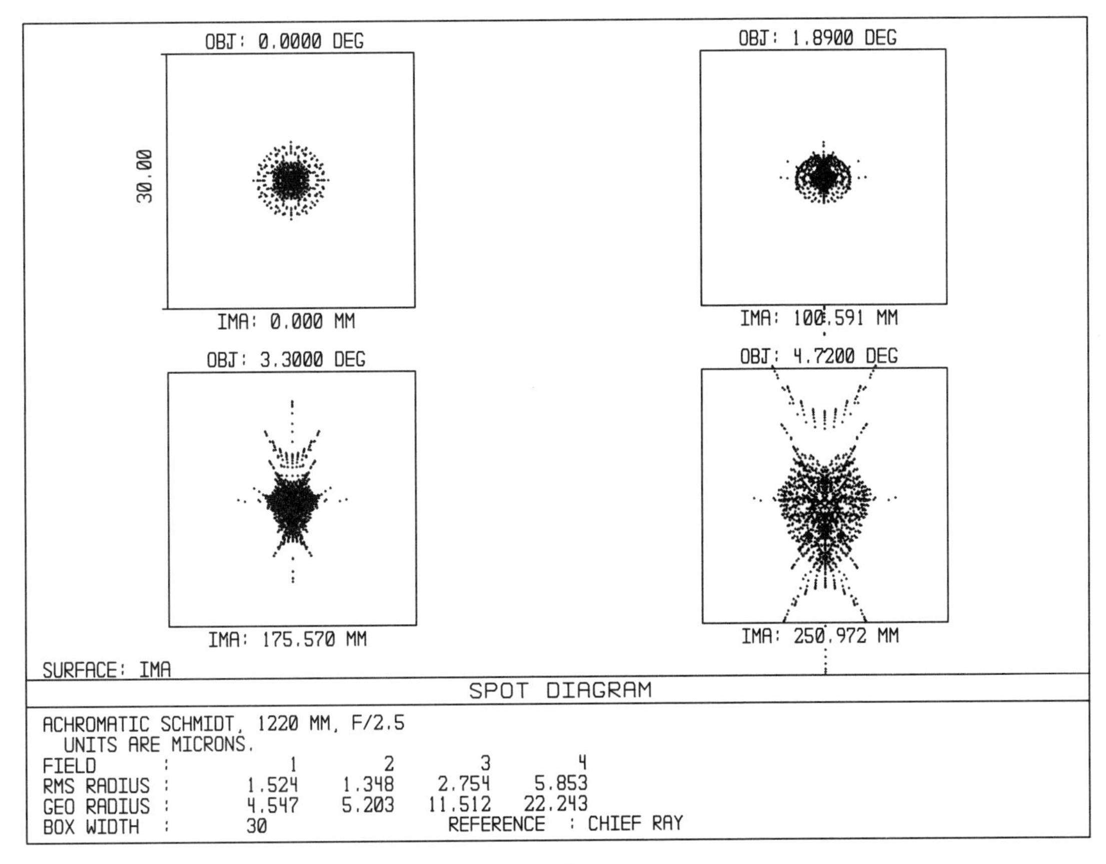

Figure B.6.3.4.

spherical. Note too that the on-axis chromatic errors (the effects of which are present all across the field) and the off-axis oblique spherical errors have opposite signs and tend to cancel each other. This balance is crucial for best overall performance. In fact, the balance could have been made even slightly better if the on-axis null had been abandoned. But given this requirement, the present on-axis curves are as asymmetrical as possible, thereby giving maximum cancellation.

Figure B.6.3.4 is the polychromatic spot diagram. Compare with Figure B.6.2.4. Figure B.6.3.5 is the matrix spot diagram. Compare with Figure B.6.2.5. As expected, the spots for the achromatic Schmidt are considerably smaller. Although spot size varies with wavelength and across the field, the main parts of the larger spots have a diameter of roughly 13 μm. A 13 μm linear spot diameter corresponds to 0.9 arc-seconds in the sky. This is a factor of 3 improvement over the classical Schmidt. Except on nights of very good seeing, the system is no longer aberration and emulsion limited.

Listing B.6.3.2 is the prescription for the achromatic Schmidt. Listings B.6.3.3 and B.6.3.4 are the sag tables for the two aspheric surfaces on the doublet corrector plate. Note that the maximum sags on these surfaces are 3 or 4 times as much as the maximum sag on the singlet corrector. Note too that the front crown element is convex in the middle surrounded by a concave outer zone, whereas the rear flint element is concave in the middle surrounded by a convex outer zone. In other words, as oriented in space, the two curves tend to follow each other.

At Palomar and Siding Spring, the achromatic corrector plates greatly improve the practical image quality of the big Schmidts. But there is a further gain. For stars (but not extended objects), a reduction in the size of the point spread function concentrates the light. Because the photographic process responds to power density, the telescopes can now record fainter stars. By these advances in resolution and sensitivity, the big Schmidts are now even more valuable as research instruments.

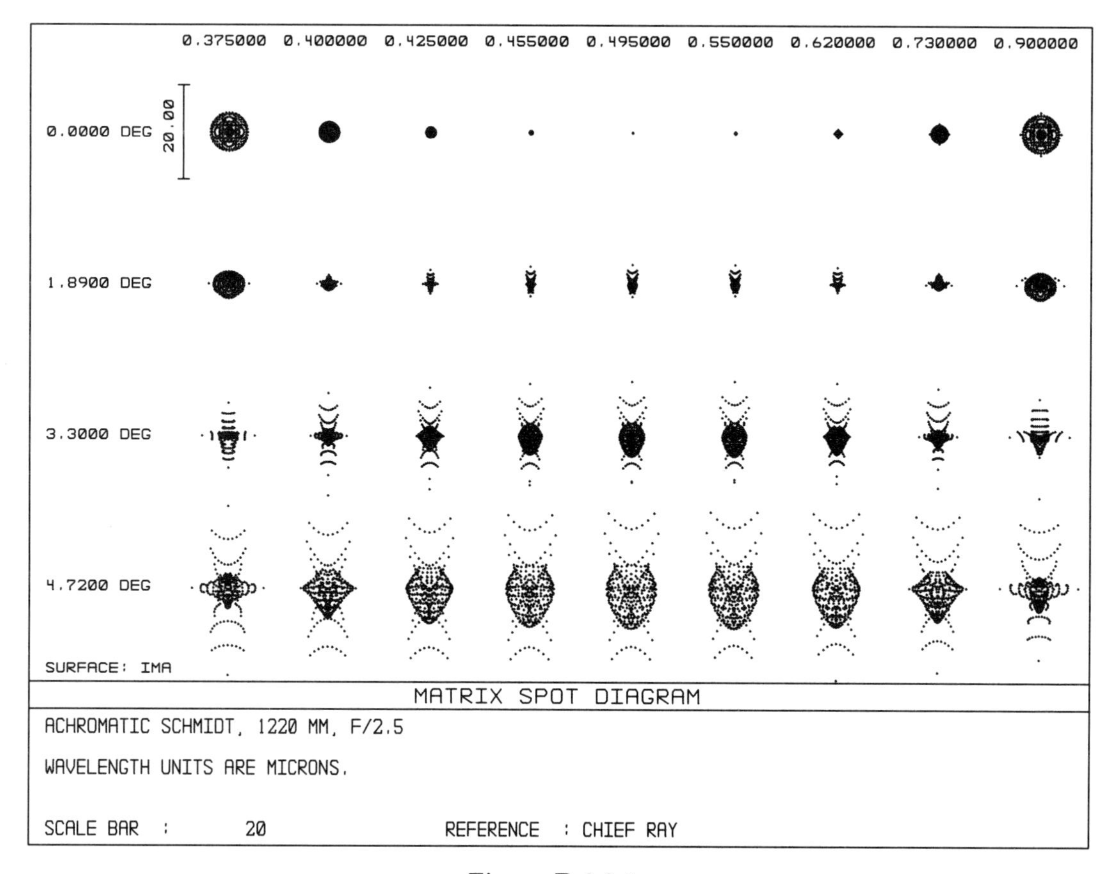

Figure B.6.3.5.

Listing B.6.3.2

```
System/Prescription Data

File : C:LENS503.ZMX
Title: ACHROMATIC SCHMIDT, 1220 MM, F/2.5

GENERAL LENS DATA:

Surfaces          :           8
Stop              :           3
System Aperture   :Entrance Pupil Diameter
Ray aiming        : Off
Apodization       :Uniform, factor =     0.000000
Eff. Focal Len.   :        3050 (in air)
Eff. Focal Len.   :        3050 (in image space)
Total Track       :     7013.82
Image Space F/#   :         2.5
Para. Wrkng F/#   :         2.5
Working F/#       :         2.5
Obj. Space N.A.   :    6.1e-008
Stop Radius       :      609.96
Parax. Ima. Hgt.  :     251.828
Parax. Mag.       :           0
Entr. Pup. Dia.   :        1220
Entr. Pup. Pos.   :     1013.14
Exit Pupil Dia.   :     1257.13
Exit Pupil Pos.   :    -3142.83
Field Type        : Angle in degrees
Maximum Field     :        4.72
Primary Wave      :    0.495000
Lens Units        : Millimeters
Angular Mag.      :    0.970462
```

```
Fields          : 4
Field Type:  Angle in degrees
#         X-Value          Y-Value          Weight
1        0.000000         0.000000         0.000000
2        0.000000         1.890000         1.000000
3        0.000000         3.300000         1.000000
4        0.000000         4.720000         1.000000

Vignetting Factors
#          VDX            VDY            VCX            VCY
1       0.000000       0.000000       0.000000       0.000000
2       0.000000       0.000000       0.000000       0.000000
3       0.000000       0.000000       0.000000       0.000000
4       0.000000       0.000000       0.000000       0.000000

Wavelengths     : 9
Units:  Microns
#         Value           Weight
1        0.375000        1.000000
2        0.400000        1.000000
3        0.425000        1.000000
4        0.455000        1.000000
5        0.495000        1.000000
6        0.550000        1.000000
7        0.620000        1.000000
8        0.730000        1.000000
9        0.900000        1.000000
SURFACE DATA SUMMARY:

Surf      Type         Radius      Thickness         Glass      Diameter          Conic
 OBJ  STANDARD       Infinity       Infinity                           0              0
   1  STANDARD       Infinity           1000                           0              0
   2  EVENASPH       105007.6             20           BK7      1222.073              0
 STO  STANDARD       Infinity             20          LLF6      1220.072              0
   4  EVENASPH         144622       2846.913                    1222.311              0
   5  STANDARD       Infinity       3126.904                         520              0
   6  STANDARD        -6076.9      -3026.905        MIRROR      2194.955              0
   7  STANDARD       Infinity   0.0006556725                           0              0
 IMA  STANDARD       -3042.17              0                    501.9878              0

SURFACE DATA DETAIL:

Surface OBJ      : STANDARD
Surface   1      : STANDARD
Surface   2      : EVENASPH
 Coeff on r  2   :              0
 Coeff on r  4   : -8.524712e-012
 Coeff on r  6   : -1.639526e-020
 Coeff on r  8   :              0
 Coeff on r 10   :              0
 Coeff on r 12   :              0
 Coeff on r 14   :              0
 Coeff on r 16   :              0
Surface STO      : STANDARD
Surface   4      : EVENASPH
 Coeff on r  2   :              0
 Coeff on r  4   : -6.226264e-012
 Coeff on r  6   : 7.563795e-020
 Coeff on r  8   :              0
 Coeff on r 10   :              0
 Coeff on r 12   :              0
 Coeff on r 14   :              0
 Coeff on r 16   :              0
Surface   5      : STANDARD
 Aperture        : Circular Obscuration
 Minimum Radius  :              0
 Maximum Radius  :            260
Surface   6      : STANDARD
Surface   7      : STANDARD
Surface IMA      : STANDARD

SOLVE AND VARIABLE DATA:

Semi Diam    1      : Fixed
Semi Diam    5      : Fixed
Semi Diam    7      : Fixed
Curvature of    8   : Variable

INDEX OF REFRACTION DATA:

Surf      Glass       0.375000     0.400000     0.425000     0.455000     0.495000
   0                1.00000000   1.00000000   1.00000000   1.00000000   1.00000000
   1                1.00000000   1.00000000   1.00000000   1.00000000   1.00000000
   2        BK7     1.53454939   1.53084854   1.52782658   1.52486897   1.52175196
   3       LLF6     1.55710079   1.55154395   1.54711436   1.54286695   1.53848201
   4                1.00000000   1.00000000   1.00000000   1.00000000   1.00000000
   5                1.00000000   1.00000000   1.00000000   1.00000000   1.00000000
   6     MIRROR     1.00000000   1.00000000   1.00000000   1.00000000   1.00000000
   7                1.00000000   1.00000000   1.00000000   1.00000000   1.00000000
   8                1.00000000   1.00000000   1.00000000   1.00000000   1.00000000

Surf      Glass       0.550000     0.620000     0.730000     0.900000
   0                1.00000000   1.00000000   1.00000000   1.00000000
   1                1.00000000   1.00000000   1.00000000   1.00000000
   2        BK7     1.51852239   1.51553950   1.51230442   1.50899686
   3       LLF6     1.53404072   1.53004392   1.52586164   1.52183901
   4                1.00000000   1.00000000   1.00000000   1.00000000
   5                1.00000000   1.00000000   1.00000000   1.00000000
   6     MIRROR     1.00000000   1.00000000   1.00000000   1.00000000
   7                1.00000000   1.00000000   1.00000000   1.00000000
   8                1.00000000   1.00000000   1.00000000   1.00000000
```

Listing B.6.3.3

Listing of surface sag

File : C:LENS503.ZMX
Title: ACHROMATIC SCHMIDT, 1220 MM, F/2.5

Units are Millimeters.

Semi diameter of surface 2: 611.036621.
Best Fit Sphere curvature : 0.000005.
Best Fit Sphere radius : 214940.429164.
Best Fit Sphere residual : 0.697583. (rms)

Y-coord	Sag	BFS Sag	Deviation	Remove
0.000000000	0.000000000	0.000000000	0.000000000	-0.279932268
10.000000000	0.000476071	0.000232623	-0.000243448	-0.280175716
20.000000000	0.001903259	0.000930490	-0.000972769	-0.280905037
30.000000000	0.004278497	0.002093603	-0.002184894	-0.282117162
40.000000000	0.007596670	0.003721962	-0.003874708	-0.283806977
50.000000000	0.011850616	0.005815565	-0.006035051	-0.285967320
60.000000000	0.017031129	0.008374414	-0.008656716	-0.288588984
70.000000000	0.023126957	0.011398507	-0.011728449	-0.291660718
80.000000000	0.030124799	0.014887846	-0.015236953	-0.295169221
90.000000000	0.038009312	0.018842431	-0.019166881	-0.299099150
100.000000000	0.046763104	0.023262261	-0.023500843	-0.303433111
110.000000000	0.056366736	0.028147336	-0.028219400	-0.308151669
120.000000000	0.066798725	0.033497656	-0.033301069	-0.313233337
130.000000000	0.078035540	0.039313222	-0.038722318	-0.318654586
140.000000000	0.090051602	0.045594033	-0.044457569	-0.324389838
150.000000000	0.102819288	0.052340090	-0.050479199	-0.330411467
160.000000000	0.116308926	0.059551392	-0.056757534	-0.336689802
170.000000000	0.130488797	0.067227940	-0.063260857	-0.343193125
180.000000000	0.145325134	0.075369733	-0.069955401	-0.349887669
190.000000000	0.160782125	0.083976773	-0.076805352	-0.356737620
200.000000000	0.176821906	0.093049057	-0.083772848	-0.363705117
210.000000000	0.193404568	0.102586588	-0.090817980	-0.370750248
220.000000000	0.210488153	0.112589365	-0.097898789	-0.377831057
230.000000000	0.228028655	0.123057387	-0.104971268	-0.384903537
240.000000000	0.245980019	0.133990655	-0.111989363	-0.391921632
250.000000000	0.264294139	0.145389170	-0.118904969	-0.398837238
260.000000000	0.282920863	0.157252930	-0.125667933	-0.405600201
270.000000000	0.301807987	0.169581937	-0.132226050	-0.412158318
280.000000000	0.320901258	0.182376190	-0.138525067	-0.418457336
290.000000000	0.340144372	0.195635690	-0.144508682	-0.424440951
300.000000000	0.359478976	0.209360436	-0.150118541	-0.430050809
310.000000000	0.378844666	0.223550428	-0.155294237	-0.435226506
320.000000000	0.398178984	0.238205667	-0.159973316	-0.439905585
330.000000000	0.417417423	0.253326153	-0.164091270	-0.444023539
340.000000000	0.436493425	0.268911886	-0.167581539	-0.447513807
350.000000000	0.455338377	0.284962865	-0.170375511	-0.450307780
360.000000000	0.473881614	0.301479092	-0.172402522	-0.452334790
370.000000000	0.492050419	0.318460566	-0.173589853	-0.453522122
380.000000000	0.509770021	0.335907287	-0.173862734	-0.453795002
390.000000000	0.526963594	0.353819255	-0.173144339	-0.453076607
400.000000000	0.543552259	0.372196471	-0.171355788	-0.451288056
410.000000000	0.559455081	0.391038935	-0.168416146	-0.448348415
420.000000000	0.574589070	0.410346646	-0.164242424	-0.444174693
430.000000000	0.588869181	0.430119605	-0.158749576	-0.438681844
440.000000000	0.602208311	0.450357812	-0.151850499	-0.431782767
450.000000000	0.614517302	0.471061267	-0.143456034	-0.423388303
460.000000000	0.625704936	0.492229971	-0.133474966	-0.413407234
470.000000000	0.635677942	0.513863922	-0.121814019	-0.401746287
480.000000000	0.644340984	0.535963123	-0.108377862	-0.388310130
490.000000000	0.651596674	0.558527572	-0.093069102	-0.373001370
500.000000000	0.657345558	0.581557270	-0.075788289	-0.355720557
510.000000000	0.661486128	0.605052216	-0.056433911	-0.336366180
520.000000000	0.663914810	0.629012412	-0.034902397	-0.314834666
530.000000000	0.664525971	0.653437857	-0.011088114	-0.291020382
540.000000000	0.663211918	0.678328552	0.015116634	-0.264815635
550.000000000	0.659862893	0.703684496	0.043821603	-0.236110665
560.000000000	0.654367075	0.729505690	0.075138615	-0.204793653
570.000000000	0.646610580	0.755792134	0.109181554	-0.170750715
580.000000000	0.636477460	0.782543828	0.146066368	-0.133865901
590.000000000	0.623849701	0.809760772	0.185911071	-0.094021198
600.000000000	0.608607224	0.837442967	0.228835743	-0.051096526
610.000000000	0.590627882	0.865590412	0.274962530	-0.004969739
611.036621338	0.588602582	0.868534850	0.279932268	0.000000000

Listing B.6.3.4

```
Listing of surface sag

File : C:LENS503.ZMX
Title: ACHROMATIC SCHMIDT, 1220 MM, F/2.5

Units are Millimeters.

Semi diameter of surface 4: 611.155390.
Best Fit Sphere curvature : 0.000003.
Best Fit Sphere radius    : 296648.326441.
Best Fit Sphere residual  : 0.506261. (rms)
```

Y-coord	Sag	BFS Sag	Deviation	Remove
0.000000000	0.000000000	0.000000000	0.000000000	0.126249104
10.000000000	0.000345667	0.000168550	-0.000177117	0.126071987
20.000000000	0.001381919	0.000674199	-0.000707720	0.125541384
30.000000000	0.003106517	0.001516948	-0.001589569	0.124659535
40.000000000	0.005515723	0.002696796	-0.002818927	0.123430177
50.000000000	0.008604309	0.004213744	-0.004390566	0.121858539
60.000000000	0.012365551	0.006067791	-0.006297760	0.119951344
70.000000000	0.016791232	0.008258938	-0.008532295	0.117716810
80.000000000	0.021871642	0.010787184	-0.011084458	0.115164646
90.000000000	0.027595577	0.013652530	-0.013943047	0.112306057
100.000000000	0.033950341	0.016854975	-0.017095366	0.109153738
110.000000000	0.040921747	0.020394520	-0.020527227	0.105721877
120.000000000	0.048494115	0.024271164	-0.024222950	0.102026154
130.000000000	0.056650274	0.028484908	-0.028165366	0.098083739
140.000000000	0.065371564	0.033035752	-0.032335812	0.093913292
150.000000000	0.074637834	0.037923695	-0.036714139	0.089534966
160.000000000	0.084427444	0.043148738	-0.041278706	0.084970398
170.000000000	0.094717268	0.048710881	-0.046006388	0.080242717
180.000000000	0.105482690	0.054610123	-0.050872567	0.075376537
190.000000000	0.116697608	0.060846464	-0.055851143	0.070397961
200.000000000	0.128334436	0.067419906	-0.060914530	0.065334574
210.000000000	0.140364103	0.074330447	-0.066033655	0.060215449
220.000000000	0.152756053	0.081578088	-0.071177965	0.055071139
230.000000000	0.165478250	0.089162829	-0.076315421	0.049933683
240.000000000	0.178497177	0.097084669	-0.081412507	0.044836597
250.000000000	0.191777834	0.105343610	-0.086434224	0.039814880
260.000000000	0.205283745	0.113939650	-0.091344095	0.034905009
270.000000000	0.218976957	0.122872790	-0.096104167	0.030144937
280.000000000	0.232818040	0.132143030	-0.100675010	0.025574095
290.000000000	0.246766089	0.141750370	-0.105015719	0.021233385
300.000000000	0.260778729	0.151694810	-0.109083919	0.017165185
310.000000000	0.274812109	0.161976350	-0.112835760	0.013413345
320.000000000	0.288820913	0.172594990	-0.116225923	0.010023181
330.000000000	0.302758353	0.183550730	-0.119207623	0.007041481
340.000000000	0.316576178	0.194843570	-0.121732608	0.004516497
350.000000000	0.330224669	0.206473510	-0.123751158	0.002497946
360.000000000	0.343652646	0.218440551	-0.125212095	0.001037009
370.000000000	0.356807469	0.230744692	-0.126062777	0.000186327
380.000000000	0.369635037	0.243385933	-0.126249104	0.000000000
390.000000000	0.382079793	0.256364274	-0.125715519	0.000533585
400.000000000	0.394084726	0.269679716	-0.124405010	0.001844094
410.000000000	0.405591369	0.283332258	-0.122259111	0.003989993
420.000000000	0.416539808	0.297321901	-0.119217908	0.007031197
430.000000000	0.426868679	0.311648644	-0.115220035	0.011029069
440.000000000	0.436515170	0.326312487	-0.110202683	0.016046422
450.000000000	0.445415028	0.341313432	-0.104101596	0.022147508
460.000000000	0.453502557	0.356651476	-0.096851080	0.029398024
470.000000000	0.460710621	0.372326622	-0.088383999	0.037865105
480.000000000	0.466970649	0.388338868	-0.078631781	0.047617324
490.000000000	0.472212635	0.404688216	-0.067524420	0.058724685
500.000000000	0.476365142	0.421374664	-0.054990479	0.071258626
510.000000000	0.479355303	0.438398213	-0.040957091	0.085292013
520.000000000	0.481108827	0.455758863	-0.025349964	0.100899140
530.000000000	0.481549995	0.473456614	-0.008093382	0.118155723
540.000000000	0.480601673	0.491491466	0.010889792	0.137138897
550.000000000	0.478185306	0.509863419	0.031678113	0.157927217
560.000000000	0.474220925	0.528572474	0.054351549	0.180600653
570.000000000	0.468627147	0.547618629	0.078991482	0.205240587
580.000000000	0.461321184	0.567001887	0.105680703	0.231929807
590.000000000	0.452218838	0.586722245	0.134503407	0.260752511
600.000000000	0.441234512	0.606779705	0.165545193	0.291794297
610.000000000	0.428281208	0.627174267	0.198893059	0.325142163
611.155389906	0.426653704	0.629552358	0.202898654	0.329147758

Chapter B.7

Tolerancing Example

In Chapter A.16, optical fabrication tolerances (allowed or expected manufacturing errors) were discussed in general. In this chapter, a specific lens is toleranced. Tolerancing is done in three steps: (1) specifying the tolerances, (2) the sensitivity analysis, and (3) estimating actual performance. In practice, this is an iterative process with earlier steps repeated.

B.7.1 Tolerancing a Tessar

The Tessar lens of Chapter B.3 is selected for demonstrating the procedures used in tolerancing a lens. Figure B.7.1, which is the same as Figure B.3.3.1, shows the layout. The Tessar is a good example for tolerancing because it is a well-known and common lens, all aberrations are active, and it includes both airspaced and cemented surfaces.

Listing B.7.1 gives the nominal Tessar lens prescription. Note that for tolerancing purposes, the conventional optical configuration is slightly modified to include object points with both positive and negative field angles. Thus, the field points are now located at 0, $\pm 70\%$, and $\pm 100\%$ of the half-field. Because this lens is rotationally symmetric about the optical axis, only object points in the y-direction are necessary. However, in a general tilted and decentered ("off-axis") system, objects arrayed in two dimensions over the field of view are required. In addition, because the Tessar has mechanical vignetting, vignetting coefficients have been added to ensure that light gets through the lens properly.

B.7.2 Specifying the Tolerances

The tolerancing procedure used here is the one in ZEMAX. In this program, the first step in tolerancing a lens is to define a set of tolerance operands in the tolerance editor. A good way to do this is by letting the program define a set of default operands, and then to modify these as required. The tolerance types (types of lens fabrication errors) to be included depend on the lens and, when more than one way is possible, on the preferences of the designer. The associated tolerance values (error limits, both plus and minus) are determined in conjunction with step two, the sensitivity analysis. For the tolerance values to be acceptable, overall optical performance must not be excessively degraded when all tolerances act together; that is, the lens must remain within its error budget and meet specs. For the Tessar example, the adopted set of tolerance operands and their values is given in Listing B.7.2 and is described below.

In operands 2 through 8, TFRN specifies a tolerance on lens surface curvature (surface power) expressed in fringes. The value in the Int1 column is the surface number. Although these errors could have been expressed directly as changes in curvature or in radius of curvature, using fringes is more convenient when the sur-

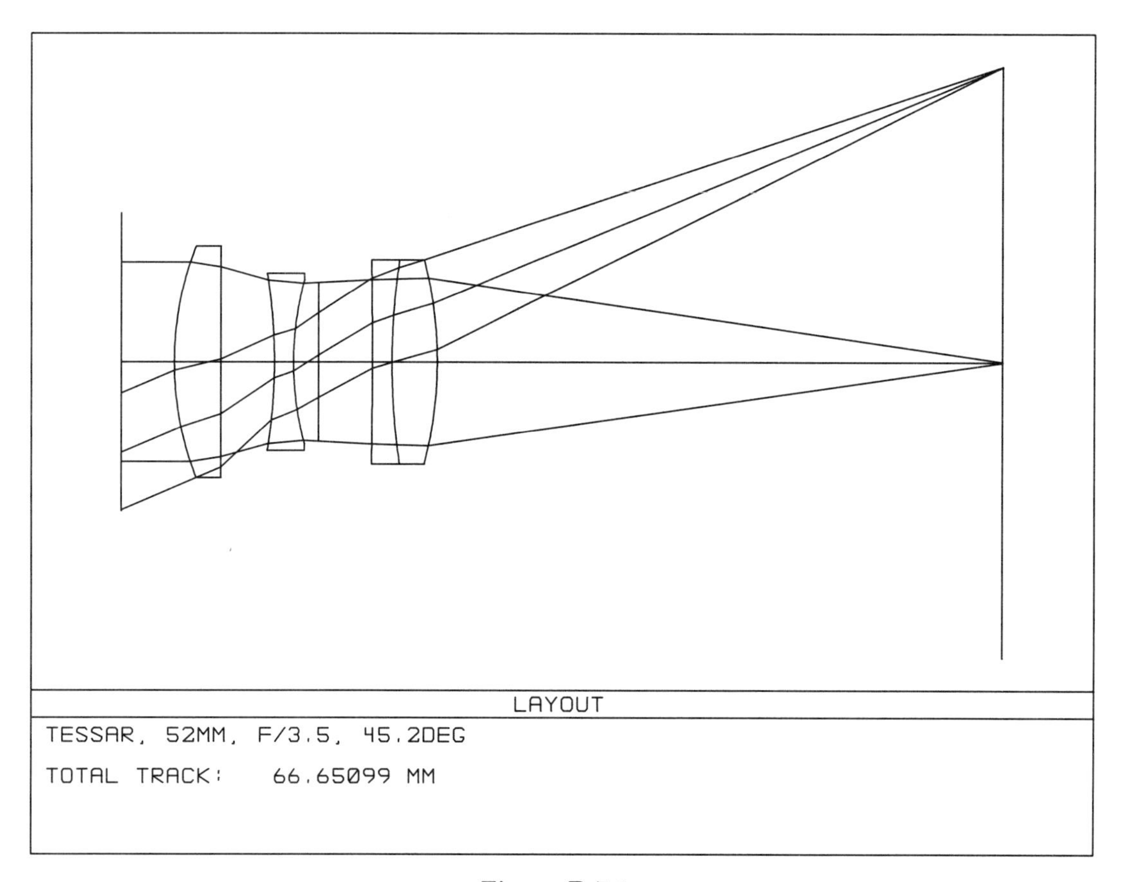

Figure B.7.1.

faces are checked by the optician with test plates and Newton's rings in the usual double-pass configuration. This test gives the departure or sag error of the lens surface relative to the test plate surface. The amount of error allowed is specified in the Min and Max columns and is ±2 double-pass fringes (±1 wavelength of sag error on the surface at its edge).

In operands 9 through 15, TTHI specifies a tolerance on air or glass thickness. The surface number is given in the Int1 column. The amount of error allowed is ±0.050 mm. Recall that in the computer, each surface location is given relative to the previous surface by local, not global, coordinates. Consequently, according to this model, an error in one of the thicknesses displaces all subsequent surfaces by the amount of the error. But usually when a lens is manufactured, a thickness error occurring toward the front of the lens is actually absorbed or compensated in the next airspace toward the rear. If this is the case, then during tolerancing, the value in the Int2 column of a TTHI operand gives the number of the surface where this compensation is to take place.

Operands 16 through 27 specify the allowed decenter and tilt tolerances on the elements and groups of elements bounded by the surface numbers given in the Int1 and Int2 columns. TEDX and TEDY are the decenter operands for transverse displacements in the x- and y-directions, respectively. The amount of decenter error is ±0.050 mm. TETX and TETY are the tilt operands for rotations about the x- and y-axes, respectively. These rotation axes pass through the vertex of the surface in the Int1 column. The amount of tilt error is ±0.050°.

In operands 28 through 35, TIRX and TIRY specify tolerances on single-surface tilt errors, which produce element wedge. These tilts are given by surface sag errors along the x- and y-directions, respectively. The sags are measured by total indicator runout (TIR) around the edge of the lens. The surface number is given in the Int1 column. Note that the sense of TIRX and TIRY is opposite that of TETX and

Listing B.7.1

```
System/Prescription Data
Title:  TESSAR, 52MM, F/3.5, 45.2DEG, Tolerancing

GENERAL LENS DATA:

Surfaces          :                 11
Stop              :                  6
System Aperture   :  Entrance Pupil Diameter = 14.86
Ray aiming        :  Paraxial Reference, cache on
 X Pupil shift    :                  0
 Y Pupil shift    :                  0
 Z Pupil shift    :                  0
Apodization       : Uniform, factor =    0.00000E+000
Eff. Focal Len.   :           52.00096 (in air)
Eff. Focal Len.   :           52.00096 (in image space)
Total Track       :             66.651
Image Space F/#   :           3.499392
Para. Wrkng F/#   :           3.499392
Working F/#       :           3.562482
Obj. Space N.A.   :          7.43e-010
Stop Radius       :           5.829161
Parax. Ima. Hgt.  :           21.64591
Parax. Mag.       :                  0
Entr. Pup. Dia.   :              14.86
Entr. Pup. Pos.   :           16.25826
Exit Pupil Dia.   :           14.65645
Exit Pupil Pos.   :          -51.28866
Field Type        :  Angle in degrees
Maximum Field     :               22.6
Primary Wave      :               0.55
Lens Units        :     Millimeters
Angular Mag.      :          0.9845164

Fields          : 5
Field Type:  Angle in degrees
#        X-Value          Y-Value           Weight
1        0.000000         0.000000          8.000000
2        0.000000       -15.800000          2.000000
3        0.000000        15.800000          2.000000
4        0.000000       -22.600000          1.000000
5        0.000000        22.600000          1.000000

Vignetting Factors
#            VDX              VDY              VCX              VCY
1       0.000000         0.000000         0.000000         0.000000
2       0.000000        -0.051634         0.001503         0.221485
3       0.000000         0.051634         0.001503         0.221485
4       0.000000        -0.085569         0.007759         0.467579
5       0.000000         0.085569         0.007759         0.467579
Wavelengths      : 5
Units:  Microns
#          Value          Weight
1       0.450000       1.000000
2       0.500000       1.000000
3       0.550000       1.000000
4       0.600000       1.000000
5       0.650000       1.000000

SURFACE DATA SUMMARY:

        Surf        Type       Radius        Thickness          Glass         Diameter        Conic
OBJ STANDARD   Infinity    Infinity                                 0                0
  1 STANDARD   Infinity           4                          21.86453                0
  2 STANDARD    22.5851         3.5           LAFN21               17                0
  3 STANDARD   3174.661    4.005808                                17                0
  4 STANDARD  -39.77737         1.5             SF15               13                0
  5 STANDARD   20.74764           2                                12                0
STO STANDARD   Infinity     4.06086                          11.65832                0
  7 STANDARD  -502.9552         1.5               F5               15                0
  8 STANDARD   47.47455         3.5           LAFN21               15                0
  9 STANDARD  -28.85977    42.58433                                15                0
 10 STANDARD   Infinity           0                          43.40646                0
IMA STANDARD   Infinity                                      43.40646                0

SURFACE DATA DETAIL:
Surface OBJ       :  STANDARD
Surface 1         :  STANDARD
Surface 2         :  STANDARD
 Aperture         :  Circular Aperture
 Minimum Radius   :               0
 Maximum Radius   :             8.5
Surface 3         :  STANDARD
Surface 4         :  STANDARD
Surface 5         :  STANDARD
Surface STO       :  STANDARD
Surface 7         :  STANDARD
Surface 8         :  STANDARD
Surface 9         :  STANDARD
 Aperture         :  Circular Aperture
 Minimum Radius   :               0
 Maximum Radius   :             7.5
Surface  10       : STANDARD
Surface IMA       : STANDARD
```

Listing B.7.2

```
Tolerance Data Listing

Title:  TESSAR, 52MM, F/3.5, 45.2DEG, Tolerancing
Num   Type   Int1  Int2             Min            Max
  1  (COMP)   10     0  -1.00000E+000   1.00000E+000
  2  (TFRN)    2        -2.00000E+000   2.00000E+000
  3  (TFRN)    3        -2.00000E+000   2.00000E+000
  4  (TFRN)    4        -2.00000E+000   2.00000E+000
  5  (TFRN)    5        -2.00000E+000   2.00000E+000
  6  (TFRN)    7        -2.00000E+000   2.00000E+000
  7  (TFRN)    8        -2.00000E+000   2.00000E+000
  8  (TFRN)    9        -2.00000E+000   2.00000E+000
  9  (TTHI)    2     3  -5.00000E-002   5.00000E-002
 10  (TTHI)    3     5  -5.00000E-002   5.00000E-002
 11  (TTHI)    4     5  -5.00000E-002   5.00000E-002
 12  (TTHI)    5     6  -5.00000E-002   5.00000E-002
 13  (TTHI)    6     9  -5.00000E-002   5.00000E-002
 14  (TTHI)    7     9  -5.00000E-002   5.00000E-002
 15  (TTHI)    8     9  -5.00000E-002   5.00000E-002
 16  (TEDX)    2     3  -5.00000E-002   5.00000E-002
 17  (TEDY)    2     3  -5.00000E-002   5.00000E-002
 18  (TETX)    2     3  -5.00000E-002   5.00000E-002
 19  (TETY)    2     3  -5.00000E-002   5.00000E-002
 20  (TEDX)    4     5  -5.00000E-002   5.00000E-002
 21  (TEDY)    4     5  -5.00000E-002   5.00000E-002
 22  (TETX)    4     5  -5.00000E-002   5.00000E-002
 23  (TETY)    4     5  -5.00000E-002   5.00000E-002
 24  (TEDX)    7     9  -5.00000E-002   5.00000E-002
 25  (TEDY)    7     9  -5.00000E-002   5.00000E-002
 26  (TETX)    7     9  -5.00000E-002   5.00000E-002
 27  (TETY)    7     9  -5.00000E-002   5.00000E-002
 28  (TIRX)    2        -2.50000E-002   2.50000E-002
 29  (TIRY)    2        -2.50000E-002   2.50000E-002
 30  (TIRX)    4        -2.50000E-002   2.50000E-002
 31  (TIRY)    4        -2.50000E-002   2.50000E-002
 32  (TIRX)    7        -2.50000E-002   2.50000E-002
 33  (TIRY)    7        -2.50000E-002   2.50000E-002
 34  (TIRX)    9        -2.50000E-002   2.50000E-002
 35  (TIRY)    9        -2.50000E-002   2.50000E-002
 36  (TIRR)    2        -5.00000E-001   5.00000E-001
 37  (TIRR)    3        -5.00000E-001   5.00000E-001
 38  (TIRR)    4        -5.00000E-001   5.00000E-001
 39  (TIRR)    5        -5.00000E-001   5.00000E-001
 40  (TIRR)    7        -5.00000E-001   5.00000E-001
 41  (TIRR)    8        -5.00000E-001   5.00000E-001
 42  (TIRR)    9        -5.00000E-001   5.00000E-001
 43  (TIND)    2        -5.00000E-004   5.00000E-004
 44  (TIND)    4        -5.00000E-004   5.00000E-004
 45  (TIND)    7        -5.00000E-004   5.00000E-004
 46  (TIND)    8        -5.00000E-004   5.00000E-004
```

TETY. For example, a TIRY error, which produces surface sag errors along the y-direction, means that the surface is rotated about the x-axis. Recall from Chapter A.16 that wedge can also be produced by a surface decenter. To prevent tolerancing the same error twice, surface decenter is not used here. In addition, surface tilt is only considered on one side of an element, again to prevent double jeopardy. Finally, note that with total indicator runout, the magnitude of each of the values in the Min and Max columns gives the total allowed runout, regardless of the sign of the tilt. In other words, the plus and minus 0.0250 mm values listed mean that for either case the dial-indicator would actually vary by ± 0.0125 mm about a central value for a total runout error of 0.0250 mm.

In operands 36 through 42, TIRR specifies a tolerance on surface figure error or irregularity; that is, the departure of a surface from the ideal (in this case, spherical) shape. In ZEMAX, TIRR is half spherical aberration error and half astigmatism (cylinder) error. As with surface power, test plates would normally be used to monitor irregularity, and thus TIRR is given in fringes (± 0.5 double-pass fringes). The value in the Int1 column is the surface number.

In operands 43 through 46, TIND specifies a tolerance on glass indices of refraction, which are assumed to be equally shifted for all wavelengths. The surface number is given in the Int1 column. The allowed error here is ± 0.00050, which corresponds to Schott Grade 3.

Finally, the first operand in Listing B.7.2 specifies that defocus (refocus) is a compensator. A compensator is a system parameter that can be adjusted during manufacture (or during use) to partially correct or control fabrication errors. As a compensator, refocus is very effective and is nearly always used. The value in the Int1 column gives the surface number (surface 10 is the dummy surface just before the image surface), the value in the Int2 column is a code for the type of compensation (0 means adjust thickness), and the values in the Min and

Listing B.7.3

```
Analysis of Tolerances

Title:  TESSAR, 52MM, F/3.5, 45.2DEG, Tolerancing

Units are Millimeters.

Fast tolerancing mode is on.  In this mode, all
compensators are ignored, except back focus error.

Mode:  Sensitivities
Merit:  RMS Wavefront Error in waves

Nominal Merit Function (MF) is 0.65302400


Fields:  User Defined Angle in degrees
#     X-Field      Y-Field      Weight     VDX     VDY    VCX    VCY
1 0.000E+000  0.000E+000 8.000E+000 0.000  0.000 0.000 0.000
2 0.000E+000 -1.580E+001 2.000E+000 0.000 -0.052 0.002 0.221
3 0.000E+000  1.580E+001 2.000E+000 0.000  0.052 0.002 0.221
4 0.000E+000 -2.260E+001 1.000E+000 0.000 -0.086 0.008 0.468
5 0.000E+000  2.260E+001 1.000E+000 0.000  0.086 0.008 0.468

Sensitivity Analysis:
                ------------ Minimum ------------  ------------ Maximum ------------
Type  Sf1 Sf2      Value       MF           Change      Value      MF           Change
TFRN        2 -2.000000 0.656770        0.003746 2.000000 0.649476        -0.003548
TFRN        3 -2.000000 0.645785       -0.007239 2.000000 0.660770         0.007746
TFRN        4 -2.000000 0.664302        0.011278 2.000000 0.642852        -0.010172
TFRN        5 -2.000000 0.645313       -0.007711 2.000000 0.661336         0.008312
TFRN        7 -2.000000 0.656878        0.003854 2.000000 0.649270        -0.003754
TFRN        8 -2.000000 0.654578        0.001554 2.000000 0.651487        -0.001537
TFRN        9 -2.000000 0.648576       -0.004448 2.000000 0.657811         0.004787
TTHI    2   3 -0.050000 0.636409       -0.016615 0.050000 0.679479         0.026455
TTHI    3   5 -0.050000 0.700940        0.047916 0.050000 0.631689        -0.021335
TTHI    4   5 -0.050000 0.637705       -0.015319 0.050000 0.670387         0.017363
TTHI    5   6 -0.050000 0.652539       -0.000485 0.050000 0.653591         0.000567
TTHI    6   9 -0.050000 0.641946       -0.011078 0.050000 0.667284         0.014260
TTHI    7   9 -0.050000 0.646342       -0.006682 0.050000 0.660818         0.007794
TTHI    8   9 -0.050000 0.645014       -0.008010 0.050000 0.662113         0.009089
TEDX    2   3 -0.050000 0.682794        0.029770 0.050000 0.682794         0.029770
TEDY    2   3 -0.050000 0.697868        0.044844 0.050000 0.697868         0.044844
TETX    2   3 -0.050000 0.663903        0.010879 0.050000 0.663903         0.010879
TETY    2   3 -0.050000 0.662711        0.009687 0.050000 0.662711         0.009687
TEDX    4   5 -0.050000 0.755038        0.102014 0.050000 0.755038         0.102014
TEDY    4   5 -0.050000 0.730404        0.077380 0.050000 0.730404         0.077380
TETX    4   5 -0.050000 0.669140        0.016116 0.050000 0.669140         0.016116
TETY    4   5 -0.050000 0.667572        0.014548 0.050000 0.667572         0.014548
TEDX    7   9 -0.050000 0.682355        0.029331 0.050000 0.682355         0.029331
TEDY    7   9 -0.050000 0.682241        0.029217 0.050000 0.682241         0.029217
TETX    7   9 -0.050000 0.654284        0.001260 0.050000 0.654284         0.001260
TETY    7   9 -0.050000 0.654754        0.001730 0.050000 0.654754         0.001730
TIRX        2 -0.025000 0.669562        0.016538 0.025000 0.669562         0.016538
TIRY        2 -0.025000 0.690119        0.037095 0.025000 0.690119         0.037095
TIRX        4 -0.025000 0.716759        0.063735 0.025000 0.716759         0.063735
TIRY        4 -0.025000 0.704050        0.051026 0.025000 0.704050         0.051026
TIRX        7 -0.025000 0.659304        0.006280 0.025000 0.659304         0.006280
TIRY        7 -0.025000 0.661297        0.008273 0.025000 0.661297         0.008273
TIRX        9 -0.025000 0.676252        0.023228 0.025000 0.676252         0.023228
TIRY        9 -0.025000 0.683680        0.030656 0.025000 0.683680         0.030656
TIRR        2 -0.500000 0.661699        0.008675 0.500000 0.645406        -0.007618
TIRR        3 -0.500000 0.645571       -0.007453 0.500000 0.661406         0.008382
TIRR        4 -0.500000 0.662448        0.009424 0.500000 0.644899        -0.008125
TIRR        5 -0.500000 0.644787       -0.008237 0.500000 0.662733         0.009709
TIRR        7 -0.500000 0.658727        0.005703 0.500000 0.647747        -0.005277
TIRR        8 -0.500000 0.654958        0.001934 0.500000 0.651135        -0.001889
TIRR        9 -0.500000 0.645210       -0.007814 0.500000 0.661700         0.008676
TIND        2 -0.000500 0.659576        0.006552 0.000500 0.646863        -0.006161
TIND        4 -0.000500 0.644402       -0.008622 0.000500 0.662463         0.009439
TIND        7 -0.000500 0.649363       -0.003661 0.000500 0.656793         0.003769
TIND        8 -0.000500 0.660255        0.007231 0.000500 0.646277        -0.006747

Worst offenders:
Type  Sf1 Sf2      Value          MF        Change
TEDX    4   5    0.050000    0.755038    0.102014
TEDX    4   5   -0.050000    0.755038    0.102014
TEDY    4   5   -0.050000    0.730404    0.077380
TEDY    4   5    0.050000    0.730404    0.077380
TIRX        4    0.025000    0.716759    0.063735
TIRX        4   -0.025000    0.716759    0.063735
TIRY        4   -0.025000    0.704050    0.051026
TIRY        4    0.025000    0.704050    0.051026
TTHI    3   5   -0.050000    0.700940    0.047916
TEDY    2   3    0.050000    0.697868    0.044844

Nominal RMS Wavefront     :      0.653024
Estimated change          :      0.186761
Estimated RMS Wavefront   :      0.839785

Merit Statistics:
Mean                 :        0.667045
Standard Deviation   :        0.024052

Compensator Statistics:
Change in back focus:
Minimum              :       -0.160033
Maximum              :        0.159992
Mean                 :        0.000006
Standard Deviation   :        0.034501


End of Run.
```

Max columns give the allowed range (±1.0 mm shift).

B.7.3 Sensitivity Analysis

The second step in tolerancing a lens is to do the sensitivity analysis. Listing B.7.3 is an example using the Tessar and its tolerance types and values just defined. A sensitivity analysis applies the several tolerances to the lens one at a time in sequence. For each tolerance, the resulting degraded optical performance is computed and compared with the unperturbed performance. Note that positive and negative tolerances are treated as two separate errors. Thus in the present example, there are 45 tolerance operands but a total of 90 individual tolerances (plus one compensator).

The performance criterion for tolerancing can usually be chosen by the lens designer. The three most common measures are: image RMS spot size, exit pupil RMS wavefront error (optical path difference), and image MTF. Some programs allow the designer to select any criterion that can be incorporated into a merit function. Note that with most lenses, first-order changes and boundary condition violations are ignored during tolerancing. There are times, however, when distortion changes and boresight (pointing) errors are important and must be considered.

Recall that for a lens optimized with a particular merit function, any subsequent change in the lens will make the revised value of the merit function worse (this is what optimization means). It follows that during a sensitivity analysis, there is an advantage in selecting a tolerancing criterion that is similar to the optimization criterion. For similar performance criteria, it is less likely that a tolerance error will actually improve rather than degrade image quality. However, this seeming anomaly can still happen because the two criteria are not identical. The present Tessar lens was optimized using RMS wavefront errors, and thus this criterion has also been selected for the sensitivity analysis in Listing B.7.3. In this analysis, lower wavefront errors are better, and negative changes are the unexpected improvements.

In any case, the idea is to look for the tolerances that cause big performance degradations (the worst offenders). By adjusting the tolerance values during a series of sensitivity runs, these changes can be brought down to acceptable levels (the values in Listing B.7.2 are the result of several such iterations). Whether these final tolerances are practical or not is a different issue that must be discussed with the people in the optical and machine shops. You may also relax the tolerances that do not matter, but avoid overdoing this. If a lens parameter is not critical, simply enter a reasonable tolerance and let its effect be negligible (or delete it from the analysis).

In Listing B.7.3, all tolerances addressing the same type of lens parameter have been given equal values. For example, all of the element tilts (TETX and TETY) are ±0.050°. This approach allows the sensitivity of similar parameters to be directly compared. In practice, the lens designer may wish to do a further, fine-tuned analysis with each tolerance value separately adjusted. In the Tessar example, this refinement would be appropriate because several of the degradations are still fairly large (in particular, the decenters of the central, negative element). However, the tolerances in this example have been kept deliberately loose to better illustrate their effects.

B.7.4 Overall Performance

After the individual tolerances have been applied and their corresponding performance changes computed, the third step in tolerancing a lens is to estimate overall performance when all tolerances are acting together. One way to do this is by computing the root-sum-square (RSS) of the individual changes. This is a standard statistical procedure for combining independent random errors. Think of an RSS analysis as like a multi-dimensional extension of the Pythagorean theorem, where the separate errors are on the triangle legs and the combined error is on the hypotenuse. Note that, because the actual value of a given tolerance cannot be both positive and negative at the same time, the average of the squares of the two corresponding changes is used in this computation.

The RSS result for the Tessar is given near the bottom of Listing B.7.3. The RMS wavefront error, weighted and averaged over the field of view, is estimated to have changed from 0.653 to 0.840 waves.

B.7.5 Monte Carlo Statistical Analysis

Another method of estimating overall optical performance when all tolerances are simultaneously active is to do a Monte Carlo statistical analysis. For each cycle of this computation, all of the toleranced lens parameters are simultaneously perturbed, each in a random way, to simulate a real situation. It is assumed that each parameter varies about the midpoint between its allowed plus and minus limits (whose absolute values need not be equal) according to a truncated Gaussian dis-

Listing B.7.4

```
Analysis of Tolerances

Title:  TESSAR, 52MM, F/3.5, 45.2DEG, Tolerancing

Units are Millimeters.

Fast tolerancing mode is on.  In this mode, all
compensators are ignored, except back focus error.

Mode:  Sensitivities
Merit:  RMS Wavefront Error in waves

Nominal Merit Function (MF) is 0.65302400

 Fields:  User Defined Angle in degrees
#    X-Field      Y-Field      Weight      VDX     VDY    VCX    VCY
1 0.000E+000  0.000E+000 8.000E+000 0.000   0.000 0.000 0.000
2 0.000E+000 -1.580E+001 2.000E+000 0.000  -0.052 0.002 0.221
3 0.000E+000  1.580E+001 2.000E+000 0.000   0.052 0.002 0.221
4 0.000E+000 -2.260E+001 1.000E+000 0.000  -0.086 0.008 0.468
5 0.000E+000  2.260E+001 1.000E+000 0.000   0.086 0.008 0.468

Monte Carlo Analysis:
Number of trials:  20

Statistics:  Normal Distribution

                             0.0,  0.0 0.0, -0.7 0.0,  0.7 0.0, -1.0 0.0,  1.0
  Trial     Merit    Change   Field 1    Field 2    Field 3    Field 4    Field 5
      1 0.735009 0.081985  0.324202  0.745250  0.940293  1.129371  1.602465
      2 0.709108 0.056084  0.404856  0.796698  0.675021  1.240612  1.417223
      3 0.942836 0.289812  0.515429  0.896589  1.359301  1.324207  1.806419
      4 0.825572 0.172548  0.523490  0.769312  0.958764  1.846088  0.958877
      5 0.709862 0.056838  0.330976  0.835260  0.748617  0.907168  1.684982
      6 0.820948 0.167924  0.317020  1.172754  0.847972  1.705391  1.238621
      7 0.917744 0.264720  0.557805  1.150596  0.982477  1.543269  1.530500
      8 0.833083 0.180059  0.603036  0.883316  0.825041  1.273603  1.504402
      9 0.756926 0.103902  0.476003  0.816396  0.867535  1.285958  1.310177
     10 1.016575 0.363551  0.717153  1.315805  1.069106  1.617589  1.410042
     11 0.792336 0.139312  0.511909  0.867032  0.807456  1.424170  1.362725
     12 0.801054 0.148030  0.457254  0.996405  0.746862  1.407551  1.492821
     13 0.828289 0.175265  0.573964  0.678499  1.019797  1.582416  1.210235
     14 0.742032 0.089008  0.324590  0.778735  0.826954  1.345809  1.572875
     15 0.802819 0.149795  0.531505  0.794125  0.883771  1.691150  1.039190
     16 0.815146 0.162122  0.514798  0.598144  1.100690  1.536220  1.297608
     17 0.708814 0.055790  0.386831  0.800259  0.726518  1.114451  1.502746
     18 0.726310 0.073286  0.376141  0.748385  0.836735  1.253495  1.470324
     19 0.724872 0.071848  0.350897  0.836103  0.849430  1.485832  1.149879
     20 0.721004 0.067980  0.333040  0.680646  0.968628  1.263209  1.411308

Nominal 0.653024           0.257862  0.706290  0.706290  1.312030  1.312030
   Best 0.708814           0.317020  0.598144  0.675021  0.907168  0.958877
  Worst 1.016575           0.717153  1.315805  1.359301  1.846088  1.806419
   Mean 0.796517           0.456545  0.858015  0.902048  1.398878  1.398671
Std Dev 0.082361           0.110543  0.173149  0.152949  0.224234  0.204834

Compensator Statistics:
Change in back focus:
Minimum            :        -0.206503
Maximum            :         0.186828
Mean               :         0.016650
Standard Deviation :         0.096382

90% of Monte Carlo lenses have a merit function below 1.602465.
50% of Monte Carlo lenses have a merit function below 1.411308.
10% of Monte Carlo lenses have a merit function below 1.039190.
End of Run.
```

tribution. The standard deviation is one fourth of the total range, with no perturbations greater than two standard deviations (the errors are kept within their allowed bounds). After the lens has been perturbed, its performance is computed. This process is then repeated for many cycles (separate simulations), each with a different set of randomly selected errors.

Listing B.7.4 gives a Monte Carlo analysis of the Tessar lens using the same tolerances and performance criterion as in Listings B.7.2 and B.7.3. The results from 20 cycles are tabulated followed by a statistical summary. In this analysis, the RMS wavefront error, weighted and averaged over the field of view, is estimated to have changed from 0.653 to 0.797 waves. But the Monte Carlo output contains much more information. The standard deviation of the 0.797 estimate is 0.082, and the extremes that were encountered are 0.709 (best) and 1.017 (worst). Furthermore, the performance degradation at various field positions can now be seen.

Note that the RSS estimate indicates greater degradation than the Monte Carlo estimate. The reason is that the RSS estimate assumes that the tolerances are all at their limits as specified in Listing B.7.2. In contrast, the Monte Carlo estimate

Listing B.7.5

```
Analysis of Tolerances

Title:  TESSAR, 52MM, F/3.5, 45.2DEG, Tolerancing

Units are Millimeters.

Fast tolerancing mode is on.  In this mode, all
compensators are ignored, except back focus error.

Mode:  Sensitivities
Merit:  MTF average S&T at 30.0000 lp/mm

Nominal Merit Function (MF) is 0.33493134

Fields:  User Defined Angle in degrees
#       X-Field        Y-Field          Weight      VDX        VDY      VCX      VCY
1    0.000E+000     0.000E+000     8.000E+000     0.000      0.000    0.000    0.000
2    0.000E+000    -1.580E+001     2.000E+000     0.000     -0.052    0.002    0.221
3    0.000E+000     1.580E+001     2.000E+000     0.000      0.052    0.002    0.221
4    0.000E+000    -2.260E+001     1.000E+000     0.000     -0.086    0.008    0.468
5    0.000E+000     2.260E+001     1.000E+000     0.000      0.086    0.008    0.468

Monte Carlo Analysis:

Monte Carlo Analysis:
Number of trials:  20

Statistics:  Normal Distribution

                               0.0,   0.0 0.0, -0.7 0.0,   0.7 0.0, -1.0 0.0,   1.0
  Trial     Merit     Change    Field 1    Field 2    Field 3    Field 4    Field 5
      1 0.302078 -0.032854   0.560885   0.273203   0.238451   0.360072   0.148555
      2 0.331611 -0.003321   0.605365   0.282904   0.245086   0.328003   0.263601
      3 0.323977 -0.010954   0.584210   0.252308   0.264537   0.263357   0.314755
      4 0.278896 -0.056035   0.573384   0.242184   0.219345   0.217898   0.210975
      5 0.299162 -0.035769   0.544402   0.224523   0.306224   0.275415   0.199633
      6 0.313828 -0.021103   0.583105   0.260522   0.282988   0.228455   0.276022
      7 0.311688 -0.023243   0.597027   0.249874   0.287387   0.265545   0.227635
      8 0.217916 -0.117015   0.402857   0.108777   0.184399   0.252139   0.173614
      9 0.307779 -0.027153   0.561392   0.263209   0.270811   0.164011   0.344241
     10 0.301767 -0.033164   0.570886   0.178317   0.307796   0.347679   0.179223
     11 0.318638 -0.016293   0.608824   0.278931   0.232300   0.304065   0.241962
     12 0.228970 -0.105961   0.547556   0.121191   0.233765   0.180988   0.141206
     13 0.208600 -0.126332   0.395543   0.172653   0.153968   0.244335   0.108428
     14 0.183533 -0.151399   0.441709   0.112685   0.130648   0.095739   0.187216
     15 0.289474 -0.045457   0.561558   0.343479   0.137456   0.190783   0.291359
     16 0.311574 -0.023357   0.569924   0.272855   0.281842   0.279227   0.212168
     17 0.270126 -0.064805   0.523526   0.205387   0.233564   0.199681   0.240259
     18 0.303904 -0.031027   0.572517   0.235686   0.302966   0.298849   0.176373
     19 0.307961 -0.026971   0.567015   0.244972   0.279605   0.185605   0.325582
     20 0.259013 -0.075918   0.484298   0.259618   0.173176   0.161344   0.262294

Nominal 0.334931             0.582329   0.298303   0.298309   0.274675   0.274543
   Best 0.331611             0.608824   0.343479   0.307796   0.360072   0.344241
  Worst 0.183533             0.395543   0.108777   0.130648   0.095739   0.108428
   Mean 0.283525             0.542799   0.229164   0.238316   0.242160   0.226255
Std Dev 0.041190             0.061161   0.060358   0.054717   0.066810   0.062754

Compensator Statistics:
Change in back focus:
Minimum            :        -0.174608
Maximum            :         0.174969
Mean               :        -0.008167
Standard Deviation :         0.096862

90% of Monte Carlo lenses have an MTF above 0.149.
50% of Monte Carlo lenses have an MTF above 0.228.
10% of Monte Carlo lenses have an MTF above 0.326.

End of Run.
```

assumes that the errors are more realistically distributed, with most errors less than their limits. Given the assumptions, both approaches are valid.

If the optical performance as described by either an RSS or Monte Carlo analysis is unsatisfactory, then the designer must go back to step one and start over with tighter tolerances. Recall that the tolerances used here are somewhat loose to show their effects.

Listing B.7.5 gives a second Monte Carlo analysis. Here the performance criterion has been changed from RMS wavefront error to MTF (average of tangential and sagittal) at 30 cycles/mm. In simulations of overall performance, MTF is often preferred because the contrast of fine image details is usually of greatest interest to the lens user. Higher MTF values are better, and positive changes are the unexpected improvements. The spatial frequency is chosen to be near the middle of the useful frequency range, not near the cutoff. In this example, weighted and averaged MTF has changed from 0.335 to 0.284±0.041; the best value is 0.332 and the worst value is 0.184.

In Listing B.7.5, it is interesting to look at the changes in MTF from one Monte Carlo cycle to the next. These variations correspond to expected variations in a real production run of lenses. Even if only one or a few lenses are made, the computed variations still indicate the range of possible outcomes. If the manufacturer has tighter tolerances, then it is likely that the performance of a given specimen will be less degraded. Similarly, tighter tolerances mean that a batch of lenses will be more consistent from one to another.

This confidence is one of the things that the consumer pays for when buying an expensive lens. Tight tolerances cost money. Relative to a similar but less expensive lens, a more expensive lens will also be expected to have mechanical parts that are more rugged, glass types that are more exotic (costly), and sometimes a system with more elements. However, within the constraints, the lens designer should have designed both the expensive and inexpensive lenses with equal care.

Bibliography

———, *CODE V Reference Manual,* latest edition, Optical Research Associates, Pasadena, California.

———, *Corning Color Filter Glasses,* latest edition, Corning Glass Works, Corning, New York.

———, *Hoya Optical Glass Catalog,* latest edition, Hoya Optics, Fremont, California, and Tokyo, Japan.

———, *Kodak Filters for Scientific and Technical Uses* (Wratten filter catalog), Publication B-3, latest edition, Eastman Kodak, Rochester, New York.

———, *Military Standardization Handbook: Optical Design,* MIL-HDBK-141, 1962, Defense Supply Agency, reprinted 1987 as *Military Handbook 141,* Sinclair Optics.

———, *Ohara Optical Glass Catalog,* latest edition, Ohara Corp., Somerville, New Jersey, Rancho Santa Margarita, California, and Kanagawa, Japan.

———, *Optics Guide,* latest edition, Melles Griot, Irvine, California.

———, *Photonics Corporate Guide, Buyers' Guide, Handbook, and Dictionary,* four volume set, latest annual edition, Laurin Publishing.

———, *Popular Photography Photo Buying Guide,* latest annual edition, Hachette Filipacchi Magazines.

———, *Schott Optical Glass Catalog,* and *Schott Optical Glass Filter Catalog,* latest editions, Schott Glass Technologies, Duryea, Pennsylvania, and Mainz, Germany.

———, *Scientific Imaging with Kodak Films and Plates,* Publication P-315, 1987, Eastman Kodak, Rochester, New York.

———, *ZEMAX Optical Design Program User's Guide,* latest edition, Focus Software, Tucson, Arizona.

Allen, C.W., *Astrophysical Quantities,* third edition, 1973, Athlone Press.

Allen, Lew *et al., The Hubble Space Telescope Optical Systems Failure Report,* NASA TM-103443, 1990, National Aeronautics and Space Administration.

Ashbrook, Joseph, *The Astronomical Scrapbook,* 1984, Cambridge University Press.

Bass, Michael *et al.*, eds., *Optical Society of America Handbook of Optics,* second edition, Vols. 1 and 2, 1995, McGraw-Hill.

Bell, Louis, *The Telescope,* 1922, reprinted 1981, Dover.

Born, Max and Emil Wolf, *Principles of Optics,* sixth edition, 1980, Pergamon.

Buchroeder, Richard A., *Design Examples of Tilted-Component Telescopes,* Technical Report 68, 1971, Optical Sciences Center, University of Arizona.

Buchroeder, Richard A., *Tilted Component Optical Systems,* Doctoral Dissertation, 1976, Optical Sciences Center, University of Arizona.

Buil, Christian, *CCD Astronomy,* 1991, Willmann-Bell.

Cagnet, Michel, Maurice Françon, and Jean Claude Thrierr, *Atlas of Optical Phenomena,* 1962, Springer-Verlag.

Cagnet, Michel, Maurice Françon, and Shamlal Mallick, *Atlas of Optical Phenomena Supplement,* 1971, Springer-Verlag.

Committee on Colorimetry, *The Science of Color,* 1953, reprinted 1963, Optical Society of America.

Conrady, A.E., *Applied Optics and Optical Design;* Part 1, 1929, reprinted 1957; Part 2, 1960; Dover.

Cox, Arthur, *A System of Optical Design,* 1964, Focal Press.

Cromwell, Richard H., and Gregory H. Smith, *Resolving Power, Signal-Induced Background, Noise, and Detective Quantum Efficiency of Image Intensifier Photographs,* Proceedings of the SPIE, Vol. 42, pp. 155–169, 1973.

Dereniak, Eustace L., and G.D. Boreman, *Infrared Detectors and Systems,* 1996, Wiley.

Dereniak, Eustace L., and Devon G. Crowe, *Optical Radiation Detectors,* 1984, Wiley.

DeVany, Arthur S., *Master Optical Techniques,* 1981, Wiley.

Dimitroff, George Z., and James G. Baker, *Telescopes and Accessories,* 1945, Blakiston.

Feder, Donald P.F, *Automatic Optical Design,* Applied Optics, Vol. 2, No. 3, pp. 1209–1226, December 1963.

Fischer, Robert E., ed., *International Lens Design Conference,* Proceedings of the SPIE, Vol. 237, 1980.

Florence, Ronald, *The Perfect Machine, Building the Palomar Telescope,* 1995, Harper Collins.

Forbes, G.W., *Optical System Assessment for Design: Numerical Ray Tracing in the Gaussian Pupil,* J. Opt. Soc. Am. A, Vol. 5, No. 11, pp. 1943–1956, November 1988.

Forbes, G.W., ed., *International Optical Design Conference,* Optical Society of America Proceedings, Vol. 22, 1994.

Gaskill, Jack D., *Linear Systems, Fourier Transforms, and Optics,* 1978, Wiley.

Goldberg, Norman, *Camera Technology,* 1992, Academic Press.

Goodman, Joseph W., *Introduction to Fourier Optics,* 1968, McGraw-Hill.

Greenleaf, Allen R., *Photographic Optics,* 1950, Macmillan.

Hardy, Arthur C., and Fred H. Perrin, *Principles of Optics,* 1932, McGraw-Hill.

Hecht, Eugene, (with Alfred Zajac), *Optics,* second edition, 1987, Addison-Wesley.

Hilbert, Robert S., *A Study of Concentric and Schmidt Type Catadioptric Systems,* Master's Thesis, 1963, Institute of Optics, University of Rochester.

Hiltner, W.A., ed., *Astronomical Techniques,* 1962, University of Chicago Press.

Ingalls, Albert G., ed., *Amateur Telescope Making;* Book 1, 1935; Book 2, 1937; Book 3, 1953; Scientific American; rearranged editions, 1996, Willmann-Bell.

James, T.H., ed., *The Theory of the Photographic Process,* fourth edition, 1977, Macmillan.

Jenkins, Francis A., and Harvey E. White, *Fundamentals of Optics,* fourth edition, 1976, McGraw-Hill.

Johnson, B.K., *Practical Optics,* second edition, 1947, reprinted 1960 as *Optics and Optical Instruments,* Dover.

King, Edward Skinner, *A Manual of Celestial Photography,* 1931, reprinted 1988, Sky Publishing.

King, Henry C., *The History of the Telescope,* 1955, reprinted 1979, Dover.

Kingslake, Rudolf, ed., *Applied Optics and Optical Engineering;* Vol. 1, 1965; Vol. 2, 1965; Vol. 3, 1965; Vol. 4, 1967; Vol. 5, 1969; Academic Press.

Kingslake, Rudolf, *Camera Optics,* in *Leica Manual,* fifteenth edition, pp. 499–521, 1973, Morgan and Morgan.

Kingslake, Rudolf, *Lens Design Fundamentals,* 1978, Academic Press.

Kingslake, Rudolf, and Brian J. Thompson, eds., *Applied Optics and Optical Engineering,* Vol. 6, 1980, Academic Press.

Kingslake, Rudolf, *Optical System Design,* 1983, Academic Press.

Kingslake, Rudolf, *A History of the Photographic Lens,* 1989, Academic Press.

Kingslake, Rudolf, *Optics in Photography,* third edition of *Lenses in Photography,* 1992, SPIE Press.

Korsch, Dietrich, *Reflective Optics,* 1991, Academic Press.

Kossel, Dierick, *Glass Compositions,* Leica Fotografie, English edition, pp. 20–25, February 1978.

Kuiper, Gerard P., and Barbara M. Middlehurst, eds., *Telescopes,* 1960, University of Chicago Press.

Laikin, Milton, *Lens Design,* second edition, 1995, Marcel Dekker.

Lang, Kenneth R., and Owen Gingerich, eds., *A Source Book in Astronomy and Astrophysics, 1900–1975,* 1979, Harvard University Press.

Lawrence, George N., ed., *International Lens Design Conference,* Proceedings of the SPIE, Vol. 1354, 1990.

Levi, Leo, *Applied Optics;* Vol. 1, 1968; Vol. 2, 1980; Wiley.

Linfoot, E.H., *Recent Advances in Optics,* 1955, Oxford University Press.

Livingston, Dorothy Michelson, *The Master of Light, A Biography of Albert A. Michelson,* 1973, University of Chicago Press.

Lloyd, J.M., *Thermal Imaging Systems,* 1975, Plenum.

Lockett, Arthur, and H.W. Lee, *Camera Lenses,* fourth edition, 1957, Pitman.

Longhurst, R.S., *Geometrical and Physical Optics,* third edition, 1973, Longman.

Macleod, H.A., *Thin-Film Optical Filters,* second edition, 1986, Macmillan.

Mahajan, Virendra N., *Aberration Theory Made Simple,* 1991, SPIE Press.

Mahajan, Virendra N., ed., *Selected Papers on Effects of Aberrations in Optical Imaging,* Vol. MS 74, 1994, SPIE Press.

Malacara, Daniel, ed., *Geometrical and Instrumental Optics,* 1988, Academic Press.

Malacara, Daniel, ed., *Optical Shop Testing,* second edition, 1992, Wiley.

Malacara, Daniel, and Zacarias Malacara, *Handbook of Lens Design,* 1994, Marcel Dekker.

Mann, Allen, ed., *Selected Papers on Zoom Lenses,* Vol. MS 85, 1993, SPIE Press.

Meinel, A.B., *Introduction to the Design of Astronomical Telescopes,* Technical Report 1, 1965, Optical Sciences Center, University of Arizona.

Meyer-Arendt, Jurgen R., *Introduction to Classical and Modern Optics,* 1972, Prentice-Hall.

Newton, Sir Isaac, *Opticks,* fourth edition, 1730, reprinted 1952, Dover.

O'Shea, Donald C., *Elements of Modern Optical Design,* 1985, Wiley.

Osterbrock, Donald E., *Pauper and Prince: Ritchey, Hale, and Big American Telescopes,* 1993, University of Arizona Press.

Osterbrock, Donald E., *James E. Keeler: Pioneer American Astrophysicist,* 1984, Cambridge University Press.

Osterbrock, Donald E., John R. Gustavson, and W. J. Shiloh Unruh, *Eye on the Sky, Lick Observatory's First Century,* 1988, University of California Press.

Osterbrock, Donald E., *Yerkes Observatory 1892-1950,* 1997, University of Chicago Press.

Pendray, G. Edward, *Men, Mirrors, and Stars,* 1935, Funk and Wagnalls.

Rancourt, James D., *Optical Thin Films Users' Handbook,* 1987, Macmillan.

Merte, W., R. Richter, and M. von Rohr, *Photographic Lenses* (*Das Photographische Objectiv*), Parts 1 and 2, 1932, translated 1949, Central Air Documents Office, Wright-Patterson Air Force Base, Dayton, Ohio.

Neblette, C.B., *Photography, Its Materials and Processes,* sixth edition, 1962, Van Nostrand.

Ray, Sidney F., *Applied Photographic Optics,* 1988, Focal Press.

Ridpath, Ian, ed., *Norton's 2000.0: Star Atlas and Reference Handbook,* eighteenth edition, 1989, Longman.

Rogliatti, G., *Leica and Leicaflex Lenses,* second edition, second printing with additions, 1984, Hove Foto Books.

Rogliatti, G., *Leica, The First Sixty Years,* 1985, Hove Foto Books.

Rotoloni, Robert, *The Nikon Rangefinder Camera,* second edition, 1983, Hove Foto Books.

Rutten, Harrie, and Martin van Venrooij, *Telescope Optics,* 1988, Willmann-Bell.

Saleh, Bahaa E.A., and Teich, Malvin Carl, *Fundamentals of Photonics,* 1991, Wiley.

Schroeder, Daniel J., *Astronomical Optics,* 1987, Academic Press.

Schroeder, Daniel J., ed., *Selected Papers on Astronomical Optics.* Vol. MS 73, 1993, SPIE Press.

Sears, Francis Weston, *Optics,* third edition, 1949, Addison-Wesley.

Selwyn, E.W.H., *Photography in Astronomy,* 1950, Eastman Kodak, Rochester, New York.

Shannon, Robert R., and James G. Wyant, *Applied Optics and Optical Engineering;* Vol. 7, 1979; Vol. 8, 1980; Vol. 9, 1983; Vol. 10, 1987; Vol. 11, 1992; Academic Press.

Shannon, Robert R., *The Art and Science of Optical Design,* 1997, Cambridge University Press.

Shapley, Harlow, ed., *Source Book in Astronomy 1900–1950,* 1960, Harvard University Press.

Smith, Gregory H., *Evaluation of Image Intensifier Tubes Using Detective Quantum Efficiency,* Doctoral Dissertation, 1972, Optical Sciences Center, University of Arizona, also Technical Report 76, 1972, Optical Sciences Center, University of Arizona.

Smith, Warren J., *Modern Optical Engineering,* second edition, 1990, McGraw-Hill.

Smith, Warren J., *Modern Lens Design,* 1992, McGraw-Hill.

Smith, Warren J., ed., *Lens Design,* Vol. CR41, 1992, SPIE Press.

Southall, James P.C., *Mirrors, Prisms, and Lenses,* third edition, 1933, reprinted 1964, Dover.

Strand, K.Aa., ed., *Basic Astronomical Data,* 1963, University of Chicago Press.

Steel, W.H., *Interferometry,* second edition, 1983, Cambridge University Press.

Stover, John C., *Optical Scattering,* 1990, McGraw-Hill.

Strong, John, *Concepts of Classical Optics,* 1958, Freeman.

Suiter, Harold Richard, *Star Testing Astronomical Telescopes,* 1994, Willmann-Bell.

Taylor, H. Dennis, *The Adjustment and Testing of Telescope Objectives,* 1891, reprinted 1983, Adam Hilger.

Taylor,William H., and Duncan T. Moore, eds., *International Lens Design Conference,* Proceedings of the SPIE, Vol. 554, 1985.

Texereau, Jean, *How To Make a Telescope,* second edition, 1984, Willmann-Bell.

Thomas, Jr., Woodlief, ed., *SPSE Handbook of Photographic Science and Engineering,* 1973, Wiley.

Thompson, Allyn J., *Making Your Own Telescope,* 1947, Sky Publishing.

Thompson, Morris M., ed., *Manual of Photogrammetry,* third edition, Vols. 1 and 2, 1966, American Society of Photogrammetry.

Twyman, F., *Prism and Lens Making,* second edition, 1952, reprinted 1988, Adam Hilger.

Walker, Bruce H., *Optical Engineering Fundamentals,* 1995, McGraw-Hill.

Walker, Gordon, *Astronomical Observations, An Optical Perspective,* 1987, Cambridge University Press.

Wallin, Walter, *Design of Optical Systems,* 1963, University of California at Los Angeles Engineering Extension.

Warner, Deborah Jean, and Robert B. Ariail, *Alvan Clark and Sons, Artists in Optics,* second edition, 1995, Willmann-Bell.

Welford, Walter T., *Useful Optics,* 1991, University of Chicago Press.

Wilson, R.N., *Reflecting Telescope Optics I: Basic Design Theory and its Historical Development,* 1996, Springer-Verlag.

Wilson, R.N., *Reflecting Telescope Optics II: Manufacture, Testing, Alignment; Modern Developments,* in press, Springer-Verlag.

Wolfe, William L., *Introduction to Infrared System Design,* Vol. TT24, 1996, SPIE Press.

Wolfe, William L., and George J. Zissis, eds., *The Infrared Handbook,* revised edition, 1985, Office of Naval Research.

Woodbury, David O., *The Glass Giant of Palomar,* 1939, Dodd, Mead, and Co.

Wright, Helen, *Palomar, The World's Largest Telescope,* 1952, Macmillan.

Wright, Helen, *Explorer of the Universe, A Biography of George Ellery Hale,* 1966, reprinted 1994, American Institute of Physics Press.

Wright, Helen, Joan N. Warnow, and Charles Weiner, eds., *The Legacy of George Ellery Hale,* 1972, MIT Press.

Wright, Helen, *James Lick's Monument,* 1987, Cambridge University Press.

Wynne, C.G., *Field Correctors,* in *Progress in Optics,* Vol. 10, pp. 137–164, 1972, North Holland.

Wynne, C.G., *A New Wide-Field Triple Lens Paraboloid Field Corrector,* Mon. Not. R. Astr. Soc., Vol. 167, pp. 189–197, 1974.

Wynne, C.G., *The Optics of the Achromatized UK Schmidt Telescope,* Q. Jl. R. Astr. Soc., Vol. 22, pp. 146–153, 1981.

Wyszecki, Günter and W.S. Stiles, *Color Science,* second edition, 1982, Wiley.

Yoder, Paul R., Jr., *Opto-Mechanical Systems Design,* second edition, 1992, Marcel Dekker.

Zügge, Hannfried, ed., *Lens and Optical Systems Design,* Proceedings of the SPIE, Vol. 1780, 1993.

Index